Springer Tracts in Civil Engineering

Series Editors

Sheng-Hong Chen, School of Water Resources and Hydropower Engineering, Wuhan University, Wuhan, China

Marco di Prisco, Politecnico di Milano, Milano, Italy

Ioannis Vayas, Institute of Steel Structures, National Technical University of Athens, Athens, Greece

Springer Tracts in Civil Engineering (STCE) publishes the latest developments in Civil Engineering - quickly, informally and in top quality. The series scope includes monographs, professional books, graduate textbooks and edited volumes, as well as outstanding PhD theses. Its goal is to cover all the main branches of civil engineering, both theoretical and applied, including:

- Construction and Structural Mechanics
- Building Materials
- Concrete, Steel and Timber Structures
- Geotechnical Engineering
- Earthquake Engineering
- Coastal Engineering; Ocean and Offshore Engineering
- Hydraulics, Hydrology and Water Resources Engineering
- Environmental Engineering and Sustainability
- Structural Health and Monitoring
- Surveying and Geographical Information Systems
- Heating, Ventilation and Air Conditioning (HVAC)
- Transportation and Traffic
- Risk Analysis
- Safety and Security

Indexed by Scopus

To submit a proposal or request further information, please contact:
Pierpaolo Riva at Pierpaolo.Riva@springer.com (Europe and Americas)Wayne Hu at wayne.hu@springer.com (China)

More information about this series at https://link.springer.com/bookseries/15088

Patrizia Gasparini · Lucio Di Cosmo ·
Antonio Floris · Davide De Laurentis
Editors

Italian National Forest Inventory—Methods and Results of the Third Survey

Inventario Nazionale delle Foreste e dei Serbatoi Forestali di Carbonio—Metodi e Risultati della Terza Indagine

Editors
Patrizia Gasparini
Centro di ricerca Foreste e Legno
CREA
Trento, Italy

Lucio Di Cosmo
Centro di ricerca Foreste e Legno
CREA
Trento, Italy

Antonio Floris
Centro di ricerca Foreste e Legno
CREA
Trento, Italy

Davide De Laurentis
Arma dei Carabinieri
Comando Unità Forestali Ambientali e Agroalimentari
Roma, Italy

ISSN 2366-259X ISSN 2366-2603 (electronic)
Springer Tracts in Civil Engineering
ISBN 978-3-030-98680-3 ISBN 978-3-030-98678-0 (eBook)
https://doi.org/10.1007/978-3-030-98678-0

© The Editor(s) (if applicable) and The Author(s) 2022. This book is an open access publication.
Open Access This book is licensed under the terms of the Creative Commons Attribution 4.0 International License (http://creativecommons.org/licenses/by/4.0/), which permits use, sharing, adaptation, distribution and reproduction in any medium or format, as long as you give appropriate credit to the original author(s) and the source, provide a link to the Creative Commons license and indicate if changes were made.
The images or other third party material in this book are included in the book's Creative Commons license, unless indicated otherwise in a credit line to the material. If material is not included in the book's Creative Commons license and your intended use is not permitted by statutory regulation or exceeds the permitted use, you will need to obtain permission directly from the copyright holder.
The use of general descriptive names, registered names, trademarks, service marks, etc. in this publication does not imply, even in the absence of a specific statement, that such names are exempt from the relevant protective laws and regulations and therefore free for general use.
The publisher, the authors, and the editors are safe to assume that the advice and information in this book are believed to be true and accurate at the date of publication. Neither the publisher nor the authors or the editors give a warranty, expressed or implied, with respect to the material contained herein or for any errors or omissions that may have been made. The publisher remains neutral with regard to jurisdictional claims in published maps and institutional affiliations.

This Springer imprint is published by the registered company Springer Nature Switzerland AG
The registered company address is: Gewerbestrasse 11, 6330 Cham, Switzerland

Foreword I

The legislative decree-19 August 2016-n. 177 containing "Provisions regarding the rationalization of police functions and absorption of the State Forestry Corps, pursuant to article 8, paragraph 1, letter a), of law no. 124, concerning the reorganization of public administrations", in article 7 paragraph 2 provides that the Carabinieri, among others, carry out the study activities related to the qualitative and quantitative survey of forest resources, to the monitoring of the phytosanitary status of forests and finally related to the controls on the level of pollution of forest ecosystems and the general monitoring of the territory collecting, processing, archiving and disseminating data. The legislator, in entrusting the Carabinieri with the qualitative and quantitative data collection activities of the forests, wanted to ensure a strategic activity for the country to a national Armed Force. As other land, air and marine resources, whose surveys are entrusted to military bodies such as the Army Geographical Institute for cartographic surveys, the Aeronautical Meteorological Service for weather surveys and the Hydrographic Institute of the Navy for those hydro-oceanographic, the surveys relating to forest resources are assigned to the Forestry, Environmental, Agri-food Unit Command of the Carabinieri.

The National Forest Inventory is a periodic sample survey which investigates the quality and quantity of the country's forest resources; it is the source of Italian forest statistics, a monitoring tool that produces concrete knowledge in support of forest and environmental policy, not addressed only to the productive aspects, but inclusive of all the functions performed by the forests and the related ecosystem services. In continuity with what happened previously during the achievement of the National Inventory of Forests and Forest Carbon Pools 2005 (INFC2005), Arma dei Carabinieri continued to make use of the scientific support of the Council for Agricultural Research and Economics—Research Centre for Forestry and Wood, thanks to a consolidated and fruitful relationship of collaboration and institutional synergy. In order to obtain updated statistics, comparable and responding to a plurality of information needs related to the management of forests and the territory, the inventory survey INFC2015—the year 2015 is assumed as the base year of reference but the surveys in the woods began in November 2017 and concluded in December 2019—strictly followed the methodology used for INFC2005 (the previous national forest inventory survey). In 1985 the first national forest inventory IFNI85 was carried out, but the different sampling design, definitions and classification system did not allow a correct comparison of the data.

The inventory makes it possible to obtain estimates as indicative as possible of the different values expressed by forest ecosystems. The quantitative and qualitative data of the country's forests are therefore available to international organizations, national and regional public administrations, civil society and research institutions. The main purpose of the third Italian forest inventory INFC2015 was to obtain estimates on the evolution of Italian forest areas ten years after the INFC2005. Now, in fact, with the results of INFC2015, realized with the same relevant methodologies, it is finally possible to obtain comparable quantitative and qualitative data that will feed a strategic historical series for the evaluation of the variations that natural capital undergoes over time.

In fact, the "photogram" of 2015 is now added to that of 2005, transforming a photograph into a film to which other frames will soon be added (the subsequent inventory surveys). It allows to observe the evolution of the Italian forest heritage in the context of global change underway triggered by the rise in the average temperature of the planet.

The inventories (INFC2005, INFC2015) produce important information, providing data necessary to describe the main characteristics of forest areas. The main qualitative parameters investigated are the biodiversity of forest formations determined by the specific composition and by the presence of dead wood; management in terms of methods and intensity; safeguards thanks to protection constraints; the state of health; structure; the evolutionary stage. The quantitative ones, on the other hand, concerned the precise quantification of the national forest heritage, both in terms of surface and amount of wood resources, understood as volume, biomass and stored carbon content.

In addition to quantifying and assessing the state of the resources at the time of observation, the repetition of the survey makes it possible to monitor changes affecting the forest heritage, in order to verify the effects of management policies, foresee evolution scenarios and plan corrective actions. INFC2015 and INFC2005 make it possible to analyse the evolution over the last ten years of the entity and quality of forest resources based on the comparison of data collected with statistical sampling. The comparison of the two inventories makes it possible, thanks to the re-proposal of the same survey methodology, to reconstruct historical series for the main characters, such as, first of all, the forest area, the consistency of the carbon stock and the annual storage capacity of CO_2, the greenhouse gas most responsible for the global rise in temperatures.

The updating of the national forest inventory is linked to the country's joining in international commitments for the conservation of terrestrial ecosystems and biodiversity, for the sustainable management of forest resources and for the reduction of greenhouse gas emissions, in line with the European Green Deal and the European Biodiversity Strategy 2030. The agreements signed provide for the monitoring of the implementation status of the commitments undertaken over time by quantifying, at set deadlines, the values of indicators (international reporting) such as the surface forestry, the number of species, the ratio of forest growth to the amount of wood used, the amount of accumulated carbon and so on.

Finally, the sampling intensity, as the number of sample units to be observed, and the selection of the sample, by regional "strata", was defined in order to ensure reliable results also at the regional level. The tables of the results produced, in fact, always report the statistics on a regional basis, thus guaranteeing the Regions and Autonomous Provinces statistical data of great utility.

The 2015 National Inventory of Forests and Carbon Pools data assume particular relevance by virtue of the functional dependence of the Forestry, Environmental and Agri-food Units Command on the Minister of Ecological Transition, sanctioned by the changes introduced by law no. 113 of conversion, with modifications, of the decree law 9 June 2021, n. 80. The aforementioned functional dependence is achieved without prejudice to the possibility of the Minister of Agricultural, Food and Forestry Policies to make use of the same Command, for the performance of the functions attributable to the attributions of that Ministry.

In this innovative institutional framework, the monitoring activity and the quantitative detection of forest resources, as well as the entire activity of the Forestry, Environmental and Agri-food Units Command, is perfectly in line with both the implementation of international Conventions on environmental matters and the relative national regulations both as regards the needs of forestry and territorial planning at the various levels of competence.

Rome, Italy Gen. Antonio Pietro Marzo

Presentazione I

Il decreto legislativo 19 agosto 2016, n. 177, recante "Disposizioni in materia di razionalizzazione delle funzioni di polizia e assorbimento del Corpo forestale dello Stato, ai sensi dell'articolo 8, comma 1, lettera (a), della legge 7 agosto 2015, n. 124, in materia di riorganizzazione delle amministrazioni pubbliche", all'articolo 7 comma 2 dispone che l'Arma dei Carabinieri, tra le altre, eserciti le attività di studio connesse alla rilevazione qualitativa e quantitativa delle risorse forestali, al monitoraggio sullo stato fitosanitario delle foreste, ai controlli sul livello di inquinamento degli ecosistemi forestali e al monitoraggio del territorio in genere con raccolta, elaborazione, archiviazione e diffusione dei dati.

Il legislatore, nell'affidare all'Arma dei Carabinieri le attività di rilevazione dei dati qualitativi e quantitativi delle foreste, ha voluto assicurare ad una Forza Armata dello Stato un'attività strategica per il Paese. Al pari di altre risorse terrestri, aeree e marine i cui rilevamenti sono affidati a organismi militari come l'Istituto geografico dell'Esercito per i rilievi cartografici, il Servizio Meteorologico dell'Aeronautica per le rilevazioni meteo e l'Istituto Idrografico della Marina per quelle idro-oceanografiche, i rilievi relativi alle risorse forestali sono assegnati al Comando Unità Forestali Ambientali e Agroalimentari dell'Arma dei Carabinieri.

L'Inventario Forestale Nazionale è un'indagine campionaria periodica finalizzata alla conoscenza della qualità e quantità delle risorse forestali del Paese ed è la fonte delle statistiche forestali italiane, uno strumento di monitoraggio che produce conoscenza concreta a supporto della politica forestale e ambientale, non rivolto solo agli aspetti produttivi, ma inclusivo di tutte le funzioni assolte dal bosco e dei relativi servizi ecosistemici che è in grado di fornire.

In continuità con quanto avvenuto in precedenza, anche per la realizzazione dell'Inventario Nazionale delle Foreste e dei serbatoi forestali di Carbonio 2015 (INFC2015), l'Arma dei Carabinieri ha continuato ad avvalersi del supporto scientifico del Consiglio per la ricerca in agricoltura e l'analisi dell'economia agraria—Centro di ricerca Foreste e Legno, grazie ad un consolidato e proficuo rapporto di collaborazione e di sinergia istituzionale.

Al fine di ottenere statistiche aggiornate, comparabili e rispondenti ad una pluralità di esigenze informative connesse alla gestione delle foreste e del territorio, l'indagine inventariale INFC2015—l'anno 2015 si assume come anno base di riferimento ma i rilievi in bosco sono iniziati nel novembre del 2017 e si sono conclusi nel dicembre del 2019—ha seguito rigorosamente la metodologia utilizzata per INFC2005 (il precedente rilievo inventariale nazionale delle foreste). Nel 1985 era stato realizzato il primo inventario forestale nazionale IFNI85 ma il disegno di campionamento, le definizioni e il sistema di classificazione differenti non permettevano una corretta comparazione dei dati.

L'inventario consente di ottenere stime il più possibile indicative dei diversi valori espressi dagli ecosistemi forestali. I dati quantitativi e qualitativi delle foreste del Paese sono quindi a disposizione delle organizzazioni internazionali, delle amministrazioni pubbliche sia nazionali che regionali, della società civile, delle istituzioni di ricerca. La finalità principale del terzo inventario forestale italiano INFC2015 è stata quella di ottenere stime sull'evoluzione delle superfici forestali italiane a dieci anni di distanza dal INFC2005. Ora, infatti, con i risultati di INFC2015, realizzato con le stesse metodologie di rilievo, finalmente si possono ottenere dati quantitativi e qualitativi comparabili che andranno ad alimentare una serie storica strategica per la valutazione delle variazioni che il capitale naturale subisce nel tempo.

Al "fotogramma" del 2005, difatti, si aggiunge ora quello del 2015 trasformando una fotografia in un film a cui a breve si aggiungeranno altri fotogrammi (le successive indagini inventariali) che consentono di osservare l'evoluzione del patrimonio forestale italiano nell'ambito del cambiamento globale in atto innescato dall'innalzamento della temperatura media del pianeta.

Gli inventari (INFC2005, INFC2015) producono informazioni importanti, fornendo dati necessari a descrivere le principali caratteristiche delle aree forestali. I principali parametri qualitativi indagati sono stati: la biodiversità delle formazioni forestali determinata dalla composizione specifica e dalla presenza di legno morto; la gestione in termini di modalità e intensità; la tutela mediante vincoli di protezione; lo stato di salute; la struttura; lo stadio evolutivo. Quelli quantitativi hanno invece riguardato la quantificazione precisa del patrimonio forestale nazionale, sia in termini di superficie che di entità delle risorse legnose, intesa come volume, biomassa e contenuti di carbonio immagazzinato.

Oltre alla quantificazione e valutazione dello stato delle risorse al momento dell'osservazione, la ripetizione dell'indagine consente di monitorare i cambiamenti che interessano il patrimonio forestale, al fine di verificare gli effetti delle politiche di gestione, prevedere scenari di evoluzione e pianificare interventi correttivi. INFC2015 e INFC2005 permettono di analizzare l'evoluzione negli ultimi dieci anni dell'entità e della qualità delle risorse forestali basandosi sul confronto di dati raccolti con campionamento statistico. La comparazione dei due inventari consente infatti, grazie alla riproposizione della stessa metodologia di indagine, di ricostruire serie storiche per i principali caratteri, quali, in primo luogo, la superficie forestale, la consistenza dello stock di carbonio e la capacità di stoccaggio annuale della CO_2, il gas serra maggiormente responsabile dell'innalzamento globale delle temperature.

L'aggiornamento dell'inventario forestale nazionale è legato all'assunzione da parte del Paese di impegni a livello internazionale per la conservazione degli ecosistemi terrestri e della biodiversità, per la gestione sostenibile delle risorse forestali e per la riduzione delle emissioni di gas ad effetto serra in linea con il Green Deal Europeo e con la Strategia Europea per la biodiversità 2030. Gli accordi sottoscritti prevedono infatti di monitorare nel tempo lo stato di attuazione degli impegni assunti quantificando, a scadenze prefissate, i valori di indicatori (reporting internazionale) quali la superficie forestale, il numero di specie, il rapporto tra accrescimento delle foreste ed entità di legname utilizzato, la quantità di carbonio accumulato e così via.

Infine, l'intensità di campionamento, cioè il numero di unità campionarie da osservare, e la selezione del campione, per "strati" regionali, è stata definita in modo da garantire risultati affidabili anche a livello di regione. Le tabelle dei risultati prodotte, infatti, riportano le statistiche sempre su base regionale garantendo quindi alle Regioni e Province autonome dati statistici di grandissima utilità.

I dati dell'Inventario Nazionale delle Foreste e dei Serbatoi di Carbonio 2015, assumono una particolare rilevanza in virtù della dipendenza funzionale del Comando Unità Forestali, Ambientali e Agroalimentari dal Ministro della transizione ecologica, sancita dalle modifiche introdotte dalla legge 6 agosto 2021, n. 113 di conversione, con modifiche, del decreto legge 9 giugno 2021, n. 80. La suddetta dipendenza funzionale si realizza ferma restando la possibilità del Ministro delle politiche agricole alimentari e forestali di avvalersi dello stesso Comando, per lo svolgimento delle funzioni riconducibili alle attribuzioni di quel Ministero.

In tale innovato quadro istituzionale, l'attività di monitoraggio e la rilevazione quali quantitativa delle risorse forestali, così come l'intera attività del Comando Unità Forestali, Ambientali e Agroalimentari, risulta perfettamente in linea sia con l'attuazione delle convenzioni internazionali in materia ambientale e delle relative norme nazionali sia per quanto riguarda le esigenze di pianificazione forestale e territoriale ai vari livelli di competenza.

Rome, Italy

Gen. Antonio Pietro Marzo

Foreword II

Forests play an essential role in guaranteeing natural and environmental balances and, at the same time, in satisfying the needs of mankind: to make forests "count" in the country's political and economic choices and strategies, it is primarily necessary to "count" forests.

Multifunctional resource par excellence, woods represent the richest ecosystems in animal and plant biodiversity and are able to perform irreplaceable services: they provides an important contribution in mitigating the effects of climate change through the removal of excess carbon dioxide present in the atmosphere, they affect in a decisive way the water cycle evaporation, absorption and regulation, the influence that the purification, the emission of the air and the influence that the oxygen purification we breathe, they favour the effect of the soils and contrast of the slopes to the phenomena of desertification and represents the vital habitat for many other species, plants and animals.

These functions, directly linked to the quality of the environment, generate a further series of a positive externalities, both in terms of production of renewable raw materials such as wood and undergrowth products, and in terms of fundamental intangible services such as touristic/recreational use, the characterization of landscape and the conservation of the traditional, cultural and even spiritual values of the territories.

The removal from the atmosphere and the storage of greenhouse gases especially of carbon dioxide is one of the most important functions recently recognized to forests which contribute to mitigate the effects of climate change regulating the climate. In fact, forests, like the whole plant kingdom, represent an irreplaceable bridge between the inorganic world and that of living beings and a formidable biological machine that captures carbon from the atmosphere storing it in its fibres and keeping it blocked for long time: a cubic metre of dry wood contains about 250 kg, a half of its weight.

In this context, the monitoring of forest ecosystems is coherently and effectively part of the achievement of the strategic objectives identified by the European Union as part of the "Green Deal", which aims to achieve the neutrality of polluting emissions by 2050. The tools of strategic planning developed are aimed at promoting a correct transition towards a sustainable economy with zero environmental impact in all member states, thanks to huge economic resources to be allocated in favour of decarbonization, circular economy and sustainable mobility. These ambitious objectives are achievable only through a radical change in current lifestyles, combining the need to ensure a better quality of life for citizens with a reinforced and essential protection of the environment through a socio-economic development that can be defined as truly sustainable.

The European Commission has set a challenging goal of reducing greenhouse gas emissions, expecting to go from 40% to at least 55% by 2030 compared to 1990 levels and to zero them in 2050. Europe therefore sets itself, towards the rest of the world, consistently as leader in the implementation of climate neutrality in the wake of the achievement of the Kyoto protocol and the Doha amendment for the reduction of climate-altering gases and the Paris Agreement for the containment of the increase in world average temperature within 1.5 °C.

Forests, as already mentioned, play a strategic role as a carbon dioxide reserve and in this context it is easy to understand the importance of forest monitoring. Forest monitoring, in fact, has a crucial meaning when it has a character of continuity over time and it allows us to know the consistency and evolution of our forests, the country's strategic resource. Furthermore, the forest ecosystem monitoring networks are an essential tool for verifying whether the limit values are respected on the territory of the State and the objectives established in order to prevent, eliminate or reduce the negative effects of global climate change in human health and in the environment.

In this context, the National Inventory of Forests and Forest Carbon Pools (INFC2015) provides a comprehensive in-depth and timely examination of the qualitative and quantitative state of Italian forests. It represents the tool used to quantify the carbon content stored in forests, in compliance with the agreements signed by Italy under the United Nations Framework Convention on Climate Change (UNFCCC).

INFC2015 was created by the Forestry Specialty of the Arma dei Carabinieri, continuing the activities carried out in the past by the Corpo forestale dello Stato with the INFC2005 and with the first National Forest Inventory of 1985 (IFNI). The Projects, Conventions and Environmental Education Office of the Forestry, Environmental and Agri-food Unit Command (CUFA) in charge of its realization, with the scientific contribution of the Council for Agricultural Research and Economics—Research Centre for Forestry and Wood, has implemented the INFC2015 project coordinating the military forces involved in the area, the financial resources allocated and the Regions and Provinces with special status, in order to effectively and efficiently organize the available resources.

In the various operational phases of the project, more than 150 soldiers, duly trained at the structures of the Carabinieri, have lavished their enthusiasm and their preparation in updating—with reference to 2015 year—the knowledge of the Italian forest heritage, highlighting the strong link that the Arma dei Carabinieri, and in particular the sector of Forestry Specialties, has with the rural and forest territory. The active participation in all phases of INFC2015 by the Forestry Corps and Services of the Regions and Autonomous Provinces also represented an important example of positive cooperation between State structures. The support of the Ministry of

Ecological Transition, responsible for the implementation of the Kyoto Protocol in Italy, also contributed to the achievement of the new Inventory.

The results of the inventory survey photograph and certify the progressive and further increase of forests in our country (10,980,000 hectares), quantified by about 520,000 hectares more than the previous inventory of 2005, with the forest occupying over a third of the national territory. The forest has, therefore, maintained the trend of expansion at national level and they continue to expand towards the mountain and hilly territory, made available by agriculture and the naturalization of the landscape.

The results of the latest Inventory also reveal further environmental aspects of great importance, such as the carbon content stored in forest soils, making us even more evident the strategic importance of our forests in contributing to compliance with the international commitments undertaken by Italy, to well-being of the environment and people and, consequently, placing ourselves in front of the responsibility of continuing in the quantitative and qualitative monitoring of forest ecosystems, with perseverance and with increasing professionalism.

The Inventory already represents, but will increasingly be in the future, a sort of "green thermometer" capable of measuring the consistency and the state of vitality of the forests, but above all it will allow to evaluate their contribution to mitigate the "planetary fever".

Rome, Italy Gen. Davide De Laurentis

Presentazione II

Le foreste svolgono un ruolo essenziale nel garantire gli equilibri naturali e ambientali globali e, contemporaneamente, nel contribuire al soddisfacimento dei bisogni del genere umano: affinché le foreste "contino" nelle scelte e nelle strategie politiche ed economiche del Paese, bisogna prima di tutto "contare" le foreste.

Risorsa multifunzionale per eccellenza, i boschi rappresentano gli ecosistemi più ricchi di biodiversità animale e vegetale ed in grado di svolgere servizi insostituibili: forniscono, infatti, un contributo importante nel mitigare gli effetti dei cambiamenti climatici attraverso la sottrazione di anidride carbonica presente in eccesso nell'atmosfera, caratterizzano in modo determinante il ciclo dell'acqua influenzandone evaporazione, assorbimento e regimazione, garantiscono la depurazione dell'aria e l'emissione dell'ossigeno che respiriamo, favoriscono il consolidamento dei suoli e dei versanti, contribuiscono al contrasto dei fenomeni di desertificazione e rappresentano l'habitat vitale per molte altre specie, vegetali ed animali.

Queste funzioni, direttamente connesse alla qualità dell'ambiente, generano una serie ulteriore di esternalità positive, sia in termini di produzione di materie prime rinnovabili come il legno e di prodotti del sottobosco, sia in termini di servizi immateriali fondamentali come la fruizione turistico-ricreativa, la caratterizzazione del paesaggio e la conservazione dei valori tradizionali, culturali e, a volte, anche spirituali dei territori.

La sottrazione dall'atmosfera e l'immagazzinamento dei gas ad effetto serra, in particolare del diossido di carbonio o anidride carbonica, è una delle funzioni più importanti di recente riconosciute alle foreste che, così, contribuiscono a mitigare gli effetti dei cambiamenti climatici e a regolare il clima. Infatti le foreste, come tutto il regno vegetale, rappresentano un ponte insostituibile tra il mondo inorganico e quello degli esseri viventi e una formidabile macchina biologica che cattura carbonio dall'atmosfera, lo immagazzina nelle sue fibre e lo tiene bloccato per tempi anche molto lunghi: un metro cubo di legno secco contiene circa 250 kg di carbonio, pari a circa la metà del suo peso.

In questo contesto l'attività di monitoraggio degli ecosistemi forestali si inserisce coerentemente ed efficacemente nella realizzazione degli obiettivi strategici individuati dall'Unione Europea nell'ambito del "Green Deal", che mira al raggiungimento della neutralità delle emissioni inquinanti entro il 2050. Gli strumenti di pianificazione strategica messi a punto hanno lo scopo di favorire una corretta transizione verso un'economia sostenibile e con impatto ambientale zero in tutti i Paesi membri, grazie ad ingenti risorse economiche da destinare in favore della decarbonizzazione, dell'economia circolare e della mobilità sostenibile. Tali ambiziosi obiettivi sono perseguibili solo attraverso un radicale mutamento degli attuali stili di vita, coniugando l'esigenza di garantire una migliore qualità della vita dei cittadini con una rafforzata ed imprescindibile tutela dell'ambiente attraverso uno sviluppo socio-economico che possa essere definito realmente sostenibile.

La Commissione Europea ha fissato una meta impegnativa di riduzione delle emissioni di gas ad effetto serra, prevedendo di passare dal 40% ad almeno il 55% per il 2030 rispetto ai livelli del 1990 e di azzerarle nel 2050. L'Europa si pone quindi, nei confronti del resto del mondo, coerentemente come capofila nella realizzazione della neutralità climatica sulla scia dell'attuazione del protocollo di Kyoto e dell'emendamento di Doha per la riduzione dei gas clima alteranti e dell'Accordo di Parigi per il contenimento dell'aumento della temperatura media mondiale entro 1.5 °C.

Le foreste, come già accennato, svolgono un ruolo strategico come serbatoio di anidride carbonica e in questo contesto è facile comprendere quale importanza assuma il monitoraggio forestale. Il monitoraggio delle foreste, infatti, ha un significato cruciale quando possiede carattere di continuità nel tempo e permette di conoscere la consistenza e l'evoluzione delle nostre foreste, risorsa strategica del Paese. Le reti di monitoraggio degli ecosistemi forestali, inoltre, sono uno strumento indispensabile per verificare se sul territorio dello Stato siano rispettati i valori limite e raggiunti gli obiettivi stabiliti al fine di prevenire, eliminare o ridurre gli effetti negativi del cambiamento climatico globale sia per la salute umana che per l'ambiente.

In tale contesto l'Inventario Nazionale delle Foreste e dei serbatoi forestali di Carbonio (INFC2015) fornisce un esame globale, approfondito e puntuale dello stato qualitativo e quantitativo delle foreste italiane. Esso rappresenta, in particolare, lo strumento con cui contabilizzare il contenuto di carbonio immagazzinato nelle foreste, in adempimento agli accordi sottoscritti dall'Italia nell'ambito della Convenzione Quadro sui Cambiamenti Climatici delle Nazioni Unite (UNFCCC).

INFC2015 è stato realizzato dall'Arma dei Carabinieri della Specialità Forestale dando continuazione alle attività svolte in passato dal Corpo Forestale dello Stato con l'INFC2005 e con il primo Inventario forestale nazionale del 1985 (IFNI). L'Ufficio Studi e Progetti del Comando Carabinieri per la tutela della biodiversità e dei parchi, incaricato della sua realizzazione, avvalendosi dell'apporto scientifico del Consiglio per la ricerca in agricoltura e l'analisi dell'economia agraria-Centro di ricerca Foreste e Legno, ha realizzato il progetto INFC2015 coordinando i militari coinvolti sul territorio, le risorse finanziarie stanziate e le Regioni e Province a statuto speciale al fine di un'efficace ed efficiente organizzazione delle risorse disponibili.

Nelle diverse fasi operative del progetto più di centocinquanta militari, debitamente formati presso le strutture dell'Arma dei Carabinieri, hanno profuso il loro entusiasmo e la loro preparazione nelle attività di aggiornamento, in riferimento all'anno 2015, della conoscenza del patrimonio forestale italiano, evidenziando il forte legame che l'Arma dei Carabinieri, ed in particolare il comparto della Specialità Forestale, ha con il territorio rurale e forestale. La partecipazione attiva a tutte le fasi dell'INFC2015 da parte dei Corpi e dei servizi forestali delle Regioni e delle Province autonome ha rappresentato, inoltre, un importante esempio di positiva cooperazione tra strutture dello Stato. Anche il supporto del Ministero della Transizione Ecologica, responsabile dell'applicazione del Protocollo di Kyoto in Italia, ha contribuito alla realizzazione del nuovo Inventario.

I risultati dell'indagine inventariale fotografano e certificano il progressivo ed ulteriore aumento delle foreste nel nostro Paese (10,980,000 ettari), quantificato da circa 520,000 ettari in più rispetto al precedente inventario del 2005, con il bosco che va ad occupare oltre un terzo del territorio nazionale. Il bosco ha, quindi, mantenuto l'andamento di espansione a livello nazionale e continua ad appropriarsi del territorio montano e collinare reso disponibile dall'agricoltura e dalla naturalizzazione del paesaggio.

I risultati dell'ultimo Inventario fanno anche emergere ulteriori aspetti ambientali di grande rilievo, quali il contenuto di carbonio stoccato nei suoli forestali, rendendoci ancor più palese l'importanza strategica delle nostre foreste nel contribuire al rispetto degli impegni internazionali assunti dall'Italia, al benessere dell'ambiente e della società e ponendoci, di conseguenza, di fronte alla responsabilità di proseguire, nell'interesse della collettività, nelle attività di monitoraggio quantitativo e qualitativo degli ecosistemi forestali, con continuità e con sempre maggiore professionalità.

L'Inventario rappresenta già, ma sempre più lo sarà in futuro, una sorta di "termometro verde" in grado di misurare la consistenza e lo stato di vitalità delle foreste, ma soprattutto permetterà di valutare il loro contributo per mitigare la "febbre planetaria".

Rome, Italy

Gen. Davide De Laurentis

Preface

This book was published under the 3rd executive plan of the cooperation agreement (Art. 15, Law 241/90) for supporting activities on the designing, the realization and the management of the third Italian National Forest Inventory INFC2015, between the Carabinieri Command of the Forestry, Environmental and Agri-food Units and the Council for Agricultural Research and Economics. Arma dei Carabinieri is entrusted with the realisation of the National Forest Inventory and with the decisions about the aims of the survey and data treatment.

The book was written to provide information on the methods used in the third Italian National Forest Inventory (INFC2015) and to present the main results on the estimated variables. Chapters 1 through 6 describe the general organization, the definitions, the methods and the estimation procedures. Chapters 7 through 12 specify the main estimates produced by INFC2015. Finally, Chap. 13 compares the estimates obtained from INFC2005 and INFC2015, the last two national surveys, to provide an understanding of the current dynamics of Italian forests. The estimates are shown in tables that are given in each chapter. The complete set of inventory estimates generated by INFC2015 is available at www.inventarioforestale.org.

The estimates are presented through texts that introduce the subject matter of the chapter, explain the way the related variables were surveyed and comment on the main outcomes. Chapters 7 through 12 introduce the topics by describing their importance in general, as well as importance of their assessment within the forest inventory. The introductions were targeted to the general public, as more and more people are now interested in the data about environmental resources. The main results are then further explained with the help of graphics.

The book intends to reach a large audience of readers interested in the Italian official statistics on forests, and for this reason, it has been written in two languages. However, the English and Italian versions differ slightly because they were not viewed as a mere translation from one language to another. In fact, much more reference material is available for Italian readers than for English readers. Moreover, some terms do not require explanation for Italian readers, because they are commonly understood through traditional usage but may need clarification for other readers. For these reasons, varying levels of descriptions were provided.

Some conventions have been adopted to simplify the writing. The categories or types that are relevant for the stratification are written in capital letters. For example, 'Forest' refers to the inventory category of Forest, and 'Chestnut forest' indicates one of the forest types adopted in the INFC. However, it is still possible to refer to 'forest' in general, without any specific reference to the inventory domain (e.g. '… forest protection is very important for human well-being'). In addition, the word '...forest' is generally omitted when indicating a forest type, especially in figures and tables (e.g. 'Fir' indicates 'Fir forest' and 'Other deciduous broadleaved' indicates 'Other deciduous broadleaved forest'. Finally, a list of acronyms and a glossary have been added at the end of the book.

Trento, Italy Patrizia Gasparini
Trento, Italy Lucio Di Cosmo
Trento, Italy Antonio Floris
Rome, Italy Davide De Laurentis

Prefazione

Il presente volume è stato prodotto nell'ambito dell'accordo di collaborazione Art. 15 Legge 241/90 per attività di supporto alla progettazione, realizzazione e gestione del terzo Inventario Forestale Nazionale Italiano, terzo piano esecutivo, sottoscritto dal Comando Unità Forestali, Ambientali e Agroalimentari Carabinieri e dal Consiglio per la ricerca in agricoltura e l'analisi dell'economia agraria. L'Arma dei Carabinieri è il soggetto titolare dell'Inventario Forestale Nazionale, a cui competono le decisioni in materia di finalità della rilevazione e trattamento dei dati.

Lo scopo di questo libro è illustrare i metodi utilizzati per il terzo Inventario Forestale Nazionale Italiano (INFC2015) e presentare i risultati ottenuti per le principali variabili rilevate. La prima parte della pubblicazione (Capitoli da 1 a 6) descrive l'organizzazione generale di INFC, le definizioni, i metodi e le procedure di stima. La seconda parte del volume (Capitoli da 7 a 12) riporta le principali statistiche forestali prodotte da INFC2015. Infine, il Capitolo 13 paragona le stime ottenute da INFC2015 con quelle di INFC2005, le ultime due indagini inventariali, per cogliere alcuni aspetti della dinamica in atto per le foreste italiane.

Le stime sono riportate in forma di tabelle, in ciascun capitolo. L'insieme di tutte le stime prodotte da INFC2015 è disponibile nel sito www.inventarioforestale.org.

Le statistiche inventariali sono presentate mediante un testo introduttivo sugli argomenti del Capitolo, spiegazioni sulle modalità di rilievo delle variabili e commenti sui principali risultati ottenuti. Nei Capitoli da 7 a 12, le variabili oggetto di stima sono introdotte da testi che ne descrivono l'importanza generale e l'importanza del loro rilievo in ambito inventariale. Le introduzioni sono state pensate per un pubblico ampio, non necessariamente specialistico, poiché l'interesse per le informazioni sulle risorse ambientali è oggigiorno abbastanza generalizzato. Infine, i risultati principali sono commentati anche con l'ausilio di grafici.

La pubblicazione è stata curata in doppia lingua, l'italiano e l'inglese, con lo scopo di raggiungere una platea ampia di lettori interessati alle statistiche ufficiali sulle foreste in Italia. Ad ogni modo, le due versioni differiscono lievemente perché non sono state pensate come una mera traduzione in una seconda lingua. Infatti, i lettori del testo in italiano hanno accesso ad una quantità di materiale bibliografico piuttosto rilevante, in parte precluso agli altri lettori. Inoltre, alcuni termini familiari, che non necessitano di spiegazioni perché di uso consolidato, non sempre trovano corrispondenza in altri Paesi. Per queste ragioni, in alcuni casi le descrizioni sono state riportate con un dettaglio diverso, nelle due lingue.

Per rendere più agevole la scrittura, sono state adottate alcune convenzioni. Le categorie rilevanti per la stratificazione del campione o delle stime sono state scritte in maiuscolo. Per esempio, con "Bosco" si intende la categoria inventariale del Bosco, e Castagneti indica una delle categorie forestali adottate da INFC. Fuori da questa accezione, rimane possibile utilizzare il termine "bosco" per il suo significato generico, senza riferimento al dominio inventariale (per esempio, "... la protezione del bosco è molto importante per il benessere dell'uomo"). Inoltre, sono stati aggiunti alla fine del volume la lista degli acronimi utilizzati e un glossario con i principali termini tecnici.

Trento, Italy Patrizia Gasparini
Trento, Italy Lucio Di Cosmo
Trento, Italy Antonio Floris
Rome, Italy Davide De Laurentis

Acknowledgments

Realization of the third Italian National Forest Inventory, the topic of this book, has been possible thanks to the involvement of many people who contributed their knowledge and expertise in different areas. Listing all of them without forgetting some names would be impossible. For this reason, we express our heartfelt appreciation to all of those who have contributed in any way in realizing this ambitious project. We are grateful to our colleagues in CREA Research Centre for Forestry and Wood in Trento, who collaborated on the planning, the field crews training and assistance, and the quality check during the data gathering. To all of the employees of the State Forestry Corps and the Carabinieri Command of Forestry, Environmental and Agri-food Units, who coordinated and carried out the surveys in the various phases, we owe them a debt of gratitude. We appreciate the contribution of the people of the autonomous regions and provinces who carried out the surveys in their districts as well as the experts of the companies who projected and developed the software for recording and storing the data. Last but not least, we want to thank our colleagues and the consultants from different disciplines who contributed to projecting the methods and the procedures of the second Italian National Forest Inventory. It was through their efforts that we were able to design the operational activities of the INFC2015.

Ringraziamenti

La realizzazione del terzo Inventario Forestale Nazionale Italiano, a cui è dedicato questo volume, è stata possibile solo grazie al contributo di moltissime persone, con ruoli e competenze diverse. Elencare i nomi di tutti coloro che hanno collaborato, senza tralasciarne qualcuno, non sarebbe possibile. Preferiamo quindi esprimere un ringraziamento collettivo, sincero e sentito, a tutti coloro che hanno contribuito a vario titolo e in momenti diversi alla realizzazione di questo ambizioso progetto. Ai colleghi del CREA Foreste e Legno di Trento, che hanno collaborato alla progettazione, alla formazione dei rilevatori e alle attività di assistenza e controllo durante i rilievi. Al personale del Corpo Forestale dello Stato e del Comando Unità Forestali Ambientali e Agroalimentari dell'Arma dei Carabinieri, per aver coordinato le operazioni ed eseguito la raccolta dati, nelle diverse fasi dell'indagine. Al personale delle Regioni a statuto speciale e delle Province autonome, che ha collaborato ai rilievi, per i territori di propria competenza. Agli esperti delle Società che hanno progettato e realizzato le componenti informatiche per la raccolta e l'archiviazione dei dati. Infine, ma non ultimi, ai colleghi e ai consulenti esterni di varie discipline che hanno contribuito all'elaborazione dei metodi e delle procedure per il secondo Inventario Forestale Nazionale, da cui abbiamo attinto per la progettazione operativa dell'INFC2015.

Contents

Indice

Acronyms/Acronimi

AdS	Sample plot, circular in INFC; the acronym is generally coupled with the length of the circle radius (AdS2, AdS4, AdS13, AdS25)/Area di saggio, circolare in INFC; l'acronimo è generalmente seguito dalla lunghezza del raggio espressa in metri (AdS2, AdS4, AdS13, AdS25).
AGEA	Agricultural Payments Agency/Agenzia per le erogazioni in agricoltura.
CBD	Convention on Biological Diversity.
CFS	Corpo Forestale dello Stato, former National Forest Service.
CRA-MPF	Former name of the CREA—Forestry and Wood Research Centre based in Trento/Unità di Ricerca per il Monitoraggio e la Pianificazione Forestale del Consiglio per la Ricerca e la Sperimentazione in Agricoltura, in Trento.
$d_{1.30}$	Stem diameter at breast height/Diametro del fusto a 1.30 m di altezza da terra.
DBH	Stem diameter at breast height/Diametro del fusto a 1.30 m di altezza da terra.
DOP	Dilution of Precision.
EGNOS	European Geostationary Navigation Overlay Service.
ES	Standard error of the estimate/Errore standard associato al valore stimato.
FRA	Global Forest Resources Assessment.
GIS	Geographic Information System.
GPS	Global Positioning System.
GNSS	Global Navigation Satellite System.
ICP forests	International Co-operative Programme on Assessment and Monitoring of Air Pollution Effects on Forests.

IFNI85	First Italian National Forest Inventory/Primo inventario forestale nazionale italiano.
INFC	Italian national forest inventory as established by the Ministerial Decree of 13 December 2001, named National Inventory of Forests and Forest Carbon Pools/Inventario Nazionale delle Foreste e dei serbatoi forestali di Carbonio istituito con Decreto del Ministro dell'Agricoltura del 13 dicembre 2001.
INFC2005	Second Italian National Forest Inventory/Secondo inventario forestale nazionale italiano.
INFC2015	Third Italian National Forest Inventory/Terzo inventario forestale nazionale italiano.
IPCC	Intergovernmental Panel on Climate Change.
ISAFA	Former name of the CREA—Forestry and Wood Research Centre based in Trento when it was established in 1969 (Experimental Institute for Forest and Range Management)/Istituto Sperimentale per l'Assestamento Forestale e per l'Alpicoltura (poi CRA-MPF e CREA—Foreste e Legno).
MCPFE	Ministerial Conference on the Protection of Forests in Europe.
nIR	Near Infrared.
NIRs	National Inventory Reports on data and methods used for the national Greenhouse Gas Inventory.
OWL/ATB	Other wooded land/Altre terre boscate.
SE	Standard error of the estimate/Errore standard associato al valore stimato.
SFM/GFS	Sustainable Forest Management/Gestione Forestale Sostenibile.
SoEF	State of Europe's Forest report/Rapporto sullo Stato delle Foreste Europee.
SDGs/OSS	Sustainable Development Goals/Obiettivi di Sviluppo sostenibile.
UNFCCC	United Nations Framework Convention on Climate Change.

Chapter 1
The Italian Forest Inventory in Brief

L'inventario forestale nazionale italiano in breve

Patrizia Gasparini and Giancarlo Papitto

Abstract Large-scale forest inventories are important sources of forest information at the national level in individual countries. These surveys have undergone strong development in recent times, driven by new information needs and by advances in statistical-mathematical theory and in survey methods and techniques. In Italy, the first national forest inventory was carried out in the mid-1980s. A thorough review of the sampling design and survey protocols was carried out in the second inventory, and the third survey has just been completed. This chapter briefly describes the history and organisational structure of the Italian National Forest Inventory and summarises its content and products.

Keywords Forest statistics · International reporting · Sustainable forest management · Climate change reporting · Natural capital · Forest policy

1.1 Introduction

The term 'forest inventory' refers to both forest information and to the process of collecting and processing the data on which information is based (Tompoo et al., 2010). Information provided by forest inventories is in the form of data tables, often accompanied by charts and thematic maps. Forest inventories are conducted on a local scale, for a forest estate or a group of forest estates, or on a broader scale for a region or a country. In the latter case, we speak of a National Forest Inventory (NFI), which aims to provide the national forest statistics of a country.

In Europe, the first large-scale forest inventories based on systematic sampling were carried out in the 1920s and 1930s (Loetsch & Haller, 1973). These were conducted in Norway from 1919 to 1930 (Tomter et al., 2010), in Finland from 1921

P. Gasparini (✉)
CREA Research Centre for Forestry and Wood, Trento, Italy
e-mail: patrizia.gasparini@crea.gov.it

G. Papitto
Carabinieri Command of the Forestry, Environmental and Agro-Food Units (CUFA), Rome, Italy

© The Author(s) 2022

P. Gasparini et al. (eds.), *Italian National Forest Inventory—Methods and Results of the Third Survey*, Springer Tracts in Civil Engineering,
https://doi.org/10.1007/978-3-030-98678-0_1

to 1924 (Kangas & Maltamo, 2007), and in Sweden from 1923 to 1929 (Axelsson et al., 2010). After the Second World War, thanks to the evolution of mathematical statistics and of sampling theory of the early twentieth century, and, in parallel, to the introduction of aerial photos as ancillary sources of data, large-scale forest inventories have seen a strong development in many European countries (Democratic Republic of Germany, France, Austria and Spain). Since the 1980s, many other countries have begun taking inventories on a sample basis, and to date, 23 NFIs have been carried out in Europe (Gschwantner et al., 2021).

Modern forest inventories are generally of a combined or multi-source type, i.e., information is collected both directly in the field through plot measurements and using remote sensing (aerial photos, satellite data) and other data sources (GIS layers, previous inventories, etc.). Furthermore, the most recent forest inventories concern a much wider spectrum of variables than the previous ones, which had as their main objective the estimation of wood production and its availability over time. The inventory protocols have been significantly expanded to include observations and measurements on the various components of forest ecosystems, such as deadwood, lower layers of vegetation, microhabitats, naturalistic and landscape emergencies, forest soils, as well as aspects related to the usability of forest resources, such as roads, recreational infrastructures, and many others. For these reasons, large-scale forest inventories have taken on the character of multi-purpose surveys.

Interest in the information produced by NFIs goes beyond national borders and increasingly concerns the international community. In fact, inventory statistics are also used to compile reports and statistics at a supra-national level, aimed at verifying compliance with the commitments undertaken by the various countries in the context of international agreements for the protection of forests, the conservation of biodiversity and the fight against climate change.

The need to update and integrate the official forest statistics available in Italy became pressing at the end of the 1970s, due to the lack of a national inventory and the lack of coherent and reliable information throughout the national territory, which is necessary to promote a modern national forest policy (Tabacchi & Tosi, 2011). The first Italian NFI, conducted in the mid-1980s, temporarily filled this lack of information on a national scale. However, the survey remained without repetition until the early 2000s when, following the emergence of new and more urgent information needs on a national and international scale, the Italian NFI took on a permanent and institutional character, with larger goals, consistent with the modern needs not only of the forestry sector, but of the entire national and international community.

This chapter briefly describes the history and organisational structure of the current Italian NFI, known as the National Inventory of Forests and Forest Carbon Pools (INFC). The text also illustrates the purpose of the survey and lists the main information produced as well as its use in evaluating the management of forest resources and the ecosystem services provided.

1.2 Historical Notes and Legal Aspects

The first Italian NFI (IFNI85) was carried out in the period 1983–87, following the activities of the Interministerial Advisory Commission dedicated to the control and testing of the Forest Map of Italy and the National Forest Inventory. The Commission identified the minimum objectives and financial resources to be used and entrusted the Experimental Institute for Forest and Range Management (ISAFA) of Trento, now merged into the Research Centre for Forestry and Wood of the Council for Agricultural Research and Economics (CREA Research Centre for Forestry and Wood), with the task of drawing up the inventory design, conducting the control of the survey activities, processing the information collected and illustrating the results obtained. The surveys were conducted by the staff of the State Forestry Corps and, for the regions with special statutes and autonomous provinces, surveys were carried out by the staff of the local forestry services. For IFNI85 a systematic sampling scheme was adopted, based on a regular 3 km × 3 km grid of observation points, with a single survey phase on the ground, sufficient to produce estimates with sampling errors acceptable at the national level. More than 33,400 points identified by the national sampling grid were first examined on the cartography and the aerial photos available or through a quick survey, to exclude survey points that were unequivocally not affected by forest cover. The remaining points, which were visited on the ground and surveyed totalled about 9000, one every 900 ha of wood (MAF-ISAFA, 1988).

At the end of the 1990s, ISAFA carried out a preparatory study for the launch of the second Italian NFI on behalf of the Ministry of Agriculture and Forestry (ISAFA, 1998). The same Ministry subsequently issued the Ministerial Decree 13/12/2001, establishing the Italian NFI as a 'permanent instrument of knowledge of the national forest heritage in support of political functions in the forest and environmental sector, including the protection and recovery of biodiversity' (Art. 1). Following the law 353/2000, the decree entrusted the implementation of the NFI to the State Forestry Corps (Art. 2) with the technical-scientific support of ISAFA (Art. 4).

The second Italian NFI was created during the period 2002–2007 and named the National Inventory of Forests and Forest Carbon Pools—INFC2005, in which the statistics refer to the year 2005. Starting from this realisation, in fact, the main purpose of the survey was the estimation of the carbon content in the forest pools identified for the activities related to the United Nations Framework Convention on Climate Change (UNFCCC) (IPCC, 2003). The carbon pools investigated by INFC2005 included aboveground biomass, deadwood, litter and soil (Di Cosmo et al., 2013; Gasparini & Di Cosmo, 2016). With INFC2005 the sampling design and the survey protocols were completely revised and modified with respect to the previous inventory. A new denser 1 km × 1 km national grid was established, with sampling points randomly selected within each grid square, and a three-phase sampling design was adopted to replace the one-phase one from the previous inventory (cf. Chap. 2). In addition, the internationally valid definition of forest, defined for the Global Forest Resources Assessment (FRA) of FAO-UNECE was adopted, and many additional observations and measures were included in order to describe the different components of forest ecosystems and the set of services they provide.

The third Italian NFI INFC2015 was launched in 2013 with the updating of the classification of land use and land cover by photointerpretation (cf. Chap. 3). Following the approval of the legislative decree of 19 August 2016, n. 177, containing 'Provisions on the rationalisation of police functions and absorption of the State Forestry Corps, pursuant to Art. 8, paragraph 1, letter a), of the law of 7 August 2015, n. 124, on the reorganisation of public administrations' the task for the carrying out of the Italian NFI was entrusted to the Forestry Specialty of the Carabinieri. This important measure involved the reorganisation of the operational structure responsible for the coordination and implementation of the NFI and consequently the postponement of the survey campaign. The collection of data in the field for INFC2015 was, therefore, started in November 2017, at the conclusion of the necessary training of the surveyors, and ended in the first months of 2020. Once the data quality control activities and the final testing of the surveys in all the Italian regions were completed, the inventory statistics presented in this volume were processed in 2021.

1.3 Institutional and Organisational Aspects

The Italian NFI is conducted by the Forestry Specialty of the Carabinieri, in continuity with the activities carried out in the past by the State Forestry Corps. The Carabinieri Command of the Forestry, Environmental and Agro-Food Units (CUFA), through its Studies and Projects Office, deals with the coordination of survey activities, the training of surveyors and organisational and logistical aspects. Surveys in the field in the 15 regions with ordinary statutes are carried out by CUFA staff located in the regional and provincial offices of the Command and coordinated by 14 regional officers. In the remaining districts, the four regions with special statutes, Valle d'Aosta, Friuli-Venezia Giulia, Sicilia and Sardegna, and the two autonomous provinces of Trento and Bolzano, the field surveys are conducted by the staff of the local forest services or by external personnel. Approximately 50 photo interpreters located in their respective offices in the area conducted the photointerpretation of the first phase of INFC2015. The surveys in the field were carried out by 35 teams of the Carabinieri, for a total of about 105 surveyors, and by 20 regional and provincial teams in the special statute regions and autonomous provinces, for a total of approximately 45 additional surveyors.

The CREA Research Centre for Forestry and Wood based in Trento was responsible for the design of the survey, data quality controls and the final data processing. The working group dedicated to INFC activities was composed of 8–10 units of research and technical personnel. The centre also took care of the training and assistance of the surveyors and collaborated in the implementation of the database and software tools for data acquisition and transfer. The latter were implemented by two external companies, Telespazio S.p.A. and AlmavivA S.p.A. For the purpose of defining methods and procedures for the INFC, during the planning of INFC2005,

CREA Research Centre for Forestry and Wood made use of the collaboration of many other subjects of universities and service companies who provided important contributions published in the documentation of the project available at www.inventarioforestale.org/documentazione di progetto.

1.4 Aims and Products

The main features of the Italian NFI were established by the Ministerial Decree 13/12/2001. These features designate the inventory as a permanent sample survey, which is periodically updated according to cycles lasting no less than five years. The inventory has the task of producing results consistent with international definitions and with other statistical surveys carried out by the National Statistical Institute (ISTAT). The Italian NFI is responsible for classifying and measuring over 50 variables, both qualitative and quantitative. These variables include those relating to the composition of the vegetation, origin and stage of development, characteristics of forest sites, ownership, forest roads, management methods, availability for wood supply, presence of planning tools and constraints, protected forests, state of health and terrain instability. The quantitative variables concern the size, in terms of diameter and height, and the growth of trees and shrubs that exceed the pre-established diameter threshold, and the count of individuals in the lower layers of vegetation. Deadwood elements are also measured, i.e., trees that have died from pathologies, senescence or broken trees, whole or in parts, standing or lying on the ground, and dead portions of living trees that have fallen to the ground, as well as stumps resulting from cuttings (cf. Chap. 4). The results are produced on a regional scale, for the 21 Italian regions, as well as on a national scale.

The INFC statistics are used in numerous national and international reporting processes. Among the formers, the national report on forests and forestry sector (RaF Italia) of the Ministry of Agricultural, Food and Forestry Policies and of Tourism (MiPAAFT, 2019), the annual inventory of greenhouse gas emissions for UNFCCC and the Kyoto Protocol prepared by the Italian Institute for Environmental Protection and Research (ISPRA, 2021) and the national report on natural capital (Comitato Capitale Naturale, 2021). Among the international reports, INFC contributes to the European report on sustainable forest management (SoEF Europe, https://foresteurope.org/foresteurope/), and to the UNECE-FAO Global FRA (https://www.fao.org/forest-resources-assessment/en/) (Table 1.1).

The products of the Italian NFI are mainly of two types: statistics on a national and regional basis and elementary data. The former are represented by area estimates (of forests, different types of forest, etc.) and by estimates of the total values and per area unit values of the main quantitative variables that describe the characteristics of forests (number of trees, growing stock volume, biomass and carbon content, annual increment in volume and biomass, deadwood volume and biomass, etc.). The elementary data are represented by the values assumed by the qualitative and quantitative variables in the single inventory plot or for the single individual (tree

Table 1.1 Use of Italian NFI information for national and international reporting on the state of forests. In the column on SFM and the one on national reports, the indicators used are listed; in the column on FAO-FRA, tables compiled through NFI data are listed / Impiego delle informazioni prodotte dall'IFN italiano nei rapporti sullo stato delle foreste di livello nazionale e internazionale. Nella colonna relativa al rapporto europeo sulla GFS e in quella relativa ai rapporti nazionali sono elencati gli indicatori valutati; nella colonna relativa al rapporto FAO-FRA sono indicate le tabelle compilate con dati dell'IFN

Forest attribute	SFM – SoEF 2015–20[1]	FAO – FRA 2015–20[2]	National reporting[3,4]
Caratteri delle foreste	GFS – Stato Foreste Europee 2015–20[1]	Rapporto FAO sullo stato delle foreste[2]	Rapporti nazionali[3,4]
Forest and OWL area by inventory category, by pure and mixed conifers and broadleaves, and by availability for wood supply Superficie Bosco e ATB per categorie inventariali e forestali, per grado di mescolanza, per disponibilità al prelievo legnoso	1.1, 1.2, 1.3; 4.3	1a, 1b, 1c, 1e; 6b; 8a, SDG 15	1.1, 1.2[3]
Growing stock volume (total, by forest type, by species, by diameter class) Volume legnoso (totale, per grado di mescolanza, per specie, per classe diametrica)	1.1, 1.2	2a, 2b	1.6[3]
Silvicultural system, development stage, age class; presence of management plan Tipo colturale, stadio di sviluppo, classe di età; presenza di piano di gestione	1.3; 4.2	1b; 3a, 3b; 8a, SDG 15	1.3[3]
Aboveground tree biomass and deadwood biomass; carbon in aboveground tree biomass, deadwood, and litter; soil carbon[5]; carbon in harvested wood	1.4	2c, 2d; 8a, SDG 15	1.7 [3]; Ecosystem service carbon storage[4]

(continued)

Table 1.1 (continued)

Forest attribute	SFM – SoEF 2015–20[1]	FAO – FRA 2015–20[2]	National reporting[3,4]
Caratteri delle foreste	GFS – Stato Foreste Europee 2015–20[1]	Rapporto FAO sullo stato delle foreste[2]	Rapporti nazionali[3,4]
Fitomassa arborea epigea e necromassa; contenuto di carbonio nella fitomassa arborea, nel legno morto e nella lettiera; contenuto di carbonio organico nel suolo[5]; carbonio nelle utilizzazioni			1.7[3]; Servizio Ecosistemico Sequestro e Stoccaggio di Carbonio[4]
Forest area by damage agents	2.4	5a	
Superficie del Bosco per causa di danno			
Annual increment of volume and biomass; annual fellings volume and biomass	3.1		1.6, 1.7[3]; Ecosystem service carbon storage[4]
Incremento annuo di volume e fitomassa; Volume e fitomassa delle utilizzazioni annue			1.6, 1.7[3]; Servizio Ecosistemico Sequestro e Stoccaggio di Carbonio[4]
Number of tree species	4.1		
Numero di specie arboree			
Stand origin	4.2, 4.3	1b, 1c	
Origine			
Area of dominated by introduced tree species forest types	4.4		

(continued)

Table 1.1 (continued)

Forest attribute	SFM – SoEF 2015–20[1]	FAO – FRA 2015–20[2]	National reporting[3,4]
Caratteri delle foreste	GFS – Stato Foreste Europee 2015–20[1]	Rapporto FAO sullo stato delle foreste[2]	Rapporti nazionali[3,4]
Superficie categorie forestali di specie non autoctone			
Volume of standing and lying deadwood	4.5		
Volume legno morto in piedi e a terra			
Protected Forest and Other wooded land area	4.9	3a, 3b; 8a, SDG 15	
Superficie Bosco e Altre terre boscate in aree protette			
Wooded area under hydrogeological constraint	5.1	3a	1.5[3]
Superficie forestale con vincolo idrogeologico			
Character of forest ownership	6.1	4a	1.4[3]
Carattere della proprietà forestale			
Forest area by primary designated management objective (protective, recreational)	5.1; 6.10	3a	
Bosco con funzione prioritaria (protettiva, ricreativa)			

[1]https://foresteurope.org/publications/#1471590853638-cbc85f9c-8e6e
[2]https://www.fao.org/forest-resources-assessment/en/
[3]National report on Forests and Forestry sector / Rapporto sullo stato delle foreste e del settore forestale in Italia (MiPAAFT, 2019)
[4]National report on natural capital / Rapporto sul capitale naturale (Comitato Capitale Naturale, 2021)
[5]Data provided by INFC2005 / Dati prodotti da INFC2005 (Gasparini et al., 2013)

or shrub) or element (stump, deadwood fragment) measured. Both types of data are fully available with open access at www.inventarioforestale.org/.

Appendix (Italian Version)

Riassunto Gli inventari forestali su ampia scala sono tra le principali fonti di statistiche forestali disponibili a livello nazionale nei singoli Paesi. Tali indagini hanno visto un forte sviluppo in tempi relativamente recenti, sulla spinta di nuove esigenze informative e dei progressi della teoria statistico-matematica e dei metodi nonché delle tecniche di rilevamento. In Italia il primo inventario forestale nazionale è stato realizzato a metà degli anni Ottanta. Dopo una profonda revisione del disegno di campionamento e dei protocolli di rilievo, messi a punto in occasione del secondo inventario, si è da poco conclusa la realizzazione della terza indagine. Il capitolo descrive brevemente la storia e la struttura organizzativa dell'inventario forestale nazionale italiano e ne illustra in sintesi i contenuti e i prodotti.

Introduzione

Il termine inventario forestale indica l'insieme delle informazioni relative alle caratteristiche delle foreste e alle operazioni necessarie per la raccolta e l'elaborazione dei dati su cui si basano (Tomppo et al., 2010). Le informazioni prodotte sono solitamente in forma tabellare, spesso accompagnate da rappresentazioni grafiche e mappe tematiche. Gli inventari forestali vengono realizzati a scala locale, per una singola proprietà forestale o per un insieme di proprietà, o a scala più ampia, di regione o nazione. In quest'ultimo caso si tratta di inventari forestali nazionali (IFN), finalizzati a produrre le statistiche forestali di un Paese.

In Europa, i primi inventari su ampie superfici basati su campionamenti sistematici sono stati realizzati negli anni Venti e Trenta del secolo scorso (Loetsch & Haller, 1973): in Norvegia nel 1919–30 (Tomter et al., 2010), Finlandia nel 1921–24 (Kangas & Maltamo, 2007), Svezia nel 1923–29 (Axelsson et al., 2010). Grazie all'evoluzione dei metodi statistico-matematici e della teoria del campionamento degli inizi del XX secolo e, parallelamente, all'introduzione delle foto aeree quale fonte ausiliaria di dati, a partire dal periodo successivo alla Seconda guerra mondiale gli inventari forestali su ampia scala hanno visto un forte sviluppo in molti Paesi europei (Repubblica Democratica di Germania, Francia, Austria, Spagna). A partire dagli anni Ottanta e Novanta, molti altri Paesi hanno avviato un proprio inventario su base campionaria, fino ad arrivare a 23 IFN attualmente realizzati in Europa (Gschwantner et al., 2021).

Gli inventari forestali moderni sono generalmente di tipo combinato o multirisorsa, ossia le informazioni vengono raccolte sia direttamente in campo, attraverso misure in aree di saggio, sia mediante l'uso di dati telerilevati (foto aeree, immagini satellitari, ecc.) o altre fonti informative (mappe tematiche o precedenti inventari). Gli inventari forestali più recenti, inoltre, riguardano uno spettro di variabili molto più ampio rispetto ai precedenti, che avevano quale obiettivo principale la stima della produzione legnosa e della sua disponibilità nel tempo. I protocolli di rilievo inventariale sono stati ampliati notevolmente, per includere osservazioni e misure sulle diverse componenti degli ecosistemi forestali come il legno morto, gli strati inferiori di vegetazione, i microhabitat, le emergenze naturalistiche e paesaggistiche, i suoli forestali, nonché su aspetti legati alla fruibilità delle risorse forestali come la viabilità, le infrastrutture ricreative e molti altri. Per queste ragioni, gli inventari forestali su ampia scala hanno assunto il carattere di indagini multi-obiettivo.

L'interesse per le informazioni prodotte dagli IFN supera i confini nazionali e riguarda sempre più spesso la comunità internazionale. Le statistiche di fonte inventariale, infatti, vengono utilizzate anche per la compilazione di report e statistiche di livello sovra-nazionale, finalizzati alla verifica del rispetto degli impegni assunti dai diversi Paesi nell'ambito degli accordi internazionali per la tutela delle foreste, la conservazione della biodiversità e la lotta al cambiamento climatico.

L'esigenza di aggiornare e integrare le statistiche forestali ufficiali disponibili in Italia è divenuta pressante alla fine degli anni Settanta, a causa della carenza di informazioni coerenti e affidabili su tutto il territorio nazionale, necessarie a promuovere una moderna politica forestale nazionale (Tabacchi & Tosi, 2011). Il primo IFN italiano, realizzato a metà degli anni Ottanta, ha temporaneamente colmato tale carenza di informazioni a scala nazionale. L'indagine è rimasta tuttavia priva di una ripetizione fino agli inizi degli anni Duemila quando, a seguito dell'emergere di nuove e più urgenti esigenze informative su scala nazionale e internazionale, l'IFN italiano ha assunto carattere permanente e istituzionale, con obiettivi più ampi, coerenti con le moderne esigenze non solo del settore forestale, bensì dell'intera comunità nazionale e internazionale.

Il capitolo descrive brevemente la storia e la struttura organizzativa dell'attuale IFN italiano, denominato Inventario Nazionale delle Foreste e dei serbatoi forestali di Carbonio (INFC). Il testo illustra inoltre le finalità dell'indagine ed elenca le principali informazioni prodotte e il loro impiego ai fini della valutazione della gestione delle risorse forestali e dei servizi ecosistemici da esse forniti.

Cenni storici e aspetti giuridici

Il primo IFN italiano (IFNI85) venne realizzato nel periodo 1983–87, a seguito delle attività della Commissione consultiva interministeriale dedicata al controllo e collaudo della Carta forestale d'Italia e dell'Inventario forestale nazionale. La Commissione individuò gli obiettivi minimi e le risorse finanziarie da impiegare e affidò all'allora Istituto Sperimentale per l'Assestamento Forestale e per l'Alpicoltura (ISAFA) di Trento, ora confluito nel Centro di ricerca Foreste e Legno del Consiglio per la ricerca in agricoltura e l'analisi dell'economia agraria (CREA Centro di ricerca Foreste e Legno), il compito di redigere la progettazione, condurre il controllo delle attività di rilievo, elaborare le informazioni raccolte e illustrare i risultati ottenuti. I rilievi furono eseguiti dal personale del Corpo Forestale dello Stato e, per le Regioni a statuto speciale e le Province autonome, dal personale dei Servizi forestali locali. Per IFNI85 fu adottato uno schema di campionamento sistematico, basato su una griglia regolare di punti di osservazione con maglie quadrangolari di 3 km di lato, con un'unica fase di rilievo al suolo, sufficiente a produrre statistiche con errori campionari accettabili a livello nazionale. Gli oltre 33,400 punti individuati dal reticolo nazionale furono dapprima esaminati sulla cartografia e le foto aree disponibili, o attraverso una ricognizione speditiva, per escludere dal rilievo i punti inequivocabilmente non interessati da copertura forestale. I punti visitati al suolo e rilevati furono circa 9000, uno ogni 900 ha circa di bosco (MAF-ISAFA, 1988).

Alla fine degli anni Novanta l'ISAFA realizzò uno studio preparatorio per l'avvio del secondo IFN italiano, su incarico del Ministero delle Politiche Agricole e Forestali (ISAFA, 1998). Lo stesso Ministero emanò successivamente il Decreto Ministeriale 13/12/2001 che istituì l'IFN italiano quale "strumento permanente di conoscenza del patrimonio forestale nazionale a supporto delle funzioni di indirizzo politico nel settore forestale e ambientale, ivi compresa la tutela e il recupero della biodiversità" (Art. 1). Il decreto, tenendo conto di quanto stabilito dalla legge 353/2000, affidò la realizzazione dell'IFN al Corpo Forestale dello Stato (Art. 2), con il supporto tecnico-scientifico dell'ISAFA (Art. 4).

Il secondo IFN italiano fu realizzato nel periodo 2002–2007 con la denominazione di Inventario Nazionale delle Foreste e dei serbatoi forestali di Carbonio—INFC2005, poiché le statistiche vengono riferite all'anno 2005. A partire da questa realizzazione, infatti, lo scopo principale dell'indagine è la stima del contenuto di carbonio nei serbatoi forestali individuati per le attività relative alla Convenzione Quadro delle Nazioni Unite sui Cambiamenti Climatici (UNFCCC) (IPCC, 2003). I serbatoi di carbonio indagati da INFC2005 sono la biomassa epigea, il legno morto, la lettiera e il suolo (Di Cosmo et al., 2013; Gasparini & Di Cosmo, 2016). Con INFC2005 il disegno di campionamento e i protocolli di rilievo furono integralmente rivisti e modificati rispetto al precedente inventario. Fu istituito un nuovo reticolo nazionale, più denso, con passo di 1 km e con un punto di campionamento posizionato casualmente all'interno di ciascuna maglia del reticolo, e fu adottato un disegno di campionamento a tre fasi (cfr. Chap. 2). Inoltre, fu adottata una definizione di foresta valida a livello internazionale, in uso per il Global Forest Resources Assessment (FRA) redatto da FAO-UNECE, e vennero incluse molte osservazioni e misure aggiuntive, allo scopo di descrivere le diverse componenti degli ecosistemi forestali e l'insieme dei servizi da essi forniti.

Il terzo IFN italiano INFC2015 fu avviato nel 2013 con l'aggiornamento della classificazione dell'uso e copertura del suolo per fotointerpretazione (cfr. Chap. 3). A seguito dell'approvazione del decreto legislativo 19 agosto 2016 n. 177, recante "Disposizioni in materia di razionalizzazione delle funzioni di polizia e assorbimento del Corpo Forestale dello Stato, ai sensi dell'articolo 8, comma 1, lettera a), della legge 7 agosto 2015, n. 124, in materia di riorganizzazione delle amministrazioni pubbliche", la competenza della realizzazione dell'IFN italiano fu assegnata alla specialità forestale dell'Arma dei Carabinieri. Questo importante provvedimento comportò la riorganizzazione della struttura operativa responsabile del coordinamento e della realizzazione dell'IFN e conseguentemente la posticipazione della campagna di rilievo al suolo. La raccolta dei dati in campo per l'INFC2015 fu quindi avviata nel novembre 2017, dopo la necessaria formazione dei rilevatori, e si concluse nei primi mesi del 2020. Completate le attività di controllo di qualità dei dati ed i collaudi dei rilievi in tutte le regioni italiane, nel corso del 2021 sono stati elaborati i dati per la produzione delle statistiche presentate in questo volume.

Aspetti istituzionali e organizzativi

L'IFN italiano viene realizzato dalla specialità forestale dell'Arma dei Carabinieri, in continuità con le attività svolte in passato dal Corpo Forestale dello Stato. Il Comando Unità Forestali, Ambientali e Agro-Alimentari (CUFA), si occupa del coordinamento delle attività di rilievo, della formazione dei rilevatori e degli aspetti organizzativi e logistici. I rilievi in campo nelle 15 regioni a statuto ordinario vengono realizzati dal personale del CUFA dislocato negli uffici regionali e provinciali del Comando e coordinato da 14 Referenti regionali dell'Arma. Nei rimanenti distretti, le quattro regioni a statuto speciale Valle d'Aosta, Friuli-Venezia Giulia, Sicilia e Sardegna, e le due province autonome di Trento e Bolzano, i rilievi in campo vengono condotti dal personale dei locali Servizi forestali o da personale esterno incaricato allo scopo. Oltre 50 fotointerpreti dislocati nei rispettivi uffici sul territorio hanno realizzato la fotointerpretazione della prima fase INFC2015. I rilievi in campo sono stati realizzati da 35 squadre dell'Arma dei Carabinieri, per un totale di circa 105 rilevatori, e da 20 squadre regionali e provinciali delle Regioni a statuto speciale e delle Province autonome, per un totale di altri 45 rilevatori circa.

Il CREA Centro di ricerca Foreste e Legno con sede a Trento è responsabile della progettazione dell'indagine, dei controlli di qualità dei dati e delle elaborazioni finali. Il gruppo di lavoro dedicato alle attività INFC è composto da 8–10 unità di personale ricercatore e tecnico. Il Centro si è occupato inoltre della formazione e assistenza dei rilevatori e ha collaborato all'implementazione del database e degli strumenti software per l'acquisizione e trasferimento dei dati. Questi ultimi sono stati realizzati da due società esterne, Telespazio S.p.A. e AlmavivA S.p.A. Ai fini della definizione di metodi e procedure per l'INFC, durante la progettazione dell'INFC2005, il CREA Centro di ricerca Foreste e Legno si è avvalso della collaborazione di molti altri soggetti di Università e Società di servizi che hanno fornito importanti contributi pubblicati nella documentazione di progetto disponibile all'indirizzo alla pagina www.inventarioforestale.org/documentazione di progetto.

Finalità e prodotti

I caratteri principali dell'IFN italiano sono stabiliti dal D.M. 13/12/2001, che definisce l'inventario come un'indagine campionaria di tipo permanente, aggiornata periodicamente secondo cicli di durata non inferiore a 5 anni. L'inventario ha il compito di produrre risultati coerenti con le definizioni internazionali e con le altre rilevazioni statistiche realizzate dall'Istituto Nazionale di Statistica (ISTAT). L'IFN italiano si occupa di classificare e misurare oltre 50 variabili, sia qualitative sia quantitative. Tra le prime, quelle relative a composizione della vegetazione, origine e stadio di sviluppo, caratteristiche delle stazioni forestali, proprietà, viabilità forestale, modalità di gestione e disponibilità per il prelievo legnoso, presenza di strumenti di pianificazione e di vincoli, eventuale grado di protezione, stato di salute e presenza di fenomeni di dissesto. Le variabili quantitative riguardano la dimensione, in termini di diametro e altezza, e l'accrescimento dei soggetti di specie arboree e arbustive che superino la soglia di misura prefissata e il conteggio degli individui degli strati inferiori di vegetazione; vengono misurati, inoltre, gli alberi morti per patologie, senescenza o schianti, in piedi o atterrati, le parti di alberi cadute a terra e le ceppaie derivanti da tagli o schianti (cfr. Chap. 4). I risultati vengono prodotti a scala regionale, per le 20 regioni italiane, oltre che a scala nazionale.

Le statistiche INFC vengono utilizzate in numerosi processi di reporting nazionale e internazionale. Tra i primi, il rapporto sullo stato delle foreste e del settore forestale (RaF Italia) del Ministero delle Politiche Agricole, Alimentari e Forestali e del Turismo (MiPAAFT, 2019), l'inventario annuale delle emissioni di gas serra per l'UNFCCC e il Protocollo di Kyoto, redatto dall'Istituto Superiore per la Protezione e la Ricerca Ambientale (ISPRA, 2021), e il rapporto sul capitale naturale (Comitato Capitale Naturale, 2021). Tra i report internazionali, l'INFC contribuisce al report europeo sulla gestione sostenibile delle foreste (SoEF Europe, https://foresteurope.org/foresteurope/) e al Global FRA di UNECE-FAO (https://www.fao.org/forest-resources-assessment/en/) (Table 1.1).

I prodotti dell'IFN italiano sono principalmente di due tipi: statistiche su base nazionale e regionale e dati elementari. Le prime sono rappresentate da stime di superficie (delle foreste, dei diversi tipi di foresta, ecc.) e da stime dei valori totali e per unità di superficie delle principali variabili quantitative che descrivono le caratteristiche delle formazioni forestali (numero di alberi, volume legnoso, biomassa e contenuto di carbonio, incremento annuo di volume e di biomassa, volume e necromassa del legno morto, ecc.). I dati elementari sono rappresentati dai valori assunti dalle variabili qualitative e quantitative nel singolo punto inventariale o per il singolo individuo (albero o arbusto) o elemento (ceppaia, frammento di legno morto) misurato. Entrambi i tipi di dati sono disponibili ad accesso aperto all'indirizzo www.inventarioforestale.org/.

References

Axelsson, A., Ståhl, G., Söderberg, U., Petersson, H., Fridman, J., & Lundström, A. (2010). National forest inventory reports—Sweden. In: E. Tomppo, T. Gschwantner, M. Lawrence, & R. E. McRoberts (Eds.), *National forest inventories—Pathways for common reporting* (pp. 541–553). Springer. ISBN 978-90-481-3232-4. https://doi.org/10.1007/978-90-481-3233-1

Comitato Capitale Naturale (2021). Quarto Rapporto sullo Stato del Capitale Naturale in Italia. Roma. Retrieved Jan 05, 2022, from https://www.mite.gov.it/pagina/il-rapporto-sullo-stato-del-capitale-naturale-italia

Di Cosmo, L., Gasparini, P., Paletto, A., & Nocetti, M. (2013). Deadwood basic density values for national-level carbon stock estimates in Italy. *Forest Ecology and Management, 295*, 51–58. https://doi.org/10.1016/j.foreco.2013.01.010

Gasparini, P., & Di Cosmo, L. (2016). National forest inventory reports—Italy. In: C. Vidal, I. Alberdi, L. Hernández, J., & Redmond (Eds.), *National forest inventories—Assessment of wood availability and use* (pp. 485–506). Springer. ISBN 978-3-319-44014-9. https://doi.org/10.1007/978-3-319-44015-6

Gasparini, P., Di Cosmo, L., & Pompei, E. (Eds.) (2013). Il contenuto di carbonio delle foreste italiane. Inventario Nazionale delle Foreste e dei serbatoi forestali di Carbonio INFC2005. Metodi e risultati dell'indagine integrativa. Consiglio per la Ricerca e la Sperimentazione in Agricoltura, Roma (267 pp.). ISBN 978-88-97081-36-4.

Gschwantner, T., Alberdi, I., Bauwens, S., Bender, S., Borota, D., Bosela, M., Bouriaud, O., Breidenbach, J., Donis, J., Fischer, C., Gasparini, P., Heffernan, L., Hervé, J., Kolozs, L., Korhonenn, K. T., Koutsias, N., Kovácsevics, P., Kucera, M., Kulbokas, G., … Tomter, S. M. (2021). Growing stock monitoring by European National Forest Inventories: Historical origins, current methods, and advances in harmonisation. *Forest Ecology and Management, 505*, 119868. https://doi.org/10.1016/j.foreco.2021.119868

IPCC (2003). Good practice guidance for land use, land-use change and forestry. IPCC National Greenhouse Gas Inventories Programme. Institute for Global Environmental Strategies, Hayama, Japan. ISBN 4-88788-003-0. Retrieved Jan 4, 2022, from https://www.ipcc-nggip.iges.or.jp/public/gpglulucf/gpglulucf_files/GPG_LULUCF_FULL.pdf

ISAFA (1998). Secondo Inventario Forestale Nazionale. Studio di fattibilità. Ministero per le Politiche Agricole e Forestali, Direzione Generale delle Risorse Forestali, Montane e Idriche. Istituto Sperimentale per l'Assestamento Forestale e per l'Alpicoltura – ISAFA. Trento (210 pp.). Retrieved Jan 05, 2022, from https://www.inventarioforestale.org/sites/default/files/datiinventario/Fattib.pdf

ISPRA (2021). Italian Greenhouse Gas Inventory 1990–2019. National Inventory Report 2021. Rapporti 341/2021. Roma (610 pp.). Retrieved Feb 03, 2022, from https://www.isprambiente.gov.it/files2021/pubblicazioni/rapporti/nir2021_italy_14apr_completo.pdf

Kangas, A., & Maltamo, M. (Eds.). (2007). *Forest inventory. Methodology and applications. Managing forest ecosystems* (Vol. 10, p. 364). Springer. ISBN 978-1-4020-4379-6

Loetsch, F., & Haller, K. E. (1973). *Forest inventory* (2nd ed., Vol. I, 436 pp.). BLV Verlagsgesellscaft.

MAF-ISAFA (1988). Inventario forestale nazionale—IFN1985. Sintesi metodologica e risultati. Ministero dell'Agricoltura e delle Foreste, Corpo Forestale dello Stato (461 pp.). Istituto Sperimentale per l'Assestamento Forestale e per l'Alpicoltura. Trento.

MiPAAFT (2019). RaF Italia 2017–18—Rapporto sullo stato delle Foreste e del settore forestale in Italia. Prodotto dalla Rete Rurale Nazionale (RRN 2014–20). Compagnia delle Foreste. Arezzo (279 pp.). ISBN: 978-88-98850-34-1. Retrieved Dec 9, 2021, from https://www.reterurale.it/flex/cm/pages/ServeBLOB.php/L/IT/IDPagina/19231.

Tabacchi, G., & Tosi, V. (2011). Breve storia degli inventari forestali in Italia. In: P. Gasparini & G. Tabacchi (Eds.), *L'Inventario Nazionale delle Foreste e dei serbatoi forestali di Carbonio INFC 2005. Secondo inventario forestale nazionale italiano. Metodi e risultati* (pp. 3–8). Edagricole-Il Sole 24 Ore, Milano. ISBN 978-88-506-5394-2.

Tompoo, E., Schadauer, K., McRoberts, R. E., Gschwantner, T., Gabler, K., & Ståhl, G. (2010). Introduction. In: E. Tomppo, T. Gschwantner, M. Lawrence, & R. E. McRoberts (Eds.), *National forest inventories—Pathways for common reporting* (pp. 1–18). Springer. ISBN 978-90-481-3232-4. https://doi.org/10.1007/978-90-481-3233-1

Tomter, S. M., Hylen, G., & Nilsen, J. (2010). Norway. In: E. Tomppo, T. Gschwantner, M. Lawrence, & R. E. McRoberts (Eds.), *National forest inventories: Pathways for common reporting* (pp. 411–424). Springer. ISBN 978-90-481-3232-4. https://doi.org/10.1007/978-90-481-3233-1

Open Access This chapter is licensed under the terms of the Creative Commons Attribution 4.0 International License (http://creativecommons.org/licenses/by/4.0/), which permits use, sharing, adaptation, distribution and reproduction in any medium or format, as long as you give appropriate credit to the original author(s) and the source, provide a link to the Creative Commons license and indicate if changes were made.

The images or other third party material in this chapter are included in the chapter's Creative Commons license, unless indicated otherwise in a credit line to the material. If material is not included in the chapter's Creative Commons license and your intended use is not permitted by statutory regulation or exceeds the permitted use, you will need to obtain permission directly from the copyright holder.

Chapter 2
Definitions and Sampling Design

Definizioni e disegno di campionamento

Patrizia Gasparini and Antonio Floris

Abstract The population studied by the INFC is made up of the Forest and Other wooded land as defined by the FAO Global Forest Resources Assessment (FRA). INFC estimates are produced for the 21 Italian administrative units (the regions), the macro-categories and inventory categories and the forest types that represent the divisions of the inventory domain. This chapter illustrates the classification system adopted for the INFC and provides descriptions of the different classes. The sampling design adopted and the construction of the national 1 km × 1 km grid, on which the identification of the sampling points is based, are also illustrated. Finally, the system of reference units or plots used to operate the classifications and carry out the measurements required by the survey protocol is presented.

Keywords Forest definition · Forest type · Stratified sampling · Inventory grid · Sample units

2.1 Introduction

The object of study in a statistical survey consists of the set of units in which one or more common characteristics are to be studied, and it is indicated by the term population. The population can be divided into study domains, i.e., its different components for which it is required to produce separate estimates (OECD, 2007). The population and its possible divisions must be identified clearly and unambiguously from the start of the survey design. The term inventory domain refers to the main object of the INFC inventory survey, which is represented by the Forest and Other wooded land as defined by the FAO Global Forest Resources Assessment (FRA) report. The divisions of the inventory domain are represented by 21 administrative units into which the Italian territory is divided. These units consist of 15 regions with ordinary statutes, 4 regions with special statutes, and the two autonomous provinces, called regions for brevity. Further divisions for which separate estimates are produced

P. Gasparini (✉) · A. Floris
CREA Research Centre for Forestry and Wood, Trento, Italy
e-mail: patrizia.gasparini@crea.gov.it

© The Author(s) 2022

P. Gasparini et al. (eds.), *Italian National Forest Inventory—Methods and Results of the Third Survey*, Springer Tracts in Civil Engineering,
https://doi.org/10.1007/978-3-030-98678-0_2

are those relating to the inventory macro-categories and inventory categories and the forest types.

Once the population of interest and its divisions have been identified, the design of a statistical survey requires the definition of a sampling plan or sampling design, i.e., the procedure to select the units to be observed: the structure of the sample, the methods for selecting the population units, their probability of inclusion in the sample itself and the fraction or sampling rate, on which the sample size depends (Fabbris, 1989). The characteristics of the study population, called parameters, are described starting with the data collected in the sample through estimation techniques, i.e., the calculation of appropriate estimators, such as the sample mean and its variance. The set consisting of the sampling design and the estimators is called the sampling strategy (Cicchitelli et al., 1997).

This chapter describes the criteria adopted to define the inventory domain of the INFC and its divisions. The sampling design and the national grid underlying the localisation of observations and measurements on the territory are then described. These correspond with the points or plots of different shapes or sizes named in the complex sampling units, which are listed and described in the concluding part of the chapter. The estimators and the procedures used to calculate the inventory estimates are the subject of Chaps. 5 and 6.

2.2 Inventory Domain and Classification System

Starting with the second Italian forest inventory INFC2005, the areas characterised by a forest cover responding to the definitions of Forest and Other wooded land elaborated on the FAO Global FRA2000 survey (FAO, 1998, 2000) were adopted as inventory domain. These definitions were then further detailed and applied in all subsequent FRAs (FAO, 2010, 2012, 2018; FAO-ITTO, 2003), becoming a reference point for the harmonisation of forest statistics on an international scale. They are based on some objective characteristics, some of which can also be evaluated from aerial photos or other remote images, given sufficient resolution, and others that require observation on the ground for checking or for accurate evaluation. Among the former, the minimum size is determined in terms of extension and width of the areas with tree-shrub cover. Among the latter, the coverage of tree species, distinct from that of shrub species, the height of the mature subjects and the land use are the determinants.

In the INFC, the two classes of Forest and Other wooded land, which together identify the inventory domain of the INFC, are called inventory macro-categories. These are divided into more detailed classes called inventory categories, some of which are distinguished based on the height of the tree species and identified to facilitate the comparison between the estimates produced by the first Italian forest inventory IFNI85 and those of the second one INFC2005. In IFNI85, in fact, the threshold for the height of the trees at maturity is 2 m. Table 2.1 shows the definitions for the inventory macro-categories and inventory categories used in the third

Table 2.1 Inventory macro-categories and inventory categories that identify the divisions of the inventory domain in INFC2015 / Macrocategorie e categorie inventariali che individuano le ripartizioni del dominio inventariale INFC2015

Inventory macro-category Macrocategoria inventariale	Description Descrizione	Inventory category Categoria inventariale	Description Descrizione
Forest	Land spanning more than 0.5 hectares with trees higher than 5 meters and a canopy cover of more than 10%, or trees able to reach these thresholds in situ. It does not include land that is predominantly under agricultural or urban land use. It does include young trees that have not yet reached but are expected to reach a canopy cover of 10% and tree height of 5 meters; temporarily unstocked areas due to clear-cutting or natural disasters; forest roads, firebreaks and small open areas; corridors of trees with an area of more than 0.5 hectares and width of more than 20 meters; cork-oak plantations	Tall trees forest	Land classified as Forest, with a cover of more than 10% of tree species with a potential height in situ of at least 5 meters at the time of the survey; it includes temporarily unstocked areas; plantations of artificial origin specialized for timber and wood production are excluded
		Plantations	Forest stands of artificial origin specialized for timber and wood production, with a cover of more than 10% of tree species with a potential height in situ of at least 5 meters
Bosco	Area di superficie maggiore di 0.5 ettari, caratterizzata da una copertura superiore a 10% di alberi in grado di superare un'altezza di 5 metri a maturità, oppure in grado di superare tali soglie in situ. Sono escluse le aree con uso prevalente di carattere agricolo o urbano. Sono inclusi: soprassuoli giovani, in grado di raggiungere le soglie di copertura e altezza; le aree temporaneamente prive di soprassuolo per effetto di tagli a raso o disastri naturali; strade forestali, viali tagliafuoco e piccole aree prive di copertura; fasce di alberi con larghezza maggiore di 20 metri; impianti di querce da sughero	Boschi alti	Area classificata come "Bosco", caratterizzata da una copertura superiore a 10% di specie con altezza potenziale in situ di almeno 5 metri al momento del rilievo; sono incluse le aree temporaneamente prive di soprassuolo; sono esclusi i soprassuoli di origine artificiale specializzati per la produzione di legna e legname
		Impianti di arboricoltura da legno	Impianti di specie forestali specializzati per la produzione legnosa, con copertura maggiore del 10% e con altezza potenziale in situ di almeno 5 m

(continued)

Table 2.1 (continued)

Inventory macro-category Macrocategoria inventariale	Description Descrizione	Inventory category Categoria inventariale	Description Descrizione
Other wooded land	Land not classified as Forest, spanning more than 0.5 hectares, with trees higher than 5 meters and a canopy cover of 5-10%, or trees able to reach these thresholds in situ; or with a combined cover of shrubs, bushes and trees above 10%. It does not include land that is predominantly under agricultural or urban land use. It includes areas with trees that will not reach a height of 5 meters in situ and with a canopy cover of 10% or more, e.g. some alpine tree vegetation types, or vegetation of arid zones	Short trees forest	Formations with cover of more than 10% of tree species with potential height in situ between 2 and 5 m
		Scrubland	Formations with cover of more than 10% of trees with potential height in situ of less than 2 meters
		Sparse forest	Formations with cover between 5% and 10% of tree species with potential height in situ higher than 5 meters
		Shrubs	Formations with cover of more than 10% of shrub species and tree crown cover less than 5%
		Not accessible or not classified wooded area	Land spanning more than 0.5 hectares and width of more than 20 meters, with a total cover of tree and shrub species greater than 10% assessed by photointerpretation, characterized by the absence of a more detailed classification from the field surveys

(continued)

Table 2.1 (continued)

Inventory macro-category Macrocategoria inventariale	Description Descrizione	Inventory category Categoria inventariale	Description Descrizione
Altre terre boscate	Area di superficie maggiore di 0.5 ettari non classificata come "Bosco", caratterizzata da una copertura pari a 5-10% di alberi con altezza superiore a 5 metri, o in grado di superare tali soglie in situ, oppure con una copertura complessiva di arbusti, cespugli e alberi superiore a 10%. Sono escluse le aree con uso prevalente di carattere agricolo o urbano. Include le aree con copertura superiore a 10% di alberi non in grado di raggiungere i 5 metri a maturità, come alcune formazioni alpine o di zone aride	Boschi bassi	Formazioni con copertura maggiore del 10% di specie arboree con altezza potenziale in situ compresa fra 2 e 5 metri
		Boscaglie	Formazioni con copertura maggiore del 10% di specie arboree con altezza potenziale in situ minore di 2 metri
		Boschi radi	Formazioni con copertura compresa fra il 5% e il 10% di specie arboree con altezza potenziale in situ superiore a 5 metri
		Arbusteti	Formazioni con copertura maggiore del 10% di specie arbustive e copertura delle chiome arboree inferiore al 5%
		Aree boscate inaccessibili o non classificate	Aree di estensione maggiore di 0.5 ettari e larghezza maggiore di 20 metri, con una copertura complessiva di specie arboree e arbustive superiore a 10% rilevata attraverso la sola fotointerpretazione, in cui manca una classificazione di dettaglio da rilievi in campo

Italian forest inventory INFC2015. In the INFC2015 results, the inventory category of Temporarily unstocked areas, originally distinguished in the classification of vegetation by the surveyors, has been merged with that of Tall trees forest. In addition, the list of categories in Table 2.1 includes the residual category of Not accessible or not classified wooded areas, which refers to all situations not otherwise classifiable included by convention in the processing of data, in the Other wooded land macro-category. These are formations characterised by a cover of trees and shrubs greater than 10% verified by photointerpretation, but for which a more detailed classification of the vegetation is lacking, mainly due to their inaccessibility by the surveyors.

The INFC classification system includes two further levels in addition to those described above, the forest types and subtypes, respectively, which identify the inventory domains in greater detail. The formations of the inventory categories Tall trees forest, Short trees forest, Sparse forest and Scrubland are classified according to 17 forest types, Plantations according to 3 other forest types and Shrubs according to 3 further forest types (Table 2.2). The forest types are therefore divided into subtypes, for a total number of 68 for the woods, 7 for the Plantations and 16 for the Shrubs, which are distinguished based on the dominant species or according to ecological criteria. For example, the subtypes Subalpine Norway spruce forests and Mountain Norway spruce forests are distinguished on the basis of an ecological criterion, while those of the forest type of Mediterranean pines (*Pinus pinaster* forests, *Pinus pinea* forests and *Pinus halepensis* forests) differ in species.

The forest type is assigned according to the dominant species or group of species in terms of crown coverage, the latter assessed in a neighbourhood of the inventory point with an area of approximately 2000 m^2 (AdS25, cf. Chap. 4). The prevailing group of species is first determined as conifers, deciduous broadleaved or evergreen broadleaved, and then the forest type based on the prevailing species of the group, again in terms of crown coverage. In INFC2015, in the absence of a classification in the field of the forest type, the inventory point was assigned to the not classified class. The classification system with the related forest types and subtypes (Pignatti, 2003) was defined during the planning of the second Italian forest inventory and also remained unchanged in the third.

2.3 Sampling Design

The design of the Italian NFI includes three sampling phases, with samples extracted according to a stratified sampling (Fattorini et al., 2006; Gasparini & Di Cosmo, 2016; Gasparini & Tabacchi, 2011). The first phase (or phase 1) consists of the preliminary classification of land use and land cover through the photointerpretation of orthophotos (cf. Chap. 3) at over 301,000 points, one for each mesh of the 1 km $\times$ 1 km grid in which the national territory has been divided (cf. Sect. 2.4). This sampling scheme is called tessellated sampling (Särndal et al., 1992). The points are positioned randomly within the meshes and therefore their distribution is of a non-aligned systematic type (Gallego, 1995). The purpose of the first phase is to

Table 2.2 Forest types and subtypes and corresponding codes CORINE Biotopes and EUNIS / Categorie e sottocategorie forestali e corrispondenti codici secondo le classificazioni CORINE Biotopes ed EUNIS

Forest type Categoria forestale	CORINE Biotopes	EUNIS	Forest subtypes Sottocategorie forestali
Larch and Swiss stone pine forest Boschi di larice e cembro	42.3	G3.2	(1) Mixed Larch and Swiss stone pine forests; (2) Dense Larch forests; (3) Sparse larch trees in Subalpine heath; (4) Other formations of larch and Swiss stone pine
			(1) Boschi di larice e cembro; (2) Lariceto in fustaia chiusa; (3) Larici isolati nella brughiera subalpina; (4) Altre formazioni di larice e cembro
Norway spruce forest Boschi di abete rosso	42.2	G3.1	(1) Subalpine Norway spruce forests; (2) Mountain Norway spruce forests; (3) Other formations with Norway spruce as the prevailing species
			(1) Pecceta subalpina; (2) Pecceta montana; (3) Altre formazioni con prevalenza di abete rosso
Fir forest Boschi di abete bianco	42.1	G3.1	(1) Fir forests or mixed fir-beech forests with *Vaccinium* and *Majanthemum*; (2) Fir forests with *Cardamine*; (3) Fir forests with *Campanula*; (4) Other formations with Fir as the prevailing species
			(1) Abetina e abeti-faggeta a mirtillo e *Majanthemum*; (2) Abetina a *Cardamine*; (3) Abetina a *Campanula*; (4) Altre formazioni di abete bianco

(continued)

Table 2.2 (continued)

Forest type Categoria forestale	CORINE Biotopes	EUNIS	Forest subtypes Sottocategorie forestali
Scots pine and Mountain pine forest Pinete di pino silvestre e montano	42.5, 42.4	G3.3, G3.4	(1) Scots pine forests with *Erica*; (2) Scots pine forests with *Carex* or with *Astragalus;* (3) Scots pine forests with *Quercus robur* and *Molina*; (4) Scots pine forests with *Quercus pubescens* and *Cytisus sessilifolius*; (5) *Pinus uncinata* forests; (6) Other formations with *Pinus sylvestris* and *Pinus uncinata*
			(1) Pineta di pino silvestre a *Erica;* (2) Pineta di pino silvestre a carici oppure astragali; (3) Pineta di pino silvestre a farnia o e molina; (4) Pineta di pino silvestre a roverella e citiso a foglie sessili; (5) Pineta di pino montano; (6) Altre formazioni a pino silvestre e montano
Black pines forest Pinete di pino nero, laricio e loricato	42.6, 42.7	G3.5, G3.6	(1) Black pine forests with *Erica* and *Fraxinus ornus*; (2) Black pine forests with *Chamaecytisus* and *Genista*; (3) *Pinus laricio* forests; (4) *Pinus leucodermis* forests; (5) Other formations with *Pinus nigra* and *Pinus laricio*
			(1) Pineta di pino nero a erica e orniello; (2) Pineta di pino nero a citiso e ginestra; (3) Pineta di pino laricio (*Pinus laricio*); (4) Pineta di pino loricato (*Pinus leucodermis*); (5) Altre formazioni di pino nero e laricio

(continued)

Table 2.2 (continued)

Forest type Categoria forestale	CORINE Biotopes	EUNIS	Forest subtypes Sottocategorie forestali
Mediterranean pines forest Pinete di pini mediterranei	42.8	G3.7	(1) *Pinus pinaster* forests; (2) *Pinus pinea* forests; (3) *Pinus halepensis* forests
			(1) Pinete di *Pinus pinaster;* (2) Pinete di *Pinus pinea;* (3) Pinete di *Pinus halepensis*
Other coniferous forest Altri boschi di conifere, pure o miste	42.A	G3.9	(1) *Cupressus* spp. forests; (2) Other coniferous forests
			(1) Formazioni a cipresso; (2) Altre formazioni di conifere
Beech forest Faggete	41.1	G1.6	(1) Mesophile beech forests; (29 Acidophilous beech-forests with *Luzula*; (3) Termophilus beech-forests with *Cephalanthera*; (4) Beech forests with *Ilex,* ferns, and *Campanula*; (5) Other formations with beech as the prevailing species
			(1) Faggete mesofile; (2) Faggete acidofile a *Luzula*; (3) Faggete termofile a *Cephalanthera*; (4) Faggete ad agrifoglio, felci e *Campanula*; (5) Altre formazioni di faggio
Temperate oaks forest Querceti a rovere, roverella e farnia	41.2, 41.7	G1.7, G1.8	(1) *Quercus petraea* forests; (2) *Quercus pubescens* forests; (3) *Quercus robur* forests; (4) Other forests of *Q. petraea*, *Q. pubescens* or *Q. robur*

(continued)

Table 2.2 (continued)

Forest type Categoria forestale	CORINE Biotopes	EUNIS	Forest subtypes Sottocategorie forestali
			(1) Boschi di rovere; (2) Boschi di roverella; (3) Boschi di farnia; (4) Altre formazioni di rovere, roverella o farnia
Mediterranean oaks forest Cerrete, boschi di farnetto, fragno, vallonea	41.7	G1.7	(1) *Quercus cerris* forests in the plain; (2) *Q. cerris* forests on hills or in mountain; (3) *Quercus frainetto* forests; (4) *Quercus trojana* or *Quercus macrolepis* forests; (5) Other formations with *Q. cerris*, *Q. frainetto*, *Q. trojana*, *Q. macrolepis*
			(1) Cerrete di pianura; (2) Cerrete collinari e montane; (3) Boschi di farnetto; (4) Boschi di fragno e nuclei di vallonea; (5) Altre formazioni di cerro, farnetto, fragno, vallonea
Chestnut forest Castagneti	41.9, 41.5	G1.7	(1) *Castanea sativa* forests for wood production; (2) *Castanea sativa* forests for chestnut production
			(1) Castagneti da legno; (2) Castagneti da frutto, selve castanili
Hornbeam and Hophornbeam forest Ostrieti, carpineti	41.8, 41.2	G1.7	(1) *Ostrya carpinifolia* and *Fraxinus ornus* forests; (2) *Carpinus orientalis* forests; (3) Pure *Carpinus betulus* forests
			(1) Boschi di carpino nero e orniello; (2) Boscaglia di carpino orientale; (3) Boschi di carpino bianco

(continued)

Table 2.2 (continued)

Forest type Categoria forestale	CORINE Biotopes	EUNIS	Forest subtypes Sottocategorie forestali
Hygrophilous forest Boschi igrofili	44.1, 44.2, 44.4, 44.6, 44.7	G1.1, G1.3	(1) *Fraxynus oxycarpa* and *Ulmus forests*; (2) *Alnus incana* forests; (3) *Alnus glutinosa* forests; (4) *Populus* forests; (5) Riparian *Salix* forests; (6) *Platanus* forests; (7) Other formations in humid environments
			(1) Boschi a frassino ossifillo e olmo; (2) Boschi a ontano bianco; (3) Boschi a ontano nero; (4) Pioppeti naturali; (5) Saliceti riparali; (6) Plataneti; (7) Altre formazioni forestali in ambienti umidi
Other deciduous broadleaved forest Altri boschi caducifogli	41.4, 41.8, 41.B, 41.C, 41.D, 41.H, 83.3	G1.9, G1.A, G1.B	(1) Mountain forests with *Acer pseudoplatanus*, *Fraxinus excelsior*, and other species; (2) Apennine *Acer* forests; (3) *Alnus cordata* forests; (4) *Cercis* forests; (5) *Betula* forests and forests of pioneer tree species; (6) *Robinia* and *Ailanthus* forests; (7) Other deciduous formations
			(1) Acero-tilieti di monte e boschi di frassino e altre specie; (2) Acereti appenninici; (3) Boschi di ontano napoletano; (4) Boscaglie di *Cercis*; (5) Betuleti, boschi montani pionieri; (6) Robinieti e ailanteti; (7) Altre formazioni caducifoglie

(continued)

Table 2.2 (continued)

Forest type Categoria forestale	CORINE Biotopes	EUNIS	Forest subtypes Sottocategorie forestali
Holm oak forest Leccete	45.3	G2.1	(1) Coastal thermophilus *Holm oak* forests; (2) *Holm oak* and *Fraxinus ornus* mixed forests; (3) Saxicolous holm oak forests, (4) *Holm oak* scrubs
			(1) Lecceta termofila costiera; (2) Bosco misto di leccio e orniello; (3) Lecceta rupicola; (4) Boscaglia di leccio
Cork oak forest Sugherete	45.2	G2.1	(1) Mediterranean cork oak forests; (2) Wood-pasture with cork oak
			(1) Sugherete mediterranee; (2) Pascolo arborato a sughera
Other evergreen broadleaved forest Altri boschi di latifoglie sempreverdi	42.4, 45.1, 45.4, 45.5, 45.8	G2.1, G2.4, G2.6	(1) Mediterranean thermophilus scrubs; (2) Evergreen forests of humid environments
			(1) Boscaglie termo mediterranee; (2) Boschi sempreverdi di ambienti umidi
Poplar plantations Pioppeti artificiali	83.3	G1.C, G2.8, G3.F	(1) Poplar plantations
			(1) Pioppeti artificiali
Other broadleaved plantations Piantagioni di altre latifoglie			(1) Broadleaves plantations; (2) *Eucalyptus* plantations
			(1) Piantagioni di latifoglie; (2) Piantagioni di eucalipti

(continued)

Table 2.2 (continued)

Forest type Categoria forestale	CORINE Biotopes	EUNIS	Forest subtypes Sottocategorie forestali
Piantagioni di conifere Coniferous plantations			(1) Indigenous conifers plantations; (2) *Pseudotsuga menziesii* plantations; (3) *Pinus radiata* plantations; (4) Other conifer species plantations
			(1) Piantagioni di conifere indigene; (2) Piantagioni di *Pseudotsuga menziesii*; (3) Piantagioni di *Pinus radiata*; (4) Altre piantagioni di conifere esotiche
Subalpine shrubs Arbusteti subalpini	31.4, 31.5, 31.6	F2.2, F2.3, F2.4	(1) Dwarf mountain pine forests; (2) Other subalpine needle-leaves shrubs; (3) subalpine heath, (4) *Alnus viridis* formations; (5) Alpine *Salix* formations
			(1) Mughete; (2) Altri arbusteti subalpini di aghifoglie; (3) Brughiera subalpina; (4) Formazione ad ontano verde; (5) Saliceti alpini
Temperate climate shrubs Arbusteti di clima temperato	31.8	F3.1, F3.2, F5.4	(1) *Prunus* and *Corylus* formations; (2) Other deciduous shrubs; (3) *Spartium junceum* formations; (4) *Genista aetnensis* formations; (5) Other formations with *Cytisus* or *Genista*; (6) *Juniperus* shrubs
			(1) Pruneti e corileti; (2) Altri arbusteti di specie decidue; (3) Arbusteti a ginestra (*Spartium junceum*); (4) Arbusteti a ginestra dell'Etna (*Genista aetnensis*); (5) Altre formazioni di ginestre; (6) Arbusteti a ginepro

(continued)

Table 2.2 (continued)

Forest type Categoria forestale	CORINE Biotopes	EUNIS	Forest subtypes Sottocategorie forestali
Mediterranean scrubs and shrubs Macchia, arbusteti mediterranei	32.1, 32.2, 32.3	F5.1, F5.2, F5.5	(1) Coastal *Juniperus* formations; (2) *Pistacia* Mediterranean scrub; (3) Coastal Mediterranean scrub; (4) *Cytisus* Mediterranean scrub; (5) Other evergreen shrubs
			(1) Formazione a ginepri sul litorale; (2) Macchia a lentisco; (3) Macchia litorale; (4) Cisteti; (5) Altri arbusteti sempreverdi

identify the sampling points of the strata of interest for the subsequent survey phase, represented by the classes of forest land use and cover (Forest formations, Sparse forest formations, Temporarily unstocked areas and Plantations; cf. Chap. 3) and from the class of points not classifiable from orthophotos. Furthermore, for INFC2015, additional strata of interest have been identified, as described below.

The second sampling phase (or phase 2) involves a subsample of the first phase points, over 30,000, selected according to a sampling stratified by region and class of land use and land cover. The points of the second phase sample are visited on the ground, to verify the preliminary classification by photointerpretation, confirming it or not, and to assign the inventory category and the forest type. During the second phase surveys, the qualitative characteristics of forests are also evaluated and classified (cf. Chap. 4), in order to produce estimates of the distribution of the wooded area according to the different characteristics (composition by species groups, degree of coverage, silvicultural system, stage of development, management methods, presence of constraints and protected areas, characteristics of forest stations, road conditions, etc.). These will be presented in Chaps. 7, 8, 9, 10 and 11 of this volume.

The third phase strata are identified by the forest type assigned in the second phase together with the land use and cover class, and the region. For each Forest stratum, a subsample of second phase points is extracted to carry out measurements envisaged for the third sampling phase (phase 3) relating to the quantitative characteristics of forests (cf. Chap. 4). The measurements performed on approximately 7000 points produce the estimates of the totals and densities, or values per unit area, of the quantitative variables presented in Chaps. 7, 9 and 12 of this volume. These include growing stock, biomass, annual volume increment, deadwood biomass, etc., which are important for assessing the state of Italian forests and their role as a carbon pool and biodiversity reservoir.

The sampling design described above, outlined for the second Italian NFI, was also applied for INFC2015 with some adaptations. In fact, with the operational planning, the indications from the beginning of the operational structure responsible for the coordination and implementation of the INFC were incorporated. They constituted organizing the surveys in a single campaign, including as many points already detected in the previous survey as possible in order to reduce the time that the survey was on the ground and to facilitate its organization, while maintaining an unchanged quantity, quality, and level of detail of the produced estimates.

The sample for the new ground survey was then constituted by including: (i) the points of the third phase INFC2005 not affected by significant changes in land use and cover, (ii) the points of the second phase sample INFC2005 affected by significant changes and (iii) an additional sample, stratified by region, for strata consisting of new points in wooded areas and plantations and new non-classifiable points. The significant changes were highlighted by comparing two photointerpretations of INFC2005 and INFC2015 and are related to the points transited from strata of inventory interest to the stratum "other land uses" or vice versa. The outcome of the comparison is shown in Fig.2.1 for the set of 301,271 points on the national territory according to the opinion of the photo interpreters, out of a total of 301,328 points identified by the national grid (cf. Sect. 2.4).

		INFC2005		
		Forest land use and cover Uso e copertura forestale	Other land use and cover Altro uso e copertura del suolo	Total Totale
INFC2015	Forest land use and cover Uso e copertura forestale	101,902	11,561	113,463
	Other land use and cover Altro uso e copertura del suolo	10,446	177,362	187,808
	Total Totale	112,348	188,923	301,271

Fig. 2.1 Comparison between the results of the photointerpretation of land cover and use in the last two Italian national forest inventories: number of photoplots / Confronto tra i risultati della fotointerpretazione dell'uso e copertura del suolo dei due inventari forestali italiani più recenti: numero di fotopunti

The actual change from forest use and cover to other land uses was verified on a subsample of 2338 second phase points. Of these, almost half belonged to other land uses in the second phase of INFC2005, and therefore it was not necessary to carry out a further verification in the field. The remaining 1303 points were instead detected on the ground during the INFC2015 campaign or classified through a photointerpretation that included more recent images, if they were inaccessible to the surveyors. Points in the second phase sample not affected by significant changes in land use and cover represent the greatest part. For them the third phase stratum attributed during the INFC2005 surveys was considered valid and the new surveys concerned 6597 points of the third phase subsample. Finally, the extracted subsample from the new points transited in strata of inventory interest consisted of 874 points. The sampling rate for these strata is lower than that adopted for the other ones, due to the limits mentioned above. However, it was considered suitable for evaluating the actual change in land use and cover and ensured a balanced subsample compared to the overall sample of points selected for the field survey.

Overall, the INFC2015 second phase sample consisted of 30,877 points of which 8774 were classified for the qualitative characteristics during the new survey campaign or, if inaccessible, evaluated remotely or through recent orthophotos, in order to verify the land use and cover and the forest type. Of these, 6993 points were measured to estimate the quantitative characteristics and represent the third phase sample of INFC2015. The composition of the sample selected for the INFC2015 field campaign is represented in Fig. 2.2; the largest group is that of the sample units already recorded both in phase 2 and in phase 3 INFC2005, which represents more than three-quarters of the total.

The number of sample units in the three phases and the resulting sampling rates by region and at a national level are shown in Table 2.3 together with the area represented

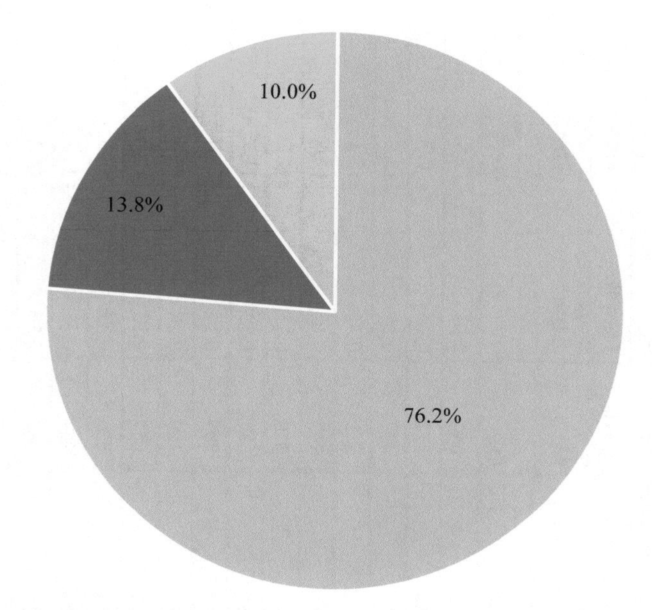

Fig. 2.2 Distribution of the sample selected for the INFC2015 field campaign in relation to the previous field surveys of INFC2005 / Distribuzione del campione selezionato per la campagna di rilievo INFC2015 in relazione ai precedenti rilievi in campo INFC2005

by each sample point or plot. At the national level, the sampling rate of phase 2 and phase 3 is equal to 0.26 and 0.30, respectively, and the area represented by each sample unit is 3.9 km^2 for phase 2 units and 13.0 km^2 for phase 3. In the second phase, the sampling rate at regional level varies little, from 0.23 for Lombardia to 0.28 for Trentino, and the area represented by each sample unit varies from 3.7 km^2 in Toscana to 4.6 km^2 in Lombardia. In the third phase, the regional sampling rate varies from 0.23 for Toscana to 0.44 for Valle d'Aosta and Molise, while the area represented by a sample unit varies from 8.7 km^2 in Valle d'Aosta to 15.8 km^2 in Toscana.

2.4 National Grid

The grid used to subdivide the national territory into portions of equal area and identify the points of the INFC sample was built during the design of the second Italian inventory INFC2005. It contains quadrangular meshes, 1 km × 1 km, geometrically

Table 2.3 Number of INFC2015 sample units selected and surveyed and their representativity, by sampling phase and region / Numero di unità campionarie INFC2015 estratte e rilevate e relativa rappresentatività, per fase di campionamento e regione

Region/Regione	Phase 1 photoplots	Of which in forest land use&cover	Phase 2 plots	Phase 2 sampling rate	Wooded area per phase 2 plot	Phase 3 plots	Phase 3 sampling rate	Forest area per phase 3 plot
	Fotopunti di fase 1	Di cui in uso&copertura forestale	Aree di saggio di fase 2	Tasso di campionamento di fase 2	Superficie forestale per area di saggio di fase 2	Aree di saggio di fase 3	Tasso di campionamento di fase 3	Superficie del Bosco per area di saggio di fase 3
	(n)	(n)	(n)		(km^2)	(n)		(km^2)
Piemonte	25,418	10,603	2642	0.25	4.2	644	0.30	13.8
Valle d'Aosta	3251	1138	306	0.27	3.8	114	0.44	8.7
Lombardia	23,877	7405	1698	0.23	4.6	401	0.29	15.5
Alto Adige	7384	3759	1030	0.27	3.8	275	0.31	12.3
Trentino	6220	4190	1184	0.28	3.6	297	0.29	12.6
Veneto	18,380	4863	1287	0.26	3.9	328	0.31	12.7
Friuli V.G	7837	3825	1012	0.26	3.9	284	0.33	11.7
Liguria	5379	4216	1135	0.27	3.8	295	0.32	11.6
Emilia Romagna	22,470	6919	1874	0.27	3.8	430	0.28	13.6
Toscana	22,982	12,665	3428	0.27	3.7	657	0.23	15.8
Umbria	8442	4298	1122	0.26	3.9	325	0.32	12.0
Marche	9387	3328	865	0.26	3.9	264	0.35	11.1
Lazio	17,201	7143	1831	0.26	3.9	369	0.25	15.2
Abruzzo	10,811	5038	1335	0.26	3.8	337	0.31	12.2
Molise	4429	1830	435	0.24	4.2	160	0.44	9.6
Campania	13,612	5561	1437	0.26	4.0	299	0.28	13.5

(continued)

Table 2.3 (continued)

Region/Regione	Phase 1 photoplots	Of which in forest land use&cover	Phase 2 plots	Phase 2 sampling rate	Wooded area per phase 2 plot	Phase 3 plots	Phase 3 sampling rate	Forest area per phase 3 plot
	Fotopunti di fase 1	Di cui in uso&copertura forestale	Aree di saggio di fase 2	Tasso di campionamento di fase 2	Superficie forestale per area di saggio di fase 2	Aree di saggio di fase 3	Tasso di campionamento di fase 3	Superficie del Bosco per area di saggio di fase 3
	(n)	(n)	(n)		(km^2)	(n)		(km^2)
Puglia	19,330	2114	553	0.26	4.1	158	0.44	9.0
Basilicata	9989	4243	1077	0.25	4.0	255	0.35	11.3
Calabria	15,062	7329	1871	0.26	3.9	317	0.25	15.6
Sicilia	25,705	4384	1125	0.26	4.3	260	0.39	11.0
Sardegna	24,105	13,718	3630	0.26	3.9	524	0.33	11.9
Italia	301,271	118,569	30,877	0.26	3.9	6993	0.30	13.0

coupled to meridians and parallels. Such meshes, although square in plane projection only near the meridian of origin, comply everywhere with the requirement of identical area (1 km^2) of each mesh (Fig. 2.3).

The WGS84 datum (DMA, 1991) was adopted for the generation of the grid nodes, due to its perfect congruence with the GNSS systems that would have been used later for finding and positioning the points in the field. After establishing a pair of geographical coordinates, φ_0, λ_0, at the South-West corner of the quadrangle which contains the whole national territory, further nodes of the grid have been generated moving 1000 m eastwards along the same parallel. Once the South-East node of the quadrangle is reached, the procedure starts again from the initial node, this time moving 1000 m North along the meridian and repeating the previous step. The procedure was repeated until the quadrangle containing the national territory was completed, ending in the last North-East node (detailed algorithms available in Floris & Scrinzi, 2011).

Intersecting the grid with the border of the national territory, the actual inventory grid was identified and consisted of 306,831 quadrilateral meshes, with side 1000 m on the ellipsoid surface. By converting the geographical coordinates of the grid nodes

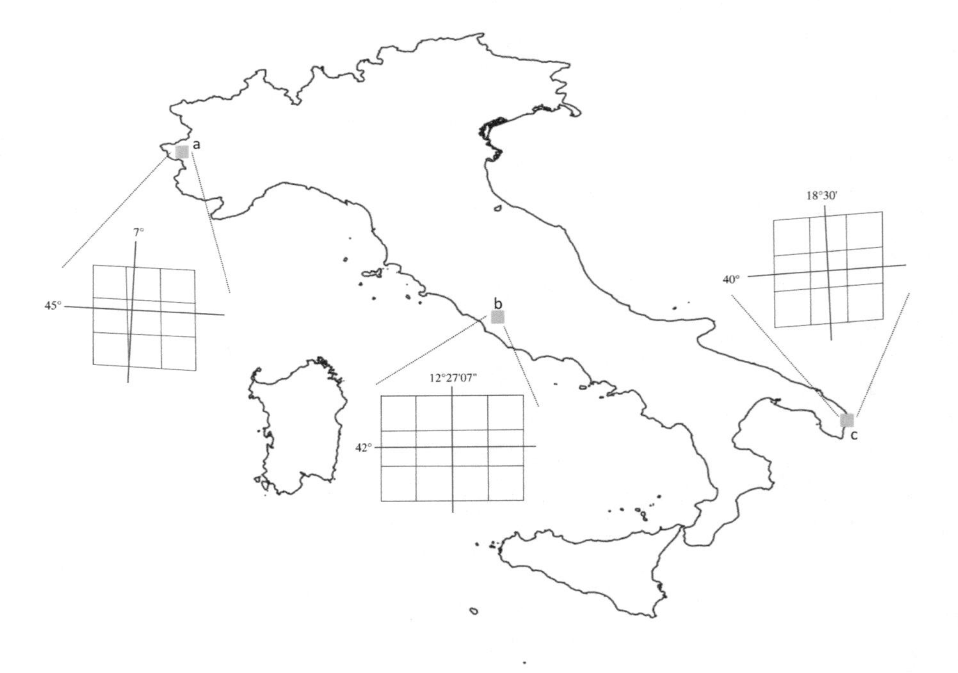

Fig. 2.3 Pieces of an ellipsoidal grid in three different zones of Italy: **a** North-Western zone, **b** along the meridian of origin, **c** South-Eastern zone. The different shape of the meshes can be observed, each having the same area / Porzioni di grigliato ellissoidico in tre zone diverse del territorio nazionale: **a** zona nord-occidentale, **b** centrale lungo il meridiano d'origine, **c** sud-orientale. Si osserva la diversa forma delle maglie, tutte aventi medesima superficie

into plane coordinates in the UTM-WGS84 reference system and recalculating the surface of each mesh with the Gauss formula for a closed polygon, deviations of less than one square metre were consistently obtained with respect to the nominal value of 1 km^2.

To randomly select a sample point for each grid mesh, a pair of angular positive values $\delta\varphi$, $\delta\lambda$ has been added to the coordinates of the South-West node of each mesh in which a portion of the national territory is included (Fig. 2.4). The pair of angular values were randomly generated within the limits of 1000 m from the node itself. The sample points have again been intersected with the official national administrative borders, obtaining the final sample of 301,328 points.

The plane coordinates of the sample points have also been transformed into the reference systems UTM-ED50 and Gauss-Boaga-Roma40 (aka Italy-Monte Mario) (Cima et al., 2003) in such a way as to make the points identifiable on any kind of

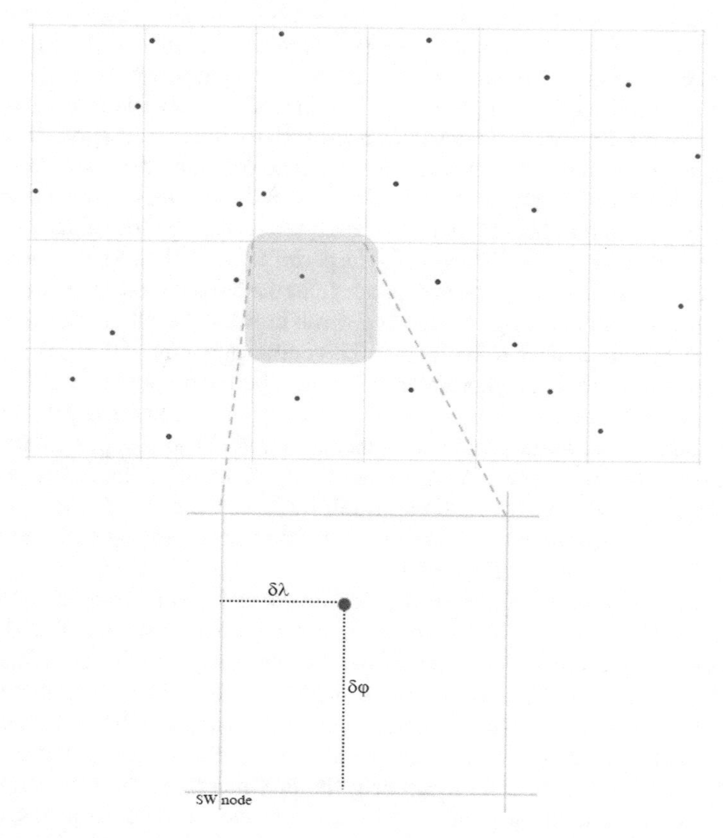

Fig. 2.4 A piece of inventory grid in plane projection, with a mesh detail, a sample point (red dot) and $\delta\varphi$, $\delta\lambda$ from the SW node / Porzione di grigliato in proiezione piana, con dettaglio di una maglia, il relativo punto di campionamento (punto rosso) e $\delta\varphi$, $\delta\lambda$ dal nodo Sud-Ovest

map or mapping tool used during the inventory surveys. As described in Chap. 3, the reference system Gauss-Boaga-Roma40 was adopted in the WebGIS platform Geoinfo and used for photointerpretation (cf. Chap. 3).

2.5 Reference Units for the Survey

In this text, the term sampling unit indicates the different reference units (i.e., points and plots) used to collect information on the numerous attributes detected by the INFC in different ways and at different phases of the survey. These comply with the expected accuracy goals and are in relation to the available financial and time resources.

In the first phase for the photointerpretation of land use and land cover, the sampling unit is the sample point with an analysis window consisting of nine contiguous squares, of 50 m side, arranged according to a 3 × 3 scheme, for a total area of 22,500 m^2 (Gasparini et al., 2014). The central square of the analysis window is centred on the sample point. The analysis window is functional to the assessment of the dimensional thresholds necessary for the correct application of the definition of Forest and Other wooded land adopted (cf. Sect. 2.2 and Chap. 3).

In the ground survey, besides the check of the classification of land use and land cover, many qualitative characteristics are assessed, including those related to a more detailed classification of the vegetation (Gasparini et al., 2016). The administrative and regulatory attributes, height and distance from the closest road are measured with reference to point C. Coverage of tree and shrub crowns is attributed by observing, both on the ground and in orthophotos, the central quadrant (50 × 50 m) of the analysis window used in the photointerpretation, called photoplot 2500 (FP2500). A circular area centred on point C, inscribed in the same central quadrant and with a radius of 25 m, known as the ground reference area (AdS25, area approximately 2000 m^2) is the sampling unit for evaluating accessibility, descriptive characteristics of the vegetation, site characteristics, and the health status of the stand. In AdS25, further observations on qualitative characteristics related to management and silvicultural practices are also conducted (Fig. 2.5).

The third phase of INFC is mainly addressed to collect the quantitative data necessary for estimating dendrometric parameters (cf. Chap. 4). In INFC2015, the third phase measurements were carried out simultaneously with the classification of the qualitative characteristics in all points visited (cf. Sect. 2.3). Measurements are performed on circular sample plots, useful in maximising the area/perimeter ratio and reducing the number of edge elements (de Vries, 1986). Moreover, they can be easily instituted in the field by measuring the distance from the plot centre. Two concentric plots centred on point C, with a radius of 4 m and 13 m respectively (AdS4, area 50.27 m^2 and AdS13, area 530.93 m^2), are used to select the standing trees of different size to be measured. AdS4 is used for measuring trees with DBH between 4.5 and 9.4 cm, and AdS13 is used for trees with DBH equal or larger than 9.5 cm. Tree heights, incremental cores, deadwood lying on the ground and stumps

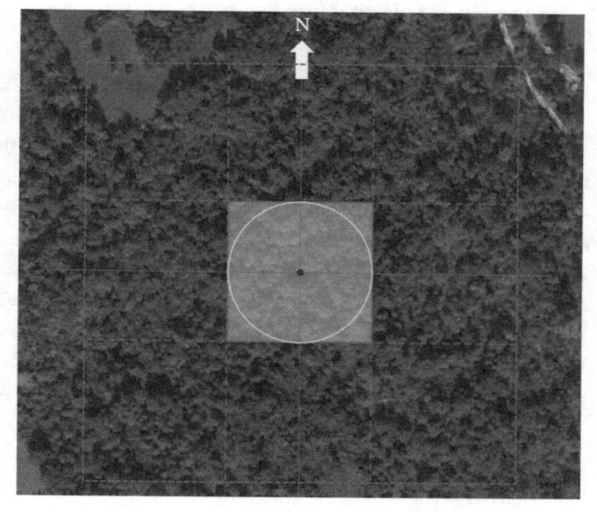

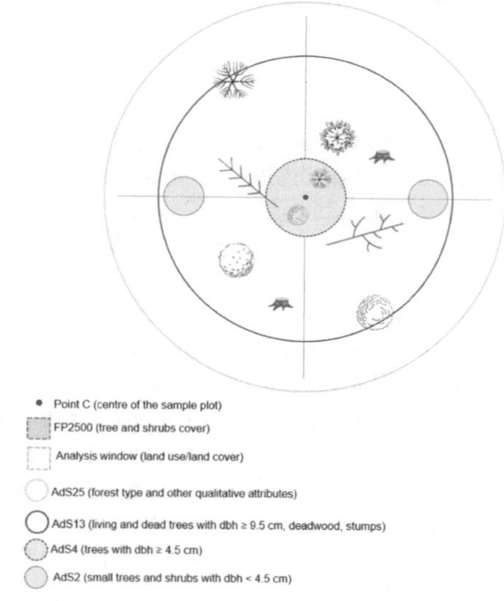

Fig. 2.5 Sampling units in INFC2015 and related attributes to be classified and measured, in brackets / Le unità di riferimento per i rilievi INFC2015 con indicazione degli attributi da rilevare per ciascuna di esse, tra parentesi

are measured and collected in AdS13. In addition, two circular sub-areas with a radius of 2 m (AdS2, area 12.57 m^2), located 10 m eastwards and westwards from point C are dedicated to tree regeneration and shrub surveys.

Appendix (Italian Version)

Riassunto La popolazione oggetto di studio dell'INFC è costituita dal Bosco e dalle Altre terre boscate così come definiti per il Global Forest Resources Assessment – FRA della FAO. Le statistiche INFC vengono prodotte per le 21 unità amministrative e territoriali italiane (regioni), le macrocategorie e categorie inventariali e le categorie forestali, che rappresentano ripartizioni del dominio inventariale. Il capitolo illustra il sistema di classificazione adottato per l'INFC e fornisce le descrizioni delle diverse classi. Vengono inoltre illustrati il disegno di campionamento adottato e la modalità di costruzione del reticolo nazionale 1 km × 1 km su cui si basa l'individuazione dei punti di campionamento. Infine, viene presentato il sistema di unità di riferimento (punti o aree di saggio) utilizzato per operare le classificazioni ed eseguire le misure previste dal protocollo di rilievo.

Introduzione

L'oggetto di studio di un'indagine statistica, costituito dall'insieme delle unità elementari di cui si intendono osservare uno o più caratteri ad esse comuni, viene indicato con il termine popolazione. La popolazione di interesse può essere suddivisa in domini di studio, ossia le sue diverse componenti per le quali è richiesto di produrre statistiche separate (OECD, 2007). La popolazione e le eventuali ripartizioni debbono essere individuate in modo chiaro e inequivocabile fin dall'avvio della progettazione dell'indagine. Con il termine dominio inventariale si intende qui l'oggetto di indagine dell'inventario forestale nazionale italiano INFC, rappresentato dal Bosco e dalle Altre terre boscate così come definiti per il rapporto sulle risorse forestali mondiali (Global Forest Resources Assessment—FRA) della FAO. Le ripartizioni del dominio inventariale sono rappresentate dalle 21 unità amministrative in cui è suddiviso il territorio italiano, 15 Regioni a statuto ordinario, 4 Regioni a statuto speciale, e due Province autonome, chiamate regioni per brevità; altre ripartizioni per le quali vengono prodotte statistiche separate sono quelle relative alle macrocategorie e categorie inventariali e alle categorie forestali.

Una volta individuata la popolazione di interesse e le sue ripartizioni, la progettazione di una indagine statistica richiede di definire la procedura di selezione delle unità da osservare attraverso un piano o disegno di campionamento. Questo riguarda la struttura del campione, le modalità per selezionare le unità della popolazione, la loro probabilità di inclusione nel campione stesso e la frazione o tasso di campionamento, da cui dipende la numerosità del campione (Fabbris, 1989). Le caratteristiche della popolazione di studio, dette parametri, vengono descritte a partire dai dati raccolti nel campione attraverso le tecniche di stima, ossia il calcolo di opportuni stimatori, quali ad esempio la media e la relativa varianza campionaria. L'insieme costituito dal piano di campionamento e dagli stimatori viene detto strategia campionaria (Cicchitelli et al., 1997).

Il capitolo descrive i criteri adottati per definire il dominio inventariale dell'INFC e le sue ripartizioni. Vengono quindi descritti il piano di campionamento e il reticolo nazionale alla base della localizzazione sul territorio di osservazioni e misure. Le stesse vengono condotte in corrispondenza di punti o di aree di forma o dimensione diversa denominate nel complesso unità di campionamento, che vengono elencate e descritte nella parte conclusiva del capitolo. Gli stimatori e le procedure utilizzate per il calcolo delle stime inventariali sono invece oggetto dei Chaps. 5 e 6.

Dominio inventariale e sistema di classificazione

A partire dal secondo inventario forestale nazionale italiano INFC2005, il dominio inventariale corrisponde all'insieme delle aree caratterizzate da una copertura forestale rispondente alle definizioni di Foresta e di Altre terre boscate elaborate per l'indagine FAO Global FRA2000 (FAO, 1998, 2000). Tali definizioni sono state in seguito ulteriormente dettagliate e applicate in tutti i successivi FRA (FAO, 2010, 2012, 2018; FAO-ITTO, 2003), divenendo un punto di riferimento per l'armonizzazione delle statistiche forestali a scala internazionale. Esse si basano su alcune caratteristiche oggettive, alcune delle quali valutabili anche su foto aeree o altre immagini da remoto, purché a risoluzione sufficiente, e altre che richiedono l'osservazione al suolo per una conferma o per la corretta valutazione. Tra le prime, la dimensione minima in termini di estensione e larghezza delle aree con copertura arboreo-arbustiva. Tra le seconde, la copertura delle specie arboree, distinta da quella delle specie arbustive, l'altezza dei soggetti a maturità e l'uso del suolo.

Nell'INFC le due classi Bosco e Altre terre boscate, che insieme individuano il dominio inventariale, vengono denominate macrocategorie inventariali. Queste sono ripartite in classi di maggiore dettaglio chiamate categorie inventariali, alcune delle quali distinte per lo sviluppo in altezza dei soggetti di specie arboree, allo scopo di facilitare il confronto tra le stime del primo inventario forestale italiano IFNI85 e quelle del secondo inventario INFC2005. In IFNI85, infatti, la soglia per l'altezza delle piante a maturità è di 2 m. La Table 2.1 riporta le definizioni per le macrocategorie e le categorie inventariali utilizzate nel terzo inventario italiano INFC2015. Ai fini della trattazione dei risultati INFC2015 la categoria inventariale delle Aree temporaneamente prive di soprassuolo, originariamente distinta in sede di classificazione della vegetazione da parte dei rilevatori, è stata accorpata a quella dei Boschi alti. Inoltre, l'elenco delle categorie in Table 2.1 include la categoria residuale delle Aree boscate inaccessibili o non classificate, che si riferisce a tutte le situazioni non altrimenti classificabili incluse per convenzione, in sede di elaborazione dei dati, nelle Altre terre boscate. Si tratta di formazioni caratterizzate da una copertura arboreo-arbustiva superiore al 10% verificata da fotointerpretazione, ma per le quali manca una classificazione di maggiore dettaglio della vegetazione, principalmente a causa della loro totale inaccessibilità per i rilievi.

Il sistema di classificazione INFC prevede ulteriori due livelli oltre a quelli sopra descritti, rispettivamente delle categorie e sottocategorie forestali, che individuano i domini inventariali di maggiore dettaglio. Le formazioni delle categorie inventariali dei Boschi alti, dei Boschi bassi, dei Boschi radi e delle Boscaglie vengono classificate secondo 17 categorie forestali, gli Impianti di arboricoltura da legno secondo 3 categorie forestali e gli Arbusteti secondo ulteriori 3 categorie (Table 2.2). Le categorie forestali si articolano quindi in sottocategorie, in numero totale di 68 per i boschi, 7 per gli Impianti e 16 per gli Arbusteti, distinte sulla base della specie dominante o secondo criteri ecologici. Ad esempio, le sottocategorie Peccete subalpine e Peccete montane per i Boschi di abete rosso sono distinte sulla base di un criterio ecologico, mentre quelle della categoria delle Pinete mediterranee si differenziano per la specie (pino marittimo, pino domestico o d'Aleppo).

La categoria forestale viene assegnata in funzione della specie o del gruppo di specie dominanti in termini di copertura, quest'ultima valutata in un intorno del punto inventariale di superficie di circa 2000 m^2 (AdS25, cfr. Chap. 4). Dapprima viene determinato il gruppo di specie prevalente, se conifere, latifoglie decidue o latifoglie sempreverdi, e successivamente la categoria forestale in base alla o alle specie prevalente/i del gruppo, sempre in termini di copertura delle chiome. In INFC2015, in mancanza di una classificazione in campo della categoria forestale, il punto inventariale è stato assegnato alla classe non classificato. Il sistema di classificazione e le relative categorie e sottocategorie forestali (Pignatti, 2003) è stato definito in occasione della progettazione del secondo inventario italiano ed è stato mantenuto invariato anche nel terzo.

Disegno di campionamento

Il disegno dell'INFC comprende tre fasi di campionamento, con campioni estratti secondo un campionamento stratificato (Fattorini et al., 2006; Gasparini & Di Cosmo, 2016; Gasparini & Tabacchi, 2011). La prima fase (o fase 1) consiste nella classificazione preliminare dell'uso e copertura del suolo attraverso la fotointerpretazione di ortofoto (cfr. Chap. 3) in corrispondenza di oltre 301,000 punti (punti di campionamento o punti inventariali), individuati in numero di uno per ciascuna maglia del reticolo 1 km × 1 km con cui è stata suddivisa la superficie nazionale (cfr. Sect. 2.4). Tale schema di campionamento viene definito tessellated sampling (Särndal et al., 1992). I punti sono posizionati entro le maglie con modalità casuale e perciò la loro distribuzione risulta di tipo sistematico non allineato (Gallego, 1995). Lo scopo della prima fase è quello di individuare i punti appartenenti agli strati di interesse per la fase di rilievo successiva, rappresentati dalle classi di uso e copertura forestale (Formazioni forestali, Formazioni forestali rade, Aree temporaneamente prive di soprassuolo e Impianti di arboricoltura da legno; cfr. Chap. 3) e dalla classe dei Punti non classificabili da ortofoto. Per l'INFC2015, inoltre, sono stati individuati ulteriori strati di interesse, come descritto nel seguito. La seconda fase di campionamento (o fase 2) interessa un sotto-campione dei punti di prima fase, oltre 30,000, selezionati secondo un campionamento stratificato per regione e classe di uso e copertura del suolo. I punti del campione di seconda fase vengono visitati al suolo, allo scopo di verificare la classificazione preliminare da fotointerpretazione, confermandola o meno, e assegnare la categoria inventariale e la categoria forestale. Durante i rilievi di seconda fase vengono inoltre valutati e classificati i caratteri qualitativi delle formazioni forestali (cfr. Chap. 4), allo scopo di produrre le stime della ripartizione della superficie forestale secondo le diverse caratteristiche (composizione per gruppi di specie, grado di copertura, tipo colturale, stadio di sviluppo, modalità di gestione, presenza di vincoli ed aree protette, caratteristiche delle stazioni forestali, condizioni della viabilità, ecc.) presentate in Chaps. 7, 8, 9, 10 e 11 del presente volume. La categoria forestale assegnata in seconda fase insieme all'uso e copertura del suolo e alla regione di appartenenza individuano gli strati di campionamento della terza fase (o fase 3). Per ciascuno degli strati appartenenti al Bosco viene estratto un sotto-campione dei punti di seconda fase su cui eseguire i rilievi di maggiore dettaglio, relativi ai caratteri quantitativi (cfr. Chap. 4). Le misure eseguite in terza fase su circa 7000 punti consentono di produrre le stime dei totali e delle densità, o valori per unità di superficie, delle variabili quantitative (volume, fitomassa, incremento annuo di volume, necromassa, ecc.) presentate in Chaps. 7, 9 e 12 del presente testo, importanti per valutare lo stato del patrimonio forestale italiano e il suo ruolo quale serbatoio di carbonio e riserva di biodiversità.

Il disegno di campionamento sopra descritto, delineato in occasione della progettazione di INFC2005, è stato applicato con alcuni adattamenti anche per INFC2015. Con la progettazione operativa, infatti, furono recepite le indicazioni provenienti sin dall'inizio dalla struttura operativa responsabile del coordinamento e della realizzaziobe dell'INFC di organizzare i rilievi in un'unica campagna, includendo il maggior numero possibile di punti già rilevati nella precedente indagine, allo scopo di ridurre i tempi del rilievo al suolo e facilitarne l'organizzazione, pur mantenendo invariati la quantità, qualità e il livello di dettaglio delle statistiche prodotte.

Il campione di punti per i nuovi rilievi al suolo è stato quindi costituito includendo: (i) tutti i punti della terza fase INFC2005 non interessati da cambiamenti significativi dell'uso e copertura del suolo, (ii) i punti del campione di seconda fase INFC2005 interessati da cambiamenti significativi e (iii) un ulteriore campione, stratificato per regione, per gli strati costituiti da nuovi punti in aree boscate e impianti di arboricoltura da legno e da nuovi punti non classificabili. I cambiamenti significativi sono stati evidenziati dal confronto tra la fotointerpretazione di INFC2005 e quella di INFC2015 e sono relativi ai punti transitati da strati di interesse inventariale allo strato "altri usi del suolo" o viceversa. L'esito del confronto tra i risultati delle due sessioni di fotointerpretazione è riportato in Fig. 2.1 per l'insieme dei 301,271 punti ricadenti sul territorio nazionale secondo il giudizio dei fotointerpreti, sul totale di 301,328 punti individuati dal reticolo nazionale (cfr. Sect. 2.4). L'effettivo cambiamento da uso e copertura forestale ad altri usi del suolo è stato verificato su un sotto-campione di punti di seconda fase costituito da 2338 punti. Di questi, quasi la metà era risultata appartenere ad altri usi del suolo già nella seconda fase INFC2005 e pertanto non è stato necessario procedere ad una ulteriore verifica in campo. I rimanenti 1303 punti sono stati invece rilevati al suolo nel corso della campagna INFC2015 oppure classificati attraverso una ulteriore fotointerpretazione, su immagini più recenti, qualora inaccessibili per i rilevatori. I punti del campione di seconda fase non interessati da cambiamenti significativi dell'uso e copertura rappresentano la parte numericamente più consistente. Per essi è stato considerato valido lo strato di terza fase attribuito nel corso dei rilievi INFC2005 e i nuovi rilievi hanno riguardato 6597 punti del sotto-campione di terza fase. Il sotto-campione estratto per i nuovi punti transitati in strati di interesse inventariale, infine, è costituito da 874 punti. Il tasso di campionamento per questi strati è inferiore a quello adottato per gli altri strati, a causa dei limiti sopra ricordati, ma è stato ritenuto idoneo per valutare l'effettivo cambiamento dell'uso e copertura del suolo e allo stesso tempo assicurare una ripartizione equilibrata del campione complessivo di punti interessato dai rilievi.

Complessivamente, il campione di seconda fase INFC2015 è costituito da 30,877 punti di cui 8774 sono stati classificati per i caratteri qualitativi durante la campagna di rilievo o, se inaccessibili, valutati a distanza o mediante ortofoto recenti, allo scopo di verificare l'uso e copertura del suolo e la categoria forestale. Di questi, 6993 punti sono stati oggetto di misure per la stima dei caratteri quantitativi e costituiscono il campione di terza fase INFC2015. La composizione del campione selezionato per la campagna di rilievo INFC2015 è illustrata in Fig. 2.2; il gruppo più numeroso è quello delle unità campionarie già rilevate sia in fase 2 sia in fase 3 INFC2005, che rappresenta oltre i tre quarti del totale.

La numerosità delle unità campionarie nelle tre fasi e i tassi di campionamento risultanti per regione e a livello nazionale sono riportati in Table 2.3 unitamente alla superficie rappresentata da ciascun punto o area di saggio inventariale. A livello nazionale, il tasso di campionamento di fase 2 e di fase 3 è rispettivamente pari a 0.26 e 0.30 e la superficie rappresentata da ciascuna unità di campionamento è di 3.9 km^2 per la fase 2 e 13.0 km^2 per la fase 3. Nella seconda fase, il tasso di campionamento a livello regionale varia di poco, da 0.23 per la Lombardia a 0.28 per il Trentino, e la superficie rappresentata da ciascuna unità di campionamento varia da 3.7 km^2 in Toscana a 4.6 km^2 in Lombardia. Nella terza fase il tasso di campionamento regionale varia da 0.23 per la Toscana a 0.44 per Valle d'Aosta e Molise, mentre la superficie rappresentata da una unità di campionamento varia da 8.7 km^2 in Valle d'Aosta a 15.8 km^2 in Toscana.

Il reticolo nazionale

Il reticolo impiegato per suddividere il territorio nazionale in porzioni di uguale superficie e individuare quindi i punti del campione INFC è stato generato in occasione della progettazione del secondo inventario forestale italiano INFC2005. Si tratta di un reticolo a maglie quadrangolari, di dimensione 1 km × 1 km, geometricamente agganciate a meridiani e paralleli; tali maglie, pur avendo forma quadrata in proiezione piana solo in prossimità del meridiano di origine, hanno ovunque una superficie identica, pari a 1 km^2 (Fig. 2.3).

Per la generazione dei nodi del reticolo è stato adottato il datum WGS84 (DMA, 1991), per la sua perfetta congruenza con i sistemi GNSS che sarebbero stati impiegati in seguito per l'identificazione e il posizionamento dei punti in campo. Stabilite delle coordinate geografiche di origine (emanazione) in detto sistema, φ_0, λ_0, nello spigolo Sud-Ovest del quadrilatero che contiene tutto il territorio nazionale, ulteriori nodi del reticolo sono stati generati muovendosi verso Est lungo lo stesso parallelo con passo di 1000 m. Giunti nel nodo a Sud-Est del quadrilatero, la procedura riparte dal nodo iniziale di emanazione, spostandosi questa volta di 1000 m a Nord lungo il meridiano e ripetendo il passo precedente. La procedura è stata reiterata fino alla completa copertura del quadrilatero di contenimento del territorio nazionale, terminando nel nodo a Nord-Est (dettagli degli algoritmi disponibili in Floris & Scrinzi, 2011). Intersecando il reticolo così ottenuto con la superficie del territorio nazionale, è stato individuato il reticolo inventariale vero e proprio, costituito da 306,831 maglie quadrangolari di lato 1000 m sulla superficie ellissoidica. Convertendo le coordinate geografiche dei nodi del reticolo in coordinate piane nel sistema di riferimento UTM-WGS84 e ricalcolando la superficie di ciascuna maglia con la formula di Gauss per una poligonale chiusa, si sono ottenuti sempre scostamenti inferiori al metro quadro, nel confronto con il valore atteso di 1 km^2.

Al fine di determinare le coordinate dei punti di campionamento, in numero di uno per maglia del reticolo, al nodo Sud-Ovest di ciascuna maglia in cui fosse compresa una porzione del territorio nazionale è stata associata una coppia di valori angolari positivi $\delta\varphi$ e $\delta\lambda$ generati casualmente entro i limiti di 1000 m dal nodo stesso (Fig. 2.4). I punti di campionamento sono stati nuovamente intersecati con i limiti amministrativi nazionali ufficiali, ottenendo il campione definitivo di 301,328 punti. Le coordinate piane dei punti di campionamento sono state trasformate anche nei sistemi di riferimento UTM-ED50 e Gauss-Boaga-Roma40 (Cima et al., 2003), in modo da rendere i punti individuabili su qualsiasi supporto cartografico, cartaceo o digitale, in uso durante i rilievi inventariali. In particolare, il sistema di riferimento Gauss-Boaga-Roma40 è quello adottato nella piattaforma GeoInfo, utilizzata per la fotointerpretazione (cfr. Chap. 3).

Le unità di riferimento per i rilievi

Con il termine unità di campionamento vengono indicate, in questo testo, le diverse unità di riferimento (punti e aree) dove acquisire le informazioni sui numerosi attributi rilevati dall'INFC, con diverse modalità di rilievo e in fasi diverse dell'indagine, rispettando i previsti obiettivi di accuratezza e in relazione alle risorse, materiali e di tempo, a disposizione.

Il riferimento della fotointerpretazione di prima fase, finalizzata alla classificazione di uso e copertura del suolo, è il punto di campionamento con relativo intorno di analisi costituito da nove quadrati contigui, ciascuno avente 50 m di lato, disposti secondo uno schema 3 × 3, per un'area totale di 22,500 m^2 (Gasparini et al., 2014). Il quadrato centrale dello schema è centrato sul punto di campionamento. L'intorno di analisi è funzionale alla valutazione dei requisiti dimensionali necessari per la corretta applicazione delle definizioni di Bosco e di Altre terre boscate adottate (cfr. Sect. 2.2 e Chap. 3).

In campo, oltre alla verifica della classificazione di uso e copertura del suolo, vengono rilevati molti caratteri di tipo qualitativo, tra cui quelli necessari alla classificazione di maggior dettaglio del tipo di vegetazione (Gasparini et al., 2016). Gli attributi di tipo amministrativo e normativo, la quota e la distanza dalla viabilità vengono rilevati con riferimento al punto C. Il grado di copertura delle chiome di alberi e arbusti viene attribuito osservando, sia sul terreno che su ortofoto, il quadrante centrale (50 $\times$ 50 m) dell'intorno di analisi utilizzato in fotointerpretazione di fase 1, chiamato photoplot 2500 (FP2500). Un'area circolare avente per centro il punto C, inscritta nel medesimo quadrante centrale e con raggio pari a 25 m, denominata AdS25 e con superficie di circa 2000 m^2, è invece l'unità di campionamento per il rilievo dell'accessibilità, dei caratteri descrittivi della vegetazione, delle caratteristiche stazionali e dello stato di salute del soprassuolo. In AdS25 vengono acquisite anche le osservazioni sui caratteri qualitativi inerenti alla gestione e alla selvicoltura (Fig. 2.5).

La terza fase INFC è deputata principalmente all'acquisizione di dati quantitativi necessari alla stima delle grandezze dendrometriche. In INFC2015 la terza fase di rilievi è stata eseguita contestualmente alla classificazione ex novo dei caratteri qualitativi in tutti i punti visitati (cfr. Sect. 2.3). I rilievi dendrometrici vengono eseguiti su aree di saggio di forma circolare utili per massimizzare il rapporto area/perimetro e ridurre il numero di elementi limite (de Vries, 1986), che possono essere facilmente materializzate in campo attraverso la semplice verifica della distanza dal punto centrale. Vengono istituite due aree concentriche aventi come centro il punto C, con raggio rispettivamente di 4 m (AdS4, superficie pari a 50.27 m^2) e 13 m (AdS13, superficie pari a 530.93 m^2). AdS4 viene utilizzata per le misure sui soggetti aventi $d_{1.30}$ compreso tra 4.5 e 9.4 cm, AdS13 per le misure sui soggetti con $d_{1.30}$ maggiore o uguale a 9.5 cm. In AdS13 vengono eseguiti anche i rilievi delle altezze degli alberi, il prelievo delle carote incrementali, le misure sul legno morto a terra e sulle ceppaie. Vengono inoltre istituite due sotto-aree circolari aventi raggio di 2 m (AdS2, superficie pari a 12.57 m^2), con centro a 10 m dal punto C in direzione Est e Ovest, dedicate ai rilievi della rinnovazione e delle specie arbustive.

References

Cicchitelli, G., Herzel, A., & Montanari, G. E. (1997). Il campionamento statistico. Il Mulino, Bologna (581 pp.). ISBN 88-15-05744-7.

Cima, V., Maseroli, R., & Surace, L. (2003). Il processo di georeferenziazione dal telerilevamento ai GIS. Atti della VII Conferenza Nazionale ASITA (pp. 49–68). Verona 28–31 ottobre 2003. ISBN 88-900943-5-4.

De Vries, P.G. (1986). Sampling theory for forest inventory. A Teach-Yourself Course (399 pp.). Springer-Verlag, Berlin.

DMA WGS84 Development Committee (1991). Department of Defense World Geodetic System 1984, its definition and relationships with local geodetic systems. The Defense Mapping Agency. Retrieved Jan 07, 2022, from https://apps.dtic.mil/sti/pdfs/ADA280358.pdf.

Fabbris, L. (1989). L'indagine campionaria: metodi, disegni e tecniche di campionamento. NIS, Roma (261 pp.).

FAO (1998). Appendix 2. Terms and definitions. In: *Global Forest Resources Assessment 2000. FRA 2000.* FRA Forest Resources Assessment Working Paper 1. Rome. Retrieved Jan 07, 2022, from https://www.fao.org/3/y1997e/y1997e00.htm#Contents.

FAO (2000). On definitions of forest and forest change. Forest resources assessment working paper 33, Rome (15 pp.). Retrieved Jan 05, 2022, from https://www.fao.org/3/ad665e/ad665e00.htm.

FAO-ITTO (2003). Global forest resources assessment update 2005. FRA2005 Draft terms and definitions. Forest resources assessment working paper 73, Rome (50 pp.).

FAO (2010). Global forest resources assessment 2010. FRA 2010 terms and definitions. Forest resources assessment programme working paper 144/E, Rome (27 pp.). Retrieved Jan 05, 2022, from https://www.fao.org/3/am665e/am665e.pdf.

FAO (2012). Global forest resources assessment 2015. FRA 2015 terms and definitions. Forest resources assessment working paper 180, Rome (31 pp.). Retrieved Jan 05, 2022, from http://www.fao.org/docrep/017/ap862e/ap862e00.pdf.

FAO (2018). Global forest resources assessment 2020. FRA2020 terms and definitions. Forest resources assessment working paper 188, Rome (32 pp.). Retrieved Jan 05, 2022, from https://www.fao.org/3/I8661EN/i8661en.pdf.

Fattorini, L., Marcheselli, M., & Pisani, C. (2006). A three-phase sampling strategy for multi-resource forest inventories. *Journal of Biological, Agricultural and Environmental Statistics, 11*(3) 296–316.

Floris, A., & Scrinzi, G. (2011). Il reticolo nazionale. In: P. Gasparini & G. Tabacchi (Eds.), *L'Inventario Nazionale delle Foreste e dei serbatoi forestali di Carbonio INFC 2005. Secondo inventario forestale nazionale italiano. Metodi e risultati* (pp. 30–36). Edagricole-Il Sole 24 Ore, Milano. ISBN 978-88-506-5394-2.

Gallego, F. J. (1995). Sampling frames of square segments. Joint Research Centre, European Commission. Office for Official Publications of the European Communities, Luxembourg (68 pp.).

Gasparini, P., & Di Cosmo, L. (2016). National forest inventory reports—Italy. In: C. Vidal, I. Alberdi, L. Hernández, & J. Redmond (Eds.), *National forest inventories—Assessment of wood availability and use* (pp. 485–506). Springer, Cham, Switzerland. ISBN 978-3-319-44014-9. https://doi.org/10.1007/978-3-319-44015-6.

Gasparini, P., Di Cosmo, L., Floris, A., Notarangelo, G., & Rizzo, M. (2016). Guida per i rilievi in campo. INFC2015—Terzo inventario forestale nazionale. Consiglio per la ricerca in agricoltura e l'analisi dell'economia agraria, Unità di Ricerca per il Monitoraggio e la Pianificazione Forestale (CREA-MPF); Corpo Forestale dello Stato, Ministero per le Politiche Agricole, Alimentari e Forestali (341 pp.). ISBN 9788899595449. Retrieved Jan 05, 2022, from https://www.inventarioforestale.org/it/node/72.

Gasparini, P., Rizzo, M., & De Natale, F. (2014). Manuale di fotointerpretazione per la classificazione delle unità di campionamento di prima fase. Inventario Nazionale delle Foreste e dei Serbatoi Forestali di Carbonio, INFC2015—Terzo inventario forestale nazionale. Consiglio per la Ricerca e la sperimentazione in Agricoltura, Unità di Ricerca per il Monitoraggio e la Pianificazione Forestale (CRA-MPF). Corpo Forestale dello Stato, Ministero per le Politiche Agricole, Alimentari e Forestali (64 pp.). Retrieved Jan 05, 2022, from https://www.inventarioforestale.org/it/node/72.

Gasparini, P., & Tabacchi, G. (2011). Il piano di campionamento. In: P. Gasparini & G. Tabacchi (Eds.), *L'Inventario Nazionale delle Foreste e dei serbatoi forestali di Carbonio INFC 2005. Secondo inventario forestale nazionale italiano. Metodi e risultati* (pp. 25–29). Edagricole-Il Sole 24 Ore, Milano. ISBN 978-88-506-5394-2.

OECD (2007). Glossary of statistical terms. Retrieved Jan 21, 2022, from https://stats.oecd.org/glossary/.

Pignatti, S. (2003). Guida alla classificazione della vegetazione forestale. Inventario Nazionale delle Foreste e dei Serbatoi Forestali di Carbonio INFC (2003), Trento (61 pp.). Retrieved Jan 05, 2022, from https://www.inventarioforestale.org/it/node/72.

Särndal, C. E., Svensson, B., & Wretman, J. (1992). Model assisted survey sampling (694 pp.). Springer-Verlag, New York.

Open Access This chapter is licensed under the terms of the Creative Commons Attribution 4.0 International License (http://creativecommons.org/licenses/by/4.0/), which permits use, sharing, adaptation, distribution and reproduction in any medium or format, as long as you give appropriate credit to the original author(s) and the source, provide a link to the Creative Commons license and indicate if changes were made.

The images or other third party material in this chapter are included in the chapter's Creative Commons license, unless indicated otherwise in a credit line to the material. If material is not included in the chapter's Creative Commons license and your intended use is not permitted by statutory regulation or exceeds the permitted use, you will need to obtain permission directly from the copyright holder.

Chapter 3
Land Use and Land Cover Photointerpretation

Fotointerpretazione dell'uso e copertura del suolo

Maria Rizzo and Patrizia Gasparini

Abstract Most national forest inventories use remote sensing data, mainly aerial photos and orthophotos, for the preliminary classification of land use and cover in the inventory points, also for the purpose of estimating the forest area. The classification of land use and land cover during the first phase of the third Italian forest inventory INFC2015 was carried out by interpreting 4-band digital orthophotos (RGB colors and near infrared) in over 301,000 points located on a grid with quadrangular meshes of 1 km^2. The classification system adopted includes three hierarchical levels, of which the first corresponds to the same level of the European CORINE Land Cover system and the subsequent ones aimed at highlighting the classes of inventory interest, for the subsequent stratification of the sample of points to be surveyed on the ground. A rigorous quality control procedure was implemented, during the photointerpretation and its conclusion, in order to assess the accuracy of the classifications and the extent of changes to and from forest land use and land cover.

Keywords Land cover classification · WebGIS · Land use categories · Visual interpretation · Quality assurance

3.1 Introduction

Aerial photographs have been and continue to be a widely used source of information in the forestry sector (Hall, 2003; Howard, 1991). Currently, remote sensing is used in most countries that conduct large-scale inventories. Many forest inventories use remote sensing images such as aerial photos, orthophotos and satellite images for the classification of land use and land cover at the sampling points in order to select the points to be later detected on the ground or for estimation of the forest area, at least preliminarily. Some inventories also use aerial photos to acquire data on the characters of forest formations (Tomppo et al., 2010).

M. Rizzo (✉) · P. Gasparini
CREA Research Centre for Forestry and Wood, Trento, Italy
e-mail: maria.rizzo@crea.gov.it

© The Author(s) 2022

P. Gasparini et al. (eds.), *Italian National Forest Inventory—Methods and Results of the Third Survey*, Springer Tracts in Civil Engineering,
https://doi.org/10.1007/978-3-030-98678-0_3

In the first inventory phase, the Italian national forest inventory INFC uses orthophotos for the preliminary classification of land use and cover to stratify the sample to be used for the field surveys of the two following sampling phases (cf. Chap. 2). The classification carried out during the first inventory phase of the third inventory INFC2015 also allowed a preliminary estimate of the changes relating to the forest use and cover that have occurred since the previous inventory survey, useful for reporting purposes at the end of the first commitment period of the Kyoto Protocol.

After some hints on the characteristics of aerial photos and orthophotos and on the photo interpretation process (Sect. 3.2), the chapter illustrates the land use and cover classification system adopted for INFC (Sect. 3.3) and the tools and procedure used in the first phase of INFC2015 (Sect. 3.4). Finally, the chapter describes the set of controls used to evaluate the quality of the data resulting from the classification of photo interpreters (Sect. 3.5).

3.2 Materials and Methods for Photointerpretation of Land Use and Cover

The quality of digital images (and the use that can be made of them) primarily depends on their resolution, which is divided into spatial or geometric, spectral, radiometric and temporal resolution (Brivio et al., 2006). Spatial resolution is linked to the size of the elementary area on the ground where the electromagnetic energy is detected, that is to the size of the pixels. The spectral resolution indicates the number and width of the spectral bands (wavelength intervals) in which an image is acquired.[1] Radiometric resolution is given by the minimum difference in electromagnetic energy detectable by the sensors on the photographed surfaces. Finally, temporal resolution indicates the time interval between two successive shots of the same area.

Orthophotos simultaneously provide a photographic and cartographic representation of the territory. They are obtained through geometric correction (orthorectification) and georeferencing of digital aerial photos or previously digitized frames. Orthorectification entails straightening and projection on the horizontal plane of the images, in order to allow a correct representation on the plane of distances, angles and surfaces. Georeferencing allows each point of the territory represented by the orthophoto to be associated with its position in space, referable to a system of geographic or plane coordinates (Gasparini et al., 2014).

The interpretative process of an orthophoto, or any remotely sensed image, is characterised by the presence of two fundamental phases: (a) examination, recognition

[1] In digital images, all visible colors are represented by additive synthesis of the three basic colors red, green and blue (R, G, B). The portions of the electromagnetic spectrum outside these regions are not visible to the human eye. To make them visible, they are assigned to one of the three basic colors. The most frequent example is that of near infrared (nIR), a particularly important region in the study of vegetation, which is assigned to the red band, transferring the red band to green and the green band to blue, thus giving up the information contribution of the blue band.

and, if necessary, measurement of the elements in the image; (b) the formulation of deductive and inductive reasoning, based on the observations made, in order to classify what is represented in the image (Dainelli, 2011). Knowledge of the landscape and its characteristic elements is crucial for effective photointerpretation. The classification is based on the analysis of spatial (localisation, association), spectral (color, tone) and geometric (shape, size) characters. Photointerpreters are recommended to proceed first with a broad observation of the territorial context, based mainly on the analysis of the shape and size of the different elements or polygons and their arrangement in space, defined as structure or pattern. The individual elements are recognizable on the image based on the brightness and intensity of the color, the texture given by micro-changes in the distribution of color tones, and their shapes and sizes. The presence of shadows on an image plays a double role, and they can act as a disturbing element or as a contribution to photointerpretation. This presents an obstacle, especially when large portions of the territory are obscured, but it can also provide important clues for identifying the vertical profile and the height of the elements to be interpreted, facilitating, for example, the distinction between trees and shrubs.

3.3 Classification System

The land use and land cover classification system used for the first INFC phase (Table 3.1) includes three hierarchical levels, the first corresponding to the same level of the CORINE Land Cover classification system (European Commission, 1993) and the following two aimed at further detailing the classes of greatest interest to produce the inventory statistics (Gasparini & Di Cosmo, 2016). The first level is divided into the five main classes of the CORINE Land Cover system (Artificial surfaces, Agricultural areas, Forest and semi-natural areas, Wetlands, Water bodies), from which it differs for the inclusion of *Castanea sativa* forests for fruit production and pastures in the class Forest and semi-natural areas rather than in Agricultural areas. The second level of INFC classification includes 12 subclasses, of which two (Plantations for timber and wood production and Woodland) are of interest for the subsequent sampling phases, together with the residual class of non-classifiable points. The third level of classification is present only for the subclass Woodland, which is divided into three further subclasses on the basis of coverage thresholds of tree and shrub species consistent with the definitions of Forest and Other wooded land and used since 2000 for the Global Forest Resources Assessment (FRA) (FAO, 2001) and adopted for INFC (cf. Chap. 2).

The classification system for the INFC2015 photointerpretation has remained unchanged compared to the second Italian forest inventory INFC2005, to allow the comparison between the results of the two photointerpretations and highlight the significant changes to and from the subclasses of inventory interest. The only difference concerns the introduction, for the Agricultural areas class, of the new subclass of Fruit plantations, which includes orchards, vineyards and olive groves.

Table 3.1 Classification scheme adopted for the INFC photointerpretation of land use and cover / Schema di classificazione adottato per la fotointerpretazione dell'uso e copertura del suolo nell'INFC

Class/Classe	Subclass/Sottoclasse
1. Artificial surfaces 1. Superfici artificiali	1.1 Urban parks 1.2 Other artificial surfaces
	1.1 Parchi urbani 1.2 Altre superfici artificiali
2. Agricultural areas 2. Superfici agricole	2.1 Plantations for timber and wood production 2.2 Fruit plantations 2.3 Other agricultural areas
	2.1 Impianti di arboricoltura da legno 2.2 Impianti di arboricoltura da frutto 2.3 Altre superfici agricole
3. Forest and semi-natural areas 3. Superfici boscate e ambienti seminaturali	3.1 Woodland 3.1.a Forest formations 3.1.b Sparse forest formations 3.1.c Temporarily unstocked areas 3.2 Grassland, pastures, uncultivated land 3.3 Open areas with little or no vegetation
	3 Aree boscate 3.1.a Formazioni forestali 3.1.b Formazioni forestali rade 3.1.c Aree temporaneamente prive di soprassuolo 3.2 Praterie, pascoli e incolti 3.3 Aree con vegetazione rada o assente
4. Wetlands/Zone umide	–
5. Water bodies/Acque	–

For a detailed description of the photointerpretation classes and instructions for their identification on the orthophotos, refer to the Photointerpretation Manual (Gasparini et al., 2014).

3.4 Classification Tools and Procedure

The classification of land use and land cover for the more than 301,000 inventory points constituting the INFC first phase sample was performed through photointerpretation of digital orthophotos on a dedicated WebGIS platform called GeoInfo, developed by the AlmavivA company with the collaboration of Telespazio (Gasparini et al., 2020). The platform allows operators to view orthophotos (Fig. 3.1) and other useful information layers, such as administrative limits, roads, toponymy, hydrographic network, altitude bands, state nature reserves and areas covered by fires, as

Fig. 3.1 Viewing of orthophotos on the WebGIS platform GeoInfo / Visualizzazione di ortofoto sulla piattaforma WebGIS GeoInfo

well as to acquire the classifications of the photo interpreters and store the data in a national central archive.

The INFC2015 photointerpretation was conducted on digital orthophotos in color and infrared-false color (RGB + nIR) with a resolution of 50 cm derived from the 2010–2012 AGEA coverage. To solve doubtful cases, the photointerpreters could also consult the digital color orthophotos of equal resolution derived from the AGEA 2007–2009 coverage, and orthophotos with a resolution of 1 m of the 1999–2005 coverage. The latter, used by photointerpreters in the first phase of the second inventory INFC2005, were used to evaluate any significant changes in use and land cover that occurred subsequently, through the visual comparison of the images referring to the two different periods.

The first phase of INFC2015 was conducted by about 50 photointerpreters, partly from the State Forestry Corps and partly from the Forest Services of the regions with special statutes, and the autonomous provinces, suitably instructed and trained through a specific training course.

The INFC classification procedure consists of assigning each sample point to the class and subclass of land use and cover of the polygon in which the point falls, after checking whether or not the minimum dimensional thresholds are exceeded. The polygon represents a homogeneous area for land use and cover, having an area greater than 0.5 ha and a width greater than 20 m. Limited to polygons characterised by a cover of trees or shrubs, the photointerpreter verifies also whether or not the minimum coverage thresholds are exceeded (if tree species, 10% for Forest and 5% for Sparse forest categories, and if shrubs species 10% for Shrubs category) in accordance with the definition adopted for the inventory domain (cf. Chap. 2). An additional 40% coverage threshold, relating to the herbaceous component, is used to distinguish the subclass Grasslands, pastures and uncultivated areas from that of Open areas with little or no vegetation. Sample points falling in smaller polygons,

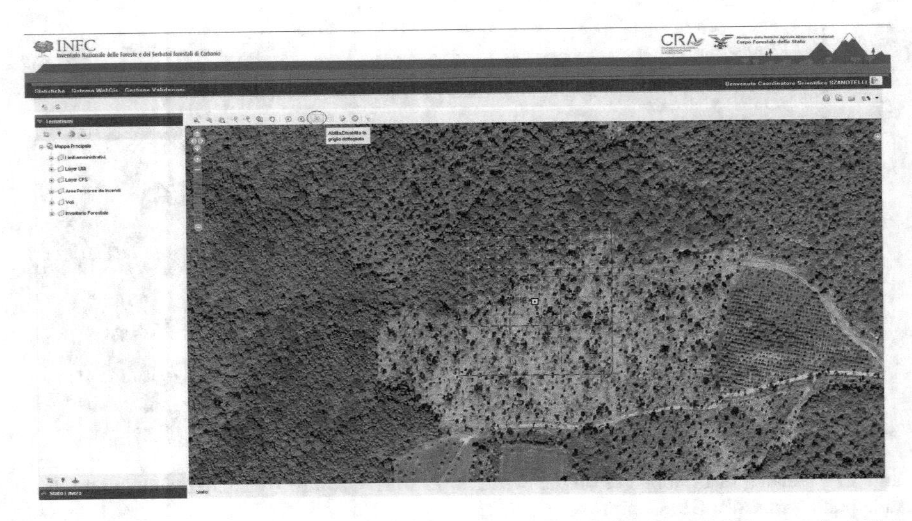

Fig. 3.2 Analysis window, grid and measuring tools on the WebGIS GeoInfo / Intorno di analisi, griglia e strumenti di misura nel WebGIS GeoInfo

with an area between 500 and 5000 m^2 or, if elongated, with a width between 3 and 20 m, are assigned to the land use and cover class of the nearest polygon that respects the minimum area and width thresholds indicated above, 0.5 ha and 20 m, respectively. In these cases, the presence of an 'included polygon' and its land use and land cover are also recorded.

The WebGIS GeoInfo automatically manages the sequence of operations for photointerpretation, proposing to the operators the points to be classified and allowing verification of the minimum thresholds through the display of an analysis window (cf. Chap. 2), a grid and tools to measure distances and surfaces (Fig. 3.2) (Gasparini et al., 2021). The analysis window, consisting of nine contiguous squares of 50 m side for a total area of 22,500 m^2 and centred on the sample point, allows for visual estimation of the extension and width of the homogeneous polygons identified. The grid of points spaced 10 m apart, on the other hand, enables quick and objective evaluation of the crown cover degree by counting the points that intercept trees or shrubs crowns.

For details on classification procedures and any specific cases, refer to the Photointerpretation Manual (Gasparini et al., 2014).

3.5 Quality Controls of the Photointerpretation

Land cover classification by visual interpretation of aerial photos always involves a certain degree of subjectivity, and the experience of the photointerpreters plays an important role in determining the quality of the result (Strand et al., 2002). Subjectivity should be limited as much as possible in order to obtain comparable classifications. The implementation of a quality control procedure during the photointerpretation activity and at its conclusion is fundamental, both to evaluate the uniformity of judgment by the photointerpreters and the reproducibility of the classification, and to evaluate the accuracy of the classification made.

The data quality assurance (QA) procedures implemented for the first INFC phase aimed at checking non-sampling errors due to measurement errors (during the verification of polygons minimum extension and width thresholds and of tree and shrub cover), or to an incorrect understanding of the rules of interpretation of the images by the photointerpreters. The control procedure is based on the identification of quality objectives (MQOs, Measurement Quality Objectives) and quality limits (DQLs, Data Quality Limits), respectively corresponding to incorrect classifications or tolerable measurement errors and the relative maximum permissible thresholds (Gasparini et al., 2009). The MQOs related to tolerable incorrect classifications for the QA of INFC2015 are reported in Table 3.2. In regard to the minimum surface and width thresholds, measurement errors of 200 m^2 and 2 m, respectively, for the polygons and of 50 m^2 and 1 m, respectively, for the included polygons were considered tolerable. For the crown coverage, errors up to 2.8% and 5.5% were considered tolerable for tree or shrub coverage and herbaceous coverage, respectively, values corresponding to one point and two points of the grid used (cf. Chap. 4). Table 3.3 shows the DQLs for the different land use and cover classes, established according to the importance of the individual classes and subclasses and the relative difficulty of recognition on orthophotos. The greater the importance of a class or subclass, the higher its DQL and the lower the percentage threshold of admissible errors; on the contrary, the greater the classification difficulty of a class or subclass, the lower its DQL and the higher the percentage threshold of admissible errors.

The QA activity during photointerpretation was carried out through the reclassification of a certain number of randomly selected points by a group of expert operators of the CREA Research Centre for Forestry and Wood, who took on the role of reference operators. The CREA operators also used the GeoInfo platform for the classification of the points. However, the results of the classification, which had been previously performed by the photointerpreters, were not made available to them. Later, the reference operator compared his own classification with that of the photointerpreter and, if there were discrepancies, assessed whether they were admissible based on measurement or classification errors deemed tolerable according to the MQOs. A specially implemented IT platform, accessible from the intranet of CREA Research Centre for Forestry and Wood, was used by internal operators in charge of periodic and final checks to record the identification code of the points concerned

Table 3.2 Admitted misclassifications of the land use and cover in the photointerpretation during INFC2015 first phase / Tolleranze di classificazione dell'uso e copertura del suolo nella fotointerpretazione della prima fase INFC2015

Class and subclass	Specific case	Other class or subclass tolerated
Classe e sottoclasse	Caso specifico	Altra classe o sottoclasse ammissibile
Urban parks Parchi urbani	Urban parks bordering urban areas	Forest formations
	Parco urbano al margine di centri urbani	Formazioni forestali
Other artificial surfaces Altre superfici artificiali	Quarry disused or in restoration	Open areas with little or no vegetation
	Cava dismessa o in via di ripristino	Zone aperte con vegetazione rada o assente
Plantations for timber and wood production Impianti di arboricoltura da legno	Young plantations	Fruit plantations
	Plantations close to the end of the cultivation cycle	Forest formations
	Plantations on forest boundary and/or without evident signs of cultivation	Forest formations
	Impianti giovani	Impianti di arboricoltura da frutto
	Impianti a fine turno	Formazioni forestali
	Impianti ai margini del bosco e/o privi di segni evidenti di coltivazione	Formazioni forestali
Fruit plantations Impianti di arboricoltura da frutto	Young plantations	Plantations for timber and wood production
	Plantations on forest boundary and/or without evident signs of cultivation	Forest formations
	Impianti giovani	Impianti di arboricoltura da legno
	Impianti ai margini del bosco e/o privi di segni evidenti di coltivazione	Formazioni forestali
Other agricultural areas Altre superfici agricole	Stable grasslands without evident mowing signs	Grassland, pastures, uncultivated land
	Prati stabili con segni sfalcio non evidenti	Praterie, pascoli e incolti
Forest formations Formazioni forestali	Close to crown coverage threshold	Sparse forest formations

(continued)

Table 3.2 (continued)

Class and subclass	Specific case	Other class or subclass tolerated
Classe e sottoclasse	Caso specifico	Altra classe o sottoclasse ammissibile
	Ai limiti della soglia di copertura	Formazioni forestali rade
Temporarily unstocked areas Aree temporaneamente prive di soprassuolo	Close to crown coverage threshold	Forest formations
	Ai limiti della soglia di copertura	Formazioni forestali
Praterie, pascoli e incolti Grassland, pastures, uncultivated land	Close to crown coverage threshold	Sparse forest formations
		Open areas with little or no vegetation
	Limits of photointerpretation, lack of auxiliary data	Wetlands
	Ai limiti della soglia di copertura	Formazioni forestali rade
		Aree con vegetazione rada o assente
	Limiti della fotointerpretazione, mancanza di dati ausiliari	Zone umide

and the results of the checks. In total, checks were performed on 9766 points during the photointerpretation, equal to 3.2% of the points of the INFC sample.

In addition to the on-going checks described above, final checks for approval were carried out on a randomly selected subsample of the inventory points per region. The final checks covered 2% of the inventory points in each region. Subsequently, for three regions which had not reached the DQLs for the subclasses of greatest interest, a partial revision of the classification was conducted and a new control was carried out on a further 2% of the sampling points. The results of the final checks for Woodland (Table 3.4) show that the classification discrepancy between photointerpreters and reference operators affects 2.1% of the points at national level, and percentages were always lower than the maximum threshold of 5% in all regions.

During the classification activity, a further blind check was conducted on a random subsample of inventory points, which were assigned simultaneously to two photo interpreters from the same region without their knowledge. The results of the blind check will be used for subsequent analyses aimed at any modification of the classification system.

Table 3.3 Data Quality Limits (DQLs) for the photointerpretation in the INFC2015 first phase / Limiti di Qualità (DQLs) per la fotointerpretazione della prima fase INFC2015

Class and subclass	Significance for INFC2015	Complexity of classification	DQLs (%)
Classe e sottoclasse	Importanza per INFC2015	Difficoltà di classificazione	
Artificial surfaces/Superfici artificiali	Quite important/Abbastanza importante	Easy/Facile	90
Urban parks/Parchi urbani	Important/Importante	Difficult/Difficile	85
Other artificial surfaces/Altre superfici artificiali	Quite important/Abbastanza importante	Easy/Facile	90
Agricultural areas/Superfici agricole	Quite important/Abbastanza importante	Quite easy/Abbastanza facile	85
Plantations for timber and wood production/ Impianti di arboricoltura da legno	Important/Importante	Difficult/Difficile	85
Fruit plantations/Impianti di arboricoltura da frutto	Quite important/Abbastanza importante	Quite easy/Abbastanza facile	85
Other agricultural areas/Altre superfici agricole	Quite important/Abbastanza importante	Quite easy/Abbastanza facile	85
Forest and semi-natural areas/Superfici boscate e ambienti seminaturali	Very important/Molto importante	Difficult/Difficile	90
Woodland/Aree boscate	Very important/Molto importante	Quite easy/Abbastanza facile	95
Forest formations/Formazioni forestali	Very important/Molto importante	Quite easy/Abbastanza facile	95
Sparse forest formations/Formazioni forestali rade	Very important/Molto importante	Difficult/Difficile	90
Temporarily unstocked areas/Aree temporaneamente prive di soprassuolo	Very important/Molto importante	Difficult/Difficile	90
Grassland, pastures, uncultivated land/Praterie, pascoli e incolti	Important/Importante	Difficult/Difficile	85
Open areas with little or no vegetation/Aree con vegetazione rada o assente	Important/Importante	Quite easy/Abbastanza facile	90

(continued)

Table 3.3 (continued)

Class and subclass	Significance for INFC2015	Complexity of classification	DQLs (%)
Classe e sottoclasse	Importanza per INFC2015	Difficoltà di classificazione	
Wetlands/Zone umide	Quite important/Abbastanza importante	Difficult/Difficile	80
Water bodies/Acque	Quite important/Abbastanza importante	Easy/Facile	90
All Class and subclass/Tutte le classi e sottoclassi	Important/Importante	Quite easy/Abbastanza facile	90

Appendix (Italian Version)

Riassunto La maggior parte degli inventari forestali nazionali utilizza dati telerilevati, principalmente foto aeree e ortofoto, per la classificazione preliminare dell'uso e copertura del suolo nei punti inventariali, anche ai fini della stima della superficie forestale. La classificazione dell'uso e copertura del suolo durante la prima fase del terzo inventario forestale italiano INFC2015 è stata effettuata mediante interpretazione di ortofoto digitali a 4 bande (colori RGB e infrarosso vicino) in oltre 301,000 punti localizzati su un reticolo a maglie quadrangolari di 1 km^2. Il sistema di classificazione adottato prevede tre livelli gerarchici, di cui il primo corrispondente all'analogo livello del sistema europeo CORINE Land Cover e i successivi finalizzati ad evidenziare le classi di interesse inventariale, per la successiva stratificazione del campione di punti da rilevare al suolo. Nel corso della fotointerpretazione e alla sua conclusione è stata attuata una rigorosa procedura di controllo della qualità, allo scopo di valutare l'accuratezza delle classificazioni e l'entità dei cambiamenti da e verso l'uso e copertura forestale.

Introduzione

Le fotografie aeree sono state e continuano ad essere una fonte di informazione ampiamente utilizzata nel settore forestale (Hall, 2003; Howard, 1991). Attualmente, il telerilevamento è utilizzato nella maggior parte dei Paesi che realizzano inventari su ampia scala. Molti inventari forestali utilizzano immagini telerilevate quali foto aeree, ortofoto e immagini satellitari per la classificazione dell'uso e copertura del suolo in corrispondenza dei punti di campionamento, allo scopo di selezionare i punti da rilevare successivamente al suolo o per la stima, almeno preliminare, della superficie forestale. Alcuni inventari utilizzano le foto aeree anche per acquisire dati sui caratteri delle formazioni forestali (Tomppo et al., 2010).

Table 3.4 Results of the final checks for approval for the subclass Woodland (Forest formations, Sparse forest formations and Temporary unstocked areas), by region; positive check means concordant classification between control operators and photointerpreters / Risultati dei controlli finali per la sottoclasse delle Aree boscate (Formazioni forestali, Formazioni forestali rade e Aree temporaneamente prive di soprassuolo), per regione; collaudo positivo in caso di classificazione concorde tra fotointerpreti e operatori di controllo

Region	Number of checked points	Positive checks	Negative checks
Regione	Numero di punti collaudati	Collaudi positivi	Collaudi negativi
	(n)	(%)	(%)
Piemonte	204	98.0	2.0
Valle d'Aosta	22	100.0	0.0
Lombardia	134	97.8	2.2
Alto Adige	74	100.0	0.0
Trentino	84	100.0	0.0
Veneto	104	98.1	1.9
Friuli V.G	82	100.0	0.0
Liguria	75	100.0	0.0
Emilia Romagna	122	99.2	0.8
Toscana	238	97.9	2.1
Umbria	84	98.8	1.2
Marche	152	98.0	2.0
Lazio	139	99.3	0.7
Abruzzo	106	99.1	0.9
Molise	43	97.7	2.3
Campania	204	94.1	5.9
Puglia	88	97.7	2.3
Basilicata	171	98.8	1.2
Calabria	286	95.8	4.2
Sicilia	164	95.7	4.3
Sardegna	271	98.5	1.5
Italia	2847	97.9	2.1

L'inventario forestale nazionale italiano INFC prevede l'uso di ortofoto per la classificazione preliminare dell'uso e copertura del suolo, nella prima fase inventariale, ai fini della stratificazione del campione da impiegare per i rilievi in campo delle due successive fasi di campionamento (cfr. Chap. 2). La classificazione effettuata durante la prima fase inventariale del terzo inventario INFC2015 ha consentito anche una stima preliminare dei cambiamenti relativi all'uso e copertura forestale intercorsi dalla precedente indagine inventariale, utile ai fini delle rendicontazioni a conclusione del primo periodo di impegno del Protocollo di Kyoto.

Dopo alcuni cenni sulle caratteristiche di foto aeree ed ortofoto e sul processo di fotointerpretazione (Sect. 3.2), il capitolo illustra il sistema di classificazione dell'uso e copertura adottato per INFC (Sect. 3.3) e gli strumenti e la procedura utilizzati nella prima fase dell'INFC2015 (Sect. 3.4). Il capitolo, infine, descrive l'insieme di controlli impiegato per valutare la qualità dei dati derivanti dalla classificazione dei fotointerpreti (Sect. 3.5).

Materiali e metodi per la fotointerpretazione dell'uso e copertura del suolo

La qualità delle immagini digitali (e l'uso che di esse di può fare) dipende principalmente dalla loro risoluzione, che si distingue in spaziale o geometrica, spettrale, radiometrica e temporale (Brivio et al., 2006). La risoluzione spaziale è legata alle dimensioni dell'area elementare al suolo di cui si rileva l'energia elettromagnetica, ossia alla dimensione dei pixel. La risoluzione spettrale indica il numero e l'ampiezza delle bande spettrali (intervalli di lunghezza d'onda) nelle quali viene acquisita un'immagine.[2] La risoluzione radiometrica è data dalla minima differenza di energia elettromagnetica rilevabile dai sensori sulle superfici fotografate. La risoluzione temporale, infine, indica l'intervallo di tempo che intercorre tra due riprese successive di una stessa area.

Le ortofoto forniscono una rappresentazione allo stesso tempo fotografica e cartografica del territorio. Esse si ottengono attraverso la correzione geometrica (ortorettifica) e la georeferenziazione di foto aeree digitali o di fotogrammi preventivamente digitalizzati. Il procedimento di ortorettifica consiste nel raddrizzamento e nella proiezione sul piano orizzontale delle immagini, in modo da consentire una corretta rappresentazione sul piano di distanze, angoli e superfici. La georeferenziazione permette di associare a ciascun punto del territorio rappresentato dall'ortofoto la sua posizione nello spazio, riferibile a un sistema di coordinate geografiche o piane (Gasparini et al., 2014).

Il processo interpretativo di un'ortofoto, o di qualsiasi immagine telerilevata, si caratterizza per la presenza due fasi fondamentali: (a) l'esame, il riconoscimento e, se necessario, la misurazione degli elementi presenti; (b) la formulazione di ragionamenti deduttivi e induttivi, basati sulle osservazioni fatte, allo scopo di classificare quanto rappresentato nell'immagine (Dainelli, 2011). Per una buona fotointerpretazione è molto importante conoscere il territorio in cui si opera e saperne riconoscere gli elementi caratteristici. La classificazione si basa sull'analisi di caratteri spaziali (localizzazione, associazione), spettrali (colore, tonalità) e geometrici (forma, dimensione). Ai fotointerpreti viene raccomandato di procedere dapprima con un'osservazione ampia del contesto territoriale, basata principalmente

[2] Nelle immagini digitali, la rappresentazione di tutti i colori visibili avviene per sintesi addittiva dei tre colori fondamentali rosso, verde e blu (R, G, B). Le porzioni di spettro elettromagnetico al di fuori di queste regioni non sono visibili all'occhio umano. Per renderle visibili, le si assegna a uno dei tre colori fondamentali. L'esempio più frequente è quello dell'infrarosso vicino (nIR), regione particolarmente importante nello studio della vegetazione, che si assegna alla banda del rosso, trasferendo la banda del rosso sul verde e la banda del verde sul blu, ottenendo la composizione denominata infrarosso-falso colore (IRFC).

sull'analisi della forma e della dimensione dei diversi elementi o poligoni e della loro disposizione nello spazio, definita come struttura o pattern. I singoli elementi sono riconoscibili sull'immagine sulla base della luminosità e intensità del colore, della tessitura, data da micro-cambiamenti nella distribuzione delle tonalità di colore, e delle loro forme e dimensioni. La presenza di ombre su un'immagine ha un doppio ruolo, come elemento di disturbo o come contributo alla fotointerpretazione. Se da una parte, infatti, esse rappresentano un ostacolo, soprattutto quando oscurano porzioni molto estese di territorio, dall'altra esse possono fornire importanti indizi nell'identificazione del profilo verticale e dell'altezza degli elementi da interpretare, facilitando ad esempio la distinzione tra le piante arboree e le piante arbustive.

Il sistema di classificazione

Il sistema di classificazione dell'uso e copertura del suolo utilizzato per la prima fase INFC (Table 3.1) prevede tre livelli gerarchici, di cui il primo corrispondente all'analogo livello del sistema di classificazione CORINE Land Cover (European Commission, 1993) e i successivi due finalizzati a dettagliare ulteriormente le classi di maggiore interesse ai fini della produzione delle statistiche inventariali (Gasparini & Di Cosmo, 2016). Il primo livello si articola nelle cinque classi principali del sistema CORINE Land Cover (Superfici artificiali, Superfici agricole, Superfici boscate e ambienti seminaturali, Zone umide, Acque), da cui si differenzia per l'inclusione dei castagneti da frutto e dei pascoli nella classe Superfici boscate e ambienti seminaturali anziché in quella delle Superfici agricole. Il secondo livello di classificazione INFC comprende 12 sottoclassi, di cui due (Impianti di arboricoltura da legno e Aree boscate) di interesse per le successive fasi di campionamento, insieme alla classe residua dei punti non classificabili. Il terzo livello di classificazione è presente solo per la sottoclasse Aree boscate, che viene distinta in ulteriori tre sottoclassi sulla base di soglie di copertura delle specie arboree e arbustive coerenti con le definizioni di Bosco e Altre terre boscate utilizzate dall'anno 2000 per il Global Forest Resources Assessment (FRA) (FAO, 2001) e adottate per INFC (cfr. Chap. 2).

Il sistema di classificazione per la fotointerpretazione INFC2015 è rimasto invariato rispetto al secondo inventario forestale italiano INFC2005, per consentire il confronto tra gli esiti delle due fotointerpretazioni ed evidenziare i cambiamenti significativi da e verso le sottoclassi di interesse inventariale. L'unica differenza riguarda l'introduzione, per la classe Superfici agricole, della nuova sottoclasse degli Impianti di arboricoltura da frutto, che include frutteti, vigneti e oliveti.

La descrizione delle singole classi e sottoclassi di fotointerpretazione e le indicazioni per la loro individuazione sulle ortofoto sono riportate nel manuale di fotointerpretazione per la classificazione delle unità di campionamento di prima fase (Gasparini et al., 2014).

Strumenti e procedure per la classificazione

La classificazione dell'uso e copertura del suolo per gli oltre 301,000 punti inventariali del campione di prima fase INFC è stata eseguita tramite fotointerpretazione di ortofoto digitali su una piattaforma WebGIS dedicata denominata GeoInfo, sviluppata dalla società AlmavivA con la collaborazione di Telespazio (Gasparini et al., 2020). La piattaforma consente di visualizzare le ortofoto (Fig. 3.1) e altri strati informativi utili, quali limiti amministrativi, viabilità, toponomastica, rete idrografica, fasce altimetriche, riserve naturali statali e aree percorse da incendi, nonché di acquisire le classificazioni dei fotointepreti e memorizzare i dati in un archivio centrale nazionale.

La fotointerpretazione INFC2015 è stata realizzata su ortofoto digitali a colori (RGB) e all'infrarosso-falso colore con risoluzione di 50 cm derivanti dalle coperture AGEA 2010–2012. Per risolvere casi dubbi, i fotointerpreti potevano consultare anche le ortofoto digitali a colori di uguale risoluzione derivanti dalle coperture AGEA 2007–2009, e ortofoto con risoluzione di 1 m derivanti dalla copertura 1999–2005. Queste ultime, utilizzate dai fotointerpreti nella prima fase del secondo inventario INFC2005, sono state impiegate per valutare eventuali cambiamenti significativi di uso e copertura del suolo intervenuti successivamente, attraverso il confronto visivo delle immagini riferite alle due diverse epoche.

La prima fase INFC2015 è stata condotta da circa 50 fotointerpreti, in parte del Corpo Forestale dello Stato e in parte dei Servizi forestali delle Regioni a statuto speciale e delle Province autonome, opportunamente formati e addestrati mediante uno specifico corso di formazione. I fotointerpreti hanno operato da remoto, presso i rispettivi uffici, utilizzando la piattaforma dedicata.

La procedura di classificazione INFC consiste nell'attribuire ciascun punto di campionamento alla classe e sottoclasse di uso e copertura del suolo del poligono in cui il punto ricade, previa verifica del superamento delle soglie minime di estensione e larghezza. Il poligono rappresenta un'area omogenea per uso e copertura del suolo, avente una superficie maggiore di 0.5 ha e una larghezza superiore a 20 m. Limitatamente ai poligoni caratterizzati da una copertura arborea o arbustiva, il fotointerprete verifica anche il superamento o meno delle soglie minime di copertura (se specie arboree, 10% per il Bosco e 5% per i Boschi radi; se specie arbustive 10% per gli Arbusteti) previste dalla definizione adottata per il dominio inventariale (cfr. Chap. 2). Un'ulteriore soglia di copertura del 40%, relativa alla componente erbacea, viene utilizzata per distinguere la sottoclasse Praterie, pascoli e incolti da quella delle Aree con vegetazione rada o assente. In presenza di poligoni di dimensioni più piccole, con superficie compresa fra 500 e 5000 m^2 oppure, se di forma allungata, con larghezza compresa fra 3 e 20 m, si registra la presenza di un "incluso" e si assegna al punto di campionamento l'uso e copertura del suolo del poligono più vicino che rispetta le soglie minime di superficie e larghezza sopra indicate, rispettivamente 0.5 ha e 20 m.

Il WebGIS GeoInfo gestisce in automatico la sequenza delle operazioni per la fotointerpretazione, proponendo all'operatore i punti da classificare e permettendo di verificare le soglie minime attraverso la visualizzazione di un intorno di analisi (cfr. Chap. 2), di una griglia di punti e di strumenti di misura delle distanze e delle superfici (Fig. 3.2) (Gasparini et al., 2021). L'intorno di analisi, costituito da una griglia di nove celle quadrate di lato pari a 50 m e superficie di 2500 m^2 centrata nel punto di campionamento, permette di stimare a vista l'estensione e la larghezza dei poligoni omogenei individuati. La griglia di punti distanti tra loro 10 m, invece, consente di valutare in modo rapido e oggettivo il grado di copertura mediante il conteggio dei punti che intercettano le chiome di alberi o arbusti.

Per dettagli sulle procedure di classificazione ed eventuali casi particolari si rimanda a Gasparini et al. (2014).

Controlli di qualità della fotointerpretazione

La classificazione della copertura del suolo mediante interpretazione visiva di foto aeree comporta sempre un certo grado di soggettività e l'esperienza dei fotointerpreti gioca un ruolo importante nel determinare la qualità del risultato (Strand et al., 2002). La soggettività dovrebbe essere limitata il più possibile allo scopo di ottenere classificazioni comparabili tra loro. L'attuazione di una procedura di controllo della qualità durante l'attività di fotointerpretazione e alla sua conclusione è fondamentale, sia per valutare l'uniformità di giudizio da parte dei fotointerpreti e la riproducibilità della classificazione, sia per valutare l'accuratezza della classificazione.

Le procedure di assicurazione della qualità dei dati (QA, Quality Assurance) attuate per la prima fase INFC sono finalizzate al controllo degli errori non campionari dovuti a errori di misura nella verifica delle soglie minime di estensione e larghezza dei poligoni e della copertura arborea e arbustiva, oppure alla non corretta interpretazione delle regole di interpretazione delle immagini da parte dei fotointerpreti. La procedura di controllo si basa sull'individuazione di obiettivi di qualità (MQOs, Measurement Quality Objectives) e di limiti di qualità (DQLs, Data Quality Limits), corrispondenti rispettivamente a errate classificazioni o errori di misurazione tollerabili e alle relative soglie massime ammissibili (Gasparini et al., 2009). I MQOs relativi a errate classificazioni tollerabili per la QA di INFC2015 sono riportati in Table 3.2. Riguardo alle soglie minime di superficie e larghezza sono stati considerati tollerabili errori di misura rispettivamente di 200 m^2 e 2 m per i poligoni e di 50 m^2 e 1 m rispettivamente per gli inclusi. Per la copertura, sono stati considerati tollerabili errori fino a 2.8% di copertura arborea o arbustiva e 5.5% di copertura erbacea, valori corrispondenti rispettivamente a un punto e a due punti della griglia utilizzata per la stima della copertura (cfr. Chap. 4). In Table 3.3 sono riportati i DQLs per le diverse classi di uso e copertura del suolo, stabiliti in funzione dell'importanza delle singole classi e sottoclassi e della relativa difficoltà di riconoscimento sulle ortofoto. Maggiore è l'importanza di una classe o sottoclasse, più elevato è il relativo DQL e minore è la soglia percentuale di errori ammissibili; al contrario, maggiore è la difficoltà di classificazione, minore è il relativo DQL e più elevata è la soglia percentuale di errori ammissibili.

L'attività di QA durante la fotointerpretazione è stata realizzata attraverso la riclassificazione di una certa quantità di punti, selezionati casualmente, da parte di un gruppo di operatori esperti del CREA Centro di ricerca Foreste e Legno, i quali hanno assunto il ruolo di operatori di riferimento. Gli operatori del CREA hanno utilizzato anch'essi la piattaforma GeoInfo per la classificazione dei punti, senza che fosse reso loro disponibile l'esito della classificazione già eseguita del fotointerprete incaricato. Successivamente, l'operatore di riferimento confrontava la propria classificazione con quella del fotointerprete incaricato e, in caso di discordanza, valutava se essa fosse ammissibile sulla base degli errori di misura o di classificazione ritenuti tollerabili secondo i MQOs. In caso di discordanza non risolvibile veniva inviata al fotointerprete una richiesta di verifica ed eventuale modifica della classificazione. Una piattaforma informatica appositamente implementata, accessibile dall'intranet del CREA Centro di ricerca Foreste e Legno, veniva utilizzata dagli operatori interni addetti ai controlli periodici e finali per registrare l'identificativo dei punti interessati e gli esiti dei controlli. In totale, durante la fotointerpretazione sono stati eseguiti controlli su 9766 punti, pari al 3.2% dei punti del campione INFC.

Oltre ai controlli in corso d'opera sopra descritti, sono stati realizzati dei controlli finali, o collaudi, su un sottocampione dei punti inventariali per regione selezionato in modo casuale. Il controllo finale ha riguardato il 2% dei punti inventariali di ciascuna regione. Successivamente, per tre regioni interessate dal mancato raggiungimento dei DQLs per le sottoclassi di maggiore interesse, si è proceduto ad una parziale revisione della classificazione e ad un nuovo controllo su un ulteriore 2% dei punti di campionamento. L'esito finale dei collaudi per le Aree boscate (Table 3.4) mostra che la discordanza di classificazione tra fotointerpreti e operatori di riferimento interessa il 2.1% dei punti a livello nazionale e percentuali sempre inferiori alla soglia massima del 5% in tutte le regioni.

Durante l'attività di classificazione è stato realizzato un ulteriore controllo di tipo blind check su un sottocampione casuale di punti inventariali, i quali venivano assegnati contemporaneamente a due fotointerpreti della stessa regione, all'insaputa degli stessi. I risultati del blind check verranno utilizzati per successive analisi finalizzate all'eventuale modifica del sistema di classificazione.

References

Brivio, P. A., Lechi, G., & Zilioli, E. (2006). Principi e metodi di Telerilevamento. Città Studi Edizioni, Torino. 525 pp. ISBN 978-88-251-7293-5.

Dainelli, N. (2011). L'osservazione della Terra—Fotointerpretazione: Metodologie di analisi a video delle immagini digitali per la creazione di cartografia tematica. Dario Flaccovio Editore s.r.l., Palermo. 235 pp.

European Commission (1993). CORINE Land Cover guide technique. Office des Pubblications Officielles des Communautés Européennes, Luxembourg. 144 pp.

FAO (2001). Global Forest Resources Assessment 2000. FRA 2000. Main report. FAO Forestry Paper 140. Rome. Retrieved Jan 25, 2022, from https://www.fao.org/forestry/fra/86624/en/.

Gasparini, P., Floris, A., Rizzo, M., Di Cosmo, L., Morelli, S., & Zanotelli, S. (2021). Il contributo della geomatica alle attività del terzo inventario forestale nazionale italiano INFC2015. In: Atti AsitaAcademy2021 (pp. 243–260). Asita, Milano. ISBN 978-88-941232-7-2.

Gasparini, P., Floris, A., Rizzo, M., Patrone, A., Credentino, L., Papitto, G., & Di Martino, D. (2020). Il terzo inventario forestale nazionale italiano INFC2015: procedure, strumenti e applicazioni. *GEOmedia, 24*(6), 6–17. ISSN 1128-8132.

Gasparini, P., & Di Cosmo, L. (2016). National Forest Inventory Reports—Italy. In: C. Vidal, I. Alberdi, L. Hernández, & J. Redmond (Eds.), *National forest inventories—Assessment of wood availability and use*. Springer. ISBN 978-3-319-44014-9. https://doi.org/10.1007/978-3-319-44015-6.

Gasparini, P., Rizzo, M., & De Natale, F. (2014). Manuale di fotointerpretazione per la classificazione delle unità di campionamento di prima fase. Inventario Nazionale delle Foreste e dei Serbatoi Forestali di Carbonio, INFC2015—Terzo inventario forestale nazionale. Consiglio per la Ricerca e la sperimentazione in Agricoltura, Unità di Ricerca per il Monitoraggio e la Pianificazione Forestale (CRA-MPF); Corpo Forestale dello Stato, Ministero per le Politiche Agricole, Alimentari e Forestali (64 pp.). ISBN 978-88-97081-73-9.

Gasparini, P., Bertani, R., De Natale, F., Di Cosmo, L., & Pompei, E. (2009). Quality control procedures in the Italian national forest inventory. *Journal of Environmental Monitoring, 11*, 761–768.

Hall, R. J. (2003). The roles of aerial photographs in forestry remote sensing image analysis. In: M. A. Wulder, & S. E. Franklin, (Eds.), *Remote Sensing of Forest Environments* (pp. 47–75). Springer. https://doi.org/10.1007/978-1-4615-0306-4.

Howard, J. A. (1991). *Remote sensing of forest resources. Theory and application* (420 pp.). Chapman & Hall. ISBN 0-412-29930-5.

Strand, G., Dramstad, W., & Engan, G. (2002). The effect of field experience on the accuracy of identifying land cover types in aerial photographs. *International Journal of Applied Earth Observation and Geoinformation, 4*, 137–146. https://doi.org/10.1016/S0303-2434(02)00011-9

Tomppo, E., Gschwantner, T., Lawrence, M., & McRoberts, R. (Eds.) (2010). National forest inventories—Pathways for common reporting (606 pp.). Springer. ISBN 978-90-481-3232-4. https://doi.org/10.1007/978-90-481-3233-1.

Open Access This chapter is licensed under the terms of the Creative Commons Attribution 4.0 International License (http://creativecommons.org/licenses/by/4.0/), which permits use, sharing, adaptation, distribution and reproduction in any medium or format, as long as you give appropriate credit to the original author(s) and the source, provide a link to the Creative Commons license and indicate if changes were made.

The images or other third party material in this chapter are included in the chapter's Creative Commons license, unless indicated otherwise in a credit line to the material. If material is not included in the chapter's Creative Commons license and your intended use is not permitted by statutory regulation or exceeds the permitted use, you will need to obtain permission directly from the copyright holder.

Chapter 4
Field Assessment—Survey Protocols and Data Collection

Modalità di rilievo in campo e archiviazione dei dati

Antonio Floris, Lucio Di Cosmo, Maria Rizzo, and Amato Patrone

Abstract The INFC2015 field campaign surveyed almost 9000 sample points and assessed and measured a relevant number of variables. This chapter describes the procedures adopted to retrieve the sample points marked in the previous NFI (INFC2005) and to reach and mark sample points never located before on the ground. The chapter also describes the protocol used for classifying and measuring the variables, either at a stand level or a single item level as well as the relevant sample unit (point or plot) for measuring each variable. Some sections of the chapter describe the electronic devices and procedures used for storing the field data and sending them to the central database. It also details the way in which crews were supported remotely and the way data quality assurance was applied during the campaign, both in the field and from a distance.

Keywords Sample plots · NFI survey · GNSS positioning · DBH measurements · Tree cores · Coarse woody debris · Sample trees

4.1 Introduction

Field surveys can be easily sketched by a sequence of basic, fundamental steps. In the great majority of cases, they consist of reaching the NFI sample point defined by its coordinates; assess the value of some variables; lay out sample plots (AdS) around or near it, within which to assess or measure other variables; and permanently mark the point, so as to be found in future field campaigns. The listed steps are also specific stages in the field campaign planning process, which are described in this chapter.

Navigation and positioning are determined through the use of global navigation satellite systems (GNSS); more specifically, special devices equipped with GNSS

A. Floris (✉) · L. Di Cosmo · M. Rizzo
CREA Research Centre for Forestry and Wood, Trento, Italy
e-mail: antonio.floris@crea.gov.it

A. Patrone
AlmavivA S.P.A. - Agriculture Operations & Solutions, Roma, Italy

© The Author(s) 2022

P. Gasparini et al. (eds.), *Italian National Forest Inventory—Methods and Results of the Third Survey*, Springer Tracts in Civil Engineering,
https://doi.org/10.1007/978-3-030-98678-0_4

that are suitable to operate under dense tree cover and still maintain good positioning accuracy are employed. Sample point position is a fundamental variable assessed by NFIs, as it allows revisiting the sample plot and coupling ground data with remote sensing data for further analysis (Dalponte et al., 2011; Gobakken & Næsset, 2009; Kitahara et al., 2010). Assuring effective positioning for future retrievals of the sample point is done by marking it with either permanent buried markers or with other visible marks that help crews find the buried one, based on recorded information on their relative positions.

Land use and land cover classifications, the first carried out as preliminary to any other assessment, classification of vegetation and other qualitative characters as well as the measuring on the quantitative variables are the essence of NFIs. In this chapter, the methods adopted to survey the variables in INFC2015 are described. These are of general validity but adaptations of the protocol to special and rare circumstances are available from the field manual (Gasparini et al., 2016) or other cited references.

Recording the information collected in the field, either what is necessary for internal needs (e.g., those about the sample point markers) or necessary to produce statistics, is supported by computer tools for many reasons. These tools can make information more accessible and useful to surveyors (e.g., to validate or update information collected in the past field survey). They may help in more specific tasks (e.g., navigation and positioning). They may suggest logical order during the different steps of the survey, or ease data input by menu lists. They also allow for automatic checks, preventing entering inconsistent data or warnings in case the information entered is possible but dubious (possible inconsistency due to typing errors). Lastly, recordings allow easy electronic storage in the INFC central database simply by data transmission.

Data quality is another important aspect of field surveys. In high complex monitoring projects developed over years, good quality data cannot be based simply on a sequence of data collection, data check, data approval or rejection, as rejection implies severe consequences, with the work to be done again. It is important to support the crews while they are conducting the survey, not only to prevent mistakes and check the data in the central database to correct them, but also to recognize errors and prevent repetition of mistakes (Gasparini et al., 2009). Supporting the crews and carrying out quality checks on the database during the field campaign are activities to be taken into consideration during the planning of the field surveys, because they imply choosing appropriate communication tools and a proper database design.

4.2 Navigation, Positioning and Marking of Sample Points

Limiting the positional uncertainties of sample points in the different phases of the NFI, from photointerpretation to ground surveys, is a very important requirement for data quality. INFC adopts a three-phase sampling design. Phase 1 is carried out by photointerpretation and the other two phases use field surveys. Phase 2 uses about 30,000 sampling points, and phase 3 is a subsample of those. Photointerpretation

Fig. 4.1 GNSS navigation, positioning and data recording devices used during the field campaign of INFC2015 / La strumentazione per la navigazione, il posizionamento e la registrazione dei dati utilizzata durante i rilievi in campo INFC2015

is conducted in a sample area (photoplot 2500—FP2500) whose centre is the NFI sample point defined by its coordinates (cf. Chap. 2). The procedure to locate an inventory point on the ground with the highest possible accuracy in respect to its position in the interpreted orthophoto was developed in INFC2005 (Floris et al., 2011). This procedure was also adopted in INFC2015. INFC2005 also established marking procedures capable of ensuring retrieval of the sample points after a few months during phase 3. INFC2015 carried out a sole field campaign on sampling points that in 90% of cases had already been surveyed and marked in INFC2005. For this reason, it has been necessary to experiment and adopt a navigation and positioning strategy appropriate to maximise retrieval of existing marked points or establish new ones on the ground.

Navigation to the sample point (called C-point after its establishment on the ground) and its positioning were done using a multi-constellation GNSS receiver Trimble R1, operated by a datalogger Trimble Juno SB. Given the technological difficulties in using a real-time differential correction in many Italian forest areas, it was decided to use, whenever possible, a satellite-based augmentation system (SBAS) provided by the EGNOS service (European Commission, 2017; Gasparini et al., 2021). Approaching a position about 10–15 m distance from the sample point was carried out in instant position navigation (NPI) mode, while reaching specific coordinates (e.g., C-point established in INFC2005) from that position was carried out by a procedure called navigation from average position (NPM) (Colle et al., 2009).

Figure 4.1 shows the devices used for navigation, positioning and data recording in the field.

The INFC sample points visited in the field during the previous inventory have a positioning uncertainty of 3–4 m, depending on the receiver used at the time (Colle et al., 2009). Those points were marked with two types of buried stakes: temporary stakes, for the points visited in INFC2005 phase 2 only, consisting of 30 cm long nail and a small aluminum head plate; and permanent stakes, with a steel tip with anchors and a larger aluminum head, for the points also visited in INFC2005 phase 3. Near the sample point, at approx. 10–15 m under the best local conditions for receiving the GNSS signal, an end of navigation point (F-point) was also marked through a buried temporary stake. Markings included nailing small aluminum plates to the base of three reference trees, whose monographic information recorded their species, DBH, distances and azimuth to C-point or F-point and photographs. In INFC2015, F-point and C-point stakes were searched by a metal detector, with the help of the monographic descriptions and the photographs taken in the previous inventory campaign. Examples of the marks are shown in Fig. 4.2. Figure 4.3 shows the regional and national rates of sample points for which at least one of the stakes (either in the F-point or in the C-point) was successfully found.

In the INFC2015 field campaign, only permanent stakes were buried in C-points, in sample points in forest land use/land cover. To mark F-points, temporary stakes were used instead, considering the successful retrieval rate. At the end of the survey for each inventory point, a stationary GNSS positioning on F-point was conducted, calculating the coordinates as the mean of 50 single positions. Table 4.1 shows some positioning uncertainty parameters calculated from the GNSS raw files on more than 8000 points. As the C-point is reached measuring the azimuth and the distance from the F-point with a compass and a rangefinder, C-point and F-point positions undergo similar uncertainty. Finally, before leaving the sample plot, the monographic descriptions of the markers were updated and new photographs were taken.

4.3 Variables, Their Classifications and Measurements

Field surveys allow definitive classification of land use and vegetation features relevant to verify correct inclusion of each sample point in the NFI domain and to classify the point in relation to a wide range of qualitative characteristics in order to measure quantitative variables. Table 4.2 lists the variables assessed or measured with the field survey and the reference sample unit (sample point or plot) used for each variable. Assessments and measurements are carried out with reference to the sample point or to one of the sample plots (AdS) that have it as the centre (AdS4, AdS13, AdS25) or close to it (AdS2) (cf. Chap. 2). This section describes the methods used for variable assessment or measurement. The classes adopted are described in the tables at the end of the chapters with the results, which show the statistics for the variables.

Fig. 4.2 The buried stakes and the external markings used to mark the sample points in the field / I picchetti interrati e le marcature esterne usati per marcare i punti di campionamento in campo

4.3.1 Land Use, Land Cover, Pure/mixed Forests, Stand Origin

Crews first assess if the inventory sample point is in Total wooded area, by observing the land use and land cover (of trees and shrubs) status. The latter is quantified based on the orthophoto already used by the photointerpreter, available in the tablet. Surveyors judge potential crown cover, in case of temporarily unstocked areas, and accurately measure land features (e.g., roads and water streams width) when relevant for the adopted definitions in Chap. 2. A software application specifically developed in INFC2015, INFC_APP (cf. Sect. 4.4) guides assignment of a sample plot to the

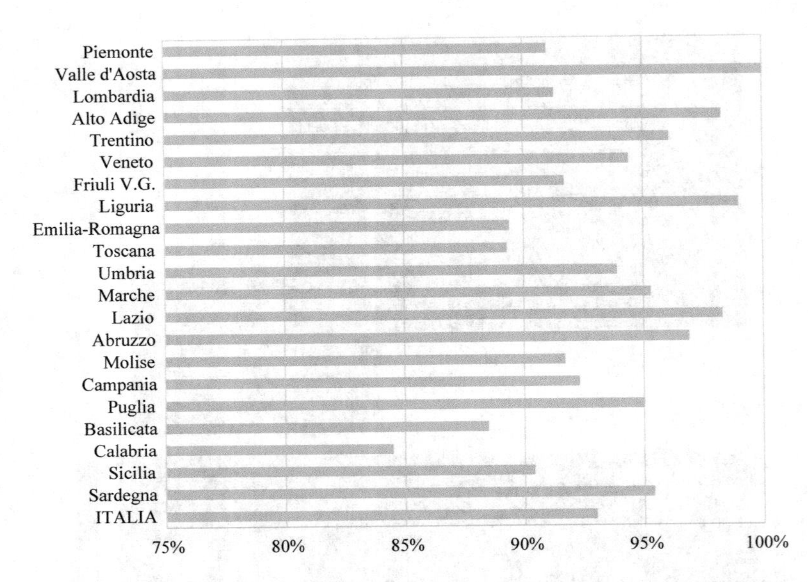

Fig. 4.3 Regional and national rates of sample points where at least one stake of INFC2005 (either point C or point F) was found in INFC2015 / Aliquote percentuali regionali e nazionale dei punti di campionamento nei quali almeno un picchetto di INFC2005 (punto C o punto F) sia stato ritrovato in INFC2015

Table 4.1 Positioning uncertainty parameters of sample points under different SBAS-EGNOS correction status / Parametri di incertezza del posizionamento dei punti di campionamento, in diverse condizioni di correzione SBAS-EGNOS

Correction status	N obs	Max PDOP	Max HDOP	Horizontal precision (m)	Vertical precision (m)	Standard deviation (m)
Stato della correzione	Num casi			Precisione orizzontale (m)	Precisione verticale (m)	Deviazione standard (m)
Correct Corretti	2792	1.99	1.06	0.71	1.06	0.67
Uncorrected Non corretti	5381	2.65	1.46	4.75	6.79	1.24
Total Totali	8173	2.43	1.33	3.37	4.83	1.05

proper crown cover class. A systematic grid is superimposed to the orthophoto in the reference area (FP2500, cf. Chap. 2). Crews count the number of sample points falling on crowns and specify if these are shrubs or trees. Besides distinguishing Forest from Other wooded land, this assessment enables producing estimates on areas by total canopy cover, and by tree canopy cover. INFC_APP can also compute

Table 4.2 Variables surveyed in the field in INFC2015, and their reference sample units / Variabili rilevate in campo in INFC2015 e rispettive unità di campionamento

Variable/Variabile	Sample unit Unità di campionamento
Land use-Land cover/Classe di uso-copertura del suolo	Analysis window Intorno di analisi
FOREST STAND ATTRIBUTES/ATTRIBUTI DELLA VEGETAZIONE FORESTALE	
Canopy cover (trees and shrubs)/Copertura delle chiome (alberi e arbusti)	FP2500
Inventory category/Categoria inventariale	Analysis window Intorno di analisi
Forest type and subtype/Categoria e sottocategoria forestale	AdS25
Conifers and broadleaves pure-mixed condition/Grado di mescolanza conifere e latifoglie	AdS25
Stand origin/Origine del soprassuolo	AdS25
Development stage/Stadio di sviluppo	AdS25
Stand age (even-aged stands)/Età media del soprassuolo (soprassuoli coetanei)	AdS25
Vertical structure/Struttura verticale	AdS25
LEGAL STATUS/STATO GIURIDICO-AMMINISTRATIVO	
Ownership/Proprietà	Point C
Constraints/Vincoli	Point C
Nature protection/Protezione naturalistica	Point C
Forest planning/Pianificazione forestale	Point C
SITE CONDITIONS/CARATTERI STAZIONALI	
Accessibility/Accessibilità	Point C
Distance of roads/Distanza da viabilità	Point C
Altitude/Altitudine	Point C
Slope/Pendenza	Point C
Aspect/Esposizione	Point C
Local land position/Giacitura locale	AdS25
Extended land position/Giacitura estesa	Large area
Terrain instability/Dissesto	AdS25
Terrain roughness/Accidentalità	AdS25
SILVICULTURE/SELVICOLTURA	
Silviculture system/Tipo colturale	AdS25
Intensity of silvicultural practices/Pratiche selvicolturali	AdS25
Primary designated management objective/Funzione prioritaria	AdS25

(continued)

Table 4.2 (continued)

Variable/Variabile	Sample unit Unità di campionamento
Non-wood products and services/Prodotti e servizi non legnosi	AdS25
Availability for wood supply/Disponibilità al prelievo legnoso	Point C
Utilisation mode/Modalità di utilizzazioni	AdS25
Logging mode/Modalità di esbosco	AdS25
FOREST HEALTH/STATO DI SALUTE	
Damaging diffusion and severity/Diffusione e intensità di patologie o danni	AdS25
Damage causes/Cause di patologie o danni	AdS25
Defoliation intensity/Intensità della defogliazione	AdS25
Defoliation localization/Localizzazione della defogliazione	AdS25
TALLY TREES/ALBERI MISURATI	
Species/Specie	AdS4/AdS13
DBH/$d_{1.30}$	AdS4/AdS13
Vitality and integrity/Vitalità e integrità	AdS4/AdS13
Type of tree/Dendrotipo	AdS4/AdS13
Height of broken trees/Altezza degli alberi troncati	AdS4/AdS13
SAMPLE TREES/ALBERI CAMPIONE	
Total tree height/Altezza totale	AdS4/AdS13
Crown base height/Altezza di inserzione della chioma	AdS4/AdS13
Social position/Posizione sociale	AdS4/AdS13
Increment cores/Prelievo di carote incrementali	AdS4/AdS13
FOREST UNDERSTOREY/RINNOVAZIONE E ARBUSTI	
Diameter and height class/Classe dimensionale	AdS2
Species/Specie	AdS2
Number/Numero	AdS2
Origin/Origine	AdS2
Damages/Patologie e danni	AdS2
STANDING DEAD TREES/ALBERI MORTI IN PIEDI	
Species/Specie	AdS4/AdS13
DBH/$d_{1.30}$	AdS4/AdS13
Decay class/Classe di decadimento	AdS4/AdS13
Tree height (if broken)/Altezza del fusto (se troncati)	AdS4/AdS13
DEADWOOD LYING ON THE GROUND/LEGNO MORTO GROSSO A TERRA	

(continued)

Table 4.2 (continued)

Variable/Variabile	Sample unit Unità di campionamento
Species/Specie	AdS13
End sections diameter/Diametro delle estremità	AdS13
Length/Lunghezza	AdS13
Decay class/Classe di decadimento	AdS13
STUMPS/CEPPAIE	
Species/Specie	AdS13
Diameter/Diametro	AdS13
Height/Altezza	AdS13
Decay class/Classe di decadimento	AdS13
Cutting age/Epoca del taglio	AdS13

the predominance of conifer or broadleaf cover or neither (in mixed forests) to classify the stand as pure or mixed conifers/broadleaves.

Stand origin is assigned considering the intensity of human actions to promote or sustain regeneration; actions may be absent (natural stands), by silviculture (semi-natural stands) or intense (plantations).

4.3.2 Inventory Categories, Forest Types and Forest Subtypes

Inventory categories of Forest and Other wooded land are classified by observing features mainly related to composition (e.g., trees or shrubs), site potential for growth (e.g., tall trees or short trees), and stand origin (e.g., seminatural or planted). The inventory categories of Forest and Other wooded land are further classified into forest types and subtypes.

Forest type is classified based on the dominant species or group of species in terms of crown coverage. First, it is assessed if dominance is due to coniferous or broadleaved species, deciduous or evergreen; forest type must be consistent with that information, i.e., if crown cover prevalence is by conifer, one of the coniferous forest types is expected. In case of mixed stands, deciding which species or group of species is predominant may require a walk along with transects and recording of the species of the upper layer tree at established points (Fig. 4.4). Such procedures and computations are assisted by INFC_APP. Forest subtype is assigned based on the species composition and stand ecological characteristics.

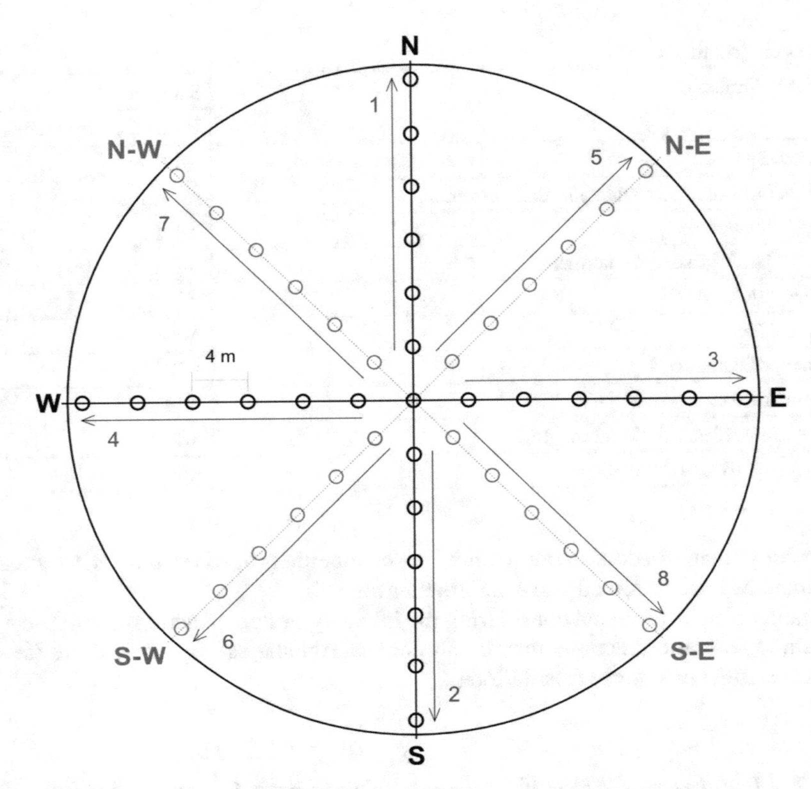

Fig. 4.4 Walking transects across AdS25 to assess tree species crown cover, by standing in the positions indicated by the circles. Red numbers show the order to follow to obtain at least 25 observations / Disposizione dei punti di osservazione della copertura delle specie entro AdS25; i numeri in rosso indicano l'ordine dei transect fino al raggiungimento di un totale di 25 classificazioni

4.3.3 Legal Status

Ownership was assessed based on the position of the inventory sample point, accessing cadastral information available to crews as a GIS layer, or by interviewing local forestry or other administrative personnel.

Limitations on the use of forest resources or the inclusion in protected areas (type and protection level) were assessed on documents or maps or given by local personnel.

Presence of forest planning was assessed based on documents or local personnel knowledge; planning may be present at various levels (e.g., regional guideline plans or property-based operative management plans).

4.3.4 Accessibility, Presence of Roads and Site Features

A sample point is accessible if the crew can reach it and can carry out the required assessment and measurements in the relative AdS. Physical conditions, prohibited access under laws or from owner's decisions and risk of severe injury for the surveyors may result in inaccessibility. Flexibility is allowed during the survey in challenging cases. For example, if it is too risky, DBH of some trees might be visually estimated or the measurements limited to AdS4. Any special adaptation is recorded to allow consistent data processing.

Simple and fast access to a sample point also depends on the presence of roads. This is first assessed on maps, orthophotos, etc. and then checked and verified in the field. Crews record distances and differences of elevation between the NFI sample point and the closest road or forest track.

The inventory sample point altitude is measured by the GNSS device, while recording the coordinates (cf. Sect. 4.2).

Aspect is measured using a compass, by reading the magnetic azimuth; the measurer stands at the sample point aiming the compass downwards. The slope angle is measured by a clinometer. The measurer stands at the sample point and aims towards two topographic poles, placed 25 m distance uphill and downhill along the maximum slopes; the recorded value is the average of the two measures.

Land position is observed by referring to both the AdS25 (local land position) and to a larger area, from a few hectares to a few tens of hectares around the plot (extended land position). Roughness, assessed in the AdS25, is the micro-morphology of the terrain, determined by the presence of obstacles such as boulders, rocks, ditches and sinkholes that could condition any logging operations (felling, concentration, removal) and, in general, the availability of the area around the sample point.

Any presence of terrain instability under way, repetitive or occasional, is surveyed in the AdS25 or the surrounding area. The instability must affect an area of at least 100 m^2 to be recorded.

4.3.5 Silviculture, Stand Characters, Production

Applied silviculture considers type and intensity of practices. These are strictly related to the products that are obtained, so the classification system has specific classes for the practices aiming at obtaining secondary, non-wood products and social services.

When the management is strongly and clearly oriented to mainly obtain one specific good or service from the forest, this is classified under the primary designated management objective condition, and the specific good or service is noticed.

Availability for wood supply is information provided based on documents (e.g., restrictions by environmental protection laws) or assessing in the field eventual economic convenience of utilisation or presence of severe constraints. Convenience of utilisation might come by considering, for example, the value of the exploitable wood in relation to the difficulties and costs in accessing remote forests.

The silvicultural system and the utilisation mode are recorded when observing stand structure and characters, while the logging mode also considers the exploitable products (e.g., timber or firewood), local tradition (e.g., cables are mainly used in the Alps and less on the Apennines) and road presence.

Development stage information is based on visual assessment as well as the age class, which is recorded only in even-aged stands; however, the information is confirmed after the tree coring described in Sect. 4.3.8, and once tree rings have been counted.

4.3.6 *Forest Health*

The survey aims to provide a general overview on forest health conditions and records the presence of diseases and damages when they affect at least 30% of AdS25. When this threshold is reached, the cause of the disease or damage is assessed as well as its severity on the trees that are sick or damaged. In case the disease or damage implies defoliation, its intensity is recorded, and an indication is given about which part of the crown is primarily affected.

4.3.7 *Tally Trees*

Tally trees are woody plants, either alive or dead, with DBH $\geq$ 4.5 cm inside the AdS4 or $\geq$ 9.5 cm in the AdS13. A tree is in or out of a sample plot depending on the distance between the plot centre and the vertical axis passing through the centre of the section at 1.30 m aboveground level. DBHs are measured using a calliper. On slope terrain (>10°) the measurer stands uphill from the tree, otherwise the graduated beam is along the plot radius. When the first measured DBH exceeds 9.5 cm, its cross-section diameter is also measured. For each callipered tree, the species, vitality and integrity (Table 4.3) and type of tree (Table 4.4) are also recorded. In case of broken trees, the height is measured. For dead trees, the decay class is recorded, based on visual assessment (Table 4.5).

4.3.8 *Sample Trees*

Ten trees among the tally trees are selected for additional measurements. These include the five closest trees to the plot centre, the three remaining largest and two 'rare' trees among the remaining. A rare tree may be a tree of species or size class (in monospecific stands) not included within the previous eight. Whenever possible,

Table 4.3 Tree vitality-integrity / Classi per la vitalità-integrità degli alberi

Tree vitality-integrity Vitalità-integrità dell'albero	Description Descrizione
Alive and intact Vivo ed integro	Living tree with no dry or broken portions, neither missing part of the crown
	Individuo vivo e privo di evidenti parti secche, danneggiate o mancanti della chioma
Little damaged Vivo, con leggere menomazioni	Living tree with dry or broken portions not exceeding 1/3 of the crown
	Individuo vivo con parti secche, danneggiate o mancanti che interessano meno di 1/3 della chioma
Moderately damaged Individuo vivo con moderate menomazioni	Living tree with portions dry, broken or missing in between 1/3 and 2/3 of the crown
	Individuo vivo con parti secche, danneggiate o mancanti comprese tra 1/3 e 2/3 della chioma
Heavy damaged Individuo vivo con forti menomazioni	Living tree with portions dry, broken or missing exceeding 2/3 of the crown
	Individuo vivo, ma fortemente compromesso, con parti secche, danneggiate o mancanti per oltre 2/3 della chioma
Alive but broken Individuo vivo ma troncato	The stem is trunked but there are still living portions
	Individuo ancora vivo per la presenza di getti o parti verdi, nonostante presenti il fusto troncato
Dead and intact Individuo morto e integro	Standing dead tree, still intact
	Individuo morto, completamente secco in piedi, ma integro
Dead and broken Individuo morto e troncato	Dead tree still standing but the stem is trunked
	Individuo morto, completamente secco, con il fusto troncato

sample trees must be free from visible faults. For each tree, the total tree height and crown base height are measured, and the tree is classified as dominant, intermediate or dominated tree. Tree height is measured using a Haglof Vertex hypsometer.

In order to estimate the annual diameter increment, the sample trees are cored with a Pressler increment borer, at 1.30 m aboveground level, along with the plot radius direction. From each sampled tree, one core is extracted and the length of the five outermost rings (excluding the one from the current year) is measured with a ruler. Each core is labelled and sent to the dendrochronology laboratory at CREA Research Centre for Forestry and Wood in Trento for a measurement check.

Table 4.4 Type of tree / Dendrotipo

Type of tree/Dendrotipo	Description/Descrizione
High tree Individuo d'alto fusto	It is a tree originated from seed germination. Trees of unknown origin but shaped as high trees and those originated as coppice shoots but shaped as a high tree after singling
	Albero proveniente dalla germinazione di un seme; in questa classe devono essere annoverati anche gli individui per i quali la determinazione dell'origine (da seme o da ceppaia) risulta piuttosto incerta ma che hanno un portamento di albero d'alto fusto; si assegnano a questa classe anche i polloni rilasciati dopo il taglio di avviamento all'alto fusto in una conversione
Shoot Pollone	Tree originated by a stool, either for silviculture operations (coppicing) or because of natural damage
	Individuo originato dal ricaccio della ceppaia a seguito di operazioni colturali (ceduazione) o per cause naturali o accidentali (traumi alla ceppaia che hanno provocato ricacci, polloni radicali, ecc.)
Standard Matricina	Standard left in a coppice stand to produce seeds to replace old stools with reduced vitality
	Pianta rilasciata dopo il taglio di un ceduo per uno o più turni successivi, allo scopo di disseminare e di sostituire, dopo il taglio, le ceppaie esaurite
Indeterminable Non determinabile	Neither of the three cases; shrub-like subjects, lianas, etc.
	Tutti i casi non rientranti in una delle tre classi precedenti; forme arbustive, liane, ecc.

4.3.9 Forest Understorey

Small trees and shrubs are distinguished based on the INFC species list. Distinction is not relevant for measurements, since what must be measured only relies on size thresholds. Small trees and shrubs are woody entities with a diameter of less than the calipering threshold of 4.5 cm but higher than 50 cm. They are assigned to one of the classes in Table 4.6. They are measured in two AdS2, 10 m distance from the sampling point, East and West positioned (cf. Chap. 2). The survey consists of counting the plants separately by species and size class. For each species surveyed, the prevalent origin is recorded (artificial, agamic, by seed), eventual damage, if affecting at least 30% of the plants, and the cause, if recognisable (by animals, e.g., by pasture or wildlife, or by weather).

4.3.10 Deadwood Lying on the Ground and Stumps

Deadwood lying on the ground (complete trees, stems, branches, etc.) must have a diameter and length of at least 9.5 cm and is measured within the AdS13. Each

Table 4.5 Decay classes adopted for deadwood assessment (tally trees, stumps, deadwood lying on the ground) / Classi di decadimento per il legno morto grosso (alberi morti, ceppaie, legno morto grosso a terra)

Decay classes Classi di decadimento	Description Descrizione
Recently dead Individuo non decomposto	Bark still attached. Small branches (diameter < 3 cm) present. Wood consistency intact. Fungus mycelium absent or poorly developed
	Corteccia intatta e attaccata al legno. Rametti con diametro inferiore ai 3 cm presenti. Legno intatto. Micelio fungino assente o poco sviluppato
Weakly decayed Individuo parzialmente decomposto	Bark is loose but not fragmented. Small branches are only partially present. Wood consistency intact but fungus mycelium under bark well developed. Presence of rotten areas but narrower than 3 cm
	Corteccia intatta ma allentata (non ancora distaccata). Rametti parzialmente presenti. Legno intatto. Micelio fungino ben sviluppato. Presenza di aree marcescenti più piccole di 3 cm
Medium decayed Individuo mediamente decomposto	Bark is fragmented. Small branches absent. Wood consistency reduced but the log still has a hard core. Rotten areas wider than 3 cm
	Corteccia presente a tratti. Rametti totalmente assenti. Consistenza del legno ridotta. Aree marcescenti ampie più di 3 cm
Very decayed Individuo fortemente decomposto	Bark and small branches absent. Rotten throughout the log. Wood consistency compromised and the log is irregularly shaped under the effects of its own weight
	Corteccia e rametti assenti. Estese aree marcescenti. Legno privo di consistenza. Forma schiacciata per l'effetto del peso
Decomposed Individuo totalmente decomposto	Bark and small branches absent. Wood consistency is lost (dust). The log is fragmented in sections and may be mossy
	Corteccia e rametti assenti. Consistenza polverosa. Legno decomposto in scaglie, spesso coperte dal terreno, da muschio o da licheni

woody piece (called an element) is ideally divided into regular fragments of length not more than two metres. For each fragment, two cross sectional diameters at both terminal sections are measured, as well as its length. Each fragment is classified as either from coniferous or broadleaved species and the decay condition is given according to the classes in Table 4.5.

Stumps are the remains of cut trees or naturally broken trees not reaching a height of 1.30 m and with diameter at least 10 cm and are measured in the AdS13. The

Table 4.6 Size classes for small trees and shrubs / Classi dimensionali per la rinnovazione e gli arbusti

Size class	Height (h)	DBH
Classe dimensionale	Altezza (h)	$d_{1.30}$
1	50 cm < h ≤ 130 cm	–
2	h > 130 cm	DBH < 2.5 cm
		$d_{1.30\ m}$ < 2.5 cm
3	h > 130 cm	2.5 cm ≤ DBH < 4.5 cm
		2.5 cm ≤ $d_{1.30\ m}$ < 4.5 cm

diameter at the cutting height (two orthogonal measurements) and the height above-ground level (two measurements, the minimum and maximum height) are measured. The species is also recorded, if recognisable, as well as the decay class (Table 4.5). Lastly, information is given on the cutting age, specifically if the cut has occurred before or after the twelve months preceding the survey.

4.4 Data Collection, Database and Field Software (INFC_APP)

The relational database designed to store in central server data collected during the field surveys was developed in an Oracle 10 g environment and consisted of 27 tables (more than 280 fields) linked by the sample point identifier as the primary key and, in the quantitative variables tables, by the item (tree, stump, etc.) identifier as the secondary key. Accessing the central server is possible through authentication and protection protocols, which vary in relation to four profiles associated with corresponding specific roles of the user. These include CUFA, which is nationally responsible; researchers/analysts of CREA Research Centre for Forestry and Wood, regional coordinators, and the head of the field crew. The database also stores and makes available the data recorded during the photointerpretation, which must be validated or updated during the field survey.

The device used for data storage in the field is a tablet with OS Android 6.0, for which the application INFC_APP was developed (Gasparini et al., 2020). Most of the preloaded data is contained in this app, except the waypoints for navigation and raw positioning files, which were recorded with Trimble Terrasync software on the Trimble Juno SB datalogger. On the client side, the database was implemented through the Sqlite library. The information flow between the tablet device (client) and the central server can occur directly from the field, using the installed G4 card, via web service Simple Object Access Protocol (SOAP) (Oracle, 2021). With this protocol, it is possible to transfer not only alphanumeric data, but also binary data as the .ssf files containing the raw measurements and the qualitative parameters of the GNSS point positioning. The client device sends a request and waits for a response from the server. Four web services were implemented for searching, downloading, uploading, and updating the status of the points, fundamental for the distinction between the points to be surveyed, those in progress or those that have been concluded. The web service receives the tablet's IMEI code and returns the list of points assigned to the entitled NFI crew. The crew team can then choose the sample point to survey from that list and download all the preloaded information. If the procedure ends correctly, the client device automatically invokes the service, and this changes the status of the point, moving it from 'assigned' to 'in progress', thus excluding it from the list of points that are still possible to download on devices. At the end of the survey, the client invokes the service for uploading data, sends all the collected data to the central database, making them visible to other user profiles authorised to access and consult. It is possible to do partial uploads, as temporary backups of a survey not yet concluded, or a final upload to transfer all data at the end of the surveys on a sample point. The final upload can even be postponed to when data collection has been accomplished, such as after measurement of incremental cores, which is usually performed in the office.

Data security during the campaign is ensured by the establishment of a list of devices authorised to install the app and access the database according to the specific IMEI code of each tablet. In regard to INFC_APP development, one of the basic requirements during the design phase was to create a user-friendly GUI application. The user interface guides the surveyor throughout the logical and chronological phases of the survey, from navigation and positioning procedures to the collection of the qualitative and quantitative variables. It is possible to move through the various sections of the application, with the only constraint being to save the data entered in any specific section by explicitly confirming the saving. From the home screen it is possible to consult the documentation (survey protocol, electronic devices manuals, etc.) remaining in the application. The GUI sections are of two different types: the first is represented by data input modules (Fig. 4.5), and the second is the result of data queries, which resume the previously input data, both within each section and in the home screen, thus facilitating the monitoring of the survey progress (Fig. 4.6).

For all qualitative variables that are categorical and assessed using the appropriate class value, look-up tables have been adopted and the chosen class is selected from a closed list of possible values. For quantitative attributes, threshold values have been established, and warning messages have been shown in case of unlikely values. Combinations of masks and sub-masks have also been created for the input of progressively more detailed data (Fig. 4.7). Several real-time automatic cross-checks in different fields have been designed to prevent errors in terms of plausibility, congruity, completeness, or input errors (Fig. 4.8).

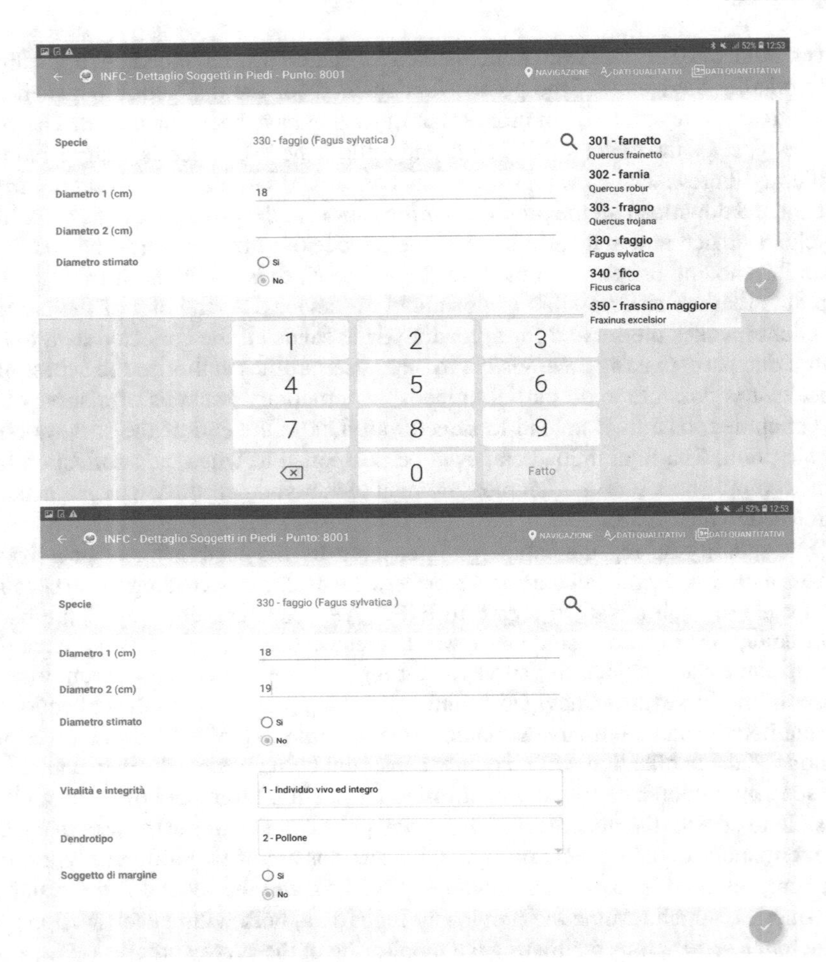

Fig. 4.5 Example of INFC_APP input module, regarding attributes referring to tree callipering / Esempio di modulo di input di INFC_APP, riguardante attributi relativi al cavallettamento degli alberi

4.5 Start-Up, Remote Assistance and Quality Check

4.5.1 Start-Up and Remote Assistance to Field Crews

As soon as the crews began their job, they were joined by the CREA Research Centre for Forestry and Wood team to be led under expert assistance. This assured appropriate reminder of the protocol procedure as learned in the training course and also provided clarification on local specific cases. However, clarification on local

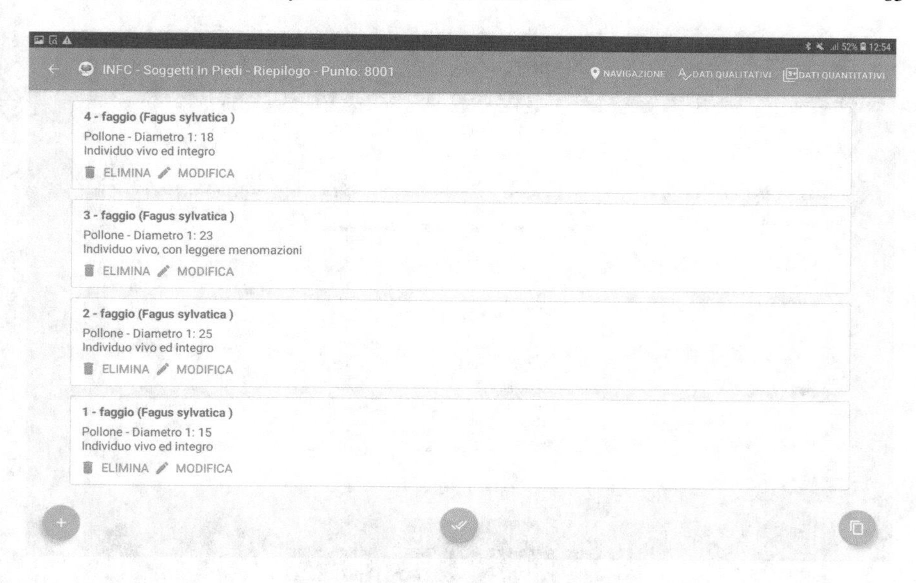

Fig. 4.6 Example of INFC_APP output module, regarding the progress status of tree callipering / Esempio di modulo di output di INFC_APP, riguardante lo stato di avanzamento del cavallettamento

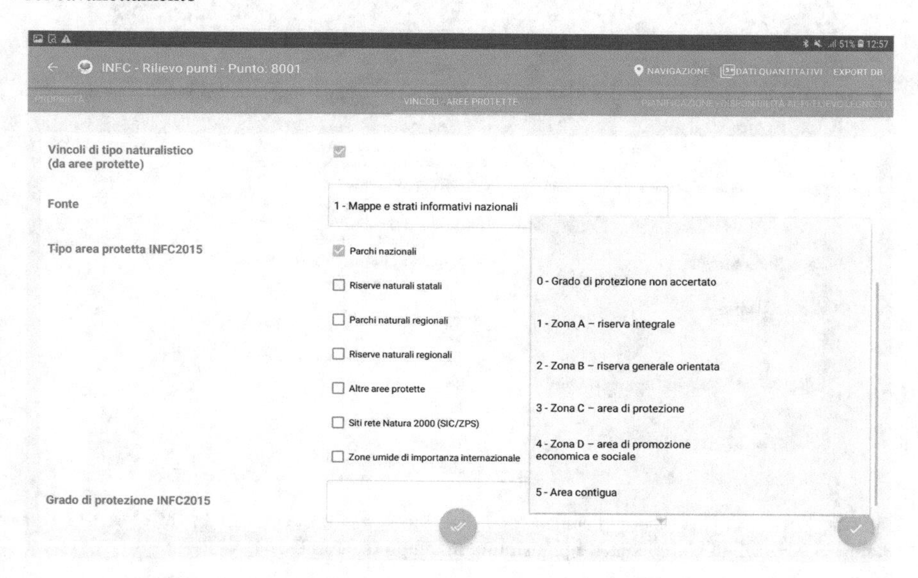

Fig. 4.7 Example of INFC_APP module, with mask and sub-mask, regarding the input of progressively more detailed data status / Esempio di modulo di INFC_APP con maschera e sotto-maschera per l'inserimento di dati a dettaglio progressivamente maggiore

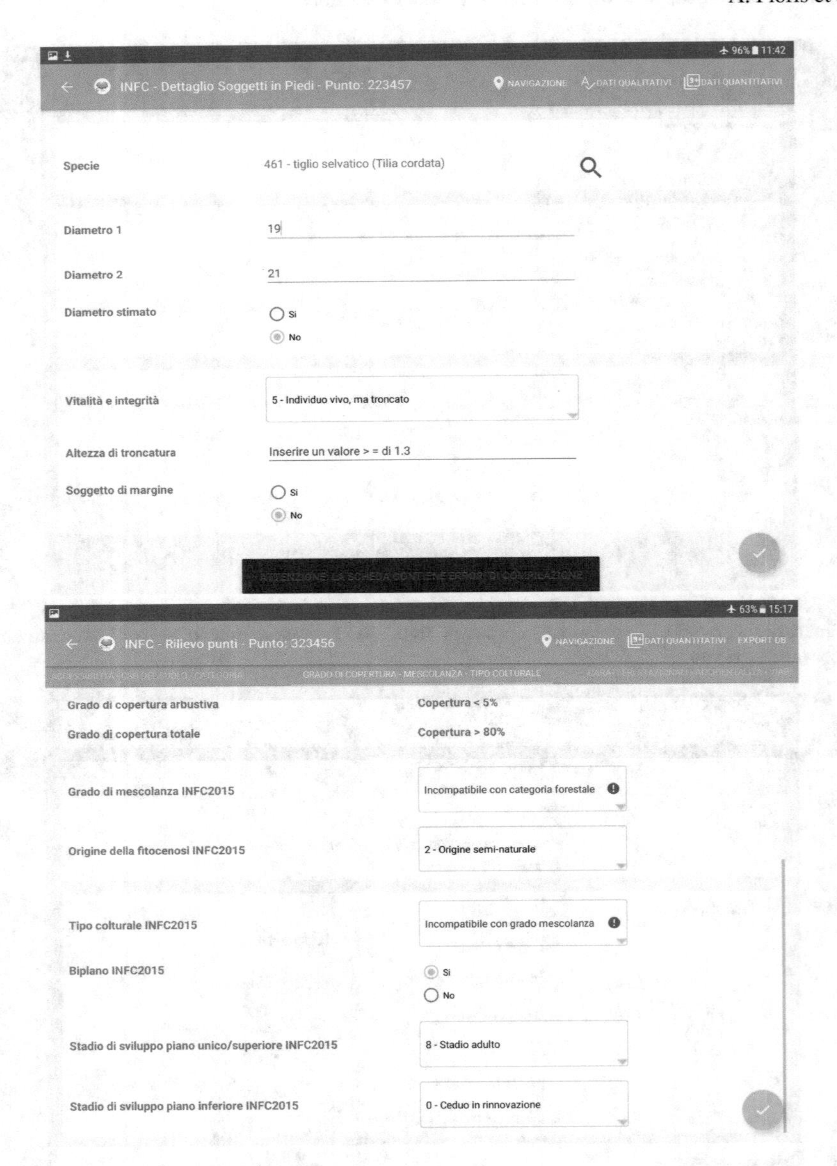

Fig. 4.8 Two examples of completeness and plausibility real-time checks in INFC_APP / Due esempi di controlli di completezza e plausibilità in tempo reale in INFC_APP

specific cases is not a secondary concern in Italy, given the highly diverse vegetation conditions.

During the field campaign, a helpdesk was active daily at CREA Research Centre for Forestry and Wood in Trento, reachable either on the phone or by email. Crew support consisted of answering questions related to any aspect of the survey protocol in an effort to help resolve doubts and uncertainties and so avoid subjective or incorrect interpretations. In fact, questions were often asked about special cases because crews were not in ordinary conditions, e.g., they were at risk of injury. All questions were entered in a database in such a way to give future consistent answers in similar cases, but especially to derive statistics on the most frequent topics of questions. Based on these statistics, the helpdesk could warn crews (all crews or only those potentially interested) to prevent reiteration or possible mistakes. The helpdesk database also allowed to verify full functioning of the automatic controls of INFC_APP, especially the first releases (cf. Sect. 4.4), but above all, they could suggest further improvements.

4.5.2 Data Quality Check and Final Field Work Checks for Approval

Regular checks on the central database are fundamental to guaranteeing high data quality. They also permit monitoring the progress of the field work, revealing possible problems in carrying out the survey which cause delays. In this case, they allowed verification of the crews' accomplishments of what was requested about data and materials. Crews had to post the woody cores to the dendrochronology laboratory at CREA Research Centre for Forestry and Wood in Trento, as well as safety backup copies of the photographs taken. Arrival of these two additional materials was expected at stated intervals after sample plots data were sent to the central database.

Checks in the database during the fieldwork aimed at assessing completeness, plausibility, and consistency of information. Completeness indicates that all data expected have been recorded. This is possibly the simplest check to be implemented through automatic checks in the crews' software; nevertheless, control is needed to verify the proper functioning of the software. In our case, controls revealed malfunctioning of early releases of INFC_APP, which caused a limited number of data loss. Checks about materials sent to CREA Research Centre for Forestry and Wood in Trento were special cases still related to completeness checks.

Plausibility means that information is reliable in absolute terms. An example of qualitative information that is not credible is the declared presence of a species in an area where it cannot be found (e.g., cork oak on the Alps). Another example involving quantitative values is a recording size that cannot be reached (e.g., exaggerated tree heights). The two examples should highlight the limits of automatic checks and importance of judgment based on data control, because there are places where a species is not expected but would still be possible to be found as well as a tree height that is possible may not be in a specific stand.

Consistency means that two pieces of information on related variables make sense; in other words, the two pieces of information are conditioned with each other. Nevertheless, the two values provided are reliable. Plausibility checks and consistency checks were carried out through cross-check data within the INFC2015 database, but they could also rely on crossing information from the INFC2005 survey.

Final field work checks for approval after the end of the field campaign was carried out by controllers. They fully surveyed some plots per region, in the presence of the crews and the data gathered were compared with those stored in the central database, previously recorded by the crews. The comparison was aimed to compute the level of reproducibility. For each variable measured, a maximum proportion of disagreement was stated, and a score was assigned, based on a level of agreement. The overall score was based on the scores obtained in each variable assessed. Crew knew from the very beginning that their job would be evaluated. For this reason, although limited in numbers, these checks were effective in maintaining a high level of quality throughout the entire field campaign.

Appendix (Italian Version)

Riassunto La fase di campagna di INFC2015 ha previsto il raggiungimento di quasi 9000 punti di campionamento e il rilievo di un numero rilevante di attributi, di tipo sia qualitativo sia quantitativo. Questo capitolo illustra le procedure di ritrovamento dei punti già istituiti nel precedente inventario (INFC2005), il posizionamento ex-novo di punti mai visitati in precedenza, le modalità di classificazione e misura delle variabili rilevate, sia a livello di popolamento sia di singolo elemento; inoltre fornisce indicazioni su quali unità di campionamento (punto o area di saggio) siano state utilizzate per il rilievo di dette variabili. Altre sezioni del capitolo sono dedicate alle modalità di registrazione informatica dei dati rilevati in campo e del loro invio al database centrale, all'assistenza fornita ai rilevatori e ai controlli per la qualità dei dati eseguiti durante i rilievi, sia in presenza sia da remoto, nonché ai controlli finali.

Introduzione

La conduzione dei rilievi inventariali in campo può essere efficacemente descritta elencando poche operazioni fondamentali. Si tratta, in concreto e nella generalità dei casi, di raggiungere un punto prestabilito (punto di campionamento) definito dalle sue coordinate, assumere quel punto come luogo di osservazione di alcune variabili di interesse, considerare quel punto come il centro di aree di saggio (AdS) entro cui condurre altre osservazioni e/o misurazioni, registrare le informazioni raccolte, e infine marcare quel punto in maniera permanente, così da poterlo ritrovare nelle successive campagne di rilievo. La schematizzazione proposta afferisce a fasi ben distinte del processo di progettazione di una campagna inventariale, che sono oggetto di questo capitolo.

Navigazione e posizionamento implicano l'utilizzo di procedure assistite da tecnologia di posizionamento satellitare globale (GNSS) con strumentazione performante nelle condizioni di copertura della volta, tipiche dei boschi, e adeguata a garantire una sufficiente accuratezza delle coordinate registrate. Queste rappresentano di per sé una variabile rilevata, che permette di poter rivisitare l'unità di campionamento e coniugare ai dati rilevati al suolo anche informazioni provenienti da remote sensing (Dalponte et al., 2011; Gobakken & Næsset, 2009; Kitahara et al., 2010). Un posizionamento utile anche ai fini di future ripetizioni dei rilievi implica l'utilizzo di marcatori permanenti ma non visibili (generalmente interrati) e altri visibili, di ausilio nel ritrovamento a rilevatori che abbiano accesso alle informazioni associate.

La classificazione dell'uso e copertura del suolo, preliminare ad ogni eventuale successiva operazione di rilievo, le classificazioni della vegetazione e degli attributi qualitativi nonché le misurazioni delle variabili quantitative, sono l'essenza stessa dell'indagine inventariale. In questo capitolo sono descritte le variabili rilevate in INFC2015 e le modalità seguite, rimandando al manuale di campagna (Gasparini et al., 2016), o ad altra bibliografia citata, per i casi particolari e di minore rilievo nell'ambito generale.

La registrazione delle informazioni raccolte, siano esse necessarie per le esigenze di progetto (ad esempio quelle relative alle marcature) o relative alle variabili oggetto di stima, è supportata per diverse motivazioni da applicativi informatici. Questi rendono più agevole consultare le informazioni necessarie ai rilevatori prima ancora di acquisirne di nuove (ad esempio quelle pregresse da verificare in virtù di possibili cambiamenti rispetto al precedente rilievo); possono supportare i rilevatori in varie fasi, per esempio se il loro uso è integrato nella procedura di navigazione e posizionamento, oppure proponendo da un menù a scelta le uniche classi coerenti con informazione di rango superiore; consentono controlli automatici, impedendo la registrazione di informazioni incongruenti o richiamando l'attenzione nel caso di combinazioni possibili ma poco frequenti; infine, consentono l'archiviazione delle informazioni registrate nel database INFC mediante semplice invio telematico.

Un ultimo aspetto relativo alle attività in campo riguarda l'importanza della qualità dei dati rilevati. In attività di rilievo complesse con durata pluriennale, non è possibile adottare una procedura di qualità basata solo su controlli a posteriori, che si risolvano in un'approvazione o rigetto dei dati, con conseguente ripetizione del rilievo. È necessario, invece, supportare le squadre di rilevatori mentre compiono le valutazioni in campo, se richiesto, per prevenire errori e controllare i dati che via via arrivano in archivio per individuare possibili errori di valutazione e/o misurazione compiuti dalle squadre, perché possano essere corretti ma soprattutto per prevenirne la ripetizione, avvisando le squadre interessate (Gasparini et al., 2009). L'attività di assistenza alle squadre e il controllo dei dati di archivio durante la fase di campo va prevista già in fase di progettazione, soprattutto per le sue implicazioni sugli strumenti di comunicazione e sulla struttura del database.

Navigazione, localizzazione e marcatura dei punti di campionamento

La massima accuratezza del posizionamento dei punti inventariali nelle diverse fasi dell'inventario, nella fotointerpretazione e al suolo, è un requisito molto importante ai fini della qualità dei dati rilevati. INFC è un inventario trifasico con una fase di fotointerpretazione a video e due fasi al suolo: la fase 2, su circa 30,000 punti, e la fase 3 su un loro sottoinsieme. La fotointerpretazione è condotta su un intorno di analisi che ha come centro il punto di campionamento inventariale, definito dalle sue coordinate (cfr. Chap. 2). Nell'INFC2015, la procedura per un'identificazione in campo più accurata possibile del punto di campionamento fotointerpretato è stata ripresa da INFC2005 (Floris et al., 2011). Sempre da INFC2005 derivano le modalità di marcatura adeguate a garantire il ritorno sulla stessa posizione, che nella fase 3 di INFC2005 avveniva a distanza di alcuni mesi. INFC2015 ha previsto una sola campagna di rilievi al suolo, su punti che per il 90% erano già stati visitati in campo in INFC2005. È stato per questo necessario sperimentare, in fase di progettazione dei rilievi, opportune strategie di navigazione e di posizionamento capaci di ridurre al minimo l'aliquota di punti non ritrovati ma efficienti anche per posizionarne di nuovi.

La navigazione per il ritrovamento del punto di campionamento (anche detto punto C, dopo essere stato materializzato al suolo) è avvenuta con ausilio di un ricevitore GNSS multicostellazione Trimble R1, pilotato da un datalogger Trimble Juno SB. Considerate le difficoltà tecnologiche ad utilizzare una vera e propria correzione differenziale in tempo reale in molte aree forestali italiane, si è stabilito di fruire, quando possibile, del sistema di aumento della precisione (SBAS) fornito dal servizio EGNOS (European Commission, 2017; Gasparini et al., 2021). L'avvicinamento fino a circa 10–15 m dal punto di campionamento è avvenuto in modalità navigazione da posizione istantanea (NPI), mentre il ritrovamento di precisione del punto, se già visitato in INFC2005, oppure il suo primo posizionamento se non visitato, sono avvenuti con procedura definita navigazione da posizione media (NPM) (Colle et al., 2009). La Fig. 4.1 mostra la strumentazione di navigazione e posizionamento in campo.

I punti di campionamento rilevati al suolo in INFC2005, la cui incertezza di posizionamento è stimata in 3–4 m a seconda del ricevitore utilizzato (Colle et al., 2009), erano stati marcati con picchetti interrati di due tipi: picchetti provvisori nei punti visitati nella sola fase 2 di INFC2005, che sono costituiti da chiodo lungo 30 cm e placca di testa in alluminio, e picchetti permanenti nei punti visitati anche in fase 3 di INFC2005, che hanno puntale in acciaio zincato con ancoraggi e testa in alluminio. Nelle migliori condizioni locali di ricezione del segnale GNSS, a 10–15 m dal punto di campionamento, era stato inoltre interrato un picchetto del tipo provvisorio sopra descritto, ad indicare il punto di fine navigazione (denominato punto F dopo la sua materializzazione al suolo). Il ritrovamento del picchetto del punto F e l'applicazione dei valori di offset verso il punto C (distanza e azimut) misurati in INFC2005 permettono di individuare la posizione del punto di campionamento. Le marcature prevedevano anche l'utilizzo di piccole placche in alluminio apposte alla base di tre alberi di riferimento, corredati di informazioni monografiche essenziali (specie, diametro a 1.30 m, distanza e azimut dal punto C o dal punto F) e alcune fotografie. La ricerca dei picchetti del punto F e del punto C è avvenuta con un metal detector, con l'aiuto delle descrizioni monografiche e delle fotografie. L'insieme delle marcature è mostrato in Fig. 4.2. La Fig. 4.3 mostra l'aliquota regionale e nazionale dei punti di campionamento in cui almeno uno dei picchetti, del punto F o del punto C, è stato ritrovato.

Nella campagna di rilievo INFC2015, per i punti ricadenti in uso del suolo forestale sono stati utilizzati solo picchetti di tipo permanente. Nel punto F sono stati invece usati ancora picchetti del tipo provvisorio, considerata anche la buona performance di ritrovamento. Al termine di tutti i rilievi, su ciascun punto inventariale è stato eseguito un posizionamento GNSS stazionando sul punto F e calcolando le coordinate medie di 50 posizionamenti istantanei. La Table 4.1 riporta i valori di alcuni parametri di incertezza del posizionamento, calcolati dai file grezzi dei rilievi GNSS su oltre 8000 punti. Essendo C raggiunto dal punto F mediante azimut e distanza misurati sul terreno, le coordinate di F e di C sono caratterizzate da incertezza di posizionamento simile. Prima di abbandonare il punto inventariale, alla fine dei rilievi, sono state aggiornate le descrizioni monografiche delle marcature ed eseguite nuove riprese fotografiche.

Le variabili rilevate: modalità di classificazione e misura

I rilievi in campo consentono di classificare in maniera definitiva l'uso e la copertura del suolo e i caratteri della vegetazione rilevanti per validare l'inclusione del punto inventariale nel dominio di INFC, ma anche di condurre le classificazioni sui caratteri qualitativi e di misurare le variabili quantitative. La Table 4.2 elenca le variabili rilevate in campo e le unità di campionamento (punto o area di saggio) adottate per ognuna; valutazioni e misurazioni sono condotte sul punto di campionamento o nelle aree di saggio (AdS) che sono centrate (AdS4, AdS13, AdS25) o vicine (AdS2) al punto di campionamento (cfr. Chap. 2). Questa sezione descrive i metodi di rilievo delle variabili qualitative e quantitative; le classi adottate sono riportate nelle tabelle in fondo ai capitoli dedicati ai risultati, insieme alle stime per le variabili.

Uso e copertura del suolo, grado di mescolanza, origine del soprassuolo

La prima valutazione riguarda la verifica che il punto di campionamento inventariale sia in un'area boscata, osservando l'uso del suolo e la copertura da alberi e/o arbusti; la copertura viene valutata sulla stessa ortofoto utilizzata in fase di fotointerpretazione, visualizzabile con il tablet in dotazione ai rilevatori (cfr. Sect. 4.4). Il rilievo in campo consente di valutare la copertura potenziale, nel caso di aree temporaneamente prive di soprassuolo, e condurre misure più accurate su elementi del territorio rilevanti per l'applicazione delle definizioni adottate (es. larghezza delle strade e dei corsi d'acqua) (cfr. Chap. 2). Un applicativo specificamente sviluppato, INFC_APP (cfr. Sect. 4.4), guida l'assegnazione del plot a una delle classi di copertura previste, anche mediante una griglia di punti sovrimpressa all'ortofoto nell'area di riferimento (FP2500, cfr. Chap. 2). I rilevatori contano il numero di punti che intercettano le chiome, specificando se si tratta di alberi o arbusti e INFC_APP calcola la copertura totale e quella arborea. Oltre a permettere di distinguere tra Bosco e Altre terre boscate, questo rilievo permette di produrre le statistiche di superficie per classe di copertura totale e di copertura arborea. INFC_APP assiste i rilevatori anche nei calcoli necessari a stabilire l'eventuale dominanza di copertura da parte di specie di conifere o di latifoglie, per classificare il soprassuolo secondo il grado di mescolanza (puro di conifere o di latifoglie, oppure misto) (cfr. Sect. 4.3.2).

L'origine dei soprassuoli viene assegnata valutando il grado di intensità delle azioni volte a favorire o sostenere la rinnovazione, che possono mancare (boschi naturali), afferire alla selvicoltura (soprassuoli seminaturali) o essere di tipo intensivo (piantagioni).

Categorie inventariali, categorie forestali e sottocategorie forestali

Le categorie inventariali del Bosco e delle Altre terre boscate sono classificate osservando caratteristiche riguardanti principalmente la composizione (es. alberi o arbusti), le potenzialità di sviluppo offerte dalla stazione (es. Boschi alti o Boschi bassi), e l'origine del soprassuolo (es. seminaturale o artificiale). Le categorie inventariali del Bosco e delle Altre terre boscate sono ulteriormente ripartite in categorie forestali e sottocategorie forestali (cfr. Chap. 2).

La categoria forestale viene classificata sulla base della specie con copertura delle chiome prevalente. Innanzitutto, viene stabilita l'eventuale prevalenza a conifere o a latifoglie, decidue o sempreverdi; la categoria forestale deve essere coerente con quella informazione; per esempio, se prevale la copertura di conifere la categoria forestale dovrà essere una tra quelle a conifere. Nel caso di boschi misti, la stima del gruppo di specie o della specie con copertura prevalente può richiedere una valutazione più accurata che prevede di camminare lungo percorsi predefiniti e indicare la specie con la chioma nel piano dominante in corrispondenza di punti di osservazione prestabiliti (Fig. 4.4); la procedura e i calcoli relativi sono guidati da INFC_APP. La sottocategoria forestale viene assegnata sulla base delle specie presenti e dell'ecologia della stazione.

Proprietà, grado di protezione e pianificazione

Per la classificazione della proprietà risulta rilevante la posizione del punto di campionamento. Le informazioni sulla proprietà derivano dal catasto, le cui informazioni sono disponibili mediante layer tematico su webGIS, oppure da conoscenza del personale forestale o amministrativo locale.

L'esistenza di restrizioni all'utilizzo delle risorse forestali e l'appartenenza ad aree protette (di vario tipo e livello) derivano da base documentale, cartografica o da conoscenza del personale forestale locale.

La presenza di strumenti di pianificazione è stata accertata su documenti oppure dichiarata dal personale forestale locale, ai diversi livelli previsti (per esempio piani di orientamento generali oppure piani di assestamento aziendali).

Accessibilità, viabilità e caratteri fisico-stazionali

Il punto di campionamento inventariale viene definito accessibile se raggiungibile e se risulta possibile condurre le rilevazioni e le misurazioni nelle AdS previste intorno ad esso. L'inaccessibilità può essere dovuta ad ostacoli fisici o a divieti di legge o imposti dal proprietario, ma anche a condizioni di pericolo per l'incolumità dei rilevatori. Nei casi di pericolo è ammessa una certa tolleranza; ad esempio, il diametro di alcuni alberi potrebbe essere solo stimato, oppure il rilievo condotto solamente nell'AdS4. Queste circostanze vengono registrate per consentire un trattamento dei dati adeguato, durante le elaborazioni.

La facilità e la velocità nel raggiungimento di un punto inventariale dipendono anche dalla presenza di viabilità. Questa viene valutata prima di tutto su mappe, ortofoto, ecc. e poi verificata in campo. I rilevatori registrano la distanza e il dislivello del punto di campionamento dalla strada, indicando anche la tipologia di viabilità.

L'altitudine del punto inventariale viene registrata dal dispositivo GNSS durante il rilievo delle coordinate (cfr. Sect. 4.2). L'esposizione viene misurata con una bussola; il rilevatore staziona sul punto di campionamento e orienta la bussola verso valle, leggendo il valore dell'azimut magnetico. L'inclinazione viene misurata con un clisimetro. Il rilevatore staziona sul punto di campionamento inventariale e traguarda due paline, una posta a monte e l'altra a valle, a 25 m di distanza. Il valore di inclinazione registrato si ottiene mediando le due misurazioni.

La giacitura viene osservata con riferimento all'AdS25 (giacitura locale) e a un'area più vasta, da pochi ettari a qualche decina di ettari intorno al punto di campionamento (giacitura estesa). L'accidentalità del terreno, valutata sempre nell'AdS25, tiene conto della micro-morfologia determinata dalla presenza di ostacoli come massi, pietre, fossi e buche che possono condizionare le utilizzazioni (taglio, concentramento, esbosco) e in generale la fruizione dell'area. Infine, nell'AdS25 viene valutata la presenza di fenomeni di dissesto in corso, ripetuti o occasionali. Si registrano e classificano solo fenomeni estesi almeno su 100 m^2 di territorio.

Pratiche selvicolturali, caratteristiche del soprassuolo, produzione

Le pratiche selvicolturali vengono descritte per tipo e intensità; poiché queste sono strettamente dipendenti dai prodotti che si vogliono ottenere, il sistema di classificazione prevede classi per le pratiche orientate ad ottenere prodotti secondari non legnosi e per l'ottenimento di beni e servizi. Quando la gestione è fortemente orientata ad ottenere un prodotto o servizio specifico, il bosco è classificato come avente una funzione prioritaria, e si registra il prodotto o il servizio oggetto di interesse.

La disponibilità al prelievo legnoso è una informazione registrata sulla base di documenti (es. limitazioni da leggi sulla protezione dell'ambiente) o valutata sulla base della convenienza alle utilizzazioni, che può essere ridotta per esempio dallo scarso valore del prodotto in relazione alle difficoltà di accesso e di lavorazione dovute alle condizioni stazionali.

Tipo colturale e modalità di utilizzazioni sono rilevate osservando la struttura del soprassuolo e le sue caratteristiche fisionomiche, mentre le modalità di esbosco sono valutate prendendo in considerazione anche il tipo di materiale esboscato (es. legname da opera o legna da ardere), usi locali (ad esempio i sistemi di esbosco a fune sono diffusi principalmente sulle Alpi e poco sugli Appennini) e la presenza di viabilità.

La valutazione sullo stadio di sviluppo viene condotta sulla base delle caratteristiche strutturali, compositive e dimensionali, e così è anche per quella sulla classe di età, che viene richiesta solo per i popolamenti coetanei; tuttavia, il giudizio si avvale anche dei carotaggi descritti nella Section 4.3.8, che consentono di contare gli anelli annuali.

Stato di salute

Il rilievo sulla presenza di malattie e danni ha l'obiettivo di fornire un quadro d'insieme sulle condizioni generali dello stato di salute delle foreste, registrando la presenza di questi quando presenti su almeno il 30% della copertura arborea nell'AdS25. Al raggiungimento di quella soglia, vengono valutati cause ed intensità della malattia o del danno. Nel caso in cui la malattia o il danno si manifestino con defogliazione, vengono registrate il suo grado di intensità e la localizzazione prevalente nella chioma.

Alberi misurati

Gli individui di cui si misura il diametro sono quelli di specie legnose (per comodità indicati genericamente come alberi), vivi o morti, con $d_{1.30} \geq 4.5$ cm entro l'AdS4 o $d_{1.30} \geq 9.5$ cm nell'AdS13. L'appartenenza ad un'AdS viene verificata misurando la distanza tra il centro della stessa e l'asse passante per il centro della sezione dell'albero a 1.30 m da terra. Il diametro viene misurato con un cavalletto dendrometrico, con l'operatore a monte dell'albero su terreni con inclinazione maggiore di 10° oppure con l'asta del cavalletto orientata lungo il raggio dell'AdS, negli altri casi. Se il primo diametro così misurato è ≥ 9.5 cm, si misura un secondo diametro ortogonale. Di ogni albero cavallettato, si registrano la specie, la vitalità e integrità (Table 4.3), il dendrotipo (Table 4.4). Degli alberi troncati si misura anche l'altezza. Per gli alberi morti, si valuta la classe di decadimento secondo le descrizioni nella Table 4.5.

Alberi campione

Tra gli alberi cavallettati, si selezionano dieci alberi campione per misurazioni aggiuntive. Si tratta dei cinque alberi più vicini al punto di campionamento inventariale, seguiti dai tre rimanenti con maggiore diametro e infine due definiti rari; questi ultimi sono alberi di specie non ancora o poco campionate tra i primi otto o di dimensioni non ancora comprese, in caso di boschi monospecifici. Quando possibile, gli alberi campione sono scelti tra quelli integri. Le misurazioni aggiuntive riguardano l'altezza totale, l'altezza di inserzione della chioma e la classificazione della posizione sociale (dominante, intermedia o sottoposta). L'altezza viene misurata con un ipsometro Haglof Vertex.

Ai fini delle stime di incremento, gli alberi campione vengono carotati con una trivella di Pressler, a 1.30 m da terra e lungo la direzione del raggio dell'AdS. Viene prelevata una sola carotina legnosa per albero e si misura con un righello l'incremento degli ultimi cinque anelli annuali (escluso quello dell'anno corrente). Le carote, opportunamente etichettate, sono inviate al laboratorio di dendrocronologia presso la sede di Trento del CREA Centro di ricerca Foreste e Legno, per le misurazioni di controllo.

Rinnovazione e arbusti

Rinnovazione e arbusti sono distinti sulla base della lista di specie adottata da INFC; tale distinzione non ha rilevanza ai fini delle misurazioni, poiché queste interessano tutti i soggetti legnosi sulla sola base dei limiti dimensionali stabiliti. Rinnovazione e arbusti sono rappresentati da individui legnosi con diametro inferiore alla soglia di cavallettamento e più alti di 50 cm, rientranti nelle classi dimensionali descritte nella Table 4.6. Le misurazioni si compiono nelle due AdS2, poste a 10 m di distanza a Est e a Ovest del punto di campionamento inventariale (cfr. Chap. 2). Il rilievo consiste nel conteggio del numero di soggetti per specie e classe dimensionale. Di ogni specie viene registrata l'origine prevalente (artificiale, agamica, da seme), eventuali danni, se riguardanti almeno il 30% dei soggetti, e la loro causa, se riconoscibile (animali, es. pascolo o animali selvatici, o fattori meteorologici).

Legno morto grosso a terra e ceppaie residue

Il legno morto grosso a terra (alberi interi, fusti, rami, ecc.) viene misurato fino alla sezione con diametro più piccolo $\geq$ 9.5 cm e se di lunghezza di almeno 9.5 cm. Ogni elemento viene idealmente diviso in frammenti regolari lunghi al massimo 2 m; di ogni frammento si misurano due diametri ortogonali tra loro delle sezioni estreme, nonché la distanza tra le due sezioni. Si registrano il gruppo specie (conifera o latifoglia) e la classe di decadimento secondo quanto indicato nella Table 4.5.

Le ceppaie residue sono le rimanenze di tagli o di schianti naturali che hanno lasciato monconi non più alti di 1.30 m e diametro di almeno 9.5 cm, entro l'AdS13. Si misurano due diametri ortogonali alla sezione di taglio e due altezze, la minima e la massima. Sono inoltre registrate la specie (se riconoscibile) e la classe di decadimento (Table 4.5); infine, si valuta se il taglio è avvenuto entro 12 mesi dal rilievo inventariale.

Registrazione informatica dei dati, database e INFC_APP

Il database di tipo relazionale progettato per archiviare nel server centrale i dati della campagna di rilievo è stato sviluppato in ambiente Oracle 10 g, ed è articolato in 27 tabelle (più di 280 campi) collegate tra loro tramite l'identificativo del punto come chiave primaria e quello dell'elemento misurato (albero, ceppaia, ecc.) come chiave secondaria, nelle tabelle dei dati quantitativi. L'accesso al server centrale avviene tramite autenticazione e protocolli di protezione variabili in relazione a quattro profili, associati a corrispondenti specifici ruoli dell'utente: referenti nazionali CUFA, ricercatore/analista CREA, coordinatore regionale, caposquadra rilievi. Il database contiene anche informazioni registrate durante la fotointerpretazione, realizzata in ambiente informatico diverso (cfr. Chap. 3), da validare o aggiornare durante i rilievi in campo.

Per i rilievi in campo è stata sviluppata l'applicazione INFC_APP, residente in un tablet con SO Android 6.0 in dotazione a ciascuna squadra di rilevatori (Gasparini et al., 2020). La maggior parte dei dati precaricati e rilevati è contenuta in questa app, ad eccezione dei waypoint per la navigazione e dei file grezzi di posizionamento registrati con software Trimble Terrasync sul datalogger Trimble Juno SB. Lato client (tablet), il database è stato implementato tramite la libreria Sqlite.

Il flusso delle informazioni tra il tablet e il server centrale avviene attraverso connessione internet direttamente dal campo, con scheda 4G. Lo scambio di informazioni tra dispositivo tablet e server centrale avviene tramite web service Simple Object Access Protocol (SOAP) (Oracle, 2021). Con questo protocollo è possibile trasferire, oltre a dati alfanumerici, anche dati binari, nello specifico il file.ssf contenente le misure grezze per il posizionamento satellitare e i parametri qualitativi dello stesso. Il client invia una richiesta e resta in attesa di una risposta dal server.

Sono stati implementati quattro servizi web per la ricerca, il download, l'upload e l'aggiornamento dello stato dei punti, fondamentale per la distinzione tra i punti da rilevare, in lavorazione oppure conclusi. Il web service riceve in input il codice IMEI del tablet e invia al dispositivo in risposta la lista dei punti assegnati alla specifica squadra. A questo punto la squadra è abilitata a scaricare sul proprio tablet tutti gli attributi precaricati del punto da rilevare. Se la procedura si conclude correttamente, il tablet invoca automaticamente il servizio che modifica lo stato del punto, portandolo da "assegnato" a "in lavorazione", escludendolo in questo modo dalla lista dei punti che è possibile scaricare sui dispositivi. A conclusione del rilievo la squadra, invocando il servizio per l'upload dei dati, invia tutti i dati raccolti al database centrale, rendendoli visibili anche agli altri profili utente autorizzati all'accesso e alla consultazione. Gli upload possono essere parziali, come backup temporanei di un rilievo non ancora concluso, oppure finali per il trasferimento di tutti i dati a conclusione del rilievo. È possibile posticipare l'invio finale dei dati di un punto di campionamento al rientro in ufficio, per completare l'inserimento di alcune informazioni, come ad esempio i valori di lettura delle carote incrementali, operazione che di solito non viene eseguita in campo ma in un momento successivo.

La riservatezza dei dati durante la campagna di rilievo è garantita dalla costituzione di una lista di dispositivi autorizzati all'installazione della app e all'accesso al database in base al codice IMEI specifico di ciascun tablet.

Un requisito fondamentale previsto sin dalla fase di progettazione è stato quello che l'applicazione fosse user-friendly. L'interfaccia utente è stata realizzata attraverso un sistema di sezioni e maschere che guidano il rilevatore nello svolgimento logico e cronologico delle diverse fasi di rilievo e inserimento dei dati, dalla procedura di navigazione e posizionamento del punto di campionamento fino alla raccolta dei dati sugli attributi qualitativi e quantitativi. È possibile spostarsi liberamente tra le varie sezioni dell'applicativo, con il vincolo che per salvare i dati inseriti in una specifica sezione è necessaria una conferma esplicita. Dalla home screen è possibile consultare la documentazione di supporto (manualistica dei rilievi, istruzioni per l'uso degli strumenti, ecc.) senza uscire dall'applicativo. Le maschere sono di due tipi diversi: il primo è rappresentato da moduli di input dei dati (Fig. 4.5), il secondo è il risultato di query di riepilogo dei dati già inseriti, visualizzabili sia all'interno di ogni singola sezione sia nella home screen, facilitando così il controllo delle operazioni di raccolta dei dati da parte del rilevatore (Fig. 4.6).

Per tutte le variabili qualitative, che sono categoriche e vengono registrate scegliendo la classe opportuna, sono stati adottati schemi di codifica (look-up tables) in base ai quali la classe prescelta viene selezionata da un menù contenente la lista delle sole modalità possibili. Per gli attributi quantitativi, sono state inserite soglie dei valori possibili e messaggi di warning in presenza di valori possibili ma che si incontrano poco frequentemente. Sono state create anche combinazioni di maschere e sotto-maschere per l'inserimento di dati a dettaglio progressivamente superiore (Fig. 4.7). Il sistema di controlli automatici immediati dei dati in termini di plausibilità, congruenza e completezza, ha previsto anche un certo numero di controlli incrociati tra campi (Fig. 4.8).

Assistenza ai rilevatori, start-up e controllo dei dati

Start-up e assistenza da remoto alle squadre

All'inizio della campagna di rilievi al suolo, le squadre hanno eseguito valutazioni e misurazioni su punti inventariali a loro assegnati sotto la guida del personale della sede di Trento del CREA Centro di ricerca Foreste e Legno. Questo ha consentito di ripassare quanto appreso nei corsi di formazione e anche di applicare le procedure avendo a che fare con eventuali particolarità della zona di competenza. Per un Paese con grande variabilità nei caratteri vegetazionali come l'Italia, questo secondo aspetto ha una rilevanza non secondaria.

Durante tutto il periodo di rilievi in campo, è stato operativo presso il CREA Centro di ricerca Foreste e Legno di Trento un centro di assistenza help-desk, raggiungibile via telefono o e-mail. L'assistenza alle squadre è consistita nel discutere di quesiti relativi a qualsiasi aspetto riguardante le modalità di rilievo, soprattutto per risolvere dubbi e incertezze da cui possono derivare interpretazioni difformi o errate. Infatti, in molti casi le domande hanno riguardato casi particolari, di situazioni non incontrate frequentemente, per esempio per condizioni di pericolo in aree difficili da percorrere. Le domande venivano registrate in un database, in modo da costituire una casistica che potesse assicurare interpretazioni uniformi a casi particolari analoghi, ma soprattutto perché la catalogazione consente di derivare statistiche sugli argomenti con maggiore frequenza di richiesta di chiarimenti. Sulla base di queste statistiche, è stato possibile richiamare l'attenzione delle squadre (tutte o solo quelle potenzialmente interessate) su errori commessi o possibili, per prevenirne la reiterazione o l'insorgenza. Il database delle domande ha consentito anche di monitorare malfunzionamenti dei controlli automatici di INFC_APP, soprattutto nelle sue prime versioni, e di suggerirne ulteriori miglioramenti.

Controllo di qualità dei dati e collaudo

La conduzione con regolarità di controlli nel database centrale è fondamentale per garantire dati di buona qualità. I controlli permettono anche di monitorare l'avanzamento dei lavori, individuando così eventuali rallentamenti dovuti a problemi nella conduzione dei rilievi. Nel caso di INFC2015, i controlli hanno consentito anche di verificare che le quadre inviassero il materiale richiesto. Queste, infatti, dovevano spedire le carotine legnose al laboratorio di dendrocronologia del CREA Centro di ricerca Foreste e Legno in Trento, nonché copie di backup delle fotografie scattate, entro tempi definiti dall'invio dei dati rilevati al database centrale.

I controlli nel database durante la fase dei rilievi sono stati finalizzati ad assicurare la completezza, la plausibilità e la coerenza delle informazioni. Con completezza si intende che tutti i dati richiesti siano stati effettivamente registrati. Si tratta della tipologia di controlli più efficacemente automatizzabile nel software utilizzato dalle squadre. Tuttavia, il controllo rimane sempre opportuno e necessario, per verificare il corretto funzionamento dei controlli automatici implementati. In un numero molto esiguo di casi, infatti, problemi legati ai controlli automatici nelle prime versioni di INFC_APP hanno determinato una perdita parziale di dati, subito evidenziata dai controlli. La verifica dell'avvenuta e regolare consegna dei materiali sopra indicati rappresenta un caso particolare di controlli di completezza. La plausibilità indica che un'informazione è credibile in senso generico. Un esempio di valutazione non plausibile di variabile qualitativa è la presenza di una specie in un contesto dove non può trovarsi (ad esempio la quercia da sughero sulle Alpi); un altro per una variabile quantitativa è la registrazione di un valore di dimensione che non può essere raggiunto (ad esempio, altezze eccessive per gli alberi). Questi due esempi evidenziano anche i limiti dell'automazione dei controlli e l'importanza della valutazione basata su controlli dei dati nel database eseguiti da operatori esperti. Infatti, possono esistere zone limite dove non ci si attende di trovare una specie, ma dove la sua presenza non può essere esclusa in senso assoluto, come pure possono esistere altezze di alberi possibili in termini assoluti, ma sospette per alcuni soprassuoli in specifiche condizioni stazionali. La coerenza, infine, riguarda la possibilità e la logicità delle informazioni relative a due variabili correlate; in altri termini, entrambe le informazioni sono possibili di per sé, ma quando fornite insieme il valore che esse possono assumere è reciprocamente condizionato. I controlli di plausibilità e di coerenza sono stati svolti mediante l'incrocio delle informazioni presenti nell'archivio INFC2015, ma anche paragonando queste ultime con le informazioni contenute nell'archivio INFC2005, quando necessario.

Il collaudo finale dopo la conclusione dei lavori in campo è stato condotto da personale non coinvolto nei rilievi stessi. Sono state nuovamente rilevate alcune AdS per regione, in presenza delle squadre di rilevatori, e i risultati ottenuti confrontati con i dati precedentemente trasmessi da questi al database centrale. Il confronto ha consentito di verificare il grado di riproducibilità dei dati. Per ogni variabile misurata, era stata stabilita una percentuale di discordanza massima ammissibile e sulla base della concordanza è stato assegnato un punteggio; il punteggio finale è stato calcolato sulla base dei punteggi parziali. Le squadre sapevano fin dall'inizio che il loro lavoro sarebbe stato valutato e per questo motivo, sebbene il numero dei punti controllati sia stato limitato, il collaudo ha costituito una motivazione per mantenere alto il livello di qualità durante tutta la campagna di rilievi al suolo.

References

Colle, G., Floris, A., Scrinzi, G., Tabacchi, G., & Cavini, L. (2009). The Italian national forest inventory: geographical and positioning aspects in relation to the different phases of the project. In: *Proceedings, 8th annual forest inventory and analysis symposium*, 2006 October 16–19, Monterey, CA (pp. 1–8). Gen. Tech. Report WO-79. Washington, DC: U.S. Department of Agriculture, Forest Service.

Dalponte, M., Martinez, C., Rodeghiero, M., & Gianelle, D. (2011). The role of ground reference data collection in the prediction of stem volume with LiDAR data in mountain areas. *ISPRS Journal of Photogrammetry and Remote Sensing, 66*, 787–797. https://doi.org/10.1016/j.isprsjprs.2011.09.003

European Commission. (2017). Galileo & EGNOS: The EU satellite navigation programmes explained. Retrieved Dec 14, 2021, from https://www.euspa.europa.eu/sites/default/files/galileo-egnos_brochure_2017_web_1.pdf.

Floris, A., Scrinzi, G., & Colle, G. (2011). Localizzazione e materializzazione dei punti inventariali. In: Gasparini, P., Tabacchi, G. (eds) L'Inventario Nazionale delle Foreste e dei serbatoi forestali di Carbonio INFC 2005. Secondo inventario forestale nazionale italiano. Metodi e risultati, pp. 32–36. Edagricole-Il Sole 24 Ore, Milano. ISBN 978–88–506–5394–2.

Gasparini, P., Bertani, R., De Natale, F., Di Cosmo, L., & Pompei, E. (2009). Quality control procedures in the Italian national forest inventory. *Journal of Environmental Monitoring, 11*, 761–768.

Gasparini, P., Di Cosmo, L., Floris, A., Notarangelo, G., & Rizzo, M. (2016). Guida per i rilievi in campo. INFC2015—Terzo inventario forestale nazionale. Consiglio per la ricerca in agricoltura e l'analisi dell'economia agraria, Unità di Ricerca per il Monitoraggio e la Pianificazione Forestale (CREA-MPF), Corpo Forestale dello Stato, Ministero per le Politiche Agricole, Alimentari e Forestali (341 pp.). ISBN 9788899595449. Retrieved Jan 05, 2022 from https://www.inventarioforestale.org/it/node/72.

Gasparini, P., Floris, A., Rizzo, M., Patrone, A., Credentino, L., Papitto, G., & Di Martino D. (2020). Il terzo inventario forestale nazionale INFC2015: procedure, strumenti e applicazioni. *GEOmedia 24*(6), 6–17. ISSN 1128-8132.

Gasparini, P., Floris, A., Rizzo, M., Di Cosmo, L., Morelli, S., & Zanotelli, S. (2021). Il contributo della geomatica alle attività del terzo inventario forestale nazionale italiano INFC2015. In: Atti AsitaAcademy2021, Asita, Milano (pp. 243–260). ISBN 978-88-941232-7-2.

Gobakken, T., & Næsset, E. (2009). Assessing effects of sample plot positioning errors on biophysical stand properties derived from airborne laser scanner data. *Canadian Journal of Forest Research, 39*(5), 1036–1052. https://doi.org/10.1139/X09-025

Kitahara, F., Mizoue, N., Kajisa, T., Murakami, T., & Yoshida, S. (2010). Positional accuracy of national forest inventory plots in Japan. *Journal of Forest Planning, 15*(2), 73–79. https://doi.org/10.20659/jfp.15.2_73.

Oracle. (2021). Application Server Oracle9iAS SOAP Developer's Guide. Retrieved Dec 14, 2021, from https://docs.oracle.com/cd/A97335_02/integrate.102/a90297/overview.htm.

Open Access This chapter is licensed under the terms of the Creative Commons Attribution 4.0 International License (http://creativecommons.org/licenses/by/4.0/), which permits use, sharing, adaptation, distribution and reproduction in any medium or format, as long as you give appropriate credit to the original author(s) and the source, provide a link to the Creative Commons license and indicate if changes were made.

The images or other third party material in this chapter are included in the chapter's Creative Commons license, unless indicated otherwise in a credit line to the material. If material is not included in the chapter's Creative Commons license and your intended use is not permitted by statutory regulation or exceeds the permitted use, you will need to obtain permission directly from the copyright holder.

Chapter 5
Procedures for the Estimation of Forest Inventory Quantities

Procedure di stima delle grandezze inventariali

Maria Michela Dickson and Diego Giuliani

Abstract This chapter aims at illustrating the statistical procedures adopted to estimate the unknown values of the parameters of interest of the forest inventory. In particular, it firstly describes how the data collected during the second phase of the sampling plan have been used to estimate the areal extents of the different land use and cover categories. Secondly, it illustrates the procedures to properly estimate the total and density values of the quantities measured during the third phase of the survey campaign. These procedures were developed for INFC2005 and, as explained in this chapter, are still valid for INFC2015.

Keywords Unbiased estimators · Areal extent estimation · Total and density estimation

5.1 Introduction

In order to estimate the unknown values of the parameters of interest of the forest inventory, using the data collected through the survey campaign, it is required to apply the correct statistical procedures. In other words, it is necessary to identify the proper unbiased estimators. This chapter describes the unbiased estimators that were developed for the survey campaign of INFC2005 and explains why they can still be used for the campaign of INFC2015.

First of all, it is important to clarify which parameters of the forest inventory need to be estimated. It should be considered that the national territory is subdivided into L territorial districts (corresponding to the 21 Italian regions and autonomous provinces, called "regions" in the following text), with areal extents equal to $A_1, A_2, \ldots, A_i, \ldots, A_L$, respectively. The total areal extent of Italy is therefore equal to $A = \sum_{l=1}^{L} A_l$. Each region is characterised by $K = NF + F + 1$ land categories, where NF is the number of non-woodland categories (set equal to one in

M. M. Dickson (✉) · D. Giuliani
Department of Economics and Management, University of Trento, Trento, Italy
e-mail: mariamichela.dickson@unitn.it

© The Author(s) 2022

P. Gasparini et al. (eds.), *Italian National Forest Inventory—Methods and Results of the Third Survey*, Springer Tracts in Civil Engineering,
https://doi.org/10.1007/978-3-030-98678-0_5

INFC to simplify the estimation), F indicates the number of woodland categories, while the residual class comprises the categories that are excluded from the inventory.

The main population parameters of interest that can be estimated with the data collected during the second phase of the sampling plan are the $K \times L$ region-level areal extents of the categories $\{a_{kl} : k = 1, \ldots, K, l = 1, \ldots, L\}$. Aggregations of these quantities provide other interesting information, such as the areal extent of category k in the entire national territory, $A_k = \sum_{l=1}^{L} a_{kl}$; the areal extent of a subset C of some categories for a given region l, $a_{Cl} = \sum_{k \in C} a_{kl}$, or for the entire country, $a_C = \sum_{l=1}^{L} a_{Cl}$.

The main population parameters that can be estimated at the end of the third phase of the sampling plan concern the measured quantitative variables, such as the number of trees, the growing stock volume and the biomass. In particular, for any measured variable, it is possible to estimate the $F \times L$ total values t_{kl} for any combination of woodland category and region. Analogously to the a_{kl} parameters, aggregations of t_{kl} may also be of particular interest. Specifically, for a given variable, the quantity $T = \sum_{k=1}^{K} \sum_{l=1}^{L} t_{kl}$ corresponds to the overall value in the entire territory; $T_l = \sum_{k=1}^{K} t_{kl}$ represents the overall value in region l; and $T_k = \sum_{l=1}^{L} t_{kl}$ indicates the overall value for the category k in the whole area. Moreover, for a given subset C of categories, it is also interesting to know the overall value of the variable in region l, $t_{Cl} = \sum_{k \in C} t_{kl}$, and the overall value over the entire territory, $t_C = \sum_{l=1}^{L} t_{Cl}$. Another set of relevant parameters is the $F \times L$ density values $d_{kl} = t_{kl}/a_{kl}$ and their aggregations, that is $d_k = t_k/a_k$, $d_l = t_l/a_l$, $d_{Cl} = t_{Cl}/a_{Cl}$ and $d_C = t_C/a_C$.

Proper estimates of all these parameters and their corresponding variances can be obtained using the estimators developed for the survey campaign of INFC2005 (Fattorini et al., 2011). Such estimators are explained in the following sections of this chapter. In particular, Sect. 5.2 provides an essential description of the estimators of areal extents, while Sect. 5.3 briefly illustrates the estimation procedures of the total and density values of quantitative variables. Section 5.4 concludes explaining why these estimators are still valid for the survey campaign of INFC2015.

5.2 Estimation of Areal Extents

According to the sampling design adopted for INFC2005, during the first phase of the survey campaign (cf. Chap. 2), a point is selected at random in each of the NQ quadrats of the sampling grid. The $N_0 \leq NQ$ selected points are then classified, according to aerial photointerpretation, into $H = NF + V + 1$ land use and cover strata, where $V < F$ strata refer to a less detailed classification of the F categories (cf. Sect. 5.1). The last residual stratum includes all the points that the aerial photointerpretation cannot properly classify.

The N_0 points are also classified according to the region, thus leading to a two-way stratification characterised by $H \times L$ strata, each denoted as U_{hl}, with size $N_{hl} (h = 1, \ldots, H, l = 1, \ldots, L)$.

For any stratum hl relative to the V categories and non-classifiable land category, that is where $h = NF + 1, \ldots, H$, if $N_{hl} > 0$, a sample S_{hl} of n_{hl} points is selected through simple random sampling without replacement. The sampled points are then observed on the field in order to correct possible classification errors arising during the aerial photointerpretation. To the contrary, points belonging to the strata relative to NF categories, those with $h = 1, \ldots, NF$, are not sampled (cf. Chap. 2).

Following Fattorini et al. (2006) an unbiased estimator of a_{kl} is

$$\hat{a}_{kl} = R\left\{w_{kl} + \sum_{h>NF} w_{hl}w_{khl}\right\}, k = 1, \ldots, NF, l = 1, \ldots, L \tag{5.1}$$

or

$$\hat{a}_{kl} = R\sum_{h>NF} w_{hl}w_{khl}, k = NF + 1, \ldots, K, l = 1, \ldots, L \tag{5.2}$$

where R is the overall areal extent of the NQ quadrats, $w_{hl} = N_{hl}/NQ$ is the weight of stratum hl, $w_{kl} = N_{kl}/NQ$ is the weight of stratum kl relative to the non-woodland categories that are not sampled during the second phase, and $w_{khl} = n_{khl}/n_{hl}$ is the share of points of S_{hl} belonging to land category k.

Fattorini et al. (2006) also show that a conservative estimator of the variance of $\hat{a}_{kl}$ is

$$\hat{v}(\hat{a}_{kl}) = \frac{R^2}{NQ-1}\left\{w_{kl} + \sum_{h>NF}\frac{N_{hl}-1}{n_{hl}-1}w_{hl}w_{khl} - \sum_{h>NF}\frac{N_{hl}-n_{hl}}{n_{hl}-1}w_{hl}w_{khl}^2 - p_{kl}^2\right\},$$
$$k = 1, \ldots, NF, l = 1, \ldots, L \tag{5.3}$$

or

$$\hat{v}(\hat{a}_{kl}) = \frac{R^2}{NQ-1}\left\{\sum_{h>NF}\frac{N_{hl}-1}{n_{hl}-1}w_{hl}w_{khl} - \sum_{h>NF}\frac{N_{hl}-n_{hl}}{n_{hl}-1}w_{hl}w_{khl}^2 - p_{kl}^2\right\},$$
$$k = NF + 1, \ldots, K, l = 1, \ldots, L \tag{5.4}$$

where $p_{kl} = \hat{a}_{kl}/R$.

In applying Eqs. 5.3 and 5.4, it is necessary that if $N_{hl} > 1$ then $n_{kl} \geq 2$ while if $N_{hl} = 1$ then $n_{kl} \geq 1$.

In order to estimate the variances of aggregations of $\hat{a}_{kl}$, such as $\widehat{A}_k = \sum_{l=1}^{L}\hat{a}_{kl}$, the covariances among the estimates involved in the aggregations are also needed. According to Fattorini et al. (2006), the covariance between $\hat{a}_{kl}$ and $\hat{a}_{k'l}$ can be properly estimated with

$$\hat{c}(\hat{a}_{kl}, \hat{a}_{k'l}) = -\frac{R^2}{NQ-1}\left\{\sum_{k>NF} \frac{N_{hl}-n_{hl}}{n_{hl}-1} w_{hl} w_{khl} w_{k'hl} + p_{kl} p_{k'l}\right\},$$
$$k' \neq k = 1, \ldots, K, l = 1, \ldots, L \tag{5.5}$$

while the covariance between $\hat{a}_{kl}$ and $\hat{a}_{k'l'}$ should be estimated using

$$\hat{c}(\hat{a}_{kl}, \hat{a}_{k'l'}) = \frac{\hat{a}_{kl}\hat{a}_{k'l'}}{NQ-1}, k, k' = 1, \ldots, K, l' \neq l, \ldots, L \tag{5.6}$$

Although Eqs. 5.1 and 5.2 represent unbiased estimators of the areal extents for all the combinations of land category and region, the sum of the $K \times L$ estimates $\hat{a}_{kl}$ is not precisely equal to A. It is indeed equal to R times the share of points collected during the first sampling phase that fall within the borders of the national territory. Analogously, for each kl stratum, $\sum_{k=1}^{K} \hat{a}_{kl}$ is not equal to A_l since it corresponds to R times the share of points falling in the region l.

Since the values of A and A_l are known, the problem can be solved by calibrating the $\hat{a}_{kl}$-values so that the territorial totals correspond to the actual values. The calibration can be done using the following calibration factor,

$$p_{kl}^{cal} = \hat{a}_{kl} / \sum_{k=1}^{K} \hat{a}_{kl}, l = 1, \ldots, L \tag{5.7}$$

which is characterised by the fact that $\sum_{k=1}^{K} p_{kl}^{cal} = 1$ for any region l.

The calibrated estimates of areal extents can then be obtained as

$$\hat{a}_{kl}^{cal} = A_l \times p_{kl}^{cal}. \tag{5.8}$$

The formulas to estimate the variances and covariances for $\hat{a}_{kl}^{cal}$ can be found in Fattorini et al. (2006).

5.3 Estimation of Total and Density Values of Quantitative Variables

During the third phase of the sampling procedure, for any of the $(V+1) \times L \times F$ second phase stratum concerning a Forest category, if $n_{khl} > 0$, a sample Q_{khl} of m_{khl} points is selected through simple random sampling without replacement. Each sampled point is then observed on the field and the amount of any variable of interest is measured within a circular area centred on it (cf. Chaps. 2 and 4). According to Fattorini et al. (2006), the obtained data can then be used to estimate t_{kl}, for any variable, through the following unbiased estimator

$$\hat{t}_{kl} = NQ \sum_{h>NF} w_{hl} w_{khl} \overline{x}_{khl}, k = NF + 1, \ldots, K, l = 1, \ldots, L \tag{5.9}$$

where $\overline{x}_{khl}$ is the sample mean of the Horvitz-Thompson total estimates of the variable observed in the sample points of stratum khl.

Fattorini et al. (2006) also provide conservative estimators of the variances and covariances of $\hat{t}_{kl}$, that is

$$\hat{v}(\hat{t}_{kl}) = \frac{NQ^2}{NQ-1} \left\{ \sum_{h>NF} w_{hl} \frac{N_{hl}-1}{n_{hl}-1} w_{khl}(n_{khl}-1) \frac{s^2_{khl}}{m_{khl}}, \right.$$
$$\left. + \sum_{h>NF} w_{hl} \frac{N_{hl}-1}{n_{hl}-1} w_{khl}(1-w_{khl}) \overline{x}^2_{khl} + \sum_{h>NF} w_{hl} w^2_{khl} \overline{x}^2_{khl} - \overline{x}^2_{hl} \right\}$$
$$k = NF + 1, \ldots, K, l = 1, \ldots, L, \tag{5.10}$$

$$\hat{c}(\hat{t}_{kl}, \hat{t}_{k'l}) = -\frac{NQ^2}{NQ-1} \left\{ \sum_{k>NF} \frac{N_{hl}-n_{hl}}{n_{hl}-1} w_{hl} w_{khl} w_{k'hl} \overline{x}_{khl} \overline{x}_{k'hl} + \overline{x}_{khl} \overline{x}_{k'hl} \right\},$$
$$k' \neq k = NF + 1, \ldots, K, l = 1, \ldots, L, \tag{5.11}$$

and

$$\hat{c}(\hat{t}_{kl}, \hat{t}_{k'l'}) = \frac{\hat{t}_{kl} \hat{t}_{k'l'}}{NQ-1}, k, k' = NF + 1, \ldots, K, l' \neq l, \ldots, L \tag{5.12}$$

where s^2_{khl} is the sample variance of the estimated values of the variable observed in the sample points of stratum khl.

In applying Eq. 5.9, it is necessary that if $n_{khl} > 1$ then $m_{khl} \geq 2$ while if $n_{khl} = 1$ then $m_{kl} \geq 1$.

The density values, d_{kl}, can be straightforwardly estimated with

$$\hat{d}_{kl} = \hat{t}_{kl} / \hat{a}^{cal}_{kl}. \tag{5.13}$$

Unfortunately, the variance of $\hat{d}_{kl}$, which is a ratio estimator, is intractable. Approximate unbiased estimates of the variances and covariances of (5.13) can however be obtained using the common approach of linearising the ratio using the first leading term of its Taylor series expansion (Särndal et al., 1992). The reliability and precision of these approximated estimates depend on the level of precision of $\hat{t}_{kl}$ and $\hat{a}^{cal}_{kl}$. For this reason, in some circumstances the error of density estimates has not been reported.

For the survey campaign of INFC2005, the estimators described by Eqs. (5.9) and (5.10) were also modified to obtain the estimates of the total values of the quantitative variables for subsets of population units (e.g., the trees) identified by relevant qualitative attributes, such as the tree species. Let us consider M non-overlapping subsets. The total value, t_{kml}, of a quantitative variable of interest for subset m, forest category k, and region l can be properly estimated with

$$\hat{t}_{kml} = NQ \sum_{h>NW} w_{hl} w_{khl} \overline{x}_{kmhl}$$
$$k = NF + 1, \ldots, K - 1, \ m = 1, \ldots, M, \ l = 1, \ldots, L. \tag{5.14}$$

The estimated variance of $\hat{t}_{kml}$ is therefore

$$\begin{aligned} \hat{v}\left(\hat{t}_{kml}\right) = \frac{NQ^2}{NQ - 1} \Bigg\{ & \sum_{h>NF} w_{hl} \frac{N_{hl} - 1}{n_{hl} - 1} w_{khl} (n_{khl} - 1) \frac{s^2_{kmhl}}{m_{khl}} \\ & + \sum_{h>NF} w_{hl} \frac{N_{hl} - 1}{n_{hl} - 1} w_{khl} (1 - w_{khl}) \overline{x}^2_{kmhl} \\ & + \sum_{h>NF} w_{hl} w^2_{khl} \overline{x}^2_{kmhl} - \overline{x}^2_{klm} \Bigg\} \end{aligned}$$
$$k = NF + 1, \ldots, K - 1, \ m = 1, \ldots, M, \ l = 1, \ldots, L, \tag{5.15}$$

where $\overline{x}_{kmhl}$ and s^2_{kmhl} are, respectively, the sample mean and sample variance of the estimated values of the variable observed in the sample points of stratum khl for the population units that belong to subset m.

Further modifications of the estimators have allowed also to provide estimates for two specific cases. On the one hand, they have been modified to estimate the areal extents of forest categories for subsets, $m = 1, \ldots, M$, of sample points identified by qualitative attributes measured during the third phase of the sampling plan. On the other hand, they have been modified to estimate a_{kml} and t_{kml} in those circumstances in which subset m is identified during the second phase of the sampling plan. See Fattorini et al. (2011) for further details about these estimators.

5.4 Comparison Between the Two Forest Inventories

The presented estimation strategy has been used during the estimation process of the parameter of interest for INFC2005. Such strategy also remained valid for the estimation process of INFC2015. Indeed, the modifications occurred among the two forest inventories have no impact on the use of the estimation strategy, as explained in the following.

The sampling plan adopted for INFC2005 consisted of a three-phase structure (cf. Chap. 2). In the first phase, carried out using aerial photointerpretation, the area was divided into polygons of equal size (1 km^2), from which a point was randomly selected (one from each polygon). Then the population of such selected points was divided into 21 strata corresponding to the territorial districts of the national territory (regions), and into strata corresponding to the land use and cover categories. In the second phase, carried out by surveys on the ground, a stratified sample of points was selected from the strata defined in the first phase only for land use and cover categories of interest for INCF2005. The selected sample points were assigned then to strata corresponding to the forest types. In the third phase, from each Forest stratum, an additional sample of points was selected. Then, plots were laid out around each of these points, in order to define the area on which the measurements of the variables were carried out (cf. Chap. 4).

In INFC2015, the adopted sampling plan did not suffer substantial changes. Indeed, the three-phase structure was maintained: the first was designed to classify the land use and cover; the second was aimed at definitively classifying the land use and cover and forest types in correspondence of sampling points; sought to define the areas for the survey of interest variables. Therefore, the sampling plan did not undergo any changes that altered its original design. The changes made in the INFC2015 have regarded exclusively the definition of strata and the sample sizes. Specifically:

i. The number of points in the national territory was equal to 301,306 for INFC2005 and equal to 301,271 for INFC2015;
ii. The classification scheme of the land use and cover used for the INFC2005 aerial photointerpretation included 12 classes/subclasses, while the scheme used for the INFC2015 included 13 classes/subclasses (including the class of non-classified points);
iii. To estimate the INFC2015 areal extent, 3 new classes of forest inventory interest must be added to the 13 classes/subclasses, which following the aerial photointerpretation of the INFC2005, belonged to the classes/subclasses not of interest for the subsequent phases of the inventory;
iv. The number of classes/subclasses from aerial photointerpretation, which were not of interest for the forest inventory went from seven (for INFC2005) to eight (for INFC2015);
v. The number of classes of forest inventory interest was equal to four in the INFC2005, which became eight in the INFC2015.

As explained, the modifications occurred as a result of slightly different systems of aerial photointerpretation adopted during the INFC2015 and of changes in the classes/subclasses of interest. All this resulted in a different number of strata among INCF2005 and INCF2015, as well changes to some units from a stratum to another in the two survey waves. Nevertheless, from a methodological point of view, the sampling plan adopted during the two waves remained the same, since it was a three-phase design, with the definition of areas at each phase based on stratification. This fact confirmed the possibility of using the same estimation techniques adopted during the INCF2005 for INCF2015, both for the estimation of areal extents and for the values of the interest variables.

The estimators used for INCF2015 were implemented in the open source software R (R Core Team, 2020), a free environment for statistical computing.

Appendix (Italian Version)

Riassunto Questo capitolo si occupa di illustrare le procedure statistiche adottate per stimare i valori incogniti dei parametri di interesse per l'inventario forestale. In particolare, viene innanzitutto descritto come i dati raccolti durante la seconda fase del piano di campionamento sono stati utilizzati per stimare le superfici delle diverse categorie di uso e copertura del suolo. In secondo luogo, vengono illustrate le procedure per stimare in maniera appropriata i valori totali e di densità delle grandezze rilevate nel corso della terza fase dell'indagine campionaria. Tali procedure sono state sviluppate in occasione dell'INFC2005 e, come spiegato in questo capitolo, sono valide anche per l'INFC2015.

Introduzione

Per stimare i valori incogniti dei parametri oggetto di interesse dell'inventario forestale, utilizzando i dati rilevati mediante l'indagine campionaria, è necessario impiegare le procedure statistiche adeguate. In altre parole, si devono identificare gli appropriati stimatori corretti. Questo capitolo descrive gli stimatori corretti che sono stati sviluppati in occasione dell'indagine campionaria per l'INFC2005 e spiega perché possono essere utilizzati anche per l'indagine riguardante l'INFC2015.

Innanzitutto è importante definire quali sono i parametri oggetto di stima dell'inventario forestale (NFI). A tale fine, si deve considerare che il territorio nazionale è suddiviso in L distretti territoriali corrispondenti alle 21 regioni e province autonome italiane, chiamate "regioni" nel testo che segue. Le superfici delle L regioni sono indicate, rispettivamente, con $A_1, A_2, \ldots, A_i, \ldots, A_L$. La superficie complessiva dell'Italia è dunque uguale a $A = \sum_{l=1}^{L} A_l$. Ciascuna regione è caratterizzata da $K = NF + F + 1$ categorie, in cui NF è il numero delle categorie non facenti parte della Superficie forestale (in INFC posto uguale a uno per semplificare il calcolo), F indica il numero di categorie facenti parte della Superficie forestale, mentre la classe residua comprende le categorie che sono state escluse dall'inventario.

I parametri principali oggetto di interesse che possono essere stimati utilizzando i dati rilevati durante la seconda fase del piano di campionamento sono le $K \times L$ superfici a livello di singola regione delle diverse categorie di copertura, qui indicate con $\{a_{kl} : k = 1, \ldots, K, l = 1, \ldots, L\}$. Differenti aggregazioni di queste grandezze—quali la superficie della categoria k nell'intero territorio nazionale, $A_k = \sum_{l=1}^{L} a_{kl}$, la superficie complessiva di un sottoinsieme C di alcune categorie in una data regione l, $a_{Cl} = \sum_{k \in C} a_{kl}$, o nell'intero paese, $a_C = \sum_{l=1}^{L} a_{Cl}$—forniscono ulteriori informazioni utili.

I parametri principali oggetto di interesse che possono essere stimati al termine della terza fase del piano di campionamento riguardano, invece, le variabili quantitative rilevate al suolo, quali il numero di alberi, il volume o la biomassa. In particolare, per ognuna di queste variabili, è possibile stimare gli $F \times L$ valori totali, t_{kl}, per tutte le combinazioni di categoria forestale e regione. Come per i parametri a_{kl}, anche le aggregazioni dei valori di t_{kl} possono essere di particolare interesse. Nello specifico, per una data variabile, la grandezza $T = \sum_{k=1}^{K} \sum_{l=1}^{L} t_{kl}$ corrisponde al valore totale nel territorio nazionale; $T_l = \sum_{k=1}^{K} t_{kl}$ corrisponde al valore totale nella regione l; e $T_k = \sum_{l=1}^{L} t_{kl}$ indica il valore totale per la categoria k nel territorio nazionale. Per un dato sottoinsieme C di categorie, è inoltre interessante conoscere il valore della variabile nella regione l, $t_{Cl} = \sum_{k \in C} t_{kl}$, e nell'intero territorio nazionale, $t_C = \sum_{l=1}^{L} t_{Cl}$. Un altro insieme di parametri rilevanti è costituito dagli $F \times L$ valori di densità $d_{kl} = t_{kl}/a_{kl}$ e da aggregazioni dei medesimi, ossia $d_k = t_k/a_k$, $d_l = t_l/a_l$, $d_{Cl} = t_{Cl}/a_{Cl}$ e $d_C = t_C/a_C$.

Stime appropriate di tutti questi parametri e delle rispettive varianze possono essere ottenute utilizzando gli stimatori sviluppati per l'indagine campionaria dell'INFC2005 (Fattorini et al., 2011). Tali stimatori sono spiegati nelle sezioni successive di questo capitolo. In particolare, la Sect. 5.2 fornisce una descrizione essenziale degli stimatori delle superfici mentre la Sect. 5.3 illustra brevemente le procedure di stima dei valori totali e di densità delle variabili quantitative. Infine, la Sect. 5.4 conclude il capitolo spiegando perché questi stimatori rimangono validi anche per l'indagine campionaria dell'INFC2015.

Stima delle superfici

In base al disegno campionario adottato per l'INFC2005 durante la prima fase dell'indagine campionaria (cfr. Chap. 2), per ognuno degli NQ quadrati del reticolo, viene selezionato un punto in maniera casuale. Gli $N_0 \leq NQ$ punti selezionati sono poi classificati, mediante fotointerpretazione, in $H = NF + V + 1$ strati di uso e copertura del suolo, in cui i $V < F$ strati si riferiscono a una classificazione meno dettagliata delle F categorie di interesse forestale di cui si intende stimare la superficie (cfr. Sect. 5.1). L'ultimo strato residuo comprende tutti i punti per i quali la fotointerpretazione non ha dato esiti sufficientemente precisi per quanto riguarda l'uso e la copertura del suolo.

Gli N_0 punti sono classificati anche in base alla regione di appartenenza. Si ha dunque una doppia stratificazione caratterizzata da $H \times L$ strati, ciascuno indicato con U_{hl} e di dimensione pari a $N_{hl} (h = 1, \ldots, H, l = 1, \ldots, L)$.

Per ogni strato hl relativo ad una delle V categorie o alla categoria residua dei punti non classificabili, ossia in cui $h = NF + 1, \ldots, H$, se $N_{hl} > 0$, viene selezionato un campione S_{hl} di n_{hl} punti mediante campionamento casuale semplice senza ripetizione. I punti campionati vengono poi visitati a terra allo scopo di correggere eventuali errori di classificazione commessi durante la fase della fotointerpretazione. I punti, invece, appartenenti agli strati con $h = 1, \ldots, NF$, relativi alle categorie NFOWL, non vengono campionati (cf. Chap. 2).

Secondo Fattorini et al. (2006) uno stimatore corretto di a_{kl} è

$$\hat{a}_{kl} = R\left\{w_{kl} + \sum_{h>NF} w_{hl}w_{khl}\right\}, k = 1, \ldots, NF, l = 1, \ldots, L \tag{5.16}$$

oppure

$$\hat{a}_{kl} = R\sum_{h>NF} w_{hl}w_{khl}, k = NF + 1, \ldots, K, l = 1, \ldots, L \tag{5.17}$$

dove R è la superficie complessiva degli NQ quadrati, $w_{hl} = N_{hl}/NQ$ è il peso dello strato hl, $w_{kl} = N_{kl}/NQ$ è il peso dello strato kl relativo alle categorie non facenti parte della Superficie forestale i cui punti non sono stati campionati durante la seconda fase e $w_{khl} = n_{khl}/n_{hl}$ è la quota di punti di S_{hl} appartenenti alla categoria k.

Fattorini et al. (2006) dimostrano, inoltre, che uno stimatore conservativo della varianza di $\widehat{a}_{kl}$ è

$$\hat{v}(\hat{a}_{kl}) = \frac{R^2}{NQ-1}\left\{w_{kl} + \sum_{h>NF}\frac{N_{hl}-1}{n_{hl}-1}w_{hl}w_{khl} - \sum_{h>NF}\frac{N_{hl}-n_{hl}}{n_{hl}-1}w_{hl}w_{khl}^2 - p_{kl}^2\right\},$$
$$k = 1, \ldots, NF, l = 1, \ldots, L \tag{5.18}$$

oppure

$$\hat{v}(\hat{a}_{kl}) = \frac{R^2}{NQ-1}\left\{\sum_{h>NF}\frac{N_{hl}-1}{n_{hl}-1}w_{hl}w_{khl} - \sum_{h>NF}\frac{N_{hl}-n_{hl}}{n_{hl}-1}w_{hl}w_{khl}^2 - p_{kl}^2\right\},$$
$$k = NF + 1, \ldots, K, l = 1, \ldots, L \tag{5.19}$$

dove $p_{kl} = \widehat{a}_{kl}/R$.

Nell'utilizzare gli stimatori descritti dalle Equazioni 5.18 e 5.19, se $N_{hl} > 1$ allora è necessario che $n_{kl} \geq 2$; mentre se $N_{hl} = 1$ allora deve valere che $n_{kl} \geq 1$.

Per poter stimare le varianze delle diverse aggregazioni di $\widehat{a}_{kl}$, come ad esempio, $\widehat{A}_k = \sum_{l=1}^{L}\widehat{a}_{kl}$, è necessario ottenere anche le covarianze delle stime coinvolte nelle aggregazioni. Fattorini et al. (2006) dimostrano che la covarianza tra $\widehat{a}_{kl}$ e $\widehat{a}_{k'l}$ può essere stimata in maniera appropriata con

$$\hat{c}(\hat{a}_{kl}, \hat{a}_{k'l}) = -\frac{R^2}{NQ-1}\left\{\sum_{k>NF} \frac{N_{hl}-n_{hl}}{n_{hl}-1} w_{hl} w_{khl} w_{k'hl} + p_{kl} p_{k'l}\right\},$$
$$k' \neq k = 1, \ldots, K, l = 1, \ldots, L \tag{5.20}$$

mentre la covarianza tra $\hat{a}_{kl}$ and $\hat{a}_{k'l'}$ dovrebbe essere stimata ricorrendo allo stimatore seguente,

$$\hat{c}(\hat{a}_{kl}, \hat{a}_{k'l'}) = \frac{\hat{a}_{kl}\hat{a}_{k'l'}}{NQ-1}, k, k' = 1, \ldots, K, l' \neq l, \ldots, L \tag{5.21}$$

Sebbene le Equazioni 5.16 e 5.17 rappresentino stimatori corretti delle superfici per tutte le combinazioni di categoria e regione, la somma delle $K \times L$ stime di $\hat{a}_{kl}$ non corrisponde esattamente ad A. Tale somma, infatti, è uguale a R volte la quota di punti, rilevati durante la prima fase di campionamento, che cadono entro i confini del territorio nazionale. In modo analogo, per ciascun strato kl, $\sum_{k=1}^{K} \hat{a}_{kl}$ non è uguale a A_l poiché corrisponde a R volte la quota di punti che cadono nella regione l.

Dato che i valori di A e A_l sono noti, il problema può essere risolto calibrando i valori di $\hat{a}_{kl}$ in modo tale che i totali per territorio corrispondano ai valori effettivi. Ciò può essere ottenuto utilizzato il fattore di calibrazione seguente,

$$p_{kl}^{cal} = \hat{a}_{kl} / \sum_{k=1}^{K} \hat{a}_{kl}, l = 1, \ldots, L \tag{5.22}$$

rispetto al quale $\sum_{k=1}^{K} p_{kl}^{cal} = 1$ per ogni regione l.

Si possono dunque ottenere le stime calibrate delle superfici con

$$\hat{a}_{kl}^{cal} = A_l \times p_{kl}^{cal}. \tag{5.23}$$

Le formule per la stima delle varianze e covarianze di $\hat{a}_{kl}^{cal}$ sono riportate in Fattorini et al. (2006).

Stima dei valori totali e di densità delle variabili quantitative

Durante la terza fase della procedura di campionamento, per gli strati riferibili ad una categoria del Bosco tra i $(V+1) \times L \times F$ strati di seconda fase, se $n_{khl} > 0$ allora viene selezionato un campione Q_{khl} di m_{khl} punti mediante campionamento casuale semplice senza ripetizione. Ciascun punto selezionato viene poi visitato a terra così da rilevare l'ammontare di ogni variabile di interesse entro un'area circolare centrata su di esso (cfr. Chaps. 2 e 4). Secondo Fattorini et al. (2006), i dati ottenuti possono essere utilizzati per stimare il parametro t_{kl}, per ogni variabile, utilizzando il seguente stimatore corretto

$$\hat{t}_{kl} = NQ \sum_{h>NF} w_{hl} w_{khl} \overline{x}_{khl}, k = NF+1, \ldots, K, l = 1, \ldots, L \tag{5.24}$$

dove $\overline{x}_{khl}$ è la media cam-pio-na-ria delle stime di Horvitz-Thompson dei totali della variabile osservata nei punti campionati dello strato khl.

Fattorini et al. (2006) indicano anche quali sono gli stimatori conservativi delle varianze e covarianze di $\hat{t}_{kl}$, ossia

$$\begin{aligned}\hat{v}(\hat{t}_{kl}) = \frac{NQ^2}{NQ-1}\Bigg\{ & \sum_{h>NF} w_{hl} \frac{N_{hl}-1}{n_{hl}-1} w_{khl}(n_{khl}-1)\frac{s^2_{khl}}{m_{khl}}, \\ & + \sum_{h>NF} w_{hl} \frac{N_{hl}-1}{n_{hl}-1} w_{khl}(1-w_{khl})\overline{x}^2_{khl} + \sum_{h>NF} w_{hl} w^2_{khl} \overline{x}^2_{khl} - \overline{x}^2_{hl} \Bigg\} \\ & k = NF+1, \ldots, K, l = 1, \ldots, L,\end{aligned} \tag{5.25}$$

$$\begin{aligned}\hat{c}(\hat{t}_{kl}, \hat{t}_{k'l}) = -\frac{NQ^2}{NQ-1}\Bigg\{ & \sum_{k>NF} \frac{N_{hl}-n_{hl}}{n_{hl}-1} w_{hl} w_{khl} w_{k'hl} \overline{x}_{khl} \overline{x}_{k'hl} + \overline{x}_{khl} \overline{x}_{k'hl} \Bigg\}, \\ & k' \neq k = NF+1, \ldots, K, l = 1, \ldots, L,\end{aligned} \tag{5.26}$$

e

$$\hat{c}(\hat{t}_{kl}, \hat{t}_{k'l'}) = \frac{\hat{t}_{kl}\hat{t}_{k'l'}}{NQ-1}, k, k' = NF+1, \ldots, K, l' \neq l, \ldots, L \tag{5.27}$$

dove s^2_{khl} è la varianza cam-pio-na-ria dei valori stimati della variabile osservata nei punti campionati dello strato khl.

Nell'applicare l'Equazione 5.24, è necessario che se $n_{khl} > 1$ allora valga che $m_{khl} \geq 2$; mentre se $n_{khl} = 1$ allora si deve avere che $m_{kl} \geq 1$.

I valori di densità, d_{kl}, posso essere stimati direttamente con

$$\hat{d}_{kl} = \hat{t}_{kl} / \hat{a}^{cal}_{kl}. \tag{5.28}$$

Sfortunatamente, la varianza di $\hat{d}_{kl}$, che costituisce lo stimatore di un rapporto, è matematicamente intrattabile. Stime approssimativamente corrette delle varianze e covarianze di (5.28) possono però essere ottenute mediante l'approccio, comune in queste circostanze, che consiste nel linearizzare il rapporto utilizzando il termine del primo ordine dell'espansione in serie di Taylor (Särndal et al., 1992). L'affidabilità e la precisione di queste stime approssimate dipendono dal livello di precisione di $\hat{t}_{kl}$

e $\hat{a}_{kl}^{cal}$. Per questa ragione, in alcuni casi, l'errore della stima di densità non è stato riportato.

Per l'indagine campionaria dell'INFC2005, gli stimatori descritti dalle Equazioni (5.24) e (5.25) sono stati inoltre modificati affinché possano fornire le stime dei valori totali delle variabili quantitative per sottoinsiemi di unità della popolazione (ad es. gli alberi). Tali sottoinsiemi sono identificati da attributi qualitativi rilevanti come, ad esempio, la specie. Si considerino M sottoinsiemi non sovrapposti. Il valore totale, t_{kml}, di una variabile quantitativa di interesse per il sottoinsieme m, la categoria del Bosco k e la regione l può essere correttamente stimato con

$$\hat{t}_{kml} = NQ \sum_{h>NW} w_{hl} w_{khl} \overline{x}_{kmhl}$$
$$k = NF + 1, \dots, K - 1, \quad m = 1, \dots, M, \quad l = 1, \dots, L. \tag{5.29}$$

La varianza stimata di $\hat{t}_{kml}$ è dunque data da

$$\begin{aligned} \hat{v}(\hat{t}_{kml}) = \frac{NQ^2}{NQ-1} \Bigg\{ & \sum_{h>NF} w_{hl} \frac{N_{hl}-1}{n_{hl}-1} w_{khl}(n_{khl}-1) \frac{s^2_{kmhl}}{m_{khl}} \\ & + \sum_{h>NF} w_{hl} \frac{N_{hl}-1}{n_{hl}-1} w_{khl}(1-w_{khl}) \overline{x}^2_{kmhl} \\ & + \sum_{h>NF} w_{hl} w^2_{khl} \overline{x}^2_{kmhl} - \overline{x}^2_{klm} \Bigg\} \end{aligned}$$
$$k = NF + 1, \dots, K - 1, \quad m = 1, \dots, M, \quad l = 1, \dots, L, \tag{5.30}$$

dove $\overline{x}_{kmhl}$ e s^2_{kmhl} sono, rispettivamente, la media campionaria e la varianza campionaria dei valori stimati della variabile osservata nei punti campionati dello strato khl per le unità della popolazione che appartengono al sottoinsieme m.

Ulteriori modifiche degli stimatori hanno permesso di ottenere le stime riguardanti due casi particolari. Da un lato, gli stimatori sono stati modificati per stimare le superfici delle categorie forestali per i sottoinsiemi, $m = 1, \dots, M$, di punti di campionamento identificati da attributi qualitativi rilevati durante la terza fase del piano di campionamento. Dall'altro lato, sono stati modificati per stimare a_{kml} e t_{kml} nelle circostanze in cui il sottoinsieme m è stato identificato durante la seconda fase del piano di campionamento. Si veda Fattorini et al. (2011) per ulteriori dettagli sulla costruzione di tali stimatori.

Confronto tra i due inventari forestali

La strategia presentata è stata utilizzata durante la stima dei parametri di interesse per l'INFC2005. Tale strategia rimane valida anche per la procedura di stima dell'INFC2015. Infatti, le modifiche occorse tra i due inventari forestali non hanno avuto impatto sull'uso della strategia di stima, come di seguito esplicitato.

Il piano di campionamento adottato per l'INFC2005 è consistito in una struttura a tre fasi (cfr. Chap. 2). In una prima fase, condotta mediante fotointerpretazione, l'area è stata suddivisa in poligoni di eguale dimensione (1 km^2), da cui veniva selezionato casualmente un punto da ogni poligono. La popolazione costituita dai punti selezionati è stata poi divisa in 21 strati, corrispondenti ai distretti territoriali del territorio nazionale (le regioni), e poi in strati corrispondenti alle diverse categorie di uso e copertura del suolo. Nella seconda fase, condotta mediante campionamento al suolo, un campione stratificato di punti è stato selezionato dagli strati definiti in prima fase, esclusivamente per le categorie di uso e copertura del suolo di interesse per l'INFC2005. I punti selezionati sono stati attribuiti a strati corrispondenti alle categorie forestali. Nella terza fase, da ognuno dei campioni selezionati in seconda fase e attribuiti a categorie del Bosco, è stato selezionato un campione aggiuntivo di punti. Dunque, attorno ad ognuno di questi punti è stato disegnato un insieme di aree di saggio allo scopo di definire le aree in cui effettuare la misurazione delle variabili di interesse (cfr. Chap. 4).

Il piano di campionamento adottato per l'INFC2015 non ha subito cambiamenti sostanziali. La struttura in tre fasi precedentemente messa in atto è stata mantenuta: la prima fase è stata disegnata per classificare in via preliminare l'uso del suolo; la seconda ha avuto l'obiettivo di classificare in via definitiva le categorie di uso e copertura del suolo in corrispondenza dei punti campionati, di attribuire la categoria forestale e classificare i caratteri qualitativi per la stima delle superfici; la terza ha definito le aree per la misura delle variabili quantitative di interesse. Quindi, il piano di campionamento non ha subito cambiamenti tali da alterarne la struttura e il disegno originali. I cambiamenti occorsi durante la pianificazione dell'INFC2015 hanno riguardato esclusivamente la definizione degli strati e delle dimensioni campionarie. Nello specifico:

i. il numero di punti nel territorio nazionale è stato pari a 301,306 durante l'INFC2005 e pari a 301,271 durante l'INFC2015;
ii. lo schema di classificazione delle categorie di uso e copertura del suolo utilizzato per la fotointerpretazione dell'INFC2005 ha incluso 12 classi/sottoclassi, mentre lo schema utilizzato per l'INFC2015 ha incluso 13 classi/sottoclassi (inclusa la classe dei punti non classificati);
iii. per la stima delle superfici durante l'INFC2015, alle 13 classi/sottoclassi originarie sono state aggiunte 3 nuove classi di interesse inventariale, che, secondo la fotointerpretazione implementata per l'INFC2005, appartenevano a classi/sottoclassi non di interesse per le fasi successive dell'inventario, ritenute invece tali a seguito della nuova fotointerpretazione messa in atto per l'INFC2015;

iv. il numero di classi/sottoclassi non forestali da fotointerpretazione è stato di 7 classi/sottoclassi nell'INFC2005, mentre è stato di 8 classi/sottoclassi nell'INFC2015;
v. il numero di classi di interesse inventariale è stato pari a 4 nell'INFC2005, a fronte delle 7 dell'INFC2015.

Come visto, le modifiche sono state dovute a differenti sistemi di fotointerpretazione adottati durante l'INFC2015 e a cambiamenti nelle classi/sottoclassi di interesse. Tutto questo ha condotto ad una differente numerosità di strati definiti rispettivamente per l'INFC2005 e l'INFC2015, così come ad alcuni cambiamenti di unità da uno strato all'altro tra le due occasioni di indagine. Ciò nonostante, da un punto di vista metodologico, il piano di campionamento adottato durante i due inventari è rimasto invariato, dato che esso resta un disegno a tre fasi, con la definizione delle aree ad ogni fase basata sulla stratificazione. Questo conferma la possibilità di utilizzare per l'INFC2015 le stesse tecniche di stima adottate durante l'INFC2005, sia per quanto riguarda la stima delle superfici che per la stima dei valori delle caratteristiche di interesse.

Gli stimatori utilizzati per l'INFC2015 sono stati implementati mediante l'utilizzo del software statistico open source R (R Core Team, 2020), un ambiente software gratuito per il calcolo statistico.

References

Fattorini, L., Marcheselli, M., & Pisani, C. (2006). A three-phase sampling strategy for large-scale multiresource forest inventories. *Journal of Agricultural, Biological, and Environmental Statistics, 11*(3), 1–21.

Fattorini, L., Gasparini, P., & De Natale, F. (2011). Descrizione generale delle procedure di stima. In P. Gasparini & G. Tabacchi (Eds.), *L'Inventario Nazionale delle Foreste e dei serbatoi forestali di Carbonio INFC 2005. Secondo inventario forestale nazionale italiano. Metodi e risultati* (pp. 75–81). Edagricole-Il Sole 24 Ore, Milano. ISBN 978-88-506-5394-2.

R Core Team. (2020). R: A language and environment for statistical computing. R Foundation for Statistical Computing, Vienna, Austria. https://www.R-project.org/.

Särndal, C. E., Svensson, B., & Wretman, J. (1992). *Model assisted survey sampling* (p. 694). Springer.

Open Access This chapter is licensed under the terms of the Creative Commons Attribution 4.0 International License (http://creativecommons.org/licenses/by/4.0/), which permits use, sharing, adaptation, distribution and reproduction in any medium or format, as long as you give appropriate credit to the original author(s) and the source, provide a link to the Creative Commons license and indicate if changes were made.

The images or other third party material in this chapter are included in the chapter's Creative Commons license, unless indicated otherwise in a credit line to the material. If material is not included in the chapter's Creative Commons license and your intended use is not permitted by statutory regulation or exceeds the permitted use, you will need to obtain permission directly from the copyright holder.

Chapter 6
Plot Level Estimation Procedures and Models

Procedure di stima e modelli per le aree di saggio

Lucio Di Cosmo

Abstract Quantitative variable raw data recorded in the sample plots require preprocessing before the NFI estimators of totals and densities can be used to produce statistics. The objective of the plot level estimates is to estimate the variables of interest for each sample point expanded to the 1 km^2 area of the cell that the point represents. The intensity and complexity of the computations vary considerably depending on the variable, the way it is obtained by the measured items (e.g., DBH measurement *vs.* basal area), whether all the items in the sample plot or only a subsample of them are measured, and the availability of models. The definitive result of the computations are tallies, volumes, biomass and carbon stocks but estimates of additional variables at intermediate steps may be needed (e.g., total tree height). This chapter describes the methods and the models used in INFC2015 for the estimation of the variables related to trees (e.g., tallies, basal area), small trees and shrubs (e.g., biomass, carbon stock), stumps (e.g., volume, biomass), stock variation (e.g., the wood annually produced by growth and that removed). Some of the models described were produced in view of the INFC needs, before and after it was established in 2001, while others were created during the NFI computation processes. Finally, the conversion factors needed to estimate the biomass of deadwood, saplings and shrubs were obtained through an additional field campaign of the second Italian NFI (INFC2005) and the following laboratory analyses.

Keywords NFI · Tree volume and biomass model · Height-diameter model · Deadwood · Volume increment · Removals

6.1 Introduction

NFI estimators of totals and densities (values per surface unit) described in Chap. 5 require that the value of the quantitative variables measured in the sample plots are expressed with reference to the square plane surface of 1 km^2. The total

L. Di Cosmo (✉)
CREA Research Centre for Forestry and Wood, Trento, Italy
e-mail: lucio.dicosmo@crea.gov.it

© The Author(s) 2022

P. Gasparini et al. (eds.), *Italian National Forest Inventory—Methods and Results of the Third Survey*, Springer Tracts in Civil Engineering,
https://doi.org/10.1007/978-3-030-98678-0_6

value of a variable in the sample plot (e.g., the number or volume of the trees) must be divided by the surface of the sample plot and multiplied by the surface SQ (1 km^2) of the square containing the sample plot. As the variables are measured in sample plots with three different sizes (as described in Chap. 3), the expansion factors are $10^6/12.566371$, for the variables measured in each of the AdS2, $10^6/50.265482$, for the variables measured in the AdS4, and $10^6/530.929159$ for those measured in the AdS13. For example, each tree with $4.5 \leq$ DBH < 9.5 cm measured in the AdS4 represents $1 \times 10^6/50.265482$ per square kilometre. In the same way, the value of each variable of that tree (e.g., its basal area, volume or biomass) must be multiplied by $10^6/50.265482$ to get the amount per square kilometre.

Among the quantities surveyed, the number of trees and the basal area of the stands are easily computed. The same is also true for the carbon stocks either in the living ecosystem components (trees, small trees and shrubs, removed trees) or the dead ones (dead trees, deadwood lying on the ground, stumps) after their biomass has been estimated, because conversion was obtained assuming 50% carbon content of woody biomass (IPCC, 2006; Matthews, 1993; Woodall et al., 2012). For this reason, the descriptions of methods below will be limited to the biomass computation steps. Timber volume, tree biomass, annual volume increment and removals required more complex preliminary processing. They entailed construction of models that used data widely representative of the conditions across the entire country so that they might be used beyond NFI computations.

This chapter describes the data analysis and the models used for estimating the quantitative variables measured in the sample plots by INFC2015. Some of the models described were developed by previous or connected research projects, in view of the INFC needs. The conversion factors needed to estimate the biomass of deadwood, saplings and shrubs were obtained through an additional field campaign of the second Italian NFI (INFC2005) and the following laboratory analyses.

6.2 Tree Height

Total tree height (h) is an explanatory variable used in the model to estimate the volume and the biomass of trees. Due to the high costs, NFIs generally limit the tree height measurement on sample trees that are a subsample of those callipered (Gschwantner et al., 2016), which involves estimating the total tree height through models. INFC2015 increased the sample trees per plot to ten, as compared to the past assessment, which ranged from five to ten (Bosela et al., 2016; Gasparini & Di Cosmo, 2016; Gasparini et al., 2010). For the sample trees, five were randomly selected, and five were selected based on representative criteria; the collected data were used to build height-diameter models. The equations described were obtained re-calibrating the model developed in INFC2005; the new calibration allowed including the wider range of growing conditions revealed by the data we collected.

The model adopted draws from the Chapman-Richards function (Liu & Li, 2003; Ratkowsky, 1990) and is shown in Eq. (6.1):

$$\hat{h}_{ij} = \left(a_1 + a_2 \cdot H_{domj}\right)\left[1 - \exp\left(\left(a_3 + a_4 \cdot H_{domj}\right)d_{ij}\right)\right]^{\left(a_5 + a_6 H_{domj}\right)} \quad (6.1)$$

where:

$\hat{h}_{ij}$ is the total tree height estimated of tree i of plot j, in m;

d_{ij} is the DBH of tree i of plot j, in cm;

H_{domj} is the dominant height (H_{dom}) of plot j, in m, obtained as the arithmetical mean of the total heights of the three widest trees in DBH.

We graphically evaluated the affinity of the height-diameter relationship for species with small samples, to create groups with higher number of data and obtain more stable estimates of the numerical coefficients. Table 6.1 shows the 52 equations obtained and their relative estimated coefficients. They concern the main species or species groups of the Italian forests and allowed for an estimation of the total tree-heigh to almost all of the undamaged trees. For the species not provided with an equation, we evaluated which could be best used through graphical and analytical analyses. A limited number of sample-plots could not be provided with the H_{dom}. For the species included, we developed models based solely on the DBH as the explanatory variable. Finally, for trees of species callipered only a few times but provided with a relevant percentage of tree height data, we adopted by convention the mean measured heights, by DBH class in case of visible trends.

Table 6.2 shows the number of observations by species or group of species used for models' calibrations and the range of values for DBH and H_{dom}. This information is useful in establishing awareness beyond the INFC computations.

6.3 Growing Stock and Biomass

6.3.1 Living Trees

Distinct procedures were adopted to estimate the volume and the biomass of unbroken or broken (truncated) trees. The volume and biomass of unbroken trees were estimated through double entry equations available in Tabacchi et al., (2011a, b). Those equations were built before and in view of the second Italian NFI (INFC2005) and partially updated afterwards. The data were collected from tree samples across wide areas of the country. In addition to being measured, the trees were cut, and samples of wood were analysed in the laboratory to determine the dry weight. At present, there are prediction equations available for 26 species or group of species; a set of 5 equations is given for each species/group whose form is shown in Eq. (6.2):

$$\hat{y}_i = b_1 + b_2 \cdot DBH_i^2 \cdot h_i + b_3 \cdot DBH_i \quad (6.2)$$

Table 6.1 Tree height model: regression coefficients for Eq. (6.1) / Modello delle altezze: coefficienti di regressione per l'Eq. (6.1)

Species/Specie	a1	a2	a3	a4	a5	a6
Abies alba Mill.	4.26135E+00	9.33059E−01	−6.24124E−02	1.94073E−04	6.04051E−01	3.51822E−02
Acer campestre L.	3.09487E+00	8.88473E−01	−6.18510E−02	6.08207E−04	2.93101E−01	2.41979E−02
Acer monspessulanum L.	2.97392E+00	7.84232E−01	−1.07716E−01	1.85691E−03	5.14414E−01	2.32827E−02
Acer opalus L.	2.95587E+00	9.81602E−01	−6.09311E−02	6.31666E−04	2.60551E−01	2.59534E−02
Acer pseudoplatanus L.	2.98716E+00	1.00878E+00	−6.29325E−02	8.62163E−04	3.68112E−01	1.75223E−02
Alnus cordata Loisel.	5.65530E−01	1.00224E+00	−1.67133E−01	2.81823E−03	−5.06403E−01	1.29358E−01
Alnus glutinosa Gaertner	2.77831E+00	9.59346E−01	−1.04083E−01	1.85268E−03	4.28540E−01	2.68161E−02
Alnus incana Moench	3.32756E+00	9.79914E−01	−7.25081E−02	3.85823E−04	2.33026E−01	4.91115E−02
Arbutus unedo L.	2.67953E+00	9.51691E−01	−2.65201E−02	1.98906E−04	2.30484E−01	2.17535E−02
Betula spp.	3.73429E+00	9.17584E−01	−8.83425E−02	9.20848E−04	3.76846E−01	4.09901E−02
Carpinus betulus L.	3.63286E−01	1.21902E+00	−3.69872E−02	3.24973E−05	1.61856E−02	3.51330E−02
Castanea sativa Mill.	1.84083E+00	9.03699E−01	−1.35746E−01	2.19929E−03	3.75171E−01	3.46528E−02
Cedrus spp.	9.18242E−01	1.05824E+00	−1.16287E−01	2.33606E−03	8.65403E−01	3.28378E−02
Cupressus spp.	9.11655E−01	1.09760E+00	−6.98720E−02	7.75528E−04	3.69800E−01	4.13962E−02
Eucaliptus spp.	3.05691E+00	9.30214E−01	−7.49562E−02	1.42239E−04	9.96347E−03	6.00463E−02
Fagus silvatica L.	1.67564E+00	9.92171E−01	−9.31516E−02	1.29696E−03	4.66628E−01	2.45655E−02
Fraxinus excelsior L.	1.61459E+00	1.09099E+00	−8.69460E−02	1.40508E−03	3.87834E−01	2.33877E−02
Fraxinus ornus L.	2.77437E+00	9.90508E−01	−5.68753E−02	1.09636E−03	2.95019E−01	1.50805E−02
Ilex aquifolium L.	1.24163E+01	5.14217E−01	−2.51951E−02	2.89476E−04	7.02279E−01	1.38719E−03
Juglans regia L.	3.26648E+00	9.18860E−01	−1.10573E−01	2.17742E−03	6.83922E−01	2.31850E−02
Larix decidua Mill.	3.64773E+00	9.02113E−01	−7.32003E−02	2.27904E−05	7.47181E−01	3.20086E−02

(continued)

Table 6.1 (continued)

Species/Specie	a1	a2	a3	a4	a5	a6
Malus sylvestris Mill.	9.58462E+00	1.54356E+00	−3.29032E−03	4.76617E−05	3.54984E−01	3.45689E−03
Ostrya carpinifolia Scop.	2.75546E+00	9.65506E−01	−6.89939E−02	1.04189E−03	2.55169E−01	2.21904E−02
Picea abies K.	4.76745E+00	9.26548E−01	−6.54796E−02	1.81008E−04	5.89445E−01	4.01716E−02
Pinus cembra L.	3.867850451	7.37967E−01	−1.76506E−01	4.79385E−03	3.44848E+00	−9.12486E−02
Pinus exotic species	−1.33393E+00	1.10036E+00	−2.21212E−01	4.94339E−03	3.17873E+00	−2.45348E−02
Pinus halepensis Mill.	2.15282E+00	9.55179E−01	−3.07472E−02	−1.64443E−03	−4.07393E−02	5.27576E−02
Pinus nigra Arn. *ssp. laricio*	1.09849E+00	9.73514E−01	−1.10940E−01	8.31008E−04	4.16956E−01	6.06677E−02
Pinus nigra Arn. *ssp nigra*	1.30255E+00	9.86795E−01	−1.06805E−01	1.47339E−03	9.78724E−01	1.31181E−02
Pinus pinaster Ait.	3.01937E+00	9.22678E−01	−8.85217E−02	1.33219E−03	9.81293E−01	5.24319E−03
Pinus pinea L.	2.96657E+00	9.01837E−01	−1.75142E−02	−2.08893E−03	1.24437E−01	3.72866E−02
Pinus sylvestris L.	1.92176E+00	9.21973E−01	−9.87814E−02	1.07924E−03	4.30631E−01	4.52480E−02
Populus alba L.	2.98795E+00	1.09978E+00	−4.61830E−02	3.84586E−04	1.37007E−01	3.05839E−02
Populus nigra L.	3.36163E+00	9.77907E−01	−1.09064E−01	1.57426E−03	3.52180E−01	4.28066E−02
Populus tremula L.	4.22174E+00	8.79486E−01	−8.71131E−02	4.00909E−04	3.39362E−01	3.19146E−02
Populus x euroamericana	1.83666E+00	9.81646E−01	−2.00210E−01	3.63498E−03	1.94503E+00	−2.00903E−03
Prunus avium L.	1.53015E+00	1.06699E+00	−9.71860E−02	2.11279E−03	6.94489E−01	1.11695E−02
Pseudotsuga menziesii Franco	3.69341E+00	9.38367E−01	−6.96406E−02	5.46225E−04	5.71862E−01	9.27649E−03
Pyrus pyraster Burgsd.	1.38422E+00	7.88080E−01	−1.03402E−01	2.03117E−03	6.32454E−01	1.94897E−02
Quecus ilex L.	1.10089E+00	1.00019E+00	−8.04397E−02	8.33419E−04	1.39662E−01	4.73192E−02
Quercus cerris L.	1.41419E+00	9.88389E−01	−1.35353E−01	1.88494E−03	4.06135E−01	4.73742E−02
Quercus frainetto Ten.	4.31204E−01	1.02235E+00	−1.52229E−01	3.39307E−03	3.83981E−01	4.67553E−02

(continued)

Table 6.1 (continued)

Species/Specie	a1	a2	a3	a4	a5	a6
Quercus petraea Liebl.	2.67001E+00	9.40558E−01	−9.80209E−02	1.40505E−03	7.74707E−01	8.68106E−03
Quercus pubescens Willd.	1.31887E+00	9.72768E−01	−1.21427E−01	2.80814E−03	7.15921E−01	2.10289E−02
Quercus robur L.	2.50547E+00	1.02357E+00	−8.46062E−02	1.16936E−03	4.01505E−01	3.41839E−02
Quercus suber L.	1.84437E+00	9.34549E−01	−5.38232E−02	2.36331E−04	4.63972E−01	3.79493E−02
Quercus troiana Webb	2.46529E+00	9.70736E−01	−3.24847E−02	−3.08190E−03	−2.13425E−02	8.70640E−02
Robinia pseudoacacia L.	2.88463E+00	9.95842E−01	−1.06113E−01	1.83741E−03	3.92575E−01	2.60253E−02
Salix spp.	3.67907E−01	1.08018E+00	−4.94171E−02	2.00460E−04	−1.37295E−01	4.68130E−02
Species group one	2.76100E+00	9.13564E−01	−1.73659E−01	4.21242E−03	9.99678E−01	1.53357E−02
Tilia spp.	3.66262E+00	9.53185E−01	−1.11424E−01	2.03822E−03	1.07106E+00	2.76841E−03
Ulmus spp.	5.46657E+00	8.99278E−01	−7.91084E−02	1.51615E−03	8.19986E−01	1.23192E−03

Pinus exotic species: *Pinus excelsa* Wall., *Pinus radiata* D.Don., *Pinus strobus* L., other species.
Betula spp.: *Betula pendula* Roth., *Betula pubescens* Ehrh.
Cupressus spp.: *Cupressus sempervirens* L., *Cupressus arizonica* Greene., other species.
Salix spp.: *Salix caprea* L., *Salix alba* L., other species.
Tilia spp.: *Tilia cordata* Mill., *Tilia platyphyllos* Scop.
Species in group one: *Ailantus altissima* Mill., *Fraxinus oxycarpa* Bieb.

Table 6.2 Number of observations and extreme values of diameter at breast height (DBH) and dominant height (H_{dom}) used for calibrating the height-diameter equations, by species / Numero di osservazioni e valori estremi di diametro ($d_{1.30}$) e altezza dominante (H_{dom}) utilizzate per la calibrazione delle equazioni di previsione dell'altezza totale degli alberi, per specie

Species/Specie	N. of observations/N. di osservazioni	DBH/$d_{1.30}$ min	DBH/$d_{1.30}$ max	H_{dom} min	H_{dom} max
Abies alba Mill.	1084	5.0	140.0	7.0	47.4
Acer campestre L.	822	5.0	70.0	6.1	32.9
Acer monspessulanum L.	199	5.0	35.5	6.1	24.8
Acer opalus L.	429	5.0	54.0	6.5	40.9
Acer pseudoplatanus L.	935	5.0	63.5	4.4	39.1
Alnus cordata Loisel.	261	5.0	56.0	4.4	33.7
Alnus glutinosa Gaertner	305	5.0	56.5	9.8	35.7
Alnus incana Moench.	115	5.0	50.5	8.1	37.1
Arbutus unedo L.	241	5.0	23.0	3.8	25.0
Betula spp.	444	6.0	69.5	6.2	38.5
Carpinus betulus L.	546	5.0	63.0	5.6	40.1
Castanea sativa Mill.	3142	5.0	140.0	5.4	38.0
Cedrus spp.	114	6.0	91.0	6.5	30.9
Cupressus spp.	307	6.0	48.0	5.7	30.9
Eucaliptus spp.	501	5.0	70.0	5.0	37.6
Fagus silvatica L.	6368	5.0	114.0	5.8	44.3
Fraxinus excelsior L.	719	5.0	70.5	5.4	37.9
Fraxinus ornus L.	2862	5.0	41.5	4.4	33.4
Ilex aquifolium L.	106	5.0	26.5	6.0	44.3
Juglans regia L.	255	5.0	57.0	4.5	35.7
Larix decidua Mill.	2046	5.0	112.0	5.4	45.0
Malus sylvestris Mill.	127	5.0	34.0	4.9	29.1
Ostrya carpinifolia Scop.	3123	5.0	49.5	4.3	37.6
Picea abies K.	3887	5.0	100.0	4.2	52.2
Pinus cembra L.	294	5.0	68.5	8.3	36.1
Pinus exotic species	120	10.0	62.0	8.4	35.3
Pinus halepensis Mill.	849	5.0	71.5	3.9	29.8
Pinus nigra Arn. *ssp. laricio*	578	5.0	93.0	4.2	42.8
Pinus nigra Arn. *ssp nigra*	1601	5.0	86.5	4.3	52.2

(continued)

Table 6.2 (continued)

Species/Specie	N. of observations/N. di osservazioni	DBH/$d_{1.30}$ min	DBH/$d_{1.30}$ max	H_{dom} min	H_{dom} max
Pinus pinaster Ait.	377	5.0	77.5	4.6	25.0
Pinus pinea L.	371	5.0	88.0	4.4	25.7
Pinus sylvestris L.	1226	5.0	80.5	5.2	42.7
Populus alba L.	169	5.0	93.0	7.8	36.3
Populus nigra L.	331	5.0	102.0	8.5	38.9
Populus tremula L.	328	5.0	59.0	8.3	44.9
Populus x euroamericana	298	5.0	93.0	8.7	41.9
Prunus avium L.	823	5.0	61.5	5.8	29.8
Pseudotsuga menziesii Franco	204	9.0	87.5	14.7	40.1
Pyrus pyraster Burgsd.	284	5.0	37.0	3.3	25.6
Quercus ilex L.	2820	5.0	88.0	2.7	29.8
Quercus cerris L.	5147	5.0	118.0	3.4	40.1
Quercus frainetto Ten.	219	5.0	61.5	8.7	26.8
Quercus petraea Liebl.	618	5.0	92.0	6.2	44.9
Quercus pubescens Willd.	5905	5.0	94.0	3.0	32.3
Quercus robur L.	221	5.0	110.0	7.7	35.7
Quercus suber L.	884	5.0	86.0	2.4	19.9
Quercus troiana Webb	191	5.0	58.0	5.0	15.1
Robinia pseudoacacia L.	1191	5.0	63.5	5.7	36.5
Salix spp.	435	5.0	57.0	4.2	38.9
Species group one	83	5.0	45.0	7.8	27.9
Tilia spp.	271	5.0	53.5	8.0	37.1
Ulmus spp.	606	5.0	68.0	4.9	38.9

where:

$\hat{y}_i$ is the volume (v in dm^3) (stem plus large branches and treetop up to 5 cm cross section diameter), the biomass (in kg) for the same component (w_1), the biomass of the small branches and the treetop (w_2), the stump biomass (w_3) or the whole aboveground biomass (w_4) of tree i;

DBH_i is the stem diameter at 1.30 m from the ground level of tree i, in cm;

h_i is the total tree height of tree i, in m.

w_4 can also be obtained by the sum of its three components (w_1, w_2 and w_3) as the model assures additivity. Table 6.3 shows the coefficients for estimating the volume of the species or groups of species.

Table 6.4 shows the coefficients for estimating the four biomass components.

For the species not provided with a function, we adopted one of the functions available, i.e., the function that was thought to be the most suitable, based on the morphologic affinity of the species, or we adopted the prediction equation for the heterogeneous group of the other broadleaved species. In the inventory survey, the size interval of the measured DBH was for some species wider than the set of sample trees used to develop the prediction equations. In a small number of trees with DBH near the

Table 6.3 Tree volume functions: regression coefficients for Eq. (6.2) / Funzioni di cubatura: coefficienti per l'Eq. (6.2)

Species/Specie	b1	b2	b3
Abies alba Mill.	−1.8381E+00	3.7836E−02	3.9934E−01
Picea abies K.	−9.1298E+00	3.4866E−02	1.4633E+00
Acer spp.	1.6905E+00	3.7082E−02	0.0000E+00
Carpinus spp.	−1.4983E+00	3.8828E−02	0.0000E+00
Castanea sativa Mill.	−2.0010E+00	3.6524E−02	7.4466E−01
Quercus cerris L.	−4.3221E−02	3.8079E−02	0.0000E+00
Cupressus spp.	−2.6735E+00	3.6590E−02	6.4725E−01
Pseudotsuga menziesii Franco	−7.9946E+00	3.3343E−02	1.2186E+00
Eucaliptus spp.	−1.3789E+00	4.5811E−02	0.0000E+00
Fagus silvatica L.	8.1151E−01	3.8965E−02	0.0000E+00
Fraxinus spp.	−1.1137E−01	3.9108E−02	0.0000E+00
Larix decidua Mill.	−1.6519E+01	2.9980E−02	3.1506E+00
Quercus ilex L.	−2.2219E+00	3.9685E−02	6.2762E−01
Alnus spp.	−2.2932E+01	3.2641E−02	2.9991E+00
Pinus cembra L.	2.8521E+00	3.9504E−02	0.0000E+00
Pinus pinea L.	−4.0404E−01	4.1113E−02	0.0000E+00
Pinus nigra Arn. *ssp. laricio*	6.4383E+00	3.8594E−02	0.0000E+00
Pinus pinaster Ait.	2.9963E+00	3.8302E−02	0.0000E+00
Pinus nigra Arn. *ssp nigra*	−2.1480E+01	3.3448E−02	2.9088E+00
Pinus sylvestris L.	3.1803E+00	3.9899E−02	0.0000E+00
Exotic pine spp.	2.6279E+00	3.3389E−02	0.0000E+00
Robinia pseudoacacia L.	−2.1214E+00	3.7123E−02	1.4296E−01
Quercus pubescens Willd.	5.1025E−01	4.5184E−02	−3.6026E−01
Salix spp.	−2.3140E+00	3.8926E−02	0.0000E+00
Other broadleaved spp.	2.3118E+00	3.1278E−02	3.7159E−01
Pinus halepensis Mill.	−1.2508E−01	3.1518E−02	2.3748E+00

Table 6.4 Tree biomass functions: regression coefficients for Eq. (6.2) / Funzioni di stima della fitomassa: coefficienti per l'Eq. (6.2)

Species/Specie	w_1b_1	w_1b_2	w_1b_3	w_2b_1	w_2b_2	w_2b_3	w_3b_1	w_3b_2	w_3b_3	w_4b_1	w_4b_2	w_4b_3
Abies alba Mill.	−1.8060E+00	1.4255E−02	4.2759E−01	−4.2685E−01	3.5168E−03	6.9321E−01	9.4271E−02	3.5357E−04	−1.1866E−02	−2.1386E+00	1.8125E−02	1.1089E+00
Picea abies K.	−5.9426E+00	1.3210E−02	7.8369E−01	5.9459E+00	4.0669E−03	−2.1054E−01	1.3818E−01	3.4318E−04	−1.1062E−02	1.4146E−01	1.7620E−02	5.6209E−01
Acer spp.	8.6876E−01	2.0421E−02	0.0000E+00	5.3322E+00	5.5554E−03	0.0000E+00	2.5852E−01	3.9206E−04	0.0000E+00	6.4595E+00	2.6368E−02	0.0000E+00
Carpinus spp.	−1.0514E+00	2.3952E−02	0.0000E+00	3.8823E+00	5.6122E−03	0.0000E+00	4.1760E−01	6.0317E−04	0.0000E+00	3.2485E+00	3.0167E−02	0.0000E+00
Castanea sativa Mill.	−9.5407E−01	1.8335E−02	1.9237E−01	−7.9938E−01	2.6769E−03	6.9544E−01	−4.2047E−01	4.2980E−04	8.2936E−02	−2.1739E+00	2.1442E−02	9.7075E−01
Quercus cerris L.	−1.3658E+00	2.5533E−02	0.0000E+00	1.7649E+00	4.5457E−03	0.0000E+00	1.1733E−01	6.7201E−04	0.0000E+00	5.1651E−01	3.0751E−02	0.0000E+00
Cupressus spp.	−2.5288E+00	1.7046E−02	6.7181E−01	−1.1740E+00	7.0221E−03	6.6022E−01	−4.3163E−01	2.9100E−04	8.3616E−02	−4.1345E+00	2.4359E−02	1.4156E+00
Pseuotsuga menziesii Franco	−2.0529E+00	1.3823E−02	2.8280E−01	−2.5484E+01	5.8230E−04	3.5145E+00	1.3884E+00	3.3300E−04	−1.2753E−01	−2.6149E+01	1.4739E−02	3.6698E+00
Eucaliptus spp.	−1.6507E+00	2.8060E−02	0.0000E+00	1.9006E+00	4.7954E−03	0.0000E+00	2.3448E−01	7.2878E−04	0.0000E+00	4.8447E−01	3.3584E−02	0.0000E+00
Fagus silvatica L.	−8.3814E−01	2.4865E−02	0.0000E+00	2.5040E+00	5.1280E−03	0.0000E+00	−2.4959E−02	7.8153E−04	0.0000E+00	1.6409E+00	3.0775E−02	0.0000E+00
Fraxinus spp.	−6.5463E−01	2.5364E−02	0.0000E+00	2.7854E+00	6.8597E−03	0.0000E+00	5.8578E−02	7.2562E−04	0.0000E+00	2.1893E+00	3.2949E−02	0.0000E+00
Larix decidua Mill.	−2.4433E+00	1.5091E−02	6.2777E−01	−1.0305E+01	−4.3457E−04	2.3549E+00	−1.3120E+00	6.8577E−06	2.4822E−01	−1.4060E+01	1.4664E−02	3.2309E+00
Quercus ilex L.	−2.6095E+00	3.1220E−02	3.9794E−01	8.3687E+00	1.1971E−02	−1.0175E+00	3.9692E−01	1.0406E−03	−9.1902E−02	6.1561E+00	4.4232E−02	−7.1143E−01
Alnus spp.	−8.8718E+00	1.6235E−02	1.0538E+00	−7.2566E+00	1.3977E−03	1.5009E+00	−6.1812E−01	2.9779E−04	1.1172E−01	−1.6747E+01	1.7930E−02	2.6664E+00
Pinus cembra L.	1.3499E+00	1.5191E−02	0.0000E+00	1.8461E+00	3.1880E−03	0.0000E+00	1.1130E−01	4.6900E−04	0.0000E+00	3.3073E+00	1.8848E−02	0.0000E+00
Pinus pinea L.	−1.4475E+00	1.8207E−02	0.0000E+00	1.7416E+00	6.5400E−03	0.0000E+00	1.6474E−01	4.2966E−04	0.0000E+00	4.5885E−01	2.5176E−02	0.0000E+00
Pinus nigra Arn. ssp. *laricio*	1.4558E+00	1.4464E−02	0.0000E+00	6.509403	1.4374E−03	0.0000E+00	7.8209E−02	3.5188E−04	0.0000E+00	1.5405E+00	1.6253E−02	0.0000E+00
Pinus pinaster Ait.	6.3157E−01	1.5840E−02	0.0000E+00	1.2558E+00	4.7109E−03	0.0000E+00	6.6497E−02	2.5933E−04	0.0000E+00	1.9539E+00	2.0810E−02	0.0000E+00
Pinus nigra Arn. ssp *nigra*	−1.2732E+01	1.4692E−02	1.5988E+00	−2.0573E+01	2.5163E−03	2.4937E+00	−6.6688E−01	1.5495E−04	9.8741E−02	−3.3972E+01	1.7363E−02	4.1912E+00

(continued)

Table 6.4 (continued)

Species/Specie	w_1b_1	w_1b_2	w_1b_3	w_2b_1	w_2b_2	w_2b_3	w_3b_1	w_3b_2	w_3b_3	w_4b_1	w_4b_2	w_4b_3
Pinus sylvestris L.	6.5786E−01	1.7176E−02	0.0000E+00	2.1336E+00	4.5864E−03	0.0000E+00	9.3354E−02	3.1809E−04	0.0000E+00	2.8848E+00	2.2080E−02	0.0000E+00
Exotic pine spp.	7.4835E−01	1.2202E−02	0.0000E+00	4.7155E+00	3.5007E−03	0.0000E+00	1.5180E−01	2.3621E−04	0.0000E+00	5.6156E+00	1.5939E−02	0.0000E+00
Robinia pseudoacacia L.	−3.1067E+00	2.1606E−02	2.4442E−01	−6.7308E+00	2.1132E−03	1.8967E+00	−2.7664E−01	3.2284E−04	6.5335E−02	−1.0114E+01	2.4042E−02	2.2065E+00
Quercus pubescens Willd.	1.0832E+00	2.9634E−02	−4.9794E−01	−8.2101E+00	3.0396E−03	1.7561E+00	−4.7534E−02	6.2568E−04	−4.1211E−03	−7.1745E+00	3.3299E−02	1.2623E+00
Salix spp.	8.9159E−01	1.6329E−02	0.0000E+00	7.4672E+00	4.4069E−03	0.0000E+00	6.9724E−01	3.5095E−04	0.0000E+00	9.0561E+00	2.1087E−02	0.0000E+00
Other broadleaved spp.	−9.1098E+00	7.3484E−03	2.3666E+00	−3.6118E+00	4.3190E−03	7.4127E−01	−1.0365E−01	3.2592E−04	4.7418E−02	−1.2825E+01	1.1993E−02	3.1553E+00
Pinus halepensis Mill.	−5.4055E+00	1.6165E−02	9.6893E−01	−2.5230E+00	5.0702E−03	1.2722E+00	−1.7268E−01	3.2346E−04	1.7969E−02	−8.1012E+00	2.1559E−02	2.2591E+00

calipering threshold, volume estimates showed negative values. Their volume was estimated using two equations, one for conifers and the other for broadleaves, built with a subset of the data previously used by Tabacchi et al., (2011a, b) for calibrating equations, i.e., the trees with DBH < 13 cm. The two equations, available in Tomter et al. (2012), are as follows (Eqs. 6.3 and 6.4):

$$\hat{y}_i = 1.2849 + 3.9579 \cdot 10^{-2} \cdot DBH_i^2\, h_i \quad \text{(conifers)} \tag{6.3}$$

$$\hat{y}_i = 0.5997 + 3.9619 \cdot 10^{-2} \cdot DBH_i^2\, h_i \quad \text{(broadleaves)} \tag{6.4}$$

where:

$\hat{y}_i$ is the volume (v in dm^3) (stem plus large branches and treetop up to 5 cm cross section diameter) of tree *i*;

DBH_i is the stem diameter at 1.30 m from the ground level of tree *i,* in cm;

h_i is the total tree height of tree *i,* in m.

For a small number of trees with DBH near the calipering threshold, the prediction equation for one or two aboveground biomass components returned negative values. These values were conventionally set to zero, with a consequent underestimation of the contribution by these trees to the total biomass of the sample unit. When all three aboveground tree components returned negative values, the biomass of the stem and large branches was estimated by multiplying the previously estimated volume of this component by the basal density value (dry weight per unit fresh volume). The basal density was found for the 26 species or group using the data of all the unbroken trees in the dataset for which no null or negative biomass estimates had occurred. Finally, the value of the other two components was conventionally set to zero. Last, to avoid estimates from extrapolations far beyond the observation limit of the maximum DBH of the sample trees used for calibrating the prediction equations, the volume of excess size trees was prudently estimated assuming a conventional maximum DBH.

The volume of truncated trees was estimated as half the cylinder volume obtained by multiplying the basal area and the tree height at breakage. The biomass of each tree was obtained multiplying its volume by the basal density of the species or group calculated as explained above.

After estimating the volume or the biomass of all the trees growing in a sample-plot, the overall values in each, by surface unit, was obtained by summing up the tree values per km^2, using the expansion factors indicated in Sect. 6.1.

6.3.2 Standing Dead Trees

The volume of dead, still standing, and unbroken trees was estimated by the same procedure described for the unbroken living trees. In the same way, the volume of dead broken trees was estimated adopting the cylinder as the reference shape and using the DBH and the tree height measured in the field. The volume was then converted to biomass using the conversion factors shown in Table 6.5. Those were determined in an addressed additional field survey campaign of INFC2005, in 2008–2009, by measuring and analysing woody samples either in the field or in the laboratory (Gasparini et al., 2013). The conversion factors were obtained by category (standing dead trees, deadwood lying on the ground and stumps), group of species (conifers and broadleaves) and decay class (Di Cosmo et al., 2013).

As done for the living trees, after estimating the volume or the biomass of all the trees in a sample plot, the overall values in each by surface unit was obtained by summing up the tree values per km^2, using the expansion factors indicated in Sect. 6.1.

6.4 Average Annual Growth

The estimate of the annual volume increment of trees was based on the increments measured on the cores taken from the sample trees in the sample plots. More specifically, the cores allowed an estimation of the annual increment of stem radius as the average of the five outermost ring width (not including the one from the survey growing season).

The estimation procedure went through four computational steps for obtaining: (a) the percent volume increment of any sample tree; (b) the plot level volume mean percent increment; (c) the volume increment of any callipered tree; (d) the overall volume increment of all the trees callipered in the plot.

(a) The percent annual volume increment of a sample tree was obtained by Eq. (6.5) (Hellrigl, 1969, 1986):

$$pv_{zj} = 100\left(2\Delta d_{zj}/d'_{zj}\right) + \left(\Delta h_{zj}/h'_{zj}\right) \tag{6.5}$$

where:

Δd_{zj} is the annual DBH increment of sample tree z of plot j;

d_{zj} is the DBH of sample tree z of sample plot j during the field survey;

d'_{zj} is the DBH of sample tree z of sample plot j one year prior to the field survey;

Δh_{zj} is the annual total tree height increment of sample tree z of plot j;

h_{zj} is the total tree height of sample tree z of plot j during the field survey;

h'_{zj} is the total tree height of sample tree z of plot j one year prior to the field survey.

Table 6.5 Basic density (Mg m^{-3}) of deadwood biomass components (standing dead trees—SDT, stump, deadwood lying on the ground—DWL) by group of species (conifers and broadleaves) and class of decay (from Di Cosmo et al., 2013, modified) / Valori di densità basale (Mg m^{-3}) del legno morto (alberi morti in piedi—SDT, ceppaie—Stump, legno morto grosso a terra—DWL) per classi di decadimento e gruppo di specie (da Di Cosmo et al., 2013, modificato)

Decay class/Classe di decadimento	Conifers/Conifere			Broadleaves/Latifoglie		
	SDT	Stump	DWL	SDT	Stump	DWL
1	4.42870E−01	4.59476E−01	4.11544E−01	5.22766E−01	5.11339E−01	5.13019E−01
2	4.23881E−01	4.50893E−01	3.89840E−01	5.25795E−01	4.94286E−01	4.68573E−01
3	3.84246E−01	3.96977E−01	3.41977E−01	5.13705E−01	4.49951E−01	4.44332E−01
4	3.54348E−01	3.44740E−01	2.95413E−01	4.38568E−01	4.10431E−01	3.44010E−01
5	3.72325E−01	2.71189E−01	2.65911E−01	–	3.06834E−01	2.55066E−01

Δh_{zj} was calculated using differences between the height in the measurement year and that corresponding to the tree DBH one year before the survey, both estimated using the height-diameter model described in Sect. 6.2.

(b) The mean percent annual volume increment for the entire sample plot was calculated as the average of the percent annual increment values of all the sample trees, each weighted with the volume of the corresponding tree, as shown in Eq. (6.6):

$$pv_j = \sum_{z=1}^{m} (pv_{zj} \cdot Vol_{zj}) / \sum_{z=1}^{m} Vol_{zj} \tag{6.6}$$

where:

Pv_j is the average percent annual volume increment for plot j;

m is the number of sample trees of plot j;

pv_{zj} is the percent annual volume increment of sample tree z of plot j;

Vol_{zj} is the volume of sample tree z of plot j.

The volume of the sample trees was estimated using the volume functions of Eq. (6.2) while the total tree height was known because recorded in the field.

(c) The annual volume increment for any callipered tree in the plot was obtained by multiplying its volume by the average weighted percent increment, as shown in Eq. (6.7):

$$\Delta v_{ij} = Vol_{ij} \cdot pv_j \tag{6.7}$$

where:

Δv_{ij} is the annual volume increment of tree i of plot j;

Vol_{ij} is the volume of tree i of plot j;

pv_j is the average percent annual volume increment for plot j.

(d) The sample plot annual volume increment was obtained by summing the annual increment values of all n trees in the plot, as shown in Eq. (6.8)

$$\Delta V_j = \sum_{i=1}^{n} v_{ij} \cdot f_i \tag{6.8}$$

where:

ΔV_j is the annual volume increment for plot j;

v_{ij} is the annual volume increment of tree i of plot j;

f_{ij} is the expansion factor of tree i, that varies with the DBH threshold value of 9.5 cm.

In a limited number of sample plots, with few cores available, an alternative procedure was used. This unfavourable condition was limited in INFC2015 by the choice to sample a fixed number of ten trees in each sample plot. The annual volume increment in plots with less than four cores available (278 in total) was estimated using a model developed by Gasparini et al., (2017) from the data of INFC2005. The model predicts the five-year periodic volume increment according to Eq. (6.9):

$$\ln(PAI) = a_0 + b_1 \ln(GSV) + b_2 \ln(N) \tag{6.9}$$

where:

ln(*PAI*) is the natural logarithm of the periodic increment (PAI, period = 5 years) (m^3 ha^{-1} 5 $years^{-1}$);

ln(*GSV*) is the is the natural logarithm of the growing stock volume (GSV, m^3 ha^{-1});

ln(*N*) is the natural logarithm of the number of trees per hectare.

Coefficient a_0 is dependent on forest type, as shown in Table 6.6. The antilogarithm of *ln*(*PAI*) so estimated must be multiplied by a correction factor equal to 1.125, before being divided by 5, to get the mean annual increment. This prevents bias in log-transformed allometric equations (Sprugel, 1983).

Conversion and expansion of volume increment to the total of aboveground biomass annual production was obtained for each tree in the sample plot, multiplying its volume increment by the ratio w_4/v, computed for the same tree.

6.5 Volume and Biomass of Stumps and Lying Deadwood

6.5.1 Stumps

In order to estimate the volume of the stumps, they were considered like cylinders. The volume of each stump was easily calculated by the mean diameter of the cut section and the mean height. Conversion of stump volume to biomass was obtained by the basal density factors shown in Table 6.5.

The overall stump volume or biomass for the entire sample unit was obtained by summing up the volumes of all the stumps in the sample plot and referring to the surface unit according to the expansion factor for AdS13.

6.5.2 *Deadwood Lying on the Ground*

The volume of the deadwood lying on the ground (which includes suspended deadwood) was estimated by assimilating each fragment to a truncated cone. The variables required by the formula (the diameters of the end sections and length of the fragment) were known from the field survey. The biomass was then calculated through the conversion factors showed in Table 6.5, which were based on the decay class as recorded by the surveyors.

The overall volume of deadwood lying on the ground for a sample plot was obtained by summing the volumes of all the fragments and, similarly to earlier descriptions, the value obtained was referred to the surface unit according to the expansion factor for AdS13.

Deadwood lying on the ground is one of the three components of deadwood surveyed by INFC, along with the standing dead trees and stumps. Together, these are referred to as coarse woody debris (e.g., Russell et al., 2015), although they have been defined in many ways and differences exist, especially in the size required

Table 6.6 Volume increment stand-level model: regression coefficients for Eq. (6.9) (from Gasparini et al., 2017, modified) / Modello per l'incremento periodico di volume: coefficienti di regressione per l'Eq. (6.9) (da Gasparini et al., 2017, modificato)

Forest category/Categoria forestale	a_0	b_1	b_2
Larch and Swiss stone pine/Boschi di larice e cembro	−3.2336	0.6269	0.182
Norway spruce/Boschi di abete rosso	−2.8536	0.6269	0.182
Fir/Boschi di abete bianco	−2.8873	0.6269	0.182
Scots pine and Mountain pine/Pinete di pino silvestre e montano	−3.2336	0.6269	0.182
Black pines/Pinete di pino nero, laricio e loricato	−2.8513	0.6269	0.182
Mediterranean pines/Pinete di pini mediterranei	−2.8819	0.6269	0.182
Other coniferous forests/Altri boschi di conifere, pure o miste	−2.7383	0.6269	0.182
Beech/Faggete	−3.0877	0.6269	0.182
Temperate oaks/Querceti a rovere, roverella e farnia	−3.2336	0.6269	0.182
Mediterranean oaks/Cerrete, boschi di farnetto, fragno, vallonea	−3.0536	0.6269	0.182
Chestnut/Castagneti	−2.7562	0.6269	0.182
Hornbeam and Hophornbeam/Ostrieti, carpineti	−3.1425	0.6269	0.182
Hygrophilous forests/Boschi igrofili	−2.7040	0.6269	0.182
Other deciduous broadleaved/Altri boschi caducifogli	−2.7604	0.6269	0.182
Holm oak/Leccete	−3.2336	0.6269	0.182
Cork oak/Sugherete	−3.2336	0.6269	0.182
Other evergreen broadleaved/Altri boschi di latifoglie sempreverdi	−2.8559	0.6269	0.182
Poplar plantations/Pioppeti artificiali	−1.8020	0.6269	0.182
Other broadleaved plantations/Piantagioni di altre latifoglie	−1.9827	0.6269	0.182
Coniferous plantations/Piantagioni di conifere	−2.3944	0.6269	0.182

Table 6.7 Biomass (dry weight—kg) of small woody individuals (trees and shrubs under the callipering DBH threshold value) in the three size classes (from Di Cosmo & Gasparini, 2013) / Fitomassa (peso secco unitario—kg) degli individui di rinnovazione e arbusti nelle tre classi dimensionali adottate nel rilievo inventariale (da Di Cosmo & Gasparini, 2013)

Size class/Classe dimensionale	Biomass/Fitomassa		
	Saplings/Rinnovazione		Shrubs/Arbusti
	Conifers/Conifere	Broadleaves/Latifoglie	–
	(kg)	(kg)	(kg)
1	0.190006	0.032766	0.029615
2	0.773613	0.331598	0.243656
3	2.216675	2.154130	2.009259

(e.g., Enrong et al., 2006). The overall coarse woody debris volume or biomass was estimated at plot level following the usual procedure of summing up the individual values, previously referred to the surface unit according to the appropriate expansion factors.

6.6 Small Trees and Shrubs

Recording the species of the woody subjects under the minimum callipering threshold (4.5 cm) allowed estimating the variables of interest explicitly for small trees (seedlings and saplings) and shrubs in the survey. For both, the total number, the biomass, and the carbon stock were estimated, by size class. Biomass was estimated by multiplying the number of subjects recorded in the field by the unit dry weight of the correspondent category (small tree or shrub) and size class (three size classes), as shown in Table 6.7. The biomass values in Table 6.7 were obtained after the integrative survey of the second Italian NFI (INFC2005).

As usual, the values per unit surface were obtained using the appropriate expansion factor, as discussed in Sect. 6.1, and the total per plot by summing up the values of all measured items.

6.7 Removed Trees

Tree volume and biomass removed in the year preceding the field survey were estimated based on the data recorded for the stumps with cutting section diameter ≥ 9.5 cm, within the sample plot AdS13. The procedure for estimating the volume and the biomass of each tree is identical with that adopted for unbroken trees, both living and standing dead, previously described in this chapter. As it is based on the knowledge of DBH, the procedure started by reconstructing this value, based on the stump diameter and height data. For this purpose, we used the prediction equations developed by Di Cosmo & Gasparini (2020). Those equations allow to estime the DBH for 16 tree species and the broader groups of conifers and broadleaves; their form is as showed in Eq. (6.10):

$$\widehat{DBH} = b_0 + b_1 \cdot D_{stump} + b_2 \cdot D^2_{stump} + b_3 \cdot H_{stump} \tag{6.10}$$

where:

$\widehat{DBH}$ is the diameter at 1.30 m above the ground line, in cm;

D_{stump} is the diameter of the stump cut section, in cm;

H_{stump} is the height of the stump, in m.

Table 6.8 shows the estimated parameters needed for applying Eq. (6.10).

As usual, the volume estimated (v) is that of the stem plus large branches. The biomass removed is that of the same component (w_1) plus that of the small branches and treetop (w_2).

The volume or biomass removed from the plot was obtained by summing up all the tree level values.

Table 6.8 Coefficients of Eq. (6.8) for estimating the diameter at breast height (DBH) of removed trees by stump diameter and height (from Di Cosmo & Gasparini, 2020, modified) / Coefficienti dell'Eq. (6.8) per la stima del diametro degli alberi a 1.30 m da terra attraverso i valori di diametro e altezza della ceppaia (da Di Cosmo & Gasparini, 2020, modificato)

Species or group	b_0	b_1	b_2	b_3
Abies alba Mil.	−2.16952E+00	9.34004E−01	−2.57564E−03	8.76087E−02
Larix decidua Mill.	−6.73056E−01	5.46847E−01	3.50720E−03	1.70564E−01
Picea abies K.	−7.11912E−01	8.10644E−01	−1.11604E−04	−1.58722E−02
Pinus cembra L.	−2.50065E+00	8.97002E−01	−1.88567E−03	7.82736E−02
Pinus halepensis Mill.	5.60632E−01	6.15013E−01	4.11331E−03	5.92471E−02
Pinus nigra Arn.	−5.40925E+00	9.70951E−01	−1.18818E−03	2.25264E−01
Pinus pinea L.	−3.02325E+00	7.04435E−01	2.08621E−03	2.60534E−01
Pinus silvestris L.	2.25022E+00	5.49529E−01	4.81511E−03	4.39391E−02
Conifers	−1.51407E+00	8.37160E−01	−4.40000E−04	2.87600E−02
Castanea sativa Mill.	−3.13929E+00	7.50178E−01	−1.98945E−03	2.53306E−01
Cupressus spp.	−1.26516E+00	7.47879E−01	1.46697E−03	1.08038E−01
Fagus silvatica L.	−5.52839E+00	8.66377E−01	−2.88027E−03	1.82113E−01
Ostria carpinifolia Scop.	6.79901E−01	6.62823E−01	−1.01725E−03	−4.16815E−03
Quecus ilex L.	−2.47746E+00	8.46540E−01	−3.38432E−03	2.68358E−01
Quercu cerris L.	−4.10939E+00	8.22435E−01	−2.65966E−03	2.63364E−01
Quercus pubescens Willd.	1.32608E+00	6.25699E−01	−1.21702E−04	3.17740E−02
Robinia pseudoacacia L.	−9.24565E−01	8.50859E−01	−2.55761E−03	2.61228E−02
Broadleaves	−2.00784E+00	7.78270E−01	−2.20000E−03	1.50360E−01

Appendix (Italian Version)

Riassunto I dati quantitativi rilevati nelle aree di saggio richiedono elaborazioni preliminari ai fini del calcolo degli stimatori delle abbondanze (valori totali) e delle densità (valori medi) impiegati per la produzione delle statistiche inventariali. Le elaborazioni preliminari hanno l'obiettivo di ottenere il valore per unità di superficie delle variabili di interesse nelle singole aree di saggio. La complessità delle elaborazioni varia in funzione della variabile, del modo in cui essa è derivata a partire dalle misurazioni che la sottendono (es. il diametro a 1.30 m dal suolo *vs.* l'area basimetrica), dal fatto che siano stati misurati tutti gli elementi nell'area di saggio pertinente o solo un sotto-campione, nonché dalla disponibilità di modelli. Le elaborazioni preliminari producono numerosità, volumi, biomassa e stock di carbonio, ma le procedure di calcolo possono implicare stime di ulteriori variabili in passaggi intermedi (es. l'altezza degli alberi). Il capitolo descrive metodi e modelli utilizzati in INFC2015 per la stima delle variabili relative agli alberi (es. numerosità, area basimetrica), a rinnovazione ed arbusti (es. biomassa, stock di carbonio), alle ceppaie (es. volume, biomassa) e alle variazioni annuali (es. l'incremento annuo di volume e le utilizzazioni). Alcuni dei modelli descritti sono stati sviluppati tenendo conto delle esigenze inventariali nell'ambito di progetti preparatori e/o collaterali ad INFC, altri sono stati sviluppati nell'ambito delle procedure di calcolo di INFC e, infine, i coefficienti necessari per stimare il peso secco di legno morto, rinnovazione e arbusti derivano dai campionamenti e dalle attività di laboratorio della fase integrativa di rilievi di INFC2005.

Introduzione

Le procedure generali di stima dei valori totali di strato e di popolazione (abbondanze) e dei rispettivi valori per unità di superficie (densità), descritte nel Chap. 5, richiedono che le grandezze rilevate attraverso le unità di campionamento siano espresse con riferimento alla superficie delle maglie del reticolo INFC. Il valore complessivo per ogni variabile di interesse (ad esempio il numero di alberi o il volume) determinato per l'area di saggio va perciò diviso per la superficie dell'area stessa e moltiplicato per la superficie SQ (1 km^2) delle maglie del reticolo inventariale contenente il punto di campionamento. Poiché la superficie dell'area varia a seconda dell'unità di campionamento (cfr. Chap. 3), vengono così a determinarsi tre fattori di espansione che assumono valore di $10^6/12.566371$ per le grandezze censite entro ciascuna AdS2, $10^6/50.265482$ per quelle censite dentro l'AdS4 e infine $10^6/530.929159$ per quelle censite entro l'AdS13. Ad esempio, ogni albero con diametro a 1.30 m dal suolo ($d_{1.30}$) compreso tra 4.5 cm e 9.4 cm, rilevato quindi all'interno dell'AdS4, contribuisce ad un numero di alberi per chilometro quadrato pari al prodotto $1 \times 10^6/50.265482$. Analogamente, il contributo per chilometro quadrato di qualsiasi grandezza associata allo stesso albero (l'area basimetrica, il volume, la fitomassa, ecc.) sarà pari al valore calcolato della grandezza (l'area basimetrica, il volume, la fitomassa, ecc.) moltiplicato per $10^6/50.265482$.

Tra gli attributi quantitativi di interesse, il numero di alberi e l'area basimetrica dei soprassuoli sono ottenuti con semplici operazioni aritmetiche direttamente dai dati raccolti in campo. In un certo senso, ciò vale anche per la quantificazione del contenuto di carbonio nelle componenti vive (alberi vivi, rinnovazione ed arbusti, alberi asportati con le utilizzati) e morte (alberi morti in piedi, necromassa a terra, ceppaie), che avviene moltiplicando per un fattore di 0.5 il peso secco, rispettivamente della fitomassa e della necromassa legnosa (IPCC, 2006; Matthews, 1993; Woodall et al., 2012); per questo motivo, nelle descrizioni che seguono si omette la conversione tra fitomassa e contenuto di carbonio organico. Le stime del volume legnoso, della fitomassa, dell'incremento annuo di volume e della massa arborea asportata con le utilizzazioni forestali necessitano, invece, di elaborazioni preliminari più complesse che implicano la costruzione di modelli che, sviluppati utilizzando dati statisticamente rappresentativi delle condizioni di crescita nell'intero contesto nazionale, possono risultare di utilità generale ed essere applicati anche fuori dall'ambito INFC.

In questo capitolo vengono descritte le procedure e i modelli utilizzati nelle elaborazioni delle variabili quantitative rilevate da INFC2015. Alcuni dei modelli descritti sono stati sviluppati nell'ambito di progetti preparatori e/o collaterali ad INFC, mentre la derivazione dei coefficienti necessari per la stima del peso secco di legno morto, rinnovazione e arbusti è stata possibile grazie ad una specifica fase di campionamento integrativo di INFC2005 e delle successive attività di laboratorio.

L'altezza degli alberi

I modelli disponibili per la stima del volume e della fitomassa degli alberi misurati nelle aree di saggio utilizzano come variabili esplicative i valori di $d_{1.30}$ e di altezza totale dell'albero (h). Per motivi di costo, in ambito inventariale la misurazione delle altezze viene generalmente limitata ad un sotto-campione degli alberi cavallettati (Gschwantner et al., 2016) e quindi il valore di h di ciascun albero deve essere stimato con appositi modelli. In ambito INFC, la selezione degli alberi campione risponde a criteri di casualità e di rappresentatività. Sulla base dell'esperienza del secondo inventario, che prevedeva un numero di alberi campione variabile da cinque a dieci (Bosela et al., 2016; Gasparini & Di Cosmo, 2016; Gasparini et al., 2010), INFC2015 ha previsto la selezione di dieci alberi campione. Le informazioni ipso-diametriche raccolte costituiscono la base dei dati per lo sviluppo di equazioni di previsione di h. Le equazioni presentate costituiscono un aggiornamento di quelle approntate secondo la stessa metodologia nel secondo inventario forestale nazionale; la nuova calibrazione è stata necessaria per la più ampia varietà dei contesti di applicazione. Il modello adottato deriva dalla funzione di Chapman-Richards (Liu & Li, 2003; Ratkowsky, 1990) e si presenta nella forma dell'Eq. (6.11):

$$\hat{h}_{ij} = \left(a_1 + a_2 \cdot H_{domj}\right)\left[1 - \exp\left(\left(a_3 + a_4 \cdot H_{domj}\right)d_{ij}\right)\right]^{\left(a_5 + a_6 H_{domj}\right)} \tag{6.11}$$

in cui

$\hat{h}_{ij}$ è l'altezza stimata del generico albero *i* della generica area di saggio *j*, in m;

d_{ij} è il $d_{1.30}$ del generico albero *i* nella generica area di saggio *j*, in cm;

H_{domj} è l'altezza dominante (H_{dom}) della *j*-esima area di saggio, in m, calcolata come media aritmetica delle altezze dei tre alberi campione più grandi in $d_{1.30}$.

Nel caso di specie con un numero limitato di osservazioni, si è preferito raggruppare quelle con un rapporto ipso-diametrico simile, valutato su base grafica, a garanzia di maggiore stabilità dei coefficienti stimati delle regressioni. I coefficienti delle equazioni sviluppate sono riportati nella Table 6.1. Le 52 equazioni presentate riguardano le principali specie o gruppi di specie diffuse nei soprassuoli italiani e pertanto hanno consentito di stimare l'altezza della quasi totalità degli alberi misurati. Per le specie restanti, è stato adottato il modello che mostrava la maggiore capacità predittiva valutata con metodo grafico e analitico. Un numero limitato di unità campionarie non disponeva di misurazioni sufficienti per il calcolo di H_{dom}; sono state perciò approntate equazioni semplificate con $d_{1.30}$ come unica variabile esplicativa. Per le specie lianose o altre misurate con estrema rarità, ma di cui si disponeva di un certo numero di altezze misurate, si è preferito assumere un valore convenzionale di altezza calcolato come media aritmetica delle osservazioni registrate, eventualmente per classi diametriche. Nella Table 6.2 sono riportati i valori relativi alla numerosità delle osservazioni di ciascuna specie o gruppo con cui sono stati calibrati i modelli, nonché i valori estremi di $d_{1.30}$ e H_{dom} che delimitano i campi di variazione; queste informazioni sono utili a valutarne l'adozione in ambiti di applicazione esterni a INFC.

Volume e massa degli alberi

Gli alberi vivi

La stima del volume e del peso secco degli alberi vivi (fitomassa) è stata condotta con procedure differenziate per gli alberi integri e per quelli troncati. Il volume e la fitomassa di ogni albero vivo integro censito sono stati stimati utilizzando le equazioni di stima a doppia entrata disponibili da Tabacchi et al., (2011a, b), costruite in preparazione del secondo inventario nazionale (INFC2005) e successivamente aggiornate. La costruzione del modello è avvenuta sulla base dei dati di alberi selezionati, misurati e campionati per determinazioni di laboratorio del peso secco su vaste aree del territorio nazionale. Dal modello discendono equazioni di previsione relative a ventisei tra le principali specie o gruppi di specie forestali in Italia e a cinque variabili di stima. La forma del modello è mostrata nell'Eq. (6.12):

$$\hat{y}_i = b_1 + b_2 \cdot d_i^2 \cdot h_i + b_3 \cdot d_i \tag{6.12}$$

in cui:

$\hat{y}_i$ è la variabile stimata che può essere il volume (v in dm^3) (fusto con rami grossi e cimale fino ad una sezione di taglio di 5 cm), la fitomassa (in kg) della stessa componente (w_1), la fitomassa di ramaglia e cimale (w_2), la fitomassa della ceppaia (w_3) o la fitomassa dell'intera porzione epigea (w_4) del generico albero *i*;

d_i è $d_{1.30}$ dell'albero *i,* in cm;

h_i è l'altezza totale dell'albero *i,* in m.

w_4 può essere ottenuto anche come sommatoria delle tre componenti w_1, w_2 e w_3, poiché il modello ne assicura l'additività.

La Table 6.3 riporta i coefficienti per la stima del volume delle specie o gruppi di specie. La Table 6.4 riporta i coefficienti di regressione utili per la stima delle quattro componenti della biomassa epigea. Le equazioni citate non esauriscono la casistica delle specie misurate in ambito INFC, così per le altre specie è stata usata l'equazione di volta in volta ritenuta idonea per affinità di portamento, oppure è stata utilizzata quella per il gruppo eterogeneo delle altre latifoglie.

L'intervallo dimensionale degli alberi cavallettati con la campagna inventariale è risultato in alcuni casi più ampio rispetto a quello degli alberi modello utilizzati per l'approntamento delle equazioni di previsione citate. In un numero contenuto di casi, ciò ha comportato che per diametri piccoli e prossimi alla soglia di cavallettamento le stime del volume avessero valore negativo. A tale inconveniente si è ovviato utilizzando due equazioni generiche, una per le conifere (Eq. 6.3) e una per le latifoglie (Eq. 6.4), disponibili in Tomter et al. (2012). Le due equazioni sono state ottenute tarando il modello con l'insieme degli alberi campione già utilizzati da Tabacchi et al., (2011a e b) ma utilizzando solo valori di diametro inferiori a 13 cm. In pochi casi, sempre relativi a diametri prossimi alla soglia inferiore di cavallettamento, anche nella stima della fitomassa le equazioni di previsione hanno restituito stime con valori negativi di una o più componenti epigee. A tale inconveniente si è ovviato adottando una procedura di stima del valore di fitomassa del fusto e rami grossi che ne garantisse valori positivi, ponendo convenzionalmente uguale a zero eventuali valori negativi della fitomassa di ramaglia e/o ceppaia, dal che consegue una leggera sottostima della fitomassa arborea epigea totale. La procedura alternativa di stima della fitomassa di fusto e rami grossi è consistita nel moltiplicare il volume di questa componente per il valore di densità basale (peso secco per unità di volume fresco). La densità basale è stata ricavata per ciascuna delle ventisei specie o gruppi dotati di modello di previsione, utilizzando i dati degli alberi vivi integri per i quali non si era verificato l'inconveniente di stime negative di fitomassa o di volume. Infine, per evitare stime derivanti da estrapolazioni molto oltre i valori dei diametri massimi con cui sono stati costruiti i modelli di previsione adottati, il volume degli alberi con diametri elevati è stato prudenzialmente stimato assumendo per essi un diametro limite convenzionale.

Il volume degli alberi vivi troncati è stato calcolato con metodo geometrico, convenzionalmente assunto come metà di quello cilindrometrico ottenuto dal prodotto dell'area basimetrica per l'altezza di troncatura. La stima della fitomassa degli alberi vivi ma troncati è stata ottenuta attraverso il prodotto tra il volume e il valore medio di densità basale, calcolato nel modo poco sopra descritto.

Noto il volume e la fitomassa di ogni singolo albero dalle procedure di stima illustrate, il contributo di un albero al volume totale e alla fitomassa totale dell'unità di superficie è stato ottenuto sulla base dei fattori di espansione indicati in Sect. 6.1.

Gli alberi morti

Il volume legnoso degli alberi morti in piedi ancora integri è stato stimato con la stessa procedura adottata per gli alberi vivi. Analogamente, il volume degli alberi morti ma troncati è stato calcolato per via geometrica, sulla base dei valori di $d_{1.30}$ e dell'altezza di troncatura, sempre misurata in campo. La conversione del volume degli alberi morti in peso secco (necromassa) è stata ottenuta mediante i fattori di conversione riportati nella Table 6.5. Tali fattori sono stati determinati in ambito INFC mediante misurazioni, prelievo di campioni e determinazioni di laboratorio condotte con una specifica campagna di rilievo integrativa negli anni 2008 e 2009 (Di Cosmo & Gasparini, 2013). I coefficienti per la conversione dei volumi in peso secco sono specifici per categoria di legno morto (alberi morti in piedi, legno morto grosso a terra e ceppaie residue), gruppo di specie (conifere o latifoglie) e classe di decadimento (Di Cosmo et al., 2013).

Il contributo di volume e necromassa di ciascun albero morto all'unità di superficie è stato ottenuto sulla base dei fattori di espansione già più volte ricordati, mentre il totale per ciascuna variabile è stato ottenuto sommando i valori di albero nell'area di saggio.

L'incremento medio annuo

L'incremento annuo di volume degli alberi vivi è stato stimato a partire dalle informazioni incrementali ottenute con le carotine legnose prelevate dagli alberi campione nelle aree di saggio. In particolare, le letture incrementali hanno consentito di calcolare l'incremento diametrico medio annuo relativo alle ultime cinque stagioni vegetative precedenti a quella del rilievo inventariale.

La procedura di stima adottata è basata su quattro passaggi: (a) stima dell'incremento percentuale annuo di volume degli alberi campione; (b) stima dell'incremento percentuale medio annuo di volume dell'area di saggio; (c) stima dell'incremento annuo di volume di ogni albero cavallettato; (d) somma dei valori di incremento annuo di volume di tutti gli alberi cavallettati nell'area di saggio.

(a) Alla stima dell'incremento percentuale di volume di ogni albero campione si è pervenuti mediante l'Eq. (6.13) (metodo combinato esplicito—Hellrigl, 1969, 1986),

$$pv_{zj} = 100\left(2\Delta d_{zj}/d'_{zj}\right) + \left(\Delta h_{zj}/h'_{zj}\right) \tag{6.13}$$

in cui:

Δd_{zj} è l'incremento diametrico annuo del z-esimo albero campione della j-esima area di saggio;

d_{zj} è il $d_{1.30}$ del z-esimo albero campione della j-esima area di saggio;

d'_{zj} è il $d_{1.30}$ del z-esimo albero campione della j-esima area di saggio un anno prima del rilievo inventariale;

Δh_{zj} è l'incremento ipsometrico annuo del z-esimo albero campione della j-esima area di saggio;

h_{zj} è l'altezza del z-esimo albero campione della j-esima area di saggio al momento del rilievo inventariale;

h'_{zj} è l'altezza del z-esimo albero campione della j-esima area di saggio un anno prima del rilievo inventariale.

L'incremento annuo in altezza è stato calcolato per differenza tra le stime ottenute con i modelli ipso-diametrici descritti nella Sect. 6.2, in corrispondenza del diametro misurato e di quello stimato per l'anno precedente al rilievo inventariale.

(b) L'incremento percentuale medio di volume dell'intera unità campionaria è stato calcolato come valore medio degli incrementi percentuali degli alberi campione, pesati ciascuno con il volume dell'albero corrispondente, come mostrato nell' Eq. (6.14):

$$pv_j = \sum_{z=1}^{m} \left(pv_{zj} \cdot Vol_{zj} \right) / \sum_{z=1}^{m} Vol_{zj} \tag{6.14}$$

in cui:

m è il numero degli alberi campione nella j-esima area di saggio;

pv_{zj} è l'incremento percentuale di volume del z-esimo albero campione della j-esima area di saggio;

Vol_{zj} è il volume del z-esimo albero campione della j-esima area di saggio.

Il volume degli alberi campione dell'accrescimento è stato calcolato con le funzioni di cubatura già descritte nella Sect. 6.3.1, con l'altezza totale nota dalle misurazioni effettuate in campo.

(c) la stima dell'incremento annuo di volume di ogni albero cavallettato è stata ottenuta come prodotto tra il volume dell'albero e l'incremento percentuale medio di volume dell'area di saggio in cui esso è stato misurato, come da Eq. (6.15):

$$\Delta v_{ij} = Vol_{ij} \cdot pv_j \tag{6.15}$$

in cui:

Vol_{ij} è il volume dell'*i*-esimo albero della *j*-esima area di saggio;

pv_j è l'incremento percentuale medio annuo di volume dell'intera area di saggio *j*.

(d) l'incremento annuo di volume per l'intera unità campionaria è stato ottenuto mediante sommatoria dei valori di incremento annuo di tutti gli n alberi nella stessa area di saggio, come da Eq. (6.16)

$$\Delta V_j = \sum_{i=1}^{n} v_{ij} \cdot f_i \tag{6.16}$$

in cui:

v_{ij} è l'incremento annuo di volume dell'*i*-esimo albero nella *j*-esima area di saggio;

f_{ij} è il fattore di espansione dell'albero *i* dell'area di saggio *j*, variabile con la soglia di $d_{1.30}$ di 9.5 cm.

A questa procedura generale di stima dell'incremento annuo di volume ne è stata affiancata un'altra da adottare nelle aree di saggio con ridotta disponibilità di informazione incrementale. Questa evenienza è stata limitata in maniera considerevole in INFC2015, data la selezione di un numero fisso, pari a dieci, di alberi campione in ogni area di saggio. Nelle aree di saggio con un numero di misure incrementali inferiore a quattro (278 casi) l'incremento annuo di volume è stato stimato a livello di area di saggio mediante il modello disponibile da Gasparini et al. (2017). Si tratta di un modello di previsione dell'incremento periodico di volume (con periodo di 5 anni) sviluppato con i dati del secondo inventario forestale italiano e che assume la forma riportata in Eq. (6.17):

$$\ln(PAI) = a_0 + b_1 \ln(GSV) + b_2 \ln(N) \tag{6.17}$$

in cui:

PAI è l'incremento periodico di cinque anni, in $m^3\ ha^{-1}$;

GSV è il volume degli alberi vivi, in $m^3\ ha^{-1}$;

N è il numero di alberi per ettaro.

Il coefficiente a_0 varia a seconda della categoria forestale secondo i valori riportati nella Table 6.6. Il modello utilizza i valori delle variabili in logaritmo naturale. La stima dell'incremento annuo di volume secondo l'Eq. (6.17) richiede che l'antilogaritmo dell'incremento periodico stimato sia moltiplicato per un fattore di correzione pari a 1.125, prima di essere diviso per cinque. Ciò è necessario per tenere conto dell'errore sistematico che si determina quando la calibrazione di un modello allometrico avviene su valori log-trasformati (Sprugel, 1983).

La conversione dell'incremento annuo di volume del fusto e rami grossi in produzione annua di fitomassa espansa all'intera porzione epigea è stata ottenuta moltiplicando l'incremento di volume di ogni albero per il rapporto fitomassa totale (w_4) su volume (v), calcolato a livello di singolo albero.

Volume e massa delle ceppaie e del legno morto grosso a terra

Le ceppaie residue

Ai fini delle stime del volume delle ceppaie residue, la loro forma è stata convenzionalmente assimilata a quella cilindrica. Il volume legnoso è stato quindi calcolato per via geometrica a partire dal diametro medio alla sezione di taglio e dall'altezza media dalla linea di terra. La conversione del volume di ogni ceppaia in peso secco è stata possibile mediante l'adozione dei coefficienti di densità basale riportati nella Table 6.5, considerando anche lo stadio di decadimento del legno registrato dai rilevatori.

Il volume e la necromassa totale dell'area di saggio sono stati ottenuti sommando i relativi valori di tutte le ceppaie nella stessa, riferiti all'unità di superficie convenzionale secondo il fattore di espansione tipico dell'AdS13.

Il legno morto grosso a terra

Il volume dei frammenti di legno morto a terra (che comprende anche quello sospeso) è stato calcolato per via geometrica, secondo la formula per il tronco di cono. I valori necessari di diametro delle sezioni estreme e lunghezza erano noti dalle misurazioni effettuate in campo. La conversione del volume in peso secco è avvenuta utilizzando i coefficienti riportati nella Table 6.5, che tengono conto anche dello stadio di decadimento del legno, anch'esso noto dalle registrazioni dei rilevatori.

I valori per l'unità di campionamento sono stati ottenuti per sommatoria dei volumi e dei pesi dei singoli frammenti, mentre il riferimento all'unità di superficie è avvenuto secondo il fattore di espansione per l'AdS13.

Il legno morto a terra rappresenta una delle tre componenti di legno morto rilevate in ambito INFC2015, insieme agli alberi morti in piedi e alle ceppaie residue. Nel complesso, queste costituiscono il legno morto totale, generalmente indicato con *coarse woody debris* in ambito internazionale (es. Russell et al., 2015), sebbene manchi tutt'ora una definizione condivisa soprattutto per quanto riguarda i requisiti dimensionali minimi (es. Enrong et al., 2006). La stima d'insieme delle tre componenti, in termini di volume o necromassa, nell'area di saggio è avvenuta per semplice sommatoria dei valori stimati per gli oggetti misurati, preventivamente riferiti all'unità di superficie secondo i fattori opportuni.

La fitomassa di rinnovazione e arbusti

Il rilievo della vegetazione legnosa degli strati inferiori per specie ha consentito di stimare le grandezze di interesse in maniera distinta per la rinnovazione e per gli arbusti. Per entrambe le categorie, sono stati stimati la numerosità, i valori di fitomassa epigea e di carbonio immagazzinato. La stima della fitomassa è avvenuta attribuendo ad ogni soggetto censito un valore in peso secco, noto dalle attività di prelievo e di laboratorio condotte nella terza fase integrativa del secondo inventario forestale (INFC2005). Tale valore è funzione della componente (rinnovazione o arbusto) e della classe dimensionale (Table 6.7).

I valori per unità di superficie sono stati ottenuti mediante la procedura generale già illustrata per le altre grandezze di natura quantitativa, adottando il fattore di espansione appropriato per le due unità di campionamento AdS2. Le stime dei valori totali e medi sono state ottenute previa sommatoria dei valori di ciascun soggetto censito.

Entità delle utilizzazioni

La stima del volume e della fitomassa degli alberi asportati con i tagli avvenuti nei dodici mesi antecedenti il rilievo inventariale si è basata sulla misurazione delle ceppaie residue con diametro uguale o maggiore di 9.5 cm, nell'area di saggio AdS13. Le procedure per la stima del volume e della fitomassa di ogni albero asportato sono le stesse di quelle descritte per gli alberi integri, in questo capitolo. Esse richiedono la conoscenza del valore del $d_{1.30}$ di ogni individuo, in maniera da poterne stimare prima l'altezza totale, secondo i modelli descritti nella Sect. 6.2, e poi il volume e la fitomassa con i modelli disponibili. La particolarità nei calcoli relativi agli alberi asportati risiede dunque nella stima preventiva dell'ipotetico $d_{1.30}$ sulla base delle misure disponibili della ceppaia. A tal fine, sono stati adottati modelli di previsione disponibili dalla letteratura (Di Cosmo & Gasparini, 2020), che hanno la forma mostrata in Eq. (6.18),

$$\hat{d}_{1.30} = b_0 + b_1 \cdot D_{ce} + b_2 \cdot D_{ce}^2 + b_3 \cdot H_{ce} \tag{6.18}$$

in cui:

$\hat{d}_{1.30}$ è il $d_{1.30}$ in cm;

D_{ce} è il diametro della ceppaia alla linea di taglio in cm;

H_{ce} è l'altezza media della ceppaia da terra in m.

Sono disponibili stime dei parametri b_0, b_1, b_2 e b_3 per sedici specie (Table 6.8); per le restanti, sono stati adottati i parametri generici per il gruppo generico delle conifere o per quello delle latifoglie.

Le statistiche prodotte sono relative al volume del fusto e rami grossi, nonché alla fitomassa epigea della stessa componente (w_1) e della ramaglia con cimale (w_2). Il volume e la fitomassa degli alberi asportati per l'intera unità di campionamento sono stati ottenuti dalla sommatoria dei volumi e delle fitomasse dei singoli alberi prelevati, come già anticipato riferiti all'unità di superficie secondo il fattore di espansione proprio dell'AdS13.

References

Bosela, M., Gasparini, P., Di Cosmo, L., Parisse, B., De Natale, F., Esposito, S., & Scheer, L. (2016). Evaluating the potential of an individual-tree sampling strategy for dendroecological investigations using the Italian National Forest Inventory data. *Dendrochronologia, 38*, 90–97. https://doi.org/10.1016/j.dendro.2016.03.011

Di Cosmo, L., & Gasparini, P. (2013). Stima delle grandezze inventariali. In: P. Gasparini, L. Di Cosmo, & E. Pompei (Eds.), *Il contenuto di carbonio delle foreste italiane. Inventario Nazionale delle Foreste e dei serbatoi forestali di Carbonio INFC2005. Metodi e risultati dell'indagine integrativa* (pp. 77–86). Consiglio per la Ricerca e la sperimentazione in Agricoltura, Roma. ISBN 978-88-97081-36-4.

Di Cosmo, L., & Gasparini, P. (2020). Predicting diameter at breast height from stump measurements of removed trees to estimate cuttings, Illegal loggings and natural disturbances. *South-east European Forestry, 11*(1), 41–49. https://doi.org/10.15177/seefor.20-08.

Di Cosmo, L., Gasparini, P., Paletto, A., & Nocetti, M. (2013). Deadwood basic density values for national-level carbon stock estimates in Italy. *Forest Ecology and Management, 295*, 51–58.

Enrong, Y., Xihua, W., & Jianjun, H. (2006). Concept and classification of coarse woody debris in forest ecosystems. *Frontiers of Biology in China, 1*, 76–84. https://doi.org/10.1007/s11515-005-0019-y

Gasparini, P., & Di Cosmo, L. (2016). National forest inventory reports—Italy. In: C. Vidal, I. Alberdi, L. Hernandez, & J. Redmond (Eds.), *National forest inventories—Assessment of wood availability and use* (pp. 485–506). Springer. ISBN 978-3-319-44014-9. https://doi.org/10.1007/978-3-319-44015-6_26.

Gasparini, P., Di Cosmo, L., & Pompei, E. (Eds.) (2013). Il contenuto di carbonio delle foreste italiane. *Inventario Nazionale delle Foreste e dei serbatoi forestali di Carbonio INFC2005. Metodi e risultati dell'indagine integrativa* (267 pp.). Consiglio per la Ricerca e la sperimentazione in Agricoltura, Roma. ISBN 978-88-97081-36-4.

Gasparini, P., Di Cosmo, L., Rizzo, M., & Giuliani, D. (2017). A stand-level model derived from National Forest Inventory data to predict periodic annual volume increment of forests in Italy. *Journal of Forest Research-JPN, 22*(4), 209–217. https://doi.org/10.1080/13416979.2017.1337260

Gasparini, P., Tosi, V., & Di Cosmo, L. (2010). National forest inventory reports: Italy. In: E. Tomppo, T. Gschwantner, M. Lawrence, & R. E. McRoberts (Eds.), *National forest inventories—Pathways for common reporting* (pp. 311–331). Springer. ISBN 978-90-481-3232-4. https://doi.org/10.1007/978-90-481-3233-1.

Gschwantner, T., Lanz, A., Vidal, C., Bosela, M., Di Cosmo, L., Fridman, J., Gasparini, P., Kuliešis, A., Tomter, S., & Schadauer, K. (2016). Comparison of methods used in European National Forest Inventories for the estimation of volume increment: Towards harmonisation. *Annals of Forest Science, 73*, 807–821. https://doi.org/10.1007/s13595-016-0554-5

Hellrigl, B. (1969). Sul calcolo dell'incremento percentuale degli alberi in piedi. *Italia Forestale e Montana, XXIV*, 187–191.

Hellrigl, B. (1986). Metodologie per la determinazione degli incrementi. *Nuove metodologie nella elaborazione dei piani di assestamento dei boschi* (pp. 615–629). ISEA.

IPCC (2006). Forest land. In Guidelines for national greenhouse gas inventories, Volume 4. Institute for Global Environmental Studies (IGES). Retrieved Oct 25, 2021, from https://www.ipcc-nggip.iges.or.jp/public/2006gl/pdf/4_Volume4/V4_04_Ch4_Forest_Land.pdf.

Liu, Z., & Li, F. (2003). The generalized Chapman-Richards function and application to tree and stand growth. *Journal of Forestry Research, 14*(1), 19–26.

Matthews, G. (1993). The carbon content of trees. Forestry commission technical paper 4. ISBN 0 85538 317 8.

Ratkowsky, D. A. (1990). *Handbook of nonlinear regression models* (241 pp.). Marcel Dekker, inc.

Russell, M. B., Fraver, S., Aakala, T., Gove, J. H., Woodall, C. W., D'Amato, A. W., & Ducey, M. J. (2015). Quantifying carbon stores and decomposition in dead wood: A review. *Forest Ecology and Management, 350*, 107–128. https://doi.org/10.1016/j.foreco.2015.04.033

Sprugel, D. G. (1983). Correcting for bias in log-transformed allometric equations. *Ecology, 64*, 209–210.

Tabacchi, G., Di Cosmo, L., & Gasparini, P. (2011a). Aboveground tree volume and phytomass prediction equations for forest species in Italy. *European Journal of Forest Research, 130*(9), 11–934. https://doi.org/10.1007/s10342-011-0481-9.

Tabacchi, G., Di Cosmo, L., Gasparini P., & Morelli, S. (2011b). Stima del volume e della fitomassa delle principali specie forestali italiane. Equazioni di previsione, tavole del volume e tavole della fitomassa arborea epigea. Consiglio per la Ricerca e la sperimentazione in Agricoltura, unità di ricerca per il monitoraggio e la pianificazione Forestale. Trento (412 pp.). Retrieved Oct 25, 2021, from https://www.inventarioforestale.org/sites/default/files/datiinventario/pubb/tavole_cubatura.pdf.

Tomter, S. M., Gasparini, P., Gschwantner, T., Hennig, P., Kulbokas, G., Kuliešis, A., Polley, H., Robert, N., Rondeux, J., Tabacchi, G., & Tomppo, E. (2012). Establishing bridging functions for harmonizing growing stock estimates: Examples from European national forest inventories. *Forestry Sciences, 58*(3), 224–235. https://doi.org/10.5849/forsci.10-068

Woodall, C. W., Domke, G. M., MacFarlane, D. W., & Oswalt, C. M. (2012). Comparing field- and model-based standing dead tree carbon stock estimates across forests of the US. *Forestry, 85*(1), 125–133. https://doi.org/10.1093/forestry/cpr065

Open Access This chapter is licensed under the terms of the Creative Commons Attribution 4.0 International License (http://creativecommons.org/licenses/by/4.0/), which permits use, sharing, adaptation, distribution and reproduction in any medium or format, as long as you give appropriate credit to the original author(s) and the source, provide a link to the Creative Commons license and indicate if changes were made.

The images or other third party material in this chapter are included in the chapter's Creative Commons license, unless indicated otherwise in a credit line to the material. If material is not included in the chapter's Creative Commons license and your intended use is not permitted by statutory regulation or exceeds the permitted use, you will need to obtain permission directly from the copyright holder.

Chapter 7
Area and Characteristics of Italian Forests

Superficie e principali caratteristiche delle foreste italiane

Patrizia Gasparini, Lucio Di Cosmo, and Antonio Floris

Abstract Awareness of exhaustible forest resources is not recent in human history; rather, it dates back to the late Middle Ages, when it became clear that some kind of planning was needed to utilise forest resources and to do so, assessment was necessary. Postponed in time, enlarged to a national scale and based on statistical sampling, compared to the inventory methods adopted at that time, modern NFIs are assigned to produce sound information necessary to support forest policies. Forest areas and composition, ownership, growing stock and increment, as well as management, silviculture and structural characters are among the variables assessed by NFIs. This chapter provides statistics on those variables. For areas, estimates are shown for Total wooded area, Forest, Other wooded land, and their distribution among inventory categories and forest types, which describe species composition. In addition, the chapter also addresses distribution by altitude classes. For stands characters, areas are shown by crown coverage, development stage and age class. Lastly, inventory statistics are given on the presence and amount of small trees and shrubs.

Keywords Forest area · Other wooded land area · Forest types · Private forests · Public forests · Growing stock

P. Gasparini (✉) · L. Di Cosmo · A. Floris
CREA Research Centre for Forestry and Wood, Trento, Italy
e-mail: patrizia.gasparini@crea.gov.it

L. Di Cosmo
e-mail: lucio.dicosmo@crea.gov.it

A. Floris
e-mail: antonio.floris@crea.gov.it

© The Author(s) 2022

P. Gasparini et al. (eds.), *Italian National Forest Inventory—Methods and Results of the Third Survey*, Springer Tracts in Civil Engineering,
https://doi.org/10.1007/978-3-030-98678-0_7

7.1 Introduction

Awareness of exhaustible forest resources is not recent in human history. In Western countries, it dates back to the late Middle Ages (Sereno, 2008). By the seventh and eighth centuries, resources had been reduced because of tillage to expand the agricultural land. In some phases from the eleventh to the fourteenth centuries, wide forest areas needed to be protected from overexploitation, and public authorities started doing so with laws enacted during the fourteenth century (Fossier, 2003). However, during the twelfth and thirteenth centuries, there had been a great increase of local statutes and agreements between communities and lords by which people intended to protect uncultivated areas (Provero, 2020). Signs of what we today would indicate as environmental damages were already apparent also in the North of Italy, but what was meant by protection was the equilibrium able to assure continuity of forest goods production, i.e., wood, acorn for grazing, and game (Delort, 1989). Loetsch and Haller (1973) claim that forest inventories started at the end of the Middle Ages when a shortage of timber supplies due to overexploitation forced people to plan the utilisation of accessible forest near towns and mines. These same concerns and information needs, postponed in time and enlarged to a national scale, led to the beginning of national forest inventories (NFIs) in the early 1900s. According to Persson and Janz (2015), there was a need for information concerning areas, topography, ownership, accessibility, volume and growth. This chapter reports INFC2015 statistics on some of the main variables traditionally considered by national forest inventories.

Measuring forest area has long been a necessary condition to estimate total values of assessed variables. Estimates were strictly based on the preliminary mapping of forests, because maps provided the forest area needed to upscale the mean values of variables estimated in different units of the forest (Loetsch & Haller, 1973). Such inventories were soon acknowledged to be inadequate for compiling national forest inventories (McRoberts et al., 2010; Tomppo et al., 2010). In modern NFIs based on probability sampling theory, areas are estimated like any other assessed variable. This is true for Total wooded area and, naturally, its components, such as Forest, Other wooded land, broadleaved or coniferous forest areas, inventory categories, forest types and subtype areas.

Correlated with air temperature, elevation also influences the distribution of vegetation indirectly, by conditioning the crumbling of minerals and organic matter decomposing in the soil formation process (Avena & Dowgiallo, 1995). Temporal statistical series on the distribution of vegetation types by altitudinal belts are also a valuable information source for evaluating adaptation to climate change. INFC estimates by altitude are produced for classes 500 m wide and for classes 300 m wide. This also allows for comparisons with both the first NFI (IFNI85) and some statistics by the Italian National Statistical Institute (ISTAT), which adopts the 600 m asl altitude as the limit for mountain territories.

Forest policy makers, especially the public bodies, certainly need information about the public and private forest areas for two reasons. First, it is reasonable to expect that by managing forests, public bodies and private owners pursue different objectives. Second, policies and regulations on forests may affect the rights on estates, and such information may help predict possible social and political consequences.

Producing statistics on growing stock volume and its increment have long been the main goal of NFIs (e.g., Breidenbach et al., 2020). In recent decades, the traditional interest regarding the economic value of timber volume has been complemented with its value as a carbon pool (cf. Chap. 12). For this reason, in addition to the growing stock volume, the aboveground tree biomass is estimated, and tally lists include trees smaller than in the past. Desired sustainable management, necessary to safeguard the productive capacity of forests, relies on regulating utilisation based on volume increment. This has also become important for monitoring programmes aimed at assessing the forests' response to environmental changes (Dobbertin, 2005; Gschwantner et al., 2016; Solberg et al., 2009). When estimated in terms of biomass, increment provides a measure of the contribution of forests in removing carbon from the atmosphere. The role of NFIs in increment estimation is unique, because it is a variable almost thoroughly obtained through field surveys, by repeated measurements in permanent plots and by tree coring, rather than by remote sensing techniques (Gasparini et al., 2017).

With special reference to the inventory statistics presented in this chapter, crown coverage of trees is fundamental for assigning the sample plots to the NFI domain, given the thresholds of 5 and 10% coverage that is relevant for the adopted classifying system (cf. Chap. 2). Moreover, crown distribution allows for the descriptions of the stand structure. INFC2015 recorded the presence of crown levels to provide statistics on the vertical structure, distinguishing one-storied from two-storied forests.

Data on the silvicultural system, development stage and age class, the latter only for even-aged forests, are essential for making planning decisions and developing forest policies at a more general level. Estimates on quantitative variables by forest areas are particularly useful in making hypotheses on future forest condition under different management scenarios.

Forest understory, according to the INFC, refers to small trees and shrubs, the two components traditionally least considered by NFIs among those described in this chapter. DBH thresholds once adopted by NFIs were strongly oriented by a willingness to estimate wood or timber with current commercial value or approaching such a value. Measuring small entities not only allows for a more accurate assessment of the carbon stock in the woody vegetation, but it also allows for recording the presence of species that seldom exceed the threshold for being callipered.

7.2 Area and Composition of Italian Forests

Estimates of Forest and Other wooded land area and forest type areas are among the main results of the forest inventory. INFC provides area estimates for two inventory macro-categories, seven inventory categories and twenty-three forest types, the latter further divided into subtypes, at regional and national levels. The classification scheme adopted by INFC and the class descriptions are given in Chaps. 2 and 3, respectively, for inventory categories and forest types and for land use and land cover.

Table 7.1 shows the statistics on area estimates for Forest, Other wooded land and Total wooded area. Tables 7.2 and 7.3 provide area estimates for the inventory categories of Forest and Other wooded land, respectively. The Total wooded area in Italy is estimated to be equal to 11,054,458 ha, of which 82.2% is classified as Forest (9,085,186 ha) and 17.8% as Other wooded land (1,969,272 ha). Total wooded area covers 36.7% of the country area; Forest covers 30.2%; and Other wooded land, 6.5% of the country's area. At the regional level, Forest cover varies considerably, going from 7.4% (Puglia) to 63.3% (Liguria), and it is above 40% in five regions (Alto Adige, Trentino, Friuli-Venezia Giulia, Toscana, and Umbria).

Table 7.1 Forest and Other wooded land area and Total wooded area / Estensione di Bosco e Altre terre boscate e Superficie forestale totale

Region/Regione	Forest		Other wooded land		Total wooded area		Regional and country area
	Bosco		Altre terre boscate		Superficie forestale totale		Superficie territoriale
	Area	ES	Area	ES	Area	ES	Area
	(ha)	(%)	(ha)	(%)	(ha)	(%)	(ha)
Piemonte	890,433	1.3	84,991	8.0	975,424	1.1	2,539,983
Valle d'Aosta	99,243	3.6	8733	24.0	107,976	3.1	326,322
Lombardia	621,968	1.6	70,252	8.7	692,220	1.3	2,386,285
Alto Adige	339,270	1.7	36,081	10.4	375,351	1.4	739,997
Trentino	373,259	1.4	33,826	10.6	407,086	1.2	620,690
Veneto	416,704	1.9	52,991	9.1	469,695	1.6	1,839,122
Friuli V.G	332,556	1.9	41,058	10.6	373,614	1.4	785,648
Liguria	343,160	1.7	44,084	10.3	387,244	1.4	542,024
Emilia Romagna	584,901	1.5	53,915	9.3	638,816	1.4	2,245,202
Toscana	1,035,448	1.1	154,275	5.2	1,189,722	0.8	2,299,018
Umbria	390,305	1.6	23,651	15.2	413,956	1.3	845,604
Marche	291,767	2.1	21,314	16.2	313,081	1.8	936,513
Lazio	560,236	1.6	87,912	7.7	648,148	1.3	1,720,768
Abruzzo	411,588	1.8	63,011	8.6	474,599	1.4	1,079,512
Molise	153,248	3.0	20,025	16.0	173,273	2.2	443,765
Campania	403,927	2.1	87,332	7.6	491,259	1.6	1,359,025
Puglia	142,349	4.0	49,389	9.9	191,738	3.0	1,936,580
Basilicata	288,020	2.7	104,392	6.2	392,412	1.7	999,461
Calabria	495,177	2.0	155,443	4.8	650,620	1.4	1,508,055
Sicilia	285,489	3.2	101,745	7.1	387,234	2.4	2,570,282
Sardegna	626,140	2.1	674,851	2.0	1,300,991	0.9	2,408,989
Italia	9,085,186	0.4	1,969,272	1.4	11,054,458	0.3	30,132,845

Table 7.2 Forest area by inventory category / Estensione delle categorie inventariali del Bosco

Region/Regione	Tall trees forest/Boschi alti						Plantations		Total forest	
	Stocked		Temporarily unstocked		Total					
	Con soprassuolo		Temporaneamente privi di soprassuolo		Totale		Impianti di arboricoltura da legno		Totale bosco	
	Area	ES	Area	ES	Area	ES	Area	ES	Area	ES
	(ha)	(%)	(ha)	(%)	(ha)	(%)	(ha)	(%)	(ha)	(%)
Piemonte	865,460	1.3	4313	41.1	869,773	1.3	20,660	16.2	890,433	1.3
Valle d'Aosta	99,243	3.6	0	–	99,243	3.6	0	–	99,243	3.6
Lombardia	595,513	1.6	1322	57.7	596,836	1.6	25,132	12.5	621,968	1.6
Alto Adige	337,758	1.8	1512	49.8	339,270	1.7	0	–	339,270	1.7
Trentino	372,539	1.5	721	70.8	373,259	1.4	0	–	373,259	1.4
Veneto	411,053	1.9	374	99.9	411,427	1.9	5277	33.7	416,704	1.9
Friuli V.G	323,362	1.9	0	–	323,362	1.9	9194	17.0	332,556	1.9
Liguria	339,545	1.7	3248	49.9	342,793	1.7	367	99.2	343,160	1.7
Emilia Romagna	577,770	1.5	1082	57.8	578,852	1.5	6049	31.6	584,901	1.5
Toscana	1,027,732	1.1	933	59.2	1,028,665	1.1	6783	30.3	1,035,448	1.1
Umbria	383,928	1.6	0	–	383,928	1.6	6377	33.7	390,305	1.6
Marche	284,904	2.1	0	–	284,904	2.1	6863	35.7	291,767	2.1
Lazio	555,251	1.6	2809	35.6	558,060	1.6	2176	39.7	560,236	1.6
Abruzzo	407,169	1.8	1448	50.0	408,616	1.8	2971	36.3	411,588	1.8
Molise	150,533	3.1	0	–	150,533	3.1	2715	27.7	153,248	3.0

(continued)

Table 7.2 (continued)

Region/Regione	Tall trees forest/Boschi alti						Plantations		Total forest	
	Stocked		Temporarily unstocked		Total					
	Con soprassuolo		Temporaneamente privi di soprassuolo		Totale		Impianti di arboricoltura da legno		Totale bosco	
	Area	ES	Area	ES	Area	ES	Area	ES	Area	ES
	(ha)	(%)	(ha)	(%)	(ha)	(%)	(ha)	(%)	(ha)	(%)
Campania	399,329	2.1	1434	50.1	400,763	2.1	3163	50.4	403,927	2.1
Puglia	140,735	4.1	1514	49.9	142,248	4.0	100	99.8	142,349	4.0
Basilicata	285,778	2.7	720	70.7	286,498	2.7	1522	93.6	288,020	2.7
Calabria	485,433	2.1	7337	32.5	492,771	2.1	2406	39.3	495,177	2.0
Sicilia	279,892	3.3	4839	37.7	284,731	3.2	758	70.7	285,489	3.2
Sardegna	598,026	2.2	2229	40.8	600,255	2.2	25,885	11.8	626,140	2.1
Italia	8,920,952	0.4	35,836	12.7	8,956,787	0.4	128,399	6.2	9,085,186	0.4

Table 7.3 Other wooded land area by inventory category / Estensione delle categorie inventariali delle Altre terre boscate

Region/Regione	Short trees forest		Sparse forest		Scrubland		Shrubs		Not accessible or not classified wooded area		Total Other wooded land	
	Boschi bassi		Boschi radi		Boscaglie		Arbusteti		Aree boscate inaccessibili o non classificate		Totale Altre terre boscate	
	Area	ES	Area	ES	Area	ES	Area	ES	Area	ES	Area	ES
	(ha)	(%)	(ha)	(%)	(ha)	(%)	(ha)	(%)	(ha)	(%)	(ha)	(%)
Piemonte	8291	32.9	17,037	18.2	7084	43.2	30,499	12.7	22,079	15.3	84,991	8.0
Valle d'Aosta	0	–	2357	62.5	0	–	2697	37.3	3680	31.2	8733	24.0
Lombardia	5731	27.7	5179	27.1	882	70.7	40,462	12.3	17,998	17.8	70,252	8.7
Alto Adige	0	–	4928	26.0	0	–	27,172	12.4	3981	30.1	36,081	10.4
Trentino	2523	37.8	4215	28.7	0	–	23,126	13.3	3963	29.2	33,826	10.6
Veneto	748	70.6	1087	57.7	1869	44.6	28,540	13.3	20,746	14.4	52,991	9.1
Friuli V.G	3680	46.4	6603	23.3	0	–	20,543	17.1	10,232	18.5	41,058	10.6
Liguria	10,843	25.6	5682	32.8	1466	49.6	9048	19.7	17,045	17.2	44,084	10.3
Emilia Romagna	9578	19.5	9687	30.3	2210	40.8	13,126	21.0	19,314	13.2	53,915	9.3
Toscana	13,369	16.4	11,871	22.6	4312	40.6	51,890	10.2	72,832	7.4	154,275	5.2
Umbria	3318	33.2	3786	43.9	0	–	10,719	23.9	5828	32.2	23,651	15.2
Marche	1082	57.9	5701	38.5	371	100.0	6772	37.1	7388	19.1	21,314	16.2
Lazio	8773	28.4	10,700	28.9	3983	53.1	49,806	10.6	14,649	17.4	87,912	7.7
Abruzzo	724	70.8	11,451	24.5	362	100.0	38,991	12.4	11,483	14.9	63,011	8.6
Molise	5433	34.2	481	68.2	1172	57.5	10,315	24.6	2624	31.8	20,025	16.0
Campania	5156	26.7	9389	29.5	1473	50.0	50,397	11.2	20,918	13.6	87,332	7.6

(continued)

Table 7.3 (continued)

Region/Regione	Short trees forest		Sparse forest		Scrubland		Shrubs		Not accessible or not classified wooded area		Total Other wooded land	
	Boschi bassi		Boschi radi		Boscaglie		Arbusteti		Aree boscate inaccessibili o non classificate		Totale Altre terre boscate	
	Area	ES	Area	ES	Area	ES	Area	ES	Area	ES	Area	ES
	(ha)	(%)	(ha)	(%)	(ha)	(%)	(ha)	(%)	(ha)	(%)	(ha)	(%)
Puglia	5826	25.5	7834	33.6	4661	28.6	27,399	14.7	3,669	30.1	49,389	9.9
Basilicata	3729	31.5	9252	23.9	5896	32.5	65,910	8.3	19,605	13.4	104,392	6.2
Calabria	15,298	15.4	6997	22.9	11,567	17.8	36,814	12.2	84,768	6.5	155,443	4.8
Sicilia	14,697	17.9	8665	28.3	4460	40.0	65,753	9.4	8170	25.5	101,745	7.1
Sardegna	30,539	11.7	44,200	10.9	11,942	17.6	558,795	2.3	29,377	12.6	674,851	2.0
Italia	149,336	5.7	187,099	5.7	63,710	9.7	1,168,776	1.9	400,350	3.2	1,969,272	1.4

Other wooded land cover is around the national value in most regions, except in Sardegna, Basilicata and Calabria, where it is considerably higher (28.0%, 10.4% and 10.3%, respectively). Sardegna hosts approximately one-third of the entire Other wooded land area. Figure 7.1 shows the percent of country area covered by Forest and Other wooded land and the distribution of related inventory sample points across Italy, respectively.

By far, the most important inventory category of Forest is that of the Tall trees forest, which accounts for 8,956,787 ha, of which 35,836 ha are temporarily unstocked (Table 7.2). Tall trees forest includes land covered by trees higher than 5 m, and with a canopy cover of more than 10%, or able to reach these thresholds in situ, and where the predominant use of land is not agricultural or urban.

The same thresholds of canopy cover and tree height are applied to Plantations. However, they are distinguished by their usage in timber and wood production, being of artificial origin and subjected to intensive management; these might grow on agricultural land. Overall, Plantations cover 128,399 ha, and the highest regional percentages of this category are in Piemonte, Lombardia, Friuli-Venezia Giulia, Marche and Sardegna (Table 7.2), where they represent approximately 2–4% of Forest area.

Other wooded land is mainly formed by Shrubs, which account for 1,168,776 ha, almost half of which is located in Sardegna. The inventory categories Short trees forest, Sparse forest, and Scrubland, overall represent 20.3% of Other wooded land area (Table 7.3 and Fig. 7.2). The presence of Short trees forest and Scrubland is often an index of difficult site conditions, due to poor soils or high winds, limiting the growth of tree, while Sparse forests may be a sign of spontaneous colonisation in progress or of degradation of denser stands. However, they might also represent potential natural vegetation at high altitude sites. By convention, all areas where the presence of tree and/or shrub cover was identified by photointerpretation but more detailed data could not be collected during the field surveys, were assigned to Other wooded land in the inventory category Not accessible or not classified wooded area. This category accounts for 20.3% of Other wooded land, at the national level.

A broad classification of forests is that based on species group composition (Table 7.4), which is widely used to compile forest statistics in European and international reporting activities. Inventory sampling points were assigned to one of the classes of Table 7.4 based on the field assessment of the crown cover percentage by species group. Tables 7.5 and 7.6 give the area estimates by conifers, broadleaves and mixed forest for the inventory macro-categories Forest and Other wooded land, respectively. INFC analogous statistics are available at inventarioforestale.org/statistiche_INFC for the inventory categories of Forest and of Other wooded land. Figures 7.3 and 7.4 show the percent of area of Forest and Other wooded land by pure broadleaves, pure conifers and mixed woods.

At the national level, pure broadleaves woods dominate both in Forest (68.5%) and in Other wooded land (53.9%); in the latter, the dominance of pure broadleaves is even greater (83.4%) if we consider just the areas actually classified for this attribute. Pure coniferous Forest accounts for 12.8% of the area and are concentrated in northern regions (Valle d'Aosta, Alto Adige and Trentino), where they characterise many Alpine landscapes, and in some southern peninsular regions and in Sicilia, due to

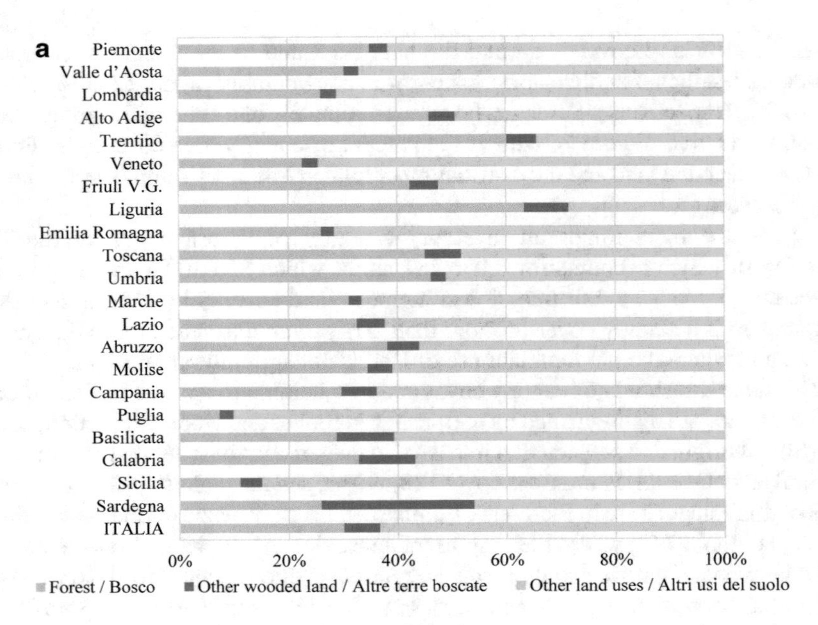

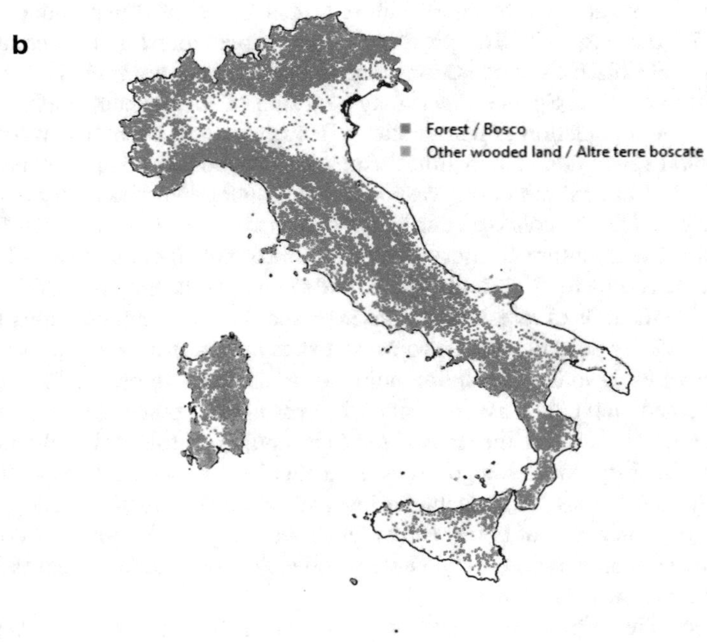

Fig. 7.1 Forest and Other wooded land in Italy: **a** percent of country area covered by Forest, Other wooded land and all other land uses, **b** distribution of inventory sample points of Forest and Other wooded land / Bosco e Altre terre boscate in Italia: **a** percentuale della superficie territoriale occupata da Bosco, Altre terre boscate e altri usi del suolo, **b** distribuzione dei punti inventariali appartenenti al Bosco e alle Altre terre boscate

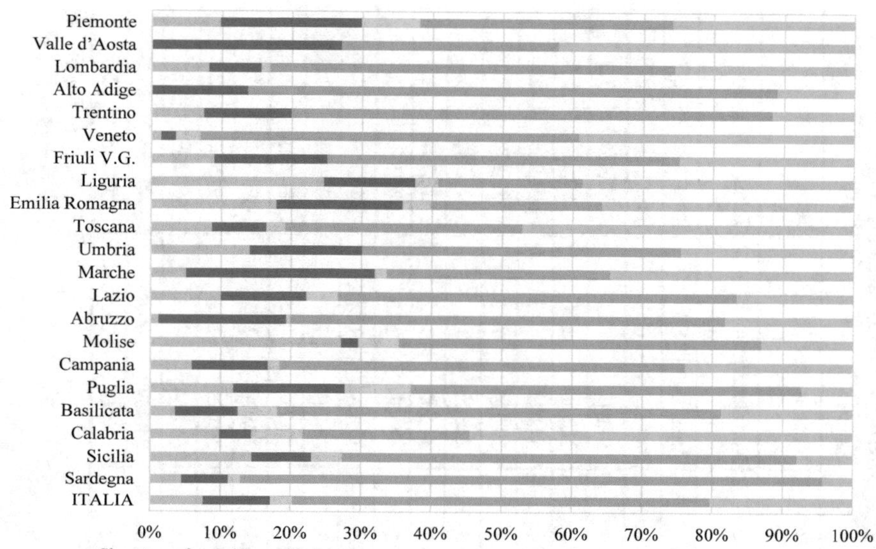

Fig. 7.2 Percent of area of Other wooded land by inventory category / Ripartizione percentuale delle Altre terre boscate per categorie inventariali

Table 7.4 Classes of pure and mixed conifers and broadleaves / Classi per il grado di mescolanza del soprassuolo

Pure and mixed conifers and broadleaves	Description
Grado di mescolanza del soprassuolo	Descrizione
Pure conifers Puro di conifere	Crown cover of coniferous species higher than 75%
	Copertura delle chiome di conifere superiore a 75%
Pure broadleaves Puro di latifoglie	Crown cover of broadleaved species, deciduous and evergreen, higher than 75%
	Copertura delle chiome di latifoglie, decidue e sempreverdi, superiore a 75%
Mixed conifers and broadleaves Misto di conifere e latifoglie	Crown cover either of coniferous species and of broadleaved species, deciduous and evergreen, lower than 75%
	Copertura delle chiome di conifere inferiore a 75% e copertura complessiva delle latifoglie, decidue e sempreverdi, inferiore a 75%

Table 7.5 Forest area by pure and mixed conifers and broadleaves / Estensione del Bosco ripartito per grado di mescolanza del soprassuolo

Region/Regione	Forest/Bosco									
	Pure conifers		Pure broadleaves		Mixed conifers and broadleaves		Not classified		Total	
	Puro di conifere		Puro di latifoglie		Misto di conifere e latifoglie		Non classificato		Totale	
	Area	ES	Area	ES	Area	ES	Area	ES	Area	ES
	(ha)	(%)	(ha)	(%)	(ha)	(%)	(ha)	(%)	(ha)	(%)
Piemonte	92,316	6.5	662,203	1.9	81,870	6.9	54,044	8.5	890,433	1.3
Valle d'Aosta	63,080	6.2	11,945	17.1	13,068	16.1	11,150	17.6	99,243	3.6
Lombardia	109,744	6.3	364,908	2.8	93,093	6.9	54,223	9.3	621,968	1.6
Alto Adige	273,731	2.4	14,219	16.1	26,595	11.6	24,724	12.2	339,270	1.7
Trentino	189,227	3.4	84,597	6.1	77,514	6.3	21,921	12.7	373,259	1.4
Veneto	97,907	5.8	197,302	4.0	87,190	6.0	34,305	11.2	416,704	1.9
Friuli V.G	47,534	8.7	179,212	3.7	66,860	7.1	38,950	10.3	332,556	1.9
Liguria	19,163	13.4	264,787	2.4	46,015	8.8	13,194	16.3	343,160	1.7
Emilia Romagna	18,050	14.1	508,355	1.9	30,218	10.8	28,278	12.0	584,901	1.5
Toscana	40,091	9.9	795,856	1.5	95,748	6.1	103,753	5.8	1,035,448	1.1
Umbria	5530	25.6	338,173	2.1	31,632	11.1	14,970	19.0	390,305	1.6
Marche	5559	25.7	226,981	2.9	29,207	12.9	30,020	10.7	291,767	2.1
Lazio	11,422	17.8	465,690	2.0	16,520	16.6	66,604	7.1	560,236	1.6
Abruzzo	12,977	16.5	336,745	2.4	27,428	11.2	34,439	9.9	411,588	1.8
Molise	4527	39.2	138,177	3.6	4296	29.8	6248	24.5	153,248	3.0

(continued)

Table 7.5 (continued)

Region/Regione	Forest/Bosco									
	Pure conifers		Pure broadleaves		Mixed conifers and broadleaves		Not classified		Total	
	Puro di conifere		Puro di latifoglie		Misto di conifere e latifoglie		Non classificato		Totale	
	Area	ES	Area	ES	Area	ES	Area	ES	Area	ES
	(ha)	(%)	(ha)	(%)	(ha)	(%)	(ha)	(%)	(ha)	(%)
Campania	8536	25.0	324,059	2.7	9939	19.1	61,392	7.6	403,927	2.1
Puglia	20,781	14.4	105,248	5.1	14,317	18.2	2003	42.6	142,349	4.0
Basilicata	11,847	20.3	215,826	3.6	8949	20.2	51,399	8.0	288,020	2.7
Calabria	63,306	7.9	286,318	3.2	77,329	7.3	68,224	8.4	495,177	2.0
Sicilia	44,580	10.7	186,405	4.5	30,977	12.8	23,528	16.1	285,489	3.2
Sardegna	26,496	11.8	513,393	2.5	44,640	9.8	41,610	9.8	626,140	2.1
Italia	1,166,403	1.6	6,220,400	0.6	913,405	2.0	784,978	2.2	9,085,186	0.4

Table 7.6 Other wooded land area by pure and mixed conifers and broadleaves / Estensione delle Altre terre boscate ripartite per grado di mescolanza del soprassuolo

Region/Regione	Other wooded land/Altre terre boscate									
	Pure conifers		Pure broadleaves		Mixed conifers and broadleaves		Not classified		Total	
	Puro di conifere		Puro di latifoglie		Misto di conifere e latifoglie		Non classificato		Totale	
	Area	ES	Area	ES	Area	ES	Area	ES	Area	ES
	(ha)	(%)	(ha)	(%)	(ha)	(%)	(ha)	(%)	(ha)	(%)
Piemonte	4742	38.8	39,783	12.4	7171	33.2	33,295	12.9	84,991	8.0
Valle d'Aosta	2136	68.5	1541	49.6	385	99.6	4671	27.2	8733	24.0
Lombardia	10,205	26.1	23,769	15.7	5250	28.9	31,029	13.7	70,252	8.7
Alto Adige	13,633	18.8	8299	21.4	3781	31.4	10,368	18.7	36,081	10.4
Trentino	9997	22.1	6750	23.6	4576	28.3	12,503	16.7	33,826	10.6
Veneto	7034	22.8	5574	25.7	848	63.4	39,535	11.1	52,991	9.1
Friuli V.G	7273	32.1	7717	21.6	1487	49.8	24,581	14.8	41,058	10.6
Liguria	1587	44.6	20,057	15.5	2881	54.9	19,560	16.6	44,084	10.3
Emilia Romagna	2511	61.5	25,824	15.2	2210	40.8	23,370	12.0	53,915	9.3
Toscana	3252	33.3	53,638	10.2	6839	29.1	90,545	6.7	154,275	5.2
Umbria	1745	44.9	9514	25.9	4720	37.7	7671	26.7	23,651	15.2
Marche	341	100.0	8536	30.7	2895	55.2	9542	21.6	21,314	16.2
Lazio	2185	40.8	52,082	10.9	3316	33.3	30,328	13.8	87,912	7.7
Abruzzo	10,402	21.9	26,930	15.4	9865	29.9	15,813	15.4	63,011	8.6
Molise	0	–	15,448	19.6	781	70.5	3796	28.2	20,025	16.0

(continued)

Table 7.6 (continued)

Region/Regione	Other wooded land/Altre terre boscate									
	Pure conifers		Pure broadleaves		Mixed conifers and broadleaves		Not classified		Total	
	Puro di conifere		Puro di latifoglie		Misto di conifere e latifoglie		Non classificato		Totale	
	Area	ES	Area	ES	Area	ES	Area	ES	Area	ES
	(ha)	(%)	(ha)	(%)	(ha)	(%)	(ha)	(%)	(ha)	(%)
Campania	0	–	41,680	12.7	1841	44.7	43,811	10.8	87,332	7.6
Puglia	3381	49.6	33,534	12.7	7151	33.0	5323	25.3	49,389	9.9
Basilicata	746	70.6	51,394	9.9	373	99.9	51,879	8.7	104,392	6.2
Calabria	1492	49.9	37,032	11.3	3731	31.5	113,188	5.7	155,443	4.8
Sicilia	1485	50.0	61,280	9.4	5887	38.6	33,093	13.6	101,745	7.1
Sardegna	7080	22.9	530,575	2.4	44,315	9.5	92,882	6.5	674,851	2.0
Italia	91,228	7.7	1,060,959	2.0	120,303	6.7	696,782	2.5	1,969,272	1.4

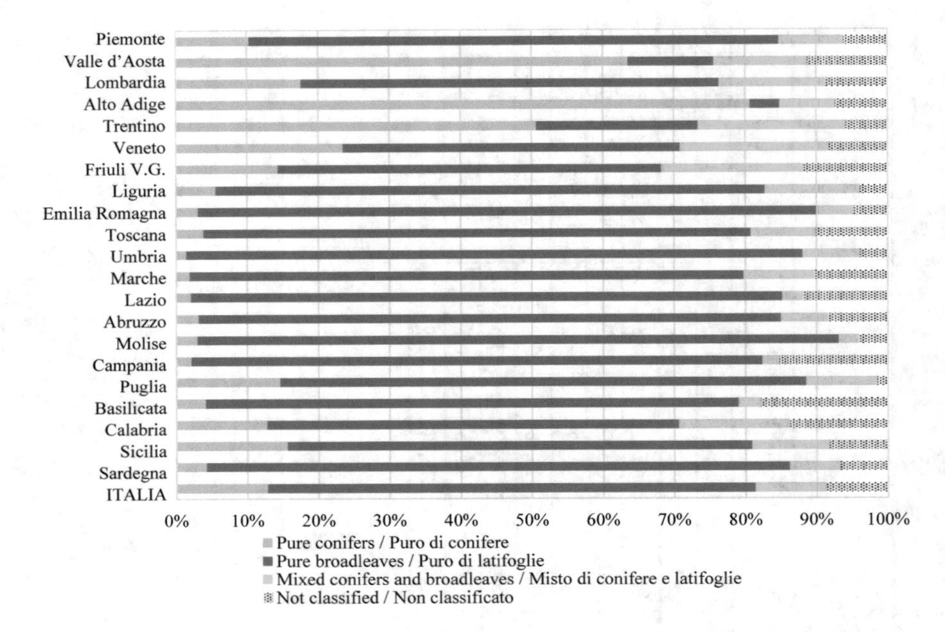

Fig. 7.3 Percent of area of Forest by pure or mixed class of conifers and broadleaves / Ripartizione percentuale della superficie del Bosco per classi del grado di mescolanza di conifere e latifoglie

coastal pine forests and to the presence of some mountain-Mediterranean conifer species. The class mixed conifers and broadleaves accounts for 10.1% of Forest area and 6.1% of Other wooded land area; it is more common in some northern regions (Lombardia, Trentino, Veneto, Friuli-Venezia Giulia) and in Calabria.

A more detailed classification of tree species composition is that adopted by INFC and described in Chap. 2. Inventory sample points are classified based on the identification of dominant species in terms of crown coverage. The identification of forest type is essential for the assignment of points to the inventory strata. For this reason, the forest type has also been attributed to the inventory points not accessible but observable from a remote location whenever it was possible to recognize the dominant species. In these cases, however, it was not possible to classify the forest subtype as well as any other information on the characteristics of the forest cover. The class 'not classified' for forest type includes the areas that were deliberately not classified during the second national forest inventory INFC2005, because the survey protocol did not prescribe classification of forest type for temporarily unstocked areas, and these could not be classified during the survey in INFC2015.

Tables 7.7 and 7.8 give the estimated area of forest types for the inventory categories Tall trees forest and Plantations, respectively. The same statistics are provided for some of the inventory categories of Other wooded land in Table 7.9, at the national level, and in Table 7.10 for the category Shrubs also at regional level. Area estimates on forest types at the provincial level are available at inventarioforestale.org/statistiche_INFC. Figure 7.5 compares the area of forest types in Tall trees forest at the

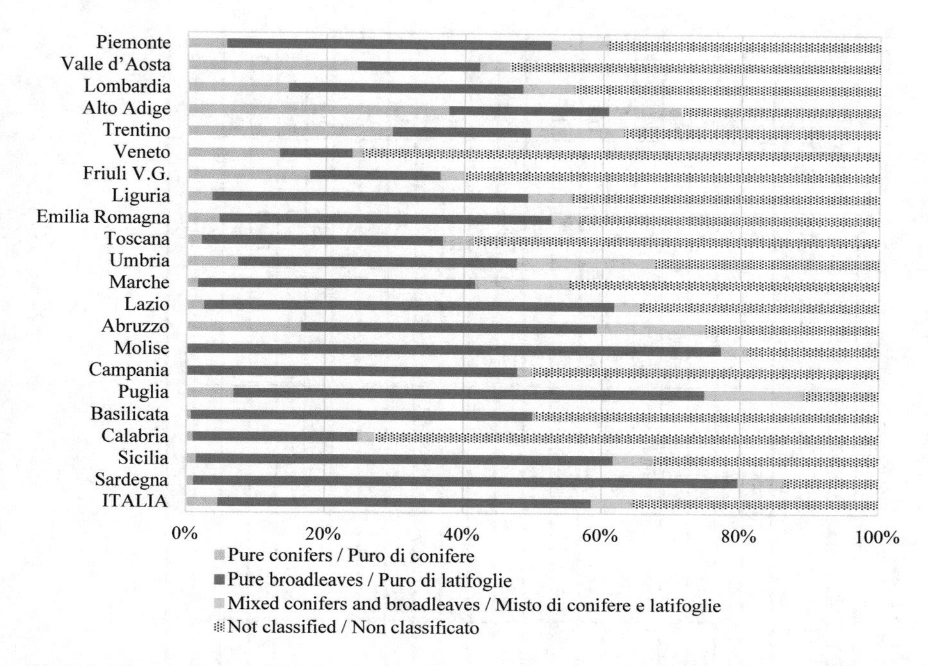

Fig. 7.4 Percent of area of Other wooded land by pure or mixed class of conifers and broadleaves / Ripartizione percentuale della superficie delle Altre terre boscate per classi del grado di mescolanza di conifere e latifoglie

national level. Four forest types account for more than one million hectares in Italy. They are, in order, Temperate oaks, Other deciduous broadleaved, Mediterranean oaks and Beech forests, which may be found in almost all Italian regions, except for Beech in Sardegna and Mediterranean oaks in the north-eastern regions. Four forest type areas are larger than one-half million hectares; they are Hornbeam and Hophornbeam, Chestnut, Holm oak and Norway spruce. The latter is found almost exclusively in the Alpine regions while the other three types can be found in most regions.

Figures in Table 7.7 show the variety of forest landscapes in Italy. On average, Italian regions host 13 forest types; only in the region of Alto Adige does one type (Norway spruce) account for more than 50% of the area of Tall trees forest. Only in three other regions are more than 40% of Tall trees forest represented by one forest type (Larch and Swiss stone pine in Valle d'Aosta, Mediterranean oaks in Basilicata and Holm oak in Sardegna). Area estimates on forest subtypes are available at national and regional level at inventarioforestale.org/statistiche_INFC. They can help to distinguish features linked to the frequency and ecology of most of the species. The list of subtypes distinguished by the classification scheme of INFC with corresponding CORINE biotopes and EUNIS codes are given in Chap. 3.

Table 7.7 Tall trees forest area by forest type / Estensione delle categorie forestali dei Boschi alti

Region/Regione	Larch and Swiss stone pine		Norway spruce		Fir		Scots pine and Mountain pine		Black pines	
	Boschi di larice e cembro		Boschi di abete rosso		Boschi di abete bianco		Pinete di pino silvestre e montano		Pinete di pino nero, laricio e loricato	
	Area	ES	Area	ES	Area	ES	Area	ES	Area	ES
	(ha)	(%)	(ha)	(%)	(ha)	(%)	(ha)	(%)	(ha)	(%)
Piemonte	80,060	7.1	18,182	14.8	14,142	16.8	20,757	13.9	3232	35.3
Valle d'Aosta	47,644	7.9	18,230	13.4	1156	57.3	11,671	17.2	385	99.6
Lombardia	53,420	9.8	87,863	6.7	9416	25.0	19,043	16.3	4408	31.6
Alto Adige	93,650	6.0	178,339	3.6	2647	37.6	37,053	9.6	378	99.8
Trentino	62,186	7.2	136,842	4.3	16,216	14.7	20,901	12.9	5766	24.9
Veneto	42,361	9.6	97,142	5.6	7104	22.8	12,251	17.5	3508	33.5
Friuli V.G	12,971	19.2	44,597	8.6	1858	44.5	12,228	20.0	31,554	11.2
Liguria	1100	57.2	367	99.2	2932	35.0	10,262	18.5	5498	25.5
Emilia Romagna	0	–	3684	31.6	2947	35.3	1842	44.7	16,208	14.9
Toscana	0	–	361	100.0	5059	26.7	1084	57.7	18,067	14.0
Umbria	0	–	0	–	0	–	737	70.6	5898	24.8
Marche	0	–	371	100.0	0	–	0	–	10,377	18.7
Lazio	0	–	368	100.0	0	–	0	–	8474	20.7
Abruzzo	0	–	362	100.0	724	70.8	1086	57.8	18,406	13.8
Molise	0	–	0	–	1172	57.5	0	–	2343	40.5
Campania	0	–	0	–	0	–	0	–	5524	25.7
Puglia	0	–	0	–	0	–	0	–	1554	49.8
Basilicata	0	–	0	–	746	70.6	0	–	2610	37.7
Calabria	0	–	0	–	3731	31.5	0	–	73,443	7.0
Sicilia	0	–	0	–	0	–	0	–	7493	22.2
Sardegna	0	–	0	–	0	–	0	–	5225	26.7
Italia	393,391	3.1	586,709	2.2	69,849	7.6	148,915	5.1	230,351	4.0

(continued)

Table 7.7 (continued)

Region/Regione	Mediterranean pines		Other coniferous forest		Beech		Temperate oaks		Mediterranean oaks	
	Pinete di pini mediterranei		Altri boschi di conifere, pure o miste		Faggete		Querceti a rovere, roverella e farnia		Cerrete, boschi di farnetto, fragno, vallonea	
	Area	ES	Area	ES	Area	ES	Area	ES	Area	ES
	(ha)	(%)	(ha)	(%)	(ha)	(%)	(ha)	(%)	(ha)	(%)
Piemonte	808	70.7	6465	24.9	117,131	5.8	76,176	7.4	6061	25.8
Valle d'Aosta	0	–	0	–	1156	57.3	4238	29.6	0	–
Lombardia	0	–	6348	36.7	66,650	8.0	41,095	10.1	2204	44.7
Alto Adige	0	–	0	–	4159	29.9	4537	28.7	0	–
Trentino	0	–	0	–	62,607	7.1	5766	24.9	0	–
Veneto	748	70.6	0	–	67,639	7.0	15,651	17.2	0	–
Friuli V.G	1115	57.5	1115	57.5	90,645	5.8	8176	21.1	0	–
Liguria	24,546	11.7	367	99.2	37,017	9.4	43,863	9.4	11,247	17.6
Emilia Romagna	2947	35.3	3315	33.3	102,507	5.6	76,993	6.6	101,263	5.6
Toscana	45,161	9.2	11,900	20.2	73,324	7.0	152,096	4.7	250,565	3.5
Umbria	8110	21.1	1843	44.6	16,517	16.4	100,886	5.7	128,595	5.0
Marche	2194	40.9	5657	33.8	17,790	14.1	61,753	7.6	24,732	12.9
Lazio	7344	22.3	1474	49.9	74,430	7.1	81,764	6.9	132,444	5.2
Abruzzo	2172	40.8	2172	40.8	124,468	4.8	87,761	6.5	32,145	11.0
Molise	2184	68.9	0	–	14,839	15.5	48,198	9.4	51,975	8.4
Campania	9146	24.0	864	62.2	56,244	8.0	60,934	8.2	74,644	7.6
Puglia	27,718	12.8	2331	40.6	4661	28.6	25,865	11.6	38,728	9.8
Basilicata	8933	20.2	5150	35.8	26,820	11.4	43,077	9.7	130,692	5.2
Calabria	17,474	16.1	12,561	21.9	79,413	6.7	53,790	9.5	43,593	9.4
Sicilia	45,327	10.3	9299	27.3	15,165	15.6	71,887	8.2	30,989	13.8
Sardegna	34,633	10.9	13,055	16.8	0	–	87,780	6.9	0	–
Italia	240,559	4.2	83,915	8.0	1,053,183	1.8	1,152,288	1.8	1,059,876	1.8

(continued)

Table 7.7 (continued)

Region/Regione	Chestnut		Hornbeam and Hophornbeam		Hygrophilous forest		Other deciduous broadleaved		Holm oak	
	Castagneti		Ostrieti, carpineti		Boschi igrofili		Altri boschi caducifogli		Leccete	
	Area	ES	Area	ES	Area	ES	Area	ES	Area	ES
	(ha)	(%)	(ha)	(%)	(ha)	(%)	(ha)	(%)	(ha)	(%)
Piemonte	163,639	4.6	16,162	15.7	22,736	15.2	323,713	3.3	0	–
Valle d'Aosta	4238	29.6	0	–	0	–	10,524	18.2	0	–
Lombardia	82,079	7.1	81,720	7.0	18,845	15.1	121,011	6.0	0	–
Alto Adige	1890	44.5	8825	20.7	2747	36.4	3532	33.3	0	–
Trentino	1441	50.0	40,841	9.0	3860	42.6	16,473	16.2	360	100.0
Veneto	17,199	14.5	82,025	6.3	10,417	22.3	52,392	10.1	2991	35.3
Friuli V.G	12,264	17.1	46,791	8.8	12,600	19.6	47,447	8.7	0	–
Liguria	109,586	4.9	46,230	8.8	3299	33.0	31,836	11.0	13,911	15.8
Emilia Romagna	40,521	9.3	111,252	5.3	28,223	13.5	86,414	7.2	737	70.7
Toscana	156,153	4.5	64,654	7.5	27,751	13.6	84,492	7.1	129,686	5.2
Umbria	2212	40.7	58,518	7.4	8038	26.0	11,284	23.1	41,289	9.0
Marche	3706	31.5	76,287	6.2	12,902	19.2	62,094	7.6	7042	22.8
Lazio	35,792	9.9	97,974	5.9	9211	19.9	54,024	9.2	48,267	8.5
Abruzzo	4705	27.7	47,924	8.3	19,866	14.6	56,371	9.0	8686	20.3
Molise	391	99.8	8982	20.3	9054	27.2	10,225	24.9	1172	57.5
Campania	55,986	8.3	53,030	8.0	11,048	18.1	34,386	11.2	37,485	9.6
Puglia	1165	57.5	4661	28.6	388	99.8	11,538	20.9	17,364	16.1
Basilicata	5955	24.8	7084	22.8	12,975	19.2	29,872	13.4	10,371	22.2
Calabria	68,966	7.0	4477	28.8	12,872	23.4	45,153	10.9	48,692	9.5
Sicilia	8720	20.7	1737	51.2	4929	27.6	13,237	16.7	19,536	16.3
Sardegna	1866	44.7	0	–	3359	33.3	7743	27.0	255,463	3.7
Italia	778,475	2.1	859,176	2.0	235,117	4.6	1,113,761	1.9	643,052	2.4

(continued)

Table 7.7 (continued)

Region/Regione	Coark oak		Other evergreen broadleaved		Not classified		Total Tall trees forest	
	Sugherete		Altri boschi di latifoglie sempreverdi		Non classificato		Totale Boschi alti	
	Area	ES	Area	ES	Area	ES	Area	ES
	(ha)	(%)	(ha)	(%)	(ha)	(%)	(ha)	(%)
Piemonte	0	–	0	–	508	100.0	869,773	1.3
Valle d'Aosta	0	–	0	–	0	–	99,243	3.6
Lombardia	0	–	2292	67.3	441	100.0	596,836	1.6
Alto Adige	0	–	0	–	1512	49.8	339,270	1.7
Trentino	0	–	0	–	0	–	373,259	1.4
Veneto	0	–	0	–	0	–	411,427	1.9
Friuli V.G	0	–	0	–	0	–	323,362	1.9
Liguria	0	–	0	–	733	70.1	342,793	1.7
Emilia Romagna	0	–	0	–	0	–	578,852	1.5
Toscana	6143	24.2	1445	50.0	723	70.7	1,028,665	1.1
Umbria	0	–	0	–	0	–	383,928	1.6
Marche	0	–	0	–	0	–	284,904	2.1
Lazio	2579	37.7	2579	37.7	1335	50.8	558,060	1.6
Abruzzo	0	–	1407	100.0	362	100.0	408,616	1.8
Molise	0	–	0	–	0	–	150,533	3.1
Campania	368	100.0	737	70.8	368	100.0	400,763	2.1
Puglia	0	–	5150	27.0	1125	57.7	142,248	4.0
Basilicata	0	–	1864	44.6	347	99.9	286,498	2.7
Calabria	5224	26.6	21,143	15.4	2239	40.7	492,771	2.1
Sicilia	17,261	17.7	39,151	12.4	0	–	284,731	3.2
Sardegna	152,755	5.1	38,004	14.0	373	100.0	600,255	2.2
Italia	184,330	4.7	113,772	7.4	10,067	19.3	8,956,787	0.4

Table 7.8 Plantations area by forest type / Estensione delle categorie forestali degli Impianti di arboricoltura da legno

Region/Regione	Poplar plantations		Other broadleaved plantations		Coniferous plantations		Total plantations	
	Pioppeti artificiali		Piantagioni di altre latifoglie		Piantagioni di conifere		Totale Impianti di arboricoltura da legno	
	Area	ES	Area	ES	Area	ES	Area	ES
	(ha)	(%)	(ha)	(%)	(ha)	(%)	(ha)	(%)
Piemonte	15,510	18.2	3999	43.0	1151	57.9	20,660	16.2
Valle d'Aosta	0	–	0	–	0	–	0	–
Lombardia	18,508	14.8	6183	38.7	441	100.0	25,132	12.5
Alto Adige	0	–	0	–	0	–	0	–
Trentino	0	–	0	–	0	–	0	–
Veneto	2332	61.0	2945	56.0	0	–	5277	33.7
Friuli V.G	6540	23.8	2654	44.7	0	–	9194	17.0
Liguria	367	99.2	0	–	0	–	367	99.2
Emilia Romagna	4252	41.2	1428	49.9	368	100.0	6049	31.6
Toscana	736	69.6	4595	41.0	1452	50.0	6783	30.3
Umbria	0	–	6377	33.7	0	–	6377	33.7
Marche	0	–	6863	35.7	0	–	6863	35.7
Lazio	789	70.8	1387	51.0	0	–	2176	39.7
Abruzzo	528	75.5	2443	41.1	0	–	2971	36.3
Molise	391	99.8	2325	27.7	0	–	2715	27.7
Campania	1539	92.3	1624	44.5	0	–	3163	50.4

(continued)

Table 7.8 (continued)

Region/Regione	Poplar plantations		Other broadleaved plantations		Coniferous plantations		Total plantations	
	Pioppeti artificiali		Piantagioni di altre latifoglie		Piantagioni di conifere		Totale Impianti di arboricoltura da legno	
	Area	ES	Area	ES	Area	ES	Area	ES
	(ha)	(%)	(ha)	(%)	(ha)	(%)	(ha)	(%)
Puglia	0	–	100	99.8	0	–	100	99.8
Basilicata	0	–	1522	93.6	0	–	1522	93.6
Calabria	100	99.9	1560	50.1	746	70.6	2406	39.3
Sicilia	0	–	758	70.7	0	–	758	70.7
Sardegna	0	–	24,392	12.2	1493	50.0	25,885	11.8
Italia	51,592	9.9	71,156	9.1	5651	25.9	128,399	6.2

Table 7.9 Other wooded land area by forest type, at the national level / Estensione delle categorie forestali delle Altre terre boscate, a livello nazionale

Forest type/Categoria forestale	Short trees forest, Sparse forest, Scrubland/Boschi bassi, Boschi radi, Boscaglie							
	Short trees forest		Sparse forest		Scrubland		Total	
	Boschi bassi		Boschi radi		Boscaglie		Totale	
	Area	ES	Area	ES	Area	ES	Area	ES
	(ha)	(%)	(ha)	(%)	(ha)	(%)	(ha)	(%)
Larch and Swiss stone pine forest/Boschi di larice e cembro	0	–	19,879	16.1	0	–	19,879	16.1
Norway spruce forest/Boschi di abete rosso	0	–	2834	35.1	0	–	2834	35.1
Fir forest/Boschi di abete bianco	361	100.0	404	100.0	0	–	765	70.8
Scots pine and Mountain pine forest/Pinete di pino silvestre e montano	1891	80.0	1417	100.0	0	–	3308	62.7
Black pines forest/Pinete di pino nero, laricio e loricato	1218	57.9	3631	46.2	373	99.9	5222	35.5
Mediterranean pines forest/Pinete di pini mediterranei	2638	37.7	7158	32.7	388	99.8	10,184	25.3
Other coniferous forest/Altri boschi di conifere, pure o miste	2628	37.8	1455	50.0	373	100.0	4456	28.9
Beech forest/Faggete	3005	35.3	6164	37.3	367	99.2	9535	26.8
Temperate oaks forest/Boschi a rovere, roverella e farnia	20,372	14.7	39,257	12.3	4855	27.7	64,483	9.0
Mediterranean oaks forest/Cerrete, boschi di farnetto, fragno, vallonea	7546	27.3	10,327	25.0	1108	57.6	18,980	17.7
Chestnut forest/Castagneti	1139	57.7	3947	53.1	1119	57.6	6,205	36.9
Hornbeam and Hophornbeam forest/Ostrieti, carpineti	18,295	16.0	13,328	26.4	3714	46.1	35,337	13.7
Hygrophilous forest/Boschi igrofili	1952	44.8	5513	34.2	3739	31.6	11,203	21.3
Other deciduous broadleaved forest/Altri boschi caducifogli	28,583	14.2	29,936	14.5	18,318	21.2	76,838	9.1
Holm oak forest/Leccete	33,112	10.6	17,063	14.7	14,847	17.8	65,022	7.7
Coark oak forest/Sugherete	3003	35.3	13,180	19.1	746	70.7	16,930	16.4
Other evergreen broadleaved forest/Altri boschi di latifoglie sempreverdi	23,594	16.1	11,608	27.7	13,763	20.8	48,965	11.7
Subalpine shrubs/Arbusteti subalpini	0	–	0	–	0	–	0	–
Temperate climate shrubs/Arbusteti di clima temperato	0	–	0	–	0	–	0	–
Mediterranean scrubs and shrubs/ Macchia, arbusteti mediterranei	0	–	0	–	0	–	0	–
Not classified/Non classificato	0	–	0	–	0	–	0	–
Total/Totale	149,336	5.7	187,099	5.7	63,710	9.7	400,146	3.7

(continued)

Table 7.9 (continued)

Forest type/Categoria forestale	Shrubs		Not accessible or not classified wooded area		Total Other wooded land	
	Arbusteti		Aree boscate inaccessibili o non classificate		Totale Altre terre boscate	
	Area	ES	Area	ES	Area	ES
	(ha)	(%)	(ha)	(%)	(ha)	(%)
Larch and Swiss stone pine forest/Boschi di larice e cembro	0	–	0	–	19,879	16.1
Norway spruce forest/Boschi di abete rosso	0	–	0	–	2834	35.1
Fir forest/Boschi di abete bianco	0	–	0	–	765	70.8
Scots pine and Mountain pine forest/Pinete di pino silvestre e montano	0	–	0	–	3308	62.7
Black pines forest/Pinete di pino nero, laricio e loricato	0	–	0	–	5222	35.5
Mediterranean pines forest/Pinete di pini mediterranei	0	–	0	–	10,184	25.3
Other coniferous forest/Altri boschi di conifere, pure o miste	0	–	0	–	4456	28.9
Beech forest/Faggete	0	–	0	–	9535	26.8
Temperate oaks forest/Boschi a rovere, roverella e farnia	0	–	0	–	64,483	9.0
Mediterranean oaks forest/Cerrete, boschi di farnetto, fragno, vallonea	0	–	0	–	18,980	17.7
Chestnut forest/Castagneti	0	–	0	–	6205	36.9
Hornbeam and Hophornbeam forest/Ostrieti, carpineti	0	–	0	–	35,337	13.7
Hygrophilous forest/Boschi igrofili	0	–	0	–	11,203	21.3
Other deciduous broadleaved forest/Altri boschi caducifogli	0	–	0	–	76,838	9.1
Holm oak forest/Leccete	0	–	0	–	65,022	7.7
Coark oak forest/Sugherete	0	–	0	–	16,930	16.4
Other evergreen broadleaved forest/Altri boschi di latifoglie sempreverdi	0	–	0	–	48,965	11.7
Subalpine shrubs/Arbusteti subalpini	149,895	6.0	0	–	149,895	6.0
Temperate climate shrubs/Arbusteti di clima temperato	220,395	5.0	0	–	220,395	5.0
Mediterranean scrubs and shrubs/ Macchia, arbusteti mediterranei	798,485	2.2	0	–	798,485	2.2
Not classified/Non classificato	0	–	400,350	3.2	400,350	3.2
Total/Totale	1,168,776	1.9	400,350	3.2	1,969,272	1.4

Table 7.10 Shrubs area by forest type / Estensione delle categorie forestali degli Arbusteti

Region/Regione	Subalpine shrubs		Temperate climate shrubs		Mediterranean scrubs and shrubs		Total Shrubs	
	Arbusteti subalpini		Arbusteti di clima temperato		Macchia, arbusteti mediterranei		Totale Arbusteti	
	Area	ES	Area	ES	Area	ES	Area	ES
	(ha)	(%)	(ha)	(%)	(ha)	(%)	(ha)	(%)
Piemonte	17,227	18.2	13,273	17.4	0	–	30,499	12.7
Valle d'Aosta	2312	40.4	385	99.6	0	–	2697	37.3
Lombardia	30,875	14.7	9588	21.2	0	–	40,462	12.3
Alto Adige	27,172	12.4	0	–	0	–	27,172	12.4
Trentino	22,766	13.5	360	100.0	0	–	23,126	13.3
Veneto	25,584	14.4	2243	40.7	713	70.7	28,540	13.3
Friuli V.G	18,314	18.6	2230	40.6	0	–	20,543	17.1
Liguria	252	99.2	5131	26.4	3665	31.3	9048	19.7
Emilia Romagna	368	100.0	12,758	21.4	0	–	13,126	21.0
Toscana	0	–	28,110	13.4	23,780	16.4	51,890	10.2
Umbria	0	–	9316	23.4	1402	99.8	10,719	23.9
Marche	0	–	6402	38.8	371	100.0	6772	37.1
Lazio	0	–	20,599	14.5	29,207	15.2	49,806	10.6
Abruzzo	5026	36.1	33,240	13.7	724	70.8	38,991	12.4
Molise	0	–	8904	27.4	1412	50.6	10,315	24.6
Campania	0	–	11,169	27.7	39,227	13.0	50,397	11.2

(continued)

Table 7.10 (continued)

Region/Regione	Subalpine shrubs		Temperate climate shrubs		Mediterranean scrubs and shrubs		Total Shrubs	
	Arbusteti subalpini		Arbusteti di clima temperato		Macchia, arbusteti mediterranei		Totale Arbusteti	
	Area	ES	Area	ES	Area	ES	Area	ES
	(ha)	(%)	(ha)	(%)	(ha)	(%)	(ha)	(%)
Puglia	0	–	3107	35.1	24,292	16.1	27,399	14.7
Basilicata	0	–	21,139	16.2	44,771	10.1	65,910	8.3
Calabria	0	–	2239	40.7	34,575	12.7	36,814	12.2
Sicilia	0	–	23,496	16.1	42,258	12.1	65,753	9.4
Sardegna	0	–	6707	23.5	552,088	2.3	558,795	2.3
Italia	149,895	6.0	220,395	5.0	798,485	2.2	1,168,776	1.9

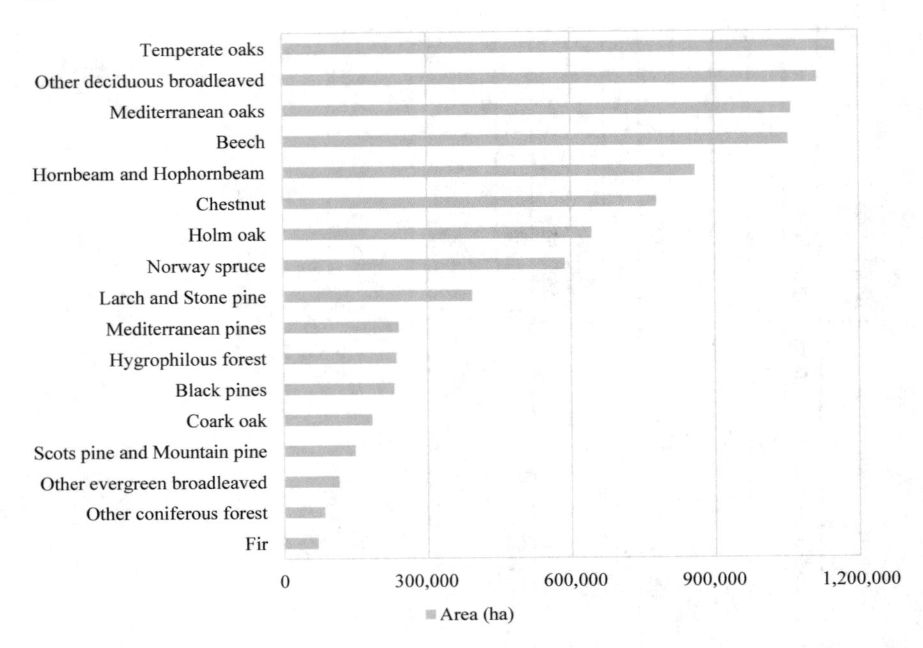

Fig. 7.5 Area of Tall trees forest types, in descending order / Superficie delle categorie forestali dei Boschi alti, in ordine decrescente

7.3 Distribution by Altitude Classes

Table 7.11 shows Forest area by altitude class, considering five altitude classes, each 500 m asl wide. Forest is distributed over a wide range of elevation, following the wide range of land elevation due to the orographic characteristics of the country. At the national level, 34.1% of Forest area is in the first class, i.e., up to 500 m asl, and 37.9% is in the second class, between 501 m and 1000 m asl. Forest area rates decrease progressively with altitude as represented by the remaining three following classes: 19.5% of Forest area is in the class 1001–1500 m asl, 7.5% is in the class 1501–2000 m, and 1.1% is above 2000 m asl.

Forest area distribution by elevation class is rather variable at the regional level. Figure 7.6 shows that some regions are marked by a more homogeneous distribution of Forest by elevation (e.g., Veneto, Lombardia, Piemonte, Abruzzo) while in others, Forest is mainly in a few classes (e.g., Toscana, Sardegna). This is also due, of course, to the distribution of regional land area along elevation gradients and explains why Puglia has 100% of its Forest below 1000 m asl. In some Alpine regions the percent of Forest area in the class 0–500 m asl is much lower than the national mean value: 0.8% in Valle d'Aosta, 1.8% in Alto Adige and 5.5% in Trentino. These are mountainous regions with a relevant part of their Forest at altitudes higher than 1500 m asl: 31.9% in Trentino, 54.8% in Alto Adige and 61.7% in Valle d'Aosta. Among non-Alpine

Table 7.11 Forest area by classes of altitude above sea level (asl), module 500 m / Estensione del Bosco ripartito per classi di altitudine sul livello del mare (slm), modulo 500 m

Region/Regione	Forest/Bosco											
	0–500 m asl/s.l.m		501–1000 m asl/s.l.m		1001–1500 m asl/s.l.m		1501–2000 m asl/s.l.m		>2000 m asl/s.l.m		Total/Totale	
	Area	ES	Area	ES	Area	ES	Area	ES	Area	ES	Area	ES
	(ha)	(%)	(ha)	(%)	(ha)	(%)	(ha)	(%)	(ha)	(%)	(ha)	(%)
Piemonte	302,173	3.5	274,164	3.5	211,318	4.1	85,761	6.9	17,017	15.3	890,433	1.3
Valle d'Aosta	771	70.3	11,945	17.1	25,277	11.0	46,929	7.7	14,321	17.5	99,243	3.6
Lombardia	164,142	4.9	218,530	4.1	154,649	5.1	73,451	7.7	11,197	24.5	621,968	1.6
Alto Adige	6049	24.8	35,392	9.9	111,844	5.2	153,812	4.0	32,173	11.8	339,270	1.7
Trentino	20,556	14.1	91,428	5.8	142,380	4.2	102,186	5.3	16,709	14.5	373,259	1.4
Veneto	101,713	6.3	123,140	5.2	123,978	5.1	63,474	7.3	4400	40.5	416,704	1.9
Friuli V.G	100,831	5.6	123,685	4.8	85,814	6.0	22,226	14.7	0	–	332,556	1.9
Liguria	146,909	4.1	151,536	4.2	42,515	8.7	2199	40.4	0	–	343,160	1.7
Emilia Romagna	220,368	3.9	259,076	3.2	96,617	5.8	8841	20.3	0	–	584,901	1.5
Toscana	592,052	2.0	371,515	2.7	70,074	7.2	1807	44.7	0	–		1.1
Umbria	164,775	4.2	185,221	3.8	39,203	10.2	1106	57.6	0	–	390,305	1.6
Marche	118,717	5.1	143,129	4.3	28,808	11.7	1112	57.8	0	–	291,767	2.1
Lazio	225,175	3.7	211,328	3.8	103,765	6.0	19,968	13.3	0	–	560,236	1.6
Abruzzo	77,084	7.1	142,073	4.8	141,115	4.7	51,316	8.0	0	–	411,588	1.8
Molise	36,551	11.1	90,302	5.8	24,052	12.9	2343	40.5	0	–	153,248	3.0
Campania	147,657	4.8	175,842	4.2	76,746	7.0	3683	31.6	0	–	403,927	2.1

(continued)

Table 7.11 (continued)

Region/Regione	Forest/Bosco											
	0–500 m asl/s.l.m		501–1000 m asl/s.l.m		1001–1500 m asl/s.l.m		1501–2000 m asl/s.l.m		>2000 m asl/s.l.m		Total/Totale	
	Area	ES	Area	ES	Area	ES	Area	ES	Area	ES	Area	ES
	(ha)	(%)	(ha)	(%)	(ha)	(%)	(ha)	(%)	(ha)	(%)	(ha)	(%)
Puglia	83,189	6.2	59,160	7.7	0	–	0	–	0	–	142,349	4.0
Basilicata	46,726	9.6	178,592	4.2	57,855	7.8	4847	27.6	0	–	288,020	2.7
Calabria	104,330	6.6	221,232	4.0	142,004	4.8	27,611	11.4	0	–	495,177	2.0
Sicilia	97,000	7.1	118,562	5.9	59,749	9.2	10,179	19.1	0	–	285,489	3.2
Sardegna	337,088	3.4	259,394	3.6	29,658	11.9	0	–	0	–	626,140	2.1
Italia	3,093,856	1.0	3,445,245	1.0	1,767,420	1.4	682,850	2.2	95,816	6.9	9,085,186	0.4

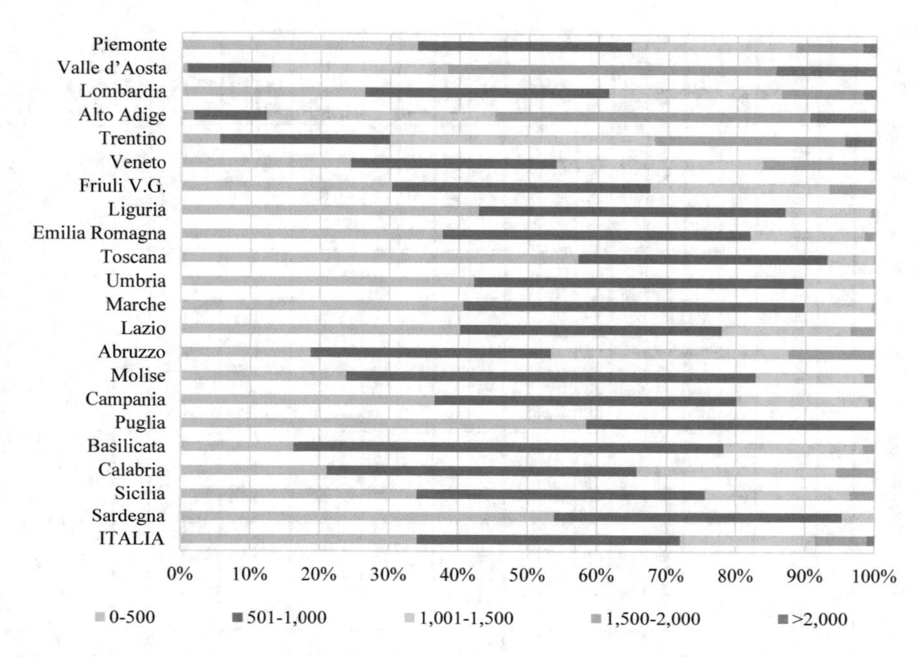

Fig. 7.6 Percent of area of Forest by elevation class (m asl) / Ripartizione percentuale dell'area del Bosco per classi di altitudine (m s.l.m.)

regions, Abruzzo is the sole region with a relevant rate (12.5%) of Forest above 1500 m asl.

Observing the distribution of Other wooded land by altitude class (Table 7.12 and Fig. 7.7), it is apparent that more than half of the national area (54.6%) falls into the first class, below 500 m asl, but that percentage is much higher in some regions of central and southern Italy: Toscana (72.5%), Sardegna (76.3%) and Puglia (83.1%). Southern regions, in general, host limited area of Other wooded land above 1500 m asl, except Abruzzo (13.5%), which is also the sole non-Alpine region with Other wooded land at above 2000 m asl (0.6% of regional value). Other wooded land above 1500 m asl is, hence, a characteristic of the Alpine regions and very high rates are found in Alto Adige (92.5%) and in Trentino (73.0%).

Additional estimates concerning the elevation distribution of the inventory categories and forest types of Tall trees forest are available at inventarioforestale.org/statistiche_INFC. All estimates produced by 300 m wide elevation classes are also available at the website.

Table 7.12 Other wooded land area by classes of altitude above sea level (asl), module 500 m / Estensione delle Altre terre boscate ripartite per classi di altitudine, modulo 500 m

Region/Regione	Other wooded land/Altre terre boscate											
	0–500 m/s.l.m		501–1000 m asl/s.l.m		1001–1500 m asl/s.l.m		1501–2000 m asl/s.l.m		>2000 m asl/s.l.m		Total/Totale	
	Area	ES	Area	ES	Area	ES	Area	ES	Area	ES	Area	ES
	(ha)	(%)	(ha)	(%)	(ha)	(%)	(ha)	(%)	(ha)	(%)	(ha)	(%)
Piemonte	18,513	20.0	8936	21.3	19,209	14.4	32,078	14.8	6256	32.4	84,991	8.0
Valle d'Aosta	0	–	486	81.7	1003	56.5	6,374	30.1	871	63.3	8733	24.0
Lombardia	7814	23.3	12,541	20.6	13,753	19.7	24,899	15.8	11,245	26.3	70,252	8.7
Alto Adige	0	–	100	99.8	2606	37.7	15,831	15.2	17,543	16.1	36,081	10.4
Trentino	721	70.8	2523	37.8	5898	25.0	11,891	17.0	12,794	19.0	33,826	10.6
Veneto	6982	29.0	2243	40.7	9973	19.5	28,154	13.4	5639	25.3	52,991	9.1
Friuli V.G	3629	31.3	8176	21.1	9539	23.4	19,713	17.6	0	–	41,058	10.6
Liguria	21,827	16.6	15,676	16.8	5848	24.7	733	70.1	0	–	44,084	10.3
Emilia Romagna	33,576	12.3	15,250	16.9	4353	40.2	737	70.7	0	–	53,915	9.3
Toscana	111,862	6.3	33,409	11.5	8643	24.8	361	100.0	0	–	154,275	5.2
Umbria	9046	24.0	12,365	21.5	2240	66.8	0	–	0	–	23,651	15.2
Marche	10,238	26.8	8581	22.8	2495	62.1	0	–	0	–	21,314	16.2
Lazio	55,583	10.3	24,137	15.0	6246	36.6	1946	38.2	0	–	87,912	7.7
Abruzzo	7452	34.9	25,260	14.5	21,827	16.2	8110	29.4	362	100.0	63,011	8.6
Molise	5167	26.6	9785	25.4	4292	38.9	781	70.5	0	–	20,025	16.0
Campania	50,459	10.9	28,229	15.3	8644	20.2	0	–	0	–	87,332	7.6

(continued)

Table 7.12 (continued)

Region/Regione	Other wooded land/Altre terre boscate											
	0–500 m/s.l.m		501–1000 m asl/s.l.m		1001–1500 m asl/s.l.m		1501–2000 m asl/s.l.m		>2000 m asl/s.l.m		Total/Totale	
	Area	ES	Area	ES	Area	ES	Area	ES	Area	ES	Area	ES
	(ha)	(%)	(ha)	(%)	(ha)	(%)	(ha)	(%)	(ha)	(%)	(ha)	(%)
Puglia	41,032	11.3	8357	21.0	0	–	0	–	0	–	49,389	9.9
Basilicata	55,553	8.8	35,486	11.9	13,353	18.9	0	–	0	–	104,392	6.2
Calabria	66,702	8.2	60,135	8.1	24,875	12.0	3731	31.5	0	–	155,443	4.8
Sicilia	53,904	10.7	37,005	12.2	10,078	22.6	758	70.7	0	–	101,745	7.1
Sardegna	515,132	2.5	147,041	5.3	12,678	17.1	0	–	0	–	674,851	2.0
Italia	1,075,191	2.0	495,722	3.1	187,551	5.1	156,098	6.0	54,711	9.9	1,969,272	1.4

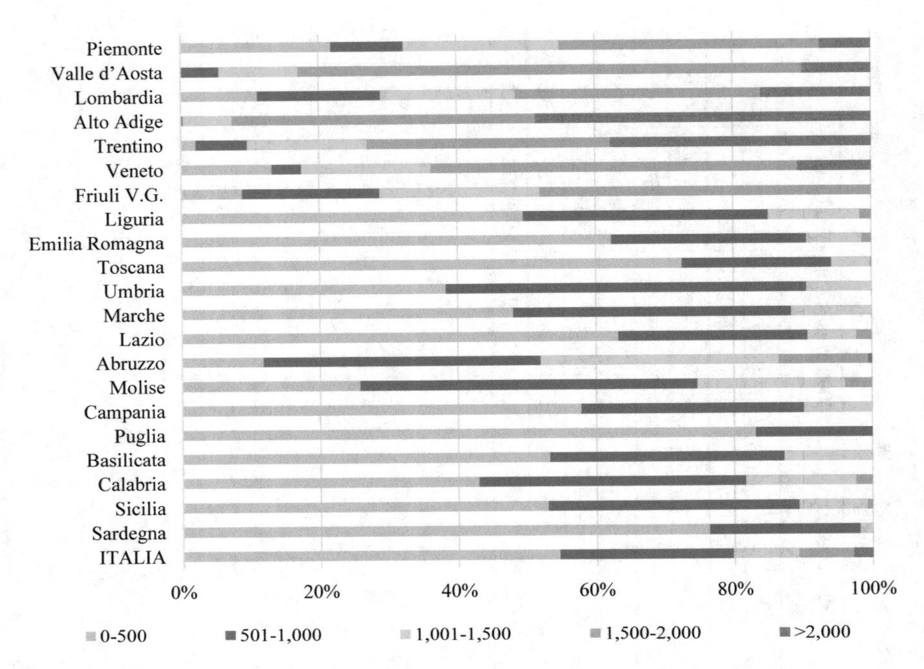

Fig. 7.7 Percent of area of Other wooded land by elevation class (m asl) / Ripartizione percentuale dell'area delle Altre terre boscate per classi di altitudine (m s.l.m.)

7.4 Ownership

Ownership was classified considering two hierarchical levels, the first relative to the character of the property, private or public, and the second to the type of owner, as reported in Table 7.13. For the first level, in case of mixed property, only private or public was indicated, evaluating case by case if the function of the area was mainly of public or private interest.

Tables 7.14 and 7.15 give the area of Forest and Other wooded land by private and public ownership, respectively; Tables 7.16 and 7.17 show the area estimates of Forest and Other wooded land by ownership type. Similar statistics are available at inventarioforestale.org/statistiche_INFC also for the inventory categories of Forest and Other wooded land and for the forest types of Forest.

In Italy, 63.5% of the Total wooded area (Forest and Other wooded land) is private property, 32.0% is public property and 4.5% of the area was not classified for this characteristic. In Forest, the prevalence of private property is even more accentuated (66.4%), but the lower percentage of private forest in Other wooded land (50.2%) might also be due to the high portion of not classified area for this macro-category. A picture of the distribution of private and public ownership of Forest and Other wooded land in Italian regions is shown in Figs. 7.8 and 7.9 respectively. Private property of Forest prevails in almost all regions, except in Trentino, Abruzzo and

Table 7.13 Character and type of ownership / Classi per il carattere e il tipo di proprietà

Ownership character and type		Description
Carattere e tipo di proprietà		Descrizione
Private ownership/ Proprietà privata	Unknown or undefined Tipo non noto o non definito	Private property in general, of unknown or undefined type
		Proprietà privata in genere, di tipo non noto o non definito
	Individual Individuale	Private property of individuals
		Proprietà privata individuale
	Companies, industries Società, imprese, industrie	Private property of companies, industries
		Proprietà privata di società, imprese, industrie
	Other private agencies Altri enti privati	Private property of other private agencies (collective property managed by private agencies, foundations, etc.)
		Proprietà privata di altri enti privati (proprietà collettive gestite da enti di diritto privato, fondazioni, ecc.)
Public ownership/ Proprietà pubblica	Unknown or undefined Tipo non noto o non definito	Public property in general, of unknown or undefined type
		Proprietà pubblica in genere, di tipo non noto o non definito
	State, region Stato, regione	Public property of the State, Regions or Autonomous provinces
		Proprietà pubblica statale o regionale: demanio statale, regionale o di Provincia autonoma
	Municipality, province Comune, provincia	Public property of a Municipality or Province
		Proprietà pubblica comunale o provinciale
	Other public agency Altro ente pubblico	Public property of other public agencies
		Proprietà pubblica di altri enti pubblici

Sicilia, and accounts for more than 80% of the Forest area in the regions Liguria, Emilia Romagna, Toscana, and Marche.

For the macro-category Forest, the most common type of private ownership is individual property (79.0%), which covers more than 90% of the Forest area in Liguria, Molise, Campania, and Basilicata. Private property belonging to companies is significant in Toscana (19.8%) and Umbria (14.9%), while in Trentino and Marche, a good percentage of private Forest belongs to other private agencies (14.9% and 16.3%, respectively) (Fig. 7.10). Concerning public Forest, municipalities and ordinary statute provinces prevail (65.4%), followed by the state, the regions, and the autonomous provinces (together they account for 23.5%), while only 8.3% of the area belongs to other public agencies (Fig. 7.11). The division by type of public

Table 7.14 Forest area by private and public ownership / Estensione del Bosco ripartito per carattere della proprietà

Region/Regione	Forest/Bosco							
	Private ownership		Public ownership		Not classified		Total	
	Proprietà privata		Proprietà pubblica		Non classificato		Totale	
	Area	ES	Area	ES	Area	ES	Area	ES
	(ha)	(%)	(ha)	(%)	(ha)	(%)	(ha)	(%)
Piemonte	645,315	1.9	244,714	3.8	404	100.0	890,433	1.3
Valle d'Aosta	61,900	6.3	37,343	8.5	0	–	99,243	3.6
Lombardia	400,797	2.6	217,483	4.2	3688	57.8	621,968	1.6
Alto Adige	238,778	2.8	99,736	5.6	756	70.5	339,270	1.7
Trentino	106,960	5.2	266,300	2.5	0	–	373,259	1.4
Veneto	280,662	2.9	132,585	5.0	3457	58.0	416,704	1.9
Friuli V.G	196,351	3.4	133,305	4.6	2901	68.1	332,556	1.9
Liguria	297,362	2.1	45,431	8.4	367	99.2	343,160	1.7
Emilia Romagna	497,037	1.9	86,458	6.3	1406	100.0	584,901	1.5
Toscana	878,730	1.3	154,935	4.8	1783	82.2	1,035,448	1.1
Umbria	285,330	2.6	102,171	5.7	2805	69.1	390,305	1.6
Marche	238,127	2.7	53,640	8.1	0	–	291,767	2.1
Lazio	289,970	3.1	269,247	3.1	1019	59.3	560,236	1.6
Abruzzo	184,256	4.1	227,332	3.2	0	–	411,588	1.8
Molise	98,505	5.2	54,743	7.8	0	–	153,248	3.0

(continued)

Table 7.14 (continued)

Region/Regione	Forest/Bosco							
	Private ownership		Public ownership		Not classified		Total	
	Proprietà privata		Proprietà pubblica		Non classificato		Totale	
	Area	ES	Area	ES	Area	ES	Area	ES
	(ha)	(%)	(ha)	(%)	(ha)	(%)	(ha)	(%)
Campania	221,980	3.7	180,166	4.0	1781	82.1	403,927	2.1
Puglia	91,156	6.0	50,067	7.9	1125	57.7	142,349	4.0
Basilicata	177,959	4.2	110,061	5.4	0	–	288,020	2.7
Calabria	289,278	3.2	193,370	4.0	12,529	29.9	495,177	2.0
Sicilia	134,458	5.5	145,322	5.4	5709	49.1	285,489	3.2
Sardegna	413,521	2.9	211,188	4.2	1431	100.0	626,140	2.1
Italia	6,028,434	0.6	3,015,594	1.0	41,158	16.8	9,085,186	0.4

Table 7.15 Other wooded land area by private and public ownership / Estensione delle Altre terre boscate ripartite per carattere della proprietà

Region/Regione	Other wooded land/Altre terre boscate							
	Private ownership		Public ownership		Not classified		Total	
	Proprietà privata		Proprietà pubblica		Non classificato		Totale	
	Area	ES	Area	ES	Area	ES	Area	ES
	(ha)	(%)	(ha)	(%)	(ha)	(%)	(ha)	(%)
Piemonte	27,717	15.2	32,361	13.6	24,913	15.6	84,991	8.0
Valle d'Aosta	4062	41.6	991	56.9	3680	31.2	8733	24.0
Lombardia	18,391	17.6	31,042	13.3	20,819	17.8	70,252	8.7
Alto Adige	17,467	14.5	14,632	18.0	3981	30.1	36,081	10.4
Trentino	1934	45.2	27,930	11.9	3963	29.2	33,826	10.6
Veneto	9895	19.3	16,525	14.9	26,571	14.5	52,991	9.1
Friuli V.G	7761	21.6	17,264	17.3	16,033	20.0	41,058	10.6
Liguria	19,891	16.5	5733	24.7	18,460	17.3	44,084	10.3
Emilia Romagna	32,391	13.3	1842	44.7	19,682	13.1	53,915	9.3
Toscana	69,550	8.5	10,110	22.4	74,615	7.5	154,275	5.2
Umbria	14,333	20.9	3490	31.6	5828	32.2	23,651	15.2
Marche	10,730	25.7	1783	82.1	8800	22.7	21,314	16.2
Lazio	29,208	13.2	38,298	13.0	20,405	18.3	87,912	7.7
Abruzzo	22,011	17.5	28,110	14.6	12,890	17.2	63,011	8.6
Molise	11,727	22.4	5673	33.1	2624	31.8	20,025	16.0

(continued)

Table 7.15 (continued)

Region/Regione	Other wooded land/Altre terre boscate							
	Private ownership		Public ownership		Not classified		Total	
	Proprietà privata		Proprietà pubblica		Non classificato		Totale	
	Area	ES	Area	ES	Area	ES	Area	ES
	(ha)	(%)	(ha)	(%)	(ha)	(%)	(ha)	(%)
Campania	36,644	13.3	22,709	16.2	27,980	14.7	87,332	7.6
Puglia	36,467	12.2	9253	27.1	3669	30.1	49,389	9.9
Basilicata	67,471	8.0	14,472	17.9	22,449	14.7	104,392	6.2
Calabria	44,121	10.1	22,668	12.6	88,654	6.7	155,443	4.8
Sicilia	64,081	8.9	22,358	15.8	15,306	24.3	101,745	7.1
Sardegna	442,743	2.8	201,301	4.2	30,808	12.8	674,851	2.0
Italia	988,595	2.1	528,545	3.0	452,132	3.3	1,969,272	1.4

Table 7.16 Forest area by ownership type / Estensione del Bosco ripartito per tipo di proprietà

Region/Regione	Private Forest/Bosco di proprietà privata									
	Individual		Companies, industries		Other private agencies		Unknown or undefined type		Total	
	Individuale		Società, imprese, industrie		Altri enti privati		Tipo non noto o non definito		Totale	
	Area	ES	Area	ES	Area	ES	Area	ES	Area	ES
	(ha)	(%)	(ha)	(%)	(ha)	(%)	(ha)	(%)	(ha)	(%)
Piemonte	403,219	2.7	17,119	19.3	12,122	18.2	212,856	4.5	645,315	1.9
Valle d'Aosta	18,439	14.9	385	99.6	6,550	23.6	36,525	9.3	61,900	6.3
Lombardia	351,920	2.9	23,151	14.5	13,736	17.8	11,990	21.4	400,797	2.6
Alto Adige	206,062	3.3	1512	49.8	30,826	10.6	378	99.8	238,778	2.8
Trentino	88,193	5.8	2418	61.1	15,988	14.9	360	100.0	106,960	5.2
Veneto	236,143	3.4	3709	31.5	31,889	10.6	8921	24.7	280,662	2.9
Friuli V.G	89,563	6.3	5096	26.5	6690	23.3	95,002	5.6	196,351	3.4
Liguria	269,141	2.4	13,927	15.8	6964	22.6	7330	22.0	297,362	2.1
Emilia Romagna	434,125	2.2	31,660	11.8	26,091	12.6	5161	26.7	497,037	1.9
Toscana	612,092	1.9	174,273	4.4	26,016	11.7	66,350	8.4	878,730	1.3
Umbria	215,963	3.3	42,637	9.7	20,535	14.4	6195	31.1	285,330	2.6
Marche	181,015	3.6	11,958	20.3	38,855	9.3	6299	36.1	238,127	2.7
Lazio	256,605	3.5	11,422	17.8	6238	24.2	15,705	15.1	289,970	3.1
Abruzzo	164,703	4.5	4343	28.8	5067	26.6	10,143	19.5	184,256	4.1
Molise	97,208	5.3	1297	52.8	0	–	0	–	98,505	5.2
Campania	206,079	3.9	4419	28.8	3314	33.3	8168	25.8	221,980	3.7
Puglia	74,723	6.7	9207	24.2	388	99.8	6837	29.6	91,156	6.0
Basilicata	172,037	4.3	2540	61.4	0	0.0	3381	48.8	177,959	4.2
Calabria	253,516	3.6	8209	21.2	2612	37.7	24,942	13.9	289,278	3.2
Sicilia	82,592	7.3	2654	37.7	1137	57.7	48,075	9.9	134,458	5.5
Sardegna	350,086	3.2	7453	22.3	1493	50.0	54,489	9.4	413,521	2.9
Italia	4,763,427	0.8	379,391	3.2	256,510	3.8	629,106	2.6	6,028,434	0.6

(continued)

Table 7.16 (continued)

Region/Regione	Public forest/Bosco di proprietà pubblica									
	State, region		Municipality, province		Other public agency		Unknown or undefined type		Total	
	Stato, regione		Comune, provincia		Altro ente pubblico		Tipo non noto o non definito		Totale	
	Area	ES	Area	ES	Area	ES	Area	ES	Area	ES
	(ha)	(%)	(ha)	(%)	(ha)	(%)	(ha)	(%)	(ha)	(%)
Piemonte	20,484	13.9	203,775	4.1	7316	23.6	13,140	23.3	244,714	3.8
Valle d'Aosta	1541	49.6	32,333	9.4	3083	34.8	385	99.6	37,343	8.5
Lombardia	22,678	14.8	185,108	4.7	5290	28.8	4408	31.6	217,483	4.2
Alto Adige	7691	22.2	48,225	8.8	43,820	8.9	0	–	99,736	5.6
Trentino	13,590	18.2	192,230	3.3	60,480	7.3	0	–	266,300	2.5
Veneto	29,332	11.9	98,818	5.8	2991	35.3	1443	99.9	132,585	5.0
Friuli V.G	40,022	10.1	88,452	5.7	2973	35.2	1858	44.5	133,305	4.6
Liguria	8430	20.5	32,969	10.1	1833	44.3	2199	40.4	45,431	8.4
Emilia Romagna	58,093	8.0	26,154	11.7	1473	50.0	737	70.7	86,458	6.3
Toscana	126,076	5.3	15,176	15.3	5059	26.7	8625	28.5	154,935	4.8
Umbria	32,001	11.0	21,382	12.8	48,051	9.0	737	70.6	102,171	5.7
Marche	29,649	10.7	20,354	13.2	1853	44.7	1783	82.1	53,640	8.1
Lazio	20,265	13.3	211,768	3.8	30,950	10.7	6264	24.1	269,247	3.1
Abruzzo	18,460	13.8	194,100	3.6	7198	28.0	7574	21.7	227,332	3.2
Molise	7420	22.4	46,542	8.8	781	70.5	0	–	54,743	7.8
Campania	11,295	23.4	160,808	4.2	6222	24.2	1841	44.7	180,166	4.0
Puglia	15,497	15.3	31,074	10.4	1554	49.8	1942	44.5	50,067	7.9
Basilicata	19,011	13.7	83,668	6.4	3723	31.5	3659	46.1	110,061	5.4
Calabria	65,203	7.5	102,795	6.0	8582	20.7	16,791	14.7	193,370	4.0
Sicilia	97,705	6.7	37,181	11.8	1517	49.9	8920	28.1	145,322	5.4
Sardegna	65,231	7.6	137,867	5.5	4731	27.8	3359	33.3	211,188	4.2
Italia	709,672	2.3	1,970,780	1.3	249,478	3.8	85,665	8.0	3,015,594	1.0

Table 7.17 Other wooded land area by ownership type / Estensione delle Altre terre boscate ripartite per tipo di proprietà

Region/Regione	Private Other wooded land/Altre terre boscate di proprietà privata									
	Individual		Companies, industries		Other private agencies		Unknown or undefined type		Total	
	Individuale		Società, imprese, industrie		Altri enti privati		Tipo non noto o non definito		Totale	
	Area	ES	Area	ES	Area	ES	Area	ES	Area	ES
	(ha)	(%)	(ha)	(%)	(ha)	(%)	(ha)	(%)	(ha)	(%)
Piemonte	11,000	22.2	2424	40.8	2225	68.7	12,067	24.9	27,717	15.2
Valle d'Aosta	385	99.6	0	–	0	–	3677	44.8	4062	41.6
Lombardia	14,392	20.6	2275	44.8	952	70.9	771	71.5	18,391	17.6
Alto Adige	8494	21.2	0	–	8595	20.9	378	99.8	17,467	14.5
Trentino	360	100.0	0	–	1573	50.6	0	–	1934	45.2
Veneto	4113	30.0	374	99.9	5408	26.3	0	–	9895	19.3
Friuli V.G	4088	29.9	1815	44.6	372	99.7	1487	49.8	7761	21.6
Liguria	15,177	18.7	2148	69.6	733	70.1	1833	44.3	19,891	16.5
Emilia Romagna	26,192	15.0	4056	30.1	2143	70.0	0	–	32,391	13.3
Toscana	41,469	11.2	16,251	14.8	361	100.0	11,468	27.2	69,550	8.5
Umbria	11,456	22.8	369	99.8	1771	81.7	737	70.6	14,333	20.9
Marche	9277	28.8	371	100.0	1082	57.9	0	–	10,730	25.7
Lazio	22,268	14.9	3291	33.3	737	70.7	2913	55.5	29,208	13.2
Abruzzo	20,952	18.2	362	100.0	0	–	697	70.9	22,011	17.5
Molise	11,727	22.4	0	–	0	–	0	–	11,727	22.4
Campania	31,242	14.3	1105	57.8	0	–	4298	49.2	36,644	13.3
Puglia	31,572	13.3	3341	49.9	388	99.8	1165	57.5	36,467	12.2
Basilicata	66,353	8.1	373	99.9	0	–	746	70.6	67,471	8.0
Calabria	35,229	11.2	1119	57.6	373	99.9	7400	27.8	44,121	10.1
Sicilia	32,903	12.2	1485	50.0	1517	49.9	28,177	14.8	64,081	8.9
Sardegna	336,781	3.3	9692	19.6	373	100.0	95,896	6.7	442,743	2.8
Italia	735,432	2.5	50,850	9.2	28,603	13.6	173,710	5.6	988,595	2.1

(continued)

Table 7.17 (continued)

Region/Regione	Public other wooded land/Altre terre boscate di proprietà pubblica									
	State, region		Municipality, province		Other public agency		Unknown or undefined type		Total	
	Stato, regione		Comune, provincia		Altro ente pubblico		Tipo non noto o non definito		Totale	
	Area	ES	Area	ES	Area	ES	Area	ES	Area	ES
	(ha)	(%)	(ha)	(%)	(ha)	(%)	(ha)	(%)	(ha)	(%)
Piemonte	404	100.0	26,652	14.4	404	100.0	4900	44.7	32,361	13.6
Valle d'Aosta	0	–	991	56.9	0	–	0	–	991	56.9
Lombardia	1763	50.0	27,445	14.4	952	70.9	882	70.7	31,042	13.3
Alto Adige	1134	57.6	4915	27.5	7102	22.7	1481	99.8	14,632	18.0
Trentino	5548	25.8	18,658	15.1	3723	30.7	0	–	27,930	11.9
Veneto	4078	30.1	12,073	17.5	374	99.9	0	–	16,525	14.9
Friuli V.G	2602	37.6	13,919	20.1	372	99.7	372	99.7	17,264	17.3
Liguria	2068	40.8	3665	31.3	0	–	0	–	5733	24.7
Emilia Romagna	1473	50.0	368	100.0	0	–	0	–	1842	44.7
Toscana	7243	22.3	723	70.7	361	100.0	1783	82.2	10,110	22.4
Umbria	369	99.8	1008	58.2	2114	40.9	0	–	3490	31.6
Marche	371	100.0	1412	100.0	0	–	0	–	1783	82.1
Lazio	3316	33.3	29,112	16.0	4421	28.8	1448	50.0	38,298	13.0
Abruzzo	3190	50.8	23,513	15.6	1407	100.0	0	–	28,110	14.6
Molise	516	79.3	4767	37.6	0	–	391	99.8	5673	33.1
Campania	1412	100.0	18,719	17.7	1473	50.0	1105	57.8	22,709	16.2
Puglia	2993	54.5	5872	32.7	388	99.8	0	–	9253	27.1
Basilicata	2983	35.2	11,116	21.2	373	99.9	0	–	14,472	17.9
Calabria	3731	31.5	11,567	17.8	0	–	7370	22.3	22,668	12.6
Sicilia	14,164	20.2	5250	26.6	1137	57.7	1806	81.8	22,358	15.8
Sardegna	49,266	8.9	138,974	5.2	2239	40.8	10,822	18.5	201,301	4.2
Italia	108,624	6.4	360,720	3.7	26,841	12.6	32,360	13.7	528,545	3.0

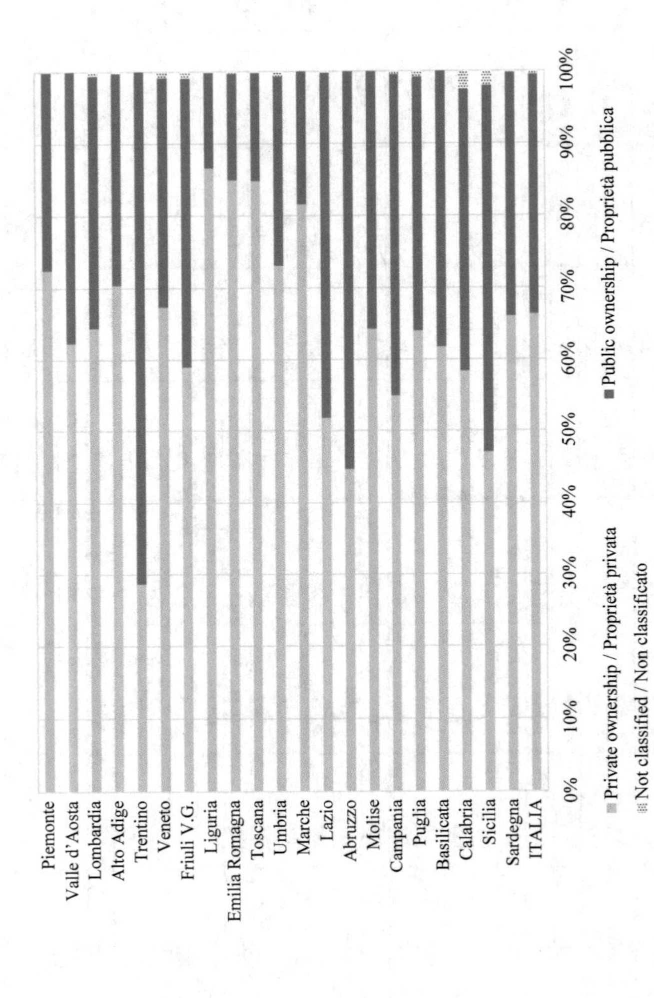

Fig. 7.8 Percent of area of Forest by private and public ownership / Ripartizione percentuale della superficie del Bosco per proprietà privata e pubblica

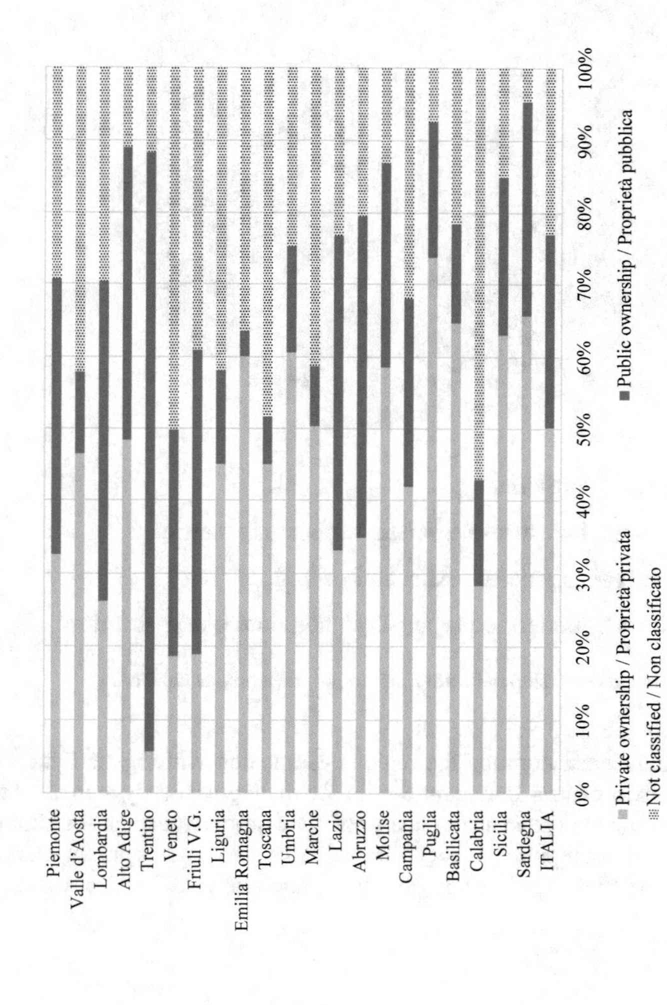

Fig. 7.9 Percent of area of Other wooded land by private and public ownership / Ripartizione percentuale della superficie del Bosco per proprietà privata e pubblica

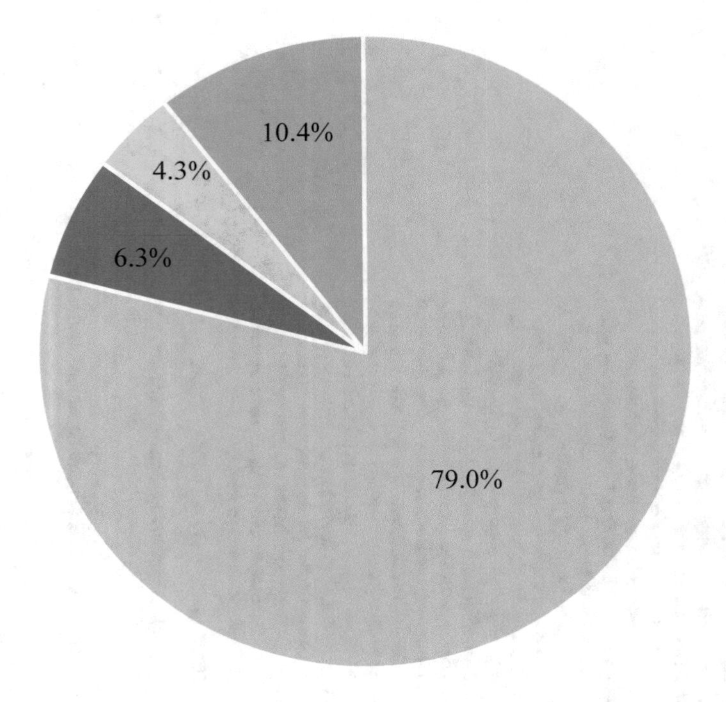

Fig. 7.10 Types of private ownership in Forest / Tipi di proprietà privata del Bosco

property varies considerably among the regions. Compared with the national situation, many regions in central Italy, and Sicilia are distinguished by a smaller rate of municipal and provincial property in favour of state property, except for Umbria, where the property of other public agencies prevails. The class other public agencies shows a considerably higher proportion than at the national level in Alto Adige and in Trentino.

7.5 Growing Stock and Increment

Table 7.18 shows the estimates on the number of living trees in the inventory categories of Forest; Tables 7.19 and 7.20 show estimates related to the forest types of Tall trees forest and Plantation. In Italian Forests there are almost 11.5 billion trees, 1264

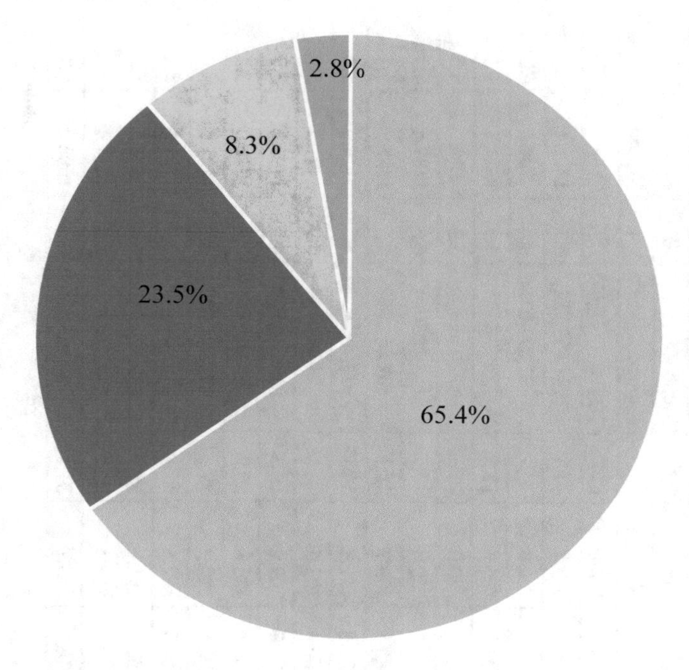

Fig. 7.11 Types of public ownership in Forest / Tipi di proprietà pubblica del Bosco

per hectare on average. Trees number per hectare ranges between 1000 and 1500 in twelve regions; in four regions (Umbria, Marche, Lazio and Molise), there are more than 1500 trees per hectare (up to 1728 in Umbria), while in the remaining five regions (Valle d'Aosta, Alto Adige, Puglia, Basilicata and Sicilia), such a density value is lower and ranges between 578 (Sicilia) and 986 (Basilicata). The values by forest type indicate a predominant contribution of the broadleaved stands, with exceptionally high values in Hornbeam and Hophornbeam forests (2160 trees per hectare) and Holm oak forests (1937 trees per hectare). To some extent, such high densities can be explained by the silviculture applied in these two forest types that are mainly coppices for firewood production. In fact, to the contrary, the lowest tree number per hectare was found in the Cork oak forests (575 trees per hectare), in which low density and relevant tree size are pursued to ensure crown complete exposure to sunlight and obtaining planks of cork of appropriate size and thickness (Gambi, 1989).

Table 7.18 Total number and number per hectare of living trees by Forest inventory category / Valori totali e per ettaro del numero di alberi per le categorie inventariali del Bosco

Region/Regione	Tall trees forest				Plantations				Total Forest			
	Boschi alti				Impianti di arboricoltura da legno				Totale Bosco			
	Number	ES	Number	ES	Number	ES	Number	ES	Number	ES	Number	ES
	Numero		Numero		Numero		Numero		Numero		Numero	
	(n)	(%)	(n ha^{-1})	(%)	(n)	(%)	(n ha^{-1})	(%)	(n)	(%)	(n ha^{-1})	(%)
Piemonte	989,883,210	3.3	1138.1	3.0	8,670,542	46.5	419.7	44.5	998,553,752	3.3	1121.4	3.0
Valle d'Aosta	72,879,071	8.0	734.4	6.8	0	–	0.0	–	72,879,071	8.0	734.4	6.8
Lombardia	743,105,421	4.6	1245.1	4.2	10,286,002	22.1	409.3	17.3	753,391,423	4.5	1211.3	4.2
Alto Adige	277,716,569	5.9	818.6	5.6	0	–	0.0	–	277,716,569	5.9	818.6	5.6
Trentino	443,222,084	6.6	1187.4	6.3	0	–	0.0	–	443,222,084	6.6	1187.4	6.3
Veneto	580,720,372	5.8	1411.5	5.4	2,140,889	45.1	405.7	28.2	582,861,261	5.8	1398.7	5.4
Friuli V.G	366,977,868	5.8	1134.9	5.4	3,787,566	27.0	411.9	19.3	370,765,434	5.7	1114.9	5.4
Liguria	494,595,827	5.9	1442.8	5.8	75,357	100.0	205.6	–	494,671,184	5.9	1441.5	5.8
Emilia Romagna	870,945,451	4.2	1504.6	3.8	1,265,138	27.7	209.2	18.7	872,210,589	4.2	1491.2	3.8
Toscana	1,359,690,069	3.4	1321.8	3.2	3,010,583	38.4	443.8	23.9	1,362,700,652	3.4	1316.0	3.2
Umbria	672,507,830	5.3	1751.6	4.9	1,786,854	32.6	280.2	17.6	674,294,685	5.2	1727.6	4.9
Marche	500,069,577	5.9	1755.2	5.4	2,455,235	50.5	357.8	35.1	502,524,812	5.9	1722.3	5.4
Lazio	879,406,902	5.4	1575.8	5.0	417,636	50.8	191.9	35.9	879,824,539	5.4	1570.5	5.0
Abruzzo	580,998,633	4.8	1421.9	4.4	1,142,353	59.2	384.4	46.2	582,140,986	4.8	1414.4	4.4
Molise	244,620,478	7.1	1625.0	6.4	1,834,648	46.2	675.7	31.1	246,455,126	7.1	1608.2	6.4

(continued)

Table 7.18 (continued)

Region/Regione	Tall trees forest				Plantations				Total Forest			
	Boschi alti				Impianti di arboricoltura da legno				Totale Bosco			
	Number	ES	Number	ES	Number	ES	Number	ES	Number	ES	Number	ES
	Numero		Numero		Numero		Numero		Numero		Numero	
	(n)	(%)	(n ha^{-1})	(%)	(n)	(%)	(n ha^{-1})	(%)	(n)	(%)	(n ha^{-1})	(%)
Campania	581,362,702	5.8	1450.6	5.4	961,228	59.1	303.9	49.2	582,323,929	5.8	1441.7	5.4
Puglia	132,234,988	7.7	929.6	6.0	34,962	100.0	349.0	–	132,269,951	7.7	929.2	6.0
Basilicata	284,004,504	6.2	991.3	5.6	48,971	100.0	32.2	–	284,053,475	6.2	986.2	5.7
Calabria	513,562,237	8.4	1042.2	8.2	1,215,818	49.2	505.3	28.7	514,778,055	8.4	1039.6	8.2
Sicilia	165,065,720	8.8	579.7	8.1	28,566	70.7	37.7	–	165,094,286	8.8	578.3	8.1
Sardegna	674,513,829	6.2	1123.7	5.9	14,317,219	22.2	553.1	20.7	688,831,047	6.1	1100.1	5.8
Italia	11,428,083,343	1.2	1275.9	1.2	53,479,566	11.6	416.5	10.4	11,481,562,909	1.2	1263.8	1.2

Table 7.19 Total number and number per hectare of living trees by Tall trees forest type / Valori totali e per ettaro del numero di alberi per le categorie forestali dei Boschi alti

Region/Regione	Larch and Swiss stone pine				Norway spruce				Fir			
	Boschi di larice e cembro				Boschi di abete rosso				Boschi di abete bianco			
	Number	ES	Number	ES	Number	ES	Number	ES	Number	ES	Number	ES
	Numero		Numero		Numero		Numero		Numero		Numero	
	(n)	(%)	(n ha^{-1})	(%)	(n)	(%)	(n ha^{-1})	(%)	(n)	(%)	(n ha^{-1})	(%)
Piemonte	32,450,757	13.0	405.3	–	11,829,452	19.3	650.6	–	10,690,514	18.8	755.9	–
Valle d'Aosta	20,792,296	16.5	436.4	–	14,611,717	18.5	801.5	–	976,977	61.1	845.2	–
Lombardia	39,367,924	15.7	736.9	–	79,471,216	14.7	904.5	–	10,646,074	34.6	1130.7	–
Alto Adige	49,046,006	11.4	523.7	–	130,910,883	8.3	734.1	–	1,532,135	43.7	578.9	–
Trentino	32,829,298	18.4	527.9	–	90,582,698	9.9	661.9	–	12,816,077	22.3	790.3	–
Veneto	48,902,663	31.3	1154.4	–	88,444,725	12.6	910.5	–	7,164,696	30.2	1008.6	–
Friuli V.G	7,201,028	26.7	555.2	–	40,054,456	21.0	898.1	–	1,740,183	49.5	936.5	–
Liguria	827,644	60.7	752.7	–	181,970	100.0	496.5	–	3,827,229	42.3	1305.3	–
Emilia Romagna	0	–	0.0	–	3,110,804	34.1	844.5	–	2,126,526	43.5	721.6	–
Toscana	0	–	0.0	–	442,208	100.0	1223.8	–	3,718,876	32.6	735.2	–
Umbria	0	–	0.0	–	0	–	0.0	–	0	–	0.0	–
Marche	0	–	0.0	–	300,864	100.0	811.8	–	0	–	0.0	–
Lazio	0	–	0.0	–	201,176	100.0	546.0	–	0	–	0.0	–
Abruzzo	0	–	0.0	–	427,118	100.0	1180.1	–	639,610	70.8	883.6	–
Molise	0	–	0.0	–	0	–	0.0	–	897,411	63.4	766.0	–
Campania	0	–	0.0	–	0	–	0.0	–	0	–	0.0	–
Puglia	0	–	0.0	–	0	–	0.0	–	0	–	0.0	–
Basilicata	0	–	0.0	–	0	–	0.0	–	621,609	72.4	833.6	–
Calabria	0	–	0.0	–	0	–	0.0	–	3,838,029	40.6	1028.6	–
Sicilia	0	–	0.0	–	0	–	0.0	–	0	–	0.0	–
Sardegna	0	–	0.0	–	0	–	0.0	–	0	–	0.0	–
Italia	231,417,616	8.4	588.3	7.8	460,569,286	5.1	785.0	4.6	61,235,946	10.3	876.7	6.5

(continued)

Table 7.19 (continued)

Region/Regione	Scots pine and Mountain pine				Black pines				Mediterranean pines			
	Pinete di pino silvestre e montano				Pinete di pino nero, laricio e loricato				Pinete di pini mediterranei			
	Number	ES	Number	ES	Number	ES	Number	ES	Number	ES	Number	ES
	Numero		Numero		Numero		Numero		Numero		Numero	
	(n)	(%)	(n ha^{-1})	(%)	(n)	(%)	(n ha^{-1})	(%)	(n)	(%)	(n ha^{-1})	(%)
Piemonte	17,704,832	22.2	852.9	–	3,617,497	41.3	1119.1	–	893,902	80.6	1106.2	–
Valle d'Aosta	12,230,172	23.9	1047.9	–	387,266	100.0	1005.1	–	0	–	0.0	–
Lombardia	30,223,557	22.7	1587.1	–	6,607,608	35.4	1499.0	–	0	–	0.0	–
Alto Adige	43,358,979	16.7	1170.2	–	42,635	100.0	112.8	–	0	–	0.0	–
Trentino	35,629,888	17.6	1704.7	–	9,315,590	27.0	1615.7	–	0	–	0.0	–
Veneto	15,357,189	24.7	1253.6	–	2,479,726	38.3	706.9	–	588,984	70.9	787.6	–
Friuli V.G	12,981,963	27.9	1061.7	–	45,338,833	20.1	1436.9	–	507,544	60.5	455.2	–
Liguria	6,495,842	24.6	633.0	–	3,402,081	31.6	618.8	–	23,370,133	24.3	952.1	–
Emilia Romagna	2,580,040	61.9	1400.8	–	23,518,714	25.0	1451.0	–	1,331,466	47.7	451.8	–
Toscana	887,818	57.7	819.0	–	16,650,565	20.4	921.6	–	37,437,551	19.4	829.0	–
Umbria	735,661	76.3	997.8	–	7,320,215	27.3	1241.0	–	11,173,336	27.6	1377.7	–
Marche	0	–	0.0	–	9,969,021	27.5	960.7	–	2,450,073	45.3	1116.7	–
Lazio	0	–	0.0	–	6,222,028	25.3	734.2	–	5,681,725	32.8	773.7	–
Abruzzo	1,451,604	60.7	1336.9	–	24,352,962	20.9	1323.1	–	1,245,085	53.1	573.3	–
Molise	0	–	0.0	–	1,723,708	43.2	735.7	–	743,544	63.2	340.5	–
Campania	0	–	0.0	–	2,631,272	36.3	476.3	–	5,472,271	39.3	598.3	–
Puglia	0	–	0.0	–	1,325,839	50.8	853.3	–	16,188,812	24.0	584.1	–
Basilicata	0	–	0.0	–	1,904,482	39.1	729.7	–	5,781,651	24.7	647.2	–
Calabria	0	–	0.0	–	45,812,903	14.8	623.8	–	12,766,398	31.0	730.6	–
Sicilia	0	–	0.0	–	5,026,154	39.0	670.8	–	15,534,991	21.3	342.7	–
Sardegna	0	–	0.0	–	3,485,939	35.7	667.2	–	21,487,427	16.8	620.4	–
Italia	179,637,544	7.8	1206.3	6.0	221,135,039	7.0	960.0	5.8	162,654,893	7.8	676.2	6.7

(continued)

Table 7.19 (continued)

	Other coniferous forests				Beech				Temperate oaks			
	Altri boschi di conifere, pure o miste				Faggete				Querceti di rovere, roverella e farnia			
	Number	ES	Number	ES	Number	ES	Number	ES	Number	ES	Number	ES
	Numero		Numero		Numero		Numero		Numero		Numero	
	(n)	(%)	(n ha^{-1})	(%)	(n)	(%)	(n ha^{-1})	(%)	(n)	(%)	(n ha^{-1})	(%)
Piemonte	5,971,359	34.2	923.7	–	129,277,246	8.8	1103.7	–	78,927,715	11.8	1036.1	–
Valle d'Aosta	0	–	0.0	–	1,857,251	62.4	1606.7	–	5,071,816	34.5	1196.6	–
Lombardia	3,060,853	38.6	482.2	–	66,433,629	12.8	996.7	–	67,965,718	19.6	1653.9	–
Alto Adige	0	–	0.0	–	4,620,411	38.0	1110.9	–	13,462,120	41.4	2967.1	–
Trentino	0	–	0.0	–	82,046,127	14.5	1310.5	–	8,039,011	35.3	1394.2	–
Veneto	0	–	0.0	–	77,401,629	12.4	1144.3	–	26,304,611	19.4	1680.7	–
Friuli V.G	693,018	74.5	621.6	–	82,042,038	12.8	905.1	–	10,941,949	25.7	1338.3	–
Liguria	191,818	100.0	523.4	–	50,657,865	18.0	1368.5	–	58,394,478	19.6	1331.3	–
Emilia Romagna	3,726,194	37.2	1123.9	–	213,980,586	10.4	2087.5	–	99,954,542	13.0	1298.2	–
Toscana	10,323,795	26.5	867.5	–	81,295,572	13.0	1108.7	–	167,751,896	9.7	1102.9	–
Umbria	1,821,822	47.6	988.4	–	31,486,399	24.4	1906.3	–	129,279,202	10.6	1281.4	–
Marche	4,799,037	39.5	848.3	–	30,381,222	20.3	1707.8	–	97,505,101	13.3	1579.0	–
Lazio	1,326,288	59.5	899.9	–	116,342,719	15.8	1563.1	–	116,446,784	15.4	1424.2	–
Abruzzo	981,817	58.5	452.1	–	190,082,347	9.5	1527.2	–	122,458,169	10.9	1395.4	–
Molise	0	–	0.0	–	14,657,454	27.8	987.8	–	81,959,708	15.1	1700.5	–
Campania	136,593	60.9	158.2	–	59,827,552	15.3	1063.7	–	67,243,894	14.4	1103.6	–
Puglia	1,910,852	45.6	819.9	–	3,676,751	33.3	788.8	–	29,883,791	19.9	1155.4	–
Basilicata	3,169,870	39.2	615.5	–	32,491,812	25.0	1211.5	–	41,160,583	17.6	955.5	–
Calabria	17,015,352	47.9	1354.6	–	100,737,450	13.6	1268.5	–	30,143,514	23.2	560.4	–
Sicilia	3,011,739	30.3	323.9	–	23,005,249	30.4	1517.0	–	31,756,196	17.3	441.8	–
Sardegna	9,421,777	25.7	721.7	–	0	–	0.0	–	47,993,469	17.2	546.7	–
Italia	67,562,184	14.5	805.1	11.2	1,392,301,309	3.5	1322.0	3.1	1,332,644,267	3.6	1156.5	3.1

(continued)

Table 7.19 (continued)

Region/Regione	Mediterranean oaks				Chestnut				Hornbeam and Hophornbeam			
	Cerrete, boschi di farnetto, fragno e vallonea				Castagneti				Ostrieti, carpineti			
	Number	ES	Number	ES	Number	ES	Number	ES	Number	ES	Number	ES
	Numero		Numero		Numero		Numero		Numero		Numero	
	(n)	(%)	(n ha^{-1})	(%)	(n)	(%)	(n ha^{-1})	(%)	(n)	(%)	(n ha^{-1})	(%)
Piemonte	7,402,382	33.0	1221.3	–	205,736,587	8.9	1257.3	–	36,131,447	25.0	2235.6	–
Valle d'Aosta	0	–	0.0	–	4,440,863	35.3	1047.8	–	0	–	0.0	–
Lombardia	2,576,895	51.1	1169.2	–	88,217,504	12.9	1074.8	–	157,701,498	13.2	1929.8	–
Alto Adige	0	–	0.0	–	1,307,189	54.9	691.5	–	23,332,576	34.4	2643.8	–
Trentino	0	–	0.0	–	2,673,662	65.5	1854.8	–	114,376,317	17.3	2800.5	–
Veneto	0	–	0.0	–	18,647,599	18.0	1084.2	–	204,335,405	12.5	2491.1	–
Friuli V.G	0	–	0.0	–	14,274,780	20.8	1163.9	–	88,905,821	14.6	1900.1	–
Liguria	15,644,731	22.9	1391.0	–	184,671,849	12.1	1685.2	–	80,835,441	14.8	1748.5	–
Emilia Romagna	127,430,756	10.6	1258.4	–	43,555,473	14.6	1074.9	–	210,946,748	9.3	1896.1	–
Toscana	335,056,339	6.7	1337.2	–	181,888,813	8.0	1164.8	–	122,446,408	13.6	1893.9	–
Umbria	229,814,718	10.7	1787.1	–	2,736,242	54.1	1237.0	–	124,400,100	13.3	2125.9	–
Marche	38,364,520	22.0	1551.2	–	4,404,503	34.1	1188.4	–	196,078,981	11.9	2570.3	–
Lazio	193,110,049	9.8	1458.1	–	47,158,777	22.8	1317.6	–	230,477,804	12.5	2352.4	–
Abruzzo	44,459,304	18.0	1383.1	–	9,427,103	32.3	2003.6	–	80,661,166	14.5	1683.1	–
Molise	90,840,221	12.9	1747.8	–	488,621	100.0	1251.3	–	26,907,855	23.3	2995.9	–
Campania	102,054,597	14.2	1367.2	–	62,926,120	17.9	1124.0	–	124,931,771	14.7	2355.9	–
Puglia	38,794,192	15.0	1001.7	–	811,478	61.0	696.4	–	5,730,364	44.3	1229.4	–
Basilicata	133,838,989	9.6	1024.1	–	6,997,782	29.6	1175.0	–	18,248,958	28.5	2575.9	–
Calabria	50,236,248	19.6	1152.4	–	71,584,179	19.2	1038.0	–	7,899,134	92.2	1764.2	–
Sicilia	13,842,093	18.5	446.7	–	6,790,342	29.5	778.7	–	1,215,825	76.4	699.8	–
Sardegna	0	–	0.0	–	1,546,870	52.0	829.0	–	0	–	0.0	–
Italia	1,423,466,034	3.5	1343.0	3.1	960,286,333	4.3	1233.5	3.8	1,855,563,618	3.8	2159.7	3.2

(continued)

Table 7.19 (continued)

Region/Regione	Hygrophilous forests				Other deciduous broadleaved forests				Holm oak			
	Boschi igrofili				Altri boschi caducifogli				Leccete			
	Number	ES	Number	ES	Number	ES	Number	ES	Number	ES	Number	ES
	Numero		Numero		Numero		Numero		Numero		Numero	
	(n)	(%)	(n ha^{-1})	(%)	(n)	(%)	(n ha^{-1})	(%)	(n)	(%)	(n ha^{-1})	(%)
Piemonte	36,379,135	23.4	1600.0	–	412,379,525	5.7	1273.9	–	0	–	0.0	–
Valle d'Aosta	0	–	0.0	–	12,510,715	25.1	1188.8	–	0	–	0.0	–
Lombardia	19,673,459	26.4	1044.0	–	167,629,197	10.6	1385.2	–	0	–	0.0	–
Alto Adige	3,685,081	47.1	1341.6	–	5,791,459	38.4	1639.7	–	0	–	0.0	–
Trentino	6,283,894	54.0	1628.1	–	47,706,181	32.0	2896.1	–	923,340	100.0	2562.2	–
Veneto	6,750,461	33.0	648.1	–	75,500,077	15.0	1441.1	–	8,842,609	39.5	2956.3	–
Friuli V.G	5,949,515	26.7	472.2	–	56,346,740	14.1	1187.6	–	0	–	0.0	–
Liguria	2,919,019	35.3	884.9	–	41,415,007	19.6	1300.9	–	21,678,511	22.0	1558.4	–
Emilia Romagna	17,771,089	23.9	629.7	–	120,172,568	13.1	1390.7	–	739,944	70.9	1004.4	–
Toscana	30,823,572	20.3	1110.7	–	108,226,675	13.0	1280.9	–	250,529,826	11.0	1931.8	–
Umbria	7,471,173	31.6	929.5	–	15,399,365	27.1	1364.7	–	110,869,597	16.2	2685.2	–
Marche	13,156,208	25.0	1019.7	–	74,113,619	16.0	1193.6	–	28,546,429	25.8	4053.9	–
Lazio	8,074,237	41.0	876.6	–	73,493,141	18.3	1360.4	–	73,452,128	24.4	1521.8	–
Abruzzo	19,093,153	20.9	961.1	–	60,431,413	18.8	1072.0	–	24,309,298	27.8	2798.5	–
Molise	7,058,443	29.6	779.6	–	16,632,012	31.5	1626.6	–	2,711,502	97.3	2314.5	–
Campania	9,333,201	58.1	844.8	–	40,426,618	22.3	1175.7	–	102,983,776	17.7	2747.3	–
Puglia	1,016,401	100.0	2616.7	–	11,797,458	29.8	1022.5	–	12,280,110	24.4	707.2	–
Basilicata	4,617,171	37.0	355.8	–	22,689,439	17.5	759.6	–	9,789,828	26.0	944.0	–
Calabria	3,947,789	35.5	306.7	–	47,121,960	21.6	1043.6	–	101,276,960	33.3	2080.0	–
Sicilia	778,971	54.3	158.1	–	8,685,465	31.6	656.1	–	31,627,064	30.3	1618.9	–
Sardegna	1,521,814	40.9	453.1	–	1,352,136	37.2	174.6	–	464,976,394	8.2	1820.1	–
Italia	206,303,785	7.9	877.4	6.5	1,419,820,769	3.5	1274.8	3.0	1,245,537,316	5.4	1936.9	5.0

(continued)

Table 7.19 (continued)

Region/Regione	Cork oak				Other evergreen broadleaved forests				Not classified				Total Tall trees forest			
	Sugherete				Altri boschi di latifoglie sempreverdi				Non classificato				Totale Boschi alti			
	Number	ES	Number	ES	Number	ES	Number	ES	Number	ES	Number	ES	Number	ES	Number	ES
	Numero		Numero		Numero		Numero		Numero		Numero		Numero		Numero	
	(n)	(%)	(n ha^{-1})	(%)	(n)	(%)	(n ha^{-1})	(%)	(n)	(%)	(n ha^{-1})	(%)	(n)	(%)	(n ha^{-1})	(%)
Piemonte	0	–	0.0	–	0	–	0.0	–	490,860	100.0	966.0	–	989,883,210	3.3	1138.1	3.0
Valle d'Aosta	0	–	0.0	–	0	–	0.0	–	0	–	0.0	–	72,879,071	8.0	734.4	6.8
Lombardia	0	–	0.0	–	2,586,344	64.8	1128.3	–	943,944	100.0	2141.4	–	743,105,421	4.6	1245.1	4.2
Alto Adige	0	–	0.0	–	0	–	0.0	–	627,095	77.4	414.6	–	277,716,569	5.9	818.6	5.6
Trentino	0	–	0.0	–	0	–	0.0	–	0	–	0.0	–	443,222,084	6.6	1187.4	6.3
Veneto	0	–	0.0	–	0	–	0.0	–	0	–	0.0	–	580,720,372	5.8	1411.5	5.4
Friuli V.G	0	–	0.0	–	0	–	0.0	–	0	–	0.0	–	366,977,868	5.8	1134.9	5.4
Liguria	0	–	0.0	–	0	–	0.0	–	82,208	70.7	112.1	–	494,595,827	5.9	1442.8	5.8
Emilia Romagna	0	–	0.0	–	0	–	0.0	–	0	–	0.0	–	870,945,451	4.2	1504.6	3.8
Toscana	7,002,873	37.6	1140.0	–	2,280,206	71.9	1577.6	–	2,927,077	70.7	4050.4	–	1,359,690,069	3.4	1321.8	3.2
Umbria	0	–	0.0	–	0	–	0.0	–	0	–	0.0	–	672,507,830	5.3	1751.6	4.9
Marche	0	–	0.0	–	0	–	0.0	–	0	–	0.0	–	500,069,577	5.9	1755.2	5.4
Lazio	1,496,680	43.9	580.3	–	998,509	87.6	387.1	–	4,924,857	53.2	3687.8	–	879,406,902	5.4	1575.8	5.0
Abruzzo	0	–	0.0	–	690,042	100.0	490.4	–	288,443	100.0	796.9	–	580,998,633	4.8	1421.9	4.4
Molise	0	–	0.0	–	0	–	0.0	–	0	–	0.0	–	244,620,478	7.1	1625.0	6.4
Campania	1,664,301	100.0	4519.4	–	236,207	70.7	320.7	–	1,494,528	100.0	4058.3	–	581,362,702	5.8	1450.6	5.4
Puglia	0	–	0.0	–	5,340,647	30.8	1037.1	–	3,478,294	69.9	3090.5	–	132,234,988	7.7	929.6	6.0
Basilicata	0	–	0.0	–	1,132,889	87.8	607.7	–	1,559,440	100.0	4490.8	–	284,004,504	6.2	991.3	5.6
Calabria	3,758,735	42.1	719.5	–	7,506,898	27.7	355.1	–	9,916,690	40.8	4429.6	–	513,562,237	8.4	1042.2	8.2
Sicilia	5,531,974	29.1	320.5	–	18,259,659	24.3	466.4	–	0	–	0.0	–	165,065,720	8.8	579.7	8.1
Sardegna	86,570,941	13.9	566.7	–	36,114,864	27.5	950.3	–	42,199	100.0	113.1	–	674,513,829	6.2	1123.7	5.9
Italia	106,025,505	11.9	575.2	11.2	75,146,265	15.3	660.5	13.3	26,775,635	23.5	2659.7	13.6	11,428,083,343	1.2	1275.9	1.2

Table 7.20 Total number and number per hectare of living trees by Plantations forest type / Valori totali e per ettaro del numero di alberi per le categorie forestali degli Impianti di arboricoltura da legno

Region/Regione	Poplar plantations				Other broadleaved plantations				Coniferous plantations				Total plantations			
	Pioppeti artificiali				Piantagioni di altre latifoglie				Piantagioni di conifere				Totale impianti di arboricoltura da legno			
	Number	ES	Number	ES	Number	ES	Number	ES	Number	ES	Number	ES	Number	ES	Number	ES
	Numero		Numero		Numero		Numero		Numero		Numero		Numero		Numero	
	(n)	(%)	(n ha^{-1})	(%)	(n)	(%)	(n ha^{-1})	(%)	(n)	(%)	(n ha^{-1})	(%)	(n)	(%)	(n ha^{-1})	(%)
Piemonte	6,169,797	63.6	397.8	–	1,680,935	47.3	420.3	–	819,810	64.8	712.4	–	8,670,542	46.5	419.7	44.5
Valle d'Aosta	0	–	0.0	–	0	–	0.0	–	0	–	0.0	–	0	–	0.0	–
Lombardia	6,447,857	24.1	348.4	–	3,588,919	50.2	580.4	–	249,226	100.0	565.4	–	10,286,002	22.1	409.3	17.3
Alto Adige	0	–	0.0	–	0	–	0.0	–	0	–	0.0	–	0	–	0.0	–
Trentino	0	–	0.0	–	0	–	0.0	–	0	–	0.0	–	0	–	0.0	–
Veneto	638,995	62.2	274.0	–	1,501,894	66.3	509.9	–	0	–	0.0	–	2,140,889	45.1	405.7	28.2
Friuli V.G	1,722,621	29.4	263.4	–	2,064,945	47.1	777.9	–	0	–	0.0	–	3,787,566	27.0	411.9	19.3
Liguria	75,357	100.0	205.6	–	0	–	0.0	–	0	–	0.0	–	75,357	100.0	205.6	–
Emilia Romagna	798,512	33.0	187.8	–	293,032	54.2	205.2	–	173,594	100.0	471.2	–	1,265,138	27.7	209.2	18.7
Toscana	298,044	71.2	404.8	–	1,832,201	55.5	398.8	–	880,338	58.9	606.2	–	3,010,583	38.4	443.8	23.9
Umbria	0	–	0.0	–	1,786,854	32.6	280.2	–	0	–	0.0	–	1,786,854	32.6	280.2	17.6
Marche	0	–	0.0	–	2,455,235	50.5	357.8	–	0	–	0.0	–	2,455,235	50.5	357.8	35.1
Lazio	44,543	70.9	56.5	–	373,093	56.8	269.0	–	0	–	0.0	–	417,636	50.8	191.9	35.9
Abruzzo	56,505	100.0	106.9	–	1,085,848	62.1	444.5	–	0	–	0.0	–	1,142,353	59.2	384.4	46.2
Molise	718,020	100.0	1838.7	–	1,116,627	40.2	480.4	–	0	–	0.0	–	1,834,648	46.2	675.7	31.1

(continued)

Table 7.20 (continued)

Region/Regione	Poplar plantations				Other broadleaved plantations				Coniferous plantations				Total plantations			
	Pioppeti artificiali				Piantagioni di altre latifoglie				Piantagioni di conifere				Totale impianti di arboricoltura da legno			
	Number	ES	Number	ES	Number	ES	Number	ES	Number	ES	Number	ES	Number	ES	Number	ES
	Numero		Numero		Numero		Numero		Numero		Numero		Numero		Numero	
	(n)	(%)	(n ha^{-1})	(%)	(n)	(%)	(n ha^{-1})	(%)	(n)	(%)	(n ha^{-1})	(%)	(n)	(%)	(n ha^{-1})	(%)
Campania	253,900	85.5	164.9	–	707,327	74.3	435.6	–	0	–	0.0	–	961,228	59.1	303.9	49.2
Puglia	0	–	0.0	–	34,962	100.0	349.0	–	0	–	0.0	–	34,962	100.0	349.0	–
Basilicata	0	–	0.0	–	48,971	100.0	32.2	–	0	–	0.0	–	48,971	100.0	32.2	–
Calabria	26,369	100.0	263.4	–	964,837	59.8	618.5	–	224,612	70.7	301.0	–	1,215,818	49.2	505.3	28.7
Sicilia	0	–	0.0	–	28,566	70.7	37.7	–	0	–	0.0	–	28,566	70.7	37.7	–
Sardegna	0	–	0.0	–	14,056,110	22.6	576.3	–	261,108	77.0	174.9	–	14,317,219	22.2	553.1	20.7
Italia	17,250,521	25.2	334.4	24.0	33,620,357	13.3	472.5	10.9	2,608,688	32.3	461.6	19.0	53,479,566	11.6	416.5	10.4

The number of trees, either total or per hectare, as an index of growing stock has limited value because size of trees may vary a lot. As a stand density index, tree numbers per hectare have little utility in natural stands and have been more frequently used in plantations (Avery & Burkhart, 1983) or in young stands (Bernetti, 1995). The number of trees per unit area is necessary but not sufficient to adequately describe stand density (Burkhart & Tomé, 2012), and must be complemented by further information, such as tree size, relative distance or stand structure (Avery, 1967).

Table 7.21 shows estimates on the basal area for the inventory categories of Forest. Table 7.22 shows the same statistics for the Tall trees forest types and Table 7.23 for the Plantations forest types. Basal area is more appropriate to give indications on the stand density because it is highly correlated to the growing stock; it has also the advantage of being calculated directly by stem diameter or DBH, a variable easy to measure with accuracy (e.g., Bueno-López & Bevilacqua, 2013; Di Cosmo & Gasparini, 2020). At forest types level, coniferous stands are marked by higher values. In fact, compared to the average national value of 22.1 m^2 ha^{-1}, all the coniferous forest types show higher densities, with basal areas ranging between 25.5 m^2 per hectare (Larch and Swiss stone pine forest) and 43.0 m^2 per hectare (Fir forest). An exception is the Mediterranean pine forest with 18.7 m^2 per hectare. Among the broadleaved types, only the Beech (31.7 m^2 ha^{-1}) and Chestnut forest (28.0 m^2 ha^{-1}) show values higher than the national average and similar to those estimated for the coniferous types. The Hornbeams and Hophornbeams and the Holm oak forest types, rich in the number of trees per hectare, have relatively lower values of the basal area (18.2 and 18.6 m^2 per hectare, respectively), confirming that trees are abundant, but the size is limited. The three forest types with highest values of total basal area are Beech (33,336,591 m^2), Norway spruce (22,314,318 m^2) and Chestnut (21,802,076 m^2), that are marked by high density values and large areas in the country.

Table 7.24 shows the growing stock volume for the inventory categories of Forest; Tables 7.25 and 7.26 shows estimates for the forest types of Forest and Plantations. Growing stock volume estimated for all Italian Forests barely exceeds 1.5 billion cubic metres, with an average value per hectare of 165.4 m^3 (Fig. 7.12). In terms of aboveground tree biomass, they amount to little more than 1 billion tons (Mg), with an average value per hectare of 114.9 Mg ha^{-1} (Table 7.27).

The regions that contribute most to the overall growing stock of Italian forests, in terms of volume and biomass, are Toscana (10.4% of total volume and 11.1% of total aboveground tree biomass), Piemonte (9.8% of volume and 9.6% of biomass) and Lombardia (8.7% of volume and 7.9% of biomass). The minimum regional values are in Valle d'Aosta, Molise and Puglia, in which growing stock volume ranges from 1.0 to 1.3% and aboveground tree biomass ranges from 1.1 to 1.5% of the national totals.

Tables 7.28 and 7.29 show estimates on the aboveground tree biomass for the forest types of Forest and Plantations. Among the forest types, the significance of Beech forests (19.6% of total growing stock volume and 21.5% of total aboveground tree biomass) and Norway spruce (16.4% of volume and 12.3% of aboveground tree biomass) has been confirmed. Chestnut forest type is still relevant but to a lesser

Table 7.21 Total basal area and basal area per hectare by Forest inventory category / Valori totali e per ettaro dell'area basimetrica per le categorie inventariali del Bosco

Region/Regione	Tall trees forest				Plantations				Total Forest			
	Boschi alti				Impianti di arboricoltura da legno				Totale Bosco			
	Basal area	ES	Basal area	ES	Basal area	ES	Basal area	ES	Basal area	ES	Basal area	ES
	Area basimetrica		Area basimetrica		Area basimetrica		Area basimetrica		Area basimetrica		Area basimetrica	
	(m^2)	(%)	(m^2 ha^{-1})	(%)	(m^2)	(%)	(m^2 ha^{-1})	(%)	(m^2)	(%)	(m^2 ha^{-1})	(%)
Piemonte	19,780,218	2.4	22.7	2.0	275,904	26.3	13.4	19.5	20,056,122	2.4	22.5	2.0
Valle d'Aosta	2,593,031	6.3	26.1	4.8	0	–	0.0	–	2,593,031	6.3	26.1	4.8
Lombardia	15,419,208	3.1	25.8	2.5	263,311	20.4	10.5	14.3	15,682,519	3.1	25.2	2.5
Alto Adige	11,720,118	3.7	34.5	3.2	0	–	0.0	–	11,720,118	3.7	34.5	3.2
Trentino	11,531,445	3.7	30.9	3.3	0	–	0.0	–	11,531,445	3.7	30.9	3.3
Veneto	11,158,997	3.3	27.1	2.5	74,198	36.1	14.1	3.2	11,233,195	3.3	27.0	2.5
Friuli V.G	8,806,092	3.8	27.2	3.2	133,953	28.7	14.6	21.9	8,940,044	3.7	26.9	3.2
Liguria	7,790,990	3.7	22.7	3.4	7161	100.0	19.5	–	7,798,151	3.7	22.7	3.4
Emilia Romagna	12,426,179	3.5	21.5	3.0	81,241	31.0	13.4	27.3	12,507,421	3.5	21.4	3.0
Toscana	21,823,936	2.5	21.2	2.2	109,090	35.6	16.1	27.9	21,933,026	2.5	21.2	2.2
Umbria	6,252,277	4.1	16.3	3.7	38,645	35.4	6.1	8.1	6,290,922	4.1	16.1	3.7
Marche	5,171,932	4.7	18.2	3.9	39,440	37.2	5.7	24.1	5,211,372	4.6	17.9	4.0
Lazio	11,105,899	3.9	19.9	3.5	18,843	42.3	8.7	18.5	11,124,742	3.9	19.9	3.5
Abruzzo	9,170,119	3.5	22.4	3.0	11,462	47.6	3.9	35.3	9,181,581	3.5	22.3	3.0
Molise	2,910,703	6.0	19.3	5.1	24,196	68.5	8.9	55.7	2,934,899	6.0	19.2	5.1

(continued)

Table 7.21 (continued)

Region/Regione	Tall trees forest				Plantations				Total Forest			
	Boschi alti				Impianti di arboricoltura da legno				Totale Bosco			
	Basal area	ES	Basal area	ES	Basal area	ES	Basal area	ES	Basal area	ES	Basal area	ES
	Area basimetrica		Area basimetrica		Area basimetrica		Area basimetrica		Area basimetrica		Area basimetrica	
	(m^2)	(%)	(m^2 ha^{-1})	(%)	(m^2)	(%)	(m^2 ha^{-1})	(%)	(m^2)	(%)	(m^2 ha^{-1})	(%)
Campania	8,138,784	4.5	20.3	3.9	9963	58.7	3.1	21.3	8,148,748	4.5	20.2	3.9
Puglia	2,280,911	6.3	16.0	4.3	281	100.0	2.8	–	2,281,192	6.3	16.0	4.3
Basilicata	4,799,165	4.8	16.8	4.2	641	100.0	0.4	–	4,799,806	4.8	16.7	4.2
Calabria	14,068,518	4.2	28.5	3.8	31,232	50.0	13.0	30.1	14,099,751	4.2	28.5	3.8
Sicilia	4,549,767	5.6	16.0	4.7	431	70.7	0.6	–	4,550,199	5.6	15.9	4.7
Sardegna	8,423,192	4.1	14.0	3.7	142,508	19.6	5.5	16.7	8,565,700	4.1	13.7	3.7
Italia	199,921,483	0.8	22.3	0.7	1,232,500	9.5	9.8	7.4	201,183,984	0.8	22.1	0.7

Table 7.22 Total basal area and basal area per hectare by Tall trees forest type / Valori totali e per ettaro dell'area basimetrica per le categorie forestali dei Boschi alti

Region/Regione	Larch and Swiss stone pine				Norway spruce				Fir			
	Boschi di larice e cembro				Boschi di abete rosso				Boschi di abete bianco			
	Basal area	ES	Basal area	ES	Basal area	ES	Basal area	ES	Basal area	ES	Basal area	ES
	Area basimetrica		Area basimetrica		Area basimetrica		Area basimetrica		Area basimetrica		Area basimetrica	
	(m^2)	(%)	(m^2 ha^{-1})	(%)	(m^2)	(%)	(m^2 ha^{-1})	(%)	(m^2)	(%)	(m^2 ha^{-1})	(%)
Piemonte	1,730,809	10.5	21.6	–	729,662	18.1	40.1	–	522,744	18.4	37.0	–
Valle d'Aosta	1,160,514	12.0	24.4	–	622,007	15.4	34.1	–	39,718	60.0	34.4	–
Lombardia	1,574,045	12.9	29.5	–	3,297,768	9.5	37.5	–	417,877	24.3	44.4	–
Alto Adige	2,825,716	9.3	30.2	–	6,750,457	5.5	37.9	–	101,880	40.9	38.5	–
Trentino	1,443,758	12.8	23.2	–	5,178,114	6.8	37.8	–	675,147	16.6	41.6	–
Veneto	1,071,570	14.3	25.3	–	3,644,186	7.5	37.5	–	318,772	23.5	44.9	–
Friuli V.G	199,411	28.7	15.4	–	1,836,112	11.2	41.2	–	105,610	47.6	56.8	–
Liguria	23,136	64.9	21.0	–	8584	100.0	23.4	–	142,471	36.8	48.6	–
Emilia Romagna	0	–	0.0	–	181,312	32.8	49.2	–	162,073	37.2	55.0	–
Toscana	0	–	0.0	–	22,366	100.0	61.9	–	264,017	31.0	52.2	–
Umbria	0	–	0.0	–	0	–	0.0	–	0	–	0.0	–
Marche	0	–	0.0	–	10,247	100.0	27.6	–	0	–	0.0	–
Lazio	0	–	0.0	–	19,549	100.0	53.1	–	0	–	0.0	–
Abruzzo	0	–	0.0	–	13,955	100.0	38.6	–	34,634	70.8	47.8	–
Molise	0	–	0.0	–	0	–	0.0	–	43,305	60.8	37.0	–
Campania	0	–	0.0	–	0	–	0.0	–	0	–	0.0	–
Puglia	0	–	0.0	–	0	–	0.0	–	0	–	0.0	–
Basilicata	0	–	0.0	–	0	–	0.0	–	30,438	74.6	40.8	–
Calabria	0	–	0.0	–	0	–	0.0	–	147,564	33.4	39.5	–
Sicilia	0	–	0.0	–	0	–	0.0	–	0	–	0.0	–
Sardegna	0	–	0.0	–	0	–	0.0	–	0	–	0.0	–
Italia	10,028,959	4.7	25.5	3.8	22,314,318	3.2	38.0	2.4	3,006,250	8.1	43.0	3.3

(continued)

Table 7.22 (continued)

Region/Regione	Scots pine and Mountain pine				Black pines				Mediterranean pines			
	Pinete di pino silvestre e montano				Pinete di pino nero, laricio e loricato				Pinete di pini mediterranei			
	Basal area	ES	Basal area	ES	Basal area	ES	Basal area	ES	Basal area	ES	Basal area	ES
	Area basimetrica		Area basimetrica		Area basimetrica		Area basimetrica		Area basimetrica		Area basimetrica	
	(m^2)	(%)	(m^2 ha^{-1})	(%)	(m^2)	(%)	(m^2 ha^{-1})	(%)	(m^2)	(%)	(m^2 ha^{-1})	(%)
Piemonte	639,779	16.5	30.8	–	79,063	39.7	24.5	–	11,818	72.8	14.6	–
Valle d'Aosta	362,187	20.6	31.0	–	7853	100.0	20.4	–	0	–	0.0	–
Lombardia	494,923	20.7	26.0	–	135,041	37.2	30.6	–	0	–	0.0	–
Alto Adige	1,395,897	12.2	37.7	–	1768	100.0	4.7	–	0	–	0.0	–
Trentino	657,838	15.5	31.5	–	195,003	28.3	33.8	–	0	–	0.0	–
Veneto	300,614	19.0	24.5	–	86,143	39.7	24.6	–	23,898	70.8	32.0	–
Friuli V.G	266,176	24.9	21.8	–	899,724	13.5	28.5	–	17,888	60.1	16.0	–
Liguria	207,909	22.0	20.3	–	163,965	33.5	29.8	–	290,915	22.4	11.9	–
Emilia Romagna	43,576	52.4	23.7	–	588,404	21.6	36.3	–	85,702	49.1	29.1	–
Toscana	43,086	59.7	39.7	–	755,177	21.4	41.8	–	1,223,441	14.4	27.1	–
Umbria	20,689	76.6	28.1	–	136,052	27.1	23.1	–	174,443	24.0	21.5	–
Marche	0	–	0.0	–	238,513	22.6	23.0	–	56,054	47.4	25.5	–
Lazio	0	–	0.0	–	246,231	23.6	29.1	–	166,096	27.3	22.6	–
Abruzzo	42,444	60.9	39.1	–	455,459	22.3	24.7	–	42,424	51.5	19.5	–
Molise	0	–	0.0	–	50,732	43.0	21.7	–	22,167	63.5	10.2	–
Campania	0	–	0.0	–	163,331	34.0	29.6	–	209,104	26.1	22.9	–
Puglia	0	–	0.0	–	58,747	53.1	37.8	–	352,510	17.4	12.7	–
Basilicata	0	–	0.0	–	56,334	40.5	21.6	–	139,976	23.6	15.7	–
Calabria	0	–	0.0	–	3,274,914	9.5	44.6	–	359,059	22.0	20.5	–
Sicilia	0	–	0.0	–	170,707	28.3	22.8	–	801,000	18.0	17.7	–
Sardegna	0	–	0.0	–	112,146	39.7	21.5	–	527,580	15.2	15.2	–
Italia	4,475,118	6.3	30.1	3.9	7,875,308	5.6	34.2	4.0	4,504,075	6.4	18.7	5.2

(continued)

Table 7.22 (continued)

Region/Regione	Other coniferous forests				Beech				Temperate oaks			
	Altri boschi di conifere, pure o miste				Faggete				Querceti di rovere, roverella e farnia			
	Basal area	ES	Basal area	ES	Basal area	ES	Basal area	ES	Basal area	ES	Basal area	ES
	Area basimetrica		Area basimetrica		Area basimetrica		Area basimetrica		Area basimetrica		Area basimetrica	
	(m^2)	(%)	(m^2 ha^{-1})	(%)	(m^2)	(%)	(m^2 ha^{-1})	(%)	(m^2)	(%)	(m^2 ha^{-1})	(%)
Piemonte	155,205	33.0	24.0	–	3,093,451	7.3	26.4	–	1,171,022	12.2	15.4	–
Valle d'Aosta	0	–	0.0	–	42,940	61.4	37.1	–	79,379	35.8	18.7	–
Lombardia	147,014	35.6	23.2	–	1,721,397	10.5	25.8	–	818,095	13.2	19.9	–
Alto Adige	0	–	0.0	–	109,776	31.6	26.4	–	103,515	31.8	22.8	–
Trentino	0	–	0.0	–	1,824,905	9.0	29.1	–	123,586	31.3	21.4	–
Veneto	0	–	0.0	–	1,974,039	9.4	29.2	–	364,522	17.7	23.3	–
Friuli V.G	29,941	60.3	26.9	–	2,645,558	8.5	29.2	–	150,727	24.7	18.4	–
Liguria	16,272	100.0	44.4	–	1,195,050	11.4	32.3	–	721,847	14.3	16.5	–
Emilia Romagna	99,228	33.7	29.9	–	3,290,657	8.2	32.1	–	1,092,818	10.8	14.2	–
Toscana	433,936	21.5	36.5	–	2,446,591	8.7	33.4	–	2,238,052	7.2	14.7	–
Umbria	67,369	46.3	36.5	–	432,736	18.6	26.2	–	1,261,538	9.9	12.5	–
Marche	87,916	31.0	15.5	–	518,515	18.4	29.1	–	986,603	12.0	16.0	–
Lazio	56,654	56.0	38.4	–	2,479,782	9.9	33.3	–	1,098,263	12.2	13.4	–
Abruzzo	37,422	56.9	17.2	–	4,172,958	6.0	33.5	–	1,400,666	9.3	16.0	–
Molise	0	–	0.0	–	525,877	18.2	35.4	–	758,267	13.9	15.7	–
Campania	7069	62.0	8.2	–	1,988,097	10.3	35.3	–	771,736	12.5	12.7	–
Puglia	61,267	46.8	26.3	–	156,285	29.6	33.5	–	385,785	15.9	14.9	–
Basilicata	96,576	34.3	18.8	–	898,718	13.7	33.5	–	500,023	14.4	11.6	–
Calabria	483,622	21.3	38.5	–	3,341,578	7.7	42.1	–	774,766	18.5	14.4	–
Sicilia	162,599	31.3	17.5	–	477,678	17.5	31.5	–	876,500	13.7	12.2	–
Sardegna	224,144	24.9	17.2	–	0	–	0.0	–	959,785	11.4	10.9	–
Italia	2,166,234	8.9	25.8	6.5	33,336,591	2.4	31.7	1.6	16,637,494	2.9	14.4	2.3

(continued)

Table 7.22 (continued)

Region/Regione	Mediterranean oaks				Chestnut				Hornbeam and Hophornbeam			
	Cerrete, boschi di farnetto, fragno e vallonea				Castagneti				Ostrieti, carpineti			
	Basal area	ES	Basal area	ES	Basal area	ES	Basal area	ES	Basal area	ES	Basal area	ES
	Area basimetrica		Area basimetrica		Area basimetrica		Area basimetrica		Area basimetrica		Area basimetrica	
	(m^2)	(%)	(m^2 ha^{-1})	(%)	(m^2)	(%)	(m^2 ha^{-1})	(%)	(m^2)	(%)	(m^2 ha^{-1})	(%)
Piemonte	177,303	30.4	29.3	–	4,833,974	6.0	29.5	–	312,034	20.1	19.3	–
Valle d'Aosta	0	–	0.0	–	112,652	31.1	26.6	–	0	–	0.0	–
Lombardia	51,177	46.0	23.2	–	2,630,247	9.3	32.0	–	1,329,754	11.2	16.3	–
Alto Adige	0	–	0.0	–	48,095	51.0	25.4	–	223,760	22.4	25.4	–
Trentino	0	–	0.0	–	49,482	52.6	34.3	–	852,124	14.8	20.9	–
Veneto	0	–	0.0	–	430,279	17.2	25.0	–	1,855,770	9.5	22.6	–
Friuli V.G	0	–	0.0	–	535,801	18.8	43.7	–	888,487	12.9	19.0	–
Liguria	265,457	20.7	23.6	–	3,098,549	7.1	28.3	–	833,018	13.0	18.0	–
Emilia Romagna	1,756,606	10.0	17.3	–	1,226,392	12.4	30.3	–	1,981,588	8.9	17.8	–
Toscana	4,000,114	5.7	16.0	–	4,578,727	6.7	29.3	–	1,291,042	12.1	20.0	–
Umbria	2,038,350	8.2	15.9	–	49,393	48.1	22.3	–	816,896	12.0	14.0	–
Marche	517,297	17.6	20.9	–	108,914	33.9	29.4	–	1,204,279	10.9	15.8	–
Lazio	2,528,420	8.7	19.1	–	995,566	18.5	27.8	–	1,717,380	9.8	17.5	–
Abruzzo	653,824	14.0	20.3	–	158,025	29.9	33.6	–	860,624	13.0	18.0	–
Molise	991,345	12.4	19.1	–	9598	100.0	24.6	–	224,477	22.0	25.0	–
Campania	1,334,546	12.5	17.9	–	1,032,238	16.3	18.4	–	988,987	13.8	18.6	–
Puglia	571,452	14.4	14.8	–	35,605	59.6	30.6	–	85,053	33.0	18.2	–
Basilicata	2,134,014	8.2	16.3	–	132,284	28.8	22.2	–	134,292	25.5	19.0	–
Calabria	1,180,745	13.6	27.1	–	1,517,659	15.2	22.0	–	35,108	82.4	7.8	–
Sicilia	562,803	16.6	18.2	–	173,467	25.5	19.9	–	12,910	64.4	7.4	–
Sardegna	0	–	0.0	–	45,129	45.7	24.2	–	0	–	0.0	–
Italia	18,763,452	2.9	17.7	2.4	21,802,076	3.1	28.0	2.3	15,647,583	3.2	18.2	2.5

(continued)

Table 7.22 (continued)

Region/Regione	Hygrophilous forests				Other deciduous broadleaved forests				Holm oak			
	Boschi igrofili				Altri boschi caducifogli				Leccete			
	Basal area	ES	Basal area	ES	Basal area	ES	Basal area	ES	Basal area	ES	Basal area	ES
	Area basimetrica		Area basimetrica		Area basimetrica		Area basimetrica		Area basimetrica		Area basimetrica	
	(m^2)	(%)	(m^2 ha^{-1})	(%)	(m^2)	(%)	(m^2 ha^{-1})	(%)	(m^2)	(%)	(m^2 ha^{-1})	(%)
Piemonte	439,907	20.3	19.3	–	5,880,392	4.9	18.2	–	0	–	0.0	–
Valle d'Aosta	0	–	0.0	–	165,782	25.9	15.8	–	0	–	0.0	–
Lombardia	361,492	19.8	19.2	–	2,409,797	8.1	19.9	–	0	–	0.0	–
Alto Adige	52,376	38.6	19.1	–	94,378	35.2	26.7	–	0	–	0.0	–
Trentino	118,658	49.1	30.7	–	409,807	20.5	24.9	–	3024	100.0	8.4	–
Veneto	147,037	27.5	14.1	–	868,097	12.1	16.6	–	74,070	38.0	24.8	–
Friuli V.G	142,565	23.1	11.3	–	1,088,091	11.7	22.9	–	0	–	0.0	–
Liguria	83,169	42.2	25.2	–	496,531	15.4	15.6	–	243,070	20.8	17.5	–
Emilia Romagna	407,105	19.5	14.4	–	1,494,127	12.4	17.3	–	16,592	72.0	22.5	–
Toscana	591,341	19.2	21.3	–	1,501,075	10.7	17.8	–	2,290,996	7.9	17.7	–
Umbria	212,561	37.7	26.4	–	146,624	27.2	13.0	–	895,626	13.3	21.7	–
Marche	368,373	24.2	28.6	–	875,000	13.4	14.1	–	200,222	26.4	28.4	–
Lazio	160,412	28.3	17.4	–	761,827	17.1	14.1	–	783,684	17.0	16.2	–
Abruzzo	384,104	20.7	19.3	–	721,233	18.3	12.8	–	173,054	28.0	19.9	–
Molise	99,847	29.8	11.0	–	173,566	28.0	17.0	–	11,522	89.3	9.8	–
Campania	299,458	27.6	27.1	–	606,546	16.6	17.6	–	711,364	17.0	19.0	–
Puglia	6757	100.0	17.4	–	272,589	23.3	23.6	–	211,379	24.2	12.2	–
Basilicata	131,253	34.1	10.1	–	396,910	17.2	13.3	–	114,537	28.3	11.0	–
Calabria	206,199	31.7	16.0	–	909,547	18.1	20.1	–	1,521,815	20.4	31.3	–
Sicilia	78,859	54.7	16.0	–	213,549	27.4	16.1	–	468,560	21.3	24.0	–
Sardegna	47,226	38.1	14.1	–	24,609	30.5	3.2	–	4,225,589	6.6	16.5	–
Italia	4,338,701	6.5	18.5	4.8	19,510,076	2.9	17.5	2.3	11,945,102	4.4	18.6	3.8

(continued)

Table 7.22 (continued)

Region/Regione	Cork oak				Other evergreen broadleaved forests				Not classified				Total Tall trees forest			
	Sugherete				Altri boschi di latifoglie sempreverdi				Non classificato				Totale Boschi alti			
	Basal area	ES	Basal area	ES	Basal area	ES	Basal area	ES	Basal area	ES	Basal area	ES	Basal area	ES	Basal area	ES
	Area basimetrica		Area basimetrica		Area basimetrica		Area basimetrica		Area basimetrica		Area basimetrica		Area basimetrica		Area basimetrica	
	(m^2)	(%)	(m^2 ha^{-1})	(%)	(m^2)	(%)	(m^2 ha^{-1})	(%)	(m^2)	(%)	(m^2 ha^{-1})	(%)	(m^2)	(%)	(m^2 ha^{-1})	(%)
Piemonte	0	–	0.0	–	0	–	0.0	–	3057	100.0	6.0	–	19,780,218	2.4	22.7	2.0
Valle d'Aosta	0	–	0.0	–	0	–	0.0	–	0	–	0.0	–	2,593,031	6.3	26.1	4.8
Lombardia	0	–	0.0	–	26,239	59.4	11.4	–	4340	100.0	9.8	–	15,419,208	3.1	25.8	2.5
Alto Adige	0	–	0.0	–	0	–	0.0	–	12,500	68.7	8.3	–	11,720,118	3.7	34.5	3.2
Trentino	0	–	0.0	–	0	–	0.0	–	0	–	0.0	–	11,531,445	3.7	30.9	3.3
Veneto	0	–	0.0	–	0	–	0.0	–	0	–	0.0	–	11,158,997	3.3	27.1	2.5
Friuli V.G	0	–	0.0	–	0	–	0.0	–	0	–	0.0	–	8,806,092	3.8	27.2	3.2
Liguria	0	–	0.0	–	0	–	0.0	–	1048	70.7	1.4	–	7,790,990	3.7	22.7	3.4
Emilia Romagna	0	–	0.0	–	0	–	0.0	–	0	–	0.0	–	12,426,179	3.5	21.5	3.0
Toscana	113,071	29.3	18.4	–	11,313	64.2	7.8	–	19,591	70.7	27.1	–	21,823,936	2.5	21.2	2.2
Umbria	0	–	0.0	–	0	–	0.0	–	0	–	0.0	–	6,252,277	4.1	16.3	3.7
Marche	0	–	0.0	–	0	–	0.0	–	0	–	0.0	–	5,171,932	4.7	18.2	3.9
Lazio	35,728	47.9	13.9	–	22,762	89.5	8.8	–	33,546	52.6	25.1	–	11,105,899	3.9	19.9	3.5
Abruzzo	0	–	0.0	–	18,245	100.0	13.0	–	1048	100.0	2.9	–	9,170,119	3.5	22.4	3.0
Molise	0	–	0.0	–	0	–	0.0	–	0	–	0.0	–	2,910,703	6.0	19.3	5.1
Campania	14,132	100.0	38.4	–	2172	70.7	2.9	–	10,003	100.0	27.2	–	8,138,784	4.5	20.3	3.9
Puglia	0	–	0.0	–	60,216	31.1	11.7	–	23,266	62.5	20.7	–	2,280,911	6.3	16.0	4.3
Basilicata	0	–	0.0	–	24,585	93.0	13.2	–	9224	100.0	26.6	–	4,799,165	4.8	16.8	4.2
Calabria	108,355	39.2	20.7	–	149,737	30.8	7.1	–	57,851	40.9	25.8	–	14,068,518	4.2	28.5	3.8
Sicilia	156,825	22.5	9.1	–	394,310	15.9	10.1	–	0	–	0.0	–	4,549,767	5.6	16.0	4.7
Sardegna	1,978,031	8.2	12.9	–	278,414	20.7	7.3	–	538	100.0	1.4	–	8,423,192	4.1	14.0	3.7
Italia	2,406,143	7.3	13.1	5.9	987,993	10.8	8.7	8.5	176,012	22.5	17.5	11.9	199,921,483	0.8	22.3	0.7

Table 7.23 Total basal area and basal area per hectare by Plantations forest type / Valori totali e per ettaro dell'area basimetrica per le categorie forestali degli Impianti di arboricoltura da legno

Region/Regione	Poplar plantations				Other broadleaved plantations				Coniferous plantations				Total plantations			
	Pioppeti artificiali				Piantagioni di altre latifoglie				Piantagioni di conifere				Totale impianti di arboricoltura da legno			
	Basal area	ES	Basal area	ES	Basal area	ES	Basal area	ES	Basal area	ES	Basal area	ES	Basal area	ES	Basal area	ES
	Area basimetrica		Area basimetrica		Area basimetrica		Area basimetrica		Area basimetrica		Area basimetrica		Area basimetrica		Area basimetrica	
	(m^2)	(%)	(m^2 ha^{-1})	(%)	(m^2)	(%)	(m^2 ha^{-1})	(%)	(m^2)	(%)	(m^2 ha^{-1})	(%)	(m^2)	(%)	(m^2 ha^{-1})	(%)
Piemonte	171,104	35.4	11.0	–	62,997	50.7	15.8	–	41,803	59.5	36.3	–	275,904	26.3	13.4	19.5
Valle d'Aosta	0	–	0.0	–	0	–	0.0	–	0	–	0.0	–	0	–	0.0	–
Lombardia	156,982	24.6	8.5	–	90,706	44.9	14.7	–	15,623	100.0	35.4	–	263,311	20.4	10.5	14.3
Alto Adige	0	–	0.0	–	0	–	0.0	–	0	–	0.0	–	0	–	0.0	–
Trentino	0	–	0.0	–	0	–	0.0	–	0	–	0.0	–	0	–	0.0	–
Veneto	32,064	72.6	13.8	–	42,134	55.9	14.3	–	0	–	0.0	–	74,198	36.1	14.1	3.2
Friuli V.G	103,662	36.1	15.9	–	30,291	42.3	11.4	–	0	–	0.0	–	133,953	28.7	14.6	21.9
Liguria	7161	100.0	19.5	–	0	–	0.0	–	0	–	0.0	–	7161	100.0	19.5	–
Emilia Romagna	40,838	37.5	9.6	–	25,825	54.3	18.1	–	14,578	100.0	39.6	–	81,241	31.0	13.4	27.3
Toscana	14,098	69.7	19.1	–	33,340	42.7	7.3	–	61,651	56.7	42.5	–	109,090	35.6	16.1	27.9
Umbria	0	–	0.0	–	38,645	35.4	6.1	–	0	–	0.0	–	38,645	35.4	6.1	8.1
Marche	0	–	0.0	–	39,440	37.2	5.7	–	0	–	0.0	–	39,440	37.2	5.7	24.1
Lazio	4554	70.9	5.8	–	14,289	52.1	10.3	–	0	–	0.0	–	18,843	42.3	8.7	18.5
Abruzzo	2446	100.0	4.6	–	9016	54.3	3.7	–	0	–	0.0	–	11,462	47.6	3.9	35.3
Molise	16,118	100.0	41.3	–	8078	47.6	3.5	–	0	–	0.0	–	24,196	68.5	8.9	55.7

(continued)

Table 7.23 (continued)

Region/Regione	Poplar plantations				Other broadleaved plantations				Coniferous plantations				Total plantations			
	Pioppeti artificiali				Piantagioni di altre latifoglie				Piantagioni di conifere				Totale impianti di arboricoltura da legno			
	Basal area	ES	Basal area	ES	Basal area	ES	Basal area	ES	Basal area	ES	Basal area	ES	Basal area	ES	Basal area	ES
	Area basimetrica		Area basimetrica		Area basimetrica		Area basimetrica		Area basimetrica		Area basimetrica		Area basimetrica		Area basimetrica	
	(m^2)	(%)	(m^2 ha^{-1})	(%)	(m^2)	(%)	(m^2 ha^{-1})	(%)	(m^2)	(%)	(m^2 ha^{-1})	(%)	(m^2)	(%)	(m^2 ha^{-1})	(%)
Campania	6686	82.5	4.3	–	3277	59.2	2.0	–	0	–	0.0	–	9963	58.7	3.1	21.3
Puglia	0	–	0.0	–	281	100.0	2.8	–	0	–	0.0	–	281	100.0	2.8	–
Basilicata	0	–	0.0	–	641	100.0	0.4	–	0	–	0.0	–	641	100.0	0.4	–
Calabria	412	100.0	4.1	–	10,885	62.0	7.0	–	19,936	70.7	26.7	–	31,232	50.0	13.0	30.1
Sicilia	0	–	0.0	–	431	70.7	0.6	–	0	–	0.0	–	431	70.7	0.6	–
Sardegna	0	–	0.0	–	124,540	21.1	5.1	–	17,968	52.7	12.0	–	142,508	19.6	5.5	16.7
Italia	556,124	15.9	10.8	12.6	534,815	13.3	7.5	9.0	171,561	29.6	30.4	14.8	1,262,500	9.5	9.8	7.4

Table 7.24 Total value and value per hectare of stem volume by Forest inventory category / Valori totali e per ettaro del volume del fusto per le categorie inventariali del Bosco

Region/Regione	Tall trees forest				Plantations				Total Forest			
	Boschi alti				Impianti di arboricoltura da legno				Totale Bosco			
	Volume	ES	Volume	ES	Volume	ES	Volume	ES	Volume	ES	Volume	ES
	(m^3)	(%)	(m^3 ha^{-1})	(%)	(m^3)	(%)	(m^3 ha^{-1})	(%)	(m^3)	(%)	(m^3 ha^{-1})	(%)
Piemonte	144,565,842	2.9	166.2	2.6	2,258,955	28.1	109.3	21.8	146,824,796	2.9	164.9	2.6
Valle d'Aosta	19,883,165	7.3	200.3	6.1	0	–	0.0	–	19,883,165	7.3	200.3	6.1
Lombardia	128,198,140	3.9	214.8	3.5	1,928,870	23.1	76.7	18.1	130,127,009	3.9	209.2	3.5
Alto Adige	116,443,137	4.4	343.2	4.0	0	–	0.0	–	116,443,137	4.4	343.2	4.0
Trentino	112,770,182	4.7	302.1	4.4	0	–	0.0	–	112,770,182	4.7	302.1	4.4
Veneto	97,882,747	4.3	237.9	3.8	582,274	40.8	110.3	18.1	98,465,021	4.3	236.3	3.7
Friuli V.G	77,420,548	4.9	239.4	4.5	1,004,823	28.8	109.3	22.6	78,425,371	4.9	235.8	4.5
Liguria	51,642,647	4.3	150.7	4.1	52,280	100.0	142.6	–	51,694,928	4.3	150.6	4.1
Emilia Romagna	83,166,929	4.1	143.7	3.7	734,463	34.3	121.4	31.7	83,901,392	4.1	143.4	3.7
Toscana	154,609,716	3.1	150.3	2.9	971,169	40.9	143.2	36.6	155,580,886	3.1	150.3	2.9
Umbria	35,546,916	5.0	92.6	4.7	178,230	33.8	28.0	10.3	35,725,146	5.0	91.5	4.7
Marche	28,706,302	5.9	100.8	5.3	198,054	42.2	28.9	33.7	28,904,356	5.9	99.1	5.3
Lazio	72,697,898	4.7	130.3	4.3	142,750	44.2	65.6	22.7	72,840,647	4.7	130.0	4.3
Abruzzo	63,392,572	4.4	155.1	4.0	47,961	51.4	16.1	42.7	63,440,533	4.4	154.1	4.0
Molise	19,614,614	7.7	130.3	7.0	151,979	82.6	56.0	70.7	19,766,594	7.7	129.0	7.0
Campania	56,704,119	5.6	141.5	5.2	58,245	65.6	18.4	27.5	56,762,364	5.6	140.5	5.2

(continued)

Table 7.24 (continued)

Region/Regione	Tall trees forest				Plantations				Total Forest			
	Boschi alti				Impianti di arboricoltura da legno				Totale Bosco			
	Volume	ES	Volume	ES	Volume	ES	Volume	ES	Volume	ES	Volume	ES
	(m^3)	(%)	(m^3 ha^{-1})	(%)	(m^3)	(%)	(m^3 ha^{-1})	(%)	(m^3)	(%)	(m^3 ha^{-1})	(%)
Puglia	15,623,974	8.9	109.8	7.7	876	100.0	8.7	–	15,624,850	8.9	109.8	7.7
Basilicata	35,610,098	6.4	124.3	5.9	2148	100.0	1.4	–	35,612,246	6.4	123.6	5.9
Calabria	111,320,185	4.7	225.9	4.3	313,623	57.8	130.3	42.3	111,633,807	4.7	225.4	4.3
Sicilia	28,411,213	6.6	99.8	5.9	1401	70.7	1.8	–	28,412,615	6.6	99.5	5.9
Sardegna	39,245,225	5.0	65.4	4.7	722,819	21.0	27.9	18.1	39,968,044	4.9	63.8	4.6
Italia	1,493,456,168	1.0	166.7	1.0	9,350,921	11.0	72.8	9.3	1,502,807,089	1.0	165.4	1.0

Table 7.25 Total value and value per hectare of stem volume by Tall trees forest type / Valori totali e per ettaro del volume del fusto per le categorie forestali dei Boschi alti

Region/Regione	Larch and Swiss stone pine				Norway spruce				Fir			
	Boschi di larice e cembro				Boschi di abete rosso				Boschi di abete bianco			
	Volume	ES	Volume	ES	Volume	ES	Volume	ES	Volume	ES	Volume	ES
	(m^3)	(%)	(m^3 ha^{-1})	(%)	(m^3)	(%)	(m^3 ha^{-1})	(%)	(m^3)	(%)	(m^3 ha^{-1})	(%)
Piemonte	13,732,717	12.1	171.5	–	7,518,340	19.5	413.5	–	5,560,238	19.4	393.2	–
Valle d'Aosta	8,759,962	13.2	183.9	–	5,714,730	16.6	313.5	–	368,290	60.5	318.6	–
Lombardia	13,718,515	14.3	256.8	–	33,631,301	10.5	382.8	–	4,619,204	24.4	490.6	–
Alto Adige	24,992,852	10.4	266.9	–	74,084,469	6.3	415.4	–	1,232,604	40.1	465.7	–
Trentino	12,713,691	14.0	204.4	–	60,540,350	7.7	442.4	–	8,303,068	18.1	512.0	–
Veneto	9,697,955	17.0	228.9	–	39,564,442	8.5	407.3	–	3,654,323	24.2	514.4	–
Friuli V.G	1,599,271	34.7	123.3	–	20,901,754	12.4	468.7	–	1,362,866	49.4	733.4	–
Liguria	182,285	65.2	165.8	–	98,767	100.0	269.5	–	1,371,646	36.6	467.8	–
Emilia Romagna	0	–	0.0	–	1,760,947	32.9	478.0	–	1,990,395	40.1	675.4	–
Toscana	0	–	0.0	–	241,602	100.0	668.6	–	3,032,981	31.6	599.6	–
Umbria	0	–	0.0	–	0	–	0.0	–	0	–	0.0	–
Marche	0	–	0.0	–	73,209	100.0	197.5	–	0	–	0.0	–
Lazio	0	–	0.0	–	354,470	100.0	962.0	–	0	–	0.0	–
Abruzzo	0	–	0.0	–	114,529	100.0	316.4	–	346,628	70.7	478.9	–
Molise	0	–	0.0	–	0	–	0.0	–	431,076	61.8	368.0	–
Campania	0	–	0.0	–	0	–	0.0	–	0	–	0.0	–
Puglia	0	–	0.0	–	0	–	0.0	–	0	–	0.0	–
Basilicata	0	–	0.0	–	0	–	0.0	–	275,828	74.6	369.9	–
Calabria	0	–	0.0	–	0	–	0.0	–	1,332,024	34.8	357.0	–
Sicilia	0	–	0.0	–	0	–	0.0	–	0	–	0.0	–
Sardegna	0	–	0.0	–	0	–	0.0	–	0	–	0.0	–
Italia	85,397,249	5.4	217.1	4.6	244,598,911	3.6	416.9	2.9	33,881,171	8.6	485.1	4.5

(continued)

Table 7.25 (continued)

Region/Regione	Scots pine and Mountain pine				Black pines				Mediterranean pines			
	Pinete di pino silvestre e montano				Pinete di pino nero, laricio e loricato				Pinete di pini mediterranei			
	Volume	ES	Volume	ES	Volume	ES	Volume	ES	Volume	ES	Volume	ES
	(m^3)	(%)	(m^3 ha^{-1})	(%)	(m^3)	(%)	(m^3 ha^{-1})	(%)	(m^3)	(%)	(m^3 ha^{-1})	(%)
Piemonte	4,376,962	18.1	210.9	–	503,811	41.9	155.9	–	54,909	71.9	67.9	–
Valle d'Aosta	2,344,424	21.6	200.9	–	44,868	100.0	116.4	–	0	–	0.0	–
Lombardia	3,709,093	24.8	194.8	–	1,063,207	44.4	241.2	–	0	–	0.0	–
Alto Adige	11,602,044	13.8	313.1	–	6985	100.0	18.5	–	0	–	0.0	–
Trentino	4,410,917	17.2	211.0	–	1,445,254	30.2	250.7	–	0	–	0.0	–
Veneto	2,119,168	20.3	173.0	–	582,374	43.9	166.0	–	206,330	70.9	275.9	–
Friuli V.G	2,047,546	28.4	167.4	–	6,164,820	14.8	195.4	–	107,678	63.3	96.6	–
Liguria	1,512,943	23.4	147.4	–	1,520,538	36.8	276.6	–	1,844,276	29.2	75.1	–
Emilia Romagna	292,739	56.2	158.9	–	4,549,760	25.4	280.7	–	675,875	54.7	229.3	–
Toscana	387,347	59.5	357.3	–	7,706,980	25.2	426.6	–	10,195,271	16.4	225.8	–
Umbria	138,442	76.6	187.8	–	917,771	28.5	155.6	–	1,093,309	24.7	134.8	–
Marche	0	–	0.0	–	1,641,021	23.9	158.1	–	319,033	50.0	145.4	–
Lazio	0	–	0.0	–	1,831,923	26.0	216.2	–	1,209,434	29.4	164.7	–
Abruzzo	365,861	63.6	336.9	–	3,091,092	26.3	167.9	–	290,614	54.1	133.8	–
Molise	0	–	0.0	–	314,979	45.0	134.4	–	145,706	64.2	66.7	–
Campania	0	–	0.0	–	1,399,605	37.8	253.4	–	1,413,292	26.0	154.5	–
Puglia	0	–	0.0	–	552,462	53.9	355.6	–	2,470,462	20.5	89.1	–
Basilicata	0	–	0.0	–	389,130	44.5	149.1	–	863,280	24.6	96.6	–
Calabria	0	–	0.0	–	27,290,996	10.3	371.6	–	2,545,993	24.3	145.7	–
Sicilia	0	–	0.0	–	1,091,658	30.8	145.7	–	5,785,644	20.5	127.6	–
Sardegna	0	–	0.0	–	782,356	45.0	149.7	–	3,034,218	17.4	87.6	–
Italia	33,307,487	7.1	223.7	5.2	62,891,589	6.5	273.0	5.2	32,255,324	7.6	134.1	6.6

(continued)

Table 7.25 (continued)

	Other coniferous forests				Beech				Temperate oaks			
	Altri boschi di conifere, pure o miste				Faggete				Querceti di rovere, roverella e farnia			
	Volume	ES	Volume	ES	Volume	ES	Volume	ES	Volume	ES	Volume	ES
	(m^3)	(%)	(m^3 ha^{-1})	(%)	(m^3)	(%)	(m^3 ha^{-1})	(%)	(m^3)	(%)	(m^3 ha^{-1})	(%)
Piemonte	1,225,558	35.9	189.6	–	24,707,388	8.0	210.9	–	7,698,574	14.7	101.1	–
Valle d'Aosta	0	–	0.0	–	369,263	70.1	319.4	–	483,202	38.8	114.0	–
Lombardia	1,530,679	39.7	241.1	–	15,299,497	12.3	229.5	–	5,562,732	16.0	135.4	–
Alto Adige	0	–	0.0	–	951,050	31.8	228.7	–	502,723	32.4	110.8	–
Trentino	0	–	0.0	–	16,013,838	11.9	255.8	–	791,230	34.4	137.2	–
Veneto	0	–	0.0	–	18,473,637	11.3	273.1	–	2,487,721	19.7	158.9	–
Friuli V.G	285,444	68.4	256.0	–	25,638,163	10.1	282.8	–	966,830	28.8	118.2	–
Liguria	169,384	100.0	462.2	–	9,663,174	11.5	261.0	–	4,342,745	16.1	99.0	–
Emilia Romagna	679,216	33.5	204.9	–	24,618,739	9.0	240.2	–	5,733,136	12.1	74.5	–
Toscana	4,311,341	25.0	362.3	–	22,205,740	9.7	302.8	–	13,467,403	8.6	88.5	–
Umbria	522,045	47.2	283.2	–	3,027,754	20.6	183.3	–	6,480,220	12.7	64.2	–
Marche	542,686	32.7	95.9	–	3,619,843	21.9	203.5	–	4,912,428	16.3	79.5	–
Lazio	576,510	61.5	391.2	–	21,667,365	10.5	291.1	–	5,527,605	14.5	67.6	–
Abruzzo	377,369	66.8	173.8	–	34,580,265	6.9	277.8	–	7,345,483	10.7	83.7	–
Molise	0	–	0.0	–	5,050,539	18.5	340.4	–	3,762,999	16.3	78.1	–
Campania	43,292	63.3	50.1	–	20,548,897	11.8	365.4	–	4,254,283	14.2	69.8	–
Puglia	474,486	53.2	203.6	–	1,817,617	31.4	390.0	–	2,078,568	22.8	80.4	–
Basilicata	727,206	39.9	141.2	–	8,874,021	16.2	330.9	–	2,597,362	16.8	60.3	–
Calabria	5,090,161	23.1	405.2	–	32,188,075	8.0	405.3	–	5,261,079	20.0	97.8	–
Sicilia	1,213,519	30.6	130.5	–	3,261,687	19.3	215.1	–	5,428,944	16.6	75.5	–
Sardegna	1,548,545	28.8	118.6	–	0	0.0	–	–	4,695,295	12.9	53.5	–
Italia	19,317,439	10.3	230.2	8.7	292,576,553	2.7	277.8	2.0	94,380,562	3.5	81.9	3.0

(continued)

Table 7.25 (continued)

Region/Regione	Mediterranean oaks				Chestnut				Hornbeam and Hophornbeam			
	Cerrete, boschi di farnetto, fragno e vallonea				Castagneti				Ostrieti, carpineti			
	Volume	ES	Volume	ES	Volume	ES	Volume	ES	Volume	ES	Volume	ES
	(m^3)	(%)	(m^3 ha^{-1})	(%)	(m^3)	(%)	(m^3 ha^{-1})	(%)	(m^3)	(%)	(m^3 ha^{-1})	(%)
Piemonte	1,216,109	28.7	200.7	–	34,341,434	6.7	209.9	–	1,573,455	21.0	97.4	–
Valle d'Aosta	0	–	0.0	–	766,833	32.3	180.9	–	0	–	0.0	–
Lombardia	412,154	48.2	187.0	–	21,389,981	10.1	260.6	–	7,075,715	13.2	86.6	–
Alto Adige	0	–	0.0	–	432,097	53.3	228.6	–	1,403,515	23.7	159.0	–
Trentino	0	–	0.0	–	452,433	58.4	313.9	–	4,553,099	19.6	111.5	–
Veneto	0	–	0.0	–	2,997,719	20.6	174.3	–	10,905,325	11.4	133.0	–
Friuli V.G	0	–	0.0	–	4,255,080	19.0	346.9	–	5,044,754	16.9	107.8	–
Liguria	1,881,147	21.5	167.3	–	19,368,980	8.1	176.7	–	4,656,158	15.0	100.7	–
Emilia Romagna	12,083,200	11.8	119.3	–	8,281,885	12.4	204.4	–	11,015,513	10.2	99.0	–
Toscana	27,342,914	7.2	109.1	–	31,160,435	7.5	199.6	–	7,560,276	12.9	116.9	–
Umbria	12,366,294	9.7	96.2	–	304,431	48.8	137.6	–	3,922,653	14.0	67.0	–
Marche	3,612,839	21.8	146.1	–	737,053	35.3	198.9	–	5,385,199	12.1	70.6	–
Lazio	17,606,398	10.8	132.9	–	6,093,107	17.6	170.2	–	8,143,267	10.6	83.1	–
Abruzzo	4,500,975	16.9	140.0	–	995,982	29.3	211.7	–	4,519,560	15.0	94.3	–
Molise	7,081,474	15.9	136.2	–	66,452	100.0	170.2	–	1,198,464	23.2	133.4	–
Campania	9,342,569	16.0	125.2	–	5,738,669	15.0	102.5	–	5,237,490	17.3	98.8	–
Puglia	3,670,563	21.6	94.8	–	190,802	58.0	163.7	–	549,921	36.0	118.0	–
Basilicata	15,781,773	10.5	120.8	–	1,128,251	34.2	189.4	–	740,375	26.4	104.5	–
Calabria	8,914,188	15.6	204.5	–	9,545,096	17.1	138.4	–	134,460	84.9	30.0	–
Sicilia	3,547,306	17.7	114.5	–	1,229,431	28.8	141.0	–	71,697	74.6	41.3	–
Sardegna	0	–	0.0	–	259,196	47.6	138.9	–	0	–	0.0	–
Italia	129,359,903	3.6	122.1	3.2	149,735,348	3.3	192.3	2.7	83,690,898	3.7	97.4	3.2

(continued)

Table 7.25 (continued)

	Hygrophilous forests				Other deciduous broadleaved forests				Holm oak			
	Boschi igrofili				Altri boschi caducifogli				Leccete			
	Volume	ES	Volume	ES	Volume	ES	Volume	ES	Volume	ES	Volume	ES
	(m^3)	(%)	($m^3\ ha^{-1}$)	(%)	(m^3)	(%)	($m^3\ ha^{-1}$)	(%)	(m^3)	(%)	($m^3\ ha^{-1}$)	(%)
Piemonte	2,731,655	22.3	120.1	–	39,312,648	5.7	121.4	–	0	–	0.0	–
Valle d'Aosta	0	–	0.0	–	1,031,592	30.1	98.0	–	0	–	0.0	–
Lombardia	2,355,028	21.1	125.0	–	17,671,061	10.0	146.0	–	0	–	0.0	–
Alto Adige	398,705	41.5	145.2	–	735,184	38.0	208.1	–	0	–	0.0	–
Trentino	876,283	49.8	227.0	–	2,660,592	21.1	161.5	–	9428	100.0	26.2	–
Veneto	957,350	28.0	91.9	–	5,830,134	13.4	111.3	–	406,270	41.9	135.8	–
Friuli V.G	850,986	26.1	67.5	–	8,195,356	13.5	172.7	–	0	–	0.0	–
Liguria	638,202	45.1	193.5	–	3,141,653	19.4	98.7	–	1,247,228	22.9	89.7	–
Emilia Romagna	2,716,086	22.9	96.2	–	8,655,037	14.1	100.2	–	114,402	73.3	155.3	–
Toscana	3,836,965	19.1	138.3	–	10,163,671	12.1	120.3	–	12,259,130	8.5	94.5	–
Umbria	1,345,714	40.6	167.4	–	824,411	32.6	73.1	–	4,603,872	14.7	111.5	–
Marche	2,109,278	23.9	163.5	–	4,981,644	15.6	80.2	–	772,070	27.9	109.6	–
Lazio	953,258	29.7	103.5	–	4,302,047	20.7	79.6	–	3,899,361	19.1	80.8	–
Abruzzo	2,256,321	23.3	113.6	–	3,788,213	21.6	67.2	–	731,498	31.7	84.2	–
Molise	526,102	29.8	58.1	–	1,003,747	28.6	98.2	–	33,077	82.1	28.2	–
Campania	1,775,664	35.2	160.7	–	3,537,316	18.2	102.9	–	3,292,291	18.3	87.8	–
Puglia	32,572	100.0	83.9	–	2,135,110	31.0	185.0	–	1,249,437	28.0	72.0	–
Basilicata	718,632	30.9	55.4	–	2,713,822	20.0	90.8	–	588,085	32.5	56.7	–
Calabria	1,594,833	31.6	123.9	–	5,928,291	19.4	131.3	–	9,768,396	27.3	200.6	–
Sicilia	290,865	32.4	59.0	–	949,122	30.1	71.7	–	2,567,639	21.9	131.4	–
Sardegna	216,010	35.8	64.3	–	75,163	33.3	9.7	–	19,660,741	8.1	77.0	–
Italia	27,180,506	7.0	115.6	5.5	127,635,813	3.4	114.6	2.9	61,202,924	5.8	95.2	5.4

(continued)

Table 7.25 (continued)

	Cork oak				Other evergreen broadleaved forests				Not classified				Total Tall trees forest			
	Sugherete				Altri boschi di latifoglie sempreverdi				Non classificato				Totale Boschi alti			
	Volume	ES	Volume	ES	Volume	ES	Volume	ES	Volume	ES	Volume	ES	Volume	ES	Volume	ES
	(m^3)	(%)	(m^3 ha^{-1})	(%)	(m^3)	(%)	(m^3 ha^{-1})	(%)	(m^3)	(%)	(m^3 ha^{-1})	(%)	(m^3)	(%)	(m^3 ha^{-1})	(%)
Piemonte	0	–	0.0	–	0	–	0.0	–	12,045	100.0	23.7	–	144,565,842	2.9	166.2	2.6
Valle d'Aosta	0	–	0.0	–	0	–	0.0	–	0	–	0.0	–	19,883,165	7.3	200.3	6.1
Lombardia	0	–	0.0	–	147,268	62.6	64.2	–	12,704	100.0	28.8	–	128,198,140	3.9	214.8	3.5
Alto Adige	0	–	0.0	–	0	–	0.0	–	100,909	85.0	66.7	–	116,443,137	4.4	343.2	4.0
Trentino	0	–	0.0	–	0	–	0.0	–	0	–	0.0	–	112,770,182	4.7	302.1	4.4
Veneto	0	–	0.0	–	0	–	0.0	–	0	–	0.0	–	97,882,747	4.3	237.9	3.8
Friuli V.G	0	–	0.0	–	0	–	0.0	–	0	–	0.0	–	77,420,548	4.9	239.4	4.5
Liguria	0	–	0.0	–	0	–	0.0	–	3521	70.7	4.8	–	51,642,647	4.3	150.7	4.1
Emilia Romagna	0	–	0.0	–	0	–	0.0	–	0	–	0.0	–	83,166,929	4.1	143.7	3.7
Toscana	576,110	30.4	93.8	–	49,914	63.7	34.5	–	111,636	70.7	154.5	–	154,609,716	3.1	150.3	2.9
Umbria	0	–	0.0	–	0	–	0.0	–	0	–	0.0	–	35,546,916	5.0	92.6	4.7
Marche	0	–	0.0	–	0	–	0.0	–	0	–	0.0	–	28,706,302	5.9	100.8	5.3
Lazio	216,206	52.3	83.8	–	129,584	93.3	50.2	–	187,363	53.3	140.3	–	72,697,898	4.7	130.3	4.3
Abruzzo	0	–	0.0	–	85,468	100.0	60.7	–	2713	100.0	7.5	–	63,392,572	4.4	155.1	4.0
Molise	0	–	0.0	–	0	–	0.0	–	0	–	0.0	–	19,614,614	7.7	130.3	7.0
Campania	58,181	100.0	158.0	–	5570	70.7	7.6	–	57,000	100.0	154.8	–	56,704,119	5.6	141.5	5.2
Puglia	0	–	0.0	–	300,681	32.4	58.4	–	101,293	60.4	90.0	–	15,623,974	8.9	109.8	7.7
Basilicata	0	–	0.0	–	172,489	97.0	92.5	–	39,845	100.0	114.7	–	35,610,098	6.4	124.3	5.9
Calabria	623,903	41.3	119.4	–	866,756	35.1	41.0	–	235,932	41.6	105.4	–	111,320,185	4.7	225.9	4.3
Sicilia	608,492	24.4	35.3	–	2,365,208	16.7	60.4	–	0	–	0.0	–	28,411,213	6.6	99.8	5.9
Sardegna	7,988,806	8.7	52.3	–	983,087	22.3	25.9	–	1808	100.0	4.8	–	39,245,225	5.0	65.4	4.7
Italia	10,071,698	7.8	54.6	6.6	5,106,025	11.9	44.9	10.4	866,767	23.7	86.1	14.1	1,493,456,168	1.0	166.7	1.0

Table 7.26 Total value and value per hectare of stem volume by Plantations forest type / Valori totali e per ettaro del volume del fusto per le categorie forestali degli Impianti di arboricoltura da legno

Region/Regione	Poplar plantations				Other broadleaved plantations				Coniferous plantations				Total Plantations			
	Pioppeti artificiali				Piantagioni di altre latifoglie				Piantagioni di conifere				Totale Impianti di arboricoltura da legno			
	Volume	ES	Volume	ES	Volume	ES	Volume	ES	Volume	ES	Volume	ES	Volume	ES	Volume	ES
	(m^3)	(%)	(m^3 ha^{-1})	(%)	(m^3)	(%)	(m^3 ha^{-1})	(%)	(m^3)	(%)	(m^3 ha^{-1})	(%)	(m^3)	(%)	(m^3 ha^{-1})	(%)
Piemonte	1,418,606	38.0	91.5	–	426,676	52.4	106.7	–	413,673	62.6	359.5	–	2,258,955	28.1	109.3	21.8
Valle d'Aosta	0	–	0.0	–	0	0.0	0.0	–	0	–	0.0	–	0	–	0.0	–
Lombardia	1,152,233	27.5	62.3	–	600,642	50.8	97.1	–	175,995	100.0	399.3	–	1,928,870	23.1	76.7	18.1
Alto Adige	0	–	0.0	–	0	0.0	0.0	–	0	–	0.0	–	0	–	0.0	–
Trentino	0	–	0.0	–	0	0.0	0.0	–	0	–	0.0	–	0	–	0.0	–
Veneto	306,608	79.3	131.5	–	275,666	55.3	93.6	–	0	–	0.0	–	582,274	40.8	110.3	18.1
Friuli V.G	829,236	34.8	126.8	–	175,587	42.4	66.1	–	0	–	0.0	–	1,004,823	28.8	109.3	22.6
Liguria	52,280	100.0	142.6	–	0	0.0	0.0	–	0	–	0.0	–	52,280	100.0	142.6	–
Emilia Romagna	355,777	42.2	83.7	–	215,795	56.9	151.1	–	162,890	100.0	442.2	–	734,463	34.3	121.4	31.7
Toscana	120,866	69.8	164.2	–	181,933	43.3	39.6	–	668,371	57.1	460.3	–	971,169	40.9	143.2	36.6
Umbria	0	–	0.0	–	178,230	33.8	28.0	–	0	–	0.0	–	178,230	33.8	28.0	10.3
Marche	0	–	0.0	–	198,054	42.2	28.9	–	0	–	0.0	–	198,054	42.2	28.9	33.7
Lazio	38,301	70.9	48.6	–	104,449	55.2	75.3	–	0	–	0.0	–	142,750	44.2	65.6	22.7
Abruzzo	16,327	100.0	30.9	–	31,634	58.7	12.9	–	0	–	0.0	–	47,961	51.4	16.1	42.7
Molise	124,764	100.0	319.5	–	27,216	49.6	11.7	–	0	–	0.0	–	151,979	82.6	56.0	70.7
Campania	45,939	81.7	29.8	–	12,306	57.8	7.6	–	0	–	0.0	–	58,245	65.6	18.4	27.5
Puglia	0	–	0.0	–	876	100.0	8.7	–	0	–	0.0	–	876	100.0	8.7	–
Basilicata	0	–	0.0	–	2148	100.0	1.4	–	0	–	0.0	–	2148	100.0	1.4	–
Calabria	2353	100.0	23.5	–	62,302	70.7	39.9	–	248,968	70.7	333.6	–	313,623	57.8	130.3	42.3
Sicilia	0	–	0.0	–	1401	70.7	1.8	–	0	–	0.0	–	1401	70.7	1.8	–
Sardegna	0	–	0.0	–	622,161	22.9	25.5	20.2	100,658	51.9	67.4	–	722,819	21.0	27.9	18.1
Italia	4,463,290	17.1	86.5	13.8	3,117,075	15.4	43.8	11.8	1,770,555	31.1	313.3	17.7	9,350,921	11.0	72.8	9.3

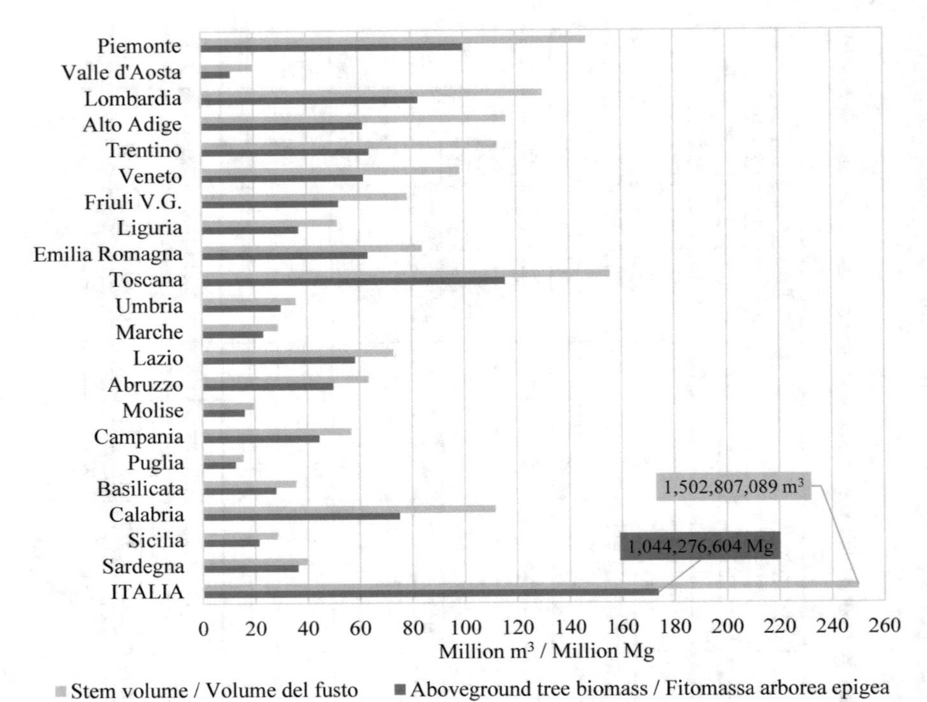

Fig. 7.12 Growing stock volume and aboveground tree biomass by region (X-axis was limited to improve readability; exceeding values are given in numbers) / Volume del fusto e fitomassa arborea epigea nelle regioni italiane (per migliorare la lettura, è stata limitata la lunghezza di barre molto lunghe e i valori reali sono indicati con numero)

extent compared with the comments on basal area; Chestnut forest is still in third position for growing stock (10.0% of total national volume) but was surpassed by the Mediterranean oaks in the biomass (10.1% of total national value), due to the higher wood basal density. Figures 7.13 and 7.14 show the growing stock volume and aboveground tree biomass values, overall and per hectare, in the Tall trees forest types.

Considerable differences between the total value and value per hectare allow for an understanding as to which forest types are the most widespread in Italy. The ratio between the growing stock volume (i.e., the volume of the stem and large branches and treetop up to 5 cm section) and the aboveground tree biomass (i.e., the volume of the stem, large branches, stump, small branches and treetop) of each forest type was previously used in the estimates provided by INFC2005 to calculate the Biomass Conversion and Expansion Factor (BCEF) (Di Cosmo & Tabacchi, 2011). This may be used to understand the biomass of stands when the only known variable is their volume. Small fluctuations of those coefficients after each NFI are expected because the sample of trees is not the same. For example, that ratio is size dependent within a species, the rate of species within a forest type is not constant,

Table 7.27 Total value and value per hectare of aboveground tree biomass by Forest inventory category / Valori totali e per ettaro della fitomassa arborea epigea per le categorie inventariali del Bosco

Region/Regione	Tall trees forest				Plantations				Total Forest			
	Boschi alti				Impianti di arboricoltura da legno				Totale Bosco			
	Biomass	ES	Biomass	ES	Biomass	ES	Biomass	ES	Biomass	ES	Biomass	ES
	Fitomassa		Fitomassa		Fitomassa		Fitomassa		Fitomassa		Fitomassa	
	(Mg)	(%)	(Mg ha^{-1})	(%)	(Mg)	(%)	(Mg ha^{-1})	(%)	(Mg)	(%)	(Mg ha^{-1})	(%)
Piemonte	98,618,491	2.8	113.4	2.4	1,272,845	26.9	61.6	20.1	99,891,336	2.8	112.2	2.4
Valle d'Aosta	11,227,949	6.9	113.1	5.6	0	–	0	–	11,227,949	6.9	113.1	5.6
Lombardia	81,605,175	3.7	136.7	3.2	1,125,370	21.2	44.8	15.4	82,730,544	3.6	133.0	3.2
Alto Adige	61,507,973	4.3	181.3	3.9	0	–	0	–	61,507,973	4.3	181.3	3.9
Trentino	63,838,069	4.3	171.0	4.0	0	–	0	–	63,838,069	4.3	171.0	4.0
Veneto	61,451,389	4.0	149.4	3.3	347,306	38.2	65.8	12.4	61,798,696	4.0	148.3	3.3
Friuli V.G	51,670,841	4.8	159.8	4.3	547,225	26.3	59.5	19.3	52,218,065	4.7	157.0	4.3
Liguria	36,848,401	4.1	107.5	3.9	25,664	100.0	70.0	–	36,874,064	4.1	107.5	3.9
Emilia Romagna	62,862,543	4.0	108.6	3.6	388,468	33.2	64.2	30.1	63,251,010	4.0	108.1	3.6
Toscana	115,055,051	2.9	111.8	2.6	562,351	37.8	82.9	32.2	115,617,402	2.9	111.7	2.6
Umbria	29,803,987	4.8	77.6	4.4	136,717	33.1	21.4	10.4	29,940,704	4.8	76.7	4.4
Marche	23,169,797	5.7	81.3	5.1	140,485	40.0	20.5	29.8	23,310,282	5.7	79.9	5.1
Lazio	58,266,635	4.6	104.4	4.2	98,442	48.1	45.2	31.0	58,365,077	4.6	104.2	4.2
Abruzzo	49,850,959	4.3	122.0	3.9	37,011	48.5	12.5	37.3	49,887,970	4.3	121.2	3.9
Molise	15,915,418	7.5	105.7	6.8	96,203	74.5	35.4	62.1	16,011,621	7.5	104.5	6.8

(continued)

Table 7.27 (continued)

Region/Regione	Tall trees forest				Plantations				Total Forest			
	Boschi alti				Impianti di arboricoltura da legno				Totale Bosco			
	Biomass Fitomassa (Mg)	ES (%)	Biomass Fitomassa (Mg ha^{-1})	ES (%)	Biomass Fitomassa (Mg)	ES (%)	Biomass Fitomassa (Mg ha^{-1})	ES (%)	Biomass Fitomassa (Mg)	ES (%)	Biomass Fitomassa (Mg ha^{-1})	ES (%)
Campania	44,388,019	5.6	110.8	5.2	42,391	66.3	13.4	26.8	44,430,410	5.6	110.0	5.2
Puglia	12,617,013	8.6	88.7	7.4	912	100.0	9.1	–	12,617,926	8.6	88.6	7.4
Basilicata	28,000,541	6.3	97.7	5.9	2039	100.0	1.3	–	28,002,580	6.3	97.2	5.9
Calabria	74,958,164	5.0	152.1	4.6	176,291	54.2	73.3	36.9	75,134,455	5.0	151.7	4.6
Sicilia	21,383,690	6.3	75.1	5.6	1195	70.7	1.6	–	21,384,884	6.3	74.9	5.6
Sardegna	35,691,220	5.2	59.5	4.9	544,367	20.9	21.0	18.0	36,235,587	5.1	57.9	4.8
Italia	1,038,731,323	1.0	116.0	0.9	5,545,281	10.1	43.2	8.1	1,044,276,604	1.0	114.9	0.9

Table 7.28 Total value and value per hectare of aboveground tree biomass by Tall trees forest type / Valori totali e per ettaro della fitomassa arborea epigea per le categorie forestali dei Boschi alti

Region/Regione	Larch and Swiss stone pine				Norway spruce				Fir			
	Boschi di larice e cembro				Boschi di abete rosso				Boschi di abete bianco			
	Biomass	ES	Biomass	ES	Biomass	ES	Biomass	ES	Biomass	ES	Biomass	ES
	Fitomassa		Fitomassa		Fitomassa		Fitomassa		Fitomassa		Fitomassa	
	(Mg)	(%)	(Mg ha^{-1})	(%)	(Mg)	(%)	(Mg ha^{-1})	(%)	(Mg)	(%)	(Mg ha^{-1})	(%)
Piemonte	7,612,541	11.9	95.1	–	4,033,665	19.2	221.8	–	3,129,679	19.4	221.3	–
Valle d'Aosta	4,824,984	12.8	101.3	–	2,999,649	16.6	164.5	–	197,069	60.1	170.5	–
Lombardia	7,535,821	13.9	141.1	–	17,911,007	10.2	203.9	–	2,383,916	24.3	253.2	–
Alto Adige	13,107,913	10.2	140.0	–	38,033,837	6.2	213.3	–	641,074	39.9	242.2	–
Trentino	6,910,278	13.8	111.1	–	31,442,275	7.6	229.8	–	4,380,044	17.8	270.1	–
Veneto	5,355,833	16.4	126.4	–	20,671,970	8.3	212.8	–	2,033,349	24.0	286.2	–
Friuli V.G	968,140	34.8	74.6	–	11,730,433	12.4	263.0	–	707,825	48.5	380.9	–
Liguria	101,927	64.8	92.7	–	50,707	100.0	138.4	–	755,523	37.1	257.7	–
Emilia Romagna	0	–	0.0	–	910,414	32.8	247.1	–	1,091,661	38.8	370.4	–
Toscana	0	–	0.0	–	127,530	100.0	352.9	–	1,600,439	30.8	316.4	–
Umbria	0	–	0.0	–	0	–	0.0	–	0	–	0.0	–
Marche	0	–	0.0	–	40,547	100.0	109.4	–	0	–	0.0	–
Lazio	0	–	0.0	–	179,957	100.0	488.4	–	0	–	0.0	–
Abruzzo	0	–	0.0	–	64,597	100.0	178.5	–	191,314	71.1	264.3	–
Molise	0	–	0.0	–	0	–	0.0	–	248,628	62.1	212.2	–
Campania	0	–	0.0	–	0	–	0.0	–	0	–	0.0	–
Puglia	0	–	0.0	–	0	–	0.0	–	0	–	0.0	–
Basilicata	0	–	0.0	–	0	–	0.0	–	179,092	75.6	240.2	–
Calabria	0	–	0.0	–	0	–	0.0	–	783,441	34.8	210.0	–
Sicilia	0	–	0.0	–	0	–	0.0	–	0	–	0.0	–
Sardegna	0	–	0.0	–	0	–	0.0	–	0	–	0.0	–
Italia	46,417,437	5.2	118.0	4.4	128,196,588	3.5	218.5	2.9	18,323,055	8.4	262.3	4.2

(continued)

Table 7.28 (continued)

Region/Regione	Scots pine and Mountain pine				Black pines				Mediterranean pines			
	Pinete di pino silvestre e montano				Pinete di pino nero, laricio e loricato				Pinete di pini mediterranei			
	Biomass Fitomassa	ES	Biomass Fitomassa	ES	Biomass Fitomassa	ES	Biomass Fitomassa	ES	Biomass Fitomassa	ES	Biomass Fitomassa	ES
	(Mg)	(%)	(Mg ha^{-1})	(%)	(Mg)	(%)	(Mg ha^{-1})	(%)	(Mg)	(%)	(Mg ha^{-1})	(%)
Piemonte	2,559,984	17.8	123.3	–	328,956	41.5	101.8	–	32,134	72.6	39.8	–
Valle d'Aosta	1,308,268	21.4	112.1	–	31,386	100.0	81.5	–	0	–	0.0	–
Lombardia	2,193,272	24.1	115.2	–	699,079	44.3	158.6	–	0	–	0.0	–
Alto Adige	6,501,145	13.4	175.5	–	5221	100.0	13.8	–	0	–	0.0	–
Trentino	2,631,383	16.7	125.9	–	912,173	29.6	158.2	–	0	–	0.0	–
Veneto	1,253,578	20.0	102.3	–	390,749	42.8	111.4	–	127,511	71.0	170.5	–
Friuli V.G	1,273,356	27.7	104.1	–	4,078,370	14.6	129.3	–	67,605	63.3	60.6	–
Liguria	906,684	23.4	88.4	–	921,169	36.2	167.6	–	1,173,999	27.3	47.8	–
Emilia Romagna	190,741	55.0	103.6	–	2,861,888	24.3	176.6	–	394,831	54.3	134.0	–
Toscana	227,192	59.1	209.6	–	4,628,974	24.1	256.2	–	6,449,350	16.0	142.8	–
Umbria	82,559	76.6	112.0	–	628,654	28.3	106.6	–	769,502	23.9	94.9	–
Marche	0	–	0.0	–	1,088,000	24.1	104.8	–	207,295	48.9	94.5	–
Lazio	0	–	0.0	–	1,168,687	25.3	137.9	–	816,132	28.6	111.1	–
Abruzzo	224,661	62.9	206.9	–	2,040,214	25.5	110.8	–	206,733	53.6	95.2	–
Molise	0	–	0.0	–	214,822	44.2	91.7	–	102,414	64.6	46.9	–
Campania	0	–	0.0	–	849,082	36.8	153.7	–	965,649	26.4	105.6	–
Puglia	0	–	0.0	–	351,609	53.4	226.3	–	1,747,462	20.6	63.0	–
Basilicata	0	–	0.0	–	257,444	43.9	98.6	–	604,763	24.7	67.7	–
Calabria	0	–	0.0	–	12,232,860	11.0	166.6	–	1,659,040	23.5	94.9	–
Sicilia	0	–	0.0	–	551,327	29.4	73.6	–	3,796,759	20.6	83.8	–
Sardegna	0	–	0.0	–	480,360	43.9	91.9	–	1,924,342	17.2	55.6	–
Italia	19,352,822	6.9	130.0	4.9	34,721,024	6.4	150.7	5.0	21,045,520	7.4	87.5	6.4

(continued)

Table 7.28 (continued)

Region/Regione	Other coniferous forests				Beech				Temperate oaks			
	Altri boschi di conifere, pure o miste				Faggete				Querceti di rovere, roverella e farnia			
	Biomass	ES	Biomass	ES	Biomass	ES	Biomass	ES	Biomass	ES	Biomass	ES
	Fitomassa		Fitomassa		Fitomassa		Fitomassa		Fitomassa		Fitomassa	
	(Mg)	(%)	(Mg ha^{-1})	(%)	(Mg)	(%)	(Mg ha^{-1})	(%)	(Mg)	(%)	(Mg ha^{-1})	(%)
Piemonte	719,586	35.0	111.3	–	18,967,409	7.9	161.9	–	6,134,043	14.6	80.5	–
Valle d'Aosta	0	–	0.0	–	266,877	67.5	230.9	–	396,678	37.9	93.6	–
Lombardia	848,596	39.1	133.7	–	11,364,965	12.4	170.5	–	4,361,171	15.5	106.1	–
Alto Adige	0	–	0.0	–	679,640	31.4	163.4	–	424,618	32.1	93.6	–
Trentino	0	–	0.0	–	10,704,907	10.6	171.0	–	599,116	33.4	103.9	–
Veneto	0	–	0.0	–	13,540,760	10.9	200.2	–	1,930,179	19.0	123.3	–
Friuli V.G	171,586	67.5	153.9	–	18,954,267	9.8	209.1	–	783,523	28.0	95.8	–
Liguria	98,281	100.0	268.2	–	7,422,867	11.4	200.5	–	3,487,239	15.1	79.5	–
Emilia Romagna	404,486	33.6	122.0	–	19,306,204	8.9	188.3	–	4,945,030	11.7	64.2	–
Toscana	2,428,906	23.6	204.1	–	17,247,367	9.5	235.2	–	11,208,658	8.3	73.7	–
Umbria	356,024	46.4	193.1	–	2,425,472	20.5	146.8	–	5,622,299	11.9	55.7	–
Marche	374,674	31.8	66.2	–	2,887,735	21.7	162.3	–	4,282,269	15.2	69.3	–
Lazio	305,601	58.4	207.4	–	17,220,397	10.5	231.4	–	4,861,315	13.9	59.5	–
Abruzzo	219,882	66.0	101.3	–	27,415,999	6.9	220.3	–	6,343,281	10.3	72.3	–
Molise	0	–	0.0	–	3,961,857	18.5	267.0	–	3,296,101	15.6	68.4	–
Campania	29,242	61.3	33.9	–	16,184,788	11.8	287.8	–	3,552,340	13.8	58.3	–
Puglia	296,167	49.7	127.1	–	1,428,803	31.4	306.5	–	1,818,720	22.3	70.3	–
Basilicata	460,077	37.5	89.3	–	6,906,068	16.3	257.5	–	2,237,102	16.2	51.9	–
Calabria	2,732,101	22.9	217.5	–	24,286,115	8.2	305.8	–	4,173,673	19.6	77.6	–
Sicilia	767,091	30.3	82.5	–	2,578,783	19.1	170.0	–	4,313,246	15.2	60.0	–
Sardegna	919,248	28.1	70.4	–	0	–	0.0	–	3,939,359	12.2	44.9	–
Italia	11,131,548	9.8	132.7	8.1	223,751,279	2.6	212.5	2.0	78,709,959	3.3	68.3	2.9

(continued)

Table 7.28 (continued)

Region/Regione	Mediterranean oaks				Chestnut				Hornbeam and Hophornbeam			
	Cerrete, boschi di farnetto, fragno e vallonea				Castagneti				Ostrieti, carpineti			
	Biomass	ES	Biomass	ES	Biomass	ES	Biomass	ES	Biomass	ES	Biomass	ES
	Fitomassa		Fitomassa		Fitomassa		Fitomassa		Fitomassa		Fitomassa	
	(Mg)	(%)	(Mg ha^{-1})	(%)	(Mg)	(%)	(Mg ha^{-1})	(%)	(Mg)	(%)	(Mg ha^{-1})	(%)
Piemonte	985,634	28.9	162.6	–	22,038,896	6.5	134.7	–	1,339,361	21.0	82.9	–
Valle d'Aosta	0	–	0.0	–	477,413	32.0	112.6	–	0	–	0.0	–
Lombardia	334,894	48.2	151.9	–	13,625,383	10.0	166.0	–	5,892,534	12.4	72.1	–
Alto Adige	0	–	0.0	–	249,973	53.8	132.2	–	1,047,927	22.7	118.7	–
Trentino	0	–	0.0	–	263,342	56.4	182.7	–	3,710,222	17.1	90.8	–
Veneto	0	–	0.0	–	1,962,322	19.4	114.1	–	8,866,997	10.5	108.1	–
Friuli V.G	0	–	0.0	–	2,706,129	18.6	220.7	–	4,078,523	16.2	87.2	–
Liguria	1,486,662	21.3	132.2	–	12,849,715	7.9	117.3	–	3,755,256	14.6	81.2	–
Emilia Romagna	9,823,304	11.6	97.0	–	5,370,966	12.4	132.5	–	9,232,902	9.8	83.0	–
Toscana	22,365,673	7.0	89.3	–	20,083,621	7.2	128.6	–	6,236,927	11.9	96.5	–
Umbria	10,368,804	9.5	80.6	–	210,480	49.4	95.2	–	3,468,604	13.4	59.3	–
Marche	2,972,795	21.5	120.2	–	481,440	34.9	129.9	–	4,907,129	11.4	64.3	–
Lazio	14,145,124	10.4	106.8	–	3,977,409	17.4	111.1	–	7,242,284	10.4	73.9	–
Abruzzo	3,643,703	16.9	113.4	–	665,937	28.9	141.5	–	3,719,661	14.4	77.6	–
Molise	5,784,411	15.7	111.3	–	45,130	100.0	115.6	–	1,050,348	23.0	116.9	–
Campania	7,596,136	15.8	101.8	–	3,799,214	14.6	67.9	–	4,502,182	16.4	84.9	–
Puglia	3,187,374	20.3	82.3	–	126,698	57.7	108.7	–	454,147	34.9	97.4	–
Basilicata	12,818,858	10.4	98.1	–	716,548	32.8	120.3	–	649,735	26.3	91.7	–
Calabria	7,002,062	15.2	160.6	–	5,897,440	16.1	85.5	–	125,157	86.9	28.0	–
Sicilia	2,764,896	17.2	89.2	–	816,570	27.9	93.6	–	57,796	69.9	33.3	–
Sardegna	0	–	0.0	–	164,175	46.9	88.0	–	0	–	0.0	–
Italia	105,280,329	3.5	99.3	3.1	96,528,800	3.2	124.0	2.5	70,337,694	3.5	81.9	2.9

(continued)

Table 7.28 (continued)

	Hygrophilous forests				Other deciduous broadleaved forests				Holm oak			
	Boschi igrofili				Altri boschi caducifogli				Leccete			
	Biomass Fitomassa	ES	Biomass Fitomassa	ES	Biomass Fitomassa	ES	Biomass Fitomassa	ES	Biomass Fitomassa	ES	Biomass Fitomassa	ES
	(Mg)	(%)	(Mg ha^{-1})	(%)	(Mg)	(%)	(Mg ha^{-1})	(%)	(Mg)	(%)	(Mg ha^{-1})	(%)
Piemonte	1,850,336	21.6	81.4	–	28,873,392	5.6	89.2	–	0	–	0.0	–
Valle d'Aosta	0	–	0.0	–	725,625	26.9	69.0	–	0	–	0.0	–
Lombardia	1,568,062	20.5	83.2	–	12,757,600	9.8	105.4	–	0	–	0.0	–
Alto Adige	267,369	39.5	97.3	–	495,790	38.5	140.4	–	0	–	0.0	–
Trentino	518,539	49.4	134.3	–	1,756,524	20.0	106.6	–	9267	100.0	25.7	–
Veneto	596,129	27.6	57.2	–	4,329,069	12.8	82.6	–	392,941	42.4	131.4	–
Friuli V.G	511,548	25.4	40.6	–	5,639,535	13.1	118.9	–	0	–	0.0	–
Liguria	412,762	43.3	125.1	–	2,267,178	17.2	71.2	–	1,155,391	23.1	83.1	–
Emilia Romagna	1,613,731	21.5	57.2	–	6,606,730	14.2	76.5	–	109,656	72.8	148.8	–
Toscana	2,576,452	17.8	92.8	–	7,558,614	11.6	89.5	–	11,614,784	8.5	89.6	–
Umbria	871,362	40.3	108.4	–	636,629	30.4	56.4	–	4,363,599	14.2	105.7	–
Marche	1,343,408	23.2	104.1	–	3,841,284	15.3	61.9	–	743,223	27.7	105.5	–
Lazio	609,234	28.5	66.1	–	3,475,716	20.6	64.3	–	3,822,040	19.5	79.2	–
Abruzzo	1,362,695	22.2	68.6	–	2,990,592	20.4	53.1	–	697,355	32.0	80.3	–
Molise	385,668	29.7	42.6	–	792,327	28.0	77.5	–	33,712	82.8	28.8	–
Campania	996,640	33.2	90.2	–	2,611,948	18.6	76.0	–	3,193,474	18.1	85.2	–
Puglia	29,224	100.0	75.2	–	1,603,410	31.0	139.0	–	1,215,708	28.2	70.0	–
Basilicata	468,746	29.5	36.1	–	1,968,268	19.3	65.9	–	571,914	33.0	55.1	–
Calabria	957,547	30.1	74.4	–	4,267,417	19.1	94.5	–	9,375,273	25.6	192.5	–
Sicilia	168,575	30.0	34.2	–	720,446	29.1	54.4	–	2,516,784	22.0	128.8	–
Sardegna	139,262	36.0	41.5	–	66,989	30.4	8.7	–	19,235,620	8.3	75.3	–
Italia	17,247,289	6.7	73.4	5.1	93,985,083	3.3	84.4	2.8	59,050,741	5.7	91.8	5.2

(continued)

Table 7.28 (continued)

	Cork oak				Other evergreen broadleaved forests				Not classified				Total Tall trees forest			
	Sugherete				Altri boschi di latifoglie sempreverdi				Non classificato				Totale Boschi alti			
	Biomass	ES	Biomass	ES	Biomass	ES	Biomass	ES	Biomass	ES	Biomass	ES	Biomass	ES	Biomass	ES
	Fitomassa		Fitomassa		Fitomassa		Fitomassa		Fitomassa		Fitomassa		Fitomassa		Fitomassa	
	(Mg)	(%)	(Mg ha^{-1})	(%)	(Mg)	(%)	(Mg ha^{-1})	(%)	(Mg)	(%)	(Mg ha^{-1})	(%)	(Mg)	(%)	(Mg ha^{-1})	(%)
Piemonte	0	–	0.0	–	0	–	0.0	–	12,876	100.0	25.3	–	98,618,491	2.8	113.4	2.4
Valle d'Aosta	0	–	0.0	–	0	–	0.0	–	0	–	0.0	–	11,227,949	6.9	113.1	5.6
Lombardia	0	–	0.0	–	116,973	62.9	51.0	–	11,902	100.0	27.0	–	81,605,175	3.7	136.7	3.2
Alto Adige	0	–	0.0	–	0	–	0.0	–	53,465	76.7	35.4	–	61,507,973	4.3	181.3	3.9
Trentino	0	–	0.0	–	0	–	0.0	–	0	–	0.0	–	63,838,069	4.3	171.0	4.0
Veneto	0	–	0.0	–	0	–	0.0	–	0	–	0.0	–	61,451,389	4.0	149.4	3.3
Friuli V.G	0	–	0.0	–	0	–	0.0	–	0	–	0.0	–	51,670,841	4.8	159.8	4.3
Liguria	0	–	0.0	–	0	–	0.0	–	3042	70.7	4.1	–	36,848,401	4.1	107.5	3.9
Emilia Romagna	0	–	0.0	–	0	–	0.0	–	0	–	0.0	–	62,862,543	4.0	108.6	3.6
Toscana	570,662	30.2	92.9	–	41,277	62.7	28.6	–	88,624	70.7	122.6	–	115,055,051	2.9	111.8	2.6
Umbria	0	–	0.0	–	0	–	0.0	–	0	–	0.0	–	29,803,987	4.8	77.6	4.4
Marche	0	–	0.0	–	0	–	0.0	–	0	–	0.0	–	23,169,797	5.7	81.3	5.1
Lazio	205,965	53.4	79.9	–	86,681	91.8	33.6	–	150,094	53.0	112.4	–	58,266,635	4.6	104.4	4.2
Abruzzo	0	–	0.0	–	60,953	100.0	43.3	–	3381	100.0	9.3	–	49,850,959	4.3	122.0	3.9
Molise	0	–	0.0	–	0	–	0.0	–	0	–	0.0	–	15,915,418	7.5	105.7	6.8
Campania	55,841	100.0	151.6	–	6233	70.7	8.5	–	45,250	100.0	122.9	–	44,388,019	5.6	110.8	5.2
Puglia	0	–	0.0	–	273,182	33.0	53.0	–	84,510	62.3	75.1	–	12,617,013	8.6	88.7	7.4
Basilicata	0	–	0.0	–	128,890	95.9	69.1	–	33,037	100.0	95.1	–	28,000,541	6.3	97.7	5.9
Calabria	594,708	42.2	113.8	–	661,901	34.2	31.3	–	209,429	40.8	93.5	–	74,958,164	5.0	152.1	4.6
Sicilia	568,864	24.3	33.0	–	1,762,552	16.6	45.0	–	0	–	0.0	–	21,383,690	6.3	75.1	5.6
Sardegna	7,910,785	8.9	51.8	–	909,519	22.2	23.9	–	1561	100.0	4.2	–	35,691,220	5.2	59.5	4.9
Italia	9,906,823	7.9	53.7	6.7	4,048,162	11.5	35.6	9.9	697,171	22.9	69.3	12.7	1,038,731,323	1.0	116.0	0.9

Table 7.29 Total value and value per hectare of aboveground tree biomass by Plantations forest type / Valori totali e per ettaro della fitomassa arborea epigea per le categorie forestali degli Impianti di arboricoltura da legno

Region/Regione	Poplar plantations				Other broadleaved plantations				Coniferous plantations				Total Plantations			
	Pioppeti artificiali				Piantagioni di altre latifoglie				Piantagioni di conifere				Totale Impianti di arboricoltura da legno			
	Biomass Fitomassa	ES	Biomass Fitomassa	ES	Biomass Fitomassa	ES	Biomass Fitomassa	ES	Biomass Fitomassa	ES	Biomass Fitomassa	ES	Biomass Fitomassa	ES	Biomass Fitomassa	ES
	(Mg)	(%)	(Mg ha^{-1})	(%)	(Mg)	(%)	(Mg ha^{-1})	(%)	(Mg)	(%)	(Mg ha^{-1})	(%)	(Mg)	(%)	(Mg ha^{-1})	(%)
Piemonte	757,850	36.2	48.9	–	299,881	53.7	75.0	–	215,113	61.2	186.9	–	1,272,845	26.9	61.6	20.1
Valle d'Aosta	0	–	0.0	–	0	–	0.0	–	0	–	0.0	–	0	–	0.0	–
Lombardia	654,093	25.7	35.3	–	373,671	44.8	60.4	–	97,606	100.0	221.4	–	1,125,370	21.2	44.8	15.4
Alto Adige	0	–	0.0	–	0	–	0.0	–	0	–	0.0	–	0	–	0.0	–
Trentino	0	–	0.0	–	0	–	0.0	–	0	–	0.0	–	0	–	0.0	–
Veneto	155,443	77.1	66.7	–	191,863	56.8	65.1	–	0	–	0.0	–	347,306	38.2	65.8	12.4
Friuli V.G	417,576	33.6	63.9	–	129,649	42.3	48.8	–	0	–	0.0	–	547,225	26.3	59.5	19.3
Liguria	25,664	100.0	70.0	–	0	–	0.0	–	0	–	0.0	–	25,664	100.0	70.0	–
Emilia Romagna	185,081	40.0	43.5	–	127,126	58.0	89.0	–	76,261	100.0	207.0	–	388,468	33.2	64.2	30.1
Toscana	72,644	70.9	98.7	–	140,168	43.4	30.5	–	349,539	56.7	240.7	–	562,351	37.8	82.9	32.2
Umbria	0	–	0.0	–	136,717	33.1	21.4	–	0	–	0.0	–	136,717	33.1	21.4	10.4
Marche	0	–	0.0	–	140,485	40.0	20.5	–	0	–	0.0	–	140,485	40.0	20.5	29.8
Lazio	18,549	70.9	23.5	–	79,894	57.3	57.6	–	0	–	0.0	–	98,442	48.1	45.2	31.0
Abruzzo	9040	100.0	17.1	–	27,971	55.7	11.4	–	0	–	0.0	–	37,011	48.5	12.5	37.3
Molise	70,596	100.0	180.8	–	25,608	47.6	11.0	–	0	–	0.0	–	96,203	74.5	35.4	62.1
Campania	32,380	84.8	21.0	–	10,011	58.8	6.2	–	0	–	0.0	–	42,391	66.3	13.4	26.8
Puglia	0	–	0.0	–	912	100.0	9.1	–	0	–	0.0	–	912	100.0	9.1	–
Basilicata	0	–	0.0	–	2039	100.0	1.3	–	0	–	0.0	–	2039	100.0	1.3	–
Calabria	1630	100.0	16.3	–	47,795	68.5	30.6	–	126,866	70.7	170.0	–	176,291	54.2	73.3	36.9
Sicilia	0	–	0.0	–	1195	70.7	1.6	–	0	–	0.0	–	1195	70.7	1.6	–
Sardegna	0	–	0.0	–	488,342	22.5	20.0	–	56,025	50.0	37.5	–	544,367	20.9	21.0	18.0
Italia	2,400,544	16.2	46.5	12.9	2,223,326	14.1	31.2	10.3	921,410	30.8	163.1	17.0	5,545,281	10.1	43.2	8.1

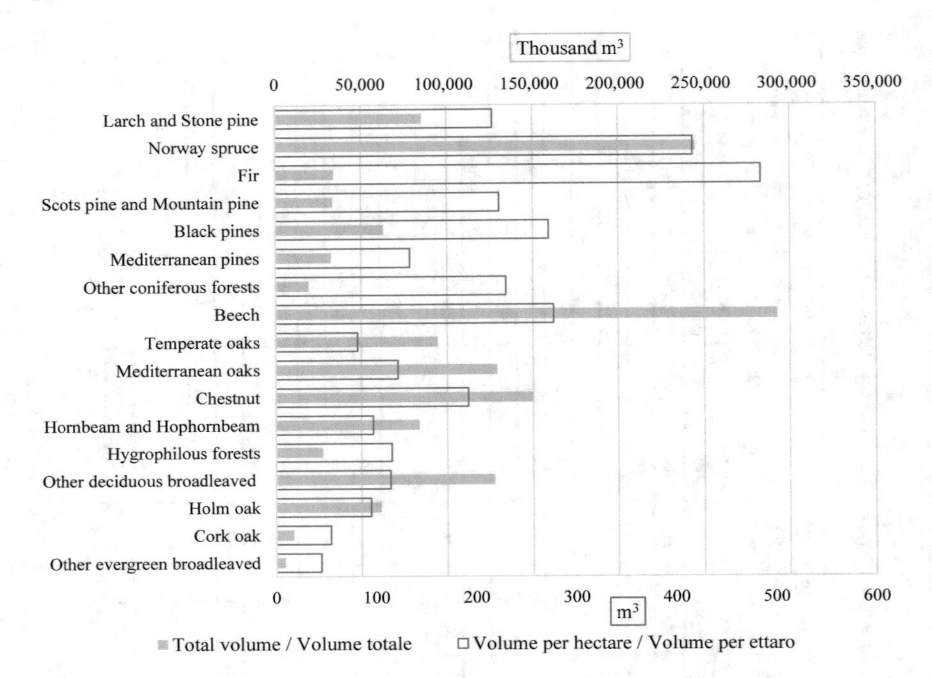

Fig. 7.13 Growing stock volume, total and per hectare, by Tall trees forest type / Volume del fusto, totale e per ettaro, per categoria forestale dei Boschi alti

the rate of truncated trees is not constant and their biomass is estimated using a simplified method (Chap. 6). For this reason, we let users compute new BCEF based on the INFC2015 statistics. However, the Mediterranean pines forest type will show a coefficient behind the small, expected difference with the former; this is due to the novel use of volume and biomass functions used for *Pinus halepensis*, that were not available in 2005.

Estimates of the annual volume increments are shown in Table 7.30 for the inventory categories of Forest; in Table 7.31, for the Tall trees forest types; and in Table 7.32, for Plantations. The mean annual volume increment of Forest in Italy amounts to 37.8 million cubic metres, 4.2 m^3 per hectare on average. Among the Tall trees forests (Fig. 7.15), the main contribution is due to Beech forest (15.5%), followed by Norway spruce (12.8%) and Other broadleaved forests (12.4%).

For totals at the national level, a main contribution is due to the broadleaved forest types, since each of these generally shows higher values that those characterised by conifers, although it is among the broadleaved that we found the minimum values (Cork oak and Other evergreen broadleaved forests, both contributing 0.6% of total volume increment). The four highest per hectare values were all in coniferous forest types (Fir, Norway spruce, Black pines and Other coniferous forests), ranging from 6.1 and 9.1 m^3 per hectare and year. Among the broadleaved types, Beech forest contributes with 5.5 m^3 per hectare and year, Chestnut with di 5.4 m^3 per hectare and year, a very similar value.

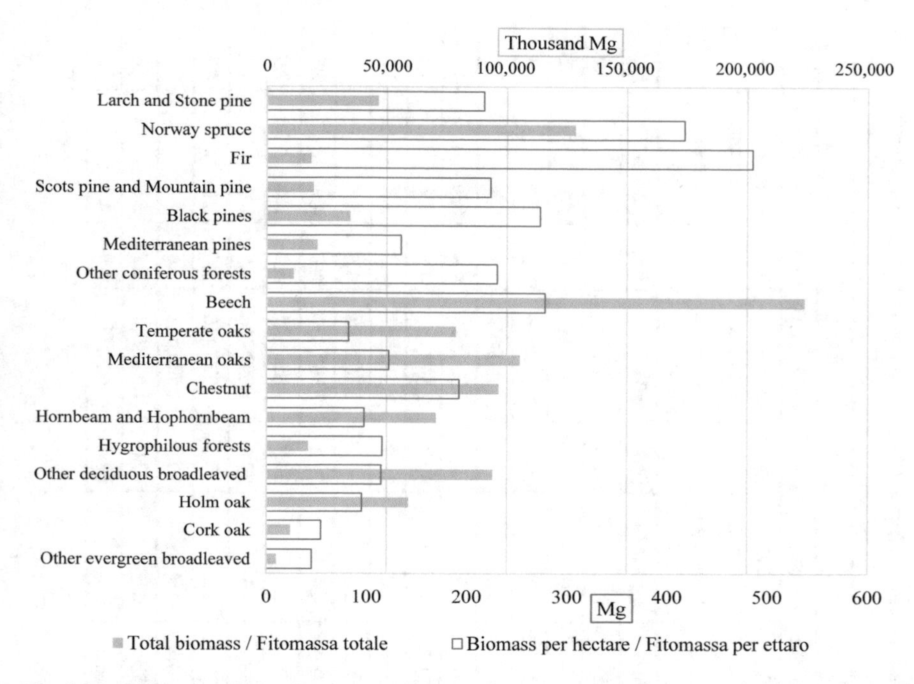

Fig. 7.14 Aboveground tree biomass, total and per hectare, by Tall trees forest type / Biomassa arborea epigea, totale e per ettaro, per le categorie forestali dei Boschi alti

7.6 Structure and Development

Stands structure and development are described through a series of variables referring to crown coverage, vertical structure, silvicultural system and applied silvicultural practices, development stage and age.

Assessment of crown coverage specifically for trees and for shrubs, on orthophotos and in the field, allowed each plot to be assigned into one of the following six classes: <5%, 5–10%, 11–20%, 21–50%, 51–80%, >80%, used for the INFC statistics.

The silvicultural system, classified by combining information on trees origin (from seed or suckers, or a mixture of the two) and adopted silvicultural practices, was recorded based on ten classes, described in Table 7.33. Even-aged coppice, coppice in transition to high forest, and even-aged high forest were classified as one-layer or two-layers stands to describe the vertical structure. The development stage was classified according to the classes shown in Table 7.34; in case of even-aged stands, age was assessed through the following year classes: less than 10 years, between years 11–20, 21–30, 31–40, 41–80, 81–120, more than 120 years. Silvicultural system and development stage were not assessed in Plantations, but the age class was considered. Data on these attributes was used to derive area estimates and estimates of the quantitative attributes of growing stock and increment by the silvicultural system, development stage and age class.

Table 7.30 Total value and value per hectare of annual volume increment by Forest inventory category / Valori totali e per ettaro dell'incremento annuo di volume per le categorie inventariali del Bosco

Region/Regione	Tall trees forest				Plantations				Total Forest			
	Boschi alti				Impianti di arboricoltura da legno				Totale Bosco			
	Annual increment	ES	Annual increment	ES	Annual increment	ES	Annual increment	ES	Annual increment	ES	Annual increment	ES
	Incremento annuo		Incremento annuo		Incremento annuo		Incremento annuo		Incremento annuo		Incremento annuo	
	(m^3)	(%)	(m^3 ha^{-1})	(%)	(m^3)	(%)	(m^3 ha^{-1})	(%)	(m3)	(%)	(m^3 ha^{-1})	(%)
Piemonte	3,663,223	3.3	4.2	3.0	44,528	28.5	2.2	21.3	3,707,751	3.2	4.2	2.9
Valle d'Aosta	356,387	7.4	3.6	6.2	0	-	0.0	-	356,387	7.4	3.6	6.2
Lombardia	3,535,499	3.7	5.9	3.2	281,578	23.7	11.2	20.4	3,817,078	3.8	6.1	3.4
Alto Adige	2,000,689	4.1	5.9	3.7	0	-	0.0	–	2,000,689	4.1	5.9	3.7
Trentino	2,256,626	4.4	6.0	4.1	0	-	0.0	–	2,256,626	4.4	6.0	4.1
Veneto	2,285,405	3.9	5.6	3.2	106,726	71.8	20.2	60.7	2,392,131	4.8	5.7	4.3
Friuli V.G	1,689,390	4.7	5.2	4.2	29,561	26.6	3.2	18.9	1,718,951	4.6	5.2	4.2
Liguria	1,161,598	4.4	3.4	4.2	1390	100.0	3.8	–	1,162,988	4.4	3.4	4.2
Emilia Romagna	2,338,546	4.3	4.0	3.8	26,887	41.5	4.4	34.8	2,365,432	4.3	4.0	3.8
Toscana	3,506,840	3.2	3.4	2.9	44,189	41.9	6.5	34.9	3,551,029	3.2	3.4	2.9
Umbria	1,014,247	4.6	2.6	4.1	8027	36.1	1.3	29.6	1,022,274	4.5	2.6	4.1
Marche	968,253	5.4	3.4	4.8	9164	39.7	1.3	37.0	977,416	5.4	3.3	4.8
Lazio	1,916,400	4.6	3.4	4.2	4263	43.2	2.0	16.2	1,920,662	4.6	3.4	4.2
Abruzzo	1,567,894	3.9	3.8	3.4	2254	52.9	0.8	47.6	1,570,148	3.9	3.8	3.4
Molise	615,150	6.0	4.1	5.0	5600	80.3	2.1	68.4	620,751	5.9	4.1	5.0

(continued)

Table 7.30 (continued)

Region/Regione	Tall trees forest				Plantations				Total Forest			
	Boschi alti				Impianti di arboricoltura da legno				Totale Bosco			
	Annual increment	ES	Annual increment	ES	Annual increment	ES	Annual increment	ES	Annual increment	ES	Annual increment	ES
	Incremento annuo		Incremento annuo		Incremento annuo		Incremento annuo		Incremento annuo		Incremento annuo	
	(m^3)	(%)	(m^3 ha^{-1})	(%)	(m^3)	(%)	(m^3 ha^{-1})	(%)	(m3)	(%)	(m^3 ha^{-1})	(%)
Campania	2,001,817	5.0	5.0	4.4	3961	62.4	1.3	24.3	2,005,778	5.0	5.0	4.4
Puglia	384,099	7.1	2.7	5.3	35	100.0	0.4	-	384,134	7.1	2.7	5.3
Basilicata	1,069,016	5.6	3.7	5.0	125	100.0	0.1	-	1,069,141	5.6	3.7	5.0
Calabria	3,039,200	5.1	6.2	4.7	12,222	52.1	5.1	36.8	3,051,422	5.1	6.2	4.7
Sicilia	754,428	6.8	2.6	6.0	67	70.7	0.1	-	754,495	6.8	2.6	6.0
Sardegna	1,027,626	5.7	1.7	5.4	54,877	20.9	2.1	18.6	1,082,502	5.5	1.7	5.2
Italia	37,152,332	1.0	4.1	1.0	635,452	16.7	4.9	15.7	37,787,784	1.0	4.2	1.0

Table 7.31 Total value and value per hectare of annual volume increment by Tall trees forest type / Valori totali e per unità di superficie dell'incremento corrente di volume per le categorie forestali dei Boschi alti

Region/Regione	Larch and Swiss stone pine				Norway spruce				Fir			
	Boschi di larice e cembro				Boschi di abete rosso				Boschi di abete bianco			
	Annual increment	ES	Annual increment	ES	Annual increment	ES	Annual increment	ES	Annual increment	ES	Annual increment	ES
	Incremento annuo		Incremento annuo		Incremento annuo		Incremento annuo		Incremento annuo		Incremento annuo	
	(m^3)	(%)	(m^3 ha^{-1})	(%)	(m^3)	(%)	(m^3 ha^{-1})	(%)	(m^3)	(%)	(m^3 ha^{-1})	(%)
Piemonte	228,132	11.8	2.8	–	131,816	18.8	7.2	–	102,623	19.8	7.3	–
Valle d'Aosta	155,824	13.7	3.3	–	93,135	17.2	5.1	–	5,673	62.9	4.9	–
Lombardia	224,657	12.5	4.2	–	770,790	10.4	8.8	–	94,182	29.9	10.0	–
Alto Adige	355,390	10.2	3.8	–	1,316,269	5.9	7.4	–	19,869	40.4	7.5	–
Trentino	209,040	13.3	3.4	–	1,160,428	7.0	8.5	–	130,804	19.1	8.1	–
Veneto	174,977	16.9	4.1	–	783,771	8.7	8.1	–	64,632	24.5	9.1	–
Friuli V.G	37,484	28.6	2.9	–	422,941	13.0	9.5	–	14,865	49.5	8.0	–
Liguria	3928	64.6	3.6	–	2809	100.0	7.7	–	35,644	38.8	12.2	–
Emilia Romagna	0	–	0.0	–	59,812	34.8	16.2	–	30,658	41.2	10.4	–
Toscana	0	–	0.0	–	9628	100.0	26.6	–	59,281	35.3	11.7	–
Umbria	0	–	0.0	–	0	–	0.0	–	0	–	0.0	–
Marche	0	–	0.0	–	3090	100.0	8.3	–	0	–	0.0	–
Lazio	0	–	0.0	–	10,593	100.0	28.7	–	0	–	0.0	–
Abruzzo	0	–	0.0	–	2226	100.0	6.1	–	7584	70.7	10.5	–
Molise	0	–	0.0	–	0	–	0.0	–	7624	62.6	6.5	–
Campania	0	–	0.0	–	0	–	0.0	–	0	–	0.0	–
Puglia	0	–	0.0	–	0	–	0.0	–	0	–	0.0	–
Basilicata	0	–	0.0	–	0	–	0.0	–	7907	75.3	10.6	–
Calabria	0	–	0.0	–	0	–	0.0	–	55,597	42.7	14.9	–
Sicilia	0	–	0.0	–	0	–	0.0	–	0	–	0.0	–
Sardegna	0	–	0.0	–	0	–	0.0	–	0	–	0.0	–
Italia	1,389,431	5.1	3.5	4.3	4,767,308	3.5	8.1	2.8	636,943	9.5	9.1	5.5

(continued)

Table 7.31 (continued)

	Scots pine and Mountain pine				Black pines				Mediterranean pines			
	Pinete di pino silvestre e montano				Pinete di pino nero, laricio e loricato				Pinete di pini mediterranei			
	Annual increment	ES	Annual increment	ES	Annual increment	ES	Annual increment	ES	Annual increment	ES	Annual increment	ES
	Incremento annuo		Incremento annuo		Incremento annuo		Incremento annuo		Incremento annuo		Incremento annuo	
	(m^3)	(%)	(m^3 ha^{-1})	(%)	(m^3)	(%)	(m^3 ha^{-1})	(%)	(m^3)	(%)	(m^3 ha^{-1})	(%)
Piemonte	66,210	19.5	3.2	–	10,672	40.6	3.3	–	897	73.1	1.1	–
Valle d'Aosta	31,084	22.2	2.7	–	1252	100.0	3.3	–	0	–	0.0	–
Lombardia	74,284	22.4	3.9	–	30,976	45.4	7.0	–	0	–	0.0	–
Alto Adige	180,623	15.0	4.9	–	313	100.0	0.8	–	0	–	0.0	–
Trentino	67,163	19.6	3.2	–	20,255	30.6	3.5	–	0	–	0.0	–
Veneto	33,747	19.8	2.8	–	12,393	40.8	3.5	–	5095	70.7	6.8	–
Friuli V.G	30,552	23.6	2.5	–	98,523	16.6	3.1	–	2072	81.7	1.9	–
Liguria	32,164	24.9	3.1	–	26,996	35.6	4.9	–	39,206	26.8	1.6	–
Emilia Romagna	8647	53.6	4.7	–	106,738	22.7	6.6	–	8893	41.5	3.0	–
Toscana	10,761	61.2	9.9	–	140,638	20.4	7.8	–	152,923	16.9	3.4	–
Umbria	3908	75.4	5.3	–	24,121	26.7	4.1	–	24,489	26.7	3.0	–
Marche	0	–	0.0	–	39,014	22.6	3.8	–	8350	43.9	3.8	–
Lazio	0	–	0.0	–	42,499	24.6	5.0	–	20,605	31.1	2.8	–
Abruzzo	5403	58.5	5.0	–	82,600	20.6	4.5	–	10,207	57.8	4.7	–
Molise	0	–	0.0	–	9961	44.8	4.3	–	8472	59.6	3.9	–
Campania	0	–	0.0	–	39,075	39.5	7.1	–	42,190	34.6	4.6	–
Puglia	0	–	0.0	–	19,274	52.0	12.4	–	77,481	19.0	2.8	–
Basilicata	0	–	0.0	–	13,580	44.6	5.2	–	29,352	23.5	3.3	–
Calabria	0	–	0.0	–	645,267	10.9	8.8	–	57,402	24.9	3.3	–
Sicilia	0	–	0.0	–	23,260	30.7	3.1	–	167,523	18.8	3.7	–
Sardegna	0	–	0.0	–	27,736	37.0	5.3	–	98,231	18.6	2.8	–
Italia	544,546	7.4	3.7	5.5	1,415,142	6.3	6.1	5.0	753,389	7.2	3.1	5.8

(continued)

Table 7.31 (continued)

	Other coniferous forests				Beech				Temperate oaks			
	Altri boschi di conifere, pure o miste				Faggete				Querceti di rovere, roverella e farnia			
	Annual increment	ES	Annual increment	ES	Annual increment	ES	Annual increment	ES	Annual increment	ES	Annual increment	ES
	Incremento annuo		Incremento annuo		Incremento annuo		Incremento annuo		Incremento annuo		Incremento annuo	
	(m^3)	(%)	(m^3 ha^{-1})	(%)	(m^3)	(%)	(m^3 ha^{-1})	(%)	(m^3)	(%)	(m^3 ha^{-1})	(%)
Piemonte	31,187	41.0	4.8	–	486,933	8.1	4.2	–	220,793	17.0	2.9	–
Valle d'Aosta	0	–	0.0	–	5207	65.1	4.5	–	12,739	40.9	3.0	–
Lombardia	30,580	35.4	4.8	–	379,207	11.6	5.7	–	201,254	14.3	4.9	–
Alto Adige	0	–	0.0	–	20,186	34.3	4.9	–	12,818	31.2	2.8	–
Trentino	0	–	0.0	–	309,883	11.4	4.9	–	19,985	35.4	3.5	–
Veneto	0	–	0.0	–	402,067	9.1	5.9	–	79,881	27.1	5.1	–
Friuli V.G	4942	64.4	4.4	–	442,999	8.9	4.9	–	25,610	27.9	3.1	–
Liguria	3522	100.0	9.6	–	183,645	12.0	5.0	–	100,336	16.4	2.3	–
Emilia Romagna	24,096	39.3	7.3	–	569,817	9.0	5.6	–	188,164	13.8	2.4	–
Toscana	87,038	25.8	7.3	–	412,712	10.9	5.6	–	316,164	8.7	2.1	–
Umbria	11,987	45.8	6.5	–	46,766	18.9	2.8	–	174,522	10.4	1.7	–
Marche	24,884	33.3	4.4	–	85,998	20.5	4.8	–	149,922	14.1	2.4	–
Lazio	11,524	67.6	7.8	–	350,942	10.3	4.7	–	164,383	13.7	2.0	–
Abruzzo	10,363	59.1	4.8	–	626,567	6.9	5.0	–	228,626	10.4	2.6	–
Molise	0	–	0.0	–	86,452	19.1	5.8	–	139,861	14.7	2.9	–
Campania	1617	74.8	1.9	–	462,736	10.8	8.2	–	157,924	12.8	2.6	–
Puglia	13,533	51.4	5.8	–	30,179	31.0	6.5	–	57,161	18.6	2.2	–
Basilicata	23,678	33.9	4.6	–	165,772	16.7	6.2	–	89,549	18.9	2.1	–
Calabria	147,103	24.0	11.7	–	623,700	8.8	7.9	–	205,938	23.2	3.8	–
Sicilia	37,668	31.7	4.1	–	64,506	20.3	4.3	–	132,376	17.5	1.8	–
Sardegna	53,885	26.3	4.1	–	0	–	0.0	–	114,030	15.6	1.3	–
Italia	517,607	10.2	6.2	8.3	5,756,273	2.6	5.5	2.0	2,792,036	3.7	2.4	3.2

(continued)

Table 7.31 (continued)

	Mediterranean oaks				Chestnut				Hornbeam and Hophornbeam			
	Cerrete, boschi di farnetto, fragno e vallonea				Castagneti				Ostrieti, carpineti			
	Annual increment	ES	Annual increment	ES	Annual increment	ES	Annual increment	ES	Annual increment	ES	Annual increment	ES
	Incremento annuo		Incremento annuo		Incremento annuo		Incremento annuo		Incremento annuo		Incremento annuo	
	(m^3)	(%)	(m^3 ha^{-1})	(%)	(m^3)	(%)	(m^3 ha^{-1})	(%)	(m^3)	(%)	(m^3 ha^{-1})	(%)
Piemonte	26,407	32.8	4.4	–	898,083	7.2	5.5	–	48,104	26.0	3.0	–
Valle d'Aosta	0	–	0.0	–	19,610	35.6	4.6	–	0	–	0.0	–
Lombardia	14,875	45.8	6.7	–	645,315	10.1	7.9	–	257,023	12.3	3.1	–
Alto Adige	0	–	0.0	–	5148	45.9	2.7	–	41,893	26.9	4.7	–
Trentino	0	–	0.0	–	9624	56.8	6.7	–	159,349	14.2	3.9	–
Veneto	0	–	0.0	–	85,975	20.5	5.0	–	357,427	9.9	4.4	–
Friuli V.G	0	–	0.0	–	126,471	20.0	10.3	–	160,470	15.1	3.4	–
Liguria	40,764	20.3	3.6	–	434,823	8.8	4.0	–	129,172	13.5	2.8	–
Emilia Romagna	319,387	10.5	3.2	–	191,020	12.4	4.7	–	357,611	9.7	3.2	–
Toscana	670,786	7.0	2.7	–	705,245	7.8	4.5	–	213,098	13.2	3.3	–
Umbria	387,904	8.3	3.0	–	6706	53.1	3.0	–	135,314	13.2	2.3	–
Marche	87,890	17.0	3.6	–	15,819	35.1	4.3	–	211,855	11.9	2.8	–
Lazio	519,329	8.8	3.9	–	204,383	21.0	5.7	–	264,245	11.2	2.7	–
Abruzzo	127,106	14.5	4.0	–	32,126	29.9	6.8	–	142,817	13.1	3.0	–
Molise	237,728	11.7	4.6	–	3329	100.0	8.5	–	43,778	23.5	4.9	–
Campania	359,874	13.1	4.8	–	313,555	16.4	5.6	–	229,689	15.4	4.3	–
Puglia	78,150	14.9	2.0	–	2479	62.0	2.1	–	12,073	37.1	2.6	–
Basilicata	500,527	8.7	3.8	–	42,575	31.1	7.1	–	20,955	27.7	3.0	–
Calabria	224,209	18.6	5.1	–	395,378	15.8	5.7	–	5727	87.4	1.3	–
Sicilia	77,765	18.4	2.5	–	45,408	31.2	5.2	–	3212	57.8	1.8	–
Sardegna	0	–	0.0	–	7839	57.4	4.2	–	0	–	0.0	–
Italia	3,672,699	3.2	3.5	2.7	4,190,912	3.7	5.4	3.0	2,793,810	3.5	3.3	2.9

(continued)

Table 7.31 (continued)

Region/Regione	Hygrophilous forests				Other deciduous broadleaved forests				Holm oak			
	Boschi igrofili				Altri boschi caducifogli				Leccete			
	Annual increment	ES	Annual increment	ES	Annual increment	ES	Annual increment	ES	Annual increment	ES	Annual increment	ES
	Incremento annuo		Incremento annuo		Incremento annuo		Incremento annuo		Incremento annuo		Incremento annuo	
	(m^3)	(%)	(m^3 ha^{-1})	(%)	(m^3)	(%)	(m^3 ha^{-1})	(%)	(m^3)	(%)	(m^3 ha^{-1})	(%)
Piemonte	131,331	33.8	5.8	–	1,279,143	6.2	4.0	–	0	–	0.0	–
Valle d'Aosta	0	–	0.0	–	31,863	24.6	3.0	–	0	–	0.0	–
Lombardia	109,095	21.2	5.8	–	696,370	10.1	5.8	–	0	–	0.0	–
Alto Adige	18,774	39.2	6.8	–	24,698	35.4	7.0	–	0	–	0.0	–
Trentino	57,934	66.8	15.0	–	111,522	22.3	6.8	–	639	100.0	1.8	–
Veneto	41,290	30.7	4.0	–	230,347	12.6	4.4	–	13,803	39.4	4.6	–
Friuli V.G	40,984	24.0	3.3	–	281,477	13.9	5.9	–	0	–	0.0	–
Liguria	12,061	39.6	3.7	–	87,623	17.0	2.8	–	28,751	20.8	2.1	–
Emilia Romagna	142,627	33.6	5.1	–	326,954	14.5	3.8	–	4123	76.4	5.6	–
Toscana	109,144	20.4	3.9	–	311,527	13.1	3.7	–	293,621	10.8	2.3	–
Umbria	52,444	38.2	6.5	–	29,413	27.8	2.6	–	116,672	15.3	2.8	–
Marche	90,574	25.5	7.0	–	219,798	14.7	3.5	–	31,059	30.3	4.4	–
Lazio	36,417	29.7	4.0	–	177,570	22.5	3.3	–	102,433	21.3	2.1	–
Abruzzo	93,109	19.6	4.7	–	155,132	17.5	2.8	–	37,241	32.6	4.3	–
Molise	30,209	29.0	3.3	–	46,402	28.9	4.5	–	1335	71.8	1.1	–
Campania	81,331	38.0	7.4	–	149,714	16.2	4.4	–	160,405	17.9	4.3	–
Puglia	2662	100.0	6.9	–	41,017	23.9	3.6	–	32,578	26.3	1.9	–
Basilicata	37,805	29.9	2.9	–	100,313	20.4	3.4	–	20,384	25.8	2.0	–
Calabria	55,555	33.3	4.3	–	300,158	24.0	6.6	–	239,636	24.6	4.9	–
Sicilia	4468	59.1	0.9	–	21,236	27.5	1.6	–	83,082	24.1	4.3	–
Sardegna	11,141	33.5	3.3	–	2713	34.7	0.4	–	500,679	9.0	2.0	–
Italia	1,158,957	8.6	4.9	7.1	4,624,991	3.6	4.2	3.1	1,666,441	5.7	2.6	5.2

(continued)

Table 7.31 (continued)

Region/Regione	Cork oak				Other evergreen broadleaved forests				Not classified				Total Tall trees forest			
	Sugherete				Altri boschi di latifoglie sempreverdi				Categoria forestale non nota				Totale Boschi alti			
	Annual increment	ES	Annual increment	ES	Annual increment	ES	Annual increment	ES	Annual increment	ES	Annual increment	ES	Annual increment	ES	Annual increment	ES
	Incremento annuo		Incremento annuo		Incremento annuo		Incremento annuo		Incremento annuo		Incremento annuo		Incremento annuo		Incremento annuo	
	(m^3)	(%)	(m^3 ha^{-1})	(%)	(m^3)	(%)	(m^3 ha^{-1})	(%)	(m^3)	(%)	(m^3 ha^{-1})	(%)	(m^3)	(%)	(m^3 ha^{-1})	(%)
Piemonte	0	–	0.0	–	0	–	0.0	–	892	100.0	1.8	–	3,663,223	3.3	4.2	3.0
Valle d'Aosta	0	–	0.0	–	0	–	0.0	–	0	–	0.0	–	356,387	7.4	3.6	6.2
Lombardia	0	–	0.0	–	5451	64.1	2.4	–	1441	100.0	3.3	–	3,535,499	3.7	5.9	3.2
Alto Adige	0	–	0.0	–	0	–	0.0	–	4710	53.3	3.1	–	2,000,689	4.1	5.9	3.7
Trentino	0	–	0.0	–	0	–	0.0	–	0	–	0.0	–	2,256,626	4.4	6.0	4.1
Veneto	0	–	0.0	–	0	–	0.0	–	0	–	0.0	–	2,285,405	3.9	5.6	3.2
Friuli V.G	0	–	0.0	–	0	–	0.0	–	0	–	0.0	–	1,689,390	4.7	5.2	4.2
Liguria	0	–	0.0	–	0	–	0.0	–	154	70.7	0.2	–	1,161,598	4.4	3.4	4.2
Emilia Romagna	0	–	0.0	–	0	–	0.0	–	0	–	0.0	–	2,338,546	4.3	4.0	3.8
Toscana	9724	40.6	1.6	–	1665	69.3	1.2	–	2884	70.7	4.0	–	3,506,840	3.2	3.4	2.9
Umbria	0	–	0.0	–	0	–	0.0	–	0	–	0.0	–	1,014,247	4.6	2.6	4.1
Marche	0	–	0.0	–	0	–	0.0	–	0	–	0.0	–	968,253	5.4	3.4	4.8
Lazio	3854	46.9	1.5	–	2472	94.3	1.0	–	5153	51.5	3.9	–	1,916,400	4.6	3.4	4.2
Abruzzo	0	–	0.0	–	6606	100.0	4.7	–	183	100.0	0.5	–	1,567,894	3.9	3.8	3.4
Molise	0	–	0.0	–	0	–	0.0	–	0	–	0.0	–	615,150	6.0	4.1	5.0
Campania	1889	100.0	5.1	–	345	70.7	0.5	–	1473	100.0	4.0	–	2,001,817	5.0	5.0	4.4
Puglia	0	–	0.0	–	9939	33.7	1.9	–	7572	62.1	6.7	–	384,099	7.1	2.7	5.3
Basilicata	0	–	0.0	–	14,272	99.3	7.7	–	2346	100.0	6.8	–	1,069,016	5.6	3.7	5.0
Calabria	16,338	45.5	3.1	–	48,969	41.6	2.3	–	18,221	43.6	8.1	–	3,039,200	5.1	6.2	4.7
Sicilia	11,328	22.7	0.7	–	82,596	20.0	2.1	–	0	–	0.0	–	754,428	6.8	2.6	6.0
Sardegna	176,769	11.5	1.2	–	34,523	33.8	0.9	–	79	100.0	0.2	–	1,027,626	5.7	1.7	5.4
Italia	219,901	10.1	1.2	9.2	206,836	16.0	1.8	14.8	45,107	23.6	4.5	13.7	37,152,332	1.0	4.1	1.0

Table 7.32 Total value and value per hectare of annual volume increment by Plantations forest type / Valori totali e per ettaro dell'incremento corrente di volume per le categorie forestali degli Impianti di arboricoltura da legno

Region/Regione	Poplar plantations				Other broadleaved plantations				Coniferous plantations				Total plantations			
	Pioppeti artificiali				Piantagioni di altre latifoglie				Piantagioni di conifere				Totale impianti di arboricoltura da legno			
	Annual increment / Incremento annuo	ES	Annual increment / Incremento annuo	ES	Annual increment / Incremento annuo	ES	Annual increment / Incremento annuo	ES	Annual increment / Incremento annuo	ES	Annual increment / Incremento annuo	ES	Annual increment / Incremento annuo	ES	Annual increment / Incremento annuo	ES
	(m^3)	(%)	(m^3 ha^{-1})	(%)	(m^3)	(%)	(m^3 ha^{-1})	(%)	(m^3)	(%)	(m^3 ha^{-1})	(%)	(m^3)	(%)	(m^3 ha^{-1})	(%)
Piemonte	24,455	33.9	1.6	–	11,450	69.5	2.9	–	8623	63.4	7.5	–	44,528	28.5	2.2	21.3
Valle d'Aosta	0	–	0.0	–	0	–	0.0	–	0	–	0.0	–	0	–	0.0	–
Lombardia	226,194	28.4	12.2	–	49,018	51.1	7.9	–	6366	100.0	14.4	–	281,578	23.7	11.2	20.4
Alto Adige	0	–	0.0	–	0	–	0.0	–	0	–	0.0	–	0	–	0.0	–
Trentino	0	–	0.0	–	0	–	0.0	–	0	–	0.0	–	0	–	0.0	–
Veneto	88,423	89.9	37.9	–	18,302	54.3	6.2	–	0	–	0.0	–	106,726	71.8	20.2	60.7
Friuli V.G	15,552	37.6	2.4	–	14,008	42.9	5.3	–	0	–	0.0	–	29,561	26.6	3.2	18.9
Liguria	1390	100.0	3.8	–	0	–	0.0	–	0	–	0.0	–	1390	100.0	3.8	–
Emilia Romagna	16,180	55.8	3.8	–	4684	58.5	3.3	–	6023	100.0	16.4	–	26,887	41.5	4.4	34.8
Toscana	14,842	92.8	20.2	–	14,770	61.7	3.2	–	14,577	59.4	10.0	–	44,189	41.9	6.5	34.9
Umbria	0	–	0.0	–	8027	36.1	1.3	–	0	–	0.0	–	8027	36.1	1.3	29.6
Marche	0	–	0.0	–	9164	39.7	1.3	–	0	–	0.0	–	9164	39.7	1.3	37.0
Lazio	1820	70.9	2.3	–	2442	56.0	1.8	–	0	–	0.0	–	4263	43.2	2.0	16.2
Abruzzo	1019	100.0	1.9	–	1235	51.6	0.5	–	0	–	0.0	–	2254	52.9	0.8	47.6
Molise	4476	100.0	11.5	–	1124	40.8	0.5	–	0	–	0.0	–	5600	80.3	2.1	68.4
Campania	2759	84.3	1.8	–	1202	70.1	0.7	-	0	–	0.0	–	3961	62.4	1.3	24.3
Puglia	0	–	0.0	–	35	100.0	0.4	–	0	–	0.0	–	35	100.0	0.4	–
Basilicata	0	–	0.0	–	125	100.0	0.1	–	0	–	0.0	–	125	100.0	0.1	–
Calabria	1089	100.0	10.9	–	2807	77.9	1.8	–	8326	70.7	11.2	–	12,222	52.1	5.1	36.8
Sicilia	0	–	0.0	–	67	70.7	0.1	–	0	–	0.0	–	67	70.7	0.1	–
Sardegna	0	–	0.0	–	52,046	21.9	2.1	–	2830	54.9	1.9	–	54,877	20.9	2.1	18.6
Italia	398,199	26.2	7.7	23.5	190,507	17.2	2.7	14.2	46,746	31.6	8.3	17.8	635,452	16.7	4.9	15.7

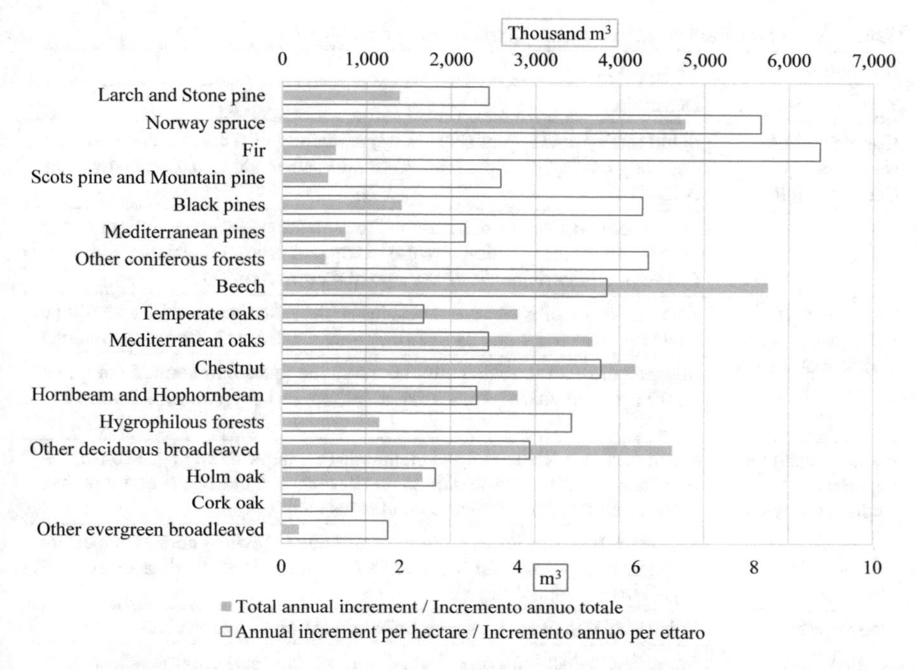

Fig. 7.15 Annual increment, total and per hectare, by Tall trees forest type / Incremento annuo, totale e per ettaro, per le categorie forestali dei Boschi alti

Tables 7.35 and 7.36 give the Forest area by classes of total canopy and tree canopy cover, respectively. Similar statistics are provided for the inventory categories Tall trees forest and Plantations and for the Tall trees forest types at national level at inventarioforestale.org/statistiche_INFC.

Figure 7.16 shows the Forest area by classes of total canopy cover at the regional level.

The largest portion of Italian Forest (74.4%) is marked by a total canopy cover higher than 80% and the total cover class of 51–80% accounts for 18.0% of Forest area. Less dense coverages are more frequent in Sicilia, Sardegna and Alto Adige (Table 7.35). The prevalence of dense or very dense crown coverages was also confirmed by considering only tree canopy cover: the percentage of Forest area with tree canopy coverage in the last two classes is, respectively, 21.8% and 62.7% (Table 7.36). Forest types marked by higher tree cover are Fir, Beech, Chestnut and Hornbeam and Hophornbeam forests; Cork oak, Other evergreen broadleaved and Larch and Swiss stone pine forests are the forest types with the lowest tree crown coverage (Fig. 7.17 and related table at inventarioforestale.org/statistiche_INFC).

The following remarks on silvicultural systems and development of Italian forests refer mainly to Tall trees forest, which is by far the largest inventory category. Additionally, most attributes discussed later in this paragraph were not assessed for Plantations. Table 7.37 gives the area estimates of Tall trees forest by silvicultural system.

Table 7.33 Silvicultural system classes / Classi per il tipo colturale

Silvicultural system	Description
Tipo colturale	Descrizione
Coppice without standards Ceduo semplice	Stand totally made up of suckers or prevalently of these if compared with trees of gamic origin (less than 20 standards/hectare or less than 4 within AdS25)
	Soprassuolo ceduo senza matricine, totalmente edificato da polloni o prevalenza di questi ultimi rispetto ai soggetti arborei di origine gamica (meno di 20 matricine/ettaro o meno di 4 entro AdS25)
Coppice with standards Ceduo matricinato	Stand made up of suckers and standards (the latter between 20 and 120 per hectare—between 4 and 24 in AdS25—and age 1 or 2 times the rotation)
	Soprassuolo costituito da polloni e matricine (queste in numero compreso tra 20 e 120 ad ettaro—tra 4 e 24 in AdS25—ed età pari a 1 o 2 volte il turno)
Mixed-management coppice Ceduo composto	Stand made up of suckers and standards (the latter more than 120 per hectare – more than 24 in AdS25 and various age classes, even more than 3 times the rotation); includes coniferous coppices
	Soprassuolo costituito da polloni e matricine (queste in numero superiore a 120 ad ettaro—oltre 24 in AdS25 e di diverse classi di età, anche superiore a 3 volte il turno)
Uneven-aged coppice Ceduo a sterzo	Stand made up of suckers with different age on the same plant
	Soprassuolo costituito da polloni di dimensioni (età) differenziate sulla stessa ceppaia
Coppice in transition to high forest Fustaia transitoria	Stand totally made up of shoots or prevalence of these against gamic origin trees; evident, recognisable signs of conversion cutting
	Soprassuolo totalmente edificato da polloni o prevalenza di questi ultimi rispetto ai soggetti arborei di origine gamica; riconoscibili segni evidenti di taglio di conversione
Even-aged high forest Fustaia coetanea	Prevalence of seed trees; presence of a single structural type (group of trees with the same development stage) on an area of at least 5000 m^2
	Prevalenza di soggetti arborei da seme; presenza di un solo tipo strutturale (gruppo di soggetti aventi lo stesso stadio di sviluppo) su una superficie di almeno 5000 m^2
Uneven-aged high forest Fustaia disetanea	Prevalence of seed trees; contemporary presence of trees at all stages of development, not aggregated in structural types or otherwise aggregated in structural types not larger than 1000 m^2
	Prevalenza di soggetti arborei da seme; presenza contemporanea di individui di tutte le fasi di sviluppo, non aggregati in tipi strutturali o altrimenti aggregati in tipi strutturali normalmente non più estesi di 1000 m^2
Irregular or structured high forest Fustaia irregolare o articolata	Presence of few structural types, normally with extension varying from 1000 to 5000 m^2 and other situations not falling under the two previous classes
	Presenza di più tipi strutturali su superfici di estensione variabile tra 1000 e 5000 m^2 e di altre situazioni non inquadrabili nelle due classi precedenti

(continued)

Table 7.33 (continued)

Silvicultural system	Description
Tipo colturale	Descrizione
Special: Chestnut forests, Walnut forests, Cork oak forests Tipo colturale speciale	Populations in forest context or at forest edges, specialised for production of so-called secondary products (chestnuts, cork, walnuts), planted widely spaced and generally subject to cultivation activity (pruning, grafting, undergrowth cleaning etc.)
	Popolamenti siti in contesto forestale o al margine di questo, specializzati per la produzione di prodotti cosiddetti secondari (castagne, sughero, tartufi, querce da ghianda), allevati a sesto d'impianto largo e di norma sottoposti ad interventi colturali (potature, innesti, ripulitura del sottobosco, ecc.)
Undefined Tipo colturale non definito	Spontaneous vegetation not subjected to any silvicultural activity or otherwise to occasional or sporadic activity (stands at inaccessible stations or with stational limits, even rupicolous, new forest stands, macchia, abandoned stands)
	Tipo colturale non identificabile con le classi sopra descritte, per esempio rinvenibile nei soprassuoli di origine spontanea, non sottoposti ad alcun intervento selvicolturale o a interventi occasionali o sporadici (formazioni su stazioni impervie o con limiti stazionali, anche rupestri, boschi di neoformazione, formazioni a macchia, soprassuoli abbandonati)

Figure 7.18 shows the distribution of Tall trees forest area by groups of silvicultural system in the Italian regions. Nationwide, the area of coppice and that of high forest are approximately the same. However, the area percentage of the two groups of silvicultural systems differs considerably at the regional level. Coppice silvicultural systems cover more than 50% of Tall trees forest area in nine regions (Lombardia, Liguria, Emilia-Romagna, Toscana, Umbria, Marche, Lazio, Molise, and Puglia). High forests have greater representation in Alpine regions (Valle d'Aosta, Piemonte, Alto Adige, Trentino, Friuli-Venezia Giulia, and Veneto) and in the two southernmost regions Calabria and Sicilia. The class undefined silvicultural system, which is marked by the absence of silvicultural activity or by only sporadic or occasional silvicultural activity, represents a relevant portion of Tall trees forest area in almost all regions.

The most common silvicultural system is coppice with standards (29.2% of Tall trees forest area), followed by uneven-aged high forest (16.1%), even-aged high forest (15.1%) and undefined silvicultural system (13.9%) (Fig. 7.19).

Tables 7.38 and 7.39 give the estimates of Tall trees forest area by development stage and by vertical structure, respectively. Additionally, Table 7.40 gives the even-aged Forest area by age class. Similar statistics are provided for the forest types at inventarioforestale.org/statistiche_INFC. Figures 7.20, 7.21, and 7.22 show the percentage of coppice, high forest and special or undefined silvicultural system area by development stage. In coppice, regenerating and young stages together account for 10.6% of the area; in even-aged high forest regenerating-empty, young growth and thicket add up to 3.0% of the area. Although a difference between those two

Table 7.34 Development stage classes for the different silvicultural systems / Classi per lo stadio di sviluppo per i diversi tipi colturali

Silvicultural system and development stage		Description
Tipo colturale e stadio di sviluppo		Descrizione
Even-aged or transitory high forest/Fustaia coetanea e fustaia transitoria	Young growth Novelleto	Even-aged (or almost even-aged) aggregation of very young trees that do not reach on average a height greater than 1/10 of the mature height and that are covered in branches down to the base; in artificial stands it may be called posticcia
		Aggregazione coetanea o quasi di alberi molto giovani che non raggiungono, nella media, un'altezza superiore ad 1/10 dell'altezza media di maturità e che sono coperti da rami fino alla base; nei soprassuoli di origine artificiale si parla anche di posticcia
	Thicket Spessina (spessaia, forteto)	Even-aged (or almost even-aged) aggregation of young trees of average height from 1/10 to 3/10 of the average mature height; phase with strong competition between species and mortality of less favoured elements; under normal density conditions the lower branches die gradually (self-pruning)
		Aggregazione coetanea o quasi di alberi giovani aventi altezza media compresa tra 1/10 e 3/10 dell'altezza media di maturità; fase di forte concorrenza intraspecifica e mortalità degli elementi meno favoriti; in condizioni di densità normale i rami più bassi disseccano gradualmente (auto potatura)
	Pole forest Perticaia	Even-aged (or so seeming) aggregation of relatively young trees of height from 3/10 to 7/10 of the average mature height; phase of strong diameter and height increment; in normal density conditions self-pruning is clearly seen; competition tends to reduce and a certain social difference is seen (dominant and dominated levels)
		Aggregazione coetanea o dall'aspetto coetaneo di alberi relativamente giovani aventi altezza media compresa tra 3/10 e 7/10 dell'altezza media di maturità; fase di forte incremento diametrico e longitudinale; in condizioni di densità normali si manifesta chiaramente e vistosamente il fenomeno dell'autopotatura; la concorrenza tende a ridursi e si manifesta una certa differenza sociale (piano dominante e piano dominato)
	Young/adult Fustaia giovane/ adulta	Even-aged (or seeming) aggregation of trees of average height from 7/10 to 9/10 of the average mature height, in which the number of trees is considerably reduced and the individual size increased; the process of social differentiation tends to reduce

(continued)

Table 7.34 (continued)

Silvicultural system and development stage		Description
Tipo colturale e stadio di sviluppo		Descrizione
		Aggregazione coetanea o dall'aspetto coetaneo di alberi aventi altezza media compresa tra 7/10 e 9/10 dell'altezza media di maturità, nella quale il numero di alberi risulta sensibilmente ridotto e la loro dimensione individuale notevolmente aumentata; il processo di differenziazione sociale tende a ridursi
	Mature and overage Fustaia matura e stramatura	Even-aged (or so seeming) aggregation of trees of average height greater than 9/10 of the average mature height, in which the reduction in number and the parallel increase in individual size are still more evident; in cases of greater aging (overmature phase), the vegetative vigour declines, as can also be seen in the general aspect
		Aggregazione coetanea o dall'aspetto coetaneo di alberi con altezza media superiore ai 9/10 dell'altezza media di maturità, nella quale la riduzione numerica ed il parallelo incremento della dimensione individuale risultano ancora più evidenti; nei casi di maggiore invecchiamento (f. stramatura) si denota un declino di vigore vegetativo riscontrabile anche nell'aspetto generale
	Regenerating (empty) Fustaia in rinnovazione (vuoto)	Phase following a clear cut at ground level or successive cuts, uniform or with reserves; regenerated trees, if present, do not reach 1.3 m height
		Fase successiva ad un taglio a raso oppure a tagli successivi uniformi o con rilascio di riserve; la rinnovazione, se presente, non raggiunge 1,3 m di altezza
Coppice/Ceduo	Young Stadio giovanile	With reference to the usual rotation as applied locally or in neighbouring areas to simple coppices or coppices with standards, of the same forest type, phase in which the (presumed) age of the suckers does not exceed half the rotation time
		Con riferimento al turno consuetudinario praticato nel contesto territoriale di riferimento ai cedui semplici o matricinati di quel tipo forestale, fase in cui l'età dei polloni non supera la metà del turno
	Adult Stadio adulto	Phase in which the age of the suckers is near the rotation time
		Fase in cui l'età dei polloni è prossima al turno
	Overage Stadio invecchiato	The age of the suckers is clearly greater than the normal rotation time
		Fase in cui l'età dei polloni è chiaramente superiore a quella del turno consuetudinario
	Regenerating Ceduo in rinnovazione	Stage immediately following a cut carried out during the current or the preceding year; the suckers, if present reach 1.3 m height

(continued)

Table 7.34 (continued)

Silvicultural system and development stage		Description
Tipo colturale e stadio di sviluppo		Descrizione
		Fase immediatamente successiva ad un intervento di taglio eseguito nell'anno in corso o in quello precedente; i ricacci, se presenti, raggiungono 1,3 m di altezza
Special or undefined silvicultural system/Tipo colturale speciale o non definito	Young Stadio giovanile	Initial settling stage of tree and/or shrub stands, with high dynamic thrust and development vitality; considered as such until the canopy closes
		Fase iniziale di insediamento di formazioni di alberi e/o arbusti, con elevato dinamismo e vitalità nello sviluppo; è considerata tale fino al raggiungimento della chiusura della copertura delle chiome
	Adult Stadio adulto	Intermediate stage of evolution with high mortality due to natural selection and accentuated social differentiation
		Fase evolutiva intermedia, con elevata mortalità per selezione naturale e accentuata differenziazione sociale
	Overage Stadio invecchiato	Phase of decadence with high mortality and evident withering, even in dominant trees, reduced vitality
		Fase di decadenza, mortalità elevata o disseccamenti evidenti anche nei soggetti dominanti, vitalità ridotta
	Not recognisable stage Stadio non riconoscibile	No evidence of any of the three stages above
		Mancata evidenza di uno dei tre stadi precedenti

percentages was expected because development stages have different durations for coppice and high forest, early stage accounts for low percentages of area in both cases. Mature and overage stage accounts for 34.3% of the area of even-aged high forest or coppice in transition to high forest, and the overage stage accounts for 32.6% of the coppice area. Figure 7.22 on special or undefined silvicultural systems, which comprises specialised stands for secondary products and spontaneous vegetation like new forest stands on abandoned land, shows a higher percentage of young stands and a lower percentage of overaged ones compared to other silvicultural systems. This was in line with expectations.

The one-layer vertical structure was found in 76.8% of the classified Tall trees forest area. The two-layers structure was found in the remaining 23.2% of the area, which accounts for 1,557,010 ha. Uneven-aged, irregular or structured high forest and uneven-aged coppice together, which are characterised by a multi-layer vertical structure or by different structural types on a small area, account for 2,219,995 ha (see Table 7.38).

Table 7.35 Forest area by classes of total canopy cover (trees and shrubs) / Estensione del Bosco ripartito per grado di copertura totale del suolo (alberi e arbusti)

Region/Regione	Forest/Bosco															
	<5%		5–10%		11–20%		21–50%		51–80%		>80%		Not classified		Total	
													Non classificato		Totale	
	Area	ES	Area	ES	Area	ES	Area	ES	Area	ES	Area	ES	Area	ES	Area	ES
	(ha)	(%)	(ha)	(%)	(ha)	(%)	(ha)	(%)	(ha)	(%)	(ha)	(%)	(ha)	(%)	(ha)	(%)
Piemonte	2064	79.7	2888	70.8	4742	51.7	49,150	15.8	216,656	7.1	613,160	2.9	1774	50.8	890,433	1.3
Valle d'Aosta	0	–	0	–	0	–	4154	56.0	20,770	21.1	74,198	7.1	120	99.6	99,243	3.6
Lombardia	0	–	3474	70.2	1034	100.0	4584	55.3	70,513	14.5	538,025	2.5	4338	41.7	621,968	1.6
Alto Adige	0	–	0	–	8874	39.1	58,285	13.9	131,251	8.2	138,840	7.7	2020	45.0	339,270	1.7
Trentino	0	–	1372	100.0	1433	100.0	19,476	25.1	66,222	12.6	284,263	3.6	492	100.0	373,259	1.4
Veneto	0	–	0	–	0	–	14,049	32.2	102,469	9.7	293,728	3.9	6459	27.1	416,704	1.9
Friuli V.G	0	–	0	–	1450	99.7	4379	43.6	28,266	21.8	295,560	3.0	2901	68.1	332,556	1.9
Liguria	753	99.2	0	–	1853	75.2	13,044	30.7	42,465	15.8	284,311	3.3	733	70.1	343,160	1.7
Emilia Romagna	1365	100.0	0	–	0	–	24,010	24.3	105,279	10.9	452,789	3.1	1458	49.6	584,901	1.5
Toscana	0	–	0	–	0	–	21,629	27.9	80,866	13.9	930,713	1.7	2240	38.1	1,035,448	1.1
Umbria	0	–	0	–	0	–	11,545	32.1	31,892	19.5	344,063	2.6	2805	69.1	390,305	1.6
Marche	0	–	0	–	847	100.0	11,845	35.0	39,297	17.5	239,778	3.8	0	–	291,767	2.1
Lazio	1275	74.1	0	–	3561	63.4	45,500	19.2	92,988	12.5	414,814	3.6	2099	41.4	560,236	1.6
Abruzzo	1660	81.3	0	–	3748	54.0	17,768	26.0	81,224	11.7	305,851	3.7	1336	57.4	411,588	1.8
Molise	0	–	0	–	0	–	0	–	0	–	153,248	3.4	0	–	153,248	3.0

(continued)

Table 7.35 (continued)

Region/Regione	Forest/Bosco															
	<5%		5–10%		11–20%		21–50%		51–80%		>80%		Not classified		Total	
													Non classificato		Totale	
	Area	ES	Area	ES	Area	ES	Area	ES	Area	ES	Area	ES	Area	ES	Area	ES
	(ha)	(%)	(ha)	(%)	(ha)	(%)	(ha)	(%)	(ha)	(%)	(ha)	(%)	(ha)	(%)	(ha)	(%)
Campania	921	70.2	659	100.0	2207	67.0	13,516	32.2	78,080	12.7	307,439	4.0	1105	57.8	403,927	2.1
Puglia	0	–	0	–	0	–	13,794	27.7	42,811	13.9	84,418	8.1	1326	50.1	142,349	4.0
Basilicata	0	–	0	–	1566	99.9	6451	44.7	57,242	13.8	219,467	4.6	3294	33.1	288,020	2.7
Calabria	2741	73.9	2503	71.3	5471	46.3	37,570	24.2	75,573	13.6	361,892	4.0	9427	35.7	495,177	2.0
Sicilia	9441	36.3	1487	78.7	9773	35.9	49,862	14.8	85,772	10.6	125,921	7.6	3233	63.1	285,489	3.2
Sardegna	0	–	3857	55.5	21,425	22.2	116,285	9.2	187,178	7.3	296,763	4.8	633	71.9	626,140	2.1
Italia	20,220	24.3	16,239	28.6	67,983	13.2	536,895	4.9	1,636,814	2.6	6,759,243	0.8	47,792	12.6	9,085,186	0.4

Table 7.36 Forest area by classes of tree canopy cover / Estensione del Bosco ripartito per grado di copertura arborea

Region/Regione	Forest/Bosco															
	<5%		5–10%		11–20%		21–50%		51–80%		>80%		Not classified		Total	
													Non classificato		Totale	
	Area	ES	Area	ES	Area	ES	Area	ES	Area	ES	Area	ES	Area	ES	Area	ES
	(ha)	(%)	(ha)	(%)	(ha)	(%)	(ha)	(%)	(ha)	(%)	(ha)	(%)	(ha)	(%)	(ha)	(%)
Piemonte	2064	79.7	7479	44.6	20,901	24.9	106,964	10.5	253,399	6.4	497,852	3.6	1774	50.8	890,433	1.3
Valle d'Aosta	0	–	0	–	650	99.6	8477	36.9	29,193	16.6	60,801	8.5	120	99.6	99,243	3.6
Lombardia	0	–	3474	70.2	1034	100.0	15,773	30.9	85,329	13.0	512,021	2.8	4338	41.7	621,968	1.6
Alto Adige	0	–	0	–	14,391	29.2	83,926	11.0	146,090	7.5	92,843	10.0	2020	45.0	339,270	1.7
Trentino	0	–	2745	70.6	6237	49.9	26,368	21.5	68,408	12.5	269,009	3.9	492	100.0	373,259	1.4
Veneto	0	–	561	99.9	5592	57.1	38,978	18.4	119,982	8.6	245,132	4.6	6,459	27.1	416,704	1.9
Friuli V.G	0	–	0	–	5623	42.6	14,324	31.3	41,729	17.1	267,980	3.6	2901	68.1	332,556	1.9
Liguria	2169	73.4	0	–	13,908	28.3	23,731	20.6	54,395	14.4	248,224	4.0	733	70.1	343,160	1.7
Emilia Romagna	1365	100.0	1637	100.0	7753	41.7	48,998	16.8	142,657	9.0	381,033	3.9	1458	49.6	584,901	1.5
Toscana	0	–	0	–	9502	40.5	66,889	15.3	112,347	11.5	844,469	2.1	2240	38.1	1,035,448	1.1
Umbria	0	–	0	–	5934	43.1	32,688	18.7	63,250	13.4	285,628	3.7	2805	69.1	390,305	1.6
Marche	0	–	0	–	4419	59.7	32,158	20.1	67,565	12.6	187,626	5.3	0	–	291,767	2.1
Lazio	1275	74.1	2412	83.8	10,761	40.4	79,295	13.9	113,768	11.0	350,626	4.4	2099	41.4	560,236	1.6
Abruzzo	1660	81.3	362	100.0	5836	42.8	34,383	18.3	92,532	10.9	275,479	4.2	1336	57.4	411,588	1.8
Molise	0	–	0	–	1401	99.8	4266	56.6	15,108	25.5	132,472	4.4	0	–	153,248	3.0

(continued)

Table 7.36 (continued)

Region/Regione	Forest/Bosco															
	<5%		5–10%		11–20%		21–50%		51–80%		>80%		Not classified		Total	
													Non classificato		Totale	
	Area	ES	Area	ES	Area	ES	Area	ES	Area	ES	Area	ES	Area	ES	Area	ES
	(ha)	(%)	(ha)	(%)	(ha)	(%)	(ha)	(%)	(ha)	(%)	(ha)	(%)	(ha)	(%)	(ha)	(%)
Campania	921	70.2	659	100.0	12,439	32.8	38,725	18.8	71,617	13.5	278,461	4.5	1105	57.8	403,927	2.1
Puglia	0	–	1225	99.8	1644	58.3	46,301	13.5	51,485	11.9	40,367	14.4	1326	50.1	142,349	4.0
Basilicata	1566	99.9	0	–	1119	99.9	23,318	22.6	73,962	11.6	184,761	5.3	3294	33.1	288,020	2.7
Calabria	5726	55.1	3933	57.8	11,480	36.3	63,123	17.0	113,562	10.7	287,927	5.1	9427	35.7	495,177	2.0
Sicilia	10,549	34.0	1487	78.7	18,789	25.2	78,182	11.3	84,340	10.2	88,909	9.7	3233	63.1	285,489	3.2
Sardegna	0	–	5033	48.4	54,047	14.6	224,331	6.4	181,385	7.4	160,711	7.9	633	71.9	626,140	2.1
Italia	27,294	21.8	31,005	21.0	213,459	7.8	1,091,199	3.3	1,982,104	2.3	5,692,333	0.9	47,792	12.6	9,085,186	0.4

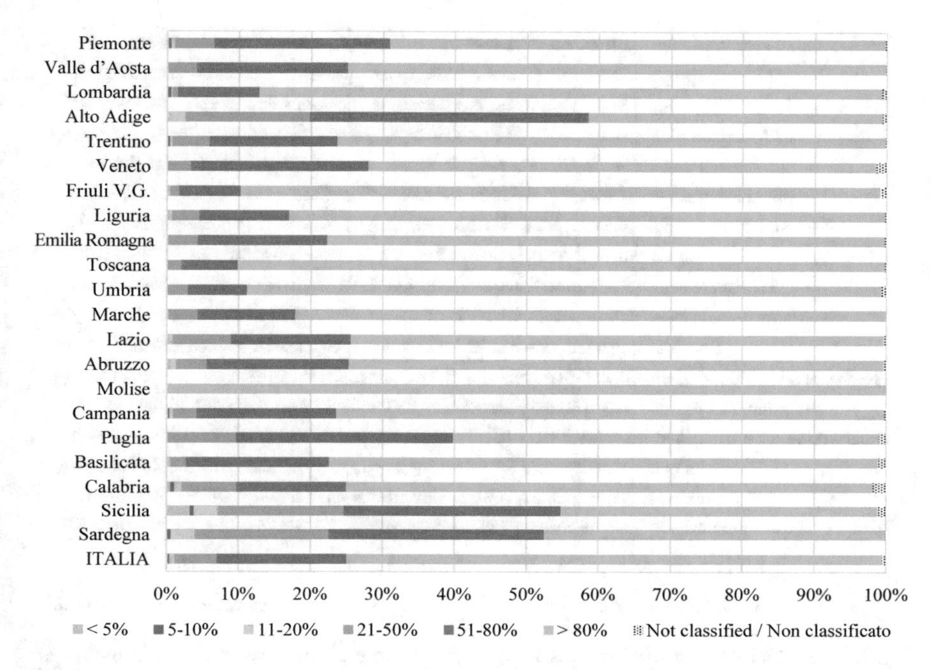

Fig. 7.16 Percent of area of Forest by total canopy cover class / Ripartizione percentuale della superficie del Bosco per classi di copertura totale

Estimates of total values and values per hectare of growing stock volume and aboveground tree biomass, and of annual volume increment of Forest by group of silvicultural systems are given in Tables 7.41, 7.42 and 7.43, respectively. In those tables, Plantations are included in the class not classified silvicultural system. Similar statistics are available by forest types, at the national level, at inventarioforestale.org/statistiche_INFC. Figures 7.23 and 7.24 illustrate the national values on growing stock volume and volume annual increment for coppice, high-forest, special and undefined silvicultural system.

Growing stock volume in high forest is about double that of coppice, both for the total volume and for the volume per hectare. The total growing stock is 905.4 million m^3 and 466.0 million m^3 for high forest and coppice, respectively, and value per hectare is respectively 241.0 m^3 and 123.0 m^3. As regards growth, the annual volume increment of high forest is much larger than that of coppice, both the total value and the value per hectare, which are 18.9 million m^3 and 5.0 m^3 per hectare, respectively, for high forest and 14.3 million m^3 and 3.8 m^3 per hectare, respectively, for coppice.

Tables 7.44, 7.45, and 7.46 provide the total values of growing stock volume and aboveground tree biomass, and the annual volume increment of Forest by age class. Similar statistics for forest types, at the national level, are available at inventarioforestale.org/statistiche_INFC. Figures 7.25, 7.26, and 7.27 show the total growing stock

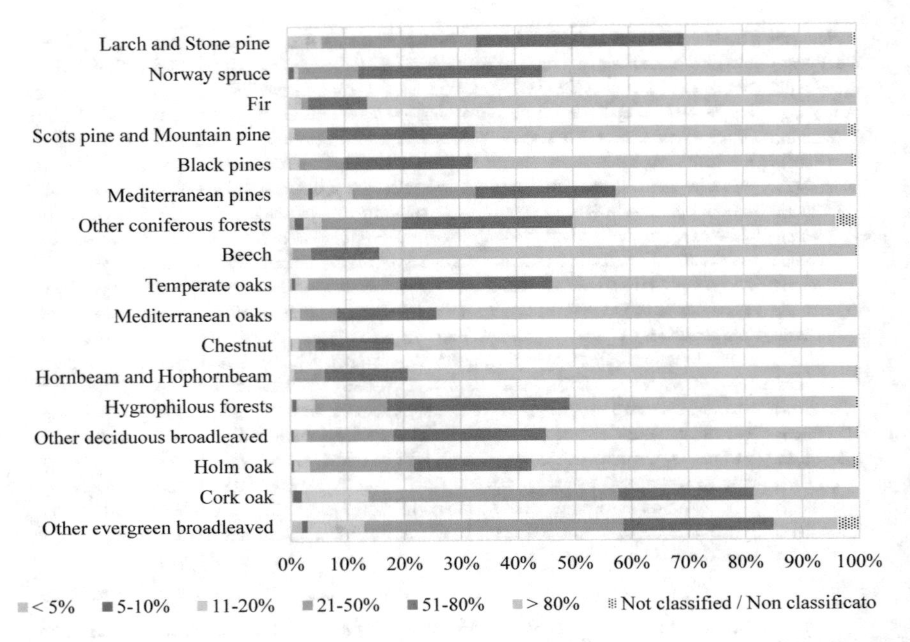

Fig. 7.17 Percent of area of Tall trees forest types by tree canopy cover class / Ripartizione percentuale della superficie delle categorie forestali dei Boschi alti per classi di copertura arborea

volume, the total annual volume increment, and the percent annual volume increment by age class, respectively, in even-aged coppice and high forest.

7.7 Forest Understory

Woody vegetation in the underlying layer was assessed by counting individual tree and shrub species growing between the forest canopy and the forest floor under the size thresholds for being callipered but higher than 50 cm. Counting was carried out using three classes of height and diameter (cf. Chap. 4). For tree species, total and per hectare estimates of number of individuals, aboveground biomass and content of organic carbon were derived by dimensional class and total. Similarly, statistics were derived for shrubs aboveground biomass and organic carbon by class of height and diameter. Counts of small trees and shrubs were processed separately and together to estimate total aboveground biomass and total organic carbon of forest understory woody vegetation (cf. Chap. 12).

Tables 7.47, 7.48, and 7.49 give the estimates of total and per hectare number of small trees by Forest inventory category and forest type. Analogue statistics by dimensional class and those on shrub species are available at inventarioforestale.org/

Table 7.37 Tall trees forest area by silvicultural system / Estensione dei Boschi alti ripartiti per tipo colturale

Region/Regione	Coppice/Cedui									
	Coppice without standards		Coppice with standards		Mixed-management coppice		Uneven-aged coppice		Total	
	Ceduo semplice		Ceduo matricinato		Ceduo composto		Ceduo a sterzo		Totale	
	Area	ES	Area	ES	Area	ES	Area	ES	Area	ES
	(ha)	(%)	(ha)	(%)	(ha)	(%)	(ha)	(%)	(ha)	(%)
Piemonte	88,599	12.4	118,118	10.0	114,387	10.7	1534	100.0	322,639	5.1
Valle d'Aosta	2572	44.8	0	–	424	99.6	0	–	2996	40.7
Lombardia	78,849	13.4	182,270	7.7	40,686	20.1	0	–	301,805	4.4
Alto Adige	5928	29.9	3757	40.4	6235	29.9	378	99.8	16,298	15.7
Trentino	14,546	31.7	17,000	30.0	11,329	37.4	1337	100.0	44,212	15.5
Veneto	24,759	23.3	101,419	8.7	21,438	23.4	7826	43.0	155,442	6.1
Friuli V.G	30,596	19.6	11,048	35.7	3552	63.8	0	–	45,196	15.3
Liguria	72,408	12.3	117,030	8.7	27,273	21.4	1423	99.2	218,134	4.5
Emilia Romagna	30,561	21.8	302,979	4.4	13,468	33.2	2731	70.4	349,738	3.8
Toscana	113,465	11.5	495,315	4.0	56,499	16.6	0	–	665,280	2.8
Umbria	4569	57.2	236,634	4.8	47,075	16.2	840	99.8	289,117	3.5
Marche	4713	48.3	145,285	6.2	4406	55.7	503	100.0	154,908	5.8
Lazio	34,495	21.9	296,335	5.2	51,379	17.8	0	–	382,209	3.8
Abruzzo	25,396	22.3	108,836	9.4	1,374	100.0	0	–	135,606	8.0
Molise	10,692	26.4	63,751	10.2	1,401	99.8	0	–	75,844	8.6
Campania	17,880	30.4	151,471	7.3	5972	49.6	0	–	175,324	6.1
Puglia	30,028	15.4	36,144	14.5	8,912	35.0	0	–	75,085	8.0
Basilicata	9089	33.9	71,064	11.9	402	99.9	0	–	80,555	10.9
Calabria	73,764	14.6	58,633	17.9	14,980	41.5	0	–	147,377	9.0
Sicilia	20,671	23.3	14,88 5	23.9	3586	50.1	0	–	39,142	15.2
Sardegna	23,727	24.2	82,367	12.1	4792	54.2	1621	100.0	112,507	9.8
Italia	717,308	4.3	2,614,342	1.8	439,570	5.7	18,193	27.4	3,789,413	1.3

(continued)

Table 7.37 (continued)

Region/Regione	High forest/Fustaie									
	Coppice in transition to high forest		Even–aged high forest		Uneven–aged high forest		Irregular or structured high forest		Totale	
	Fustaia transitoria		Fustaia coetanea		Fustaia disetanea		Fustaia irregolare o articolata		Totale	
	Area	ES	Area	ES	Area	ES	Area	ES	Area	ES
	(ha)	(%)	(ha)	(%)	(ha)	(%)	(ha)	(%)	(ha)	(%)
Piemonte	18,120	28.7	95,901	10.5	205,740	7.1	149,445	8.9	469,206	3.6
Valle d'Aosta	0	–	3706	42.6	650	99.6	83,389	5.1	87,745	4.7
Lombardia	0	–	52,846	16.4	163,540	7.0	29,712	22.9	246,098	5.1
Alto Adige	0	–	236,250	4.4	33,566	19.2	49,620	15.4	319,437	2.2
Trentino	18,471	25.9	114,564	8.8	61,989	13.1	67,179	12.6	262,204	3.7
Veneto	3103	70.6	73,384	12.5	81,358	10.8	48,558	15.8	206,403	5.0
Friuli V.G	11,811	35.3	108,187	9.0	34,185	19.2	22,536	23.4	176,719	5.6
Liguria	0	–	52,430	13.5	38,349	16.3	4301	53.6	95,080	9.2
Emilia Romagna	10,022	37.4	33,173	16.0	42,805	17.6	21,134	25.7	107,134	9.8
Toscana	60,794	15.4	93,101	10.5	43,379	19.4	57,943	16.6	255,217	6.3
Umbria	1414	99.8	14,486	22.7	21,927	21.7	9138	36.5	46,964	13.9
Marche	5815	50.0	20,645	17.6	40,648	16.7	17,348	27.0	84,456	10.1
Lazio	3331	70.3	28,125	18.9	85,849	12.0	12,569	36.4	129,874	9.2
Abruzzo	32,090	19.8	58,023	12.7	29,247	20.3	38,177	18.1	157,538	6.9
Molise	1401	99.8	13,550	24.5	7816	36.2	13,805	27.8	36,572	14.7
Campania	9281	40.5	44,951	14.3	45,725	16.2	9335	39.0	109,293	8.4
Puglia	1225	99.8	18,090	21.9	36,542	14.2	4347	48.8	60,205	9.3
Basilicata	8424	40.4	63,010	12.6	42,671	16.5	52,759	14.8	166,865	6.3
Calabria	4112	58.4	80,276	10.6	229,610	6.4	6586	47.7	320,583	4.5
Sicilia	2148	72.3	74,032	10.7	94,047	9.9	16,952	26.6	187,180	5.5
Sardegna	8106	44.3	76,223	10.0	98,648	10.3	48,676	14.5	231,653	5.8
Italia	199,668	8.4	1,354,955	2.5	1,438,291	2.7	763,511	3.8	3,756,425	1.3

(continued)

Table 7.37 (continued)

Region/Regione	Other silvicultural systems/Altri tipi colturali						Total Tall trees forest	
	Special		Undefined		Not classified			
	Tipo colturale speciale		Non definito		Non classificato		Totale Boschi alti	
	Area	ES	Area	ES	Area	ES	Area	ES
	(ha)	(%)	(ha)	(%)	(ha)	(%)	(ha)	(%)
Piemonte	15,521	31.4	61,140	15.1	1266	59.7	869,773	1.3
Valle d'Aosta	0	–	8381	23.9	120	99.6	99,243	3.6
Lombardia	1411	100.0	43,184	18.2	4338	41.7	596,836	1.6
Alto Adige	0	–	1516	56.6	2020	45.0	339,270	1.7
Trentino	0	–	66,351	12.6	492	100.2	373,259	1.4
Veneto	0	–	43,984	16.1	5598	30.7	411,427	1.9
Friuli V.G	0	–	98,546	9.8	2901	68.1	323,362	1.9
Liguria	2846	69.8	25,999	20.8	733	70.1	342,793	1.7
Emilia Romagna	11,476	33.7	109,786	9.9	718	70.8	578,852	1.5
Toscana	25,633	22.6	80,654	13.6	1881	41.7	1,028,665	1.1
Umbria	442	99.8	46,002	15.1	1402	99.8	383,928	1.6
Marche	371	100.0	45,170	15.9	0	–	284,904	2.1
Lazio	3737	63.2	40,905	19.4	1335	50.8	558,060	1.6
Abruzzo	941	69.9	113,196	8.8	1336	57.4	408,616	1.8
Molise	0	–	38,116	14.4	0	–	150,533	3.1
Campania	20,372	23.3	94,670	10.1	1105	57.8	400,763	2.1
Puglia	0	–	5733	35.2	1226	53.6	142,248	4.0
Basilicata	0	–	37,083	17.4	1994	55.7	286,498	2.7
Calabria	0	–	15,757	41.1	9054	37.0	492,771	2.1
Sicilia	4802	40.6	50,753	14.8	2854	70.3	284,731	3.2
Sardegna	34,477	16.8	220,985	6.6	633	71.9	600,255	2.2
Italia	122,030	9.8	1,247,913	3.0	41,006	14.1	8,956,787	0.4

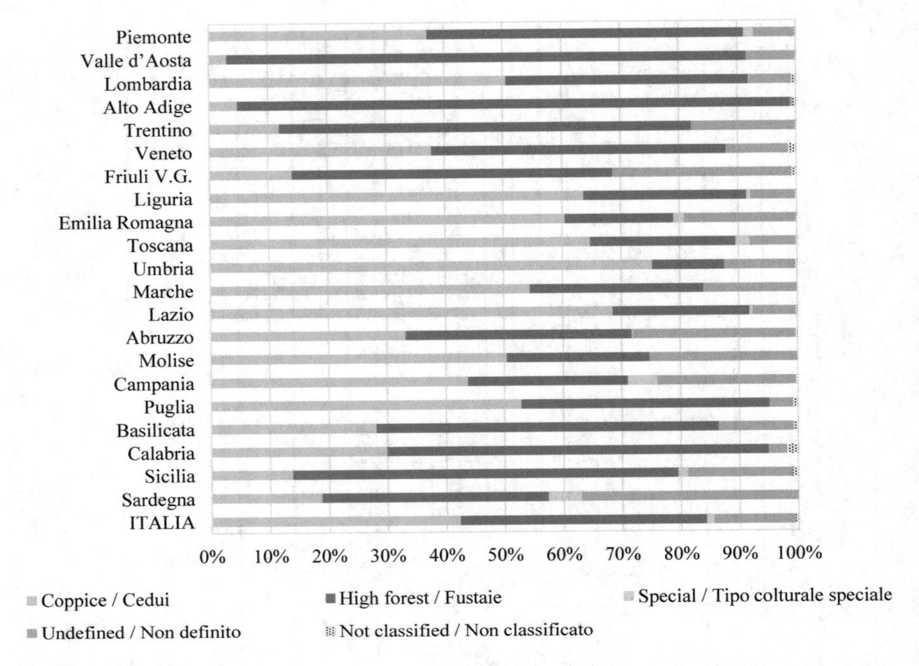

Fig. 7.18 Percent of area of Tall trees forest by groups of silvicultural system / Ripartizione percentuale della superficie dei Boschi alti per gruppi di tipi colturali

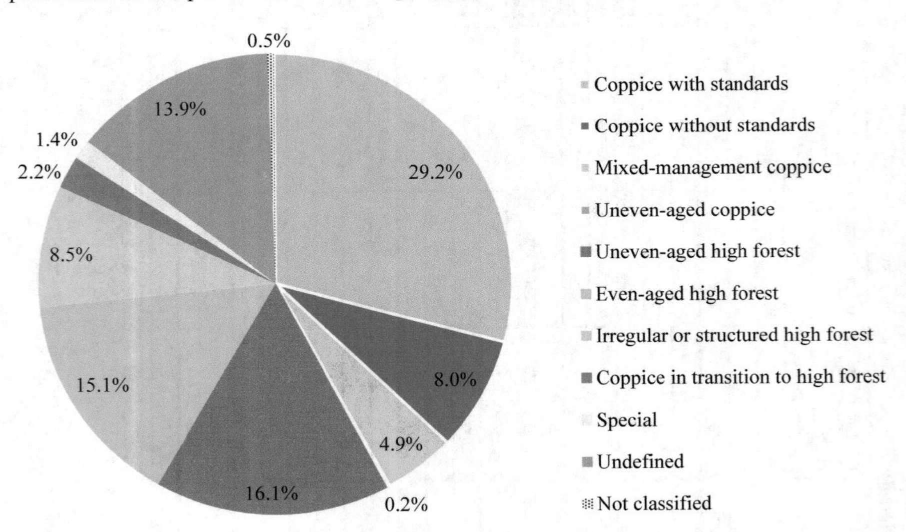

Fig. 7.19 Percent of area of Tall trees forest by silvicultural system, at the national level / Ripartizione percentuale della superficie dei Boschi alti per tipo colturale, a livello nazionale

Table 7.38 Tall trees forest area by development stage / Estensione dei Boschi alti ripartiti per stadio di sviluppo

Region/Regione	Even-aged coppice—Coppice without standards, coppice with standards, mixed-management coppice											
	Cedui coetanei—ceduo semplice, ceduo matricinato, ceduo composto											
	Young		Adult		Overage		Regenerating		Not classified		Total	
	Stadio giovanile		Stadio adulto		Stadio invecchiato		In rinnovazione		Non classificato		Totale	
	Area	ES	Area	ES	Area	ES	Area	ES	Area	ES	Area	ES
	(ha)	(%)	(ha)	(%)	(ha)	(%)	(ha)	(%)	(ha)	(%)	(ha)	(%)
Piemonte	26,262	24.0	206,629	7.3	63,648	14.5	1542	100.0	23,023	24.8	321,105	5.1
Valle d'Aosta	650	99.6	2346	44.7	0	–	0	–	0	–	2996	40.7
Lombardia	16,309	32.8	188,932	7.5	96,563	11.9	0	–	0	–	301,805	4.4
Alto Adige	756	99.8	7566	26.3	7598	26.5	0	–	0	–	15,920	15.9
Trentino	2059	94.6	25,793	21.9	15,023	31.2	0	–	0	–	42,875	15.7
Veneto	16,615	28.8	75,027	11.5	53,892	14.0	0	–	2082	79.1	147,616	6.4
Friuli V.G	8742	38.9	24,616	21.8	10,187	38.7	0	–	1652	99.7	45,196	15.3
Liguria	29,852	20.7	113,768	8.9	73,091	11.5	0	–	0	–	216,711	4.6
Emilia Romagna	19,886	27.4	207,150	6.5	116,969	9.8	0	–	3002	71.0	347,007	3.9
Toscana	17,909	31.7	325,606	5.8	321,407	5.8	0	–	358	100.0	665,280	2.8
Umbria	22,205	24.9	196,880	5.8	66,414	13.0	0	–	2779	70.6	288,278	3.5
Marche	7245	44.8	104,196	8.6	42,964	14.9	0	–	0	–	154,405	5.8
Lazio	27,426	24.9	182,455	8.0	167,029	8.5	1665	100.0	3633	72.2	382,209	3.8
Abruzzo	7149	39.4	52,831	14.8	75,625	12.1	0	–	0	–	135,606	8.0
Molise	1401	99.8	58,982	11.2	15,461	27.4	0	–	0	–	75,844	8.6
Campania	35,855	20.2	91,8 27	11.2	44,656	17.8	0	–	2986	70.9	175,324	6.1
Puglia	14,730	25.2	55,968	10.2	2948	52.6	0	–	1439	99.8	75,085	8.0
Basilicata	7905	40.6	69,736	12.1	2914	60.0	0	–	0	–	80,555	10.9
Calabria	68,203	14.5	55,594	18.7	21,435	36.0	0	–	2145	99.9	147,377	9.0
Sicilia	8062	37.8	19,824	20.9	11,255	32.1	0	–	0	–	39,142	15.2
Sardegna	17,336	29.0	74,850	12.8	18,699	28.4	0	–	0	–	110,886	9.9
Italia	356,557	6.4	2,140,578	2.2	1,227,778	3.1	3208	70.8	43,099	18.6	3,771,219	1.3

(continued)

Table 7.38 (continued)

Region/Regione	Coppice in transition to high forest and even–aged high forest											
	Fustaia transitoria e fustaia coetanea											
	Young growth		Thicket		Pole forest		Young/adult		Mature and overage		Regenerating, empty	
	Novelleto		Spessina		Perticaia		Fustaia giovane – adulta		Fustaia matura – stramatura		Fustaia in rinnovazione, vuoto	
	Area	ES	Area	ES	Area	ES	Area	ES	Area	ES	Area	ES
	(ha)	(%)	(ha)	(%)	(ha)	(%)	(ha)	(%)	(ha)	(%)	(ha)	(%)
Piemonte	1542	100.0	0	–	5198	53.0	64,369	13.9	42,912	16.0	0	–
Valle d'Aosta	0	–	0	–	0	–	3706	42.6	0	–	0	–
Lombardia	100	100.0	1812	100.0	1737	100.0	30,258	22.1	18,939	28.3	0	–
Alto Adige	1403	99.8	3890	52.2	8516	39.6	69,600	12.5	145,924	7.4	6918	42.4
Trentino	4970	50.5	0	–	1372	100.0	73,920	11.9	50,027	15.1	2745	70.6
Veneto	0	–	2937	70.7	4239	57.5	51,347	15.6	17,964	26.7	0	–
Friuli V.G	0	–	539	99.8	389	99.8	88,847	10.7	30,223	20.0	0	–
Liguria	350	99.2	0	–	0	–	17,330	24.9	34,499	17.5	252	99.2
Emilia Romagna	0	–	0	–	1406	100.0	34,148	17.1	7641	37.0	0	–
Toscana	0	–	460	100.0	0	–	91,813	12.2	60,200	14.6	1422	100.0
Umbria	0	–	0	–	0	–	14,656	23.9	1244	58.3	0	–
Marche	577	100.0	1783	82.1	0	–	23,562	18.4	0	–	539	100.0
Lazio	0	–	0	–	0	–	21,745	22.8	9711	32.2	0	–
Abruzzo	0	–	0	–	1298	100.0	57,088	13.5	31,004	19.3	724	70.8
Molise	0	–	0	–	1403	99.8	6854	36.4	6695	34.4	0	–
Campania	0	–	0	–	2869	70.1	40,970	15.9	10,393	36.9	0	–
Puglia	0	–	0	–	1869	81.7	12,461	29.7	4985	48.3	0	–
Basilicata	347	99.9	0	–	461	99.9	43,883	15.3	25,812	22.4	0	–
Calabria	0	–	1430	99.9	466	99.9	74,020	11.5	7528	42.5	944	99.9
Sicilia	4154	58.5	1760	81.4	1371	77.4	52,851	13.2	16,044	27.5	0	–
Sardegna	1621	100.0	3052	70.9	9090	35.2	58,988	11.5	11,027	33.8	552	100.0
Italia	15,065	29.5	17,662	27.2	41,684	17.6	932,414	3.4	532,772	4.5	14,094	28.5

(continued)

Table 7.38 (continued)

Region/Regione	Coppice in transition to high forest and even–aged high forest				Uneven–aged high forest, irregular or structured high forest, uneven–aged coppice	
	Fustaia transitoria e fustaia coetanea					
	Not classified		Total		Fustaia disetanea, fustaia irregolare o articolata, ceduo a sterzo	
	Non classificato		Totale		Not classified/Non classificato	
	Area	ES	Area	ES	Area	ES
	(ha)	(%)	(ha)	(%)	(ha)	(%)
Piemonte	0	–	114,021	9.8	356,720	4.7
Valle d'Aosta	0	–	3706	42.6	84,039	5.1
Lombardia	0	–	52,846	16.4	193,253	6.6
Alto Adige	0	–	236,250	4.4	83,564	11.2
Trentino	0	–	133,035	8.1	130,506	7.9
Veneto	0	–	76,487	12.3	137,742	7.3
Friuli V.G	0	–	119,998	8.3	56,721	14.0
Liguria	0	–	52,430	13.5	44,073	15.2
Emilia Romagna	0	–	43,195	14.9	66,670	13.6
Toscana	0	–	153,895	8.6	101,322	12.2
Umbria	0	–	15,900	22.4	31,904	18.0
Marche	0	–	26,460	17.4	58,499	13.3
Lazio	0	–	31,456	18.3	98,417	11.2
Abruzzo	0	–	90,113	10.0	67,424	13.0
Molise	0	–	14,951	23.9	21,621	21.2
Campania	0	–	54,232	13.4	55,061	14.6
Puglia	0	–	19,315	21.4	40,889	13.5
Basilicata	932	99.9	71,434	11.7	95,431	10.1
Calabria	0	–	84,388	10.4	236,195	6.3
Sicilia	0	–	76,181	10.5	110,999	8.8
Sardegna	0	–	84,329	9.9	148,945	7.8
Italia	932	99.9	1,554,623	2.4	2,219,995	2.0

(continued)

Table 7.38 (continued)

Region/Regione	Special and undefined silvicultural systems											
	Tipi colturali speciale e non definito											
	Young		Adult		Overage		Not recognisable		Not classified		Totale	
	Stadio giovanile		Stadio adulto		Stadio invecchiato		Non riconoscibile		Non classificato		Totale	
	Area	ES	Area	ES	Area	ES	Area	ES	Area	ES	Area	ES
	(ha)	(%)	(ha)	(%)	(ha)	(%)	(ha)	(%)	(ha)	(%)	(ha)	(%)
Piemonte	13,704	33.1	36,733	20.0	18,548	28.9	7676	44.5	0	–	76,661	13.5
Valle d'Aosta	2062	53.0	4325	35.1	1570	59.3	424	99.6	0	–	8381	23.9
Lombardia	7982	45.1	26,348	24.0	8342	41.9	1923	100.0	0	–	44,595	17.8
Alto Adige	507	99.8	1008	68.7	0	–	0	–	0	–	1516	56.6
Trentino	16,292	29.1	31,109	19.6	8597	38.9	10,353	36.4	0	–	66,351	12.6
Veneto	18,610	26.8	15,976	27.3	2520	63.6	6878	43.4	0	–	43,984	16.1
Friuli V.G	22,401	24.0	69,635	12.6	2625	76.9	3886	53.8	0	–	98,546	9.8
Liguria	7188	40.0	11,245	33.9	9659	36.7	753	99.2	0	–	28,846	20.0
Emilia Romagna	14,500	31.7	64,752	14.3	31,814	20.8	9823	39.2	372	100.0	121,263	9.4
Toscana	4527	69.8	45,471	19.1	48,812	17.4	7477	47.8	0	–	106,287	11.6
Umbria	5740	48.4	24,011	21.6	14,106	29.3	2588	62.2	0	–	46,444	15.0
Marche	4253	59.8	32,808	19.2	3878	60.8	4601	58.2	0	–	45,541	15.8
Lazio	10,531	44.2	11,594	38.8	13,759	36.2	8759	41.2	0	–	44,642	18.5
Abruzzo	39,223	17.8	48,236	15.6	19,062	26.1	7616	43.8	0	–	114,137	8.7
Molise	3956	52.2	28,727	17.4	5433	47.0	0	–	0	–	38,116	14.4
Campania	7882	40.5	57,036	14.7	37,284	18.9	12,839	30.6	0	–	115,042	9.0
Puglia	0	–	5267	37.5	0	–	466	99.8	0	–	5733	35.2
Basilicata	8001	41.8	19,101	23.9	1506	99.9	8107	38.7	368	99.9	37,083	17.4
Calabria	2332	99.9	6840	52.3	2145	99.9	4440	99.9	0	–	15,757	41.1
Sicilia	4050	58.6	42,567	16.0	6118	45.6	2820	61.7	0	–	55,555	13.9
Sardegna	60,204	14.6	108,582	10.2	6103	48.3	80,574	11.8	0	–	255,462	5.8
Italia	253,945	7.3	691,372	4.2	241,882	7.6	182,004	8.6	740	70.7	1,369,943	2.8

(continued)

Table 7.38 (continued)

Region/Regione	Not classified silvicultural system and development stage		Total Tall trees forest	
	Non classificato per tipo colturale e stadio di sviluppo		Totale Boschi alti	
	Area	ES	Area	ES
	(ha)	(%)	(ha)	(%)
Piemonte	1266	59.7	869,773	1.3
Valle d'Aosta	120	99.6	99,243	3.6
Lombardia	4338	41.7	596,836	1.6
Alto Adige	2020	45.0	339,270	1.7
Trentino	492	100.0	373,259	1.4
Veneto	5598	30.7	411,427	1.9
Friuli V.G	2901	68.1	323,362	1.9
Liguria	733	70.1	342,793	1.7
Emilia Romagna	718	70.8	578,852	1.5
Toscana	1881	41.7	1,028,665	1.1
Umbria	1402	99.8	383,928	1.6
Marche	0	–	284,904	2.1
Lazio	1335	50.8	558,060	1.6
Abruzzo	1336	57.4	408,616	1.8
Molise	0	–	150,533	3.1
Campania	1105	57.8	400,763	2.1
Puglia	1226	53.6	142,248	4.0
Basilicata	1994	55.7	286,498	2.7
Calabria	9054	37.0	492,771	2.1
Sicilia	2854	70.3	284,731	3.2
Sardegna	633	71.9	600,255	2.2
Italia	41,006	14.1	8,956,787	0.4

Table 7.39 Tall trees forest area by one–layer and two-layer structure / Estensione dei Boschi alti ripartiti per struttura monoplana e biplana

Region/Regione	One-layer		Two-layer		Not classified		Total Tall trees forest	
	Struttura monoplana		Struttura biplana		Non classificato		Totale Boschi alti	
	Area	ES	Area	ES	Area	ES	Area	ES
	(ha)	(%)	(ha)	(%)	(ha)	(%)	(ha)	(%)
Piemonte	362,047	4.8	149,232	9.2	358,494	4.7	869,773	1.3
Valle d'Aosta	9076	24.4	6007	29.5	84,160	5.1	99,243	3.6
Lombardia	318,323	4.8	80,922	13.5	197,591	6.5	596,836	1.6
Alto Adige	208,932	5.4	44,753	16.3	85,584	11.0	339,270	1.7
Trentino	215,007	5.3	27,254	21.9	130,998	7.9	373,259	1.4
Veneto	231,134	5.1	44,262	16.4	136,031	7.2	411,427	1.9
Friuli V.G	228,998	4.5	34,743	18.3	59,621	13.6	323,362	1.9
Liguria	265,472	3.7	33,938	18.5	43,383	15.2	342,793	1.7
Emilia Romagna	389,190	3.8	121,902	9.9	67,761	13.4	578,852	1.5
Toscana	624,710	3.4	300,393	6.2	103,562	11.9	1,028,665	1.1
Umbria	173,541	6.8	177,920	6.6	32,467	18.1	383,928	1.6
Marche	177,542	5.6	48,863	14.7	58,499	13.3	284,904	2.1
Lazio	364,477	4.2	93,830	12.3	99,753	11.0	558,060	1.6
Abruzzo	295,380	4.0	44,477	16.3	68,760	12.8	408,616	1.8
Molise	70,229	10.2	58,683	11.6	21,621	21.2	150,533	3.1
Campania	300,969	4.1	43,629	17.9	56,165	14.3	400,763	2.1
Puglia	48,807	12.7	51,327	11.2	42,115	13.2	142,248	4.0
Basilicata	134,652	7.7	53,121	14.3	98,725	9.8	286,498	2.7
Calabria	185,698	7.6	61,823	18.2	245,249	6.2	492,771	2.1
Sicilia	143,730	7.1	27,148	20.6	113,853	8.7	284,731	3.2
Sardegna	399,514	3.9	52,784	14.8	147,957	7.8	600,255	2.2
Italia	5,147,430	1.1	1,557,010	2.7	2,252,348	2.0	8,956,787	0.4

Table 7.40 Even-aged Forest by age class / Estensione dei tipi colturali coetanei del Bosco ripartiti per classi di età

Region/Regione	Even-aged coppice—Coppice without standards, coppice with standards, mixed-management coppice													
	Cedui coetanei—ceduo semplice, ceduo matricinato, ceduo composto													
	0–10 years		11–20 years		21–30 years		31–40 years		41–80 years		Not classified		Total	
	0–10 anni		11–20 anni		21–30 anni		31–40 anni		41–80 anni		Non classificato		Totale	
	Area	ES	Area	ES	Area	ES	Area	ES	Area	ES	Area	ES	Area	ES
	(ha)	(%)	(ha)	(%)	(ha)	(%)	(ha)	(%)	(ha)	(%)	(ha)	(%)	(ha)	(%)
Piemonte	9083	40.8	34,458	20.9	56,285	16.0	98,407	11.5	45,781	17.6	77,091	13.4	321,105	5.1
Valle d'Aosta	0	–	650	99.6	650	99.6	1695	48.8	0	–	0	–	2,996	40.7
Lombardia	1923	100.0	26,350	24.7	65,754	15.3	98,231	12.0	90,118	12.4	19,429	29.7	301,805	4.4
Alto Adige	0	–	1512	68.7	2765	44.7	2893	45.0	7220	27.4	1530	58.9	15,920	15.9
Trentino	0	–	2059	94.6	5817	55.4	12,220	35.4	11,543	35.4	11,236	37.9	42,875	15.7
Veneto	1538	99.9	16,642	28.9	16,665	28.8	48,646	15.5	52,184	14.3	11,940	31.7	147,616	6.4
Friuli V.G	0	–	10,810	34.4	4994	51.9	13,850	32.0	10,187	38.7	5356	49.2	45,196	15.3
Liguria	5015	50.7	30,004	20.4	21,173	25.7	78,308	11.5	68,796	11.9	13,415	31.2	216,711	4.6
Emilia Romagna	6137	49.9	20,232	27.5	56,885	15.1	129,164	9.2	109,509	10.2	25,079	24.2	347,007	3.9
Toscana	1586	100.0	29,271	24.7	78,443	14.1	187,005	8.5	273,810	6.6	95,164	12.6	665,280	2.8
Umbria	7367	44.4	28,020	21.9	75,787	12.0	86,936	11.3	52,009	15.0	38,159	18.3	288,278	3.5
Marche	0	–	14,456	30.1	39,771	17.1	54,561	13.8	42,117	15.2	3500	58.6	154,405	5.8
Lazio	14,844	35.1	22,650	26.2	53,475	17.6	115,679	11.0	155,826	8.8	19,735	29.8	382,209	3.8
Abruzzo	0	–	7149	39.4	14,216	30.9	38,615	17.4	75,625	12.1	0	–	135,606	8.0
Molise	0	–	1401	99.8	7235	40.9	48,286	13.0	15,461	27.4	3461	60.8	75,844	8.6
Campania	20,731	27.4	47,955	16.8	45,091	17.6	13,904	32.8	40,198	18.9	7445	42.7	175,324	6.1
Puglia	2225	69.4	13,618	26.5	10,970	29.2	36,921	14.0	2482	59.8	8869	33.8	75,085	8.0
Basilicata	1506	99.9	12,660	32.2	21,380	24.3	37,419	17.3	1968	82.1	5623	43.6	80,555	10.9
Calabria	38,420	18.8	37,874	22.4	29,062	27.1	12,116	48.7	13,401	48.5	16,503	37.6	147,377	9.0
Sicilia	1108	100.0	8589	35.8	3255	59.4	10,028	32.2	9813	35.3	6348	33.7	39,142	15.2
Sardegna	866	100.0	16,470	30.1	7647	41.5	23,991	24.4	12,970	34.7	48,942	16.7	110,886	9.9
Italia	112,349	11.5	382,832	6.2	617,322	4.8	1,148,873	3.3	1,091,018	3.3	418,825	5.9	3,771,219	1.3

(continued)

Table 7.40 (continued)

Region/Regione	Coppice in transition to high forest and even–aged high forest									
	Fustaia transitoria e fustaia coetanea									
	0–10 years		11–20 years		21–30 years		31–40 years		41–80 years	
	0–10 anni		11–20 anni		21–30 anni		31–40 anni		41–80 anni	
	Area	ES	Area	ES	Area	ES	Area	ES	Area	ES
	(ha)	(%)	(ha)	(%)	(ha)	(%)	(ha)	(%)	(ha)	(%)
Piemonte	1542	100.0	0	–	3085	70.6	15,268	30.0	43,730	16.7
Valle d'Aosta	0	–	0	–	0	–	1121	71.2	2585	53.0
Lombardia	0	–	1912	95.0	0	–	6850	43.9	25,144	24.7
Alto Adige	6072	46.2	4896	46.6	4047	56.3	5711	48.5	69,600	12.5
Trentino	4970	50.5	2745	70.6	0	–	11,306	32.2	60,905	13.5
Veneto	0	–	0	–	4380	57.7	16,590	27.0	36,155	19.2
Friuli V.G	0	–	539	99.8	389	99.8	14,937	29.3	67,002	13.0
Liguria	0	–	602	71.1	0	–	4107	56.4	10,088	32.4
Emilia Romagna	0	–	0	–	1406	100.0	12,974	30.5	18,310	22.1
Toscana	0	–	1422	100.0	460	100.0	13,349	35.2	77,715	13.3
Umbria	0	–	0	–	0	–	3349	49.4	11,306	27.5
Marche	1115	70.9	1783	82.1	0	–	6406	32.3	17,156	22.9
Lazio	0	–	0	–	0	–	4548	58.5	17,197	24.6
Abruzzo	362	100.0	362	100.0	0	–	5809	44.7	52,577	14.2
Molise	0	–	0	–	1403	99.8	1172	57.5	5682	42.4
Campania	0	–	0	–	1434	100.0	7852	38.2	33,001	17.9
Puglia	0	–	0	–	388	99.8	9408	33.7	3094	63.2
Basilicata	347	99.9	0	–	461	99.9	7269	40.5	34,101	17.4
Calabria	944	99.9	0	–	1430	99.9	29,181	21.0	39,814	16.2
Sicilia	2489	71.1	2044	83.5	2373	71.6	29,016	18.9	20,298	22.9
Sardegna	552	100.0	1621	100.0	10,531	34.9	27,675	16.1	28,060	18.5
Italia	18,394	25.4	17,926	26.5	31,786	20.3	233,898	7.0	673,522	4.1

(continued)

Table 7.40 (continued)

Region/Regione	Coppice in transition to high forest and even–aged high forest							
	Fustaia transitoria e fustaia coetanea							
	81–120 years		more than 120 years		Not classified		Total	
	81–120 anni		più di 120 anni		Non classificato		Totale	
	Area	ES	Area	ES	Area	ES	Area	ES
	(ha)	(%)	(ha)	(%)	(ha)	(%)	(ha)	(%)
Piemonte	36,008	17.9	2451	70.3	11,937	33.1	114,021	9.8
Valle d'Aosta	0	–	0	–	0	–	3706	42.6
Lombardia	15,827	32.1	0	–	3112	55.6	52,846	16.4
Alto Adige	57,522	14.3	88,402	10.9	0	–	236,250	4.4
Trentino	41,673	16.7	8355	40.4	3081	70.1	133,035	8.1
Veneto	13,638	31.5	395	99.9	5329	47.5	76,487	12.3
Friuli V.G	26,499	21.5	720	99.7	9911	39.1	119,998	8.3
Liguria	32,273	18.4	0	–	5361	42.6	52,430	13.5
Emilia Romagna	5945	38.0	0	–	4560	58.0	43,195	14.9
Toscana	42,269	17.2	0	–	18,679	28.2	153,895	8.6
Umbria	1244	58.3	0	–	0	–	15,900	22.4
Marche	0	–	0	–	0	–	26,460	17.4
Lazio	6361	32.2	3350	70.7	0	–	31,456	18.3
Abruzzo	25,509	21.6	5495	49.4	0	–	90,113	10.0
Molise	6695	34.4	0	–	0	–	14,951	23.9
Campania	8959	40.1	1434	100.0	1552	100.0	54,232	13.4
Puglia	3505	55.9	1481	99.8	1439	99.8	19,315	21.4
Basilicata	22,847	23.8	0	–	6408	45.3	71,434	11.7
Calabria	7528	42.5	0	–	5491	50.1	84,388	10.4
Sicilia	9614	36.0	0	–	10,347	34.8	76,181	10.5
Sardegna	7412	39.5	0	–	8479	42.5	84,329	9.9
Italia	371,329	5.7	112,081	10.0	95,687	12.0	1,554,623	2.4

(continued)

Table 7.40 (continued)

Region/Regione	Plantations													
	Impianti di arboricoltura da legno													
	0–10 years		11–20 years		21–30 years		31–40 years		41–80 years		Not classified		Total	
	0–10 anni		11–20 anni		21–30 anni		31–40 anni		41–80 anni		Non classificato		Totale	
	Area	ES	Area	ES	Area	ES	Area	ES	Area	ES	Area	ES	Area	ES
	(ha)	(%)	(ha)	(%)	(ha)	(%)	(ha)	(%)	(ha)	(%)	(ha)	(%)	(ha)	(%)
Piemonte	5119	49.9	9483	32.4	1924	49.2	1417	100.0	747	71.0	1970	82.1	20,660	16.2
Valle d'Aosta	0	–	0	–	0	–	0	–	0	–	0	–	0	–
Lombardia	13,728	25.2	10,289	32.6	337	100.0	337	100.0	0	–	441	100.0	25,132	12.5
Alto Adige	0	–	0	–	0	–	0	–	0	–	0	–	0	–
Trentino	0	–	0	–	0	–	0	–	0	–	0	–	0	–
Veneto	1988	69.7	1644	81.8	1301	99.9	0	–	0	–	344	99.9	5277	33.7
Friuli V.G	5148	28.1	2283	49.4	1052	57.2	372	99.8	0	–	340	99.8	9194	17.0
Liguria	0	–	0	–	367	99.2	0	–	0	–	0	–	367	99.2
Emilia Romagna	1383	53.9	2408	67.1	1521	49.9	0	–	368	100.0	368	100.0	6049	31.6
Toscana	736	69.6	361	100.0	4234	43.7	368	100.0	361	100.0	723	70.7	6783	30.3
Umbria	0	–	1893	36.8	3081	52.0	0	–	0	–	1402	99.8	6377	33.7
Marche	0	–	3902	52.9	2490	63.1	100	100.0	371	100.0	0	–	6863	35.7
Lazio	0	–	368	100.0	650	73.6	737	70.7	0	–	420	100.0	2176	39.7
Abruzzo	362	100.0	759	100.0	1684	57.3	166	100.0	0	–	0	–	2971	36.3
Molise	0	–	962	63.2	1363	47.0	391	99.8	0	–	0	–	2715	27.7
Campania	1105	57.8	1799	80.3	260	100.0	0	–	0	–	0	–	3163	50.4
Puglia	0	–	0	–	0	–	0	–	0	–	100	99.8	100	99.8
Basilicata	1422	99.9	100	99.9	0	–	0	–	0	–	0	–	1522	93.6
Calabria	660	86.1	441	99.9	560	99.9	373	99.9	0	–	373	99.9	2406	39.3
Sicilia	0	–	0	–	379	100.0	0	–	0	–	379	100.0	758	70.7
Sardegna	12,924	19.8	5780	40.5	3190	53.7	2411	56.9	1579	70.8	0	–	25,885	11.8
Italia	44,575	12.8	42,474	14.9	24,391	17.0	6672	33.1	3426	40.6	6861	35.3	128,399	6.2

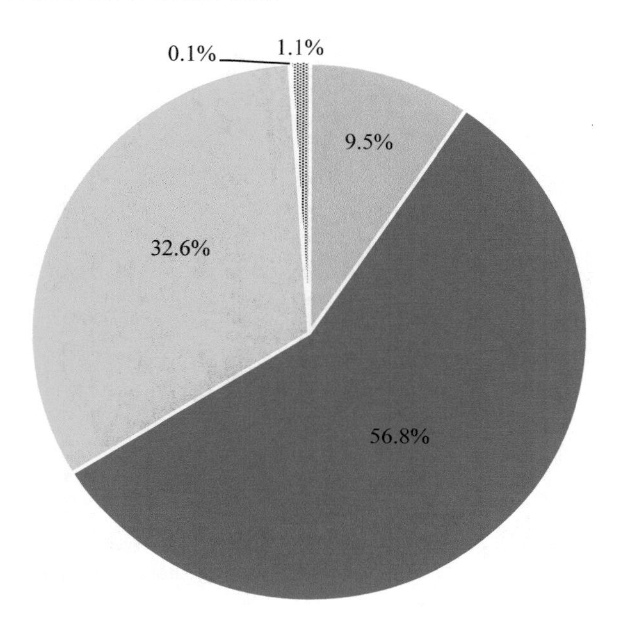

Fig. 7.20 Percent of area of coppice by development stage / Ripartizione percentuale della superficie dei cedui per stadio di sviluppo

statistiche_INFC, together with the statistics on understory woody biomass and its stored organic carbon by understory component.

Figure 7.28 shows the total number of small trees per hectare in the Italian regions. Values vary from 1506 small trees per ha in Valle d'Aosta to 9539 small trees per ha in Molise. Differences are also due to the uneven distribution of forest types in the regions. Broadleaved forests are generally characterised by a higher number of small trees than coniferous ones. Among coniferous forests, pine forests show the highest values, which are close to those of the broadleaved types with limited number of small trees like Beech and Hygrophilous forests, but higher than the number in Cork oak and Other evergreen broadleaved types.

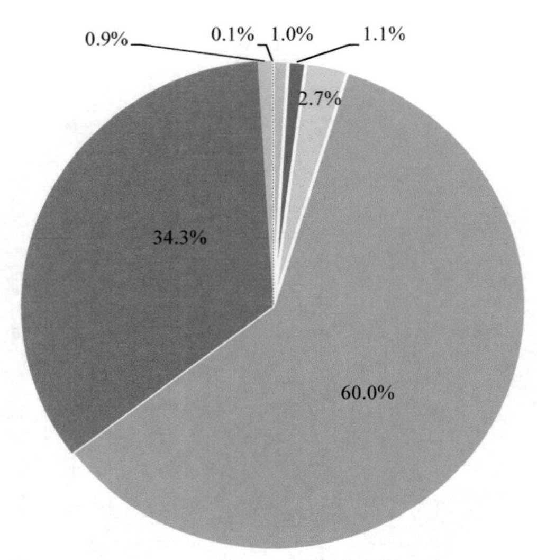

Fig. 7.21 Percent of area of even-aged high forest by development stage / Ripartizione percentuale della superficie delle fustaie coetanee per stadio di sviluppo

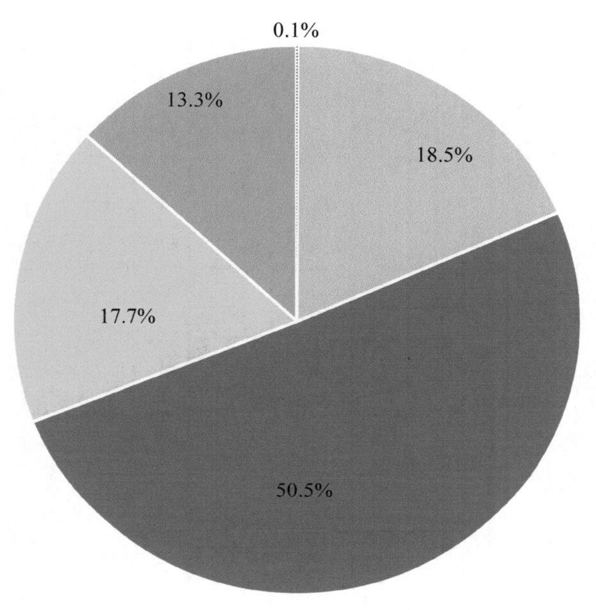

Fig. 7.22 Percent of area under special or undefined silvicultural system, by development stage / Ripartizione percentuale della superficie con tipo colturale speciale o non definito, per stadio di sviluppo

Table 7.41 Total value and value per hectare of stem volume of Forest by silvicultural system / Valori totali e per ettaro del volume del fusto del Bosco ripartito per tipo colturale

Region/Regione	Coppice			High forest			Special		
	Ceduo			Fustaia			Tipo colturale speciale		
	Volume	ES	Volume	Volume	ES	Volume	Volume	ES	Volume
	(m^3)	(%)	(m^3 ha^{-1})	(m^3)	(%)	(m^3 ha^{-1})	(m^3)	(%)	(m^3 ha^{-1})
Piemonte	49,692,519	6.4	154.0	86,303,531	4.9	183.9	3,227,336	34.9	207.9
Valle d'Aosta	450,499	44.1	150.4	18,521,062	7.9	211.1	0	–	0.0
Lombardia	51,235,889	6.4	169.8	71,687,463	6.8	291.3	152,532	100.0	108.1
Alto Adige	2,263,981	17.9	138.9	113,678,433	4.5	355.9	0	–	0.0
Trentino	6,165,548	19.2	139.5	98,758,704	5.6	376.6	0	–	0.0
Veneto	23,699,496	8.8	152.5	68,912,913	6.4	333.9	0	–	0.0
Friuli V.G	7,825,680	19.7	173.1	57,052,758	7.3	322.8	0	–	0.0
Liguria	32,556,604	6.7	149.3	16,595,937	11.7	174.5	508,434	71.9	178.6
Emilia Romagna	48,387,736	6.3	138.4	23,948,151	11.5	223.5	1,339,397	42.4	116.7
Toscana	77,247,730	4.7	116.1	63,828,497	7.4	250.1	4,065,472	26.8	158.6
Umbria	23,406,631	6.1	81.0	8,938,584	16.3	190.3	78,624	100.0	177.7
Marche	13,523,271	9.8	87.3	12,044,504	12.6	142.6	79,759	100.0	215.2
Lazio	41,655,442	6.8	109.0	27,809,933	11.5	214.1	591,618	66.6	158.3
Abruzzo	19,839,871	10.6	146.3	35,603,099	8.5	226.0	185,240	70.2	196.8
Molise	7,909,097	12.2	104.3	9,052,686	16.8	247.5	0	–	0.0
Campania	16,867,092	10.1	96.2	29,755,553	10.1	272.3	1,994,225	28.0	97.9
Puglia	5,555,130	12.5	78.8	9,542,816	14.1	158.5	0	–	0.0
Basilicata	4,824,362	14.8	59.9	28,227,900	8.5	169.2	0	–	0.0
Calabria	18,919,604	14.6	128.4	89,889,809	6.0	280.4	0	–	0.0
Sicilia	4,280,879	18.8	109.4	19,446,150	9.3	103.9	243,045	49.9	50.6
Sardegna	9,710,550	13.1	86.3	15,818,442	8.9	68.3	2,353,746	21.8	68.3
Italia	466,017,612	2.0	123.0	905,416,925	1.7	241.0	14,819,429	13.0	121.4

(continued)

Table 7.41 (continued)

Region/Regione	Undefined			Not classified			Total forest		
	Tipo colturale non definito			Non classificato			Totale bosco		
	Volume	ES	Volume	Volume	ES	Volume	Volume	ES	Volume
	(m^3)	(%)	(m^3 ha^{-1})	(m^3)	(%)	(m^3 h^{-1})	(m^3)	(%)	(m^3 ha^{-1})
Piemonte	5,274,113	21.0	86.3	2,327,298	27.4	106.1	146,824,796	2.9	164.9
Valle d'Aosta	855,882	39.0	102.1	55,722	100.0	462.6	19,883,165	7.3	200.3
Lombardia	4,425,377	22.8	102.5	2,625,747	21.3	89.1	130,127,009	3.9	209.2
Alto Adige	266,113	65.9	175.6	234,610	67.7	116.2	116,443,137	4.4	343.2
Trentino	7,700,026	17.7	116.0	145,905	100.0	296.4	112,770,182	4.7	302.1
Veneto	3,981,443	24.8	90.5	1,871,169	21.9	172.1	98,465,021	4.3	236.3
Friuli V.G	12,036,272	13.1	122.1	1,510,660	29.8	124.9	78,425,371	4.9	235.8
Liguria	1,978,151	28.7	76.1	55,802	93.8	50.8	51,694,928	4.3	150.6
Emilia Romagna	9,472,739	14.3	86.3	753,370	33.5	111.3	83,901,392	4.1	143.4
Toscana	9,118,304	18.0	113.1	1,320,883	32.6	152.4	155,580,886	3.1	150.3
Umbria	3,095,259	23.1	67.3	206,047	31.6	26.5	35,725,146	5.0	91.5
Marche	3,058,767	21.3	67.7	198,054	42.2	28.9	28,904,356	5.9	99.1
Lazio	2,453,541	28.1	60.0	330,113	35.7	94.0	72,840,647	4.7	130.0
Abruzzo	7,469,925	14.7	66.0	342,397	75.6	79.5	63,440,533	4.4	154.1
Molise	2,652,831	22.5	69.6	151,979	82.6	56.0	19,766,594	7.7	129.0
Campania	8,017,608	15.3	84.7	127,886	54.2	30.0	56,762,364	5.6	140.5
Puglia	422,479	36.4	73.7	104,424	58.7	78.8	15,624,850	8.9	109.8
Basilicata	2,368,587	34.2	63.9	191,398	78.3	54.4	35,612,246	6.4	123.6
Calabria	1,656,592	59.1	105.1	1,167,803	42.9	101.9	111,633,807	4.7	225.4
Sicilia	4,188,004	20.2	82.5	254,537	71.4	70.5	28,412,615	6.6	99.5
Sardegna	11,356,997	11.6	51.4	728,310	20.8	27.5	39,968,044	4.9	63.8
Italia	101,849,011	4.5	81.6	14,704,112	9.2	86.8	1,502,807,089	1.0	165.4

Table 7.42 Total value and value per hectare of aboveground tree biomass of Forest by silvicultural system / Valori totali e per ettaro della fitomassa arborea epigea del Bosco ripartito per tipo colturale

Region/Regione	Coppice			High forest			Special			Undefined			Not classified			Total Forest		
	Ceduo			Fustaia			Tipo colturale speciale			Tipo colturale non definito			Non classificato			Totale Bosco		
	Biomass		Biomass	Biomass		Biomass	Biomass		Biomass	Biomass		Biomass	Biomass		Biomass	Biomass		Biomass
	Fitomassa	ES	Fitomassa	Fitomassa	ES	Fitomassa	Fitomassa	ES	Fitomassa	Fitomassa	ES	Fitomassa	Fitomassa	ES	Fitomassa	Fitomassa	ES	Fitomassa
	(Mg)	(%)	(Mg ha^{-1})	(Mg)	(%)	(Mg ha^{-1})	(Mg)	(%)	(Mg ha^{-1})	(Mg)	(%)	(Mg ha^{-1})	(Mg)	(%)	(Mg ha^{-1})	(Mg)	(%)	(Mg ha^{-1})
Piemonte	35,199,884	6.3	109.1	57,578,997	5.0	122.7	1,952,680	34.7	125.8	3,834,739	20.7	62.7	1,325,036	26.0	60.4	99,891,336	2.8	112.2
Valle d'Aosta	289,200	43.2	96.5	10,344,175	7.6	117.9	0	–	0.0	565,898	35.1	67.5	28,675	100.0	238.1	11,227,949	6.9	113.1
Lombardia	36,571,976	6.3	121.2	41,337,948	6.8	168.0	92,019	100.0	65.2	3,213,610	22.9	74.4	1,514,991	19.8	51.4	82,730,544	3.6	133.0
Alto Adige	1,793,471	17.7	110.0	59,421,037	4.4	186.0	0	–	0.0	162,110	63.6	107.0	131,355	67.0	65.0	61,507,973	4.3	181.3
Trentino	4,861,270	18.5	110.0	53,805,166	5.4	205.2	0	–	0.0	5,071,048	17.1	76.4	100,586	100.0	204.3	63,838,069	4.3	171.0
Veneto	18,331,147	8.4	117.9	39,648,953	6.5	192.1	0	–	0.0	2,734,241	23.8	62.2	1,084,354	21.0	99.7	61,798,696	4.0	148.3
Friuli V.G	5,844,810	18.9	129.3	36,922,666	7.4	208.9	0	–	0.0	8,569,096	13.0	87.0	881,494	30.7	72.9	52,218,065	4.7	157.0
Liguria	23,887,400	6.6	109.5	11,064,908	11.8	116.4	337,253	72.3	118.5	1,555,798	28.1	59.8	28,706	89.7	26.1	36,874,064	4.1	107.5
Emilia Romagna	38,538,512	6.2	110.2	16,218,834	11.7	151.4	859,220	41.7	74.9	7,230,837	13.8	65.9	403,608	32.0	59.7	63,251,010	4.0	108.1
Toscana	60,803,337	4.4	91.4	44,561,468	7.4	174.6	2,729,742	25.5	106.5	6,678,808	17.5	82.8	844,048	29.8	97.4	115,617,402	2.9	111.7
Umbria	20,482,184	6.0	70.8	6,825,689	16.8	145.3	47,457	100.0	107.3	2,421,509	21.8	52.6	163,866	31.5	21.1	29,940,704	4.8	76.7
Marche	11,691,825	9.2	75.5	9,086,114	13.0	107.6	52,947	100.0	142.9	2,338,912	20.9	51.8	140,485	40.0	20.5	23,310,282	5.7	79.9
Lazio	34,734,806	6.6	90.9	21,129,955	11.6	162.7	363,530	65.6	97.3	1,888,250	26.8	46.2	248,536	37.1	70.8	58,365,077	4.6	104.2
Abruzzo	16,158,957	10.4	119.2	27,379,944	8.6	173.8	123,061	70.7	130.8	5,954,324	14.2	52.6	271,685	74.6	63.1	49,887,970	4.3	121.2
Molise	6,750,809	12.0	89.0	6,993,251	17.0	191.2	0	–	0.0	2,171,358	22.1	57.0	96,203	74.5	35.4	16,011,621	7.5	104.5
Campania	14,002,022	9.7	79.9	22,818,516	10.2	208.8	1,264,240	27.7	62.1	6,247,574	15.4	66.0	98,058	54.9	23.0	44,430,410	5.6	110.0
Puglia	4,948,829	12.4	65.9	7,255,873	14.4	120.5	0	–	0.0	326,567	35.8	57.0	86,657	60.8	65.4	12,617,926	8.6	88.6
Basilicata	3,981,863	14.4	49.4	22,020,291	8.6	132.0	0	–	0.0	1,846,576	34.5	49.8	153,850	77.2	43.8	28,002,580	6.3	97.2
Calabria	15,147,262	15.7	102.8	57,883,463	6.7	180.6	0	–	0.0	1,218,292	59.7	77.3	885,439	43.5	77.3	75,134,455	5.0	151.7
Sicilia	3,528,270	19.8	90.1	14,115,629	9.1	75.4	239,645	50.1	49.9	3,249,624	20.2	64.0	251,717	71.2	69.7	21,384,884	6.3	74.9
Sardegna	9,402,886	13.5	83.6	13,145,125	9.5	56.7	2,366,928	21.9	68.7	10,771,938	11.8	48.7	548,710	20.7	20.7	36,235,587	5.1	57.9
Italia	366,950,719	2.0	96.8	579,558,001	1.8	154.3	10,428,721	12.2	85.5	78,051,108	4.4	62.5	9,288,055	9.1	54.8	1,044,276,604	1.0	114.9

Table 7.43 Total value and value per hectare of annual volume increment of Forest by silvicultural system / Valori totali e per ettaro dell'incremento annuo di volume del Bosco ripartito per tipo colturale

Region/Regione	Coppice			High forest			Special		
	Ceduo			Fustaia			Tipo colturale speciale		
	Annual increment	ES	Annual increment	Annual increment	ES	Annual increment	Annual increment	ES	Annual increment
	Incremento annuo		Incremento annuo	Incremento annuo		Incremento annuo	Incremento annuo		Incremento annuo
	(m^3)	(%)	(m^3 ha^{-1})	(m^3)	(%)	(m^3 ha^{-1})	(m^3)	(%)	(m^3 ha^{-1})
Piemonte	1,386,005	6.6	4.3	2,022,216	5.6	4.3	67,388	35.9	4.3
Valle d'Aosta	9772	41.5	3.3	321,843	8.2	3.7	0	–	0.0
Lombardia	1,703,579	6.2	5.6	1,627,444	7.2	6.6	1601	100.0	1.1
Alto Adige	71,467	19.2	4.4	1,910,172	4.3	6.0	0	–	0.0
Trentino	199,675	18.4	4.5	1,829,712	5.7	7.0	0	–	0.0
Veneto	764,485	7.9	4.9	1,370,224	6.5	6.6	0	–	0.0
Friuli V.G	249,421	19.7	5.5	1,097,259	7.4	6.2	0	–	0.0
Liguria	772,066	6.7	3.5	333,787	11.7	3.5	10,629	76.6	3.7
Emilia Romagna	1,412,648	6.6	4.0	594,947	11.6	5.6	44,742	46.0	3.9
Toscana	1,956,368	5.0	2.9	1,212,866	7.7	4.8	63,587	30.5	2.5
Umbria	719,301	5.9	2.5	195,636	15.4	4.2	384	100.0	0.9
Marche	492,512	9.7	3.2	346,559	11.8	4.1	1460	100.0	3.9
Lazio	1,285,099	6.9	3.4	527,042	10.7	4.1	9468	65.2	2.5
Abruzzo	524,820	10.0	3.9	715,533	8.1	4.5	3977	70.0	4.2
Molise	313,933	10.8	4.1	192,668	16.5	5.3	0	–	0.0
Campania	862,922	9.4	4.9	722,616	10.1	6.6	9957	27.6	2.9
Puglia	153,863	11.1	2.0	202,058	12.4	3.4	0	–	0.0
Basilicata	273,881	14.9	3.4	697,679	8.2	4.2	0	–	0.0
Calabria	792,629	14.0	5.4	2,116,122	6.2	6.6	0	–	0.0
Sicilia	117,123	18.5	3.0	522,136	9.4	2.8	4563	47.6	1.0
Sardegna	279,663	14.5	2.5	391,238	8.4	1.7	43,470	25.1	1.3
Italia	14,341,231	2.0	3.8	18,949,756	1.8	5.0	311,226	14.0	2.6

(continued)

Table 7.43 (continued)

Region/Regione	Undefined			Not classified			Total Forest		
	Tipo colturale non definito			Non classificato			Totale Bosco		
	Annual increment	ES	Annual increment	Annual increment	ES	Annual increment	Annual increment	ES	Annual increment
	Incremento annuo		Incremento annuo	Incremento annuo		Incremento annuo	Incremento annuo		Incremento annuo
	(m^3)	(%)	(m^3 ha^{-1})	(m^3)	(%)	(m^3 ha^{-1})	(m^3)	(%)	(m^3 ha^{-1})
Piemonte	185,942	19.5	3.0	46,201	27.6	2.1	3,707,751	3.2	4.2
Valle d'Aosta	24,056	30.4	2.9	716	100.0	5.9	356,387	7.4	3.6
Lombardia	182,330b	22.8	4.2	302,123	22.3	10.3	3,817,078	3.8	6.1
Alto Adige	12,135	60.6	8.0	6916	48.3	3.4	2,000,689	4.1	5.9
Trentino	223,428	15.6	3.4	3811	100.0	7.7	2,256,626	4.4	6.0
Veneto	128,637	21.0	2.9	128,785	58.8	11.8	2,392,131	4.8	5.7
Friuli V.G	334,014	12.2	3.4	38,256	25.9	3.2	1,718,951	4.6	5.2
Liguria	44,962	29.1	1.7	1544	90.3	1.4	1,162,988	4.4	3.4
Emilia Romagna	285,183	13.8	2.6	27,912	40.0	4.1	2,365,432	4.3	4.0
Toscana	264,977	17.5	3.3	53,231	35.5	6.1	3,551,029	3.2	3.4
Umbria	96,375	23.3	2.1	10,578	36.0	1.4	1,022,274	4.5	2.6
Marche	127,722	21.4	2.8	9164	39.7	1.3	977,416	5.4	3.3
Lazio	89,639	29.2	2.2	9415	34.2	2.7	1,920,662	4.6	3.4
Abruzzo	318,074	12.9	2.8	7744	59.4	1.8	1,570,148	3.9	3.8
Molise	108,550	18.6	2.8	5601	80.3	2.1	620,751	5.9	4.1
Campania	354,452	15.3	3.7	5831	49.6	1.4	2,005,778	5.0	5.0
Puglia	20,258	37.2	3.5	7955	59.2	6.0	384,134	7.1	2.7
Basilicata	92,953	25.4	2.5	4628	66.5	1.3	1,069,141	5.6	3.7
Calabria	94,484	64.3	6.0	48,187	27.8	4.2	3,051,422	5.1	6.2
Sicilia	92,824	20.0	1.8	17,850	71.6	4.9	754,495	6.8	2.6
Sardegna	312,925	13.0	1.4	55,207	20.8	2.1	1,082,502	5.5	1.7
Italia	3,393,919	4.6	2.7	791,652	13.6	4.7	37,787,784	1.0	4.2

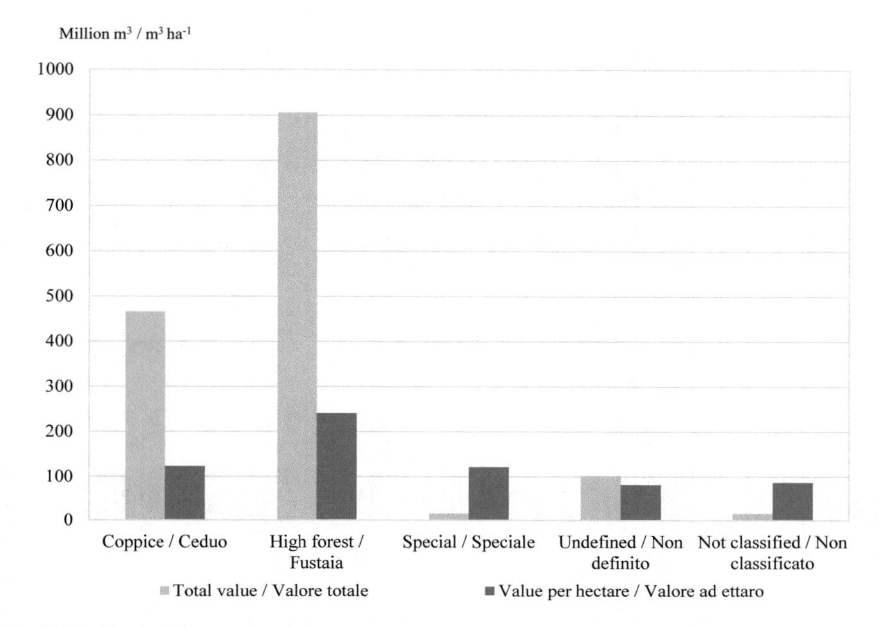

Fig. 7.23 Total value and value per hectare of growing stock volume by the silvicultural system / Valori totali e per ettaro del volume per tipo colturale

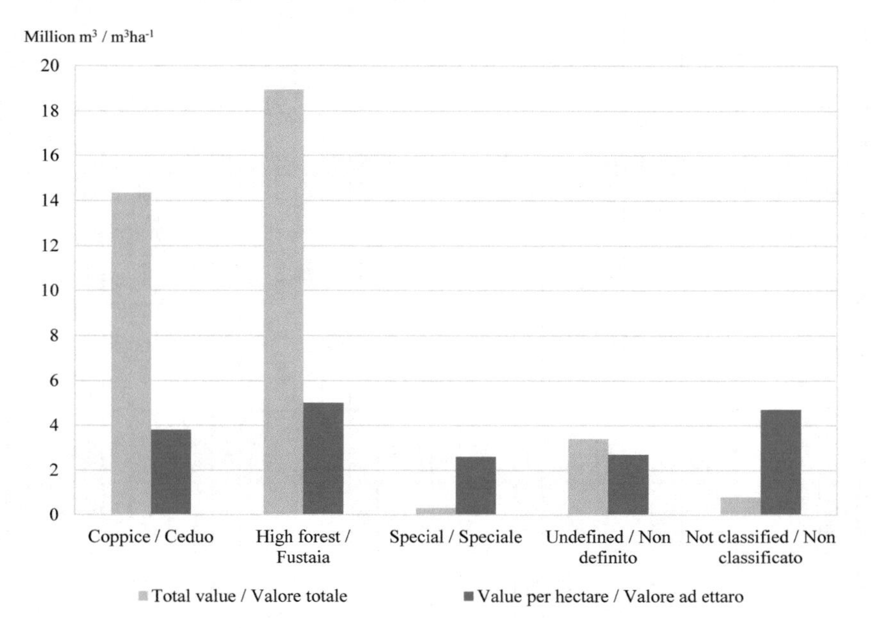

Fig. 7.24 Total value and value per hectare of the annual volume increment by the silvicultural system / Valori totali e per ettaro dell'incremento annuo di volume per tipo colturale

Table 7.44 Total value of Forest stem volume by age class / Valori totali del volume del fusto del Bosco ripartito per classe di età

Region/Regione	Even-aged coppice—Coppice without standards, coppice with standards, mixed-management coppice													
	Cedui coetanei—ceduo semplice, ceduo matricinato, ceduo composto													
	0–10 years		11–20 years		21–30 years		31–40 years		41–80 years		Not classified		Total	
	0–10 anni		11–20 anni		21–30 anni		31–40 anni		41–80 anni		Non classificato		Totale	
	Volume	ES	Volume	ES	Volume	ES	Volume	ES	Volume	ES	Volume	ES	Volume	ES
	m^3	(%)	m^3	(%)	m^3	(%)	m^3	(%)	m^3	(%)	m^3	(%)	m^3	(%)
Piemonte	753,576	55.8	3,322,924	25.6	5,483,839	19.0	16,804,655	13.5	10,643,444	19.6	12,296,069	16.4	49,304,508	6.4
Valle d'Aosta	0	–	20,823	100.0	45,810	100.0	383,866	50.2	0	–	0	–	450,499	44.1
Lombardia	16,037	100.0	1,796,093	35.7	9,123,744	21.6	16,672,873	14.5	20,662,219	14.1	2,964,925	35.7	51,235,889	6.4
Alto Adige	0	–	180,166	69.0	333,411	48.7	347,546	48.1	1,094,005	29.4	258,682	69.4	2,213,811	18.2
Trentino	0	–	157,042	94.1	548,288	64.9	1,565,175	35.6	2,023,736	38.6	1,669,570	46.3	5,963,809	19.6
Veneto	7443	100.0	1,046,241	32.3	1,721,826	36.3	7,680,683	18.7	8,270,781	17.3	2,485,593	35.3	21,212,566	9.1
Friuli V.G	0	–	868,215	37.4	733,775	53.4	2,385,233	38.5	2,185,419	42.0	1,653,038	51.0	7,825,680	19.7
Liguria	402,722	77.5	1,871,113	23.7	2,027,656	29.0	11,121,129	14.1	14,485,658	13.9	2,538,067	35.5	32,446,345	6.7
Emilia Romagna	207,318	59.7	1,398,232	42.5	4,852,349	19.0	17,627,289	11.8	22,125,148	12.3	1,851,675	36.2	48,062,010	6.4
Toscana	222,464	100.0	1,839,951	29.6	6,719,942	17.9	19,183,761	11.0	37,929,064	8.6	11,352,547	16.6	77,247,730	4.7
Umbria	124,871	58.9	852,785	30.3	5,435,931	15.8	6,182,963	13.7	5,728,514	17.1	4,982,049	20.7	23,307,113	6.2
Marche	0	–	583,497	34.4	3,624,348	20.3	4,729,734	18.9	4,207,762	21.6	353,989	67.6	13,499,331	9.8
Lazio	415,936	59.1	2,088,620	30.0	5,684,364	23.6	11,650,352	14.6	19,300,936	11.7	2,515,235	42.0	41,655,442	6.8
Abruzzo	0	–	509,326	40.0	625,168	40.8	5,479,817	22.7	13,225,560	14.2	0	–	19,839,871	10.6
Molise	0	–	150,318	100.0	410,880	45.0	4,556,873	17.0	2,340,882	31.2	450,144	66.2	7,909,097	12.2
Campania	602,838	34.3	3,551,208	22.6	4,279,822	22.9	1,889,799	35.6	5,686,537	22.9	856,890	55.3	16,867,092	10.1
Puglia	69,862	84.2	427,802	29.9	430,397	34.4	3,277,821	19.0	454,408	58.8	894,842	37.0	5,555,130	12.5
Basilicata	13,404	100.0	360,752	36.1	970,438	33.6	2,545,229	21.8	246,685	72.6	687,853	47.0	4,824,362	14.8
Calabria	1,202,108	34.2	4,388,211	37.2	5,030,707	33.4	2,592,107	52.7	2,225,808	55.8	3,480,664	39.7	18,919,604	14.6
Sicilia	719	100.0	473,422	45.9	270,625	88.2	1,233,665	40.9	1,184,428	41.6	1,118,021	37.8	4,280,879	18.8
Sardegna	59,071	100.0	721,408	38.4	570,693	59.0	2,390,514	28.4	911,614	38.8	4,381,562	19.0	9,034,861	12.2
Italia	4,098,366	19.3	26,608,149	9.4	58,924,012	6.7	140,301,081	4.3	174,932,608	4.2	56,791,414	7.4	461,655,629	2.0

(continued)

Table 7.44 (continued)

Region/Regione	Coppice in transition to high forest and even–aged high forest									
	Fustaia transitoria e fustaia coetanea									
	0–10 years		11–20 years		21–30 years		31–40 years		41–80 years	
	0–10 anni		11–20 anni		21–30 anni		31–40 anni		41–80 anni	
	Volume	ES	Volume	ES	Volume	ES	Volume	ES	Volume	ES
	m^3	(%)	m^3	(%)	m^3	(%)	m^3	(%)	m^3	(%)
Piemonte	9390	100.0	0	–	250,330	80.3	2,729,888	34.4	10,535,984	19.8
Valle d'Aosta	0	–	0	–	0	–	94,141	75.4	598,194	59.0
Lombardia	0	–	151,233	98.2	0	–	1,313,924	72.2	10,112,846	26.8
Alto Adige	746,744	47.9	236,626	53.6	149,514	87.0	1,741,277	49.2	26,494,437	14.6
Trentino	276,432	65.4	20,206	100.0	0	–	3,934,166	37.0	20,613,718	15.5
Veneto	0	–	0	–	357,262	60.8	4,565,820	33.7	12,557,338	21.0
Friuli V.G	0	–	1148	100.0	43,996	100.0	2,160,778	36.2	21,217,913	15.6
Liguria	0	–	816	100.0	0	–	713,910	56.1	1,815,032	39.4
Emilia Romagna	0	–	0	–	147,508	100.0	2,388,599	32.1	5,157,768	22.8
Toscana	0	–	4415	100.0	214,783	100.0	2,004,435	36.7	20,608,485	14.0
Umbria	0	–	0	–	0	–	474,532	48.2	2,645,510	34.7
Marche	27,619	99.0	44,335	79.9	0	–	739,172	32.2	3,052,533	25.0
Lazio	0	–	0	–	0	–	608,001	71.2	4,195,212	27.1
Abruzzo	0	–	159	100.0	0	–	280,117	43.1	12,197,743	17.0
Molise	0	–	0	–	18,971	100.0	128,208	59.7	1,213,428	46.9
Campania	0	–	0	–	83,801	100.0	1,286,855	57.3	11,628,414	21.4
Puglia	0	–	0	–	58,166	100.0	944,171	38.8	633,979	59.1
Basilicata	2857	100.0	0	–	33,317	100.0	667,925	55.0	6,926,868	22.0
Calabria	0	–	0	–	40,441	100.0	7,904,528	25.3	16,655,531	18.4
Sicilia	0	–	0	–	323,153	93.0	2,257,189	26.0	3,374,297	25.0
Sardegna	0	–	1474	100.0	313,711	43.5	2,219,414	21.6	3,078,343	22.8
Italia	1,063,041	37.8	460,412	43.3	2,034,954	26.8	39,157,051	9.8	195,313,574	4.9

(continued)

Table 7.44 (continued)

Region/Regione	Coppice in transition to high forest and even–aged high forest							
	Fustaia transitoria e fustaia coetanea							
	81–120 years		More than 120 years		Not classified		Total	
	81–120 anni		Più di 120 anni		Non classificato		Totale	
	Volume	ES	Volume	ES	Volume	ES	Volume	ES
	m^3	(%)	m^3	(%)	m^3	(%)	m^3	(%)
Piemonte	8,179,741	20.4	171,237	70.6	1,564,788	36.8	23,441,356	11.8
Valle d'Aosta	0	–	0	–	0	–	692,334	51.9
Lombardia	7,075,746	36.5	0	–	952,467	58.9	19,606,216	19.3
Alto Adige	25,373,357	16.3	38,408,892	12.7	0	–	93,150,847	6.2
Trentino	18,779,889	19.4	5,749,474	53.4	836,127	70.1	50,210,012	10.7
Veneto	6,198,862	35.6	239,299	100.0	878,599	50.5	24,797,180	14.5
Friuli V.G	12,573,920	23.2	178,047	100.0	3,692,126	46.0	39,867,928	10.4
Liguria	6,795,160	19.9	0	–	109,393	52.6	9,434,311	16.4
Emilia Romagna	2,645,922	42.9	0	–	956,157	66.4	11,295,955	16.1
Toscana	16,405,254	19.5	0	–	5,175,899	31.9	44,413,271	9.6
Umbria	290,792	67.6	0	–	0	–	3,410,834	28.2
Marche	0	–	0	–	0	–	3,863,660	20.4
Lazio	1,358,910	41.7	846,515	73.2	0	–	7,008,639	20.8
Abruzzo	8,266,375	23.6	1,908,018	51.2	0	–	22,652,412	12.1
Molise	2,738,438	38.7	0	–	0	–	4,099,044	28.7
Campania	3,105,272	41.9	437,674	100.0	46,699	100.0	16,588,715	16.5
Puglia	778,116	59.4	153,724	100.0	100,947	100.0	2,669,103	24.3
Basilicata	4,206,969	28.0	0	–	1,136,461	57.6	12,974,397	14.8
Calabria	3,153,130	43.2	0	–	2,335,742	62.6	30,089,371	12.4
Sicilia	1,500,147	47.0	0	–	1,782,491	52.5	9,237,278	15.7
Sardegna	901,197	60.2	0	–	534,639	46.7	7,048,778	13.9
Italia	130,327,196	6.8	48,092,879	12.3	20,102,533	16.3	436,551,640	3.1

(continued)

Table 7.44 (continued)

Region/Regione	Plantations													
	Impianti di arboricoltura da legno													
	0–10 years		11–20 years		21–30 years		31–40 years		41–80 years		Not classified		Total	
	0–10 anni		11–20 anni		21–30 anni		31–40 anni		41–80 anni		Non classificato		Totale	
	Volume	ES	Volume	ES	Volume	ES	Volume	ES	Volume	ES	Volume	ES	Volume	ES
	m^3	(%)	m^3	(%)	m^3	(%)	m^3	(%)	m^3	(%)	m^3	(%)	m^3	(%)
Piemonte	205,147	87.8	1,202,019	44.7	248,260	51.6	189,856	100.0	282,607	79.0	131,066	100.0	2,258,955	28.1
Valle d'Aosta	0	–	0	–	0	–	0	–	0	–	0	–	0	–
Lombardia	430,526	42.4	1,225,866	36.6	72,380	100.0	24,103	100.0	0	–	175,995	100.0	1,928,870	23.1
Alto Adige	0	–	0	–	0	–	0	–	0	–	0	–	0	–
Trentino	0	–	0	–	0	–	0	–	0	–	0	–	0	–
Veneto	282,523	85.7	126,448	83.2	130,721	100.0	0	–	0	–	42,583	100.0	582,274	40.8
Friuli V.G	326,887	37.1	419,726	58.0	116,832	57.6	114,798	100.0	0	–	26,580	100.0	1,004,823	28.8
Liguria	0	–	0	–	52,280	100.0	0	–	0	–	0	–	52,280	100.0
Emilia Romagna	77,188	57.9	155,826	65.4	255,496	59.4	0	–	162,890	100.0	83,062	100.0	734,463	34.3
Toscana	120,866	69.8	4,908	100.0	179,763	44.2	21,967	100.0	225,458	100.0	418,208	73.3	971,169	40.9
Umbria	0	–	51,975	48.2	89,746	49.4	0	–	0	–	36,509	100.0	178,230	33.8
Marche	0	–	72,163	57.0	48,149	55.6	8,537	100.0	69,206	100.0	0	–	198,054	42.2
Lazio	0	–	17,494	100.0	50,326	81.4	54,523	74.8	0	–	20,406	100.0	142,750	44.2
Abruzzo	0	–	968	100.0	42,723	57.5	4,270	100.0	0	–	0	–	47,961	51.4
Molise	0	–	6,041	64.3	21,175	68.0	124,764	100.0	0	–	0	–	151,979	82.6
Campania	5,860	92.6	47,921	78.4	4,464	100.0	0	–	0	–	0	–	58,245	65.6
Puglia	0	–	0	–	0	–	0	–	0	–	876	100.0	876	100.0
Basilicata	0	–	2,148	100.0	0	–	0	–	0	–	0	–	2,148	100.0
Calabria	5,000	70.8	19,247	100.0	40,408	100.0	124,484	100.0	0	–	124,484	100.0	313,623	57.8
Sicilia	0	–	0	–	701	100.0	0	–	0	–	701	100.0	1,401	70.7
Sardegna	224,633	43.2	99,739	46.4	162,914	53.9	136,100	61.0	99,434	71.5	0	–	722,819	21.0
Italia	1,678,628	23.7	3,452,489	22.0	1,516,337	19.8	803,401	37.3	839,595	44.1	1,060,470	38.8	9,350,921	11.0

(continued)

Table 7.44 (continued)

Region/Regione	Uneven–aged high forest, irregular or structured high forest, uneven–aged coppice		Special and undefined silvicultural systems		Not classified		Total Forest	
	Fustaia disetanea, fustaia irregolare o articolata, ceduo a sterzo		Tipi colturali speciale e non definito		Non classificato		Totale Bosco	
	Volume	ES	Volume	ES	Volume	ES	Volume	ES
	m^3	(%)	m^3	(%)	m^3	(%)	m^3	(%)
Piemonte	63,250,186	6.2	8,501,449	18.4	68,343	78.7	146,824,796	2.9
Valle d'Aosta	17,828,728	8.3	855,882	39.0	55,722	100.0	19,883,165	7.3
Lombardia	52,081,247	8.4	4,577,910	22.3	696,877	48.4	130,127,009	3.9
Alto Adige	20,577,756	14.5	266,113	65.9	234,610	67.7	116,443,137	4.4
Trentino	48,750,431	9.6	7,700,026	17.7	145,905	100.0	112,770,182	4.7
Veneto	46,602,663	9.0	3,981,443	24.8	1,288,895	28.7	98,465,021	4.3
Friuli V.G	17,184,830	17.1	12,036,272	13.1	505,837	68.3	78,425,371	4.9
Liguria	7,271,885	18.3	2,486,585	27.1	3,521	70.7	51,694,928	4.3
Emilia Romagna	12,977,922	17.2	10,812,136	13.4	18,907	70.7	83,901,392	4.1
Toscana	19,415,226	15.4	13,183,776	14.8	349,714	47.7	155,580,886	3.1
Umbria	5,627,269	20.2	3,173,883	22.7	27,817	100.0	35,725,146	5.0
Marche	8,204,785	16.8	3,138,526	20.9	0	–	28,904,356	5.9
Lazio	20,801,295	14.3	3,045,159	25.7	187,363	53.3	72,840,647	4.7
Abruzzo	12,950,687	15.7	7,655,165	14.4	294,436	87.5	63,440,533	4.4
Molise	4,953,642	23.7	2,652,831	22.5	0	–	19,766,594	7.7
Campania	13,166,837	17.9	10,011,833	13.3	69,641	83.1	56,762,364	5.6
Puglia	6,873,714	18.1	422,479	36.4	103,548	59.2	15,624,850	8.9
Basilicata	15,253,503	13.6	2,368,587	34.2	189,250	79.1	35,612,246	6.4
Calabria	59,800,437	8.1	1,656,592	59.1	854,180	54.7	111,633,807	4.7
Sicilia	10,208,872	12.8	4,431,049	19.2	253,135	71.8	28,412,615	6.6
Sardegna	9,445,353	13.2	13,710,742	10.1	5,491	74.7	39,968,044	4.9
Italia	473,227,267	2.7	116,668,440	4.2	5,353,192	16.8	1,502,807,089	1.0

Table 7.45 Total value of Forest aboveground tree biomass, by age class / Valori totali della fitomassa arborea epigea del Bosco ripartito per classe di età

Region/Regione	Even–aged coppice—Coppice without standards, coppice with standards, mixed-management coppice													
	Cedui coetanei—ceduo semplice, ceduo matricinato, ceduo composto													
	0–10 years		11–20 years		21–30 years		31–40 years		41–80 years		Not classified		Total	
	0–10 anni		11–20 anni		21–30 anni		31–40 anni		41–80 anni		Non classificato		Totale	
	Biomass		Biomass		Biomass		Biomass		Biomass		Biomass		Biomass	
	Fitomassa	ES	Fitomassa	ES	Fitomassa	ES	Fitomassa	ES	Fitomassa	ES	Fitomassa	ES	Fitomassa	ES
	Mg	(%)	Mg	(%)	Mg	(%)	Mg	(%)	Mg	(%)	Mg	(%)	Mg	(%)
Piemonte	559,482	55.2	2,548,974	24.5	4,060,316	18.5	11,924,712	13.3	7,056,961	19.7	8,741,732	15.9	34,892,176	6.3
Valle d'Aosta	0	–	15,817	100.0	38,109	100.0	235,274	50.3	0	–	0	–	289,200	43.2
Lombardia	16,120	100.0	1,426,276	32.4	6,774,616	19.8	12,069,121	14.2	14,124,840	14.2	2,161,004	35.4	36,571,976	6.3
Alto Adige	0	–	145,111	69.1	265,731	49.3	280,235	47.8	885,784	29.1	186,070	69.2	1,762,932	17.9
Trentino	0	–	155,578	93.8	505,938	63.2	1,231,464	35.5	1,543,377	38.6	1,261,364	44.7	4,697,720	18.8
Veneto	7175	100.0	858,482	32.2	1,389,877	36.0	5,967,116	17.9	6,664,149	16.9	1,769,148	35.0	16,655,947	8.6
Friuli V.G	0	–	670,608	35.5	541,822	52.3	1,800,634	35.1	1,590,036	41.8	1,241,711	52.0	5,844,810	18.9
Liguria	315,166	75.3	1,539,776	22.8	1,573,323	29.1	8,552,678	14.0	9,976,729	13.7	1,857,858	36.1	23,815,530	6.6
Emilia Romagna	174,986	60.5	1,051,574	39.1	4,147,110	18.5	14,231,488	11.6	17,235,980	12.4	1,435,404	35.5	38,276,542	6.3
Toscana	171,291	100.0	1,497,372	29.2	5,451,029	17.1	15,318,338	10.7	29,762,460	8.2	8,602,846	15.8	60,803,337	4.4
Umbria	109,673	56.4	830,803	29.5	4,718,021	15.8	5,595,378	13.7	4,916,414	17.1	4,232,786	20.6	20,403,075	6.0
Marche	0	–	593,150	34.6	3,154,298	19.7	4,096,190	18.1	3,510,641	20.9	313,509	66.3	11,667,789	9.2
Lazio	390,082	63.4	1,633,954	29.6	4,503,422	22.7	9,826,277	14.2	16,305,478	11.6	2,075,594	41.0	34,734,806	6.6
Abruzzo	0	–	38 4,194	39.7	540,697	40.1	4,569,004	22.1	10,665,062	14.0	0	–	16,158,957	10.4
Molise	0	–	123,665	100.0	366,760	46.1	3,930,818	16.9	1,944,486	31.1	385,080	65.1	6,750,809	12.0
Campania	530,152	33.7	2,874,177	22.2	3,652,507	22.6	1,402,457	35.3	4,847,159	22.4	695,569	53.4	14,002,022	9.7
Puglia	68,227	81.7	412,243	29.9	383,759	33.5	2,967,265	19.0	366,346	58.9	750,988	37.7	4,948,829	12.4
Basilicata	10,918	100.0	306,044	36.2	805,374	33.4	2,151,987	21.3	199,856	74.9	507,685	44.3	3,981,863	14.4
Calabria	831,247	32.9	3,665,903	43.6	3,544,928	34.5	2,161,974	58.2	2,154,394	58.2	2,788,816	40.4	15,147,262	15.7
Sicilia	558	100.0	386,865	45.4	219,950	86.8	1,041,788	42.5	1,039,216	43.9	839,893	38.7	3,528,270	19.8
Sardegna	47,586	100.0	720,684	38.4	555,056	60.4	2,279,798	29.0	888,682	38.8	4,186,485	19.0	8,678,290	12.3
Italia	3,232,664	18.8	21,841,249	9.9	47,192,641	6.3	111,633,994	4.2	135,678,050	4.1	44,033,543	7.2	363,612,142	2.0

(continued)

Table 7.45 (continued)

Region/Regione	Coppice in transition to high forest and even-aged high forest									
	Fustaia transitoria e fustaia coetanea									
	0–10 years		11–20 years		21–30 years		31–40 years		41–80 years	
	0–10 anni		11–20 anni		21–30 anni		31–40 anni		41–80 anni	
	Biomass		Biomass		Biomass		Biomass		Biomass	
	Fitomassa	ES	Fitomassa	ES	Fitomassa	ES	Fitomassa	ES	Fitomassa	ES
	Mg	(%)	Mg	(%)	Mg	(%)	Mg	(%)	Mg	(%)
Piemonte	6511	100.0	0	–	179,716	79.9	1,886,317	35.9	6,721,585	19.6
Valle d'Aosta	0	–	0	–	0	–	81,545	75.6	346,108	55.7
Lombardia	0	–	88,877	97.6	0	–	767,720	67.3	5,681,917	26.2
Alto Adige	408,688	46.7	144,892	50.8	81,953	86.1	929,445	49.1	14,046,799	14.4
Trentino	153,273	62.2	20,296	100.0	0	–	2,149,036	36.0	11,826,411	15.0
Veneto	0	–	0	–	284,749	59.3	2,662,565	31.5	7,763,796	20.7
Friuli V.G	0	–	835	100.0	35,651	100.0	1,526,902	37.0	13,957,547	15.4
Liguria	0	–	476	100.0	0	–	462,196	56.9	1,201,904	39.4
Emilia Romagna	0	–	0	–	96,429	100.0	1,680,341	33.8	3,324,486	24.2
Toscana	0	–	3190	100.0	130,847	100.0	1,508,860	37.1	14,830,029	14.0
Umbria	0	–	0	–	0	–	358,977	49.1	1,953,421	36.8
Marche	20,063	98.8	41,534	84.7	0	–	507,602	32.3	2,241,829	26.2
Lazio	0	–	0	–	0	–	491,569	70.3	3,004,164	28.6
Abruzzo	0	–	180	100.0	0	–	206,884	44.1	9,263,900	17.1
Molise	0	–	0	–	12,595	100.0	93,144	60.0	941,364	47.8
Campania	0	–	0	–	64,361	100.0	964,689	59.9	8,871,186	21.9
Puglia	0	–	0	–	40,849	100.0	671,924	39.1	479,604	58.5
Basilicata	3819	100.0	0	–	22,342	100.0	525,441	56.7	5,274,639	22.3
Calabria	0	–	0	–	29,768	100.0	4,480,936	26.1	8,417,813	18.6
Sicilia	0	–	0	–	205,544	92.4	1,496,634	23.7	2,276,835	24.8
Sardegna	0	–	1215	100.0	245,825	44.2	1,373,894	21.2	2,029,316	21.9
Italia	592,353	36.2	301,493	40.1	1,430,627	25.9	24,826,621	9.5	124,454,651	4.8

(continued)

Table 7.45 (continued)

Region/Regione	Coppice in transition to high forest and even–aged high forest							
	Fustaia transitoria e fustaia coetanea							
	81–120 years		More than 120 years		Not classified		Total	
	81–120 anni		Più di 120 anni		Non classificato		Totale	
	Biomass		Biomass		Biomass		Biomass	
	Fitomassa	ES	Fitomassa	ES	Fitomassa	ES	Fitomassa	ES
	Mg	(%)	Mg	(%)	Mg	(%)	Mg	(%)
Piemonte	5,098,273	20.7	102,854	70.3	1,044,425	35.8	15,039,682	11.8
Valle d'Aosta	0	–	0	–	0	–	427,653	47.2
Lombardia	3,921,213	35.6	0	–	581,401	58.7	11,041,127	18.7
Alto Adige	13,134,636	16.2	19,793,429	12.7	0	–	48,539,842	6.1
Trentino	9,840,071	19.1	2,934,139	53.3	617,509	70.0	27,540,735	10.3
Veneto	3,855,054	35.0	139,243	100.0	621,700	51.4	15,327,106	14.1
Friuli V.G	8,053,603	23.5	98,253	100.0	2,229,191	44.2	25,901,982	10.3
Liguria	4,434,715	20.1	0	–	69,201	52.0	6,168,491	16.6
Emilia Romagna	1,599,435	42.7	0	–	737,483	67.0	7,438,174	16.6
Toscana	10,238,084	19.4	0	–	3,586,302	32.6	30,297,312	9.6
Umbria	179,173	65.1	0	–	0	–	2,491,570	29.9
Marche	0	–	0	–	0	–	2,811,028	21.5
Lazio	920,654	44.1	543,772	71.9	0	–	4,960,159	21.5
Abruzzo	6,445,970	23.8	1,509,244	51.2	0	–	17,426,177	12.2
Molise	2,111,909	39.3	0	–	0	–	3,159,011	29.2
Campania	2,482,932	42.0	346,024	100.0	37,744	100.0	12,766,935	16.9
Puglia	554,088	58.7	108,329	100.0	69,539	100.0	1,924,332	24.3
Basilicata	3,313,936	28.1	0	–	869,635	59.1	10,009,812	15.0
Calabria	1,626,634	43.2	0	–	1,160,512	57.9	15,715,663	12.7
Sicilia	1,065,312	45.7	0	–	1,281,703	52.1	6,326,028	15.6
Sardegna	827,895	67.9	0	–	519,451	47.9	4,997,596	15.7
Italia	79,703,584	6.6	25,575,288	12.2	13,425,796	15.5	270,310,412	3.0

(continued)

Table 7.45 (continued)

Region/Regione	Plantations													
	Impianti di arboricoltura da legno													
	0–10 years		11–20 years		21–30 years		31–40 years		41–80 years		Not classified		Total	
	0–10 anni		11–20 anni		21–30 anni		31–40 anni		41–80 anni		Non classificato		Totale	
	Biomass		Biomass		Biomass		Biomass		Biomass		Biomass		Biomass	
	Fitomassa	ES	Fitomassa	ES	Fitomassa	ES	Fitomassa	ES	Fitomassa	ES	Fitomassa	ES	Fitomassa	ES
	Mg	(%)	Mg	(%)	Mg	(%)	Mg	(%)	Mg	(%)	Mg	(%)	Mg	(%)
Piemonte	120,964	83.1	643,326	42.7	151,419	53.6	142,022	100.0	141,924	77.0	73,190	100.0	1,272,845	26.9
Valle d'Aosta	0	–	0	–	0	–	0	–	0	–	0	–	0	–
Lombardia	306,051	44.0	661,757	36.3	42,202	100.0	17,754	100.0	0	–	97,606	100.0	1,125,370	21.2
Alto Adige	0	–	0	–	0	–	0	–	0	–	0	–	0	–
Trentino	0	–	0	–	0	–	0	–	0	–	0	–	0	–
Veneto	142,257	83.8	71,053	83.5	100,337	100.0	0	–	0	–	33,659	100.0	347,306	38.2
Friuli V.G	180,043	34.9	206,238	57.0	85,268	57.6	57,148	100.0	0	–	18,528	100.0	547,225	26.3
Liguria	0	–	0	–	25,664	100.0	0	–	0	–	0	–	25,664	100.0
Emilia Romagna	44,644	56.4	81,920	63.9	135,161	58.6	0	–	76,261	100.0	50,483	100.0	388,468	33.2
Toscana	72,644	70.9	3637	100.0	133,287	44.9	18,411	100.0	118,020	100.0	216,352	73.3	562,351	37.8
Umbria	0	–	39,989	46.9	69,765	48.5	0	–	0	–	26,963	100.0	136,717	33.1
Marche	0	–	57,031	55.4	34,192	55.1	5047	100.0	44,215	100.0	0	–	140,485	40.0
Lazio	0	–	12,837	100.0	40,041	86.5	35,682	79.5	0	–	9883	100.0	98,442	48.1
Abruzzo	0	–	843	100.0	32,447	55.0	3720	100.0	0	–	0	–	37,011	48.5
Molise	0	–	6016	64.4	19,592	66.6	70,596	100.0	0	–	0	–	96,203	74.5
Campania	5006	94.7	34,072	80.8	3313	100.0	0	–	0	–	0	–	42,391	66.3
Puglia	0	–	0	–	0	–	0	–	0	–	912	100.0	912	100.0
Basilicata	0	–	2039	100.0	0	–	0	–	0	–	0	–	2039	100.0
Calabria	4402	73.1	15,182	100.0	29,842	100.0	63,433	100.0	0	–	63,433	100.0	176,291	54.2
Sicilia	0	–	0	–	597	100.0	0	–	0	–	597	100.0	1195	70.7
Sardegna	179,416	41.8	79,607	47.0	129,467	54.6	85,386	60.7	70,493	72.2	0	–	544,367	20.9
Italia	1,055,426	22.3	1,915,547	20.7	1,032,593	19.5	499,199	38.3	450,911	42.2	591,606	37.4	5,545,281	10.1

(continued)

Table 7.45 (continued)

Region/Regione	Uneven–aged high forest, irregular or structured high forest, uneven–aged coppice		Special and undefined silvicultural systems		Not classified		Total Forest	
	Fustaia disetanea, fustaia irregolare o articolata, ceduo a sterzo		Tipi colturali speciale e non definito		Non classificato		Totale Bosco	
	Biomass		Biomass		Biomass		Biomass	
	Fitomassa	ES	Fitomassa	ES	Fitomassa	ES	Fitomassa	ES
	Mg	(%)	Mg	(%)	Mg	(%)	Mg	(%)
Piemonte	42,847,024	6.2	5,787,419	17.9	52,191	74.0	99,891,336	2.8
Valle d'Aosta	9,916,522	8.1	565,898	35.1	28,675	100.0	11,227,949	6.9
Lombardia	30,296,821	8.5	3,305,629	22.4	389,622	46.6	82,730,544	3.6
Alto Adige	10,911,734	14.2	162,110	63.6	131,355	67.0	61,507,973	4.3
Trentino	26,427,981	9.4	5,071,048	17.1	100,586	100.0	63,838,069	4.3
Veneto	25,997,047	9.0	2,734,241	23.8	737,047	28.9	61,798,696	4.0
Friuli V.G	11,020,684	17.6	8,569,096	13.0	334,269	68.4	52,218,065	4.7
Liguria	4,968,287	18.9	1,893,051	26.4	3042	70.7	36,874,064	4.1
Emilia Romagna	9,042,631	17.3	8,090,056	13.0	15,140	71.9	63,251,010	4.0
Toscana	14,264,156	15.3	9,408,550	14.4	281,697	47.8	115,617,402	2.9
Umbria	4,413,228	20.5	2,468,966	21.5	27,148	100.0	29,940,704	4.8
Marche	6,299,122	17.1	2,391,859	20.5	0	–	23,310,282	5.7
Lazio	16,169,797	14.3	2,251,780	24.6	150,094	53.0	58,365,077	4.6
Abruzzo	9,953,767	15.9	6,077,384	13.9	234,674	86.0	49,887,970	4.3
Molise	3,834,240	24.0	2,171,358	22.1	0	–	16,011,621	7.5
Campania	10,051,581	18.1	7,511,814	13.5	55,667	82.5	44,430,410	5.6
Puglia	5,331,541	18.2	326,567	35.8	85,745	61.5	12,617,926	8.6
Basilicata	12,010,480	13.7	1,846,576	34.5	151,811	78.2	28,002,580	6.3
Calabria	42,167,800	8.8	1,218,292	59.7	709,147	52.6	75,134,455	5.0
Sicilia	7,789,602	12.5	3,489,268	19.1	250,523	71.5	21,384,884	6.3
Sardegna	8,872,125	13.9	13,138,866	10.2	4343	73.4	36,235,587	5.1
Italia	312,586,167	2.7	88,479,828	4.1	3,742,775	17.1	1,044,276,604	1.0

Table 7.46 Total value of Forest annual volume increment by age class / Valori totali dell'incremento annuo di volume del Bosco ripartito per classe di età

Region/Regione	Even-aged coppice—Coppice without standards, coppice with standards, mixed-management coppice													
	Cedui coetanei—ceduo semplice, ceduo matricinato, ceduo composto													
	0–10 years		11–20 years		21–30 years		31–40 years		41–80 years		Not classified		Total	
	0–10 anni		11–20 anni		21–30 anni		31–40 anni		41–80 anni		Non classificato		Totale	
	Annual increment	ES	Annual increment	ES	Annual increment	ES	Annual increment	ES	Annual increment	ES	Annual increment	ES	Annual increment	ES
	Incremento annuo		Incremento annuo		Incremento annuo		Incremento annuo		Incremento annuo		Incremento annuo		Incremento annuo	
	m^3	(%)	m^3	(%)	m^3	(%)	m^3	(%)	m^3	(%)	m^3	(%)	m^3	(%)
Piemonte	22,114	50.8	172,192	24.8	206,437	21.4	450,209	14.9	253,072	19.5	274,124	16.7	1,378,148	6.7
Valle d'Aosta	0	–	1219	100.0	1748	100.0	6805	51.3	0	–	0	–	9772	41.5
Lombardia	2019	100.0	111,786	34.0	331,465	19.3	580,592	13.5	589,988	14.8	87,729	33.8	1,703,579	6.2
Alto Adige	0	–	9196	70.8	13,263	42.5	9588	49.6	30,041	35.9	6905	63.3	68,993	19.6
Trentino	0	–	12,049	98.6	14,896	56.6	58,535	39.3	50,686	40.0	49,633	39.3	185,798	18.3
Veneto	1675	100.0	61,798	33.0	72,168	32.6	261,739	17.7	245,527	16.6	73,222	36.1	716,129	8.3
Friuli V.G	0	–	39,626	48.7	24,411	51.7	86,843	40.3	51,805	42.2	46,737	48.6	249,421	19.7
Liguria	15,212	83.4	78,225	25.7	51,388	29.6	281,997	14.4	293,545	14.1	48,009	36.0	768,375	6.7
Emilia Romagna	8054	54.8	78,672	50.4	209,890	19.2	500,698	12.2	553,240	12.6	47,363	35.8	1,397,916	6.6
Toscana	8351	100.0	75,318	29.7	234,162	19.4	529,645	12.0	855,895	8.9	252,998	16.2	1,956,368	5.0
Umbria	4894	53.1	31,127	28.5	194,059	14.8	212,062	14.1	129,901	17.6	145,813	20.7	717,855	5.9
Marche	0	–	35,390	34.4	150,727	20.6	174,514	21.0	119,521	18.6	11,479	74.2	491,631	9.7
Lazio	10,766	46.7	123,569	34.8	226,471	24.4	383,548	13.9	467,791	11.4	72,955	37.2	1,285,099	6.9
Abruzzo	0	–	30,908	43.4	25,703	40.7	156,172	20.6	312,038	14.1	0	–	524,820	10.0
Molise	0	–	15,328	100.0	29,409	43.2	188,608	15.2	63,927	30.8	16,662	59.2	313,933	10.8
Campania	45,643	33.5	264,466	21.9	228,702	22.8	98,414	38.3	194,902	21.4	30,795	50.6	862,922	9.4
Puglia	2239	85.6	18,403	30.0	21,759	36.7	78,662	17.7	9677	58.7	23,124	37.3	153,863	11.1
Basilicata	1685	100.0	25,262	36.7	78,742	35.5	137,635	21.0	9063	73.5	21,495	54.2	273,881	14.9
Calabria	108,239	38.8	290,697	29.2	168,415	31.1	92,830	49.1	48,152	61.4	84,296	42.4	792,629	14.0
Sicilia	19	100.0	15,656	39.1	3,426	74.3	29,939	38.5	33,659	41.3	34,425	38.6	117,123	18.5
Sardegna	3089	100.0	20,723	37.5	23,336	55.3	86,727	35.0	18,513	48.1	113,506	19.1	265,894	14.5
Italia	233,999	21.1	1,511,607	9.3	2,310,574	6.5	4,405,760	4.3	4,330,940	4.3	1,441,269	7.3	14,234,148	2.1

(continued)

Table 7.46 (continued)

Region/Regione	Coppice in transition to high forest and even–aged high forest									
	Fustaia transitoria e fustaia coetanea									
	0–10 years		11–20 years		21–30 years		31–40 years		41–80 years	
	0–10 anni		11–20 anni		21–30 anni		31–40 anni		41–80 anni	
	Annual increment	ES	Annual increment	ES	Annual increment	ES	Annual increment	ES	Annual increment	ES
	Incremento annuo		Incremento annuo		Incremento annuo		Incremento annuo		Incremento annuo	
	m^3	(%)	m^3	(%)	m^3	(%)	m^3	(%)	m^3	(%)
Piemonte	1597	100.0	0	–	9218	76.7	54,475	36.0	218,851	20.9
Valle d'Aosta	0	–	0	–	0	–	3302	70.6	12,695	58.3
Lombardia	0	–	10,183	96.0	0	–	61,509	60.9	213,507	29.7
Alto Adige	25,247	54.0	14,356	47.1	11,127	88.2	64,268	53.9	575,284	14.4
Trentino	8444	55.0	4574	100.0	0	–	122,103	39.7	417,360	15.6
Veneto	0	–	0	–	27,426	70.1	135,885	30.3	249,958	22.0
Friuli V.G	0	–	146	100.0	1416	100.0	69,334	38.7	450,727	15.9
Liguria	0	–	44	100.0	0	–	14,589	56.3	39,521	42.1
Emilia Romagna	0	–	0	–	31,029	100.0	72,381	32.5	144,591	22.3
Toscana	0	–	101	100.0	1001	100.0	58,026	41.7	409,613	14.6
Umbria	0	–	0	–	0	–	11,789	47.2	49,030	30.6
Marche	1536	95.4	6935	98.0	0	–	30,937	32.8	70,719	25.3
Lazio	0	–	0	–	0	–	17,262	65.3	91,467	27.7
Abruzzo	0	–	32	100.0	0	–	11,645	49.8	265,994	16.2
Molise	0	–	0	–	3287	100.0	4292	61.3	35,092	47.4
Campania	0	–	0	–	12,414	100.0	43,735	53.6	271,589	20.5
Puglia	0	–	0	–	4420	100.0	31,038	36.9	14,734	60.5
Basilicata	503	100.0	0	–	2086	100.0	24,001	51.8	156,606	20.8
Calabria	0	–	0	–	1769	100.0	235,712	27.2	384,712	18.6
Sicilia	0	–	0	–	6322	84.5	77,190	26.6	86,842	24.4
Sardegna	0	–	193	100.0	22,087	45.3	95,408	21.1	74,341	22.7
Italia	37,327	39.1	36,563	39.5	133,601	31.8	1,238,880	9.9	4,233,231	4.9

(continued)

Table 7.46 (continued)

Region/Regione	Coppice in transition to high forest and even–aged high forest							
	Fustaia transitoria e fustaia coetanea							
	81–120 years		More than 120 years		Not classified		Total	
	81–120 anni		Più di 120 anni		Non classificato		Totale	
	Annual increment	ES	Annual increment	ES	Annual increment	ES	Annual increment	ES
	Incremento annuo		Incremento annuo		Incremento annuo		Incremento annuo	
	m^3	(%)	m^3	(%)	m^3	(%)	m^3	(%)
Piemonte	109,483	21.8	2929	83.5	40,420	37.7	436,973	12.7
Valle d'Aosta	0	–	0	–	0	–	15,996	48.4
Lombardia	153,145	36.9	0	–	17,076	56.0	455,420	20.0
Alto Adige	395,180	16.1	421,886	12.6	0	–	1507,348	6.2
Trentino	269,240	19.9	53,230	51.4	16,513	76.6	891,463	10.3
Veneto	106,785	36.7	4869	100.0	21,695	55.9	546,618	14.3
Friuli V.G	184,170	24.3	3395	100.0	53,770	55.3	762,958	10.8
Liguria	137,473	20.7	0	–	4539	47.9	196,165	17.0
Emilia Romagna	42,528	43.5	0	–	21,938	70.6	312,467	17.4
Toscana	227,737	19.2	0	–	76,273	30.7	772,750	9.7
Umbria	6460	63.9	0	–	0	–	67,280	24.4
Marche	0	–	0	–	0	–	110,126	19.3
Lazio	21,952	42.6	10,871	70.9	0	–	141,552	21.1
Abruzzo	124,119	23.1	15,620	49.8	0	–	417,410	11.7
Molise	39,121	38.0	0	–	0	–	81,792	27.2
Campania	47,385	40.3	4377	100.0	1941	100.0	381,440	15.8
Puglia	13,546	62.2	1514	100.0	9587	100.0	74,838	25.1
Basilicata	86,269	26.8	0	–	35,465	51.4	304,930	14.1
Calabria	44,602	46.3	0	–	47,825	55.6	714,619	12.8
Sicilia	32,139	43.7	0	–	44,646	53.1	247,139	15.1
Sardegna	9645	56.4	0	–	11,618	58.1	213,291	12.6
Italia	2,050,976	6.9	518,690	11.8	403,305	15.8	8,652,574	3.1

(continued)

Table 7.46 (continued)

Region/Regione	Plantations													
	Impianti di arboricoltura da legno													
	0–10 years		11–20 years		21–30 years		31–40 years		41–80 years		Not classified		Total	
	0–10 anni		11–20 anni		21–30 anni		31–40 anni		41–80 anni		Non classificato		Totale	
	Annual increment	ES	Annual increment	ES	Annual increment	ES	Annual increment	ES	Annual increment	ES	Annual increment	ES	Annual increment	ES
	Incremento annuo		Incremento annuo		Incremento annuo		Incremento annuo		Incremento annuo		Incremento annuo		Incremento annuo	
	m^3	(%)	m^3	(%)	m^3	(%)	m^3	(%)	m^3	(%)	m^3	(%)	m^3	(%)
Piemonte	3884	84.9	20,701	40.3	3521	49.7	7798	100.0	4309	77.8	4314	100.0	44,528	28.5
Valle d'Aosta	0	–	0	–	0	–	0	–	0	–	0	–	0	–
Lombardia	59,106	37.8	206,012	34.1	8140	100.0	1954	100.0	0	–	6366	100.0	281,578	23.7
Alto Adige	0	–	0	–	0	–	0	–	0	–	0	–	0	–
Trentino	0	–	0	–	0	–	0	–	0	–	0	–	0	–
Veneto	79,975	98.9	15,807	70.9	7889	100.0	0	–	0	–	3055	100.0	106,726	71.8
Friuli V.G	8785	43.4	7817	50.3	5578	64.8	4587	100.0	0	–	2794	100.0	29,561	26.6
Liguria	0	–	0	–	1390	100.0	0	–	0	–	0	–	1390	100.0
Emilia Romagna	1672	55.4	12,905	70.5	5335	56.4	0	–	6023	100.0	952	100.0	26,887	41.5
Toscana	14,842	92.8	338	100.0	13,845	65.2	1456	100.0	5033	100.0	8676	80.6	44,189	41.9
Umbria	0	–	3362	52.9	3787	57.8	0	–	0	–	878	100.0	8027	36.1
Marche	0	–	4029	54.0	2019	54.2	144	100.0	2972	100.0	0	–	9164	39.7
Lazio	0	–	355	100.0	967	72.5	1972	71.4	0	–	970	100.0	4263	43.2
Abruzzo	0	–	41	100.0	1899	61.4	314	100.0	0	–	0	–	2254	52.9
Molise	0	–	284	63.8	840	58.1	4476	100.0	0	–	0	–	5600	80.3
Campania	837	96.0	3003	77.9	122	100.0	0	–	0	–	0	–	3961	62.4
Puglia	0	–	0	–	0	–	0	–	0	–	35	100.0	35	100.0
Basilicata	0	–	125	100.0	0	–	0	–	0	–	0	–	125	100.0
Calabria	1250	88.1	459	100.0	2187	100.0	4163	100.0	0	–	4163	100.0	12,222	52.1
Sicilia	0	–	0	–	33	100.0	0	–	0	–	33	100.0	67	70.7
Sardegna	24,676	35.4	9230	50.0	14,920	56.3	3793	59.8	2259	70.9	0	–	54,877	20.9
Italia	195,027	43.0	284,465	25.5	72,471	24.7	30,656	37.5	20,596	44.6	32,237	37.4	635,452	16.7

(continued)

Table 7.46 (continued)

Region/Regione	Uneven–aged high forest, irregular or structured high forest, uneven–aged coppice		Special and undefined silvicultural systems		Not classified		Total Forest	
	Fustaia disetanea, fustaia irregolare o articolata, ceduo a sterzo		Tipi colturali speciale e non definito		Non classificato		Totale Bosco	
	Annual increment	ES	Annual increment	ES	Annual increment	ES	Annual increment	ES
	Incremento annuo		Incremento annuo		Incremento annuo		Incremento annuo	
	m^3	(%)	m^3	(%)	m^3	(%)	m^3	(%)
Piemonte	1,593,101	6.7	253,329	17.1	1673	66.9	3,707,751	3.2
Valle d'Aosta	305,847	8.7	24,056	30.4	716	100.0	356,387	7.4
Lombardia	1,172,024	8.6	183,931	22.7	20,545	47.8	3,817,078	3.8
Alto Adige	405,299	13.8	12,135	60.6	6916	48.3	2,000,689	4.1
Trentino	952,125	10.0	223,428	15.6	3811	100.0	2,256,626	4.4
Veneto	871,962	9.0	128,637	21.0	22,059	39.1	2,392,131	4.8
Friuli V.G	334,302	16.4	334,014	12.2	8695	69.3	1,718,951	4.6
Liguria	141,314	17.7	55,591	27.7	154	70.7	1,162,988	4.4
Emilia Romagna	297,212	16.1	329,925	13.4	1026	74.4	2,365,432	4.3
Toscana	440,116	15.2	328,564	15.3	9042	41.7	3,551,029	3.2
Umbria	129,802	20.1	96,759	23.2	2551	100.0	1,022,274	4.5
Marche	237,314	15.4	129,182	21.2	0	–	977,416	5.4
Lazio	385,490	13.1	99,107	27.1	5153	51.5	1,920,662	4.6
Abruzzo	298,123	14.3	322,051	12.7	5490	80.9	1,570,148	3.9
Molise	110,876	23.2	108,550	18.6	0	–	620,751	5.9
Campania	341,176	17.7	414,409	13.5	1870	80.2	2,005,778	5.0
Puglia	127,220	15.7	20,258	37.2	7920	59.5	384,134	7.1
Basilicata	392,749	12.5	92,953	25.4	4503	68.3	1,069,141	5.6
Calabria	1,401,502	8.2	94,484	64.3	35,965	32.8	3,051,422	5.1
Sicilia	274,997	13.8	97,387	19.2	17,783	71.9	754,495	6.8
Sardegna	191,717	12.7	356,395	11.7	330	79.7	1,082,502	5.5
Italia	10,404,265	2.7	3,705,145	4.3	156,200	15.9	37,787,784	1.0

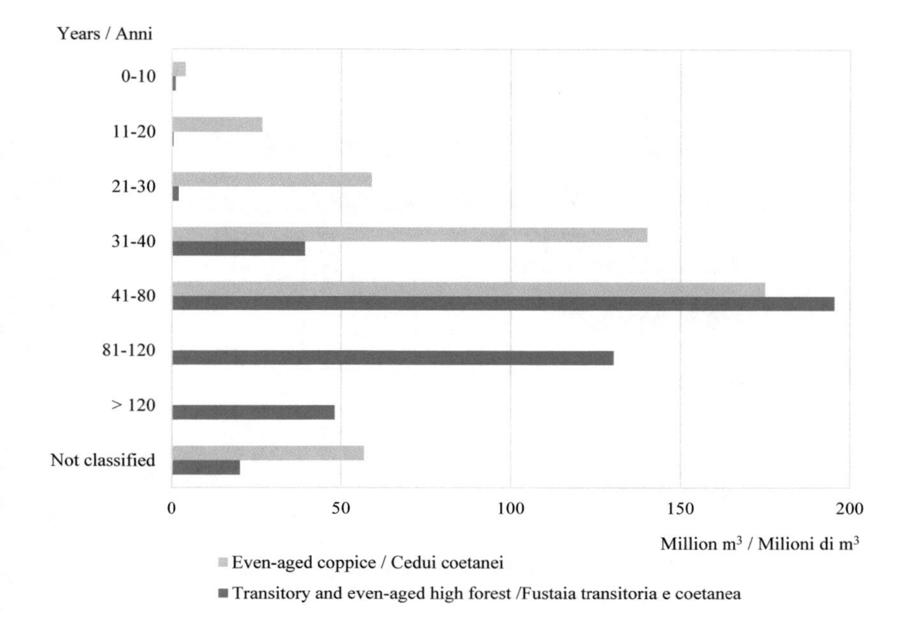

Fig. 7.25 Total growing stock volume by age class in even-aged coppice and high forest / Volume totale per classe di età, in cedui e fustaie coetanee

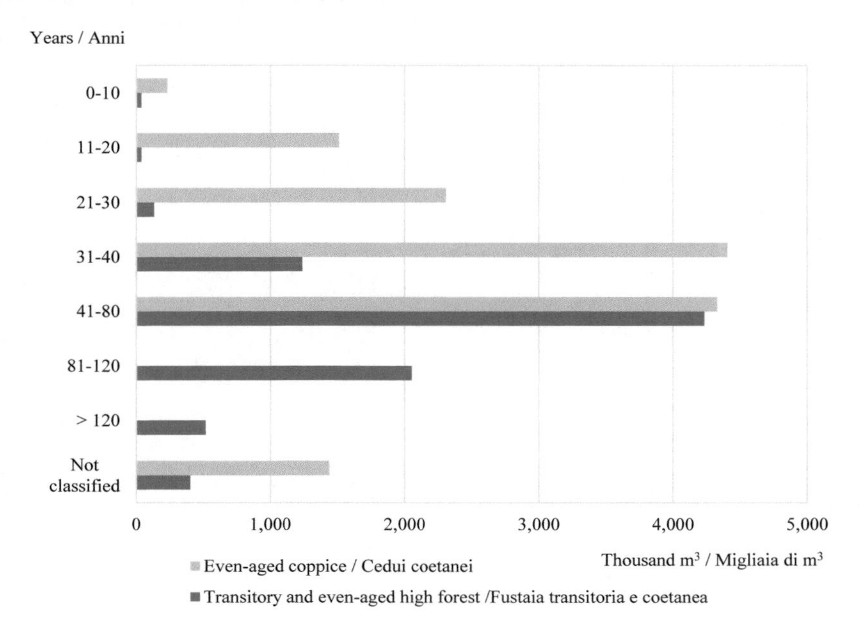

Fig. 7.26 Total annual volume increment by age class in even-aged coppice and high forest / Incremento annuo totale di volume per classe di età, in cedui e fustaie coetanee

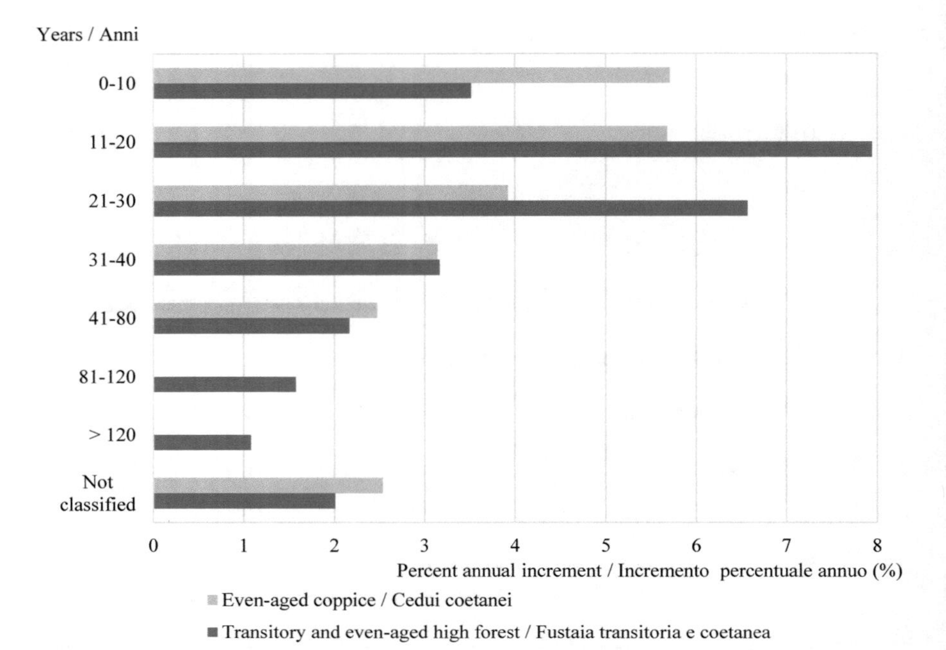

Fig. 7.27 Percent annual volume increment by age class in even-aged coppice and high forest / Incremento annuo percentuale di volume per classe di età, in cedui e fustaie coetanee

Table 7.47 Total value and value per hectare of small tree number by Forest inventory category / Valori totali e per ettaro del numero totale di soggetti della rinnovazione per le categorie inventariali del Bosco

Region/Regione	Tall trees forest				Plantations				Total Forest			
	Boschi alti				Impianti di arboricoltura da legno				Totale Bosco			
	Number	ES	Number	ES	Number	ES	Number	ES	Number	ES	Number	ES
	Numero		Numero		Numero		Numero		Numero		Numero	
	(n)	(%)	(n ha^{-1})	(%)	(n)	(%)	(n ha^{-1})	(%)	(n)	(%)	(n ha^{-1})	(%)
Piemonte	2,859,416,205	15.8	3287.5	15.8	6,349,469	40.7	307.3	35.1	2,865,765,674	15.8	3218.4	15.7
Valle d'Aosta	149,437,414	21.3	1505.8	20.9	0	–	0.0	–	149,437,414	21.3	1505.8	20.9
Lombardia	2,059,981,266	15.6	3451.5	15.5	9,299,472	56.0	370.0	54.1	2,069,280,738	15.6	3327.0	15.5
Alto Adige	629,884,152	12.5	1856.6	12.4	0	–	0.0	–	629,884,152	12.5	1856.6	12.4
Trentino	1,095,943,300	11.8	2936.1	11.7	0	–	0.0	–	1,095,943,300	11.8	2936.1	11.7
Veneto	2,381,864,047	13.3	5789.3	13.0	71,668,163	83.2	13,580.9	80.9	2,453,532,209	13.1	5887.9	12.9
Friuli V.G	1,364,289,967	11.6	4219.1	11.4	62,365,806	92.2	6783.1	90.6	1,426,655,773	11.8	4290.0	11.6
Liguria	906,938,513	11.7	2645.7	11.6	0	–	0.0	–	906,938,513	11.7	2642.9	11.6
Emilia Romagna	2,695,342,874	13.3	4656.4	13.1	1,968,548	76.1	325.5	67.7	2,697,311,421	13.3	4611.6	13.1
Toscana	6,542,543,389	10.4	6360.2	10.3	12,001,239	41.1	1769.3	21.7	6,554,544,628	10.4	6330.2	10.3
Umbria	3,707,447,601	8.0	9656.6	7.8	13,497,039	47.5	2116.6	52.2	3,720,944,640	8.0	9533.4	7.8
Marche	2,380,979,187	8.2	8357.1	7.9	24,053,486	60.8	3504.9	56.4	2,405,032,674	8.2	8243.0	7.8
Lazio	4,054,662,052	16.3	7265.6	16.2	293,093	100.0	134.7	90.5	4,054,955,145	16.3	7237.9	16.2
Abruzzo	1,770,330,190	7.0	4332.5	6.7	6,761,432	53.4	2275.5	42.2	1,777,091,622	7.0	4317.6	6.7
Molise	1,457,451,622	9.5	9682.0	9.0	4,326,974	73.8	1593.7	62.6	1,461,778,596	9.5	9538.7	9.0
Campania	2,016,751,430	9.3	5032.3	9.0	4,941,469	58.2	1562.1	45.8	2,021,692,899	9.3	5005.1	9.0
Puglia	628,381,272	11.6	4417.5	10.9	119,366	100.0	1191.5	0.8	628,500,639	11.6	4415.2	10.9
Basilicata	1,632,226,643	15.6	5697.2	15.4	923,517	72.5	606.8	45.6	1,633,150,161	15.6	5670.3	15.4

(continued)

Table 7.47 (continued)

Region/Regione	Tall trees forest				Plantations				Total Forest			
	Boschi alti				Impianti di arboricoltura da legno				Totale Bosco			
	Number	ES	Number	ES	Number	ES	Number	ES	Number	ES	Number	ES
	Numero		Numero		Numero		Numero		Numero		Numero	
	(n)	(%)	(n ha^{-1})	(%)	(n)	(%)	(n ha^{-1})	(%)	(n)	(%)	(n ha^{-1})	(%)
Calabria	2,147,475,062	16.3	4358.0	16.2	14,457,178	77.4	6008.1	66.7	2,161,932,240	16.2	4366.0	16.1
Sicilia	755,617,285	56.5	2653.8	56.4	0	–	0.0	–	755,617,285	56.5	2646.7	56.4
Sardegna	1,506,433,884	9.3	2509.7	9.2	82,825,488	31.5	3199.8	30.3	1,589,259,372	9.0	2538.2	8.9
Italia	42,743,397,357	3.4	4772.2	3.4	315,851,740	28.3	2459.9	28.0	43,059,249,097	3.4	4739.5	3.3

Table 7.48 Total value and value per hectare of small tree number by Tall trees forest type / Valori totali e per ettaro del numero totale di soggetti della rinnovazione per le categorie forestali dei Boschi alti

Region/Regione	Larch and Swiss stone pine				Norway spruce				Fir			
	Boschi di larice e cembro				Boschi di abete rosso				Boschi di abete bianco			
	Number	ES	Number	ES	Number	ES	Number	ES	Number	ES	Number	ES
	Numero		Numero		Numero		Numero		Numero		Numero	
	(n)	(%)	(n ha^{-1})	(%)	(n)	(%)	(n ha^{-1})	(%)	(n)	(%)	(n ha^{-1})	(%)
Piemonte	40,238,522	29.9	502.6	–	2,815,459	70.2	154.8	–	6,293,380	47.1	445.0	–
Valle d'Aosta	34,588,845	52.7	726.0	–	19,286,328	38.4	1057.9	–	1,832,849	71.6	1585.6	–
Lombardia	53,135,867	39.1	994.7	–	71,996,464	21.4	819.4	–	3,545,034	41.8	376.5	–
Alto Adige	40,622,045	20.5	433.8	–	284,639,383	21.0	1596.1	–	2,276,699	92.4	860.2	–
Trentino	43,085,085	39.1	692.8	–	156,884,102	23.3	1146.5	–	24,162,167	39.1	1490.0	–
Veneto	45,134,742	34.5	1065.5	–	118,781,716	20.6	1222.8	–	10,985,465	43.2	1546.4	–
Friuli V.G	19,433,969	45.6	1498.2	–	40,229,019	33.8	902.0	–	553,149	100.0	297.7	–
Liguria	434,160	74.5	394.9	–	144,720	100.0	394.9	–	6,174,718	76.9	2105.9	–
Emilia Romagna	0	–	0.0	–	880,121	100.0	238.9	–	733,434	60.0	248.9	–
Toscana	0	–	0.0	–	0	–	0.0	–	14,267,237	76.0	2820.4	–
Umbria	0	–	0.0	–	0	–	0.0	–	0	–	0.0	–
Marche	0	–	0.0	–	1,773,696	100.0	4785.8	–	0	–	0.0	–
Lazio	0	–	0.0	–	0	–	0.0	–	0	–	0.0	–
Abruzzo	0	–	0.0	–	1,442,216	100.0	3984.7	–	1,297,995	80.9	1793.1	–
Molise	0	–	0.0	–	0	–	0.0	–	2,481,185	71.2	2117.9	–
Campania	0	–	0.0	–	0	–	0.0	–	0	–	0.0	–
Puglia	0	–	0.0	–	0	–	0.0	–	0	–	0.0	–
Basilicata	0	–	0.0	–	0	–	0.0	–	1,037,915	76.9	1391.8	–
Calabria	0	–	0.0	–	0	–	0.0	–	4,448,363	58.2	1192.2	–
Sicilia	0	–	0.0	–	0	–	0.0	–	0	–	0.0	–
Sardegna	0	–	0.0	–	0	–	0.0	–	0	–	0.0	–
Italia	276,673,235	14.4	703.3	14.1	698,873,225	11.1	1191.2	10.9	80,089,591	20.9	1146.6	19.7

(continued)

Table 7.48 (continued)

Region/Regione	Scots pine and Mountain pine				Black pines				Mediterranean pines			
	Pinete di pino silvestre e montano				Pinete di pino nero, laricio e loricato				Pinete di pini mediterranei			
	Number	ES	Number	ES	Number	ES	Number	ES	Number	ES	Number	ES
	Numero		Numero		Numero		Numero		Numero		Numero	
	(n)	(%)	(n ha^{-1})	(%)	(n)	(%)	(n ha^{-1})	(%)	(n)	(%)	(n ha^{-1})	(%)
Piemonte	12,770,212	49.5	615.2	–	23,810,743	82.3	7366.2	–	1,126,184	76.9	1393.6	–
Valle d'Aosta	7,469,034	40.5	640.0	–	1,069,162	100.0	2774.8	–	0	–	0.0	–
Lombardia	96,989,825	29.2	5093.2	–	42,996,703	43.0	9754.0	–	0	–	0.0	–
Alto Adige	83,149,022	33.4	2244.1	–	150,112	100.0	397.0	–	0	–	0.0	–
Trentino	124,082,034	22.3	5936.6	–	36,975,552	58.1	6412.9	–	0	–	0.0	–
Veneto	93,573,703	28.7	7638.2	–	24,889,362	50.4	7094.8	–	14,718,871	100.0	19,683.4	–
Friuli V.G	80,156,322	34.8	6555.2	–	186,827,592	30.0	5921.0	–	3,687,660	100.0	3307.5	–
Liguria	6,753,598	46.3	658.1	–	4,496,656	45.9	817.9	–	20,220,485	35.1	823.8	–
Emilia Romagna	3,227,112	87.0	1752.1	–	68,718,497	32.8	4239.7	–	8,018,884	69.6	2721.1	–
Toscana	7,329,597	92.3	6761.7	–	127,549,358	85.1	7060.0	–	112,444,640	25.7	2489.8	–
Umbria	878,634	74.5	1191.7	–	40,856,469	38.1	6926.6	–	26,175,963	34.3	3227.5	–
Marche	0	–	0.0	–	165,544,925	37.9	15,952.7	–	10,803,207	70.7	4,924.1	–
Lazio	0	–	0.0	–	16,665,574	33.2	1966.6	–	7,831,069	63.1	1066.4	–
Abruzzo	2,451,768	76.5	2,258.0	–	90,990,409	38.2	4943.6	–	13,412,612	52.4	6176.3	–
Molise	0	–	0.0	–	20,624,848	60.1	8802.6	–	930,444	100.0	426.1	–
Campania	0	–	0.0	–	8,805,644	55.9	1594.1	–	12,475,449	42.0	1364.1	–
Puglia	0	–	0.0	–	6,633,304	82.2	4269.4	–	18,934,614	45.4	683.1	–
Basilicata	0	–	0.0	–	10,171,569	56.1	3897.1	–	7,434,218	44.5	832.2	–
Calabria	0	–	0.0	–	119,240,829	21.3	1623.6	–	28,129,351	52.8	1609.8	–
Sicilia	0	–	0.0	–	3,609,195	44.1	481.7	–	29,846,329	47.8	658.5	–
Sardegna	0	–	0.0	–	6,933,586	58.6	1327.1	–	38,254,624	58.8	1104.6	–
Italia	518,830,859	12.1	3,484.1	11.0	1,007,560,090	15.1	4374.0	14.5	354,444,604	13.8	1473.4	13.3

(continued)

Table 7.48 (continued)

Region/Regione	Other coniferous forests				Beech				Temperate oaks			
	Altri boschi di conifere, pure o miste				Faggete				Querceti di rovere, roverella e farnia			
	Number	ES	Number	ES	Number	ES	Number	ES	Number	ES	Number	ES
	Numero		Numero		Numero		Numero		Numero		Numero	
	(n)	(%)	(n ha^{-1})	(%)	(n)	(%)	(n ha^{-1})	(%)	(n)	(%)	(n ha^{-1})	(%)
Piemonte	4,554,238	55.9	704.5	–	315,591,418	55.4	2694.3	–	237,800,279	30.7	3121.7	–
Valle d'Aosta	0	–	0.0	–	229,106	100.0	198.2	–	11,760,780	48.4	2774.8	–
Lombardia	9,346,455	64.0	1472.4	–	380,465,244	71.3	5708.4	–	245,537,798	28.3	5974.8	–
Alto Adige	0	–	0.0	–	10,641,276	60.6	2558.6	–	58,157,697	41.7	12,818.4	–
Trentino	0	–	0.0	–	215,163,141	26.0	3436.7	–	35,174,670	43.1	6100.5	–
Veneto	0	–	0.0	–	425,073,537	36.5	6284.4	–	161,373,949	22.3	10,310.8	–
Friuli V.G	1,327,558	89.6	1190.7	–	189,095,503	23.7	2086.1	–	40,177,935	30.7	4914.0	–
Liguria	0	–	0.0	–	39,352,697	29.9	1063.1	–	78,311,588	29.5	1785.4	–
Emilia Romagna	18,702,579	55.0	5641.3	–	318,619,588	27.5	3108.3	–	378,085,468	21.1	4910.6	–
Toscana	22,327,554	32.6	1876.2	–	283,095,220	45.6	3860.9	–	897,060,969	20.4	5898.0	–
Umbria	6,772,802	56.9	3674.3	–	77,611,689	30.3	4698.9	–	742,430,429	14.0	7359.1	–
Marche	44,182,274	46.7	7809.6	–	72,637,059	29.2	4083.1	–	517,058,636	20.1	8373.0	–
Lazio	439,639	74.5	298.3	–	236,360,831	25.6	3175.6	–	688,637,906	54.3	8422.2	–
Abruzzo	5,480,422	71.6	2523.7	–	323,775,886	14.4	2601.3	–	436,523,881	12.3	4974.0	–
Molise	0	–	0.0	–	47,797,267	39.1	3221.0	–	548,461,219	17.0	11,379.3	–
Campania	880,564	70.7	1019.7	–	142,890,766	25.3	2540.5	–	401,613,708	21.9	6591.0	–
Puglia	9,255,773	73.2	3971.5	–	14,346,448	55.1	3077.9	–	142,046,883	30.6	5491.8	–
Basilicata	9,267,100	55.7	1799.3	–	172,317,343	45.9	6424.9	–	171,120,990	35.5	3972.5	–
Calabria	8,296,246	39.2	660.5	–	309,950,501	17.6	3903.0	–	143,645,040	35.4	2670.5	–
Sicilia	5,280,178	50.5	567.8	–	30,172,444	72.9	1989.6	–	533,494,952	79.5	7421.3	–
Sardegna	8,051,170	38.1	616.7	–	0	–	0.0	–	154,247,787	22.5	1757.2	–
Italia	154,164,552	17.9	1837.1	16.9	3,605,186,964	11.6	3,423.1	11.5	6,622,722,564	9.8	5747.5	9.7

(continued)

Table 7.48 (continued)

Region/Regione	Mediterranean oaks				Chestnut				Hornbeam and Hophornbeam			
	Cerrete, boschi di farnetto, fragno e vallonea				Castagneti				Ostrieti, carpineti			
	Number	ES	Number	ES	Number	ES	Number	ES	Number	ES	Number	ES
	Numero		Numero		Numero		Numero		Numero		Numero	
	(n)	(%)	(n ha^{-1})	(%)	(n)	(%)	(n ha^{-1})	(%)	(n)	(%)	(n ha^{-1})	(%)
Piemonte	23,596,232	51.8	3893.2	–	379,909,202	18.2	2321.6	–	27,120,345b	34.6	1678.0	–
Valle d'Aosta	0	–	0.0	–	18,817,248	36.0	4439.7	–	0	–	0.0	–
Lombardia	8,248,347	63.8	3742.4	–	291,337,133	23.7	3549.5	–	507,664,755	24.7	6212.2	–
Alto Adige	0	–	0.0	–	12,008,964	64.5	6352.5	–	87,206,057	33.3	9881.4	–
Trentino	0	–	0.0	–	5,172,746	68.9	3588.5	–	318,386,373	18.6	7795.8	–
Veneto	0	–	0.0	–	161,402,088	22.5	9384.4	–	876,748,408	28.9	10,688.8	–
Friuli V.G	0	–	0.0	–	56,140,938	34.0	4577.6	–	397,361,321	17.5	8492.3	–
Liguria	30,135,925	37.3	2679.4	–	385,507,660	16.5	3517.8	–	77,879,232	26.0	1684.6	–
Emilia Romagna	519,190,684	26.7	5127.1	–	113,505,310	22.4	2801.2	–	637,500,717	27.4	5730.3	–
Toscana	2,066,846,502	25.0	8248.8	–	1,236,804,138	23.9	7920.4	–	321,622,538	21.4	4974.5	–
Umbria	1,280,448,443	15.1	9957.2	–	7,907,704	61.8	3575.0	–	730,044,436	20.0	12,475.6	–
Marche	187,603,826	30.3	7585.6	–	28,231,322	46.7	7617.4	–	974,515,841	13.3	12,774.3	–
Lazio	805,490,319	30.3	6081.7	–	147,383,732	46.1	4117.8	–	1,003,418,737	31.8	10,241.7	–
Abruzzo	207,043,758	28.0	6440.9	–	15,374,027	38.8	3267.5	–	225,229,575	15.1	4699.7	–
Molise	579,908,266	17.8	11,157.4	–	2,015,963	100.0	5162.5	–	106,763,312	26.1	11,886.9	–
Campania	486,786,628	20.4	6521.4	–	292,254,541	31.1	5220.1	–	400,128,473	19.7	7545.4	–
Puglia	165,703,355	24.7	4278.7	–	7,558,881	83.6	6486.8	–	51,462,099	34.2	11,040.8	–
Basilicata	991,934,636	23.1	7589.9	–	36,663,466	35.5	6156.3	–	103,369,515	38.1	14,591.2	–
Calabria	100,532,993	25.8	2b306.2	–	866,102,043	28.8	12,558.4	–	3,558,690	100.0	794.8	–
Sicilia	14,791,109	39.0	477.3	–	19,557,229	48.2	2242.8	–	2,715,520	89.2	1563.0	–
Sardegna	0	–	0.0	–	9,211,765	60.6	4936.9	–	0	–	0.0	–
Italia	7,468,261,024	9.1	7046.4	9.0	4,092,866,098	10.4	5257.5	10.2	6,852,695,944	7.6	7975.9	7.4

(continued)

Table 7.48 (continued)

Region/Regione	Hygrophilous forests				Other deciduous broadleaved				Holm oak			
	Boschi igrofili				Altri boschi caducifogli				Leccete			
	Number	ES	Number	ES	Number	ES	Number	ES	Number	ES	Number	ES
	Numero		Numero		Numero		Numero		Numero		Numero	
	(n)	(%)	(n ha^{-1})	(%)	(n)	(%)	(n ha^{-1})	(%)	(n)	(%)	(n ha^{-1})	(%)
Piemonte	121,774,791	39.8	5355.9	–	1,658,170,930	24.3	5122.4	–	0	–	0.0	–
Valle d'Aosta	0	–	0.0	–	54,384,062	44.2	5167.7	–	0	–	0.0	–
Lombardia	18,836,651	71.8	999.6	–	327,950,527	20.1	2710.1	–	0	–	0.0	–
Alto Adige	13,709,028	54.5	4990.8	–	22,312,664	70.5	6317.2	–	0	–	0.0	–
Trentino	1,005,812	55.3	260.6	–	134,558,432	64.0	8168.7	–	1,293,186	100.0	3588.5	–
Veneto	44,263,355	43.6	4249.3	–	371,955,374	25.8	7099.5	–	32,963,475	38.5	11,020.4	–
Friuli V.G	14,160,615	61.8	1123.9	–	335,138,387	36.1	7063.4	–	0	–	0.0	–
Liguria	4,775,759	59.3	1447.8	–	181,215,227	40.8	5692.2	–	36,803,298	38.6	2,645.6	–
Emilia Romagna	95,455,531	40.2	3382.2	–	532,704,949	47.4	6164.5	–	0	–	0.0	–
Toscana	74,945,291	41.0	2700.6	–	547,832,966	30.4	6483.9	–	810,565,192	16.3	6250.2	–
Umbria	22,522,886	41.9	2802.0	–	27,185,346	31.5	2409.3	–	744,612,800	21.0	18,034.1	–
Marche	44,360,517	43.2	3438.4	–	225,745,162	17.2	3635.5	–	108,522,722	62.0	15,411.4	–
Lazio	13,988,513	51.1	1518.6	–	288,438,443	40.0	5339.1	–	835,094,230	41.1	17,301.4	–
Abruzzo	65,707,379	37.8	3307.6	–	289,373,316	22.6	5133.3	–	82,756,999	41.9	9527.1	–
Molise	59,195,437	35.7	6538.4	–	57,638,575	39.3	5637.0	–	31,635,106	61.1	27,003.6	–
Campania	1,467,607	64.8	132.8	–	66,338,339	32.1	1929.2	–	196,358,717	28.1	5238.3	–
Puglia	0	–	0.0	–	96,794,818	31.6	8389.2	–	92,439,067	26.4	5323.5	–
Basilicata	47,891,533	80.6	3690.9	–	30,892,812	33.2	1034.2	–	39,855,945	36.6	3843.2	–
Calabria	112,542,663	90.4	8743.5	–	251,847,769	80.1	5577.6	–	109,101,233	45.3	2240.7	–
Sicilia	0	–	0.0	–	28,803,079	85.9	2175.9	–	33,418,821	48.7	1710.6	–
Sardegna	445,731	100.0	132.7	–	12,046,665	46.0	1555.8	–	1,015,001,429	12.7	3973.2	–
Italia	757,049,096	18.2	3219.9	17.3	5,541,327,842	10.8	4975.3	10.6	4,170,422,222	10.4	6485.4	10.2

(continued)

Table 7.48 (continued)

Region/Regione	Cork oak				Other evergreen broadleaved				Not classified				Total Tall trees forest			
	Sugherete				Altri boschi di latifoglie sempreverdi				Categoria forestale non nota				Totale Boschi alti			
	Number	ES	Number	ES	Number	ES	Number	ES	Number	ES	Number	ES	Number	ES	Number	ES
	Numero		Numero		Numero		Numero		Numero		Numero		Numero		Numero	
	(n)	(%)	(n ha^{-1})	(%)	(n)	(%)	(n ha^{-1})	(%)	(n)	(%)	(n ha^{-1})	(%)	(n)	(%)	(n ha^{-1})	(%)
Piemonte	0	–	0.0	–	0	–	0.0	–	3,844,269	100.0	7565.3	–	2,859,416,205	15.8	3287.5	15.8
Valle d'Aosta	0	–	0.0	–	0	–	0.0	–	0	–	0.0	–	149,437,414	21.3	1505.8	20.9
Lombardia	0	–	0.0	–	0	–	0.0	–	1,930,464	100.0	4379.4	–	2,059,981,266	15.6	3451.5	15.5
Alto Adige	0	–	0.0	–	0	–	0.0	–	15,011,205	91.1	9925.7	–	629,884,152	12.5	1856.6	12.4
Trentino	0	–	0.0	–	0	–	0.0	–	0	–	0.0	–	1,095,943,300	11.8	2936.1	11.7
Veneto	0	–	0.0	–	0	–	0.0	–	0	–	0.0	–	2,381,864,047	13.3	5789.3	13.0
Friuli V.G	0	–	0.0	–	0	–	0.0	–	0	–	0.0	–	1,364,289,967	11.6	4219.1	11.4
Liguria	0	–	0.0	–	0	–	0.0	–	34,732,790	70.7	47,383.2	–	906,938,513	11.7	2645.7	11.6
Emilia Romagna	0	–	0.0	–	0	–	0.0	–	0	–	0.0	–	2,695,342,874	13.3	4656.4	13.1
Toscana	17,265,273	41.4	2810.7	–	0	–	0.0	–	2,586,917	70.7	3579.7	–	6,542,543,389	10.4	6360.2	10.3
Umbria	0	–	0.0	–	0	–	0.0	–	0	–	0.0	–	3,707,447,601	8.0	9656.6	7.8
Marche	0	–	0.0	–	0	–	0.0	–	0	–	0.0	–	2,380,979,187	8.2	8357.1	7.9
Lazio	1,465,463	66.3	568.2	–	0	–	0.0	–	9,447,596	62.9	7074.5	–	4,054,662,052	16.3	7265.6	16.2
Abruzzo	0	–	0.0	–	1,681,978	100.0	1,195.4	–	7,787,969	100.0	21,517.5	–	1,770,330,190	7.0	4332.5	6.7
Molise	0	–	0.0	–	0	–	0.0	–	0	–	0.0	–	1,457,451,622	9.5	9682.0	9.0
Campania	146,761	100.0	398.5	–	5,283,387	70.7	7,173.4	–	1,320,847	100.0	3586.7	–	2,016,751,430	9.3	5032.3	9.0
Puglia	0	–	0.0	–	9,785,159	47.1	1,900.1	–	13,420,871	81.9	11,924.5	–	628,381,272	11.6	4417.5	10.9
Basilicata	0	–	0.0	–	741,368	100.0	397.7	–	9,528,232	100.0	27,438.8	–	1,632,226,643	15.6	5697.2	15.4
Calabria	26,467,757	66.7	5066.8	–	24,910,830	56.5	1,178.2	–	38,700,754	67.3	17,286.8	–	2,147,475,062	16.3	4358.0	16.2
Sicilia	41,442,781	59.3	2400.9	–	12,485,648	54.5	318.9	–	0	–	0.0	–	755,617,285	56.5	2653.8	56.4
Sardegna	218,218,825	15.6	1428.6	–	26,193,081	40.5	689.2	–	17,829,222	100.0	47,776.4	–	1,506,433,884	9.3	2509.7	9.2
Italia	305,006,860	15.1	1654.7	14.4	81,081,451	24.5	712.7	23.7	156,141,135	29.5	15,509.7	22.6	42,743,397,357	3.4	4772.2	3.4

Table 7.49 Total value and value per hectare of small tree number by Plantations forest type / Valori totali e per ettaro del numero totale di soggetti della rinnovazione per le categorie forestali degli Impianti di arboricoltura da legno

Region/Regione	Poplar				Other broadleaved plantations				Coniferous plantations				Total Plantations			
	Pioppeti artificiali				Piantagioni di altre latifoglie				Piantagioni di conifere				Totale Impianti di arboricoltura da legno			
	Number Numero	ES	Number Numero	ES	Number Numero	ES	Number Numero	ES	Number Numero	ES	Number Numero	ES	Number Numero	ES	Number Numero	ES
	(n)	(%)	(n ha^{-1})	(%)	(n)	(%)	(n ha^{-1})	(%)	(n)	(%)	(n ha^{-1})	(%)	(n)	(%)	(n ha^{-1})	(%)
Piemonte	1,656,086	58.0	106.8	–	4,049,850	57.1	1012.6	–	643,534	100.0	559.2	–	6,349,469	40.7	307.3	35.1
Valle d'Aosta	0	–	0.0	–	0	–	0.0	–	0	–	0.0	–	0	–	0.0	–
Lombardia	3,704,934	89.0	200.2	–	5,594,537	72.3	904.8	–	0	–	0.0	–	9,299,472	56.0	370.0	54.1
Alto Adige	0	–	0.0	–	0	–	0.0	–	0	–	0.0	–	0	–	0.0	–
Trentino	0	–	0.0	–	0	–	0.0	–	0	–	0.0	–	0	–	0.0	–
Veneto	136,666	100.0	58.6	–	71,531,497	83.3	24,286.9	–	0	–	0.0	–	71,668,163	83.2	13,580.9	80.9
Friuli V.G	0	–	0.0	–	62,365,806	92.2	23,494.8	–	0	–	0.0	–	62,365,806	92.2	6783.1	90.6
Liguria	0	–	0.0	–	0	–	0.0	–	0	–	0.0	–	0	–	0.0	–
Emilia Romagna	591,857	100.0	139.2	–	1,376,690	100.0	963.9	–	0	–	0.0	–	1,968,548	76.1	325.5	67.7
Toscana	2,196,338	100.0	2983.1	–	8,053,240	52.5	1752.7	–	1,751,661	84.4	1206.3	–	12,001,239	41.1	1769.3	21.7
Umbria	0	–	0.0	–	13,497,039	47.5	2116.6	–	0	–	0.0	–	13,497,039	47.5	2116.6	52.2
Marche	0	–	0.0	–	24,053,486	60.8	3504.9	–	0	–	0.0	–	24,053,486	60.8	3504.9	56.4
Lazio	0	–	0.0	–	293,093	100.0	211.3	–	0	–	0.0	–	293,093	100.0	134.7	90.5
Abruzzo	132,629	100.0	251.0	–	6,628,803	54.5	2713.3	–	0	–	0.0	–	6,761,432	53.4	2275.5	42.2
Molise	3,101,481	100.0	7942.2	–	1,225,493	61.3	527.2	–	0	–	0.0	–	4,326,974	73.8	1593.7	62.6
Campania	1,125,690	100.0	731.3	–	3,815,779	69.4	2349.7	–	0	–	0.0	–	4,941,469	58.2	1562.1	45.8
Puglia	0	–	0.0	–	119,366	100.0	1191.5	–	0	–	0.0	–	119,366	100.0	1191.5	0.8
Basilicata	0	–	0.0	–	923,517	72.5	606.8	–	0	–	0.0	–	923,517	72.5	606.8	45.6

(continued)

Table 7.49 (continued)

Region/Regione	Poplar				Other broadleaved plantations				Coniferous plantations				Total Plantations			
	Pioppeti artificiali				Piantagioni di altre latifoglie				Piantagioni di conifere				Totale Impianti di arboricoltura da legno			
	Number Numero	ES	Number Numero	ES	Number Numero	ES	Number Numero	ES	Number Numero	ES	Number Numero	ES	Number Numero	ES	Number Numero	ES
	(n)	(%)	(n ha^{-1})	(%)	(n)	(%)	(n ha^{-1})	(%)	(n)	(%)	(n ha^{-1})	(%)	(n)	(%)	(n ha^{-1})	(%)
Calabria	0	–	0.0	–	10,898,488	100.0	6986.6	–	3,558,690	70.7	4768.8	–	14,457,178	77.4	6008.1	66.7
Sicilia	0	–	0.0	–	0	–	0.0	–	0	–	0.0	–	0	–	0.0	–
Sardegna	0	–	0.0	–	82,825,488	31.5	3395.6	–	0	–	0.0	–	82,825,488	31.5	3199.8	30.3
Italia	12,645,681	41.8	245.1	40.6	297,252,174	30.0	4177.5	29.5	5,953,884	50.2	1053.6	43.2	315,851,740	28.3	2459.9	28.0

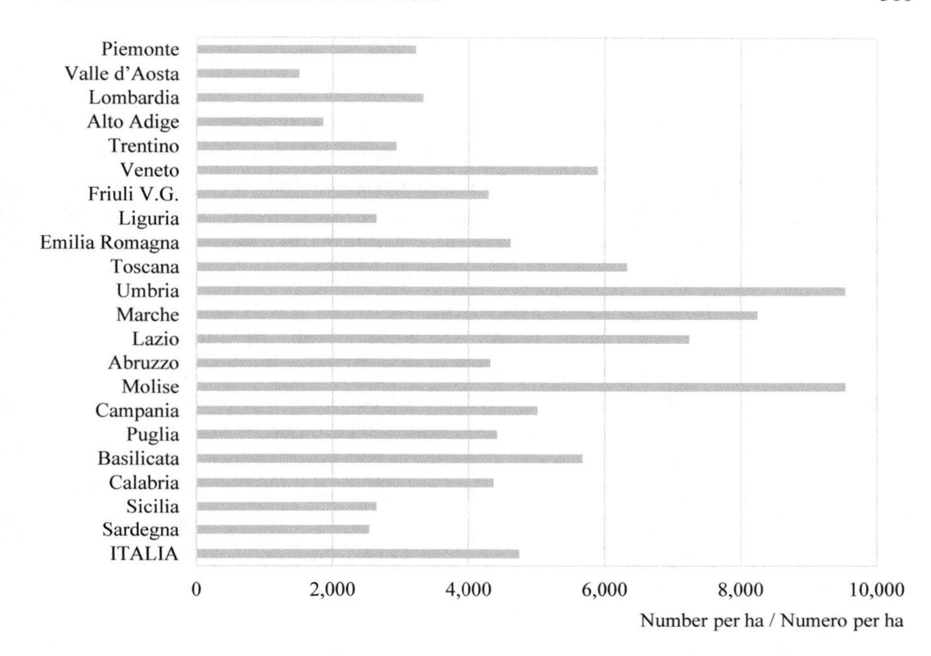

Fig. 7.28 Number per hectare of small trees in Forest, by region / Numero per ettaro di soggetti della rinnovazione per il Bosco, nelle regioni italiane

Appendix (Italian Version)

Riassunto La consapevolezza che il patrimonio forestale non è inesauribile non è acquisizione recente, ma risale al basso medioevo, quando si capì l'importanza di una qualche forma di pianificazione dello sfruttamento delle risorse forestali e la necessità di conoscerle per poterle gestire. Proiettati su scala più ampia e fondati su tecniche di campionamento statistico, rispetto alle tecniche di inventariazione ideate a quel tempo, gli inventari forestali nazionali sono deputati a produrre informazione scientificamente attendibile per supportare le decisioni di politica forestale. Tra le principali informazioni prodotte dagli inventari forestali nazionali si annoverano l'estensione delle superfici e la composizione delle foreste, la proprietà, la consistenza del capitale legnoso e il suo incremento, ma anche altre inerenti alle forme di gestione selvicolturale e alle caratteristiche colturali e strutturali del bosco. Questo capitolo riporta le statistiche prodotte nell'ambito degli attributi sopra elencati. Per l'estensione delle superfici, sono riportate quelle della Superficie forestale totale, del Bosco e delle Altre terre boscate, variamente ripartite in categorie inventariali e forestali, che ne descrivono la composizione, nonché la distribuzione per classi di quota. Tra i caratteri del soprassuolo, sono mostrati e commentati anche quelli relativi al grado di copertura, allo stadio di sviluppo, alla distribuzione per classi di età. Infine, sono riportate le statistiche sulla presenza e consistenza della rinnovazione e degli arbusti.

Introduzione

La consapevolezza che il patrimonio forestale non sia inesauribile non è acquisizione recente. Nel mondo occidentale, essa risale al basso medioevo (Sereno, 2008). Già ridotte in estensione dai cospicui dissodamenti per la conquista di suoli lavorabili, avvenuti fin dai secoli VII e VIII ma soprattutto in più fasi dall'XI al XIV, interi settori del bosco dovettero essere messi sotto tutela dall'eccessivo sfruttamento, cosa che avvenne per provvedimenti delle autorità pubbliche a partire dal XIV secolo (Fossier, 2003). Tuttavia, già tra il XII e XIII secolo gli statuti locali e gli accordi con i signori si erano moltiplicati, promossi dalle comunità per tutelare gli incolti (Provero, 2020). Si erano già manifestati, anche nell'Italia del nord, problemi che oggi definiremmo ambientali, ma l'equilibrio da preservare era principalmente quello capace di assicurare la continuità della produzione dei beni richiesti alla foresta: legna e legname, ghianda per gli animali al pascolo, selvaggina (Delort, 1989). Loetsch and Haller (1973) collocano la nascita delle tecniche di inventariazione forestale alla fine del medioevo, quando si rese necessaria una qualche forma di pianificazione dello sfruttamento delle foreste vicino alle città e alle miniere, a seguito della ridotta capacità di fornire legno. Stessi timori e stessa esigenza informativa, traslati nel tempo e proiettati su scala nazionale, hanno dato l'avvio agli inventari forestali nazionali, nei primi anni del Novecento. Secondo Persson and Janz (2015), le principali esigenze informative riguardano l'estensione delle superfici, le caratteristiche topografiche, la proprietà, l'accessibilità delle aree, la consistenza del capitale legnoso e il suo incremento. Questo capitolo riporta le statistiche prodotte da INFC2015 riguardo ad alcune delle principali variabili tradizionalmente stimate dagli inventari forestali nazionali.

La misurazione dell'estensione delle foreste ha rappresentato a lungo una condizione indispensabile per poterne stimare i valori totali. Le stime erano quindi fortemente ancorate alla preliminare operazione di mappatura, perché sulle mappe si poteva misurare la superficie del bosco cui rapportare i valori medi di massa stimati in varie sue parti (Loetsch & Haller, 1973). Inventari così condotti si sono presto rilevati inadeguati per fornire informazioni a carattere nazionale (McRoberts et al., 2010; Tomppo et al., 2010). Con gli inventari forestali moderni, basati su tecniche di campionamento statistico, le superfici sono divenute oggetto di stima al pari delle altre variabili. Ciò è vero per la Superficie forestale totale e, naturalmente, anche per le sue ripartizioni, come ad esempio il Bosco e le Altre terre boscate, le superfici per grado di mescolanza, categorie e sottocategorie forestali, che meglio descrivono la composizione delle foreste, commentate in questo capitolo.

Per la correlazione con la temperatura dell'aria, la quota esplica sulla distribuzione della vegetazione non solo un'azione diretta ma anche indiretta, condizionando i processi di pedogenesi attraverso l'alterazione dei minerali e la decomposizione della sostanza organica (Avena & Dowgiallo, 1995). Serie temporali di statistiche di distribuzione per fasce altitudinali dei tipi di vegetazione, inoltre, rappresentano un contributo fondamentale nell'ambito delle valutazioni di adattamento ai cambiamenti climatici. INFC produce stime di superficie secondo due tipi di moduli per le classi di quota, uno con intervallo di 500 m e l'altro di 300 m, onde consentire il confronto sia con le statistiche del primo inventario forestale nazionale (IFNI85) sia con le ripartizioni territoriali adottate dall'Istituto nazionale di statistica (ISTAT), in cui la quota di 600 m s.l.m. delimita i territori montani.

Informazioni utili ad orientare le politiche forestali delle istituzioni preposte sono senza dubbio rappresentate dalle stime della superficie forestale di proprietà pubblica e di proprietà privata, almeno per due ordini di motivi. Nel primo rientra l'attesa che le finalità perseguite con la gestione forestale da parte di proprietari pubblici, soprattutto quando si tratti di enti a carattere territoriale, possano differire da quelle dei privati; all'altro ordine attengono le implicazioni di carattere socio-politico che possono scaturire da provvedimenti legislativi che abbiano effetto sulla proprietà fondiaria.

Statistiche sul volume e sull'incremento legnoso sono state a lungo la motivazione principale degli inventari forestali nazionali (es. Breidenbach et al., 2020). Al tradizionale interesse per il valore economico del capitale legnoso della foresta, si è aggiunto negli ultimi decenni quello per l'insostituibile ruolo svolto dalle foreste quale serbatoio forestale di carbonio (cfr. Chap. 12). Come conseguenza, oltre al volume ne viene stimata la massa e vengono inclusi nelle stime anche alberi con diametri inferiori, rispetto al passato. La desiderata sostenibilità della gestione, necessaria per preservare la capacità produttiva delle foreste, si basa sulla regolazione dei prelievi rispetto all'incremento di volume. La sua stima ha assunto rilevanza anche nei programmi di monitoraggio volti a valutare la risposta delle foreste ai cambiamenti ambientali (Dobbertin, 2005; Gschwantner et al., 2016; Solberg et al., 2009). Espresso in termini di fitomassa, l'incremento fornisce una misura del contributo delle foreste alla fissazione di carbonio atmosferico. Il ruolo degli inventari nella stima degli incrementi è insostituibile, poiché si tratta di un attributo ancora poco valutato con le tecniche di remote sensing, che viene stimato attraverso misurazioni di campo ripetute su aree di saggio permanenti o mediante carotaggi (Gasparini et al., 2017).

Con particolare rilievo per le statistiche oggetto del presente capitolo, la valutazione del grado di copertura arborea è fondamentale per definire l'appartenenza o meno dei punti inventariali al dominio inventariale, in funzione delle soglie del 5% e del 10% rilevanti ai fini del sistema di classificazione adottato (cfr. Chap. 2). La distribuzione delle chiome, inoltre, consente di descrivere la struttura dei popolamenti; INFC2015 ha registrato la presenza di eventuali piani delle chiome per poter fornire statistiche sulla struttura verticale, mediante la differenziazione tra soprassuoli a struttura monoplana o biplana.

Informazioni relative alla forma di governo e di trattamento dei soprassuoli, nonché allo stadio di sviluppo e all'età, quest'ultima di interesse per i popolamenti coetanei, sono indispensabili per la pianificazione della gestione e per lo sviluppo di politiche forestali a livello più generale. Le stime relative alle variabili quantitative secondo le ripartizioni della superficie forestale individuate dai caratteri qualitativi sopra elencati sono inoltre particolarmente utili per ipotizzare l'evoluzione futura dei soprassuoli nei diversi scenari di intervento.

Rinnovazione e arbusti rappresentano le componenti storicamente meno considerate dagli inventari forestali, tra quelle descritte in questo capitolo. Come è stato sopra ricordato, le soglie diametriche adottate in origine dagli inventari erano fortemente orientate alla stima del capitale legnoso di interesse commerciale o prossimo a diventarlo. Nell'accezione di INFC, per rinnovazione si intendono piante di piccola dimensione, inferiore alla soglia per il cavallettamento, e non necessariamente con l'esplicita funzione riconosciuta al termine in senso selvicolturale, di ricostituzione del popolamento. La misurazione della componente di piccole dimensioni contribuisce ad una più accurata valutazione del carbonio immagazzinato dalla vegetazione legnosa, ma al contempo consente di registrare la presenza di elementi botanici che solo raramente raggiungono dimensioni utili alla misurazione dei diametri.

Superficie e composizione

Le stime della superficie del Bosco e delle Altre terre boscate sono tra i risultati principali dell'inventario forestale nazionale. INFC produce stime di superficie per due macrocategorie inventariali, sette categorie inventariali e ventitré categorie forestali, queste ultime ulteriormente ripartite in sottocategorie, a livello regionale e nazionale. Lo schema di classificazione adottato da INFC, e le descrizioni delle relative classi, sono riportati nel Chap. 2 e nel Chap. 3, rispettivamente per le categorie inventariali e forestali e per le classi di uso e di copertura del suolo.

La Table 7.1 mostra le statistiche sulla superficie del Bosco, delle Altre terre boscate e della Superficie forestale totale; le Tables 7.2 e 7.3 riportano le stime di superficie per le categorie inventariali del Bosco e delle Altre terre boscate, rispettivamente. La Superficie forestale totale in Italia ammonta a 11,054,458 ha, dei quali l'82.2% è classificata come Bosco (9,085,186 ha) e il 17.8% come Altre terre boscate (1,969,272 ha). La Superficie forestale totale copre il 36.7% della superficie territoriale nazionale; il Bosco interessa il 30.2% e le Altre terre boscate il 6.5% della superficie nazionale. A livello regionale, la superficie forestale varia considerevolmente, spaziando dal 7.4% (Puglia) al 63.3% (Liguria), e rimanendo sopra il 40% in cinque regioni (Alto Adige, Trentino, Friuli-Venezia Giulia, Toscana e Umbria). La superficie delle Altre terre boscate approssima generalmente la quantità stimata a livello nazionale anche nella maggior parte delle regioni, eccetto la Sardegna, la Basilicata e la Calabria, dove mostra valori considerevolmente più elevati (28.0%, 10.4% e 10.3%, rispettivamente). Alla Sardegna, in particolare, si deve circa un terzo della superficie delle Altre terre boscate riportate per l'Italia intera. La Fig. 7.1 mostra la percentuale della superficie territoriale italiana coperta dal Bosco e dalle Altre terre boscate e la distribuzione dei relativi punti inventariali.

La categoria inventariale dei Boschi alti è di gran lunga la più importante, con 8,956,787 ha, di cui 35,836 ha sono temporaneamente privi di copertura (Table 7.2). La categoria dei Boschi alti è rappresentata dalle superfici con una copertura delle chiome superiore al 10%, assicurata da alberi alti almeno 5 m, o capaci di raggiungere queste soglie in situ, e con uso del suolo prevalente non agricolo o urbano.

Le stesse soglie per l'altezza degli alberi e il grado di copertura valgono anche per gli Impianti di arboricoltura da legno, che però sono caratterizzati dalla destinazione produttiva specializzata di legname o di legno, sono sempre di origine artificiale e sottoposti a pratiche di gestione intensiva e possono trovarsi anche su suoli ad uso agricolo. La superficie degli Impianti di arboricoltura da legno è pari a 128,399 ha, e le percentuali maggiori si trovano in Piemonte, Lombardia, Friuli-Venezia Giulia, Marche e Sardegna (Table 7.2), dove rappresentano dal 2 al 4% della superficie del Bosco della regione.

Le Altre terre boscate sono composte principalmente da Arbusteti, per un totale di 1,168,776 ha, dei quali circa la metà sono in Sardegna. Le categorie inventariali Boschi bassi, Boschi radi e Boscaglie rappresentano nel complesso il 20.3% delle Altre terre boscate (Table 7.3 e Fig. 7.2). La presenza di Boschi bassi e Boscaglie indica spesso limitazioni stazionali importanti, dovute a fattori edafici o di ventosità, che possono ostacolare lo sviluppo di soprassuoli arborei più evoluti, mentre i Boschi radi possono rappresentare fasi di serie evolutive o regressive, legate quindi alla colonizzazione o al degrado di soprassuoli all'origine più densi; in contesti particolari ad elevate altitudini, possono però rappresentare anche condizioni di equilibrio stabile. Per convenzione, le aree in cui la presenza di copertura arborea e/o arbustiva è stata accertata per fotointerpretazione ma non è stato possibile registrare informazioni più accurate mediante i rilievi in campo, sono state assegnate alle Altre terre boscate, nella categoria inventariale Aree boscate inaccessibili o non classificate.

Una prima classificazione generale del bosco è quella basata sulla composizione del gruppo di specie (Table 7.4), ampiamente utilizzata nella compilazione di statistiche per le attività di reporting internazionale ed europeo. I punti di campionamento inventariale sono stati per questo assegnati alle classi indicate, sulla base della percentuale di copertura delle chiome relativa al gruppo di specie registrata in campo. Coerentemente, le Tables 7.5 e 7.6 riportano le statistiche sulle superfici afferenti ai boschi di conifere, a quelli di latifoglie e ai boschi misti, rispettivamente per le macrocategorie inventariali del Bosco e delle Altre terre boscate. All'indirizzo inventarioforestale.org/statistiche_INFC sono disponibili statistiche analoghe per le categorie inventariali del Bosco e per quelle delle Altre terre boscate. Le Figs. 7.3 e 7.4 mostrano la percentuale di superficie del Bosco e delle Altre terre boscate per boschi puri di latifoglie, puri di conifere e per i boschi misti. A livello nazionale, i boschi puri di latifoglie predominano sia nel Bosco (68.5%) sia nelle Altre terre boscate (53.9%); in queste ultime, la prevalenza di questa classe diventa ancora più marcata (83.4%) se si concentra l'attenzione alla superficie effettivamente classificata per questo attributo. Soprassuoli puri di conifere occupano il 12.8% della superficie del Bosco e si trovano prevalentemente nelle regioni del nord (Valle d'Aosta, Alto Adige e Trentino), dove caratterizzano vari paesaggi di tipo alpino, ma anche in alcune regioni della penisola e in Sicilia, per la presenza di pinete costiere e di alcune conifere mediterraneo-montane. La classe dei boschi misti di conifere e latifoglie occupa il 10.1% della superficie del Bosco e il 6.1% di quella delle Altre terre boscate e si rinviene maggiormente in alcune regioni del nord (Lombardia, Trentino, Veneto, Friuli-Venezia Giulia) e in Calabria.

Classificazioni di maggiore dettaglio sulla composizione specifica adottata da INFC sono descritte nel Chap. 2. I punti inventariali sono stati classificati sulla base della specie predominante in termini di copertura delle chiome. La classificazione della categoria forestale è essenziale per l'attribuzione dei punti agli strati inventariali. Per questo motivo, la categoria forestale viene attribuita anche ai punti inventariali non accessibili ma solo osservabili a distanza, quando è possibile riconoscere e valutare la predominanza della specie. In questi casi, comunque, non è stato possibile registrare informazioni sulla sottocategoria forestale e altre sulla copertura del suolo. La classe 'non classificato' include le superfici non assegnate ad alcuna categoria forestale nel secondo inventario INFC2005, in quanto Aree temporaneamente prive di soprassuolo per le quali non era previsto di indicare la categoria forestale, e che non è stato possibile classificare durante la campagna di rilievi INFC2015.

Le Tables 7.7 e 7.8 riportano le superfici forestali stimate per le categorie forestali dei Boschi alti e degli Impianti di arboricoltura da legno, rispettivamente. Statistiche analoghe sono pubblicate nella Table 7.9, a livello nazionale, per le categorie inventariali delle Altre terre boscate, e nella Table 7.10 per la categoria Arbusteti anche a livello regionale. La stima delle superfici a livello provinciale è disponibile all'indirizzo inventarioforestale.org/statistiche_INFC. La Fig. 7.5 pone a confronto la superficie delle categorie forestali dei Boschi alti, a livello nazionale. Quattro categorie forestali presentano un'estensione superiore al milione di ettari; in ordine, queste sono i Querceti di rovere, roverella e farnia, gli Altri boschi caducifogli, le Cerrete, boschi di farnetto, fragno e vallonea e le Faggete, che sono presenti in quasi tutte le regioni italiane, eccetto le Faggete per la Sardegna e le Cerrete, boschi di farnetto, fragno e vallonea per le regioni del nord-est. Altre quattro categorie forestali sono estese per più di mezzo milione di ettari: gli Ostrieti, carpineti, i Castagneti, le Leccete e i Boschi di abete rosso. Quest'ultima categoria caratterizza principalmente le regioni alpine, mentre le altre sono presenti in quasi tutte le regioni. Le statistiche nella Table 7.7 forniscono anche un'indicazione della varietà del paesaggio forestale in Italia: in media, le regioni italiane vedono la presenza di 13 categorie forestali. Solo in una regione (Alto Adige) una categoria (Boschi di abete rosso) rappresenta più del 50% della superficie dei Boschi alti, e solo in tre altre regioni più del 40% della superficie dei Boschi alti è coperta da una sola categoria (Boschi di larice e cembro in Valle d'Aosta, Cerrete, boschi di farnetto, fragno e vallonea in Basilicata e Leccete in Sardegna).

Le statistiche sull'estensione delle superfici delle sottocategorie forestali a livello nazionale e regionale sono disponibili all'indirizzo inventarioforestale.org/statistiche_INFC. Esse possono essere di aiuto per distinguere i diversi aspetti legati alla frequenza e all'ecologia della maggior parte delle specie forestali. Nel Chap. 2 è riportata la corrispondenza delle sottocategorie forestali adottate da INFC con i codici delle classificazioni CORINE ed EUNIS.

Distribuzione per classi di quota

La Table 7.11 riporta le statistiche sulla superficie del Bosco per classe di altitudine con ampiezza 500 m. Il Bosco è presente lungo un gradiente molto ampio di altitudine, rispecchiando l'ampia escursione in quota del territorio italiano dovuta alle caratteristiche orografiche. A livello nazionale, il 34.1% della superficie del Bosco si trova nella prima classe di quota, vale a dire fino a 500 m s.l.m., mentre il 37.9% si trova nella successiva fascia di quota fino a 1000 m. La percentuale di superficie forestale decresce progressivamente con l'elevazione delle quote, nelle tre classi di altitudine rimanenti: il 19.5% della superficie si colloca nella classe 1001–1500 m, il 7.5% nella classe 1501–2000 m e infine l'1.1% è sopra i 2000 m di altitudine.

La ripartizione in fasce di quota a livello regionale, comunque, risulta piuttosto variabile. In Fig. 7.6 si possono osservare regioni con una distribuzione piuttosto omogenea per ripartizione del Bosco in classi altitudinali (es. Veneto, Lombardia, Piemonte, Abruzzo) e altre in cui il Bosco si distribuisce lungo poche classi di altitudine (es. Toscana, Sardegna). Naturalmente, a questo concorre l'aliquota della superficie regionale lungo il gradiente altitudinale e a ciò si deve che il 100% della superficie del Bosco in Puglia ricada entro i 1000 m di quota. In alcune regioni alpine, la percentuale di superficie del Bosco fino a 500 m s.l.m. mostra valori molto al di sotto della media nazionale: 0.8% in Valle d'Aosta, 1.8% in Alto Adige e 5.5% in Trentino. Si tratta di regioni montuose con una parte rilevante di superficie del Bosco a quote superiori a 1500 m s.l.m.: 31.9% in Trentino, 54.8% in Alto Adige e 61.7% in Valle d'Aosta. Tra le regioni non alpine, l'Abruzzo è la sola con una percentuale rilevante di superficie del Bosco (12.5%) sopra i 1500 m.

Osservando la distribuzione delle Altre terre boscate per classi di altitudine (Table 7.12 e Fig. 7.7), si può notare che a livello nazionale più della metà della superficie (54.6%) ricade nella prima classe e dunque si colloca sotto i 500 m di quota, con percentuali maggiori, anche in misura consistente, in alcune regioni del centro e del sud Italia: Toscana (72.5%), Sardegna (76.3%) e Puglia (83.1%). Nelle regioni del sud la percentuale di Altre terre boscate sopra i 1500 m di quota è limitata, ad eccezione dell'Abruzzo (13.5%), che è anche l'unica regione non alpina ad avere Altre terre boscate sopra i 2000 m di quota (lo 0.6% della superficie totale). La presenza di Altre terre boscate sopra i 1500 m di quota è dunque una caratteristica delle regioni Alpine, e valori particolarmente elevati si riscontrano in Alto Adige (il 92.5% della superficie totale delle Altre terre boscate) e in Trentino (il 73.0%).

Statistiche ulteriori sulla distribuzione per classi di quota delle categorie inventariali e di quelle forestali dei Boschi alti sono disponibili all'indirizzo inventariofo restale.org/statistiche_INFC. Tutte le statistiche descritte per le classi di quota con modulo 500 m sono state prodotte, e disponibili all'indirizzo citato, anche per classi di altitudine con modulo 300 m.

Proprietà

La proprietà è stata valutata sulla base di due livelli gerarchici, con il primo relativo al carattere della proprietà, che può essere privato o pubblico, e il secondo al tipo di proprietà, come mostrato nella Table 7.13. Per il primo livello, i casi di comproprietà sono stati trattati di volta in volta in base alla prevalenza di un interesse privato o pubblico.

Le Tables 7.14 e 7.15 riportano la superficie del Bosco e delle Altre terre boscate, rispettivamente, per proprietà pubblica o privata; le Tables 7.16 e 7.17 mostrano le stime delle superfici del Bosco e delle Altre terre boscate ripartite per tipo di proprietà. Statistiche analoghe sono disponibili all'indirizzo inventarioforestale.org/statistiche_INFC per le categorie inventariali del Bosco e delle Altre terre boscate, e per le categorie forestali del Bosco.

In Italia, il 63.5% della Superficie forestale totale è di proprietà privata mentre il 32.0% è pubblica; il restante 4.5% della superficie risulta non classificata. Per il Bosco, la prevalenza di proprietà privata è molto accentuata (66.4%), ma la percentuale più contenuta per le Altre terre boscate (50.2%) va valutata alla luce della porzione piuttosto elevata di superficie non classificata per questo attributo. Un quadro sulla distribuzione della proprietà tra pubblico e privato del Bosco e delle Altre terre boscate nelle regioni italiane è mostrato nelle Figs. 7.8 e 7.9, rispettivamente. Nel Bosco, prevale la proprietà privata in quasi tutte le regioni, ma non in Trentino, Abruzzo e Sicilia, e rappresenta più dell'80% in Liguria, Emilia-Romagna, Toscana e Marche.

Per la macrocategoria Bosco, il tipo di proprietà privata più comune è quella individuale (79.0%), che riguarda più del 90% della superficie del Bosco in Liguria, Molise, Campania e Basilicata. La proprietà privata di società, imprese o industrie ha una certa rilevanza per la Toscana (19.8%) e l'Umbria (14.9%), mentre in Trentino e nelle Marche una percentuale apprezzabile di proprietà privata appartiene ad altri enti (14.9% e 16.3%, rispettivamente) (Fig. 7.10). Per quanto concerne la superficie pubblica del Bosco, prevale quella dei comuni e delle province a statuto ordinario (65.4%), seguita da quelle statali, regionali e delle province autonome (23.5%), mentre solo l'8.3% della superficie appartiene ad altri enti pubblici (Fig. 7.11). La ripartizione della proprietà pubblica per tipo varia considerevolmente tra le varie regioni; paragonate alla situazione nazionale, molte regioni dell'Italia centrale e la Sicilia si distinguono per la minore diffusione di proprietà comunale e provinciale rispetto a quella statale, ad eccezione dell'Umbria, dove prevale la proprietà di altri enti pubblici. Alla classe degli altri enti pubblici appartiene una percentuale considerevolmente più alta della media nazionale in Alto Adige e in Trentino.

Consistenza e accrescimento dei boschi

La Table 7.18 riporta le statistiche sul numero di alberi vivi nelle categorie inventariali del Bosco mentre le Tables 7.19 e 7.20 riportano quelle relative alle categorie forestali dei Boschi alti e degli Impianti di arboricoltura da legno. In totale, nel Bosco ci sono quasi 11.5 miliardi di alberi, per un valore medio ad ettaro pari a 1264. Il numero di alberi per ettaro si colloca tra 1000 e 1500 in dodici regioni; in quattro (Umbria, Marche, Lazio e Molise) se ne contano più di 1500 (fino a 1728 in Umbria) mentre nelle restanti cinque (Valle d'Aosta, Alto Adige, Puglia, Basilicata e Sicilia) la densità è più contenuta, con valori compresi tra 578 (Sicilia) e 986 (Basilicata). I valori per categoria forestale dei Boschi alti indicano un contributo predominante dovuto alle formazioni a latifoglie, particolarmente elevato per la categoria degli Ostrieti e carpineti (2160 alberi per ettaro) e delle Leccete (1937 alberi per ettaro). Contribuiscono a densità così elevate il prevalente utilizzo di queste formazioni per la produzione di legna da ardere e il conseguente governo a ceduo. Per confronto, la categoria con la densità più bassa in assoluto è quella delle Sugherete (575 piante per ettaro) in cui basse densità e presenza di alberi di dimensioni adeguate sono caratteristiche ricercate per una conduzione razionale della coltura, per assicurare la massima illuminazione delle chiome e plance di sughero di certe dimensioni e spessore (Gambi, 1989).

Il numero di alberi come indicatore della consistenza delle risorse forestali presenta alcuni limiti derivanti dalle dimensioni molto variabili dei soggetti censiti. Come indice di densità, il numero di alberi è una variabile poco utile nei soprassuoli naturali, che ha trovato maggiore applicazione negli impianti artificiali (Avery & Burkhart, 1983), ma anche nei popolamenti giovani (Bernetti, 1995). Il numero di alberi è un'indicazione necessaria ma non sufficiente a descrivere adeguatamente lo stato dei soprassuoli (Burkhart & Tomé, 2012), che deve essere accompagnato da informazioni aggiuntive, per esempio sulle dimensioni degli alberi, la distanza relativa o la struttura del soprassuolo (Avery, 1967).

La Table 7.21 riporta i valori di area basimetrica per le categorie inventariali del Bosco; la Table 7.22 ne riporta i valori per le categorie forestali dei Boschi alti, mentre la Table 7.23 quelli per le categorie forestali degli Impianti di arboricoltura. L'area basimetrica è un indicatore più adatto a descrivere la consistenza dei soprassuoli perché altamente correlato al volume; inoltre, ha il vantaggio di essere calcolato da valori direttamente misurati con semplicità e accuratezza del diametro a 1.30 m da terra (es. Bueno-López & Bevilacqua, 2013; Di Cosmo & Gasparini, 2020). A livello di categorie forestali, i boschi di conifere mostrano le densità più elevate. Tra le categorie a conifere solo le Pinete di pini mediterranei mostrano un valore (18.7 m^2 per ettaro) inferiore alla media nazionale, pari a 22.1 m^2 per ettaro, mentre tutte le altre hanno valori superiori compresi tra 25.5 m^2 per ettaro (Boschi di larice e cembro) e 43.0 m^2 per ettaro (Boschi di abete bianco). Tra le categorie forestali caratterizzate da specie di latifoglie, solo le Faggete (31.7 m^2 per ettaro) e i Castagneti (28.0 m^2 per ettaro) mostrano valori superiori al dato medio nazionale e paragonabili a quelli delle categorie a conifere. La categoria degli Ostrieti e carpineti e quella delle Leccete mostrano, a fronte dei valori massimi commentati per il numero di alberi ad ettaro, valori di area basimetrica contenuti (rispettivamente 18.2 e 18.6 m^2 per ettaro), a conferma del fatto che sono ricche di alberi di piccole dimensioni. Per quanto riguarda i valori totali di area basimetrica, quelli più elevati sono stati stimati per le Faggete (33,336,591 m^2), i Boschi di abete rosso (22,314,318 m^2) e i Castagneti (21,802,076 m^2), categorie che associano ad elevati valori di densità anche un'elevata superficie forestale.

Nella Table 7.24 sono mostrati i valori del volume per le categorie inventariali del Bosco. Nelle Tables 7.25 e 7.26 sono riportati i valori della variabile per le categorie forestali rispettivamente dei Boschi alti e degli Impianti di arboricoltura da legno. Il volume stimato per tutti i Boschi italiani supera di poco 1.5 miliardi di metri cubi, per un valore medio per ettaro pari a 165.4 m^3 (Fig. 7.12). Questi corrispondono ad un valore della fitomassa arborea epigea di poco superiore a 1 miliardo di tonnellate (Mg), 114.9 Mg per ettaro in media, come riportato nella Table 7.27. Le regioni che maggiormente contribuiscono alla massa complessiva dei Boschi italiani, sia in termini di volume sia di fitomassa, sono la Toscana (10.4% del volume totale e 11.1% della fitomassa arborea epigea), il Piemonte (9.8% del volume e 9.6% della fitomassa) e la Lombardia (8.7% del volume e 7.9% della fitomassa). I valori minimi regionali sono stati stimati per la Valle d'Aosta, il Molise e la Puglia, con percentuali del volume totale compresi tra l'1.0 e l'1.3% e valori della fitomassa tra l'1.1 e l'1.5%. Le Tables 7.28 e 7.29 riportano le stime della fitomassa arborea epigea per le categorie forestali dei Boschi alti e degli Impianti di arboricoltura da legno. Se si considera il contributo delle categorie forestali dei Boschi alti, si conferma il ruolo preponderante delle Faggete (19.6% del volume totale e 21.5% della fitomassa arborea epigea totale), dei Boschi di abete rosso (16.4% del volume e 12.3% della fitomassa) e, in misura minore rispetto a quanto dedotto dai valori di area basimetrica, dei Castagneti che rimangono al terzo posto per il contributo in volume (10.0% del totale) ma sono superati per la massa arborea epigea dalla categoria delle Cerrete, boschi di farnetto, fragno e vallonea (cui spetta il 10.1% del valore totale), evidentemente per le diverse caratteristiche di densità basale del legno. Le Figs. 7.13 e 7.14 mostrano i valori totali e per ettaro rispettivamente del volume del fusto e della fitomassa arborea epigea per le categorie forestali dei Boschi alti. Il divario molto evidente tra i valori totali e quelli riferiti all'ettaro che è possibile cogliere nei grafici fornisce un'indicazione indiretta dell'estensione delle superfici delle varie categorie forestali sul territorio nazionale.

Il rapporto a livello di categoria forestale tra il volume del fusto, comprendente fusto, rami grossi e cimale fino alla sezione di 5 cm, e la fitomassa arborea epigea, che include, oltre alle componenti citate, anche ceppaia, ramaglia e cimale intero, è già stato utilizzato in ambito INFC2005 per calcolare il Biomass Conversion and Expansion Factor (BCEF) (Di Cosmo & Tabacchi, 2011). Tale rapporto è utile a stimare in termini di peso secco la massa legnosa e il relativo contenuto di carbonio quando si disponga solamente di stime del volume. Piccole variazioni dei coefficienti a seguito delle ripetizioni delle indagini inventariali possono essere attese in virtù della differenza del campione degli alberi, ad esempio perché quel rapporto varia in proporzione diversa a seconda della dimensione degli alberi per ciascuna specie, per la diversa percentuale delle specie presenti in una categoria forestale oppure per la diversa percentuale di alberi troncati, per i quali la stima della densità è ottenuta con una procedura convenzionale e non mediante le funzioni di stima disponibili (cfr. Chap. 6). Pertanto, non si è ritenuto necessario presentare in questo volume i BCEF che è possibile calcolare sulla base delle statistiche di INFC2015. Risulta però opportuno evidenziare che per le Pinete di pini mediterranei il valore di BCEF ottenuto con i dati attuali si discosta in maniera apprezzabile a seguito all'adozione delle nuove funzioni per la stima del volume e della fitomassa del pino d'Aleppo, non disponibili nel 2005.

Le stime sull'incremento annuo di volume sono riportate nella Table 7.30 per le categorie inventariali del Bosco, nella Table 7.31 per le categorie forestali dei Boschi alti e nella Table 7.32 per quelle degli Impianti di arboricoltura da legno. L'incremento annuo di volume del Bosco in Italia ammonta a quasi 37.8 milioni di metri cubi, 4.2 metri cubi in media per ettaro. Nell'ambito dei Boschi alti (Fig. 7.15), il maggior contributo all'incremento annuo totale si deve alle Faggete (15.5%), seguite dai Boschi di abete rosso (12.8%) e dagli Altri boschi di latifoglie (12.4%). Per i totali a livello nazionale, si osserva un contributo rilevante delle categorie forestali a latifoglie, con valori in linea generale maggiori di quelle a conifere, anche se è tra le latifoglie che si trovano le categorie con valori minimi (Sugherete e Altri boschi di latifoglie sempreverdi, entrambe con lo 0.6% dell'incremento in volume totale). Passando invece alla produttività dei soprassuoli, i valori per ettaro più elevati si riscontrano in quattro categorie a conifere (Boschi di abete bianco, Boschi di abete rosso, Pinete di pino nero e Altri boschi di conifere), varabili tra 6.1 e 9.1 m^3 per anno. Tra le categorie a latifoglie, le Faggete contribuiscono con 5.5 m^3 annui per ettaro, i Castagneti con un valore molto simile, pari a 5.4 m^3.

Struttura e sviluppo

La struttura e lo sviluppo dei soprassuoli sono descritti mediante una serie di variabili relative al grado di copertura, alla struttura verticale, al tipo colturale e alle pratiche colturali adottate, allo stadio di sviluppo e all'età.

La valutazione della copertura delle chiome separatamente per gli alberi e gli arbusti, su ortofoto e in campo, ha consentito di assegnare ogni plot ad una delle sei seguenti classi di copertura del suolo: minore di 5, 5–10, 11–20, 21–50, 51–80, oltre 80%, secondo cui sono fornite le statistiche di INFC2015. Il tipo colturale, come definito dalla combinazione della forma di governo con quella di trattamento, è stato classificato mediante le classi descritte nella Table 7.33. Nel caso di cedui coetanei, fustaie coetanee e fustaie transitorie, la struttura verticale è stata classificata distinguendo la copertura monoplana da quella biplana. Per quanto riguarda lo stadio di sviluppo, la classificazione si è avvalsa delle descrizioni riportate nella Table 7.34; per i soprassuoli coetanei, inoltre, sì è proceduto alla classificazione per classi di età: fino a 10 anni, 11–20, 21–30, 31–40, 41–80, 81–120, oltre 120 anni. Tipo colturale e stadio di sviluppo non sono stati valutati negli Impianti di arboricoltura da legno, per i quali si è valutata invece la classe di età. Le informazioni relative a questi attributi hanno consentito di produrre statistiche di superficie e stime delle variabili quantitative relative al volume, alla massa e all'incremento dei soprassuoli per tipo colturale, stadio di sviluppo e classe di età.

Tables 7.35 e 7.36 riportano le statistiche sulla superficie del Bosco ripartita, rispettivamente, per classi di copertura totale e di copertura arborea. Statistiche analoghe per le categorie inventariali dei Boschi alti e degli Impianti di arboricoltura da legno, nonché per le categorie forestali dei Boschi alti a livello nazionale, sono disponibili all'indirizzo inventarioforestale.org/statistiche_INFC. La Fig. 7.16 mostra la superficie del Bosco ripartita per classi di copertura totale a livello regionale. L'aliquota maggiore per il Bosco italiano (74.4%) compete alla classe di copertura totale maggiore dell'80%, mentre la classe di copertura totale 51–80% occupa il 18.0%. Classi di copertura meno densa sono più frequenti in Sicilia, Sardegna e Alto Adige. La prevalenza di superfici a copertura totale densa o molto densa delle chiome è confermata anche dall'osservazione della copertura delle sole chiome arboree: la percentuale della superficie del Bosco con copertura arborea nelle ultime due classi è, rispettivamente, il 21.8 e il 62.7%. Tra le categorie forestali, una maggiore copertura delle chiome arboree si osserva nei Boschi di abete bianco, nelle Faggete, nei Castagneti e negli Ostrieti e carpineti; le Sugherete, gli Altri boschi di latifoglie sempreverdi e i Boschi di larice e cembro sono le categorie che si caratterizzano maggiormente per la più limitata copertura delle chiome degli alberi (Fig. 7.17 e tabelle in inventarioforestale.org/statistiche_INFC).

Le osservazioni che seguono sul tipo colturale e lo stadio di sviluppo del Bosco italiano si riferiscono principalmente ai Boschi alti, la categoria inventariale largamente predominante per estensione di superficie. Inoltre, la maggior parte degli attributi di questa sezione non sono oggetto di valutazione per gli Impianti di arboricoltura da legno. La Table 7.37 mostra le statistiche di superficie dei Boschi alti per tipo colturale. La Fig. 7.18 riporta la distribuzione della superficie dei Boschi alti per gruppi di tipi colturali a livello regionale. A livello nazionale, cedui e fustaie occupano approssimativamente la stessa superficie. Tuttavia, la percentuale di superficie occupata dalle due forme di governo varia sensibilmente nelle regioni. I tipi colturali del ceduo coprono più del 50% della superficie dei Boschi alti in nove regioni, (Lombardia, Liguria, Emilia-Romagna, Toscana, Umbria, Marche, Lazio, Molise e Puglia); le fustaie prevalgono nelle regioni alpine (Valle d'Aosta, Piemonte, Alto Adige, Trentino, Friuli-Venezia Giulia e Veneto) e in due regioni del sud: Calabria e Sicilia. La classe del tipo colturale non definito, che denota assenza di selvicoltura o solo attività sporadiche e occasionali, rappresenta un'aliquota importante della superficie dei Boschi alti in quasi tutte le regioni. Il tipo colturale più diffuso (29.2% della superficie dei Boschi alti) è il ceduo matricinato, seguito dalle fustaie disetanee (16.1%), dalle fustaie coetanee (15.1%) e dal tipo colturale non definito (13.9%) (Fig. 7.19).

Le Tables 7.38 e 7.39 riportano le stime della superficie dei Boschi alti per stadio di sviluppo e per struttura verticale, rispettivamente. Inoltre, la Table 7.40 mostra le stime della superficie dei Boschi alti coetanei per classe di età. Statistiche analoghe per le categorie forestali sono disponibili su inventarioforestale.org/statistiche_INFC. Le Figs. 7.20, 7.21 e 7.22 mostrano la ripartizione per stadio di sviluppo della superficie dei cedui, delle fustaie coetanee e del tipo colturale speciale o non definito. Nel ceduo, gli stadi giovanile e in rinnovazione rappresentano insieme il 10.6% della superficie totale; nelle fustaie coetanee, le classi della fustaia in rinnovazione, novelleto e spessina occupano insieme il 3.0% della superficie totale. Sebbene una differenza tra queste due percentuali fosse attesa, in relazione alla diversa durata degli stadi di sviluppo nei cedui e nelle fustaie, le classi giovanili rappresentano una percentuale bassa in entrambi i casi. Lo stadio maturo e invecchiato occupa il 34.3% della superficie delle fustaie coetanee o transitorie, lo stadio invecchiato il 32.6% della superficie dei cedui coetanei. La ripartizione per stadio di sviluppo del tipo colturale speciale o non definito, che comprende i soprassuoli specializzati per la produzione di prodotti secondari e quelli ad evoluzione naturale, come le nuove formazioni su terreni abbandonati, mostra una percentuale più alta di superficie di popolamenti giovani rispetto a quelli invecchiati, in linea con le aspettative.

Una struttura monoplana è stata stimata per il 76.8% della superficie dei Boschi alti; una struttura di tipo biplano si è osservata nel restante 23.2% della superficie, pari a 1,557,010 ha. Fustaie disetanee, irregolari o articolate e cedui disetanei, che sono caratterizzati dalla presenza di diversi tipi strutturali su superfici limitate, nel complesso occupano 2,219,995 ha (vedi Table 7.38).

Le stime dei totali e delle densità del volume, della fitomassa arborea epigea e dell'incremento annuo di volume per tipo colturale sono riportate nelle Tables 7.41, 7.42 e 7.43, rispettivamente. Nelle tabelle citate, gli Impianti di arboricoltura da legno sono stati inclusi nella classe del tipo colturale non classificato. Statistiche analoghe per le categorie forestali, a livello nazionale, sono disponibili all'indirizzo inventarioforestale.org/statistiche_INFC. Le Figs. 7.23 e 7.24 mostrano i valori di volume e incremento annuo per i cedui, le fustaie, i tipi colturali speciali e non definiti, a livello nazionale. Nelle fustaie risiede circa il doppio del volume legnoso che si trova nei cedui, sia per il valore totale sia per quello ad ettaro: il volume totale ammonta a 905.4 e 466.0 milioni di m^3, rispettivamente, per le fustaie e per i cedui; il dato per ettaro ammonta a 241.0 e 123.0 m^3, rispettivamente. Per quanto riguarda l'incremento annuo, esso è molto più sostenuto nelle fustaie, sia nei valori totali sia per ettaro, che ammontano rispettivamente a 18.9 milioni di m^3 (5.0 m^3 per ettaro) nelle fustaie e 14.3 milioni di m^3 (3.8 m^3 per ettaro) nei cedui.

Le Tables 7.44, 7.45 e 7.46 mostrano le statistiche sui valori totali del volume legnoso, la biomassa arborea epigea e l'incremento annuo di volume dei Boschi alti per classi di età. Analoghe statistiche per le categorie forestali sono disponibili su inventarioforestale.org/statistiche_INFC, per il livello nazionale. Le Figs. 7.25, 7.26 e 7.27, infine, mostrano il volume legnoso totale, l'incremento annuo di volume e l'incremento percentuale annuo di volume per classe di età nei cedui coetanei e nelle fustaie coetanee.

La rinnovazione e gli arbusti

La consistenza della vegetazione legnosa del sottobosco è stata stimata mediante il conteggio degli individui di specie arboree (genericamente indicati come rinnovazione) e arbustive sotto la soglia di cavallettamento e più alti di 50 cm. Il conteggio è stato condotto separatamente per tre classi dimensionali, definite da soglie di altezza o diametro (cfr. Chap. 4). Per le specie legnose, le statistiche sono relative ai valori totali e per ettaro del numero di individui, biomassa epigea e contenuto di carbonio, per ognuna delle tre classi dimensionali e totale. In maniera analoga, per gli arbusti sono state prodotte statistiche per la biomassa epigea e per il carbonio organico, per classe dimensionale. Infine, i dati della rinnovazione e degli arbusti sono stati trattati congiuntamente per ottenere la stima della biomassa e del carbonio organico della vegetazione legnosa del sottobosco nel suo complesso (cfr. Chap. 12).

Le Tables 7.47, 7.48 e 7.49 riportano le stime dei valori totali e per ettaro del numero di soggetti della rinnovazione per le categorie inventariali e forestali. Statistiche analoghe per classe dimensionale e quelle per gli arbusti sono disponibili all'indirizzo inventarioforestale.org/statistiche_INFC, dove sono disponibili anche le statistiche sulla biomassa della vegetazione legnosa del sottobosco e il loro contenuto di carbonio organico.

La Fig. 7.28 mostra il numero totale di soggetti della rinnovazione per ettaro nelle regioni italiane. I valori per ettaro variano da 1506 in Valle d'Aosta a 9539 in Molise. Alle differenze concorre la distribuzione delle categorie forestali tra le regioni. I boschi di latifoglie si caratterizzano generalmente per un numero più elevato di alberi di piccole dimensioni rispetto alle categorie dei boschi di conifere. Tra i boschi di conifere, le pinete mostrano i valori più elevati, simili a quelli riscontrati nelle Faggete e nei Boschi igrofili, categorie a latifoglie con numero di soggetti limitato, ma più alti che nelle Sugherete e negli Altri boschi di latifoglie sempreverdi.

References

Avena, G., & Dowgiallo, G. (1995). Substrato. In: S. Pignatti (Ed.), *Ecologia vegetale* (p. 37). UTET.

Avery, T. E. (1967). *Forest measurements* (p. 215). McGraw-Hill.

Avery, T. E., & Burkhart, H. (1983). *Forest measurements* (3rd edn., pp. 245–259). McGraw-Hill.

Bernetti, G. (1995). La descrizione particellare nella foresta. In: AA.VV. *Nuove metodologie nella elaborazione dei piani di assestamento dei boschi* (pp. 386–389). I.S.E.A.

Breidenbach, J., Granhus, A., Hylen, G., Eriksen, R., & Astrup, R. (2020). A century of National Forest Inventory in Norway—Informing past, present, and future decisions. *Forest Ecosystems, 7*(46), 1–19. https://doi.org/10.1186/s40663-020-00261-0

Bueno-López, S., & Bevilacqua, E. (2013). Diameter growth prediction for individual Pinus occidentalis Sw.trees. *iForest, 6*, 209–216. https://doi.org/10.3832/ifor0843-006

Burkhart, H. E., & Tomé, M. (2012). *Modeling forest trees and stands* (pp. 175–200). Springer Science+Business Media. https://doi.org/10.1007/978-90-481-3170-9_14.

Delort, R. (1989). *La vita quotidiana nel medioevo* (pp. 15–36). Laterza, Roma-Bari.

Di Cosmo, L., & Gasparini, P. (2020). Predicting diameter at breast height from stump measurements of removed trees to estimate cuttings, illegal loggings and natural disturbances. *South-east European Forestry, 11*(1), 41–49. https://doi.org/10.15177/seefor.20-08.

Di Cosmo, L., & Tabacchi, G. (2011). Consistenza e accrescimento dei boschi. In: P. Gasparini & G. Tabacchi (Eds.), *L'Inventario Nazionale delle Foreste e dei serbatoi forestali di Carbonio INFC 2005. Secondo inventario forestale nazionale italiano. Metodi e risultati* (pp. 117–120). Edagricole-Il Sole 24 Ore. ISBN 978-88-506-5394-2.

Dobbertin, M. (2005). Tree growth as indicator of tree vitality and of tree reaction to environmental stress: a review. *European Journal of Forest Research, 124,* 319–333. https://doi.org/10.1007/s10342-006-0110-1.

Fossier, R. (2003). Terra. *Dizionario dell'occidente medievale* (Vol. 2, pp. 1158–1167). Einaudi.

Gambi, G. (1989). Le sugherete (parte II). *Monti e Boschi, 2*, 27–38.

Gasparini, P., Di Cosmo, L., Rizzo, M., & Giuliani, D. (2017). A stand-level model derived from National Forest Inventory data to predict periodic annual volume increment of forests in Italy. *Journal of Forest Research-JPN, 22*(4), 209–217. https://doi.org/10.1080/13416979.2017.1337260

Gschwantner, T., Lanz, A., Vidal, C., Bosela, M., Di Cosmo, L., Fridman, J., Gasparini, P., Kuliešis, A., Tomter, S., & Schadauer, K. (2016). Comparison of methods used in European National Forest Inventories for the estimation of volume increment: towards harmonisation. *Annals of Forest Science, 73,* 807–821. https://doi.org/10.1007/s13595-016-0554-5.

Loetsch, F., & Haller, K. E. (1973). *Forest inventory* (2nd edn. Vol. 1, pp. 1–15). BLV Verlagsgesellschaft.

McRoberts, R. E., Tomppo, E. O., & Næsset, E. (2010). Advances and emerging issues in national forest inventories. *Scandinavian Journal of Forest Research, 25*, 368–381. https://doi.org/10.1080/02827581.2010.496739

Persson, R., & Janz, K. (2015). National forest assessments and policy influence. In: *Knowledge reference for national forest assessments* (pp. 1–12). FAO. Retrieved Nov 15, 2021, from www.fao.org/publications/card/en/c/8fd3b298-e843-4d3f-9ee0-cdb0e41739fd.

Provero, L. (2020). *Contadini e potere nel medioevo* (pp. 39–51). Carrocci.

Sereno, C. (2008). La foresta nel medioevo: temi e direzioni d'indagine. *Dendronatura, 1,* 28–34.

Solberg, S., Dobbertin, M., Reinds, G. J., Lange, H., Andreassen, K., Fernandez, P. G., Hildingsson, A., & de Vries, W. (2009). Analyses of the impact of changes in atmospheric deposition and climate on forest growth in European monitoring plots: a stand growth approach. *Forest Ecology and Management, 258*(8), 1735–1750. https://doi.org/10.1016/j.foreco.2008.09.050.

Tomppo, E., Schadauer, K., McRoberts, R. E., Gschwantner, T., Glabre, K., & Ståhl, G. (2010). Introduction. In: E. Tomppo, T. Gschwantner, M. Lawrence, & R. E. McRoberts (Eds.), *National forest inventories—Pathways for common reporting* (pp. 1–17). Springer. ISBN 978-90-481-3232-4. https://doi.org/10.1007/978-90-481-3233-1.

Open Access This chapter is licensed under the terms of the Creative Commons Attribution 4.0 International License (http://creativecommons.org/licenses/by/4.0/), which permits use, sharing, adaptation, distribution and reproduction in any medium or format, as long as you give appropriate credit to the original author(s) and the source, provide a link to the Creative Commons license and indicate if changes were made.

The images or other third party material in this chapter are included in the chapter's Creative Commons license, unless indicated otherwise in a credit line to the material. If material is not included in the chapter's Creative Commons license and your intended use is not permitted by statutory regulation or exceeds the permitted use, you will need to obtain permission directly from the copyright holder.

Chapter 8
Forest Management and Productive Function

Produzione e gestione selvicolturale

Lucio Di Cosmo and Patrizia Gasparini

Abstract Forestry originated as a science under the need to manage forests for timber production in a sustainable way. Since then, sustainability has been broadened to include a variety of values and services recognised as important for human well-being, while at maintaining a relevant productive function. Hence, silviculture exists as a need of humankind and not as a need of forests for their continued existence. Sustainability is obtained by adopting correct silvicultural practices and adjusting the planning in view of the most suitable management able to ensure the goals, for example, carbon balance under climate warming scenarios. National forest inventories provide reliable information that is needed for effective forest policies. This chapter shows area estimates on some variables important for production, such as forest accessibility, presence of roads or tracks, terrain roughness and availability for wood supply (Sect. 8.2). Silviculture practices are actualised by forest utilisation modes, which are also related to timber logging modes. Section 8.3 shows the statistics on these variables as well as those on the growing stock and the biomass annually removed from the forest. Section 8.4 shows the estimates on presence of forest planning or regulations that influence productive aspects of forests. The type of planning may vary consistently due to overlapping authorities protecting forests in Italy, so the presence of planning at different levels was emphasised.

Keywords Sustainability · Forest management · Cutting · Exploitation · Skidding road · Terrain roughness · Timber logging

L. Di Cosmo (✉) · P. Gasparini
CREA Research Centre for Forestry and Wood, Trento, Italy
e-mail: lucio.dicosmo@crea.gov.it

P. Gasparini
e-mail: patrizia.gasparini@crea.gov.it

© The Author(s) 2022

P. Gasparini et al. (eds.), *Italian National Forest Inventory—Methods and Results of the Third Survey*, Springer Tracts in Civil Engineering,
https://doi.org/10.1007/978-3-030-98678-0_8

8.1 Introduction

Managing forest is needed to utilise timber as a sustainable resource (Dorren et al., 2004), although since the 1980s, sustainability has been widened to include a variety of values and services over a large scale (O'Hara, 2016). It was in response to the threat to sustainability due to the poor conditions of Europe's remaining forests at the end of the eighteenth century that forestry originated as a science; this was the first science to explicitly acknowledge the need to safeguard finite natural resources for future generations (Perry, 1998). Hence, silviculture is a need of humankind that aims to obtain usefulness from forests, and not a need of forests for their continued existence (Piussi, 1994). Silviculture has also been described as the need to find a balance between the use of forests and their assessed capacity to provide goods and benefits, considering our undeniable need for human well-being (De Philippis, 1983). Sustainability is obtained by adopting correct silvicultural practices within appropriate forest management plans, effective at different territorial levels and compiled in compliance with the regulatory constrains in force in the different territories.

From this point of view, national forest inventories may offer an important contribution to adjust planning criteria on a national and regional scale to reach the desired level of sustainable forest management to ensure carbon balance under climate warming scenarios, because well-established decision makings should be based on reliable information (Bosela et al., 2016). The operational level of forest planning is known as detailed planning or management planning and its aim is to guarantee that forests can provide goods and benefits continuously without deteriorating (Bernetti, 1989).

Productive aspects of forests depend on many variables, some of which are described in this chapter. They are related to the possibility to access a forest, considering possible physical impediments due for example to the orography or forbidden access on legal grounds. Time and effort to reach a forest should be also reasonably limited and this is achievable only in the presence of an adequate road network (Hippoliti, 1990). INFC2015 assessed the presence of roads in terms of horizontal distance and the difference in altitude between the closest roads or tracks and the sample points. Another important characteristic conditioning the availability of forest areas for wood supply is the micromorphology of the terrain due to the presence of obstacles (boulders, rock, ditches, ravines) that might make it difficult or impossible to use tractors and harvesters. The variables cited may also interact, leading to specific features site by site. However, with equal site conditions, productive importance may be consistently different based on the value of the stand, for example, due to the species composition or timber quality. For this reason, INFC field teams are additionally requested to explicitly assess availability for wood supply based on their expert judgment.

The silvicultural practices in use are strictly related to the utilisation system, as silviculture is actualised by utilisations (Hippoliti, 2005). Section 8.3 shows the silvicultural practices adopted, the utilisation systems and the logging modes observed, i.e., the way and the tools used to cut trees and remove the wood or timber from forests.

The overlapping of state and regional authorities on forests and that of institutions entrusted with protection of the different goods and services provided by forests has brought about complexity in planning. INFC surveys the presence of plans at the various levels where they may be present and also considers the presence of laws that affect availability for wood supply. For each sample plot, all the plans and regulations are recorded. As those plans and regulations are not mutually exclusive and may be present on the same area, the sum of the areas with the presence of different planning and regulations does not equal the total area of Forest or Other wooded land.

8.2 Accessibility and Availability for Wood Supply

The extent to which forests are reachable is assessed by variables that describe the possibility to access the NFI sample points (degree of accessibility) and the ease of reaching them in relation to the horizontal distance and to the difference in altitude between the sample point and the closest road or forest track. Once in the forest, silviculture operations may be difficult in relation to ground roughness, which may also prevent the use of forestry machines (Hippoliti, 2004).

The degree of accessibility is recorded by designating the sample points as 'accessible' or 'not accessible'. Table 8.1 shows the classes for not accessible conditions, that are due to physical orographic impediments impossible to overcome or discouraging due to health or safety hazards, very intricate vegetation or forbidden access on

Table 8.1 Causes of inaccessibility / Classi relative ai motivi di non accessibilità del punto inventariale

Causes of inaccessibility	Description
Motivi di inaccessibilità	Descrizione
Physical impediment to access due to orography Ostacoli fisico-orografici	Cliffs, waterways, presence of boulders
	Presenza di massi, pareti di roccia, forti dislivelli, corsi d'acqua
Forbidden access or enclosure Divieti e/o recinzioni	Access is forbidden on legal basis or by explicit private owner decision
	Presenza di recinzioni, divieti di accesso o mancanza di permesso di accesso a fondi privati
Physical impediment due to vegetation Caratteri della vegetazione	Physical impediment to access due to very intricate vegetation
	Presenza di vegetazione molto densa che impedisce la percorribilità dell'area

legal grounds or explicit private owner decision (e.g., enclosure). Table 8.2 shows the estimates for the accessible and not accessible area of Forest and of Other wooded land. At inventarioforestale.org/statistiche_INFC, similar statistics are given for the inventory categories of Forest and Other wooded land and for the forest types of the Tall trees forest. The Italian Forest is marked by a high percentage of accessible area (90.3%) and this is also true at the regional level (Fig. 8.1), where the percentage always exceeds 80% with a rather limited variability among the regions. The same considerations can be taken for the forest types in the Tall trees forest; the lowest percentage was found in the Hygrophilous forests (78.6%).

Horizontal distance and difference in altitude between the closest road or track and the NFI sample point were assessed on maps, orthophotographs and other thematic layers (also available from WebGIS) before the field trip. This information was verified and complemented by the crew once it was in the NFI plot. This way to assess the presence of roads was introduced with INFC2015 to avoid some problems experienced in the past that caused absence of data on a large portion of the wooded territory. The previous protocol was too demanding, also considering the time needed to record the stated number of GPS positions in those years (Di Cosmo et al., 2011).

The change was substantial, since all sample plots reached in INFC2015 hold data, but as the current estimates also use information from the past NFI (cf. Chap. 5), general information is still lacking for a relevant number of plots used to derive the inventory statistics. For this reason, the comments on presence of roads should be better evaluated keeping in mind that 30.6% of Forest area is not classified for this variable.

Table 8.3 shows the estimates of Forest area by horizontal distance class (0–500, 501–1000, 1001–2000, >2000 m) to the nearest road or forest track. At inventariofo restale.org/statistiche_INFC, similar estimates are available about the Other wooded land, the inventory categories of Forest and the forest types of the Tall trees forest. A total of 47.9% of the Forest area is served by roads within a radius of 500 m. The first distance class is also the most frequent at the regional level, with values spreading from 34.3% of Basilicata and 38.9% of Valle d'Aosta to 57.1% of Alto Adige (Fig. 8.2).

The Forest area of the first two classes, i.e., distance within 1000 m, reaches 63.4% at the national level but the relative rank of the regions varies. For example, the Forest area percentage of Valle d'Aosta (61.2%) is similar to that of other regions, while Basilicata remains the region with the lowest percentage of Forest area served (45.6%); Alto Adige is still the region with the highest portion of served Forest area (74.7%), but other regions are marked by very close percentages (always over 70% in Veneto, Molise and Sardegna).

Table 8.4 shows the estimates of Forest area by difference in altitude class (0–100, 101–400, >400 m, positive when walking uphill from the road to the sample point and negative when walking downhill). At inventarioforestale.org/statistiche_INFC similar inventory statistics are available about the Other wooded land, the inventory categories of Forest and the forest types of the Tall trees forest. A total of 58.3% of the Forest area is served by roads at an altitude difference of less than 100 m. At a regional level, the Forest area with a road within 100 m of altitude difference is

Table 8.2 Forest and Other wooded land area by accessibility / Estensione del Bosco e delle Altre terre boscate ripartiti per grado di accessibilità

Region/Regione	Forest/Bosco								Other wooded land/Altre terre boscate							
	Accessible		Not accessible		Not classified		Total		Accessible		Not accessible		Not classified		Total	
	Accessibile		Non accessibile		Non classificato		Totale		Accessibile		Non accessibile		Non classificato		Totale	
	Area	ES	Area	ES	Area	ES	Area	ES	Area	ES	Area	ES	Area	ES	Area	ES
	(ha)	(%)	(ha)	(%)	(ha)	(%)	(ha)	(%)	(ha)	(%)	(ha)	(%)	(ha)	(%)	(ha)	(%)
Piemonte	832,649	1.4	52,832	8.6	4953	28.9	890,433	1.3	59,985	9.7	18,385	17.4	6621	35.6	84,991	8.0
Valle d'Aosta	87,707	4.3	11,535	17.2	0	–	99,243	3.6	4660	37.3	3588	31.9	486	81.7	8733	24.0
Lombardia	566,864	1.8	51,241	9.6	3864	33.4	621,968	1.6	40,095	11.8	24,151	15.0	6007	38.1	70,252	8.7
Alto Adige	316,436	1.9	21,949	13.0	885	71.3	339,270	1.7	23,894	13.3	9919	19.4	2269	40.6	36,081	10.4
Trentino	350,257	1.6	21,789	12.8	1213	58.5	373,259	1.4	21,683	13.7	10,931	17.9	1213	58.5	33,826	10.6
Veneto	381,508	2.1	32,274	11.5	2922	35.2	416,704	1.9	14,473	15.5	32,249	12.3	6269	31.3	52,991	9.1
Friuli V.G	294,349	2.2	38,207	10.4	0	–	332,556	1.9	17,877	16.8	22,438	15.8	743	70.5	41,058	10.6
Liguria	325,934	1.8	13,927	15.8	3299	33.0	343,160	1.7	25,928	13.7	14,960	17.3	3197	61.8	44,084	10.3
Emilia Romagna	544,832	1.7	36,017	10.5	4052	30.1	584,901	1.5	35,124	12.4	17,317	14.4	1473	50.0	53,915	9.3
Toscana	915,438	1.3	116,035	5.4	3975	30.1	1,035,448	1.1	66,534	8.9	81,634	6.6	6106	41.7	154,275	5.2
Umbria	367,630	1.8	21,569	14.9	1106	57.6	390,305	1.6	17,186	18.2	5063	26.6	1402	99.8	23,651	15.2
Marche	260,265	2.5	29,279	10.8	2224	40.8	291,767	2.1	16,566	18.6	4007	43.0	741	70.8	21,314	16.2
Lazio	478,183	2.0	72,473	6.8	9580	19.5	560,236	1.6	61,004	9.6	24,732	15.4	2176	70.3	87,912	7.7
Abruzzo	374,588	2.1	34,439	9.9	2561	37.8	411,588	1.8	56,898	9.2	5751	32.8	362	100.0	63,011	8.6
Molise	146,609	3.2	6639	23.8	0	–	153,248	3.0	17,682	17.4	2343	40.5	0	–	20,025	16.0
Campania	339,588	2.6	61,764	7.6	2574	37.8	403,927	2.1	50,518	10.9	33,197	12.9	3618	46.3	87,332	7.6
Puglia	133,643	4.3	1654	47.2	7052	23.0	142,349	4.0	42,337	11.0	3984	30.7	3068	35.2	49,389	9.9
Basilicata	232,152	3.4	52,517	7.9	3351	33.2	288,020	2.7	53,180	9.7	49,720	8.9	1491	49.9	104,392	6.2
Calabria	417,998	2.4	73,448	8.0	3731	31.5	495,177	2.0	42,512	10.3	103,914	5.8	9017	27.7	155,443	4.8
Sicilia	255,517	3.6	25,044	15.4	4929	27.6	285,489	3.2	68,250	8.9	27,140	15.2	6356	31.0	101,745	7.1
Sardegna	578,227	2.3	41,713	9.8	6200	24.2	626,140	2.1	577,938	2.3	81,407	6.8	15,507	19.0	674,851	2.0
Italia	8,200,372	0.5	816,345	2.2	68,469	7.4	9,085,186	0.4	1,314,320	1.8	576,829	2.7	78,123	9.5	1,969,272	1.4

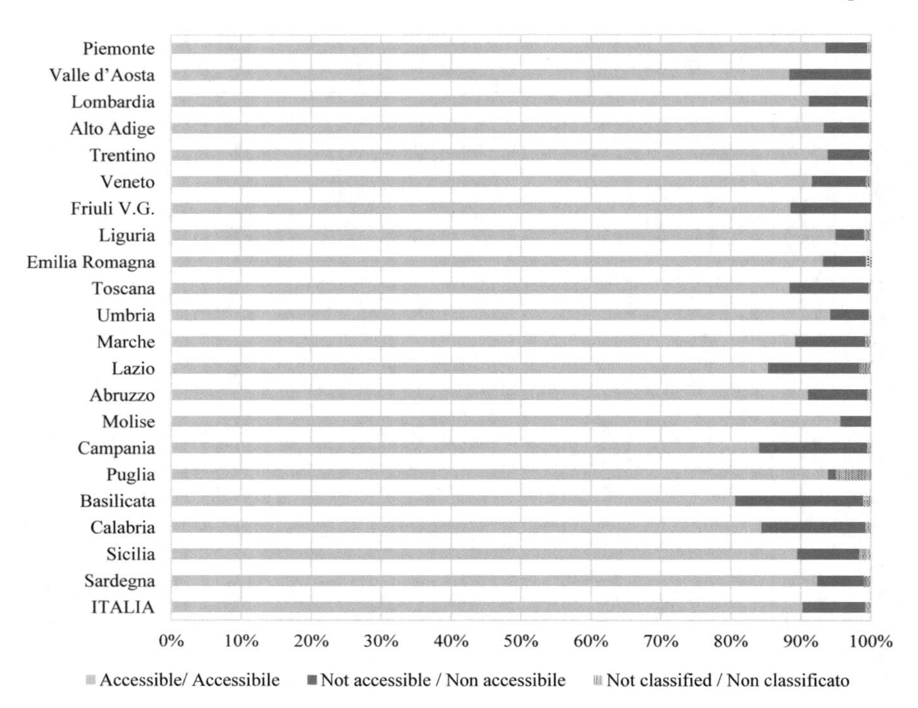

Fig. 8.1 Forest area percentage by accessibility / Ripartizione percentuale della superficie del Bosco per grado di accessibilità

at least 50% (Fig. 8.3) except Valle d'Aosta (42.1%) and Basilicata (44.2%), two regions already designated as having a low percentage of Forest area in the first distance class.

For the Tall trees forest types, the first class, i.e., 0 +/−100 m, remains the most frequent, but in the Larch and Swiss stone pine type the percentage is only 35.0% compared to 25.8% in the second class (101–400 m). Beech is the only other type where the area in the first class does not reach 50% of the total area (42.8%), since in all the other types more than 50% of the area is within a 100 m difference level. Cork oak is the forest type easiest to reach in terms of difference level, as more than 80% of its area is in the first class.

Ground roughness was classified distinguishing the three large classes in Table 8.5. Tables 8.6 and 8.7 show the area of Forest and of Other wooded land by roughness class. At inventarioforestale.org/statistiche_INFC similar estimates are available for the inventory categories of Forest and for the forest types of the Tall trees forest. A total of 62.6% of the Forest area is on smooth terrain. In the regions, at least 50% of forests grow in similar ground conditions except Valle d'Aosta, whose Forest percent area in the first class is 30.0% and 44.1% of Forest area is on rough terrain. The inventory statistics on the Other wooded land are less interesting, because the percentage of area not classified is relevant. In the regions, Other wooded land area

Table 8.3 Forest area by horizontal distance between sampling point and the closest road / Estensione del Bosco ripartito per classi di distanza tra punto di campionamento e punto più vicino di viabilità ordinaria o forestale

Region/Regione	Forest/Bosco											
	0 ÷ 500 m		501 ÷ 1000 m		1001 ÷ 2000 m		>2000 m		Not classified		Total	
	Tra 0 e 500 m		Tra 501 e 1000 m		Tra 1001 e 2000 m		Oltre 2000 m		Non classificato		Totale	
	Area	ES	Area	ES	Area	ES	Area	ES	Area	ES	Area	ES
	(ha)	(%)	(ha)	(%)	(ha)	(%)	(ha)	(%)	(ha)	(%)	(ha)	(%)
Piemonte	430,362	2.7	134,855	5.7	57,054	8.5	20,450	15.2	247,713	3.8	890,433	1.3
Valle d'Aosta	38,652	8.3	22,074	12.0	7706	21.6	1662	46.5	29,149	11.3	99,243	3.6
Lombardia	266,266	3.5	105,113	6.7	37,246	11.1	18,587	17.6	194,756	4.5	621,968	1.6
Alto Adige	193,758	3.4	59,790	7.4	17,039	17.6	2647	37.6	66,036	7.0	339,270	1.7
Trentino	180,330	3.6	55,524	7.9	19,952	13.2	4096	30.3	113,357	4.9	373,259	1.4
Veneto	230,964	3.4	76,335	7.6	20,190	13.4	5608	25.7	83,606	6.7	416,704	1.9
Friuli V.G	158,842	4.1	50,066	8.1	25,236	12.8	9663	19.3	88,751	6.1	332,556	1.9
Liguria	164,364	3.9	47,114	8.7	14,644	15.4	3665	31.3	113,372	4.8	343,160	1.7
Emilia Romagna	235,621	3.6	87,006	7.0	18,418	14.0	3684	31.6	240,171	3.3	584,901	1.5
Toscana	542,856	2.2	120,637	5.4	21,318	12.9	6504	23.5	344,133	2.9	1035,448	1.1
Umbria	196,206	3.6	56,602	8.1	9144	23.8	0	–	128,353	5.0	390,305	1.6
Marche	134,942	4.6	54,635	8.8	8154	21.1	1112	57.8	92,925	5.7	291,767	2.1
Lazio	236,172	3.6	76,600	7.2	34,635	10.1	12,527	17.0	200,302	4.0	560,236	1.6
Abruzzo	181,604	4.0	96,307	6.2	27,936	11.1	9827	19.1	95,914	6.0	411,588	1.8
Molise	71,235	6.4	39,623	10.9	7420	22.4	1953	44.4	33,018	12.4	153,248	3.0
Campania	174,919	4.3	68,154	8.0	19,825	14.8	6529	30.1	134,499	4.8	403,927	2.1
Puglia	59,341	7.7	27,772	12.0	7729	22.0	777	70.5	46,730	9.1	142,349	4.0
Basilicata	98,919	6.4	32,545	11.6	13,025	16.7	2983	35.2	140,549	4.9	288,020	2.7
Calabria	266,196	3.4	42,695	10.7	16,355	16.7	3358	33.2	166,573	4.7	495,177	2.0
Sicilia	139,860	5.3	42,827	11.9	4897	27.6	3323	49.9	94,582	7.0	285,489	3.2
Sardegna	346,194	3.2	115,474	6.4	30,953	10.9	10,449	18.8	123,070	5.8	626,140	2.1
Italia	4,347,601	0.8	1,411,747	1.7	418,875	3.1	129,403	5.8	2,777,560	1.1	9,085,186	0.4

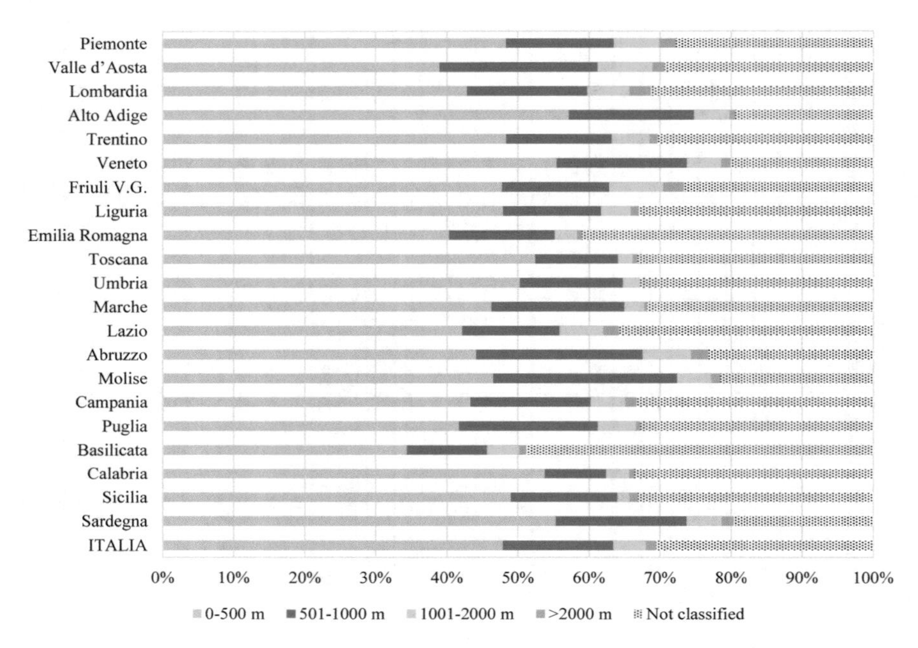

Fig. 8.2 Forest area percentage by horizontal distance between sampling point and closest road / Ripartizione percentuale della superficie del Bosco per classe di distanza orizzontale rispetto alla viabilità

is mainly in the first class but in three regions (Valle d'Aosta, Trentino and Lazio), the most frequent class is the third (ground very rough). At forest type level, only four types (Larch and Swiss stone pine, Hornbeam and Hophornbeam, Holm oak and Other evergreen broadleaved forest) have less than 50% of their forest area in the first class (Fig. 8.4), as this class is generally the most frequent.

Availability for wood supply is based on a synthetic evaluation of any significant limitations on forestry activities due to regulations or to physical features implying high costs for logging (Table 8.8). Regulatory restrictions include those deriving from protection of values, whether naturalistic (such as for integral reserves within parks), historic, cultural or of other types. Economical limitations may result from site features that make access to the area difficult (also due to lack of roads and tracks), steep slopes, ground roughness and economic constraints linked to the value of the forest products in relation to the market and local habits. Table 8.9 shows the area of Forest and of Other wooded land by availability for wood supply. Similar estimates about the inventory categories of Forest and of Other wooded land, as well as the forest types of the Tall trees forest are available at inventarioforestale.org/statistiche_INFC. At the national level, 88.5% of the Forest area is available for wood supply and the percentage is generally over 80% in the regions; exceptions are Valle d'Aosta (63.1%), Trentino (74.6%) and Friuli-Venezia Giulia (61.2%) (Fig. 8.5).

Table 8.4 Forest area by difference in altitude between sampling point and the closest road / Estensione del Bosco ripartito per classi di dislivello tra punto di campionamento e punto più vicino di viabilità ordinaria o forestale

Region/Regione	Forest/Bosco															
	0 ÷ + 100 m		0 ÷ −100 m		+101 ÷ + 400 m		−101 ÷ −400 m		> 400 m		>−400 m		Not classified		Total	
	Tra 0 e 100 m		Tra 0 e −100 m		Tra 101 e 400 m		Tra −101 e −400 m		Oltre 400 m		Oltre −400 m		Non classificato		Totale	
	Area	ES	Area	ES	Area	ES	Area	ES	Area	ES	Area	ES	Area	ES	Area	ES
	(ha)	(%)	(ha)	(%)	(ha)	(%)	(ha)	(%)	(ha)	(%)	(ha)	(%)	(ha)	(%)	(ha)	(%)
Piemonte	329,312	3.3	190,062	4.4	86,877	6.8	14,546	16.6	21,519	13.6	404	100.0	247,713	3.8	890,433	1.3
Valle d'Aosta	20,542	12.5	21,192	12.3	16,295	14.3	5009	27.1	7056	22.5	0	–	29,149	11.3	99,243	3.6
Lombardia	216,044	4.2	114,806	5.9	68,590	8.2	11,461	19.5	16,310	16.3	0	–	194,756	4.5	621,968	1.6
Alto Adige	121,167	4.9	83,756	6.1	48,086	8.8	15,937	17.2	3910	31.6	378	99.8	66,036	7.0	339,270	1.7
Trentino	118,091	4.8	74,159	6.7	53,238	7.8	6487	23.5	7568	21.7	360	100.0	113,357	4.9	373,259	1.4
Veneto	176,008	4.3	95,629	5.9	40,897	9.2	9721	19.4	10,469	18.7	374	99.9	83,606	6.7	416,704	1.9
Friuli V.G	113,547	5.0	63,027	7.8	46,419	8.9	9291	19.7	11,149	18.0	372	99.7	88,751	6.1	332,556	1.9
Liguria	128,281	4.8	63,390	7.0	27,855	11.0	8796	20.1	1466	49.6	0	–	113,372	4.8	343,160	1.7
Emilia Romagna	149,463	5.0	152,536	4.8	25,417	11.9	16,577	14.8	737	70.7	0	–	240,171	3.3	584,901	1.5
Toscana	347,223	3.0	285,219	3.3	33,941	10.8	21,318	12.9	3613	31.6	0	–	344,133	2.9	1,035,448	1.1
Umbria	117,199	5.2	124,476	5.2	12,166	17.2	8110	21.1	0	–	0	–	128,353	5.0	390,305	1.6
Marche	81,134	6.5	97,527	6.0	12,398	17.2	7042	22.8	741	70.8	0	–	92,925	5.7	291,767	2.1
Lazio	167,792	4.6	121,802	5.3	53,023	8.4	10,685	18.4	6632	23.5	0	–	200,302	4.0	560,236	1.6
Abruzzo	133,102	4.8	106,444	6.0	58,366	7.4	9772	19.1	7628	21.7	362	100.0	95,914	6.0	411,588	1.8
Molise	50,347	8.3	61,682	7.8	7029	23.1	1172	57.5	0	–	0	–	33,018	12.4	153,248	3.0
Campania	132,161	5.4	85,234	6.4	37,793	10.1	11,295	23.4	2946	35.3	0	–	134,499	4.8	403,927	2.1
Puglia	61,477	7.8	31,811	10.3	1165	57.5	1165	57.5	0	–	0	–	46,730	9.1	142,349	4.0
Basilicata	71,524	7.9	55,863	8.8	12,999	16.7	6712	23.4	373	99.9	0	–	140,549	4.9	288,020	2.7
Calabria	132,899	5.3	162,119	4.8	8955	20.3	20,589	13.3	3669	46.2	373	99.9	166,573	4.7	495,177	2.0

(continued)

Table 8.4 (continued)

Region/Regione	Forest/Bosco															
	0 ÷ + 100 m		0 ÷ −100 m		+101 ÷ + 400 m		−101 ÷ −400 m		> 400 m		>−400 m		Not classified		Total	
	Tra 0 e 100 m		Tra 0 e −100 m		Tra 101 e 400 m		Tra −101 e −400 m		Oltre 400 m		Oltre −400 m		Non classificato		Totale	
	Area	ES	Area	ES	Area	ES	Area	ES	Area	ES	Area	ES	Area	ES	Area	ES
	(ha)	(%)	(ha)	(%)	(ha)	(%)	(ha)	(%)	(ha)	(%)	(ha)	(%)	(ha)	(%)	(ha)	(%)
Sicilia	105,704	6.8	72,312	7.7	8720	20.7	4170	30.0	0	–	0	–	94,582	7.0	285,489	3.2
Sardegna	298,188	3.6	161,354	5.0	27,865	11.5	14,170	16.2	1120	57.8	373	100.0	123,070	5.8	626,140	2.1
Italia	3,071,206	1.1	2,224,399	1.3	688,094	2.3	214,025	4.3	106,906	6.1	2996	35.4	2,777,560	1.1	9,085,186	0.4

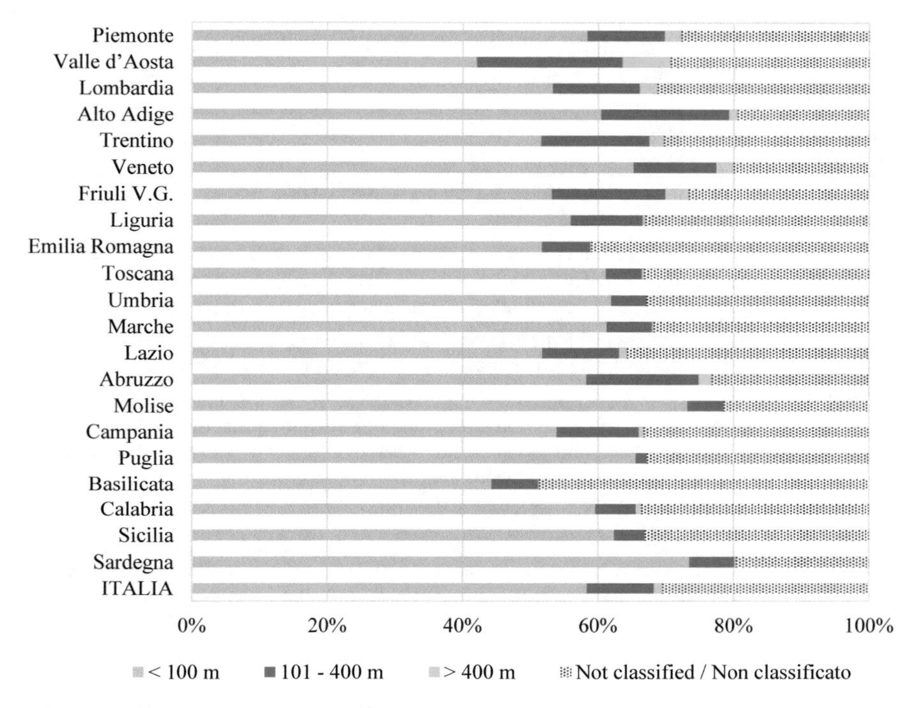

Fig. 8.3 Forest area percentage by difference in altitude between sampling point and closest road / Ripartizione percentuale della superficie del Bosco per classe di dislivello rispetto alla viabilità

The percentage of available area also exceeds 80% in the forest types, but it is 63.9% in the Larch and Swiss stone pine forest and just below 80% (79.1%) in the Scots pine and Mountain pine forest and in the Black pines forest (Fig. 8.6).

Data collected allows to estimate the areas not available for wood supply by cause of unavailability, as shown in Fig. 8.7. Unavailability is due to economic constraints in 67.8% of cases, to nature or territory protection in 27.3% of cases and to historic and cultural protection in a small percentage of cases (3.0%).

8.3 Silviculture and Forest Utilisations

The silvicultural practices were classified considering type and intensity of the applied interventions as shown in Table 8.10. Table 8.11 shows the area of Forest by silvicultural practice and Table 8.12 shows similar inventory statistics for the Tall trees forest and the Plantations. At the national level, 62.1% of the Forest area is involved in some types of silvicultural practices while 37.4% is not. Figure 8.8 shows that those percentages are variable among the regions.

Table 8.5 Terrain roughness classes / Classi per l'accidentalità del terreno

Terrain roughness	Description
Accidentalità del terreno	Descrizione
Smooth terrain	No obstacles or small obstacles but not exceeding one-quarter of the AdS25 area, with no influence on logging
Non accidentato	Assenza di ostacoli o presenza continua o discontinua di ostacoli per lo più piccoli, di dimensioni minori di 0.5 m, su una superficie non superiore a un quarto di AdS25 e normalmente non influenzanti le operazioni di prelievo
Rough terrain	Small obstacles (<0.5 m) between one and three-quarters of the AdS25 area, or large obstacles (≥0.5 m) on less than one-quarter of the AdS25 area
Accidentato	Ostacoli piccoli, di dimensioni minori di 0.5 m, presenti in modo continuo o discontinuo su una superficie estesa da un quarto a tre quarti di AdS25, oppure ostacoli grandi, di dimensioni uguali o maggiori di 0.5 m, che occupano meno di un quarto della superficie medesima
Very rough terrain	Small obstacles (<0.5 m) on more than three quarters of the AdS25 area, or large obstacles (≥0.5 m) on more than one-quarter of the AdS25 area
Molto accidentato	Ostacoli piccoli, di dimensioni minori di 0.5 m, presenti su oltre tre quarti della superficie di AdS25 oppure ostacoli grandi, di dimensioni uguali o maggiori di 0.5 m, che occupano più di un quarto della superficie medesima

Absence of practices, which mainly concern stands growing on awkward terrains or areas where utilisation is not cost effective, marks regional Forest areas from 12.0% (Alto Adige) to 72.0% (Sicilia). At the regional level, the absence of practices over a percentage of Forest area exceeding 50% was also observed in Valle d'Aosta, Liguria and Abruzzo, but other regions showed values only slightly lower than 50% (Piemonte, Friuli-Venezia Giulia, Puglia, Basilicata and Calabria). The practices more frequently applied in any region were ordinarily minimal, except Alto Adige and Trentino, where the ordinary classic denotes more than 50% of the Forest area. Hence, interventions are typically limited to harvesting wood or timber when mature or at the end of rotation.

The classic ordinary silvicultural practices, which include not only logging when the production cycle ends, but also actions aimed at improving the stands, involve 14.6% of Forest area at the national level. In the regions, only Alto Adige (53.1%), Trentino (55.6%), Veneto (30.8%), Toscana (18%) and Basilicata (19.7%) exceed that percentage. Intensive and special practices are infrequently applied and involve a limited percentage of Forest area. Ordinary intensive practices involve at least 3% of Forest area only in Piemonte (3.3%), Lombardia (4.0%) and Sardegna (4.1%) and in the last two regions they are due exclusively to presence of Plantations. Special practices for secondary products in Campania involve 4.7% of Forest area and in Sardegna 16.0%. The inventory statistics on Forest area by secondary products, shown in Table 8.13, allow for the evaluation that the special practices in Campania are related to chestnut production, while in Sardegna they are mainly due to cork production and to a lesser extent to forage and chestnut.

Table 8.6 Forest area by ground roughness / Estensione del Bosco ripartito per classi di accidentalità del terreno

Region/Regione	Forest/Bosco									
	Smooth		Rough		Very rough		Not classified		Total	
	Non accidentato		Accidentato		Molto accidentato		Non classificato		Totale	
	Area	ES	Area	ES	Area	ES	Area	ES	Area	ES
	(ha)	(%)	(ha)	(%)	(ha)	(%)	(ha)	(%)	(ha)	(%)
Piemonte	610,269	2.0	169,655	4.8	58,081	8.2	52,428	8.6	890,433	1.3
Valle d'Aosta	29,725	10.8	43,726	8.2	14,642	15.2	11,150	17.6	99,243	3.6
Lombardia	349,938	3.0	151,104	5.0	68,026	7.7	52,900	9.4	621,968	1.6
Alto Adige	230,194	2.9	65,607	7.3	19,878	13.6	23,590	12.5	339,270	1.7
Trentino	194,412	3.4	90,716	5.7	66,211	6.9	21,921	12.7	373,259	1.4
Veneto	241,387	3.4	91,568	5.8	49,818	8.7	33,931	11.3	416,704	1.9
Friuli V.G	214,489	3.2	50,129	8.1	29,732	10.8	38,207	10.4	332,556	1.9
Liguria	255,043	2.6	65,874	7.1	9048	19.7	13,194	16.3	343,160	1.7
Emilia Romagna	394,240	2.4	135,124	4.9	27,973	11.3	27,564	12.2	584,901	1.5
Toscana	703,389	1.7	182,627	4.3	46,250	8.7	103,181	5.8	1,035,448	1.1
Umbria	312,169	2.3	52,843	8.5	10,322	18.7	14,970	19.0	390,305	1.6
Marche	213,256	3.1	41,450	8.9	7042	22.8	30,020	10.7	291,767	2.1
Lazio	283,182	3.3	152,188	4.5	58,630	7.6	66,235	7.1	560,236	1.6
Abruzzo	279,278	2.9	70,216	6.9	28,016	11.1	34,077	10.0	411,588	1.8
Molise	124,510	4.1	17,963	14.0	4527	39.2	6248	24.5	153,248	3.0
Campania	204,301	3.9	93,796	6.1	44,806	9.6	61,024	7.7	403,927	2.1
Puglia	92,476	5.8	40,202	9.6	8157	21.4	1514	49.9	142,349	4.0
Basilicata	183,137	4.2	38,374	9.4	13,721	18.5	52,788	7.9	288,020	2.7
Calabria	268,690	3.3	119,583	5.9	39,426	10.5	67,478	8.5	495,177	2.0
Sicilia	180,654	4.6	68,828	8.5	12,858	16.9	23,148	16.3	285,489	3.2
Sardegna	322,657	3.4	161,301	4.9	100,831	6.4	41,350	9.8	626,140	2.1
Italia	5,687,396	0.7	1,902,874	1.4	717,996	2.3	776,920	2.3	9,085,186	0.4

Table 8.7 Other wooded land area by ground roughness / Estensione delle Altre terre boscate ripartite per classi di accidentalità del terreno

Region/Regione	Other wooded land/Altre terre boscate									
	Smooth		Rough		Very rough		Not classified		Total	
	Non accidentato		Accidentato		Molto accidentato		Non classificato		Totale	
	Area	ES	Area	ES	Area	ES	Area	ES	Area	ES
	(ha)	(%)	(ha)	(%)	(ha)	(%)	(ha)	(%)	(ha)	(%)
Piemonte	26,851	15.5	16,420	22.1	8424	21.8	33,295	12.9	84,991	8.0
Valle d'Aosta	0	–	1927	44.3	2136	68.5	4671	27.2	8733	24.0
Lombardia	11,130	19.5	15,614	20.8	12,920	22.3	30,588	13.8	70,252	8.7
Alto Adige	12,650	16.6	9401	24.3	3661	33.3	10,368	18.7	36,081	10.4
Trentino	3483	31.4	8436	24.8	9405	19.9	12,503	16.7	33,826	10.6
Veneto	7782	21.7	3430	32.4	2243	40.7	39,535	11.1	52,991	9.1
Friuli V.G	9911	22.8	4336	40.7	2230	40.6	24,581	14.8	41,058	10.6
Liguria	14,394	20.8	8047	21.0	2084	40.7	19,560	16.6	44,084	10.3
Emilia Romagna	20,734	17.1	6931	28.9	2879	55.2	23,370	12.0	53,915	9.3
Toscana	44,657	10.9	14,401	21.4	4671	38.2	90,545	6.7	154,275	5.2
Umbria	13,866	21.3	1376	50.3	1106	57.6	7303	27.6	23,651	15.2
Marche	9307	28.8	1423	49.8	1412	100.0	9171	22.2	21,314	16.2
Lazio	18,224	15.6	17,556	18.7	22,172	19.3	29,960	13.9	87,912	7.7
Abruzzo	33,738	13.7	10,256	24.8	3204	33.3	15,813	15.4	63,011	8.6
Molise	14,277	20.7	1953	44.4	0	–	3796	28.2	20,025	16.0
Campania	25,784	16.9	9393	29.4	8345	29.2	43,811	10.8	87,332	7.6
Puglia	25,572	15.2	14,610	22.1	3884	31.3	5323	25.3	49,389	9.9
Basilicata	41,412	11.3	9242	23.9	1859	44.6	51,879	8.7	104,392	6.2
Calabria	18,065	17.1	16,728	16.5	7462	22.2	113,188	5.7	155,443	4.8
Sicilia	33,726	14.6	23,674	13.4	11,252	21.2	33,093	13.6	101,745	7.1
Sardegna	246,829	4.0	205,586	4.4	130,675	5.4	91,762	6.6	674,851	2.0
Italia	632,392	2.8	400,740	3.5	242,026	4.4	694,114	2.5	1,969,272	1.4

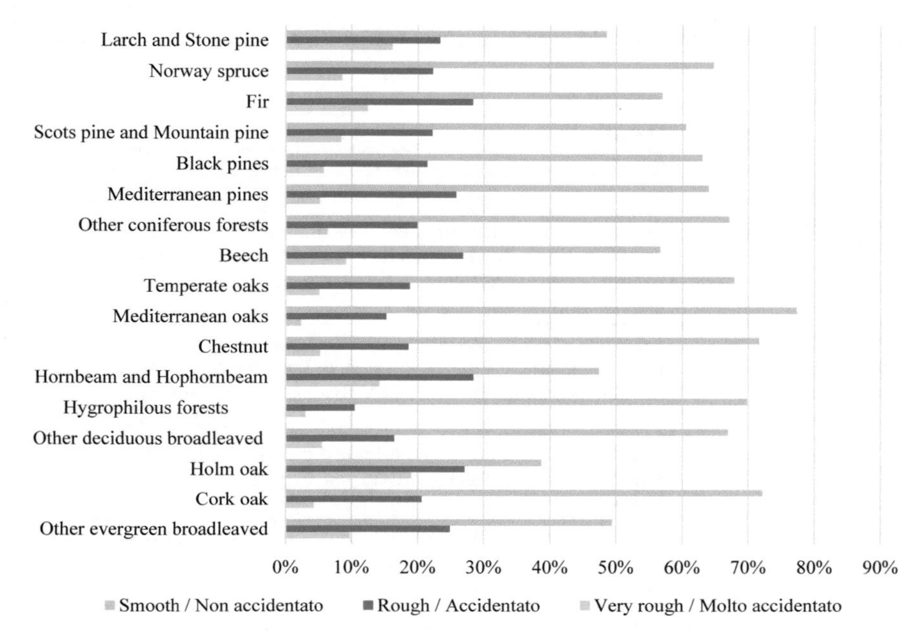

Fig. 8.4 Tall trees forest types percent of area by ground roughness / Ripartizione percentuale della superficie delle categorie forestali dei Boschi alti per accidentalità del terreno

Utilisation modes were surveyed in the case of ordinary practices. The utilisation modes are categorised based on the class system in Table 8.14. The estimates obtained are in Table 8.15 for Forest, and Table 8.16 for the inventory categories of Forest. Clearcutting with reserves for high forests, and with standards for coppice are the modes more frequently used at the national level (27.3%). This is also generally the case at the regional level (Fig. 8.9), with exceptions especially for the Alpine regions.

A survey of timber logging modes was made only for forests managed by ordinary silvicultural practices. Timber logging modes fell into the classes described in Table 8.17. Table 8.18 shows the estimates on Forest area by logging mode, while Table 8.19 shows those for the inventory categories of Forest. For Italy in general, wood/timber is extracted mainly by direct loading at the felling site (29.3%), but among the regions this mode prevails only in the southern-central regions (Islands included) except Calabria, where 32.4% of Forest area uses the skidding mode. In the Alpine regions, skidding is the predominant mode, except Valle d'Aosta, where logs are channelled downhill in 25.5% of the Forest area. Cable systems are used only in the North (down to Emilia-Romagna, but only in 1.0% of its Forest area), and helicopters are used only in Valle d'Aosta (1.4% of Forest area) and Alto Adige (0.8%) (Fig. 8.10).

Table 8.20 shows the inventory statistics on the growing stock annually removed through utilisations. During the twelve months before the inventory survey, growing stock was felled for an overall volume of almost 9.6 Mm^3, corresponding to 1.1 m^3 per hectare. The regions that report the highest absolute volumes harvested were

Table 8.8 Causes of unavailability for wood supply / Classi relative ai motivi di indisponibilità al prelievo legnoso

Causes of unavailability	Description
Motivi di indisponibilità	Descrizione
Land or nature protection Tutela delle risorse naturali e del territorio	Regulatory restrictions for protection of naturalistic value (e.g. integral reserves within parks), or for safeguarding protection forests
	Localizzazione in aree protette, zone di protezione di habitat e specie; boschi di protezione indiretta (per la tutela delle risorse idriche e del suolo) e diretta (protezione di abitati e manufatti da frane, smottamenti, valanghe, ecc.)
Social and cultural heritage protection Interesse sociale e culturale	Regulatory restrictions are applied to protect the historical, spiritual or the recreational value of the area, or the landscape; in other cases, the aim is to guarantee secondary goods or services production
	Presenza di restrizioni dovute all'esigenza di tutelare il valore estetico, storico, culturale, spirituale o ricreativo dell'area; restrizioni motivate dalla necessità di privilegiare la produzione di beni non legnosi o servizi
Economic reasons Limitazioni di carattere economico	Utilisation is not productive because of high costs due to difficult site conditions (accessibility, terrain steepness, etc.) or low value of the yield (unproductive stands)
	Presenza di condizioni che limitano in maniera significativa la convenienza economica delle utilizzazioni quali l'accessibilità, la pendenza e le condizioni del suolo; la convenienza economica considera le condizioni di mercato nel lungo periodo e non le fluttuazioni di breve periodo (es. situazioni contingenti sfavorevoli dovute a eccesso di prodotto sul mercato locale a causa di utilizzazioni consistenti a seguito di schianti)

Veneto, Piemonte, Trentino and Emilia-Romagna (all over 800,000 m^3) (Fig. 8.11). In terms of intensity, Veneto and Trentino are the only two regions that harvest more than 2 m^3 ha^{-1} (3.2 m^3 ha^{-1} and 2.9 m^3 ha^{-1}, respectively). Logging is very low in Molise (1923 m^3) and Puglia (2235 m^3). In terms of biomass, 6.3 MMg are annually removed from Forest, 0.7 Mg ha^{-1} on average, as shown in Table 8.21 and in Fig. 8.12.

8.4 Planning

Forest planning has been regulated at national level since Royal Decree No. 3267/23 was enforced, which dictated rules for forest utilisation in conformity with economic plans and Overall and Forest Police Regulations (PMPF) aiming principally at maintaining the hydro-geological equilibrium in mountain territories. Following a transfer to the regions of competence over forests (Presidential Decree No. 616/1977), the regulation of forest planning was shifted to these administrative agencies, except for the guideline and coordination functions, which remained under state rule. As a result,

Table 8.9 Forest and Other wooded land area by availability for wood supply / Estensione del Bosco e delle Altre terre boscate ripartiti per disponibilità al prelievo legnoso

Region/Regione	Forest /Bosco								Other wooded area /Altre terre boscate							
	Available		Not available		Not classified		Total		Available		Not available		Not classified		Total	
	Disponibile		Non disponibile		Non classificato		Totale		Disponibile		Non disponibile		Non classificato		Totale	
	Area	ES	Area	ES	Area	ES	Area	ES	Area	ES	Area	ES	Area	ES	Area	ES
	(ha)	(%)	(ha)	(%)	(ha)	(%)	(ha)	(%)	(ha)	(%)	(ha)	(%)	(ha)	(%)	(ha)	(%)
Piemonte	819,732	1.4	69,893	7.8	808	70.7	890,433	1.3	26,242	15.2	33,835	13.6	24,913	15.6	84,991	8.0
Valle d'Aosta	62,597	6.0	36,646	9.3	0	–	99,243	3.6	2521	60.0	2533	37.3	3680	31.2	8733	24.0
Lombardia	542,325	1.9	75,955	7.7	3688	57.8	621,968	1.6	25,047	12.9	24,386	16.6	20,819	17.8	70,252	8.7
Alto Adige	303,266	2.0	35,248	10.7	756	70.5	339,270	1.7	5221	36.2	26,878	11.5	3981	30.1	36,081	10.4
Trentino	278,280	2.3	94,979	5.5	0	–	373,259	1.4	2162	40.8	27,701	11.9	3963	29.2	33,826	10.6
Veneto	378,025	2.2	35,222	10.0	3457	58.0	416,704	1.9	5982	24.9	20,437	13.3	26,571	14.5	52,991	9.1
Friuli V.G	203,449	3.3	126,207	4.8	2901	68.1	332,556	1.9	1115	57.5	23,910	14.0	16,033	20.0	41,058	10.6
Liguria	323,133	1.9	19,660	13.2	367	99.2	343,160	1.7	12,828	16.5	12,797	22.6	18,460	17.3	44,084	10.3
Emilia Romagna	532,294	1.7	50,833	8.6	1774	82.0	584,901	1.5	12,393	21.8	21,839	16.5	19,682	13.1	53,915	9.3
Toscana	986,768	1.1	46,536	10.1	2144	70.4	1,035,448	1.1	41,453	10.8	37,846	12.2	74,976	7.5	154,275	5.2
Umbria	375,543	1.7	11,626	20.4	3136	62.7	390,305	1.6	10,324	21.8	7499	31.1	5828	32.2	23,651	15.2
Marche	286,578	2.1	5189	26.6	0	–	291,767	2.1	7554	31.0	4959	43.0	8800	22.7	21,314	16.2
Lazio	504,721	1.9	50,443	8.6	5072	26.8	560,236	1.6	44,140	12.3	22,998	12.4	20,774	18.1	87,912	7.7
Abruzzo	332,148	2.3	68,916	7.8	10,524	18.4	411,588	1.8	5605	24.7	44,516	11.4	12,890	17.2	63,011	8.6
Molise	148,562	3.2	4296	29.8	391	99.8	153,248	3.0	9650	27.8	7750	21.5	2624	31.8	20,025	16.0
Campania	329,434	2.6	55,036	8.4	19,457	15.0	403,927	2.1	17,677	14.2	39,835	13.1	29,821	14.1	87,332	7.6
Puglia	136,616	4.1	4607	39.0	1125	57.7	142,349	4.0	21,115	15.3	24,605	16.2	3669	30.1	49,389	9.9
Basilicata	275,715	2.8	12,304	17.2	0	–	288,020	2.7	55,486	9.2	26,457	12.4	22,449	14.7	104,392	6.2

(continued)

Table 8.9 (continued)

Region/Regione	Forest /Bosco								Other wooded area /Altre terre boscate							
	Available		Not available		Not classified		Total		Available		Not available		Not classified		Total	
	Disponibile		Non disponibile		Non classificato		Totale		Disponibile		Non disponibile		Non classificato		Totale	
	Area	ES	Area	ES	Area	ES	Area	ES	Area	ES	Area	ES	Area	ES	Area	ES
	(ha)	(%)	(ha)	(%)	(ha)	(%)	(ha)	(%)	(ha)	(%)	(ha)	(%)	(ha)	(%)	(ha)	(%)
Calabria	404,630	2.4	63,093	8.9	27,454	16.1	495,177	2.0	34,981	10.1	31,062	12.7	89,401	6.6	155,443	4.8
Sicilia	246,233	3.6	33,548	12.6	5709	49.1	285,489	3.2	48,900	9.8	37,539	12.8	15,306	24.3	101,745	7.1
Sardegna	567,611	2.3	56,724	9.0	1804	82.0	626,140	2.1	453,909	2.7	189,388	4.6	31,554	12.6	674,851	2.0
Italia	8,037,660	0.5	956,962	2.0	90,565	8.9	9,085,186	0.4	844,305	2.2	668,771	2.7	456,195	3.3	1,969,272	1.4

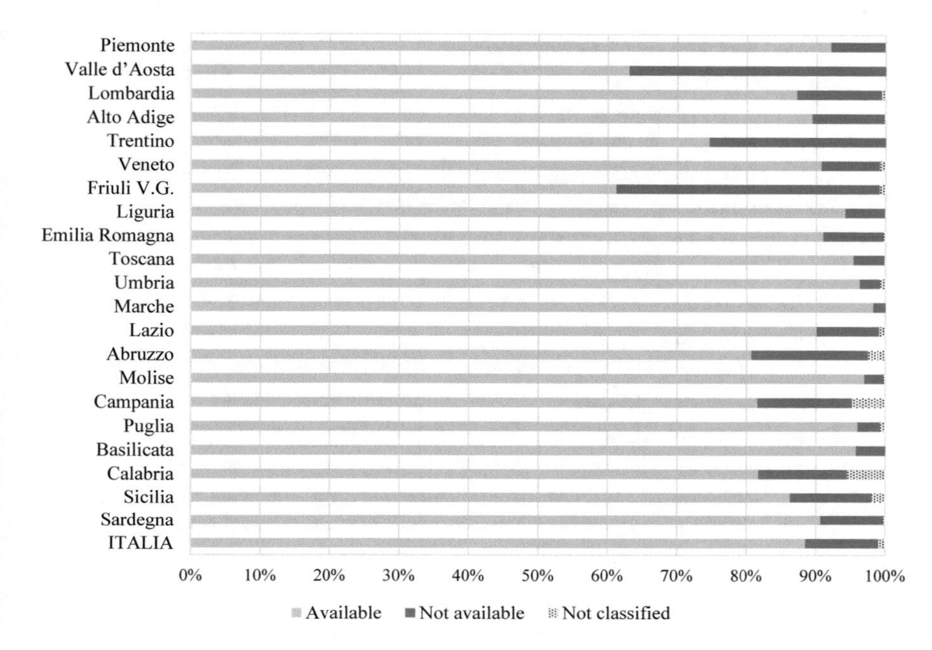

Fig. 8.5 Forest area percentage by availability for wood supply / Percentuale della superficie del Bosco per disponibilità al prelievo legnoso

current planning varies considerably in terms of both organisation and diffusion, according to the territorial district.

INFC records planning at three levels, as shown in Table 8.22. The PMPF are prescriptive rules for silvicultural activities and forest management in the administrative provinces and consequently can delineate the contours of forest planning. The presence of PMPF is recorded at the broadest level of planning. At the intermediate level, guideline planning typically relates to wider areas than single company properties, e.g., mountainous areas and natural parks even when land use is not forest.

Forest planning in its classic meaning refers to detailed management plans for forests falling under a single ownership, and this is the most detailed level recorded by INFC. In planning forms at different levels, more than one mode is recorded. Table 8.23 shows the inventory statistics on presence of planning for Forest and for Other wooded land; Table 8.24 shows those for Tall trees forest and Plantations, and Table 8.25 shows those for the inventory categories of Other wooded land. Table 8.26 shows the area of Forest and of Other wooded land with detailed plans, and the two that follow (Tables 8.27 and 8.28) show analogue statistics for the inventory categories of Forest and of Other wooded land. The estimates for Forest and Other wooded land by PMPF and those for the forest types of Forest and of Other wooded land by PMPF are available at inventarioforestale.org/statistiche_INFC. Estimates are also available for Forest and Other wooded land and their inventory categories, by

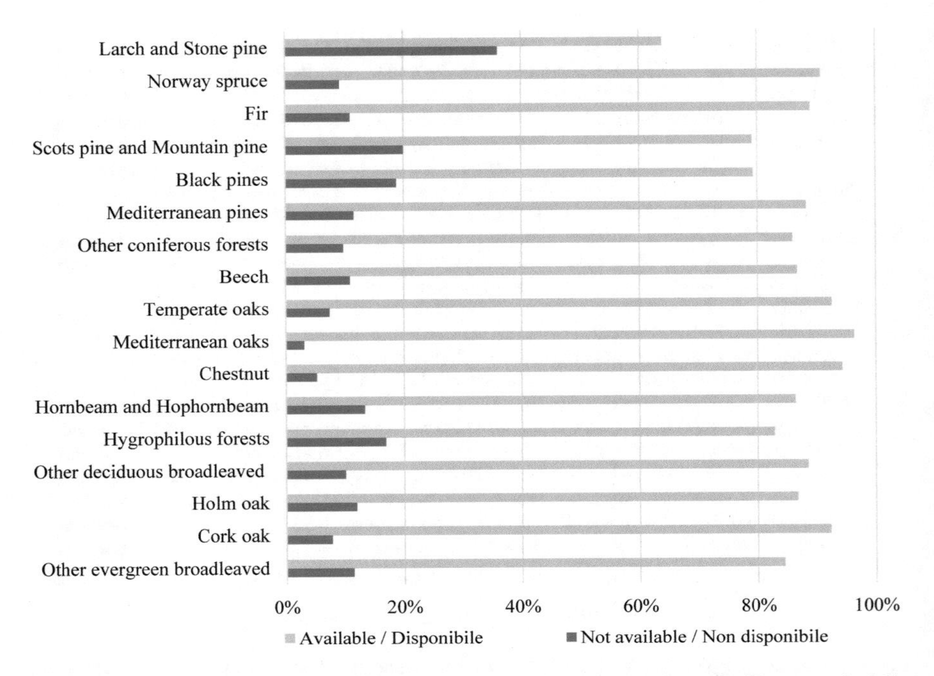

Fig. 8.6 Tall trees forest types percent of area by availability for wood supply / Percentuale della superficie delle categorie forestali dei Boschi alti, per disponibilità al prelievo legnoso

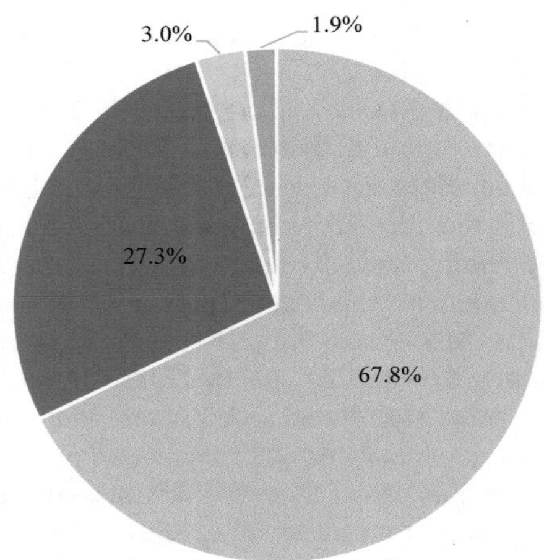

Fig. 8.7 Causes of unavailability for wood supply in Forest / Motivo di indisponibilità al prelievo legnoso nel Bosco

Table 8.10 Silvicultural practices applied to forest stands / Pratiche colturali applicate ai soprassuoli

Management methods	Description
Pratiche colturali	Descrizione
Ordinary silvicultural practices, minimal Pratiche colturali ordinarie, minimali	The product is harvested, but no cultivation or cultural action is taken (e.g. coppices cut at rotation time where no thinning is done); includes non-programmed utilisations after events, such as high winds and avalanches
	L'intervento si limita al taglio a fine turno, per la raccolta del prodotto, senza che vengano eseguiti interventi di coltivazione e/o cure colturali (es. sfolli e/o diradamenti); sono comprese le utilizzazioni non programmate, per esempio quelle effettuate in seguito a schianti e valanghe
Ordinary silvicultural practices, classic Pratiche colturali ordinarie, classiche	Traditional silviculture practices, i.e. harvesting at the end of the production cycle (rotation) but also cultural maintenance. Cultivation actions typical of agronomic practices usually applied in specialised plantations are excluded. Includes all cases where operations that favour regeneration and selection of future trees are associated with the final cut, such as cutting for seeding, secondary cutting, etc
	Pratiche improntate alla selvicoltura tradizionale; oltre alla raccolta del prodotto alla fine del ciclo produttivo (turno) vengono eseguite cure colturali, ma non sono previsti quegli interventi di coltivazione tipici delle pratiche agronomiche, di solito applicati nelle piantagioni specializzate. Rientrano nella sottoclasse tutti quei casi in cui in previsione del taglio finale si associano interventi atti a favorire l'insediamento della rinnovazione e la selezione delle piante portaseme, ad es. tagli di sementazione, secondari, ecc
Ordinary silvicultural practices, intensive Pratiche colturali ordinarie, intensive	Full cultivation procedures are followed. Practice is generally identified in plantations for timber production. During the production cycle, intensive care is taken such as soil treatment, fertilisation, pruning, etc
	Sono previsti interventi di coltivazione vera e propria, propri dell'arboricoltura da legno. Durante il ciclo di produzione si effettuano cure intensive, come ad esempio lavorazioni del terreno, concimazioni, potature, ecc
Special cultural practices for secondary products Pratiche colturali speciali per prodotti secondari	Concentrated on non-timber products, including chestnuts, cork, manna, etc
	La gestione è finalizzata alla produzione di prodotti non legnosi, tra cui castagne, sughero, manna, ecc

(continued)

Table 8.10 (continued)

Management methods	Description
Pratiche colturali	Descrizione
Special cultural practices for services Pratiche colturali speciali per servigi vari	Interventions target maximisation for tourist and recreational usage, landscape, naturalistic, faunistic and historic-cultural features of forests
	Gli interventi sono finalizzati a massimizzare le funzioni turistico-ricreative, paesaggistiche, naturalistiche, faunistiche e storico-culturali dei boschi. La finalità della gestione deve risultare evidente, nella reale pratica colturale: ad esempio un bosco in un parco naturale non è automaticamente attribuibile alla sottoclasse per la sola esistenza del parco; lo diventa però nel caso in cui venga attivamente destinato a tali servigi, ad esempio con la creazione di apposite strutture (aree pic-nic, punti panoramici, percorsi didattico-naturalistici, ecc). Solitamente si tratta di aree di estensione limitata
Other cultural practices Altre pratiche colturali	Includes all situations in which interventions do not fall explicitly within the above classes
	Comprende tutte quelle situazioni nelle quali gli interventi non rientrano esplicitamente in una delle due classi precedenti
No cultural practice Pratiche colturali assenti	Codifies all situations where no silviculture intervention of any type is foreseen. Includes cases of natural evolution stands, due to non-economic reasons (such as forests in integral natural reserves) and when utilizations are impossible (inaccessible or awkward terrains) or uneconomic (forests with negative harvesting return)
	Situazioni in cui non sono previsti interventi selvicolturali di nessun tipo, neppure le utilizzazioni finali. Comprende i casi dei soprassuoli ad evoluzione naturale, sia quando questa è dettata da motivazioni di natura giuridica (ad esempio boschi in riserve naturali integrali) sia nel caso in cui le utilizzazioni siano impraticabili (terreni inaccessibili o scomodi) o economicamente non convenienti (boschi a macchiatico negativo)

guideline planning. Finally, for the Tall trees forest types the estimates by presence of planning, PMPF, guideline planning and detailed plans are available.

In Italy, 92.2% of Forest area is regulated by at least one of the three planning levels considered; the percentage is 50.7% for Other wooded land, although this comparison should be evaluated considering that the not-classified area for Other wooded land is rather high (23.0%). Planning marks a consistent percentage of Forest area at the

Table 8.11 Forest area by management method / Estensione del Bosco ripartito per pratiche colturali

Region/Regione	Forest/Bosco																	
	Absent		Ordinary, minimal		Ordinary, classic		Ordinary, intensive		Special for secondary products		Special for services		Others		Not classified		Total	
	Assenti		Ordinarie minimali		Ordinarie classiche		Ordinarie intensive		Speciali per produzioni secondarie		Speciali per servigi vari		Altre		Non classificato		Totale	
	Area	ES	Area	ES	Area	ES	Area	ES	Area	ES	Area	ES	Area	ES	Area	ES	Area	ES
	(ha)	(%)	(ha)	(%)	(ha)	(%)	(ha)	(%)	(ha)	(%)	(ha)	(%)	(ha)	(%)	(ha)	(%)	(ha)	(%)
Piemonte	433,341	4.2	344,809	5.2	57,479	15.7	29,664	16.1	18,738	28.5	1542	100.0	3085	70.6	1774	50.8	890,433	1.3
Valle d'Aosta	53,178	10.6	41,716	12.5	2928	47.5	0	–	0	–	650	99.6	650	99.6	120	99.6	99,243	3.6
Lombardia	258,440	6.3	244,182	6.6	84,524	13.3	25,132	12.5	0	–	0	–	5351	57.7	4338	41.7	621,968	1.6
Alto Adige	40,595	16.7	98,082	9.7	180,016	5.8	0	–	9238	37.1	2810	61.4	5367	47.7	3161	45.7	339,270	1.7
Trentino	98,425	9.8	56,652	13.4	207,569	4.8	0	–	4177	57.7	1372	100.0	4571	46.1	492	100.0	373,259	1.4
Veneto	67,294	12.7	200,842	6.0	128,263	8.3	4933	35.5	0	–	0	–	8914	38.9	6459	27.1	416,704	1.9
Friuli V.G	161,348	6.7	107,096	9.3	39,651	16.9	9194	17.1	0	–	743	70.5	11,624	35.5	2901	68.1	332,556	1.9
Liguria	181,512	6.2	125,646	8.4	23,671	22.8	367	99.2	1423	99.2	1283	99.2	8132	40.9	1126	57.3	343,160	1.7
Emilia Romagna	189,459	7.1	315,490	4.6	53,020	15.5	8682	32.7	10,229	36.6	1936	60.4	4254	56.9	1831	43.8	584,901	1.5
Toscana	250,655	7.1	544,373	3.9	186,005	8.4	6783	30.3	21,957	25.0	3172	70.4	20,263	28.7	2240	38.1	1,035,448	1.1
Umbria	62,531	13.1	297,132	3.5	15,845	26.0	4974	33.7	442	99.8	1577	70.7	4998	47.4	2805	69.1	390,305	1.6
Marche	46,709	15.3	214,286	4.2	4661	51.7	6863	35.7	1112	57.8	0	–	18,137	25.1	0	–	291,767	2.1
Lazio	213,502	7.1	259,908	6.1	68,900	14.5	1756	45.4	5422	52.1	0	–	8651	43.1	2099	41.4	560,236	1.6
Abruzzo	223,591	5.2	112,095	9.2	57,216	13.3	4270	39.6	471	100.0	5817	50.5	6793	47.8	1336	57.4	411,588	1.8
Molise	29,514	17.1	106,321	6.3	3194	63.1	2715	27.7	0	–	0	–	11,503	31.2	0	–	153,248	3.0
Campania	139,638	8.3	187,976	6.7	46,925	15.8	3163	50.4	18,877	24.5	0	–	6242	49.0	1105	57.8	403,927	2.1
Puglia	69,190	9.4	56,015	11.4	14,706	24.7	0	–	0	–	0	–	1112	99.8	1326	50.1	142,349	4.0
Basilicata	140,791	7.3	83,858	10.7	56,646	14.0	1522	93.6	0	–	0	–	1908	81.6	3294	33.1	288,020	2.7

(continued)

Table 8.11 (continued)

Region/Regione	Forest/Bosco																
	Absent		Ordinary, minimal		Ordinary, classic		Ordinary, intensive		Special for secondary products		Special for services		Others		Not classified		Total	
	Assenti		Ordinarie minimali		Ordinarie classiche		Ordinarie intensive		Speciali per produzioni secondarie		Speciali per servigi vari		Altre		Non classificato		Totale	
	Area	ES	Area	ES	Area	ES	Area	ES	Area	ES	Area	ES	Area	ES	Area	ES	Area	ES
	(ha)	(%)	(ha)	(%)	(ha)	(%)	(ha)	(%)	(ha)	(%)	(ha)	(%)	(ha)	(%)	(ha)	(%)	(ha)	(%)
Calabria	247,220	6.4	181,505	8.3	48,098	16.6	4186	44.2	653	99.9	0	–	4088	63.0	9427	35.7	495,177	2.0
Sicilia	205,481	5.2	56,576	12.9	11,295	34.1	379	100.0	4802	40.6	0	–	3723	57.8	3233	63.1	285,489	3.2
Sardegna	283,144	5.4	128,223	9.1	33,279	19.6	25,885	11.8	100,176	8.3	4799	50.4	50,001	14.9	633	71.9	626,140	2.1
Italia	3,395,559	1.6	3,762,782	1.5	1,323,892	2.8	140,469	6.4	197,718	7.1	25,702	21.9	189,366	8.3	49,699	12.4	9,085,186	0.4

Table 8.12 Area of the inventory categories of Forest by management method / Estensione delle categorie inventariali del Bosco ripartite per pratiche colturali

Region/Regione	Tall trees forest/Boschi alti																	
	Absent		Ordinary, minimal		Ordinary, classic		Ordinary, intensive		Special for secondary products		Special for services		Others		Not classified		Total	
	Assenti		Ordinarie minimali		Ordinarie classiche		Ordinarie intensive		Speciali per produzioni secondarie		Speciali per servigi vari		Altre		Non classificato		Totale	
	Area	ES	Area	ES	Area	ES	Area	ES	Area	ES	Area	ES	Area	ES	Area	ES	Area	ES
	(ha)	(%)	(ha)	(%)	(ha)	(%)	(ha)	(%)	(ha)	(%)	(ha)	(%)	(ha)	(%)	(ha)	(%)	(ha)	(%)
Piemonte	433,341	4.2	344,809	5.2	57,479	15.7	9004	39.6	18,738	28.5	1542	100.0	3085	70.6	1774	50.8	869,773	1.3
Valle d'Aosta	53,178	10.6	41,716	12.5	2928	47.5	0	–	0	–	650	99.6	650	99.6	120	99.6	99,243	3.6
Lombardia	258,440	6.3	244,182	6.6	84,524	13.3	0	–	0	–	0	–	5351	57.7	4338	41.7	596,836	1.6
Alto Adige	40,595	16.7	98,082	9.7	180,016	5.8	0	–	9238	37.1	2810	61.4	5367	47.7	3161	45.7	339,270	1.7
Trentino	98,425	9.8	56,652	13.4	207,569	4.8	0	–	4177	57.7	1372	100.0	4571	46.1	492	100.0	373,259	1.4
Veneto	67,294	12.7	200,842	6.0	128,263	8.3	0	–	0	–	0	–	8914	38.9	6115	28.1	411,427	1.9
Friuli V.G.	161,348	6.7	107,096	9.3	39,651	16.9	0	–	0	–	743	70.5	11,624	35.5	2901	68.1	323,362	1.9
Liguria	181,512	6.2	125,646	8.4	23,671	22.8	0	–	1423	99.2	1283	99.2	8132	40.9	1126	57.3	342,793	1.7
Emilia Romagna	189,459	7.1	315,490	4.6	53,020	15.5	3002	15.5	10,229	36.6	1936	60.4	4254	56.9	1462	48.7	578,852	1.5
Toscana	250,655	7.1	544,373	3.9	186,005	8.4	0	–	21,957	25.0	3172	70.4	20,263	28.7	2240	38.1	1,028,665	1.1
Umbria	62,531	13.1	297,132	3.5	15,845	26.0	0	–	442	99.8	1577	70.7	4998	47.4	1402	99.8	383,928	1.6
Marche	46,709	15.3	214,286	4.2	4661	51.7	0	–	1112	57.8	0	–	18,137	25.1	0	–	284,904	2.1
Lazio	213,502	7.1	259,908	6.1	68,900	14.5	0	–	5422	52.1	0	–	8651	43.1	1678	45.3	558,060	1.6
Abruzzo	223,591	5.2	112,095	9.2	57,216	13.3	1298	100.0	471	100.0	5817	50.5	6793	47.8	1336	57.4	408,616	1.8
Molise	29,514	17.1	106,321	6.3	3194	63.1	0	–	0	–	0	–	11,503	31.2	0	–	150,533	3.1
Campania	139,638	8.3	187,976	6.7	46,925	15.8	0	–	18,877	24.5	0	–	6242	49.0	1105	57.8	400,763	2.1
Puglia	69,190	9.4	56,015	11.4	14,706	24.7	0	–	0	–	0	–	1112	99.8	1226	53.6	142,248	4.0
Basilicata	140,791	7.3	83,858	10.7	56,646	14.0	0	–	0	–	0	–	1908	81.6	3294	33.1	286,498	2.7
Calabria	247,220	6.4	181,505	8.3	48,098	16.6	2153	75.9	653	99.9	0	–	4088	63.0	9054	37.0	492,771	2.1
Sicilia	205,481	5.2	56,576	12.9	11,295	34.1	0	–	4802	40.6	0	–	3723	57.8	2854	70.3	284,731	3.2
Sardegna	283,144	5.4	128,223	9.1	33,279	19.6	0	–	100,176	8.3	4799	50.4	50,001	14.9	633	71.9	600,255	2.2
Italia	3,395,559	1.6	3,762,782	1.5	1,323,892	2.8	15,457	30.1	197,718	7.1	25,702	21.9	189,366	8.3	46,312	12.9	8,956,787	0.4

(continued)

Table 8.12 (continued)

Region/Regione	Plantations/Impianti di arboricoltura da legno																	
	Absent		Ordinary, minimal		Ordinary, classic		Ordinary, intensive		Special for secondary products		Special for services		Others		Not classified		Total	
	Assenti		Ordinarie minimali		Ordinarie classiche		Ordinarie intensive		Speciali per produzioni secondarie		Speciali per servigi vari		Altre		Non classificato		Totale	
	Area	ES	Area	ES	Area	ES	Area	ES	Area	ES	Area	ES	Area	ES	Area	ES	Area	ES
	(ha)	(%)	(ha)	(%)	(ha)	(%)	(ha)	(%)	(ha)	(%)	(ha)	(%)	(ha)	(%)	(ha)	(%)	(ha)	(%)
Piemonte	0	–	0	–	0	–	20,660	16.2	0	–	0	–	0	–	0	–	20,660	16.2
Valle d'Aosta	0	–	0	–	0	–	0	–	0	–	0	–	0	–	0	–	0	–
Lombardia	0	–	0	–	0	–	25,132	12.5	0	–	0	–	0	–	0	–	25,132	12.5
Alto Adige	0	–	0	–	0	–	0	–	0	–	0	–	0	–	0	–	0	–
Trentino	0	–	0	–	0	–	0	–	0	–	0	–	0	–	0	–	0	–
Veneto	0	–	0	–	0	–	4933	35.5	0	–	0	–	0	–	344	100.0	5277.0	33.7
Friuli V.G.	0	–	0	–	0	–	9194	17.1	0	–	0	–	0	–	0	–	9194	17.0
Liguria	0	–	0	–	0	–	367	99.2	0	–	0	–	0	–	0	–	367	99.2
Emilia Romagna	0	–	0	–	0	–	5680	33.0	0	–	0	–	0	–	368	100.0	6049	31.6
Toscana	0	–	0	–	0	–	6783	30.3	0	–	0	–	0	–	0	–	6783	30.3
Umbria	0	–	0	–	0	–	4974	33.7	0	–	0	–	0	–	1402	99.8	6377	33.7
Marche	0	–	0	–	0	–	6863	35.7	0	–	0	–	0	–	0	–	6863	35.7
Lazio	0	–	0	–	0	–	1756	45.4	0	–	0	–	0	–	420	100.0	2176	39.7
Abruzzo	0	–	0	–	0	–	2971	36.3	0	–	0	–	0	–	0	–	2971	36.3
Molise	0	–	0	–	0	–	2715	27.7	0	–	0	–	0	–	0	–	2715	27.7
Campania	0	–	0	–	0	–	3163	50.4	0	–	0	–	0	–	0	–	3163	50.4
Puglia	0	–	0	–	0	–	0	–	0	–	0	–	0	–	100	99.8	100	99.8
Basilicata	0	–	0	–	0	–	1522	93.6	0	–	0	–	0	–	0	–	1522	93.6
Calabria	0	–	0	–	0	–	2033	42.8	0	–	0	–	0	–	373	99.9	2406	39.3
Sicilia	0	–	0	–	0	–	379	100.0	0	–	0	–	0	–	379	100.0	758	70.7
Sardegna	0	–	0	–	0	–	25,885	11.8	0	–	0	–	0	–	0	–	25,885	11.8
Italia	0	–	0	–	0	–	125,012	6.2	0	–	0	–	0	–	3387	48.4	128,399	6.2

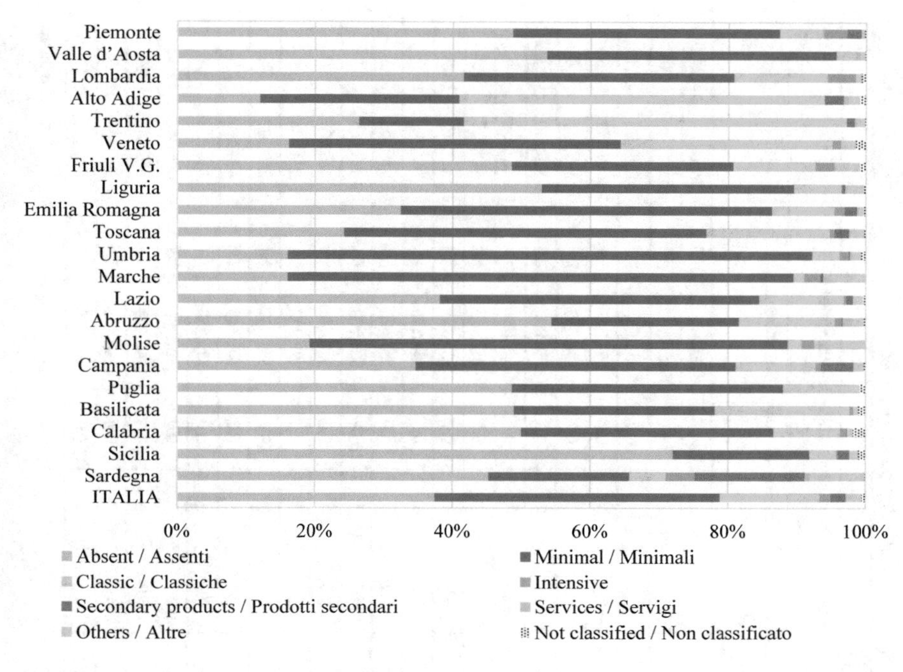

Fig. 8.8 Forest area percentage by management method / Ripartizione percentuale della superficie del Bosco per pratiche colturali

regional level also, as it generally exceeds 80%. Figure 8.13 shows the percentage of Forest area with planning at the three levels considered by INFC.

The role of PMPF is relevant in all regions as Forest area falls under their regulation; there is also little variability among the regions, compared to the two other levels of planning. In Trentino, Forest area under PMPF is apparently low because PMPF is actually replaced by analogue regulations (Trentino is an autonomous administration) and/or by the forest management plans, which have precedence over general regulations. Guideline planning only occurs in 2.3% of the Forest area and that percentage is exceeded in only seven regions: Valle d'Aosta, Lombardia, Veneto, Emilia-Romagna, Marche, Lazio and Sicilia. Detailed planning only occurs on 15.5% of the Italian Forest area. Only eight regions exceed that percentage (Valle d'Aosta, Lombardia, Alto Adige, Trentino, Veneto, Friuli-Venezia Giulia, Molise and Campania), generally in the North of the country. Alto Adige (94.6%) and Trentino (75.8%) are two exceptional cases in the percentage of Forest area with detailed plans, and they are the only two regions where the percentage exceeds 40%.

Table 8.13 Tall trees forest area by secondary products / Estensione dei Boschi alti ripartiti per tipo di prodotti secondari

Region/Regione	Tall trees forest/Boschi alti													
	Grassland (pasture)		Cork		Chestnut		Other products, edible		Other products, not edible		Not assessed		Total	
	Pascolo (foraggio)		Sughero		Castagne		Altri prodotti eduli del bosco e del sottobosco		Altri prodotti (non eduli)		Non valutato		Totale	
	Area	ES	Area	ES	Area	ES	Area	ES	Area	ES	Area	ES	Area	ES
	(ha)	(%)	(ha)	(%)	(ha)	(%)	(ha)	(%)	(ha)	(%)	(ha)	(%)	(ha)	(%)
Piemonte	0	0.0	0	0.0	17,196	29.8	0	0.0	1542	100.0	851,035	1.5	869,773	1.3
Valle d'Aosta	0	0.0	0	0.0	0	0.0	0	0.0	0	0.0	99,243	3.9	99,243	3.6
Lombardia	0	0.0	0	0.0	0	0.0	0	0.0	0	0.0	596,836	1.7	596,836	1.6
Alto Adige	9238	37.1	0	0.0	0	0.0	0	0.0	0	0.0	330,032	2.3	339,270	1.7
Trentino	4177	57.7	0	0.0	0	0.0	0	0.0	0	0.0	369,082	2.0	373,259	1.4
Veneto	0	0.0	0	0.0	0	0.0	0	0.0	0	0.0	411,427	2.0	411,427	1.9
Friuli V.G	0	0.0	0	0.0	0	0.0	0	0.0	0	0.0	323,362	2.2	323,362	1.9
Liguria	0	0.0	0	0.0	1423	99.2	0	0.0	0	0.0	341,370	2.2	342,793	1.7
Emilia Romagna	0	0.0	0	0.0	5589	48.0	4640	57.9	0	0.0	568,623	1.7	578,852	1.5
Toscana	4204	58.1	4095	30.4	9940	39.9	1807	100.0	1911	100.0	1,006,708	1.3	1,028,665	1.1
Umbria	0	0.0	0	0.0	442	99.8	0	0.0	0	0.0	383,486	1.9	383,928	1.6
Marche	0	0.0	0	0.0	1112	57.8	0	0.0	0	0.0	283,792	2.3	284,904	2.1
Lazio	0	0.0	368	100.0	5053	55.4	0	0.0	0	0.0	552,639	1.8	558,060	1.6
Abruzzo	0	0.0	0	0.0	471	100.0	0	0.0	0	0.0	408,146	2.0	408,616	1.8
Molise	0	0.0	0	0.0	0	0.0	0	0.0	0	0.0	150,533	3.4	150,533	3.1
Campania	0	0.0	0	0.0	18,877	24.5	0	0.0	0	0.0	381,886	2.5	400,763	2.1
Puglia	0	0.0	0	0.0	0	0.0	0	0.0	0	0.0	142,248	4.1	142,248	4.0
Basilicata	0	0.0	0	0.0	0	0.0	0	0.0	0	0.0	286,498	2.9	286,498	2.7
Calabria	0	0.0	653	99.9	0	0.0	0	0.0	0	0.0	492,118	2.2	492,771	2.1

(continued)

Table 8.13 (continued)

Region/Regione	Tall trees forest/Boschi alti													
	Grassland (pasture)		Cork		Chestnut		Other products, edible		Other products, not edible		Not assessed		Total	
	Pascolo (foraggio)		Sughero		Castagne		Altri prodotti eduli del bosco e del sottobosco		Altri prodotti (non eduli)		Non valutato		Totale	
	Area	ES	Area	ES	Area	ES	Area	ES	Area	ES	Area	ES	Area	ES
	(ha)	(%)	(ha)	(%)	(ha)	(%)	(ha)	(%)	(ha)	(%)	(ha)	(%)	(ha)	(%)
Sicilia	0	0.0	4802	40.6	0	0.0	0	0.0	0	0.0	279,929	3.4	284,731	3.2
Sardegna	4473	49.9	95,329	8.6	373	100.0	0	0.0	0	0.0	500,079	2.7	600,255	2.2
Italia	22,093	24.2	105,248	8.1	60,477	14.9	6447	50.2	3453	71.1	8,759,069	0.4	8,956,787	0.4

Table 8.14 Type of timber utilisation / Classi per le modalità delle utilizzazioni legnose

Type of timber utilisation Modalità di utilizzazione	Description Descrizione
Clearcutting or coppicing Taglio raso o ceduazione	On rotation, all trees are felled in a certain area
	Alla scadenza del turno vengono abbattuti tutti gli alberi su una determinata superficie
Clearcutting with reserves or coppicing with standards Taglio raso con riserve o ceduazione con rilascio di matricine	As above but on cutting some trees are left for seed production
	Come sopra, ma al momento del taglio viene rilasciato un certo numero di piante, per la produzione di seme
Patch cutting Taglio a buche	All trees are felled but the cutting area is usually circular and very limited (500–1000 m^2)
	Vengono abbattuti tutti gli alberi, ma la superficie di taglio, di solito tondeggiante, è molto limitata (500–1000 m^2)
Shelterwood cutting Tagli successivi	Cuts of various intensity and mode follow in time on the same woodland to favour birth and development of seed trees. Chronologically, these interventions include preparatory cutting, cutting for seeding, secondary cutting and removal cutting
	Tagli di intensità e modalità diversa che si susseguono nel tempo su una stessa superficie boscata, al fine di favorire nascita e sviluppo dei semenzali. Cronologicamente tali interventi si configurano in taglio di preparazione, taglio di sementazione, tagli secondari e taglio di sgombero
Selective, occasional, and uneven-aged cutting Tagli a scelta, saltuario, a sterzo	The stand consists of uneven-aged trees, also on a stump in the case of coppices. Not all the wood is used at once, but at each single intervention, mature trees (the desired economic product) and excess trees of smaller diameter classes are removed
	Il soprassuolo è costituito da alberi di diverse età, anche sulla stessa ceppaia nel caso dei cedui a sterzo. Il soprassuolo non viene utilizzato in un'unica occasione, ma con interventi periodici si effettua contemporaneamente il prelievo di alberi maturi (prodotto principale, economicamente perseguito) e di quelli eccedentari in classi diametriche più piccole (prodotto intercalare)
Other Altro	Includes cutting modes that are not classified among the above-mentioned types, and unclear situations. For instance, doubtful and not clearly classifiable cases in the silviculture practices could derive from occasional cutting on small surfaces (family cutting)
	Comprende i tipi di taglio non codificati nelle voci precedenti e le situazioni poco chiare. Ad esempio, casi dubbi o non bene codificabili nelle pratiche selvicolturali potrebbero derivare da tagli occasionali su piccole superfici (tagli famigliari)

Table 8.15 Forest area by utilisation system / Estensione del Bosco ripartito per modalità di utilizzazione del soprassuolo

Region/Regione	Forest/Bosco															
	Clearcutting or coppicing		Clearcutting with reserves or coppicing with standards		Patch cutting		Shelterwood cutting		Selective, occasional, and uneven-aged cutting		Others		Not assessed		Total	
	Taglio raso o ceduazione		Taglio raso con riserve o ceduazione con rilascio di matricine		Taglio a buche		Tagli successivi		Taglio a scelta, saltuario, a sterzo		Altro		Non valutato		Totale	
	Area	ES	Area	ES	Area	ES	Area	ES	Area	ES	Area	ES	Area	ES	Area	ES
	(ha)	(%)	(ha)	(%)	(ha)	(%)	(ha)	(%)	(ha)	(%)	(ha)	(%)	(ha)	(%)	(ha)	(%)
Piemonte	89,961	11.5	129,872	9.8	5314	47.1	5924	47.0	170,514	8.2	30,367	21.8	458,481	4.1	890,433	1.3
Valle d'Aosta	1074	71.9	0	–	1006	99.6	0	–	42,564	12.3	0	–	54,599	10.3	99,243	3.6
Lombardia	44,946	15.1	122,209	10.5	22,361	26.4	3215	64.4	121,526	10.1	39,581	20.2	268,129	6.1	621,968	1.6
Alto Adige	17,213	24.8	10,307	30.1	183,773	5.7	2805	70.3	36,142	18.0	27,857	20.6	61,172	12.8	339,270	1.7
Trentino	14,033	30.9	45,380	16.8	136,550	7.1	9820	37.0	33,586	19.6	24,852	20.9	109,038	9.1	373,259	1.4
Veneto	30,892	19.6	121,245	7.7	16,709	27.0	24,889	23.5	105,640	9.2	34,663	19.2	82,666	11.1	416,704	1.9
Friuli V.G	17,983	21.3	8229	40.6	20,110	23.5	31,651	19.6	20,752	24.6	57,216	13.5	176,616	6.1	332,556	1.9
Liguria	21,433	24.6	93,038	10.4	3893	45.9	2208	69.4	16,517	27.0	12,595	32.1	193,476	5.8	343,160	1.7
Emilia Romagna	29,979	20.1	268,277	5.2	3532	38.6	11,790	31.0	22,913	24.6	40,702	18.8	207,709	6.7	584,901	1.5
Toscana	66,082	15.3	503,583	3.9	14,508	28.8	31,478	21.5	46,241	18.4	75,270	14.0	298,287	6.3	1,035,448	1.1
Umbria	20,384	23.4	263,638	4.2	0	–	5430	35.1	20,483	23.7	8018	34.6	72,353	11.8	390,305	1.6
Marche	7353	34.7	137,245	6.8	5199	29.9	6009	45.7	65,808	12.2	4196	51.4	65,957	12.3	291,767	2.1
Lazio	13,320	34.9	280,938	5.6	1244	57.6	13,589	32.7	6276	45.7	15,196	30.7	229,673	6.8	560,236	1.6
Abruzzo	5719	38.8	69,250	12.2	1298	100.0	68,865	11.5	11,885	32.5	16,564	27.7	238,007	4.9	411,588	1.8
Molise	4905	36.2	65,857	10.1	0	–	20,637	24.0	11,247	26.6	9584	34.2	41,018	14.2	153,248	3.0
Campania	19,492	27.9	131,656	8.4	0	–	45,166	14.4	33,476	19.8	8274	40.2	165,863	7.3	403,927	2.1
Puglia	2713	57.9	37,601	13.7	0	–	4358	52.8	7564	34.8	18,485	22.8	71,628	9.1	142,349	4.0
Basilicata	3407	56.5	63,143	12.9	466	99.9	31,834	19.2	33,005	19.0	10,171	36.9	145,993	7.1	288,020	2.7
Calabria	52,526	16.3	53,574	19.0	0	–	1500	99.9	75,177	13.0	51,012	17.7	261,387	6.1	495,177	2.0

(continued)

Table 8.15 (continued)

Region/Regione	Forest/Bosco															
	Clearcutting or coppicing		Clearcutting with reserves or coppicing with standards		Patch cutting		Shelterwood cutting		Selective, occasional, and uneven-aged cutting		Others		Not assessed		Total	
	Taglio raso o ceduazione		Taglio raso con riserve o ceduazione con rilascio di matricine		Taglio a buche		Tagli successivi		Taglio a scelta, saltuario, a sterzo		Altro		Non valutato		Totale	
	Area	ES	Area	ES	Area	ES	Area	ES	Area	ES	Area	ES	Area	ES	Area	ES
	(ha)	(%)	(ha)	(%)	(ha)	(%)	(ha)	(%)	(ha)	(%)	(ha)	(%)	(ha)	(%)	(ha)	(%)
Sicilia	14,072	28.4	9508	34.1	2710	70.8	5699	42.0	25,711	19.7	10,549	34.8	217,239	4.9	285,489	3.2
Sardegna	38,678	14.4	67,790	13.6	552	100.0	2753	65.7	49,239	15.5	28,375	21.1	438,752	3.6	626,140	2.1
Italia	516,163	4.8	2,482,343	1.9	419,224	4.3	329,620	5.9	956,266	3.5	523,526	5.0	3,858,044	1.4	9,085,186	0.4

Table 8.16 Area of the inventory categories of Forest by utilisation system / Categorie inventariali del Bosco ripartite per modalità di utilizzazione del soprassuolo

Region/Regione	Tall trees forest/Boschi alti															
	Clearcutting or coppicing		Clearcutting with reserves or coppicing with standards		Patch cutting		Shelterwood cutting		Selective, occasional, and uneven–aged cutting		Others		Not assessed		Total	
	Taglio raso o ceduazione		Taglio raso con riserve o ceduazione con rilascio di matricine		Taglio a buche		Tagli successivi		Taglio a scelta, saltuario, a sterzo		Altro		Non valutato		Totale	
	Area	ES	Area	ES	Area	ES	Area	ES	Area	ES	Area	ES	Area	ES	Area	ES
	(ha)	(%)	(ha)	(%)	(ha)	(%)	(ha)	(%)	(ha)	(%)	(ha)	(%)	(ha)	(%)	(ha)	(%)
Piemonte	69,300	14.3	129,872	9.8	5314	47.1	5924	47.0	170,514	8.2	30,367	21.8	458,481	4.1	869,773	1.3
Valle d'Aosta	1074	71.9	0	–	1006	99.6	0	–	42,564	12.3	0	–	54,599	10.3	99,243	3.6
Lombardia	20,848	28.6	121,175	10.5	22,361	26.4	3215	64.4	121,526	10.1	39,581	20.2	268,129	6.1	596,836	1.6
Alto Adige	17,213	24.8	10,307	30.1	183,773	5.7	2805	70.3	36,142	18.0	27,857	20.6	61,172	12.8	339,270	1.7
Trentino	14,033	30.9	45,380	16.8	136,550	7.1	9820	37.0	33,586	19.6	24,852	20.9	109,038	9.1	373,259	1.4
Veneto	25,958	22.3	121,245	7.7	16,709	27.0	24,889	23.5	105,640	9.2	34,663	19.2	82,323	11.2	411,427	1.9
Friuli V.G.	9532	37.2	7858	42.3	20,110	23.5	31,651	19.6	20,752	24.6	56,844	13.6	176,616	6.1	323,362	1.9
Liguria	21,433	24.6	93,038	10.4	3893	45.9	2208	69.4	16,517	27.0	12,228	32.9	193,476	5.8	342,793	1.7
Emilia Romagna	24,298	23.5	268,277	5.2	3532	38.6	11,790	31.0	22,913	24.6	40,702	18.8	207,340	6.7	578,852	1.5
Toscana	59,667	16.6	503,583	3.9	14,508	28.8	31,478	21.5	46,241	18.4	74,901	14.0	298,287	6.3	1,028,665	1.1
Umbria	15,409	29.1	263,638	4.2	0	–	5430	35.1	20,483	23.7	8018	34.6	70,951	12.0	383,928	1.6
Marche	1903	76.2	137,245	6.8	5199	29.9	6009	45.7	65,808	12.2	2784	58.4	65,957	12.3	284,904	2.1
Lazio	11,984	38.4	280,938	5.6	1244	57.6	13,589	32.7	6276	45.7	14,776	31.5	229,253	6.8	558,060	1.6
Abruzzo	2747	70.5	69,250	12.2	1298	100.0	68,865	11.5	11,885	32.5	16,564	27.7	238,007	4.9	408,616	1.8
Molise	2580	64.2	65,857	10.1	0	–	20,637	24.0	11,247	26.6	9193	35.4	41,018	14.2	150,533	3.1
Campania	16,329	31.9	131,656	8.4	0	–	45,166	14.4	33,476	19.8	8274	40.2	165,863	7.3	400,763	2.1
Puglia	2713	57.9	37,601	13.7	0	–	4358	52.8	7564	34.8	18,485	22.8	71,528	9.2	142,248	4.0
Basilicata	3407	56.5	63,143	12.9	466	99.9	31,834	19.2	33,005	19.0	8649	40.2	145,993	7.1	286,498	2.7
Calabria	50,866	16.8	53,574	19.0	0	–	1500	99.9	75,177	13.0	50,639	17.9	261,014	6.1	492,771	2.1
Sicilia	13,693	29.1	9508	34.1	2710	70.8	5699	42.0	25,711	19.7	10,549	34.8	216,860	4.9	284,731	3.2
Sardegna	14,645	31.6	67,790	13.6	552	100.0	2753	65.7	49,239	15.5	26,523	22.0	438,752	3.6	600,255	2.2
Italia	399,633	6.0	2,480,937	1.9	419,224	4.3	329,620	5.9	956,266	3.5	516,450	5.1	3,854,657	1.4	8,956,787	0.4

(continued)

Table 8.16 (continued)

Region/Regione	Plantations/Impianti di arboricoltura da legno															
	Clearcutting or coppicing		Clearcutting with reserves or coppicing with standards		Patch cutting		Shelterwood cutting		Selective, occasional, and uneven–aged cutting		Others		Not assessed		Total	
	Taglio raso o ceduazione		Taglio raso con riserve o ceduazione con rilascio di matricine		Taglio a buche		Tagli successivi		Taglio a scelta, saltuario, a sterzo		Altro		Non valutato		Totale	
	Area	ES	Area	ES	Area	ES	Area	ES	Area	ES	Area	ES	Area	ES	Area	ES
	(ha)	(%)	(ha)	(%)	(ha)	(%)	(ha)	(%)	(ha)	(%)	(ha)	(%)	(ha)	(%)	(ha)	(%)
Piemonte	20,660	16.2	0	–	0	–	0	–	0	–	0	–	0	–	20,660	16.2
Valle d'Aosta	0	–	0	–	0	–	0	–	0	–	0	–	0	–	0	–
Lombardia	24,098	13.5	1034	100.0	0	–	0	–	0	–	0	–	0	–	25,132	12.5
Alto Adige	0	–	0	–	0	–	0	–	0	–	0	–	0	–	0	–
Trentino	0	–	0	–	0	–	0	–	0	–	0	–	0	–	0	–
Veneto	4933	35.5	0	–	0	–	0	–	0	–	0	–	344	99.9	5277	33.7
Friuli V.G.	8451	17.5	372	99.8	0	–	0	–	0	–	372	99.8	0	–	9194	17.0
Liguria	0	–	0	–	0	–	0	–	0	–	367	99.2	0	–	367	99.2
Emilia Romagna	5680	33.0	0	–	0	–	0	–	0	–	0	–	368	100.0	6049	31.6
Toscana	6415	31.7	0	–	0	–	0	–	0	–	368	100.0	0	–	6783	30.3
Umbria	4974	33.7	0	–	0	–	0	–	0	–	0	–	1402	99.8	6377	33.7
Marche	5450	38.5	0	–	0	–	0	–	0	–	1412	100.0	0	–	6863	35.7
Lazio	1335	50.8	0	–	0	–	0	–	0	–	420	100.0	420	100.0	2176	39.7
Abruzzo	2971	36.3	0	–	0	–	0	–	0	–	0	–	0	–	2971	36.3
Molise	2325	27.7	0	–	0	–	0	–	0	–	391	99.8	0	–	2715	27.7
Campania	3163	50.4	0	–	0	–	0	–	0	–	0	–	0	–	3163	50.4
Puglia	0	–	0	–	0	–	0	–	0	–	0	–	100	99.8	100	99.8
Basilicata	0	–	0	–	0	–	0	–	0	–	1522	93.6	0	–	1522	93.6
Calabria	1660	47.4	0	–	0	–	0	–	0	–	373	99.9	373	99.9	2406	39.3
Sicilia	379	100.0	0	–	0	–	0	–	0	–	0	–	379	100.0	758	70.7
Sardegna	24,033	13.0	0	–	0	–	0	–	0	–	1852	71.1	0	–	25,885	11.8
Italia	116,530	6.5	1406	78.2	0	–	0	–	0	–	7076	36.4	3387	48.4	128,399	6.2

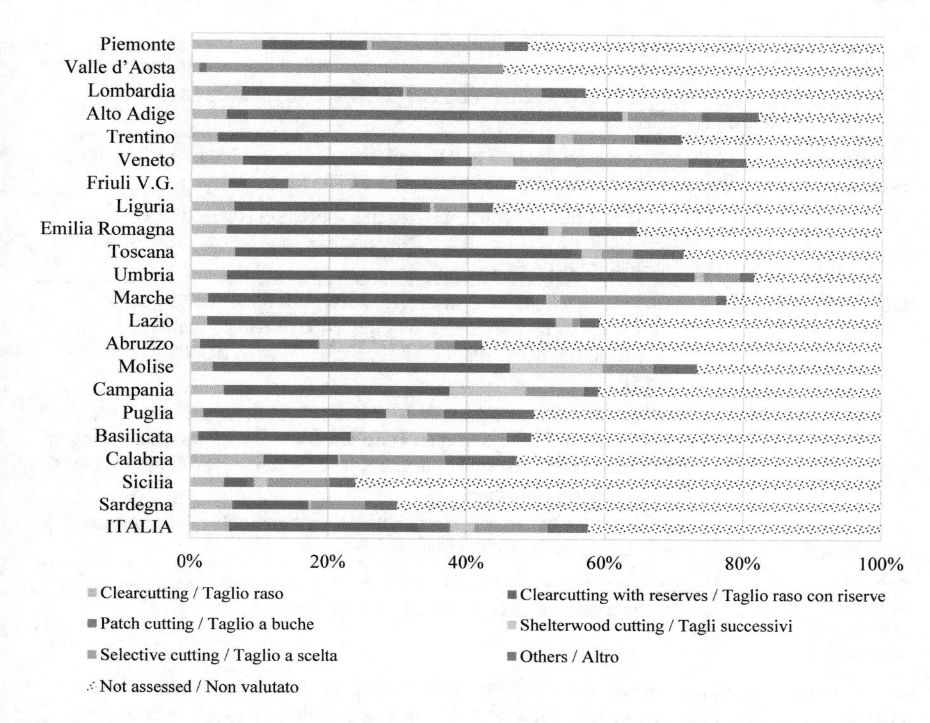

Fig. 8.9 Forest area by utilization mode / Ripartizione percentuale della superficie del Bosco per modalità di utilizzazioni forestali

Table 8.17 Logging modes / Classi per le modalità di esbosco

Logging modes Modalità di esbosco	Description Descrizione
Not classified Non classificato	For some reason, though in the presence of stands subject to silvicultural practices, the logging mode in use cannot be identified Per qualche motivo, pur in presenza di soprassuoli soggetti a pratiche colturali ordinarie o di altre pratiche, non si è in grado di valutare la modalità di esbosco in uso
Downhill Avvallamento	Logging mode where the timber slides downhill under gravity only. Includes cases of free sliding (along the slope or natural lines) and controlled sliding using artificial lines (polyethylene, masonry or timber channels) Modalità di esbosco che prevede lo scivolamento del legname a valle per sola forza di gravità. Vi rientrano i casi sia di avvallamento libero (lungo il pendio o linee naturali) sia controllato, mediante linee artificiali (risine in polietilene, muratura, legno)
Animal power or tractor without skidder A soma o con trattore senza strascico	Logging is done by loading the timber on pack animals or mechanical means directly at the felling site L'esbosco avviene mediante il carico del legname su animali da soma o mezzi meccanici, direttamente dal letto di caduta
Direct or indirect skidding by winch, sled, tractor or animals Strascico diretto o indiretto, mediante verricello, slitte, trattori o animali	Logging is carried out by skidding the timber over the terrain. Includes both direct skidding (the timber is dragged directly by animals or mechanical means) or indirect skidding (a stationary means of dragging the timber by winding a cable, generally with a winch) Modalità di esbosco che prevede il trascinamento del legname sul terreno. Vi rientrano i casi sia di strascico diretto (il legname è trascinato direttamente da animali o mezzi meccanici) sia indiretto (il mezzo è fermo e trascina il legname riavvolgendo la fune, in genere mediante verricello)
Cable systems Sistemi a fune	The timber is logged by overhead means (using cable systems such as cableways, draglines, cable cranes) Il legname viene esboscato per via aerea, mediante sistemi a fune, come teleferiche, fili a sbalzo, gru a cavo
Helicopter Elicottero	The timber is lifted out by helicopter Il legname viene esboscato mediante trasporto con elicottero

Table 8.18 Forest area by logging mode / Estensione del Bosco ripartito per modalità di esbosco

Region/Regione	Forest/Bosco													
	Downhill		Animal power or tractor without skidder		Direct or indirect skidding		Cable systems		Helicopter		Not assessed or not classified		Total	
	Avvallamento		A soma o con trattore senza strascico		Strascico diretto o indiretto		Sistemi a fune		Elicottero		Non valutato o non classificato		Totale	
	Area	ES	Area	ES	Area	ES	Area	ES	Area	ES	Area	ES	Area	ES
	(ha)	(%)	(ha)	(%)	(ha)	(%)	(ha)	(%)	(ha)	(%)	(ha)	(%)	(ha)	(%)
Piemonte	45,996	17.7	134,494	9.7	197,676	7.7	47,661	16.7	0	–	464,607	4.0	890,433	1.3
Valle d'Aosta	25,291	18.2	2080	60.7	10,588	28.9	4578	51.6	1385	99.6	55,321	10.2	99,243	3.6
Lombardia	34,715	21.3	76,551	12.4	190,141	8.0	47,767	18.0	0	–	272,794	6.0	621,968	1.6
Alto Adige	40,020	17.2	16,731	27.8	144,072	7.4	71,434	12.3	2827	70.5	64,186	12.5	339,270	1.7
Trentino	61,389	13.6	6064	42.7	124,585	8.5	71,607	12.0	0	–	109,614	9.1	373,259	1.4
Veneto	71,958	12.3	44,711	15.5	158,176	7.4	59,192	14.3	0	–	82,666	11.1	416,704	1.9
Friuli V.G	11,634	35.1	26,292	19.8	67,634	12.4	50,381	14.6	0	–	176,616	6.1	332,556	1.9
Liguria	24,766	23.1	9871	35.7	106,547	9.5	7746	42.2	0	–	194,230	5.8	343,160	1.7
Emilia Romagna	67,030	13.6	250,309	5.7	53,886	15.7	5967	47.0	0	–	207,709	6.7	584,901	1.5
Toscana	70,808	14.7	591,863	3.5	74,490	13.9	0	–	0	–	298,287	6.3	1,035,448	1.1
Umbria	11,429	34.4	287,205	3.8	18,810	26.2	0	–	0	–	72,860	11.8	390,305	1.6
Marche	80,019	10.9	95,392	9.7	48,534	15.3	0	–	0	–	67,822	12.1	291,767	2.1
Lazio	5608	58.5	293,956	5.4	30,108	21.4	0	–	0	-	230,564	6.8	560,236	1.6
Abruzzo	8670	38.6	164,791	6.5	120	100.0	0	–	0	–	238,007	4.9	411,588	1.8
Molise	4323	57.0	103,397	6.7	4511	51.3	0	–	0	–	41,018	14.2	153,248	3.0
Campania	1495	100.2	133,262	8.8	103,307	10.4	0	–	0	–	165,863	7.3	403,927	2.1
Puglia	1901	62.5	65,008	10.1	3345	57.7	0	–	0	–	72,094	9.1	142,349	4.0
Basilicata	1486	79.2	133,455	7.6	4170	54.0	0	–	0	–	148,908	7.0	288,020	2.7
Calabria	4440	99.9	64,635	16.9	160,543	8.2	0	–	0	–	265,558	6.0	495,177	2.0
Sicilia	843	100.0	47,043	14.6	17,950	25.1	0	–	0	–	219,653	4.8	285,489	3.2
Sardegna	41,688	17.9	116,338	9.4	22,822	22.5	0	–	0	–	445,292	3.5	626,140	2.1
Italia	615,508	4.5	2,663,451	1.7	1,542,014	2.6	366,332	5.7	4211	57.6	3,893,669	1.4	9,085,186	0.4

Table 8.19 Area of the inventory categories of Forest by logging mode / Estensione delle categorie inventariali del Bosco ripartite per modalità di esbosco

Region/Regione	Tall trees forest/Boschi alti													
	Downhill		Animal power or tractor without skidder		Direct or indirect skidding		Cable systems		Helicopter		Not assessed or not classified		Total	
	Avvallamento		A soma o con trattore senza strascico		Strascico diretto o indiretto		Sistemi a fune		Elicottero		Non valutato o non classificato		Totale	
	Area	ES	Area	ES	Area	ES	Area	ES	Area	ES	Area	ES	Area	ES
	(ha)	(%)	(ha)	(%)	(ha)	(%)	(ha)	(%)	(ha)	(%)	(ha)	(%)	(ha)	(%)
Piemonte	45,996	17.7	125,241	10.1	186,269	8.0	47,661	16.7	0	–	464,607	4.0	869,773	1.3
Valle d'Aosta	25,291	18.2	2080	60.7	10,588	28.9	4578	51.6	1385	99.6	55,321	10.2	99,243	3.6
Lombardia	34,715	21.3	52,453	17.0	190,141	8.0	47,767	18.0	0	–	271,760	6.1	596,836	1.6
Alto Adige	40,020	17.2	16,731	27.8	144,072	7.4	71,434	12.3	2827	70.5	64,186	12.5	339,270	1.7
Trentino	61,389	13.6	6064	42.7	124,585	8.5	71,607	12.0	0	–	109,614	9.1	373,259	1.4
Veneto	71,958	12.3	39,778	16.9	158,176	7.4	59,192	14.3	0	–	82,323	11.2	411,427	1.9
Friuli V.G.	11,634	35.1	19,170	26.2	65,561	12.7	50,381	14.6	0	–	176,616	6.1	323,362	1.9
Liguria	24,766	23.1	9505	36.9	106,547	9.5	7746	42.2	0	–	194,230	5.8	342,793	1.7
Emilia Romagna	67,030	13.6	244,629	5.8	53,886	15.7	5967	47.0	0	–	207,340	6.7	578,852	1.5
Toscana	70,808	14.7	586,821	3.5	72,748	14.1	0	–	0	–	298,287	6.3	1,028,665	1.1
Umbria	11,429	34.4	282,231	3.8	18,810	26.2	0	–	0	–	71,458	11.9	383,928	1.6
Marche	78,607	11.0	94,922	9.8	44,093	16.2	0	–	0	–	67,283	12.2	284,904	2.1
Lazio	5608	58.5	292,989	5.4	29,740	21.7	0	–	0	–	229,724	6.8	558,060	1.6
Abruzzo	8670	38.6	161,820	6.6	120	100.0	0	–	0	–	238,007	4.9	408,616	1.8
Molise	4323	57.0	100,682	6.9	4511	51.3	0	–	0	–	41,018	14.2	150,533	3.1
Campania	1495	100.0	132,876	8.8	100,530	10.6	0	–	0	–	165,863	7.3	400,763	2.1
Puglia	1901	62.5	65,008	10.1	3345	57.7	0	–	0	–	71,994	9.1	142,248	4.0
Basilicata	1486	79.2	133,455	7.6	4170	54.0	0	–	0	–	147,387	7.0	286,498	2.7
Calabria	4440	99.9	64,635	16.9	158,510	8.3	0	–	0	–	265,185	6.0	492,771	2.1
Sicilia	843	100.0	46,664	14.7	17,950	25.1	0	–	0	–	219,274	4.8	284,731	3.2
Sardegna	40,023	18.5	92,951	11.3	21,989	23.1	0	–	0	–	445,292	3.5	600,255	2.2
Italia	612,430	4.5	2,570,705	1.8	1,516,341	2.7	366,332	5.7	4211	57.6	3,886,767	1.4	8,956,787	0.4

(continued)

Table 8.19 (continued)

Region/Regione	Plantations/Impianti di arboricoltura da legno													
	Downhill		Animal power or tractor without skidder		Direct or indirect skidding		Cable systems		Helicopter		Not assessed or not classified		Total	
	Avvallamento		A soma o con trattore senza strascico		Strascico diretto o indiretto		Sistemi a fune		Elicottero		Non valutato o non classificato		Totale	
	Area	ES	Area	ES	Area	ES	Area	ES	Area	ES	Area	ES	Area	ES
	(ha)	(%)	(ha)	(%)	(ha)	(%)	(ha)	(%)	(ha)	(%)	(ha)	(%)	(ha)	(%)
Piemonte	0	–	9253	31.2	11407	30.1	0	–	0	–	0	–	20,660	16.2
Valle d'Aosta	0	–	0	–	0	–	0	–	0	–	0	–	0	–
Lombardia	0	–	24,098	13.5	0	–	0	–	0	–	1034	100.0	25,132	12.5
Alto Adige	0	–	0	–	0	–	0	–	0	–	0	–	0	–
Trentino	0	–	0	–	0	–	0	–	0	–	0	–	0	–
Veneto	0	–	4933	35.5	0	–	0	–	0	–	344	99.9	5277.0	33.7
Friuli V.G.	0	–	7122	20.2	2072	39.0	0	–	0	–	0	–	9194	17.0
Liguria	0	–	367	99.2	0	–	0	–	0	–	0	–	367	99.2
Emilia Romagna	0	–	5680	33.0	0	–	0	–	0	–	368	100.0	6049	31.6
Toscana	0	–	5041	34.8	1742	81.9	0	–	0	–	0	–	6783	30.3
Umbria	0	–	4974	33.8	0	–	0	–	0	–	1402	99.8	6377	33.7
Marche	1412	100.0	470	81.8	4441	46.8	0	–	0	–	539	100.0	6863	35.7
Lazio	0	–	967	58.9	368	100.0	0	–	0	–	840	63.6	2176	39.7
Abruzzo	0	–	2971	36.3	0	–	0	–	0	–	0	–	2971	36.3
Molise	0	–	2715	27.7	0	–	0	–	0	–	0	–	2715	27.7
Campania	0	–	387	74.9	2777	56.7	0	–	0	–	0	–	3163	50.4
Puglia	0	–	0	–	0	–	0	–	0	–	100	99.8	100	99.8
Basilicata	0	–	0	–	0	–	0	–	0	–	1522	93.6	1522	93.6
Calabria	0	–	0	–	2033	42.8	0	–	0	–	373	99.9	2406	39.3
Sicilia	0	–	379	100.0	0	–	0	–	0	–	379	100.0	758	70.7
Sardegna	1665	69.1	23,387	13.1	832	100.0	0	–	0	–	0	–	25,885	11.8
Italia	3077	59.3	92,746	7.3	25,673	18.6	0	–	0	–	6902	36.0	128,399	6.2

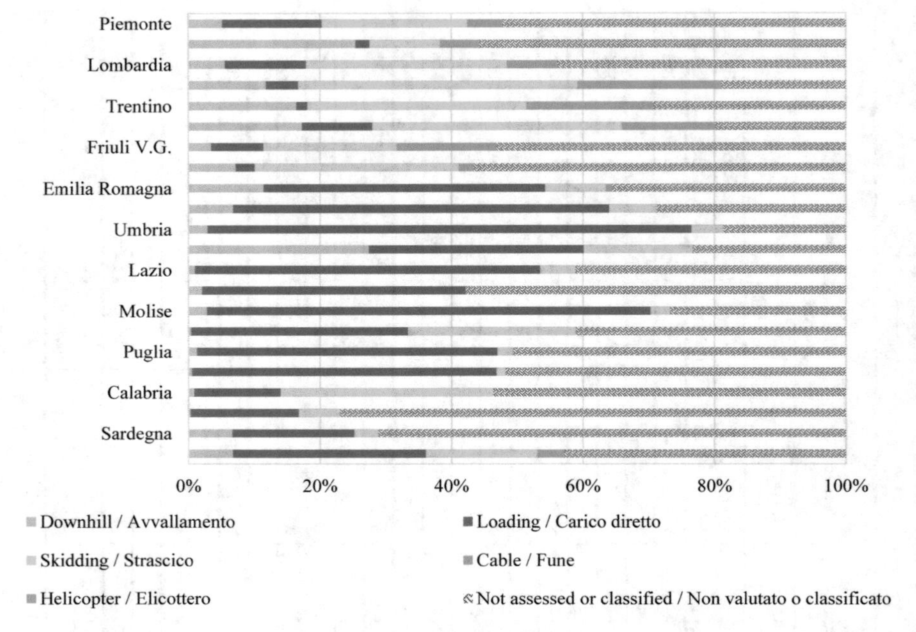

Fig. 8.10 Forest area by logging mode / Ripartizione percentuale della superficie del Bosco per modalità di esbosco

Table 8.20 Growing stock removed annually from the inventory categories of Forest, total and per hectare / Valori totali e per ettaro del volume legnoso utilizzato annualmente nelle categorie inventariali del Bosco

Region/Regione	Tall trees forest				Plantations				Total Forest			
	Boschi alti				Impianti di arboricoltura da legno				Totale Bosco			
	Volume	ES	Volume	ES	Volume	ES	Volume	ES	Volume	ES	Volume	ES
	(m^3)	(%)	(m^3ha^{-1})	(%)	(m^3)	(%)	(m^3ha^{-1})	(%)	(m^3)	(%)	(m^3ha^{-1})	(%)
Piemonte	1,075,839	45.6	1.2	45.6	168	100.0	0.0	–	1,076,007	45.6	1.2	45.6
Valle d'Aosta	42,910	49.5	0.4	49.3	0	–	0.0	–	42,910	49.5	0.4	49.3
Lombardia	110,055	47.3	0.2	47.2	0	–	0.0	–	110,055	47.3	0.2	47.3
Alto Adige	554,315	40.0	1.6	40.0	0	–	0.0	–	554,315	40.0	1.6	40.0
Trentino	1,066,069	70.4	2.9	70.4	0	–	0.0	–	1,066,069	70.4	2.9	70.4
Veneto	1,314,538	45.8	3.2	45.7	1337	92.9	0.3	92.6	1,315,875	45.7	3.2	45.7
Friuli V.G	219,472	60.8	0.7	60.7	2424	85.6	0.3	84.2	221,896	60.1	0.7	60.1
Liguria	652,739	63.2	1.9	63.1	0	–	0.0	–	652,739	63.2	1.9	63.1
Emilia Romagna	951,999	69.2	1.6	69.2	0	–	0.0	–	951,999	69.2	1.6	69.2
Toscana	645,699	55.2	0.6	55.2	14,500	100.0	2.1	–	660,199	54.0	0.6	54.0
Umbria	607,485	59.7	1.6	59.7	0	–	0.0	–	607,485	59.7	1.6	59.7
Marche	385,688	48.3	1.4	48.3	2114	100.0	0.3	–	387,802	48.1	1.3	48.0
Lazio	449,380	74.4	0.8	74.4	0	–	0.0	–	449,380	74.4	0.8	74.4
Abruzzo	208,144	64.5	0.5	64.5	0	–	0.0	–	208,144	64.5	0.5	64.5
Molise	1923	100.0	0.0	–	0	–	0.0	–	1923	100.0	0.0	–
Campania	397,081	37.2	1.0	37.1	0	–	0.0	–	397,081	37.2	1.0	37.1
Puglia	2235	100.0	0.0	–	0	–	0.0	–	2235	100.0	0.0	–
Basilicata	154,670	60.4	0.5	60.3	0	–	0.0	–	154,670	60.4	0.5	60.3
Calabria	421,515	50.1	0.9	50.1	0	–	0.0	–	421,515	50.1	0.9	50.1
Sicilia	26,514	77.7	0.1	77.7	0	–	0.0	–	26,514	77.7	0.1	77.7
Sardegna	215,643	65.6	0.4	65.6	41,802	77.1	1.6	75.2	257,445	56.4	0.4	56.3
Italia	9,503,913	16.2	1.1	16.2	62,344	56.9	0.5	56.6	9,566,257	16.1	1.1	16.1

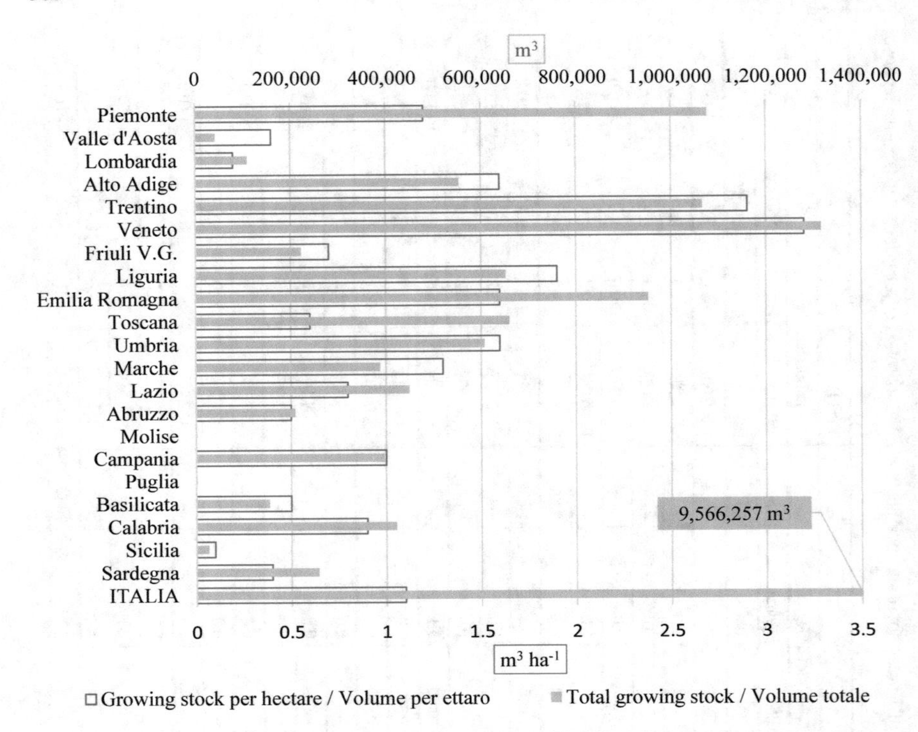

Fig. 8.11 Growing stock removed annually, total and per hectare (to improve readability, very long bars have been limited and true values are given in numbers) / Volume totale e per ettaro utilizzato annualmente (per migliorare la lettura, è stata limitata la lunghezza di barre molto lunghe e i valori reali sono indicati con numero)

Table 8.21 Biomass removed annually from the inventory categories of Forest, total and per hectare / Valori totali e per ettaro della fitomassa utilizzata annualmente per le categorie inventariali del Bosco

Region/Regione	Tall trees forest				Plantations				Total Forest			
	Boschi alti				Impianti di arboricoltura da legno				Totale Bosco			
	Biomass Fitomassa	ES	Biomass Fitomassa	ES	Biomass Fitomassa	ES	Biomass Fitomassa	ES	Biomass Fitomassa	ES	Biomass Fitomassa	ES
	(Mg)	(%)	(Mg ha^{-1})	(%)	(Mg)	(%)	(Mg ha^{-1})	(%)	(Mg)	(%)	(Mg ha^{-1})	(%)
Piemonte	714,444	43.1	0.8	43.1	169	100.0	0.0	–	714,613	43.1	0.8	43.1
Valle d'Aosta	28,489	48.8	0.3	48.7	0	–	0.0	–	28,489	48.8	0.3	48.7
Lombardia	80,277	49.6	0.1	49.6	0	–	0.0	–	80,277	49.6	0.1	49.6
Alto Adige	290,692	39.6	0.9	39.6	0	–	0.0	–	290,692	39.6	0.9	39.6
Trentino	543,674	68.4	1.5	68.3	0	–	0.0	–	543,674	68.4	1.5	68.3
Veneto	762,655	40.5	1.9	40.4	825	88.5	0.2	88.2	763,480	40.5	1.8	40.4
Friuli V.G	125,628	55.5	0.4	55.4	1510	81.1	0.2	79.6	127,138	54.8	0.4	54.8
Liguria	372,088	61.0	1.1	61.0	0	–	0.0	–	372,088	61.0	1.1	61.0
Emilia Romagna	752,847	68.6	1.3	68.6	0	–	0.0	–	752,847	68.6	1.3	68.6
Toscana	436,410	51.7	0.4	51.7	11,197	100.0	1.7	–	447,607	50.5	0.4	50.5
Umbria	474,724	59.3	1.2	59.3	0	–	0.0	–	474,724	59.3	1.2	59.3
Marche	292,039	48.4	1.0	48.3	933	100.0	0.1	–	292,972	48.2	1.0	48.2
Lazio	352,168	75.2	0.6	75.2	0	–	0.0	–	352,168	75.2	0.6	75.2
Abruzzo	161,618	67.3	0.4	67.3	0	–	0.0	–	161,618	67.3	0.4	67.3
Molise	1469	100.0	0.0	–	0	–	0.0	–	1469	100.0	0.0	–
Campania	279,928	37.6	0.7	37.5	0	–	0.0	–	279,928	37.6	0.7	37.5
Puglia	1728	100.0	0.0	–	0	–	0.0	–	1728	100.0	0.0	–
Basilicata	128,416	61.8	0.4	61.8	0	–	0.0	–	128,416	61.8	0.4	61.8
Calabria	294,167	54.4	0.6	54.4	0	–	0.0	–	294,167	54.4	0.6	54.4
Sicilia	15,090	76.4	0.1	76.4	0	–	0.0	–	15,090	76.4	0.1	76.4
Sardegna	175,289	63.3	0.3	63.3	30,605	77.2	1.2	75.4	205,894	55.1	0.3	55.1
Italia	6,283,839	15.5	0.7	15.5	45,239	57.9	0.4	57.6	6,329,078	15.4	0.7	15.4

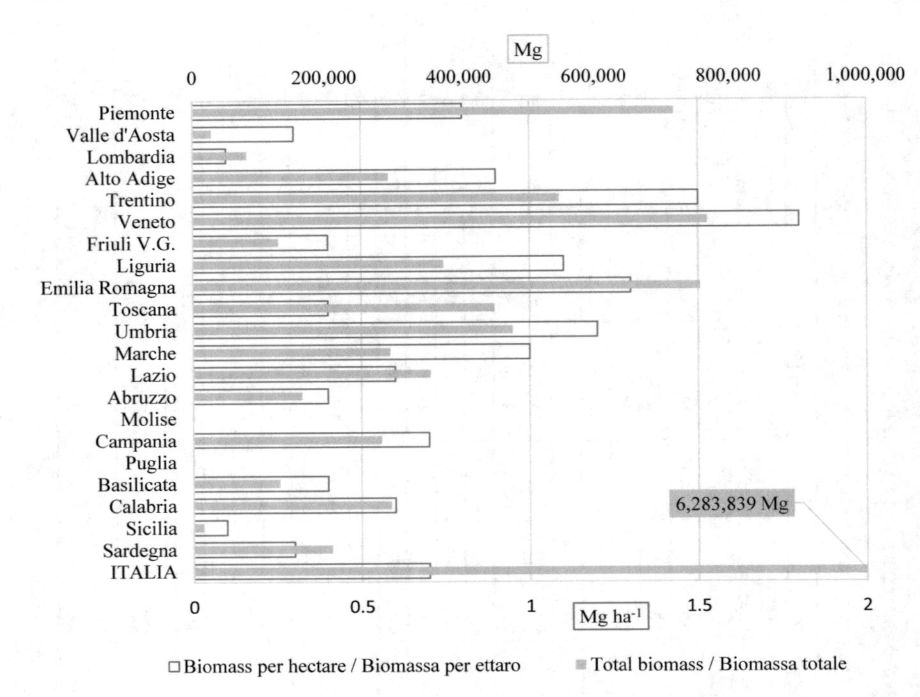

Fig. 8.12 Biomass removed annually, total and per hectare (to improve readability, very long bars have been limited and true values are given in numbers) / Biomassa totale e per ettaro utilizzata annualmente (per migliorare la lettura, è stata limitata la lunghezza di barre molto lunghe e i valori reali sono indicati con numero)

Table 8.22 Classes of the forest planning / Classi per la pianificazione forestale

Status of forest planning Pianificazione forestale	Description Descrizione
No planning Assenza di pianificazione	No forest management guidelines
	Assenza di strumenti di indirizzo della gestione forestale
Presence of PMPF Presenza di PMPF	Regulations deriving from Overall and Forest Police Regulations (PMPF)
	Regolamentazione derivante da Prescrizioni di Massima e di Polizia Forestale (PMPF)
Guideline planning Pianificazione di orientamento presente	Presence of guideline planning (multi- and inter-company planning, reorganisation planning, park planning)
	Presenza di pianificazione di orientamento (piani sovraziendali, piani di riordino, piani parco per la parte relativa alle aree forestali ecc.)
Detailed plans Pianificazione di dettaglio presente	Presence of detail planning (company plans or forest management plans)
	Presenza di piani aziendali o di assestamento forestale

Table 8.23 Forest and Other wooded land area by presence of forest planning / Estensione del Bosco e delle Altre terre boscate ripartite per stato della pianificazione forestale

Region/Regione	Forest/Bosco								Other wooded land/Altre terre boscate							
	Present		Absent		Not classified		Total		Present		Absent		Not classified		Total	
	Presente		Assente		Non classificato		Totale		Presente		Assente		Non classificato		Totale	
	Area	ES	Area	ES	Area	ES	Area	ES	Area	ES	Area	ES	Area	ES	Area	ES
	(ha)	(%)	(ha)	(%)	(ha)	(%)	(ha)	(%)	(ha)	(%)	(ha)	(%)	(ha)	(%)	(ha)	(%)
Piemonte	830,967	1.4	59,062	8.7	404	100.0	890,433	1.3	53,207	10.4	6871	34.5	24,913	15.6	84,991	8.0
Valle d'Aosta	85,757	4.4	13,486	16.0	0	–	99,243	3.6	4668	37.3	385	99.6	3680	31.2	8733	24.0
Lombardia	598,121	1.7	20,159	16.7	3688	57.8	621,968	1.6	48,661	10.5	771	71.5	20,819	17.8	70,252	8.7
Alto Adige	336,245	1.8	2b269	40.6	756	70.5	339,270	1.7	29,862	10.8	2237	70.2	3981	30.1	36,081	10.4
Trentino	361,084	1.5	12,176	20.4	0	–	373,259	1.4	27,313	11.2	2550	59.5	3963	29.2	33,826	10.6
Veneto	394,837	2.0	18,410	19.0	3457	58.0	416,704	1.9	26,419	11.6	0	–	26,571	14.5	52,991	9.1
Friuli V.G	311,588	1.9	18,068	17.2	2901	68.1	332,556	1.9	24,282	13.8	743	70.5	16,033	20.0	41,058	10.6
Liguria	339,545	1.7	3248	49.9	367	99.2	343,160	1.7	23,843	13.9	1782	81.4	18,460	17.3	44,084	10.3
Emilia Romagna	550,319	1.6	33,176	12.1	1406	100.0	584,901	1.5	31,722	13.1	2511	61.5	19,682	13.1	53,915	9.3
Toscana	1,027,617	1.1	6048	46.1	1783	82.2	1,035,448	1.1	73,973	7.8	5687	48.6	74,615	7.5	154,275	5.2
Umbria	377,764	1.7	9737	27.9	2805	69.1	390,305	1.6	12,266	19.6	5557	39.0	5828	32.2	23,651	15.2
Marche	268,362	2.3	23,405	15.5	0	–	291,767	2.1	9707	27.9	2806	56.0	8800	22.7	21,314	16.2
Lazio	533,145	1.7	26,073	14.9	1019	59.3	560,236	1.6	59,608	9.6	7898	30.7	20,405	18.3	87,912	7.7
Abruzzo	376,389	2.0	35,199	12.5	0	–	411,588	1.8	40,203	11.8	9918	29.9	12,890	17.2	63,011	8.6
Molise	145,207	3.3	8041	26.2	0	–	153,248	3.0	16,770	18.5	631	72.5	2624	31.8	20,025	16.0
Campania	392,822	2.1	9324	26.7	1781	82.1	403,927	2.1	42,532	10.2	16,820	24.8	27,980	14.7	87,332	7.6
Puglia	122,748	4.5	18,475	16.8	1125	57.7	142,349	4.0	32,753	12.0	12,967	25.9	3669	30.1	49,389	9.9
Basilicata	265,907	2.8	22,113	19.4	0	–	288,020	2.7	75,759	7.3	6184	36.7	22,449	14.7	104,392	6.2
Calabria	466,188	2.1	16,460	18.2	12,529	29.9	495,177	2.0	61,255	8.0	5534	34.1	88,654	6.7	155,443	4.8
Sicilia	239,919	3.6	39,862	12.9	5709	49.1	285,489	3.2	58,936	8.1	27,503	17.0	15,306	24.3	101,745	7.1
Sardegna	348,815	3.0	275,894	3.8	1431	100.0	626,140	2.1	245,294	3.7	398,750	3.0	30,808	12.8	674,851	2.0
Italia	8,373,344	0.4	670,684	2.7	41,158	16.8	9,085,186	0.4	999,034	2.1	518,106	3.0	452,132	3.3	1,969,272	1.4

Table 8.24 Area of the inventory categories of Forest and Other wooded land by presence of forest planning / Estensione delle categorie inventariali del Bosco ripartite per stato della pianificazione forestale

Region/Regione	Tall trees forest/Boschi alti								Plantations/Impianti di arboricoltura da legno							
	Present		Absent		Not classified		Total		Present		Absent		Not classified		Total	
	Presente		Assente		Non classificato		Totale		Presente		Assente		Non classificato		Totale	
	Area	ES	Area	ES	Area	ES	Area	ES	Area	ES	Area	ES	Area	ES	Area	ES
	(ha)	(%)	(ha)	(%)	(ha)	(%)	(ha)	(%)	(ha)	(%)	(ha)	(%)	(ha)	(%)	(ha)	(%)
Piemonte	822,089	1.4	47,280	9.7	404	100.0	869,773	1.3	8878	30.1	11,782	19.1	0	–	20,660	16.2
Valle d'Aosta	85,757	4.4	13,486	16.0	0	–	99,243	3.6	0	–	0	–	0	–	0	–
Lombardia	586,102	1.6	8882	28.7	1851	79.9	596,836	1.6	12,019	22.6	11,277	19.6	1836	83.7	25,132	12.5
Alto Adige	336,245	1.8	2269	40.6	756	70.5	339,270	1.7	0	–	0	–	0	–	0	–
Trentino	361,084	1.5	12,176	20.4	0	–	373,259	1.4	0	–	0	–	0	–	0	–
Veneto	394,150	2.0	13,820	22.1	3457	58.0	411,427	1.9	687	69.5	4590	37.5	0	–	5277	33.7
Friuli V.G	310,164	1.9	10,297	26.8	2901	68.1	323,362	1.9	1424	49.6	7771	18.6	0	–	9194	17.0
Liguria	339,545	1.7	2881	54.9	367	99.2	342,793	1.7	0	–	367	99.2	0	–	367	99.2
Emilia Romagna	549,237	1.6	28,210	12.7	1406	100.0	578,852	1.5	1082	57.8	4966	36.6	0	–	6049	31.6
Toscana	1,020,834	1.1	6048	46.1	1783	82.2	1,028,665	1.1	6783	30.3	0	–	0	–	6783	30.3
Umbria	372,789	1.7	9737	27.9	1402	99.8	383,928	1.6	4974	33.7	0	–	1402	99.8	6377	33.7
Marche	265,402	2.3	19,503	16.0	0	–	284,904	2.1	2960	54.6	3902	52.9	0	–	6863	35.7
Lazio	532,178	1.7	25,284	15.2	599	72.5	558,060	1.6	967	58.9	789	70.8	420	100.0	2176	39.7
Abruzzo	373,779	2.0	34,837	12.6	0	–	408,616	1.8	2609	39.0	362	100.0	0	–	2971	36.3
Molise	142,491	3.4	8041	26.2	0	–	150,533	3.1	2715	27.7	0	–	0	–	2715	27.7
Campania	390,045	2.1	8938	27.7	1781	82.1	400,763	2.1	2777	56.7	387	74.9	0	–	3163	50.4
Puglia	122,648	4.5	18,475	16.8	1125	57.7	142,248	4.0	100	99.8	0	–	0	–	100	99.8
Basilicata	265,907	2.8	20,591	20.2	0	–	286,498	2.7	0	–	1522	93.6	0	–	1522	93.6
Calabria	464,696	2.1	15,546	18.9	12,529	29.9	492,771	2.1	1492	49.9	914	64.0	0	–	2406	39.3
Sicilia	239,161	3.6	39,862	12.9	5709	49.1	284,731	3.2	758	70.7	0	–	0	–	758	70.7
Sardegna	340,334	3.1	258,490	4.0	1431	100.0	600,255	2.2	8481	25.2	17,404	13.4	0	–	25,885	11.8
Italia	8,314,636	0.4	604,652	2.9	37,499	17.6	8,956,787	0.4	58,708	10.1	66,032	8.4	3659	58.0	128,399	6.2

Table 8.25 Area of the inventory categories of Other wooded land by presence of planning / Estensione delle categorie inventariali delle Altre terre boscate ripartite per stato della pianificazione forestale

Region/regione	Short trees forest, Sparse forest and Scrubland							
	Boschi bassi, Boschi radi e Boscaglie							
	Present		Absent		Not classified		Total	
	Presente		Assente		Non classificato		Totale	
	Area	ES	Area	ES	Area	ES	Area	ES
	(ha)	(%)	(ha)	(%)	(ha)	(%)	(ha)	(%)
Piemonte	25,486	16.6	4092	51.7	2834	70.2	32,412	15.2
Valle d'Aosta	2357	62.5	0	–	0	–	2357	62.5
Lombardia	11,791	18.7	0	–	0	–	11,791	18.7
Alto Adige	4550	27.0	378	99.8	0	–	4928	26.0
Trentino	6737	22.8	0	–	0	–	6737	22.8
Veneto	3704	31.5	0	–	0	–	3704	31.5
Friuli V.G.	8832	20.1	0	–	1450	99.7	10,283	22.3
Liguria	14,794	19.0	1782	81.4	1415	99.2	17,992	18.5
Emilia Romagna	20,069	16.7	1406	100.0	0	–	21,475	16.7
Toscana	26,709	12.4	2843	70.0	0	–	29,552	13.0
Umbria	6834	28.9	270	99.8	0	–	7104	28.0
Marche	4689	38.1	2465	62.4	0	–	7153	32.2
Lazio	17,804	20.6	967	58.9	4686	52.8	23,456	18.4
Abruzzo	9052	23.4	2078	71.5	1407	100.0	12,537	22.9
Molise	7085	28.2	0	–	0	–	7085	28.2
Campania	14,241	19.9	1777	82.2	0	–	16,017	19.8
Puglia	16,707	19.1	1614	47.2	0	–	18,321	17.9
Basilicata	17,455	16.0	0	–	1422	99.9	18,877	16.5
Calabria	31,623	10.6	1866	44.6	373	99.9	33,862	10.2
Sicilia	19,813	14.9	6582	35.3	1427	100.0	27,822	14.2
Sardegna	39,069	10.2	47,611	10.4	0	–	86,680	7.2
Italia	309,401	4.0	75,731	9.4	15,014	30.0	400,146	3.7

Table 8.25 (continued)

Region/regione	Shrubs								Not accessible or not classified wooded area					
	Arbusteti								Aree boscate inaccessibili o non classificate					
	Present		Absent		Not classified		Total		Absent		Not classified		Total	
	Presente		Assente		Non classificato		Totale		Assente		Non classificato		Totale	
	Area	ES	Area	ES	Area	ES	Area	ES	Area	ES	Area	ES	Area	ES
	(ha)	(%)	(ha)	(%)	(ha)	(%)	(ha)	(%)	(ha)	(%)	(ha)	(%)	(ha)	(%)
Piemonte	27,721	13.5	2778	38.4	0	–	30,499	12.7	0	–	22,079	15.3	22,079	15.3
Valle d'Aosta	2312	40.4	385	99.6	0	–	2697	37.3	0	–	3680	31.2	3680	31.2
Lombardia	36,870	12.6	771	71.5	2821	69.8	40,462	12.3	0	–	17,998	17.8	17,998	17.8
Alto Adige	25,313	11.9	1859	82.0	0	–	27,172	12.4	0	–	3981	30.1	3981	30.1
Trentino	20,576	13.1	2550	59.5	0	–	23,126	13.3	0	–	3963	29.2	3963	29.2
Veneto	22,715	12.6	0	–	5825	43.2	28,540	13.3	0	–	20,746	14.4	20,746	14.4
Friuli V.G.	15,449	18.7	743	70.5	4351	53.6	20,543	17.1	0	–	10,232	18.5	10,232	18.5
Liguria	9048	19.7	0	–	0	–	9048	19.7	0	–	17,045	17.2	17,045	17.2
Emilia Romagna	11,652	22.8	1105	57.8	368	100.0	13,126	21.0	0	–	19,314	13.2	19,314	13.2
Toscana	47,264	10.2	2843	70.0	1783	82.2	51,890	10.2	0	–	72,832	7.4	72,832	7.4
Umbria	5432	25.7	5287	40.7	0	–	10,719	23.9	0	–	5828	32.2	5828	32.2
Marche	5019	42.7	341	100.0	1412	100.0	6772	37.1	0	–	7388	19.1	7388	19.1
Lazio	41,804	11.2	6563	35.4	1439	100.0	49,806	10.6	368	100.0	14,281	17.7	14,649	17.4
Abruzzo	31,150	13.9	7840	33.5	0	–	38,991	12.4	0	–	11,483	14.9	11,483	14.9
Molise	9685	25.8	631	72.5	0	–	10,315	24.6	0	–	2624	31.8	2624	31.8
Campania	28,292	12.0	15,044	26.6	7062	43.0	50,397	11.2	0	–	20,918	13.6	20,918	13.6
Puglia	16,046	15.0	11,353	28.8	0	–	27,399	14.7	0	–	3669	30.1	3669	30.1
Basilicata	58,304	8.5	6184	36.7	1422	99.9	65,910	8.3	0	–	19,605	13.4	19,605	13.4
Calabria	29,632	12.5	3669	46.2	3513	58.6	36,814	12.2	0	–	84,768	6.5	84,768	6.5
Sicilia	39,124	10.0	20,921	19.8	5709	49.1	65,753	9.4	0	–	8170	25.5	8170	25.5
Sardegna	206,225	4.1	351,139	3.2	1431	100.0	558,795	2.3	0	–	29,377	12.6	29,377	12.6
Italia	689,633	2.5	442,007	3.3	37,136	18.5	1,168,776	1.9	368	100.0	399,982	3.2	400,350	3.2

Table 8.26 Forest and Other wooded land area by presence of detailed plans / Estensione del Bosco e delle Altre terre boscate ripartiti per presenza di "Pianificazione di dettaglio"

Region/Regione	Forest/Bosco								Other wooded land/Altre terre boscate							
	Present		Absent		Not classified		Total		Present		Absent		Not classified		Total	
	Presente		Assente		Non classificato		Totale		Presente		Assente		Non classificato		Totale	
	Area	ES	Area	ES	Area	ES	Area	ES	Area	ES	Area	ES	Area	ES	Area	ES
	(ha)	(%)	(ha)	(%)	(ha)	(%)	(ha)	(%)	(ha)	(%)	(ha)	(%)	(ha)	(%)	(ha)	(%)
Piemonte	16,909	15.4	873,120	1.3	404	100.0	890,433	1.3	1212	57.8	58,866	10.0	24,913	15.6	84,991	8.0
Valle d'Aosta	35,690	8.8	63,553	6.1	0	–	99,243	3.6	991	56.9	4062	41.6	3680	31.2	8733	24.0
Lombardia	122,353	5.6	495,927	2.1	3688	57.8	621,968	1.6	14,988	17.0	34,445	13.1	20,819	17.8	70,252	8.7
Alto Adige	321,024	1.9	17,490	16.2	756	70.5	339,270	1.7	24,859	12.0	7241	28.9	3981	30.1	36,081	10.4
Trentino	283,113	2.2	90,147	5.9	0	–	373,259	1.4	23,338	12.2	6525	29.5	3963	29.2	33,826	10.6
Veneto	125,987	4.8	287,260	2.9	3457	58.0	416,704	1.9	12,477	17.1	13,942	16.3	26,571	14.5	52,991	9.1
Friuli V.G	112,609	4.9	217,047	3.2	2901	68.1	332,556	1.9	11,441	20.8	13,584	18.6	16,033	20.0	41,058	10.6
Liguria	12,828	16.5	329,965	1.8	367	99.2	343,160	1.7	1100	57.2	24,525	14.3	18,460	17.3	44,084	10.3
Emilia Romagna	39,393	9.4	544,102	1.7	1406	100.0	584,901	1.5	737	70.7	33,496	13.0	19,682	13.1	53,915	9.3
Toscana	101,186	5.8	932,479	1.3	1783	82.2	1,035,448	1.1	4333	28.8	75,326	8.2	74,615	7.5	154,275	5.2
Umbria	11,722	17.4	375,778	1.7	2805	69.1	390,305	1.6	0	–	17,823	17.8	5828	32.2	23,651	15.2
Marche	1482	50.0	290,285	2.1	0	–	291,767	2.1	0	–	12,513	24.1	8800	22.7	21,314	16.2
Lazio	55,429	8.5	503,789	1.8	1019	59.3	560,236	1.6	1842	44.7	65,664	9.2	20,405	18.3	87,912	7.7
Abruzzo	28,594	11.0	382,994	2.0	0	–	411,588	1.8	2507	37.8	47,614	10.8	12,890	17.2	63,011	8.6
Molise	24,674	13.6	128,574	4.0	0	–	153,248	3.0	516	79.3	16,885	18.4	2624	31.8	20,025	16.0
Campania	87,646	6.0	314,500	2.8	1781	82.1	403,927	2.1	5156	26.7	54,197	10.5	27,980	14.7	87,332	7.6
Puglia	0	–	141,223	4.1	1125	57.7	142,349	4.0	0	–	45,720	10.5	3669	30.1	49,389	9.9
Basilicata	2610	37.7	285,410	2.7	0	–	288,020	2.7	0	–	81,943	7.1	22,449	14.7	104,392	6.2
Calabria	2239	40.7	480,409	2.0	12,529	29.9	495,177	2.0	0	–	66,789	7.8	88,654	6.7	155,443	4.8
Sicilia	0	–	279,781	3.3	5709	49.1	285,489	3.2	0	–	86,439	7.6	15,306	24.3	101,745	7.1
Sardegna	4105	30.1	620,604	2.1	1431	100.0	626,140	2.1	373	100.0	643,671	2.1	30,808	12.8	674,851	2.0
Italia	1,389,591	1.3	7,654,437	0.5	41,158	16.8	9,085,186	0.4	105,870	6.0	1,411,270	1.7	452,132	3.3	1,969,272	1.4

Table 8.27 Area of the inventory categories of Forest by presence of detailed plans / Estensione delle categorie inventariali del Bosco, ripartite per presenza di "Pianificazione di dettaglio"

Region/Regione	Tall trees forest/Boschi alti								Plantations/Impianti di arboricoltura da legno							
	Present		Absent		Not classified		Total		Present		Absent		Not classified		Total	
	Presente		Assente		Non classificato		Totale		Presente		Assente		Non classificato		Totale	
	Area	ES	Area	ES	Area	ES	Area	ES	Area	ES	Area	ES	Area	ES	Area	ES
	(ha)	(%)	(ha)	(%)	(ha)	(%)	(ha)	(%)	(ha)	(%)	(ha)	(%)	(ha)	(%)	(ha)	(%)
Piemonte	16,909	15.4	852,460	1.3	404	100.0	869,773	1.3	0	–	20,660	16.2	0	–	20,660	16.2
Valle d'Aosta	35,690	8.8	63,553	6.1	0	–	99,243	3.6	0	–	0	–	0	–	0	–
Lombardia	120,330	5.7	474,655	2.2	1851	79.9	596,836	1.6	2023	40.1	21,273	14.5	1836	83.7	25,132	12.5
Alto Adige	321,024	1.9	17,490	16.2	756	70.5	339,270	1.7	0	–	0	–	0	–	0	–
Trentino	283,113	2.2	90,147	5.9	0	–	373,259	1.4	0	–	0	–	0	–	0	–
Veneto	125,987	4.8	281,983	2.9	3457	58.0	411,427	1.9	0	–	5277	33.7	0	–	5277	33.7
Friuli V.G	112,609	4.9	207,853	3.3	2901	68.1	323,362	1.9	0	–	9194	17.0	0	–	9194	17.0
Liguria	12,828	16.5	329,599	1.8	367	99.2	342,793	1.7	0	–	367	99.2	0	–	367	99.2
Emilia Romagna	38,679	9.5	538,768	1.7	1406	100.0	578,852	1.5	714	70.8	5335	34.7	0	–	6049	31.6
Toscana	99,727	5.8	927,154	1.3	1783	82.2	1,028,665	1.1	1459	49.6	5324	36.5	0	–	6783	30.3
Umbria	11,060	18.0	371,466	1.7	1402	99.8	383,928	1.6	663	68.5	4312	38.6	1402	99.8	6377	33.7
Marche	1482	50.0	283,422	2.1	0	–	284,904	2.1	0	–	6863	35.7	0	–	6863	35.7
Lazio	55,429	8.5	502,033	1.8	599	72.5	558,060	1.6	0	–	1756	45.4	420	100.0	2176	39.7
Abruzzo	28,594	11.0	380,022	2.0	0	–	408,616	1.8	0	–	2971	36.3	0	–	2971	36.3
Molise	24,674	13.6	125,859	4.0	0	–	150,533	3.1	0	–	2715	27.7	0	–	2715	27.7
Campania	87,278	6.0	311,705	2.8	1781	82.1	400,763	2.1	368	100.0	2795	55.5	0	–	3163	50.4
Puglia	0	–	141,123	4.1	1125	57.7	142,248	4.0	0	–	100	99.8	0	–	100	99.8
Basilicata	2610	37.7	283,888	2.7	0	–	286,498	2.7	0	–	1522	93.6	0	–	1522	93.6
Calabria	2239	40.7	478,003	2.1	12,529	29.9	492,771	2.1	0	–	2406	39.3	0	–	2406	39.3
Sicilia	0	–	279,022	3.3	5709	49.1	284,731	3.2	0	–	758	70.7	0	–	758	70.7
Sardegna	3732	31.6	595,092	2.2	1431	100.0	600,255	2.2	373	100.0	25,511	11.9	0	–	25,885	11.8
Italia	1,383,991	1.3	7,535,297	0.5	37,499	17.6	8,956,787	0.4	5600	24.7	119,140	6.5	3659	58.0	128,399	6.2

Table 8.28 Area of the inventory categories of Other wooded land by presence of detailed plans / Estensione delle categorie inventariali delle Altre terre boscate, ripartite per presenza di "Pianificazione di dettaglio"

Region/Regione	Short trees forest, Sparse forest, Scrubland							
	Boschi bassi, Boschi radi e Boscaglie							
	Present		Absent		Not classified		Total	
	Presente		Assente		Non classificato		Totale	
	Area	ES	Area	ES	Area	ES	Area	ES
	(ha)	(%)	(ha)	(%)	(ha)	(%)	(ha)	(%)
Piemonte	808	70.7	28,770	16.0	2834	70.2	32,412	15.2
Valle d'Aosta	606	68.4	1750	80.7	0	–	2,357	62.5
Lombardia	2,204	44.7	9587	20.6	0	–	11,791	18.7
Alto Adige	3,834	30.1	1093	57.7	0	–	4,928	26.0
Trentino	5,777	24.9	960	56.0	0	–	6737	22.8
Veneto	339	99.9	3365	33.2	0	–	3704	31.5
Friuli V.G.	4831	27.5	4001	29.9	1450	99.7	10,283	22.3
Liguria	733	70.1	15,843	19.4	1415	99.2	17,992	18.5
Emilia Romagna	737	70.7	20,738	17.1	0	–	21,475	16.7
Toscana	1804	44.7	27,749	13.6	0	–	29,552	13.0
Umbria	0	–	7104	28.0	0	–	7104	28.0
Marche	0	–	7153	32.2	0	–	7153	32.2
Lazio	0	–	18,771	19.7	4686	52.8	23,456	18.4
Abruzzo	0	–	11,130	22.9	1407	100.0	12,537	22.9
Molise	516	79.3	6570	29.8	0	–	7085	28.2
Campania	2946	35.3	13,071	22.9	0	–	16,017	19.8
Puglia	0	–	18,321	17.9	0	–	18,321	17.9
Basilicata	0	–	17,455	16.0	1422	99.9	18,877	16.5
Calabria	0	–	33,488	10.3	373	99.9	33,862	10.2
Sicilia	0	–	26,394	14.1	1427	100.0	27,822	14.2
Sardegna	0	–	86,680	7.2	0	–	86,680	7.2
Italia	25,135	12.1	359,996	3.9	15,014	30.0	400,146	3.7

Table 8.28 (continued)

Region/Regione	Shrubs								Not accessible or not classified wooded area					
	Arbusteti								Aree boscate inaccessibili o non classificate					
	Present		Absent		Not classified		Total		Absent		Not classified		Total	
	Presente		Assente		Non classificato		Totale		Assente		Non classificato		Totale	
	Area	ES	Area	ES	Area	ES	Area	ES	Area	ES	Area	ES	Area	ES
	(ha)	(%)	(ha)	(%)	(ha)	(%)	(ha)	(%)	(ha)	(%)	(ha)	(%)	(ha)	(%)
Piemonte	404	100.0	30,095	12.8	0	–	30,499	12.7	0	–	22,079	15.3	22,079	15.3
Valle d'Aosta	385	99.6	2312	40.4	0	–	2697	37.3	0	–	3680	31.2	3680	31.2
Lombardia	12,783	18.4	24,858	16.4	2821	69.8	40,462	12.3	0	–	17,998	17.8	17,998	17.8
Alto Adige	21,024	13.1	6147	32.5	0	–	27,172	12.4	0	–	3981	30.1	3981	30.1
Trentino	17,561	14.2	5565	33.3	0	–	23,126	13.3	0	–	3963	29.2	3963	29.2
Veneto	12,138	17.4	10,577	18.8	5825	43.2	28,540	13.3	0	–	20,746	14.4	20,746	14.4
Friuli V.G.	6610	30.1	9583	23.4	4351	53.6	20,543	17.1	0	–	10,232	18.5	10,232	18.5
Liguria	367	99.2	8682	20.1	0	–	9048	19.7	0	–	17,045	17.2	17,045	17.2
Emilia Romagna	0	–	12,758	21.4	368	100.0	13,126	21.0	0	–	19,314	13.2	19,314	13.2
Toscana	2529	37.8	47,578	10.7	1783	82.2	51,890	10.2	0	–	72,832	7.4	72,832	7.4
Umbria	0	–	10,719	23.9	0	–	10,719	23.9	0	–	5828	32.2	5828	32.2
Marche	0	–	5360	40.5	1412	100.0	6772	37.1	0	–	7388	19.1	7388	19.1
Lazio	1842	44.7	46,525	10.9	1439	100.0	49,806	10.6	368	100.0	14,281	17.7	14,649	17.4
Abruzzo	2507	37.8	36,484	13.0	0	–	38,991	12.4	0	–	11,483	14.9	11,483	14.9
Molise	0	–	10,315	24.6	0	–	10,315	24.6	0	–	2624	31.8	2624	31.8
Campania	2210	40.8	41,126	12.4	7062	43.0	50,397	11.2	0	–	20,918	13.6	20,918	13.6
Puglia	0	–	27,399	14.7	0	–	27,399	14.7	0	–	3669	30.1	3669	30.1
Basilicata	0	–	64,488	8.3	1422	99.9	65,910	8.3	0	–	19,605	13.4	19,605	13.4
Calabria	0	–	33,301	12.1	3513	58.6	36,814	12.2	0	–	84,768	6.5	84,768	6.5
Sicilia	0	–	60,045	9.4	5709	49.1	65,753	9.4	0	–	8170	25.5	8170	25.5
Sardegna	373	100.0	556,991	2.3	1431	100.0	558,795	2.3	0	–	29,377	12.6	29,377	12.6
Italia	80,734	7.0	1,050,906	2.0	37,136	18.5	1,168,776	1.9	368	100.0	399,982	3.2	400,350	3.2

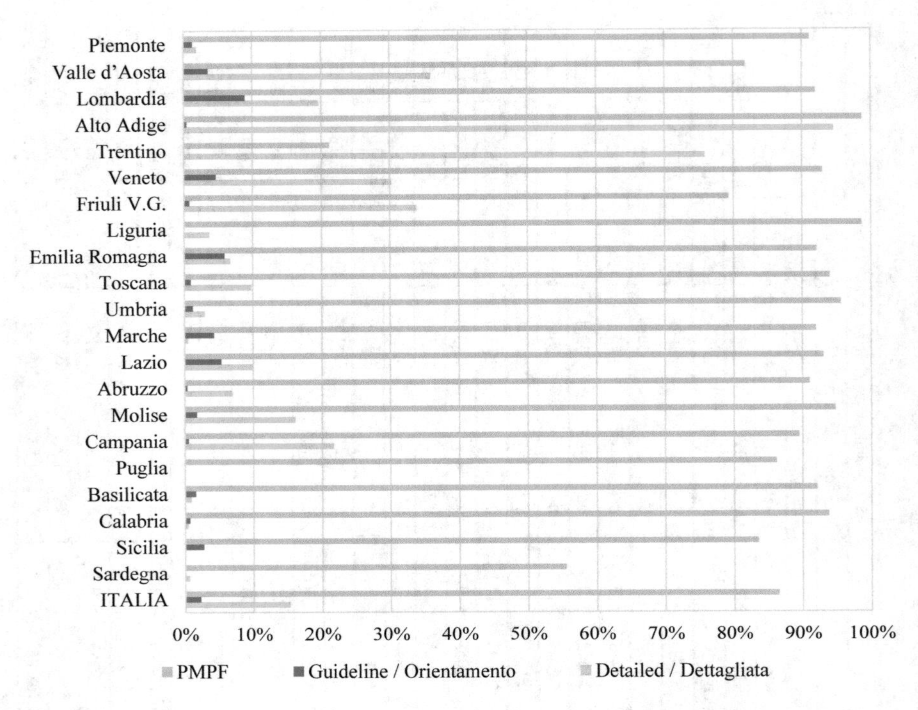

Fig. 8.13 Forest area percentage by planning level / Percentuale di superficie del Bosco per livello di dettaglio della pianificazione forestale

Appendix (Italian Version)

Riassunto La selvicoltura come scienza è nata con la necessità di gestire le foreste in modo sostenibile, per preservarne la funzione produttiva. Da allora, il concetto di sostenibilità è stato allargato ad altri valori e servizi riconosciuti importanti per il benessere del genere umano. La selvicoltura è quindi una necessità dell'uomo e non della foresta per la propria perpetuazione. La sostenibilità può essere perseguita adottando pratiche colturali corrette e adattando la pianificazione alle forme di gestione idonee ad assicurare il raggiungimento degli obiettivi, come ad esempio quelli legati al bilancio del carbonio per la mitigazione dei cambiamenti climatici. Gli inventari forestali nazionali producono le informazioni necessarie, di qualità idonea, per le scelte di politica forestale. In questo capitolo sono descritte le statistiche su alcune variabili importanti per gli aspetti produttivi, quali l'accessibilità, la presenza di viabilità forestale, l'accidentalità del terreno e la disponibilità al prelievo legnoso (Sect. 8.2). Le pratiche selvicolturali trovano concreta applicazione mediante i sistemi di utilizzazione adottati; questi sono anche correlati con le modalità di esbosco; la Sect. 8.3 descrive le statistiche su queste variabili nonché il volume e la biomassa annualmente asportati dal bosco in Italia. La Sect. 8.4 riporta le statistiche sulla presenza di pianificazione e regolamenti che influenzano gli aspetti produttivi delle foreste. I tipi di pianificazione possono variare considerevolmente in relazione alle competenze dei vari Enti preposti, quindi il rilievo ha dato particolare importanza alla presenza di pianificazione ai diversi livelli.

Introduzione

La necessità di gestire le foreste nasce principalmente dall'esigenza di ricavarne legna o legname in maniera sostenibile (Dorren et al., 2004), sebbene a partire dagli anni Ottanta dello scorso secolo il concetto di sostenibilità sia stato allargato ad altri valori e servizi ecosistemici (O'Hara, 2016). Proprio la minaccia alla sostenibilità dovuta alle condizioni di impoverimento dei soprassuoli europei a cavallo tra diciottesimo e diciannovesimo secolo ha stimolato la nascita della selvicoltura come disciplina scientifica, la prima del mondo occidentale a riconoscere esplicitamente la necessità di salvaguardare un bene naturale per le future generazioni (Perry, 1998). La selvicoltura è quindi una necessità dell'uomo, che dalla foresta vuole ottenere determinati vantaggi, e non della foresta per la sua esistenza (Piussi, 1994). Essa è anche stata indicata come una necessità, nell'intento di trovare un punto di equilibrio tra la fruizione delle diverse funzioni del bosco e l'accertata potenzialità delle sue prestazioni, dato che l'uso dei boschi è irrinunciabile per il benessere dell'umanità (De Philippis, 1983). La sostenibilità si fonda, quindi, sull'adozione di pratiche selviculturali appropriate nell'ambito di una gestione forestale correttamente pianificata ai vari livelli, sulla base dei vincoli normativi anch'essi di validità territoriale più o meno ampia. Da questo punto di vista, le statistiche di INFC2015 possono rappresentare un contributo importante per ridefinire politiche e criteri di pianificazione a livello nazionale e regionale. Nel quadro attuale di riscaldamento globale, infatti, le scelte orientate ad assicurare la sostenibilità della gestione forestale, in relazione al bilancio del carbonio e alle altre funzioni del bosco, non possono prescindere dalla disponibilità di dati affidabili (Bosela et al., 2016). Il livello di dettaglio applicativo della pianificazione è rappresentato dalla pianificazione forestale particolareggiata, anche nota come assestamento forestale, che si prefigge di garantire la continuità dei prodotti e servizi richiesti al bosco senza pericoli di deterioramento (Bernetti, 1989).

Nella gestione delle foreste, gli aspetti produttivi dipendono da vari fattori, alcuni dei quali illustrati in questo capitolo. Tra di essi, un primo gruppo riguarda la possibilità di accesso, in relazione ad ostacoli fisico-orografici o normativi. La possibilità di operare in bosco è legata anche alla possibilità di accedervi senza dispendio eccessivo di tempo ed energia, condizione possibile solo in presenza di una rete stradale adeguata (Hippoliti, 1990). INFC rileva la presenza di viabilità utile all'accesso in termini di distanza orizzontale e verticale rispetto ai punti di campionamento. Le caratteristiche micromorfologiche del terreno, dettate dalla presenza di fossi, avvallamenti o ostacoli come rocce e massi, possono impedire o ostacolare in varia misura la percorribilità del terreno con i mezzi meccanici, con ripercussioni sugli aspetti produttivi anch'esse variabili caso per caso. Oltre all'interazione delle variabili elencate, altre caratteristiche, come ad esempio il valore del soprassuolo e le condizioni del mercato, influenzano gli aspetti produttivi; per questo, la disponibilità al prelievo legnoso viene valutata anche attraverso un giudizio di tipo sintetico. La Sect. 8.2 è dedicata a tutti questi aspetti.

Le pratiche selvicolturali adottate sono strettamente connesse alle modalità delle utilizzazioni del soprassuolo, poiché la selvicoltura si concretizza con esse (Hippoliti, 2005). La Sect. 8.3 è dedicata alle pratiche selvicolturali applicate e alle modalità di utilizzazione e mezzi di esbosco adottati, cioè al modo e ai mezzi con cui gli alberi vengono tagliati e i prodotti legnosi ottenuti allontanati dal bosco.

La coesistenza di diverse competenze sulle foreste, dello Stato e delle Regioni, ma anche dei diversi Enti preposti alla salvaguardia dei molteplici valori riconosciuti al bosco, determina una certa complessità che non rende agevole valutare il grado della pianificazione in specifiche aree boscate. INFC rileva la presenza di pianificazione ai vari livelli possibili, inclusa la presenza di eventuali vincoli normativi che possono condizionare la produzione legnosa. Questa modalità di rilievo prevede che di un'area vengano indicati tutti gli strumenti e vincoli eventualmente presenti. In virtù delle possibili sovrapposizioni, sommando le aree sottoposte ai diversi strumenti pianificatori è possibile che si ottenga una superficie superiore alla superficie totale della categoria considerata, ad esempio del Bosco o delle Altre terre boscate.

Accessibilità e disponibilità al prelievo

La raggiungibilità delle aree forestali viene valutata mediante diverse caratteristiche che descrivono la possibilità di arrivare sul punto di campionamento (grado di accessibilità) e la facilità di accesso in relazione alla distanza orizzontale e al dislivello tra il punto di campionamento e la viabilità ordinaria o forestale. È stata presa in considerazione solo la viabilità ordinaria o forestale in buone condizioni, cioè concretamente, fruibile secondo la categoria di appartenenza. Sul posto, le condizioni di lavoro per le varie operazioni necessarie in occasione dei tagli possono risultare più o meno agevoli in funzione dell'accidentalità del terreno, dalle quali può dipendere la possibilità o la preclusione all'uso di macchine specializzate (Hippoliti, 2004).

Il grado di accessibilità viene espresso secondo la semplice attribuzione alla condizione di accessibile o non accessibile. I motivi di inaccessibilità, descritti nella Table 8.1, sono riconducibili a tre tipi di cause, legate ad ostacoli fisico-orografici che ne rendono impossibile il raggiungimento o il cui superamento espone a seri pericoli per l'incolumità, alla presenza di vegetazione molto densa ed intricata ed infine alla presenza di recinzioni o divieti normativi. La Table 8.2 riporta le stime di superficie per il grado di accessibilità del Bosco e delle Altre terre boscate. All'indirizzo inventarioforestale.org/statistiche_INFC sono disponibili analoghe statistiche per le categorie inventariali del Bosco e delle Altre terre boscate, nonché quelle per le categorie forestali dei Boschi alti. Il Bosco italiano è caratterizzato da un grado di accessibilità piuttosto elevato (90.3%) e questo rimane vero in generale anche a livello regionale (Fig. 8.1), poiché tra le regioni si osservano variazioni contenute e percentuali di superficie accessibile sempre sopra l'80%. Anche nelle categorie forestali dei Boschi alti la percentuale di superficie accessibile è generalmente molto elevata, con variazioni contenute; il valore più basso è stato stimato per i Boschi igrofili (78.6%).

Il rilievo della viabilità prevede di attribuire la classe di distanza orizzontale e di dislivello tra il punto inventariale e la viabilità utilizzabile per l'esbosco con mezzi meccanici attraverso la consultazione di ortofoto, carte tecniche e altra cartografia tematica disponibile (anche su WebGIS) e la verifica o l'integrazione delle informazioni raccolte con l'osservazione in campo, durante l'avvicinamento al punto di campionamento. Questa modalità di rilievo, introdotta da INFC2015, ha consentito di superare le problematiche incontrate nel passato ciclo inventariale, principalmente legate all'inattesa onerosità del rilievo secondo le specifiche allora adottate, anche per le maggiori difficoltà di allora nella ricezione del segnale GPS (Di Cosmo et al., 2011). Le modifiche introdotte nel protocollo di rilievo hanno consentito di registrare il dato per tutti i punti della campagna di INFC2015, ma essendo le stime attuali legate anche alle informazioni pregresse (cfr. Chap. 2), l'informazione rimane mancante per una porzione importante del territorio boscato. Pertanto, il commento delle stime relative alle condizioni di viabilità deve necessariamente considerare che la percentuale di punti non classificati per il Bosco rappresenta il 30.6% della superficie, aspetto che limita le possibilità di descrivere la situazione reale.

La Table 8.3 in fondo a questo capitolo mostra le statistiche sulla superficie del Bosco nelle varie classi di distanza orizzontale adottate (0–500, 501–1000, 1001–2000, >2000 m) dalla viabilità ordinaria o forestale al punto di campionamento. Sul sito inventarioforestale.org/statistiche_INFC sono disponibili le statistiche per le classi di distanza delle Altre terre boscate, per le categorie inventariali del Bosco e per le categorie forestali dei Boschi alti. Il 47.9% della superficie del Bosco dista non più di 500 m da strade ordinarie o forestali. La prima classe di distanza è la più frequente anche a livello regionale, con valori che variano dal 34.3% della Basilicata e il 38.9% della Valle d'Aosta al 57.1% dell'Alto Adige (Fig. 8.2). Se si considera la superficie entro i 1000 m di distanza, cioè nelle prime due classi, a livello nazionale la percentuale passa al 63.4% ed in parte cambia la posizione di alcune regioni. Ad esempio, la Valle d'Aosta si allinea ad altre regioni con valori prossimi al 60%, mentre la Basilicata rimane quella con percentuale di superficie del Bosco minore nelle due classi (45.6%); l'Alto Adige resta la regione con superficie del Bosco più alta entro i 1000 m di distanza (74.7%), ma anche altre regioni si approssimano a quel valore (Veneto, Molise e Sardegna superano il 70%).

La Table 8.4 mostra le statistiche sulla superficie del Bosco nelle varie classi di dislivello (0–100, 101–400, >400 m, con segno positivo per tragitto in salita dalla strada al punto di campionamento e con segno negativo in caso di tragitto in discesa). All'indirizzo inventarioforestale.org/statistiche_INFC sono disponibili le stime delle superfici per le classi di dislivello delle Altre terre boscate, delle categorie inventariali del Bosco e delle categorie forestali dei Boschi alti. Il 58.3% della superficie del Bosco è raggiungibile percorrendo un dislivello (in salita o in discesa) entro i 100 m. Anche a livello regionale più del 50% della superficie del Bosco si trova entro i 100 m di dislivello dalla viabilità ordinaria o forestale (Fig. 8.3), ad eccezione della Valle d'Aosta (42.1%) e della Basilicata (44.2%), due regioni già caratterizzate da una percentuale bassa della superficie del Bosco nella prima classe di distanza dalla viabilità. Per le categorie forestali dei Boschi alti, la classe di dislivello tra 0 e ± 100 m rimane la più frequente, ma per i Boschi di larice e cembro la percentuale di superficie rappresentata è solo del 35.0% contro il 25.8% della seconda classe (tra ±101 e ±400 m). Anche per le Faggete la superficie nella prima classe di dislivello non arriva al 50% di quella totale (42.8%), mentre in tutte le altre categorie più della metà della superficie ricade entro i 100 m di dislivello. Le Sugherete rappresentano la categoria raggiungibile con il minor dislivello da percorrere, poiché più dell'80% della superficie ricade nella prima classe.

L'accidentalità del terreno è stata classificata secondo l'attribuzione alle tre classi descritte nella Table 8.5. Le Tables 8.6 e 8.7 riportano l'estensione del Bosco e delle Altre terre boscate ripartite per le classi di accidentalità del terreno. Al link inventarioforestale.org/statistiche_INFC sono disponibili le statistiche analoghe per le categorie inventariali del Bosco e delle Altre terre boscate nonché quelle per le categorie forestali dei Boschi alti. Il 62.6% della superficie del Bosco risulta nella classe di terreno non accidentata. Almeno la metà della superficie del Bosco in tutte le regioni si trova nella stessa classe di accidentalità, ad eccezione della Valle d'Aosta, dove la percentuale è solo del 30.0% e il Bosco si trova prevalentemente su terreno accidentato (44.1% della superficie). La statistica per le Altre terre boscate risulta meno interessante, poiché affetta da una quota rilevante di superficie non classificata. Si nota una generale prevalenza di superficie nella prima classe di accidentalità a livello regionale, ma in tre regioni (Valle d'Aosta, Trentino e Lazio) la classe con maggiore estensione percentuale è la terza (molto accidentato). A livello delle categorie forestali dei Boschi alti, solo quattro categorie (Boschi di larice e cembro, Ostrieti e carpineti, Leccete e Altri boschi di latifoglie sempreverdi) hanno una quota inferiore al 50% della superficie nella prima classe (Fig. 8.4), che rimane la più estesa come casistica generale.

La disponibilità al prelievo legnoso viene valutata in maniera sintetica, considerando la presenza di limitazioni significative alle attività selvicolturali per motivi normativi o di carattere economico in misura tale da compromettere la possibilità o la remunerativitià del taglio (Table 8.8). Le restrizioni di tipo normativo possono derivare da esigenze di tutela del valore naturalistico, storico, culturale o di altro tipo. Limitazioni di carattere economico possono discendere da condizioni stazionali difficili come l'accessibilità dell'area (anche legata allo stato della viabilità forestale), la pendenza del terreno, l'accidentalità, ma anche la produttività e il valore del soprassuolo in relazione al mercato e alle consuetudini locali. La Table 8.9 mostra la ripartizione della superficie del Bosco e delle Altre terre boscate per disponibilità al prelievo legnoso. Le stime relative alle categorie inventariali del Bosco e delle Altre terre boscate, nonché quelle relative alle categorie forestali dei Boschi alti sono disponibili all'indirizzo inventarioforestale.org/statistiche_INFC. A livello nazionale l'88.5% della superficie del Bosco risulta disponibile al prelievo legnoso (Fig. 8.5) e la percentuale si mantiene elevata in tutte le regioni, generalmente sopra l'80% ad eccezione di Valle d'Aosta (63.1%), Trentino (74.6%) e Friuli-Venezia Giulia (61.2%). Anche a livello di categorie forestali dei Boschi alti si nota una netta prevalenza della superficie disponibile al prelievo, con valori superiori nella generalità dei casi all'80%, salvo per i Boschi di larice e cembro (63.9%), le Pinete di pino silvestre e montano e le Pinete di pino nero, laricio e loricato, ma queste ultime due categorie con valori molto prossimi (più del 79%) (Fig. 8.6). Le informazioni raccolte permettono di ripartire la superficie non disponibile per motivo di indisponibilità, come mostrato nella Fig. 8.7. Nel 67.8% dei casi, le limitazioni sono di carattere economico, nel 27.3% sono legate alla tutela delle risorse naturali e del territorio e solo una parte molto ridotta (3.0%) è esclusa dalle utilizzazioni per restrizioni legate ad interesse socioculturale.

Selvicoltura e utilizzazioni

Le pratiche selvicolturali applicate ai boschi italiani sono state valutate considerando la tipologia e l'intensità degli interventi, secondo quanto descritto nella Table 8.10. La Table 8.11 riporta le stime di superficie per il Bosco ripartita per pratiche colturali, mentre la Table 8.12 mostra statistiche analoghe per i Boschi alti e gli Impianti di arboricoltura da legno. A livello nazionale il 62.1% della superficie del Bosco risulta interessato da pratiche colturali di qualche tipo e intensità mentre è assente qualsiasi tipo di pratica sul 37.4% del Bosco in Italia. A livello regionale si osserva una variabilità piuttosto elevata (Fig. 8.8). L'assenza di pratiche, che caratterizza principalmente i soprassuoli di aree scomode o poco convenienti da utilizzare, interessa una percentuale del Bosco che va dal 12.0% (Alto Adige) al 72.0% (Sicilia). Aliquote superiori al 50% di Bosco non interessato da pratiche colturali si riscontrano anche per la Valle d'Aosta, la Liguria e l'Abruzzo, ma altre cinque regioni si approssimano a quella percentuale (Piemonte, Friuli-Venezia Giulia, Puglia, Basilicata, Calabria). Le pratiche ordinarie minimali sono la forma più frequente tra quelle applicate in tutte le regioni, con l'eccezione dell'Alto Adige e del Trentino, dove quelle ordinarie classiche interessano più del 50% della superficie del Bosco. In linea generale, quindi, gli interventi più diffusamente applicati si limitano al prelievo della massa legnosa a maturità o fine turno. Le pratiche colturali ordinarie classiche, che oltre alla raccolta del prodotto alla fine del ciclo produttivo prevedono interventi di miglioramento dei soprassuoli, interessano una porzione di Bosco a livello nazionale del 14.6%. Nelle regioni questa aliquota è superata solo in Alto Adige (53.1%), Trentino (55.6%), Veneto (30.8%), Toscana (18.0%) e Basilicata (19.7%). Pratiche colturali intensive, speciali o di altro tipo si riscontrano poco frequentemente e di solito per superfici limitate. Le pratiche ordinarie intensive interessano più del 3% della superficie solo in Piemonte (3.3%), Lombardia (4.0%) e Sardegna (4.1%) e nelle ultime due sono dovute esclusivamente alla presenza di Impianti di arboricoltura da legno. Pratiche colturali speciali per produzioni secondarie sono applicate in Campania sul 4.7% della superficie mentre in Sardegna sul 16.0% della superficie. Le statistiche sulle produzioni secondarie, mostrate nella Table 8.13, permettono di osservare che per la Campania quella percentuale si deve alla produzione di castagne, mentre per la Sardegna quasi esclusivamente alla produzione di sughero e, in minima parte, di foraggio e castagne. Altre pratiche interessano percentuali del Bosco regionale sopra il 3% in Friuli-Venezia Giulia (3.5%), Marche (6.2%), Molise (7.5%) e Sardegna (8.0%).

Nel caso di pratiche colturali ordinarie è stata rilevata la modalità di utilizzazione del soprassuolo, basata sul sistema di classi riportate nella Table 8.14. Le relative stime di superficie sono riportate per il Bosco nella Table 8.15 e per le categorie inventariali del Bosco nella Table 8.16. Il taglio raso con riserve, nel caso delle fustaie, e con rilascio di matricine, in quello dei cedui, è la forma di utilizzazione più diffusa per il Bosco a livello nazionale (27.3%); generalmente essa è anche la più frequente nelle varie regioni (Fig. 8.9), con alcune eccezioni soprattutto nelle regioni alpine.

Analogamente alle pratiche colturali, le modalità di esbosco sono state valutate solo per i Boschi interessati da pratiche colturali ordinarie; la Table 8.17 riporta le classi adottate. Le stime di superficie ottenute sono riportate per il Bosco nella Table 8.18 e per le sue diverse categorie inventariali nella Table 8.19. A livello nazionale l'esbosco avviene principalmente caricando la legna o il legname sul mezzo di trasporto direttamente sul letto di caduta (29.3%), ma a livello regionale la situazione è diversificata. Nelle regioni alpine prevale lo strascico (con l'eccezione della Valle d'Aosta, con avvallamento sul 25.5% della superficie del Bosco), e in quelle appenniniche e nelle isole prevale il carico diretto (con l'eccezione della Calabria, con strascico sul 32.4% della superficie). Sistemi a fune sono adoperati solo nelle regioni del nord (fino all'Emilia-Romagna ma per una quota di solo l'1.0%) e l'elicottero solo in Valle d'Aosta (14.0% della superficie) e Alto Adige (0.8%) (Fig. 8.10).

Le stime sul volume annualmente utilizzato nel Bosco italiano sono riportate nella Table 8.20. Nei dodici mesi precedenti il rilevamento inventariale sono stati abbattuti alberi per 9.6 Mm^3, per un valore di 1.1 m^3 per ettaro. Le regioni con valori assoluti più alti sono Veneto, Piemonte, Trentino ed Emilia-Romagna (superiori a 800,000 m^3) (Fig. 8.11). Per i valori di densità, la figura mostra che Veneto e Trentino sono le due uniche regioni che utilizzano un volume legnoso superiore a 2 m^3 ha^{-1}, con rispettivamente 3.2 m^3 ha^{-1} e 2.9 m^3 ha^{-1}. Le utilizzazioni in Molise e Puglia interessano volumi molto bassi. In termini di fitomassa, le utilizzazioni a livello nazionale corrispondono a 6.3 MMg, per un valore di 0.7 Mg ha^{-1}, come mostrato nella Table 8.21 e nella Fig. 8.12.

Pianificazione

La pianificazione forestale è oggetto di normative nazionali e regionali, a seguito del trasferimento alle Regioni delle competenze in materia di foreste, avvenuto con il D.P.R. n. 616/1977, e per il mantenimento di quelle relative alle funzioni di indirizzo e di coordinamento presso lo Stato. Sul piano pratico, oltre ad una certa complessità, sono derivate differenze abbastanza marcate tra le regioni. INFC rileva la presenza di tre livelli della pianificazione forestale, come descritto nella Table 8.22. A livello nazionale, il R.D.L. n. 3267/23 detta le regole per l'esecuzione delle utilizzazioni boschive in conformità a piani economici e a Prescrizioni di Massima e di Polizia

Forestale (PMPF), al fine di tutelare l'equilibrio idrogeologico dei territori montani. Come strumento prescrittivo per le attività selvicolturali e di gestione dei boschi a livello delle Province, in qualche misura le PMPF possono delineare gli ambiti della pianificazione forestale e rappresentano il piano più generale della rilevazione di INFC. La pianificazione di orientamento comprende strumenti a carattere generale validi su estensioni territoriali più o meno ampie, come ad esempio le aree montane, i parchi e le riserve naturali, validi anche per usi del suolo diversi dal bosco. Il livello di maggiore dettaglio è quello tipico della pianificazione forestale nella sua accezione classica, programmata mediante piani economici, anche noti come piani di assestamento o piani aziendali, generalmente relativi ai boschi di una singola proprietà forestale. Nel caso di coesistenza, INFC2015 ha registrato tutti i livelli di pianificazione presenti.

La Table 8.23 riporta le stime di superficie relative alla presenza di almeno una delle forme di pianificazione previste, per il Bosco e le Altre terre boscate; la Table 8.24 riporta stime analoghe per i Boschi alti e gli Impianti di arboricoltura e la Table 8.25 quelle per le categorie inventariali delle Altre terre boscate. La Table 8.26 mostra l'estensione della superficie del Bosco e delle Altre terre boscate per presenza di pianificazione di dettaglio, mentre le due tabelle successive (Tables 8.27 e 8.28) mostrano la ripartizione delle superfici per presenza di pianificazione di dettaglio per le categorie inventariali rispettivamente del Bosco e delle Altre terre boscate. All'indirizzo inventarioforestale.org/statistiche_INFC sono inoltre pubblicate le stime di superficie per presenza di PMPF per il Bosco e le Altre terre boscate, nonché quelle per le categorie inventariali del Bosco e delle Altre terre boscate. Analogamente, per la pianificazione di orientamento la stessa fonte fornisce le stime di superficie per il Bosco e le Altre terre boscate e per le loro categorie inventariali. Infine, solo per i Boschi alti, sono disponibili le stime di superficie relative allo stato della pianificazione forestale per le categorie forestali, sempre secondo i tre livelli di PMPF, pianificazione di orientamento e pianificazione di dettaglio. A livello nazionale il 92.2% della superficie del Bosco risulta regolamentata in almeno uno dei tre livelli della pianificazione considerati, valore che scende al 50.7% per la superficie delle Altre terre boscate; va però segnalato che il confronto potrebbe risentire della quota piuttosto elevata di superficie non classificata per queste ultime (23.0%). Se si osserva soltanto il Bosco, la percentuale di superficie con pianificazione forestale ad almeno uno dei tre livelli rimane piuttosto elevata in tutte le regioni e generalmente superiore all'80%. La Fig. 8.13 mostra la percentuale di superficie del Bosco, per regione e per l'Italia, in cui è presente pianificazione nei tre livelli considerati. Si può notare che le PMPF contribuiscono in maniera prevalente ad assicurare una qualche forma di pianificazione e che, dei tre livelli considerati, è quello con minore variabilità di superficie interessata tra le diverse regioni. Le stime per il Trentino in merito alla ridotta presenza di PMPF sono dovute alla presenza di regolamenti alternativi analoghi, in virtù dell'autonomia amministrativa speciale e della elevata percentuale di superficie forestale gestita secondo piani di assestamento che, secondo la normativa locale, sostituiscono e superano le regolamentazioni di carattere generale. La pianificazione di orientamento riguarda un'aliquota molto limitata della superficie del Bosco (2.3%), e il dato nazionale è superato solo in sette regioni: Valle d'Aosta, Lombardia, Veneto, Emilia-Romagna, Marche, Lazio e Sicilia. Anche la pianificazione di dettaglio risulta interessare un'aliquota del Bosco piuttosto limitata, pari al 15.5% a livello nazionale. Otto regioni superano tale valore nazionale (Valle d'Aosta, Lombardia, Alto Adige, Trentino, Veneto, Friuli-Venezia Giulia, Molise, Campania) generalmente nel nord Italia. Le uniche due regioni in cui la percentuale di superficie con piani di assestamento supera il 40% sono l'Alto Adige (94.6%) e il Trentino (75.8%), che si contraddistinguono per percentuali di superficie del Bosco dotate di un piano di gestione aziendale molto elevate.

References

Bernetti, G. (1989). Assestamento forestale. D.R.E.A.M. Firenze (261 pp.).

Bosela, M., Gasparini, P., Di Cosmo, L., Parisse, B., De Natale, F., Esposito, S., & Scheer, L. (2016). Evaluating the potential of an individual-tree sampling strategy for dendroecological investigations using the Italian National Forest Inventory data. *Dendrochronologia, 38*, 90–97. https://doi.org/10.1016/j.dendro.2016.03.011

De Philippis, A. (1983). Uso e conservazione degli ecosistemi forestali. *Italia Forestale e Montana, 48*(1), 1–9.

Di Cosmo, L., Gasparini, P., & Floris, A. (2011). Accessibilità e disponibilità al prelievo. In: P. Gasparini, & G. Tabacchi (Eds.), *L'Inventario Nazionale delle Foreste e dei serbatoi forestali di Carbonio INFC 2005. Secondo inventario forestale nazionale italiano. Metodi e risultati* (pp. 299–304). Edagricole-Il Sole 24 Ore, Milano. ISBN 978-88-506-5394-2.

Dorren, K. A., Berger, F., Imeson, A. C., Maier, B., & Rey, F. (2004). Integrity, stability and management of protection forests in the European Alps. *Forest Ecology and Management, 195*, 165–176. https://doi.org/10.1016/j.foreco.2004.02.057

Hippoliti, G. (1990). Esigenze di accessibilità nei boschi. In: *Le utilizzazioni forestali* (pp. 97–98). CUSL.

Hippoliti, G. (2004). La gestione delle utilizzazioni forestali. *Italia Forestale e Montana, 59*(6), 415–420.

Hippoliti, G. (2005). I problemi delle utilizzazioni. *Italia Forestale e Montana, 60*(2), 167–169.

O'Hara, K. L. (2016). What is close-to-nature silviculture in a changing world? *Forestry, 89*, 1–6. https://doi.org/10.1093/forestry/cpv043

Perry, D. A. (1998). The scientific basis of forestry. *Annual Review of Ecology, Evolution, and Systematics, 29*, 435–66. 0066-4162/98/1120-0435$08.00.

Piussi, P. (1994). *Selvicoltura generale* (245 pp.). UTET.

Open Access This chapter is licensed under the terms of the Creative Commons Attribution 4.0 International License (http://creativecommons.org/licenses/by/4.0/), which permits use, sharing, adaptation, distribution and reproduction in any medium or format, as long as you give appropriate credit to the original author(s) and the source, provide a link to the Creative Commons license and indicate if changes were made.

The images or other third party material in this chapter are included in the chapter's Creative Commons license, unless indicated otherwise in a credit line to the material. If material is not included in the chapter's Creative Commons license and your intended use is not permitted by statutory regulation or exceeds the permitted use, you will need to obtain permission directly from the copyright holder.

Chapter 9
Biodiversity and Protected Wooded Lands

Biodiversità e protezione delle foreste

Lucio Di Cosmo and Antonio Floris

Abstract The importance of forests for their functions other than timber and wood production has dramatically increased in the last decades with the increased awareness of the risks deriving from deforestation and the acknowledgment of the great amount of goods and benefits forests provide. Consequently, national forest inventories have widened their objectives and nowadays include variables related to environmental aspects. Among these aspects, biodiversity plays a key role for forest ecosystems' adaptation to climate change. This chapter details the INFC2015 estimates regarding tree species diversity. It also shows the estimates on the naturalness of the stands' regeneration processes and those on the presence and type of deadwood in forests. In addition to carbon storage, standing dead trees, stumps and lying deadwood also have a great potential for biodiversity. Forest protection is also pursued through laws and policies that allow for the creation of protected areas of various type and protection degree. The main inventory statistics on wooded lands in protected areas are given in the last section of this chapter.

Keywords Species richness · Regeneration · Coarse woody debris · Volume increment · Protected areas · Natura2000

9.1 Introduction

Biodiversity contributes to the functioning of the forest ecosystem and the provision of ecosystem services (e.g., Brockerhoff et al., 2017) that are important for human well-being. Since 1992, the international community has committed to conserve biodiversity under the Rio de Janeiro declaration. Biodiversity as a concept seems to be intuitive and easy to understand, but this is not the case. Furthermore, it does not

L. Di Cosmo (✉) · A. Floris
CREA Research Centre for Forestry and Wood, Trento, Italy
e-mail: lucio.dicosmo@crea.gov.it

A. Floris
e-mail: antonio.floris@crea.gov.it

© The Author(s) 2022

P. Gasparini et al. (eds.), *Italian National Forest Inventory—Methods and Results of the Third Survey*, Springer Tracts in Civil Engineering,
https://doi.org/10.1007/978-3-030-98678-0_9

necessarily imply naturalness (Pignatti, 1995a). Generally, biodiversity is referred to five levels, ranging from genetics to the geographical region passing through species, ecosystem and landscape biodiversity (Bernetti, 2005). The survey of variables specifically related to biodiversity is a recent addition to national forest inventories (Gasparini et al., 2013). INFC contributes by providing useful information for the assessment of biodiversity on four out of five of these levels. Estimates in Chap. 7 on the inventory categories and the forest types that characterise the Italian territories, such as the regions and the macro-regions, are related to the first three broad levels, i.e., the regional, landscape and community diversity. Further information is available at inventarioforestale.org/statistiche_INFC%20link, with special reference to the forest subtypes. This chapter describes INFC2015 results on the tree species diversity (Sect. 9.2), stand origin (Sect. 9.3), deadwood (Sect. 9.4) and the wooded lands in protected areas (Sect. 9.5).

Italian forest vegetation is marked by a high degree of diversity (Pedrotti, 1995). This is also due to the great variety of environmental conditions and climates that contribute to the general floristic richness of Italy (Pignatti, 1994). Although the number of species is not exhaustive in showing the biodiversity of a studied area (e.g., Pignatti, 1995b), it is considered a natural index that forms the basis of many ecological models (Gotelli & Colwell, 2001). Species richness is also one of the most important features for the naturalistic value assessment of forests (Pignatti & De Natale, 2011).

Naturalness of forests may also be evaluated based on regeneration, which INFC considers under the stand origin assessment. The most natural conditions occur when regeneration is not assisted or favoured by any human activity; to the contrary, a seeded or planted stand is marked by an artificial origin. In such a case, the impact on the environment may vary considerably depending on the material used, e.g., seeds collected in the same area or exotic species for timber production. Regeneration assisted or favoured through silvicultural operations, which creates semi-natural stands, generally guarantees conservation of local ecotypes, unless what is favoured is an exotic species or provenance previously introduced.

Deadwood is a key component of forest ecosystems. It contributes to biodiversity conservation, to carbon, nitrogen and phosphorus cycles. In addition, deadwood influences stand dynamics and regeneration of natural and semi-natural forests, plays a role in soil stabilisation on deep slopes and along the rivers (e.g., Paletto et al., 2012) and may hinder avalanche release (Berretti et al., 2006). Being combustible, as well as a resource for energy, deadwood in forests plays a role in the risk and propagation model of fires. Countries that have signed agreements on climate change mitigation need to estimate the carbon stored in deadwood, because it is one of the five forest pools recognised by IPCC (IPCC, 2003). Inclusion of deadwood among the variables measured by national forest inventories resulted in an unprecedented availability of data collected over wide areas that also showed severe limits in the modelling (e.g., Woodall et al., 2019) and demonstrated the importance of direct measurement of deadwood.

Protected areas play a crucial role in biodiversity conservation and in the safeguard of various ecosystems, including forests (Dudley & Phillips, 2006). National forest inventories may provide suitable information on protected forests. Forests are one of the essential components of the protected areas in Italy.

9.2 Tree Species

Richness of tree species is a prerequisite for having forests under diverse environmental conditions, whose diversified ecology and functioning is also important to assure the many ecosystem services that are required nowadays. INFC2015 has measured (DBH $\geq$ 4.5 cm) approximately 180 woody species.

Table 9.1 lists the 45 main species, in terms of growing stock in the Italian Forest. It shows estimates for the number of trees, basal area, growing stock volume and aboveground tree biomass, aboveground carbon stock and annual volume increment. Forty-five species have been selected from a broad list containing all the main species in each region, i.e., the species that constitute 90% of the growing stock in the region. Similar inventory statistics for the regions are available at inventarioforestale.org/statistiche_INFC%20link. Estimates on the same variables for the inventory categories and for forest types at the national and regional levels are also available. All tables share the same criterion in listing the species, i.e., according to decreasing volume values. However, the residual class other species is always at the bottom and may consist of different species depending on the variable. Figures 9.1, 9.2 and 9.3 show the cumulative distribution of the number of trees (Fig. 9.1), the growing stock (Fig. 9.2) and the aboveground tree biomass (Fig. 9.3) in the Italian Forest for the main 45 species.

The figures show that 85% of growing stock and aboveground tree biomass is due to 17 species; except for *Pinus laricio* Poiret, they are all on the list of 27 species that constitutes 85% of the number of trees. The same species are also the most important in terms of annual volume increment, and this is consistent with studies that showed volume as the main predictive variable in the increment estimation (e.g., Bevilacqua, 1999; Gasparini et al., 2017); altogether, the 17 species make up 80% of the annual increment.

Beech is the most highly represented species of the Italian Forest. For the considered characteristics, basal area is 15.2% of the total, growing stock volume is 18.1%, aboveground tree biomass is 20.7% and annual volume increment is 14.2%. Finally, numerosity of subjects is 10.8% of the total, and in second position of the list. The second most represented species in terms of basal area, growing stock volume, aboveground tree biomass and annual volume increment is the Norway spruce. It represents 3.4% of the total number of trees, 10.2% of basal area, 15.3% of growing stock, 11.1% of aboveground tree biomass and 11.8% of annual volume increment. Chestnut and Turkey oak are the two following species and their relative position changes depending on the variables considered. Chestnuts account for 6.2% of Italian Forest trees, 9.6% of basal area, 9.0% of growing stock volume, 8.1% of aboveground tree biomass and 9.9% of annual volume increment. Turkey oak trees constitute 7.6% of the total, 8.2% of basal area, 8.1% of growing stock volume, 9.4% of aboveground

Table 9.1 Total number of trees, basal area, growing stock and aboveground tree biomass, carbon stock and annual increment by species in Forest for Italy / Valori totali del numero di alberi, area basimetrica, volume, fitomassa epigea, carbonio organico e incremento annuo per specie, per il Bosco in Italia

Species/Specie	Number		Basal area		Volume		Biomass		Carbon stock		Annual volume increment	
	Numero		Area basimetrica				Fitomassa		Carbonio organico		Incremento annuo di volume	
	(n)	ES (%)	(m^2)	ES (%)	(m^3)	ES (%)	(Mg)	ES (%)	(Mg)	ES (%)	(m^3)	ES (%)
Fagus sylvatica	1,243,763,816	3.7	30,509,358	2.4	271,997,316	2.7	215,963,927	2.7	107,981,964	2.7	5,356,682	2.6
Picea abies	389,248,983	4.2	20,605,946	3.1	229,843,756	3.5	116,196,540	3.5	58,098,270	3.5	4,453,988	3.5
Castanea sativa	715,898,238	5.0	19,414,129	3.4	135,980,129	3.5	84,694,702	3.4	42,347,351	3.4	3,724,482	3.8
Quercus cerris	874,354,050	3.8	16,465,872	2.9	121,502,429	3.5	98,513,868	3.5	49,256,934	3.5	3,468,967	3.4
Larix decidua	130,299,538	5.2	10,019,908	4.1	90,196,939	4.6	49,037,263	4.5	24,518,632	4.5	1,517,416	4.3
Quercus pubescens	890,622,781	3.5	14,850,392	2.7	86,889,856	3.0	73,846,717	2.9	36,923,359	2.9	2,515,774	3.3
Ostrya carpinifolia	1,285,104,341	4.4	10,360,412	3.6	54,797,596	3.8	48,125,130	3.7	24,062,565	3.7	1,775,369	4.0
Quercus ilex	793,473,615	6.1	9,189,723	4.6	48,935,748	5.5	49,716,493	5.6	24,858,247	5.6	1,332,377	5.8
Abies alba	75,617,563	8.6	3,717,556	7.0	40,190,267	7.2	20,462,253	7.1	10,231,127	7.1	794,934	7.7
Pinus nigra	109,694,970	8.4	4,677,035	7.1	38,322,551	8.8	23,361,340	8.4	11,680,670	8.4	836,470	7.2
Pinus sylvestris	100,703,445	8.3	4,462,495	5.8	34,753,875	6.3	19,288,759	6.3	9,644,380	6.3	631,908	6.5
Pinus laricio	43,577,969	11.0	3,707,799	9.4	28,603,042	9.8	11,987,070	9.8	5,993,535	9.8	699,828	10.1
Fraxinus ornus	1,064,717,238	3.9	5,583,372	3.5	25,824,661	3.6	24,152,525	3.6	12,076,262	3.6	835,370	3.8
Robinia pseudacacia	285,399,551	8.2	3,572,949	7.4	24,709,096	7.9	20,028,573	7.7	10,014,286	7.7	1,038,460	8.7
Fraxinus excelsior	106,466,790	10.2	1,855,146	8.1	14,611,441	8.7	12,538,561	8.6	6,269,281	8.6	464,309	8.6
Quercus petraea	63,744,258	10.7	1,895,197	8.7	14,427,974	9.7	11,680,086	9.7	5,840,043	9.7	398,547	10.0
Acer pseudoplatanus	125,926,766	8.9	2,034,156	6.9	14,166,082	7.2	10,722,421	7.1	5,361,211	7.1	463,111	7.7
Pinus pinea	16,800,835	17.5	1,386,544	13.7	11,400,272	15.6	6,992,551	15.6	3,496,276	15.6	203,299	15.1
Pinus halepensis	39,589,366	11.4	1,649,562	10.2	11,249,444	11.7	7,871,675	11.7	3,935,837	11.7	370,675	11.1
Pinus pinaster	23,692,087	14.5	1,397,528	12.3	11,065,100	13.9	5,998,932	14.0	2,999,466	14.0	227,393	12.9
Pinus cembra	22,794,295	14.1	1,320,527	14.6	10,529,804	15.2	5,064,542	15.2	2,532,271	15.2	144,407	14.7
Prunus avium	96,334,726	8.3	1,788,141	6.6	10,073,429	6.8	6,132,114	6.6	3,066,057	6.6	354,515	7.5

(continued)

Table 9.1 (continued)

Species/Specie	Number		Basal area		Volume		Biomass		Carbon stock		Annual volume increment	
	Numero		Area basimetrica				Fitomassa		Carbonio organico		Incremento annuo di volume	
	(n)	ES (%)	(m^2)	ES (%)	(m^3)	ES (%)	(Mg)	ES (%)	(Mg)	ES (%)	(m^3)	ES (%)
Quercus suber	47,061,195	9.0	2,212,841	7.3	9,580,532	7.6	9,718,904	7.7	4,859,452	7.7	207,061	9.7
Acer campestre	155,822,504	9.1	1,648,549	9.1	8,633,146	9.7	6,931,589	9.2	3,465,795	9.2	312,154	9.9
Pseudotsuga menziesii	7,415,983	22.7	708,865	17.9	8,444,770	18.0	4,275,985	18.0	2,137,992	18.0	238,869	18.6
Quercus robur	24,685,758	21.3	974,252	13.5	7,877,007	14.0	6,370,784	14.0	3,185,392	14.0	234,826	13.2
Alnus glutinosa	27,733,707	12.8	998,469	12.8	7,813,947	13.3	4,670,865	13.1	2,335,433	13.1	240,307	13.8
Populus nigra	19,634,503	18.6	1,409,763	11.4	7,468,733	12.0	3,797,154	11.9	1,898,577	11.9	333,311	16.8
Carpinus betulus	128,636,578	11.8	1,230,302	9.2	7,396,373	9.8	6,302,545	9.6	3,151,273	9.6	227,145	10.3
Corylus sp.	441,722,494	7.9	1,616,560	8.2	7,138,632	8.1	5,371,848	8.3	2,685,924	8.3	255,262	10.1
Betula spp.	61,681,003	14.3	1,181,058	10.0	6,962,033	9.7	4,319,460	9.9	2,159,730	9.9	214,770	10.7
Populus tremula	46,368,780	15.9	955,221	11.8	6,753,453	12.6	3,781,674	12.0	1,890,837	12.0	233,259	13.9
Acer gr. Opalus	81,663,283	12.0	886,653	9.9	5,374,845	10.6	4,249,497	10.3	2,124,749	10.3	167,367	12.0
Alnus cordata	28,107,144	18.8	743,592	17.2	5,188,850	17.8	3,184,178	17.7	1,592,089	17.7	217,992	20.3
Tilia cordata	31,245,163	18.6	700,376	16.4	4,860,446	17.5	2,703,555	17.0	1,351,777	17.0	141,212	16.5
Eucalyptus sp.	38,549,159	14.6	771,292	13.1	4,683,027	13.8	3,488,070	13.7	1,744,035	13.7	224,566	16.0
Arbutus unedo	265,331,818	10.9	1,228,505	11.3	4,153,848	10.8	3,763,845	11.4	1,881,922	11.4	113,354	12.2
Ulmus minor	94,704,608	13.3	751,918	13.5	3,700,156	13.1	2,631,563	13.5	1,315,782	13.5	188,230	14.8
Quercus frainetto	41,484,603	21.6	575,767	17.9	3,606,476	19.3	2,935,286	19.2	1,467,643	19.2	121,123	19.5
Populus xcanadensis	8,399,035	25.8	485,780	20.9	3,604,239	22.0	1,825,221	21.2	912,610	21.2	234,953	38.2
Salix caprea	39,283,243	14.3	579,465	10.8	3,509,695	11.2	2,293,089	10.8	1,146,545	10.8	136,650	12.5
Sorbus aria	105,701,803	11.4	705,835	9.1	3,115,514	9.1	2,379,766	9.0	1,189,883	9.0	89,873	10.2
Populus alba	18,382,395	25.6	534,293	18.9	3,039,734	19.1	1,663,544	18.1	831,772	18.1	144,237	21.7
Alnus incana	14,099,655	26.4	275,602	24.2	2,011,056	28.0	1,237,445	27.0	618,723	27.0	74,455	24.3
Quercus trojana	22,511,408	21.6	238,119	20.3	1,096,541	23.5	1,088,295	23.9	544,147	23.9	33,753	21.7
Other species	1,233,511,873	3.9	9,307,162	3.5	46,723,230	4.5	32,916,402	4.3	16,458,201	4.3	1,794,777	6.1
Totals/Totali	11,481,562,909	1.2	201,183,984	0.8	1,502,807,089	1.0	1,044,276,604	1.0	522,138,302	1.0	37,787,784	1.0

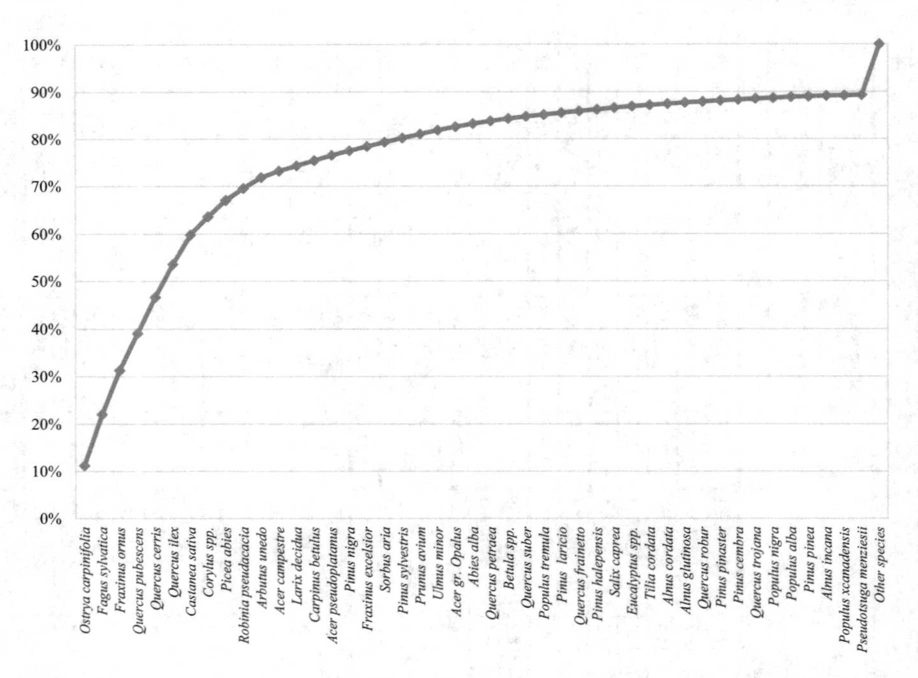

Fig. 9.1 Cumulative frequency distribution of the number of trees for the 45 main species in the Italian Forest / Distribuzione di frequenza cumulata del numero di alberi per le 45 specie più abbondanti nel Bosco in Italia

tree biomass and 9.2% of annual volume increment. Beech, Norway spruce, chestnut and Turkey oak together account for slightly more than 50% of growing stock volume and aboveground tree biomass of the Italian Forest, and 45.5% of its annual volume increment.

These four species characterise four forest types. The abundance of a species in the proper forest type is variable but always consistent: basal area ranges from 76.6% for Turkey oak in the Mediterranean oaks type to 89.0% for beech in the Beech forest type. Figure 9.4. shows the percentage of the basal area of the four species in the Tall trees forest types, and groups all others in the class other types; the figure has been drawn in such a way as to improve readability by limiting the X-axis.

The group of species that constitutes 85% of the total number of trees is wider and lists 27 species. This is due to the presence of small-sized species, often found as stools in pure or mixed coppice stands.

Some of the 45 main species described are exotic species. Black locust (*Robinia pseudacacia* L.) is a species long time naturalised with a relevant number of trees (2.5% of the total), 1.6% of growing stock volume, 1.9% of aboveground tree biomass and 2.7% of annual volume increment. *Eucaliptus* spp. trees (0.3% of the total growing stock and 0.6% of the annual volume increment) and Douglas fir (0.6% of the total growing stock and 0.6% of the annual volume increment), are planted species with a more limited occurrence.

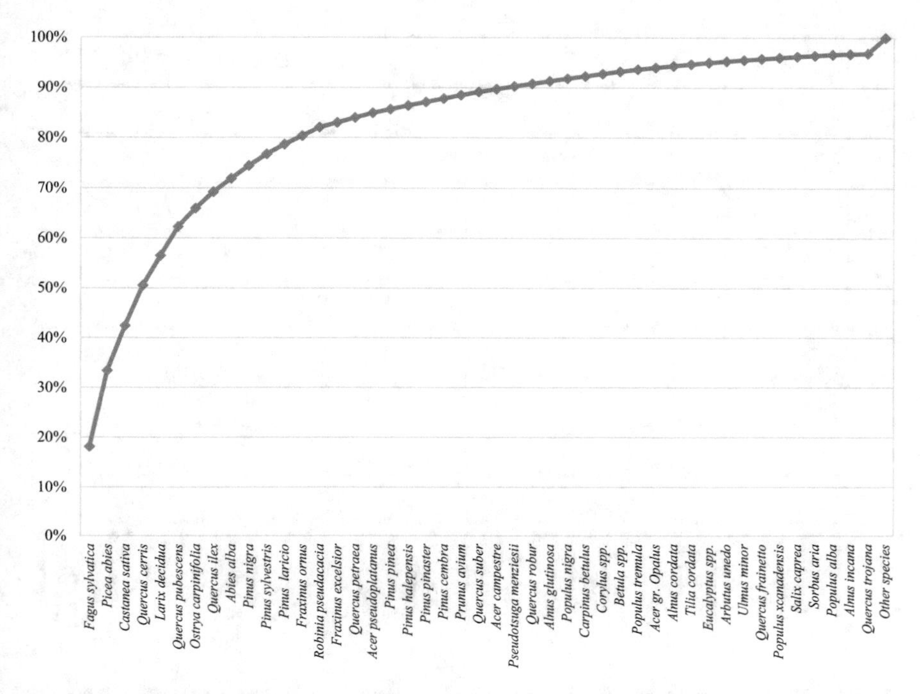

Fig. 9.2 Cumulative frequency distribution of the growing stock volume for the 45 main species in the Italian Forest / Distribuzione di frequenza cumulata del volume legnoso per le 45 specie più abbondanti nel Bosco italiano

9.3 Stands Origin

By stand origin INFC means the naturality of the regeneration process, i.e., the intensity of human intervention to favour or assist regeneration. Table 9.2 shows the three classes adopted. Tables 9.3 and 9.4 show the estimates on Forest and on Other wooded land area by stand origin. Tables 9.5 and 9.6 show corresponding statistics for Tall trees forest and Plantations. Tables 9.7 and 9.8 show the estimates for the inventory categories of Other wooded land. The inventory statistics on the Tall trees forest types area by stand origin are available at inventarioforestale.org/statistiche_INFC.

Forests are mainly semi-natural in Italy, that is regeneration is obtained or guided by silvicultural activity. Semi-natural forests amount to 68.0% at the national level but this is also the prevailing origin for stands in all the regions except Sicilia (27.7%), where it is in the second position after the natural origin class (36.8%). Calabria and Puglia are the only two remaining regions where the percentage of semi-natural stands does not reach 50%, although it is rather close (48.8% and 48.4%, respectively). Natural stands make 17.1% of Forest area. Variability among the regions is also rather high in this class, but natural stands remain the second widest area in all regions

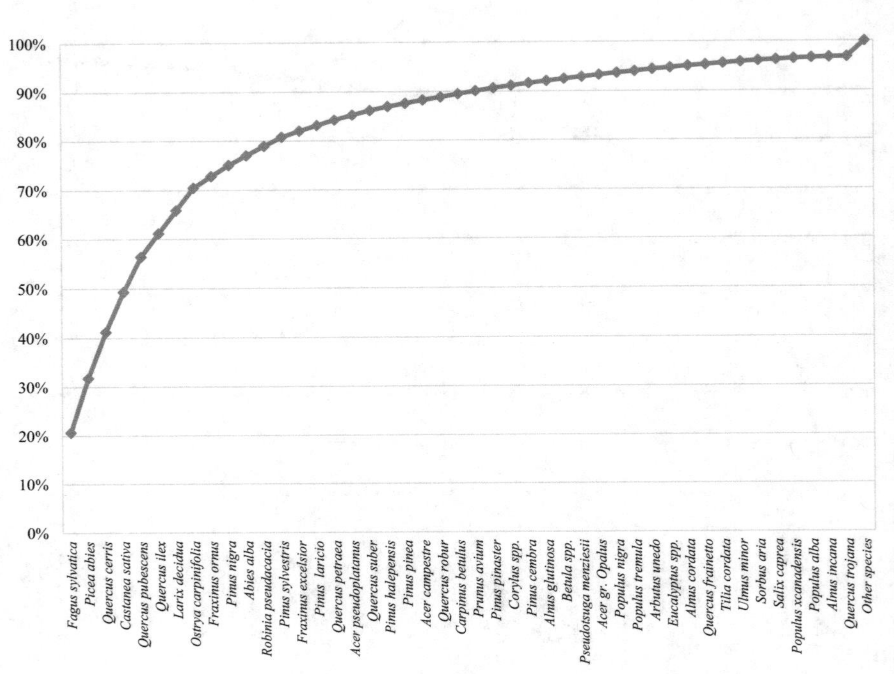

Fig. 9.3 Cumulative frequency distribution of the aboveground tree biomass for the 45 main species in the Italian Forest / Distribuzione di frequenza cumulata della biomassa arborea epigea per le 45 specie più abbondanti nel Bosco italiano

except Molise, where the percentage of planted forests is slightly higher. Natural stands are those originating after a natural disaster, e.g., forest fires, and without human intervention, but they also begin from primary or secondary successions, which are very important for forest expansion.

Comparisons between Fig. 9.5, which shows the percentage of the three stand origin classes for Forest, and Fig. 9.6, which is the equivalent for Other wooded land, show that importance of semi-natural and natural origin is inverted between Forest and Other wooded land, although the relevant unclassified percent area in the Other wooded land must be considered. However, the predominance of natural stands in Other wooded land may be mainly ascribed to tree colonisation of former agricultural lands where human interventions to guide regeneration are still limited (exception for Molise and Sardegna). Some of the stands on Other wooded land will develop into true forests and this dynamic explains in part the naturally originated forest area. To some extent, naturally originated forests may be considered an intermediate condition that lasts from the time when they transit from the Other wooded land to Forest (by reaching the threshold values of the parameters that are relevant under the adopted definitions) to the time when silviculture becomes an option under an economical point of view. Planted forests are 6.3% of the Forest area at the national level and this percentage is not reached in most regions. Sicilia is an exception because that

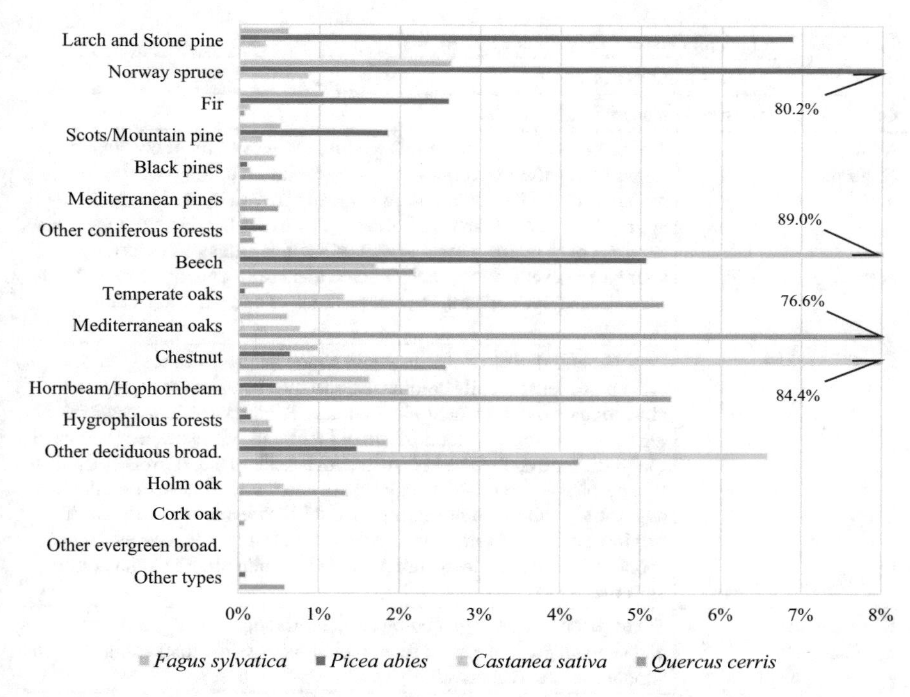

Fig. 9.4 Percentage breakdown of the main four species basal area in the forest types. X-axis was limited to 8% to improve readability; exceeding values are given in numbers / Ripartizione percentuale dell'area basimetrica delle quattro specie principali nelle categorie forestali. L'asse delle ascisse è stato limitato al valore 8% per migliorarne la leggibilità; i valori eccedenti sono mostrati in numero

percentage is the highest among all regions (27.4% of Forest area), but it is also rather close to the semi-natural Forest area (27.7%). As expected, in the Other wooded land the artificial origin class is absent in many regions and very limited in the others.

9.4 Deadwood

INFC2015 has estimated the three components of deadwood as standing dead trees, deadwood lying on the ground and stumps (Fig. 9.7) according to size thresholds generally consistent with those adopted in the international reporting (e.g., FAO, 2018). Under these thresholds, the three components combined are usually referred to as coarse woody debris. For each component, INFC2015 provides information on volume, deadwood biomass and organic carbon; standing dead trees and stumps are also estimated in terms of number. Tables with the estimates on the number and volume of standing dead trees (Tables 9.9 and 9.10) and on the number and volume of stumps (Tables 9.11 and 9.12) are shown, as well as those on the volume

Table 9.2 Stand origin classes / Origine delle fitocenosi

Stand origin	Description
Origine del soprassuolo	Descrizione
Natural Naturale	The stand derives from natural seeding by trees of the preceding cycle; no past silviculture action, such as regeneration cutting, can be recognised; the tree distribution is mostly irregular, random or partly aggregated. The crowns are normally at various heights, with gaps and canopy cover interruptions. The class includes areas colonised by spontaneous vegetation, such as scree and burnt land, and new invasive formations after abandonment of pastures or extensive agricultural practice
	Popolamenti derivati dalla disseminazione naturale dei soggetti del ciclo precedente o dalla colonizzazione spontanea di aree agricole abbandonate o di aree naturali (es. macereti) da parte della vegetazione legnosa. Non sono evidenti tracce di interventi selvicolturali (compresi i tagli di rinnovazione) e la distribuzione delle piante è irregolare, di tipo casuale o parzialmente aggregato, con vuoti e interruzioni della copertura e chiome di norma su più livelli. Non sono considerati di origine naturale oliveti, vigneti e frutteti invasi da vegetazione spontanea, in cui siano ancora presenti le piante arboree della coltura agricola
Semi-natural Seminaturale	Stand regeneration is guided by silvicultural activity; signs of cutting and/or management can be seen, including possible thickening to support natural regeneration
	Popolamenti la cui costituzione è stata guidata da interventi selvicolturali, quali tagli di rinnovazione e/o cure colturali, o da eventuali rinfoltimenti a sostegno della rinnovazione naturale. Le tracce di tali interventi risultano ancora riconoscibili
Artificial Artificiale	Afforestation or reforestation (including overage), plantations of indigenous or foreign species from seeding or planting, including abandoned plantations
	Rimboschimenti e imboschimenti, anche se invecchiati, e piantagioni derivate da semina o da impianto di specie indigene o introdotte, incluse le piantagioni abbandonate

of the lying deadwood (lying coarse woody debris, Table 9.13) and coarse woody debris total volume (Table 9.14) for Forest. Analogue statistics for the forest types are available at inventarioforestale.org/statistiche_INFC. Estimates on the deadwood biomass and the carbon stocked by the three deadwood components for either the inventory categories or the forest types of Forest are also available.

There are 1.37 billion standing dead trees in the Italian Forest, 150.3 ha^{-1} on average. The number of standing dead trees is 11.9% of the living trees; however, that percentage is variable among the regions (from 2.5% of Puglia to 20.7% of Liguria). The regional data of standing dead trees plotted on the living ones are well interpolated through a power function (Fig. 9.8).

Standing dead trees amount to 67.3 million m^3, 7.4 m^3 ha^{-1} on average. Lying deadwood is the second component in forests, with a total value of 51.8 million m^3,

Table 9.3 Forest area by stand origin / Estensione del Bosco ripartito per l'origine dei soprassuoli

Region/Regione	Forest/Bosco									
	Natural		Semi-natural		Planted		Not classified		Total	
	Origine naturale		Origine seminaturale		Origine artificiale		Non classificata		Totale	
	Area	ES	Area	ES	Area	ES	Area	ES	Area	ES
	(ha)	(%)	(ha)	(%)	(ha)	(%)	(ha)	(%)	(ha)	(%)
Piemonte	223,908	4.3	580,804	1.9	33,294	12.1	52,428	8.6	890,433	1.3
Valle d'Aosta	31,266	10.4	55,285	6.7	1541	49.6	11,150	17.6	99,243	3.6
Lombardia	125,907	5.9	407,897	2.5	35,264	11.9	52,900	9.4	621,968	1.6
Alto Adige	23,720	13.5	289,313	2.2	1890	44.5	24,346	12.3	339,270	1.7
Trentino	32,565	10.2	311,206	2.0	7568	21.7	21,921	12.7	373,259	1.4
Veneto	36,753	11.9	329,922	2.4	16,098	18.3	33,931	11.3	416,704	1.9
Friuli V.G	80,567	6.3	193,067	3.5	20,715	12.4	38,207	10.4	332,556	1.9
Liguria	63,113	7.7	259,522	2.4	7330	22.0	13,194	16.3	343,160	1.7
Emilia Romagna	112,267	5.9	407,843	2.1	36,514	11.9	28,278	12.0	584,901	1.5
Toscana	67,764	8.4	813,811	1.4	50,481	9.1	103,392	5.8	1,035,448	1.1
Umbria	27,722	13.8	335,265	2.0	12,347	19.0	14,970	19.0	390,305	1.6
Marche	40,568	10.0	198,079	3.2	23,100	15.4	30,020	10.7	291,767	2.1
Lazio	53,113	10.3	424,061	2.1	16,459	16.6	66,604	7.1	560,236	1.6
Abruzzo	138,276	5.1	218,970	3.3	20,265	13.2	34,077	10.0	411,588	1.8
Molise	7563	35.7	131,413	3.8	8023	24.9	6248	24.5	153,248	3.0
Campania	49,412	9.9	274,058	3.0	19,433	16.1	61,024	7.7	403,927	2.1
Puglia	52,676	8.9	69,408	6.4	18,262	15.6	2003	42.6	142,349	4.0
Basilicata	64,792	8.3	160,628	4.5	11,201	21.0	51,399	8.0	288,020	2.7
Calabria	116,723	5.9	239,604	3.7	70,999	7.1	67,851	8.5	495,177	2.0
Sicilia	105,187	6.7	79,022	7.6	78,132	7.5	23,148	16.3	285,489	3.2
Sardegna	102,367	6.7	401,251	2.9	80,912	7.1	41,610	9.8	626,140	2.1
Italia	1,556,230	1.7	6,180,428	0.6	569,827	2.8	778,701	2.2	9,085,186	0.4

Table 9.4 Other wooded land area by stand origin / Estensione delle Altre terre boscate ripartite per l'origine dei soprassuoli

Region/Regione	Other wooded land/Altre terre boscate									
	Natural		Semi-natural		Planted		Not classified		Total	
	Origine naturale		Origine seminaturale		Origine artificiale		Superficie non classificata per l'origine		Totale	
	Area	ES	Area	ES	Area	ES	Area	ES	Area	ES
	(ha)	(%)	(ha)	(%)	(ha)	(%)	(ha)	(%)	(ha)	(%)
Piemonte	40,131	12.8	11,161	21.8	404	100.0	33,295	12.9	84,991	8.0
Valle d'Aosta	4062	41.6	0	–	0	–	4671	27.2	8733	24.0
Lombardia	34,485	13.1	5179	28.9	0	–	30,588	13.8	70,252	8.7
Alto Adige	18,274	15.9	7439	22.0	0	–	10,368	18.7	36,081	10.4
Trentino	18,922	15.0	2402	37.4	0	–	12,503	16.7	33,826	10.6
Veneto	9717	19.2	3739	31.5	0	–	39,535	11.1	52,991	9.1
Friuli V.G	15,362	18.7	1487	49.8	0	–	24,209	14.9	41,058	10.6
Liguria	15,595	16.8	6782	28.9	2148	69.6	19,560	16.6	44,084	10.3
Emilia Romagna	21,166	16.1	9379	29.1	0	–	23,370	12.0	53,915	9.3
Toscana	41,380	12.0	21,627	15.1	723	70.7	90,545	6.7	154,275	5.2
Umbria	11,384	24.7	4596	27.6	0	–	7671	26.7	23,651	15.2
Marche	9978	28.8	1794	44.6	0	–	9542	21.6	21,314	16.2
Lazio	41,630	12.5	15,610	18.8	343	100.0	30,328	13.8	87,912	7.7
Abruzzo	45,468	11.2	1730	44.6	0	–	15,813	15.4	63,011	8.6
Molise	2159	68.5	14,071	19.7	0	–	3796	28.2	20,025	16.0
Campania	30,627	15.7	12,895	21.3	0	–	43,811	10.8	87,332	7.6
Puglia	37,966	11.9	4546	39.7	1554	49.8	5323	25.3	49,389	9.9
Basilicata	39,363	11.9	12,777	16.8	373	99.9	51,879	8.7	104,392	6.2
Calabria	30,689	12.8	7836	21.7	3731	31.5	113,188	5.7	155,443	4.8
Sicilia	55,457	9.7	8856	28.2	4339	49.3	33,093	13.6	101,745	7.1
Sardegna	297,980	3.7	282,497	3.4	1493	50.0	92,882	6.5	674,851	2.0
Italia	821,794	2.5	436,401	3.0	15,107	21.0	695,970	2.5	1,969,272	1.4

Table 9.5 Tall trees forest area by stand origin / Estensione dei Boschi alti ripartiti per l'origine dei soprassuoli

Region/Regione	Tall trees forest/Boschi alti									
	Natural		Semi-natural		Planted		Not classified		Total	
	Origine naturale		Origine seminaturale		Origine artificiale		Non classificata		Totale	
	Area	ES	Area	ES	Area	ES	Area	ES	Area	ES
	(ha)	(%)	(ha)	(%)	(ha)	(%)	(ha)	(%)	(ha)	(%)
Piemonte	223,908	4.3	580,804	1.9	12,976	17.6	52,085	8.7	869,773	1.3
Valle d'Aosta	31,266	10.4	55,285	6.7	1541	49.6	11,150	17.6	99,243	3.6
Lombardia	125,907	5.9	407,897	2.5	11,968	23.4	51,064	9.3	596,836	1.6
Alto Adige	23,720	13.5	289,313	2.2	1890	44.5	24,346	12.3	339,270	1.7
Trentino	32,565	10.2	311,206	2.0	7568	21.7	21,921	12.7	373,259	1.4
Veneto	36,753	11.9	329,922	2.4	11,164	21.3	33,587	11.3	411,427	1.9
Friuli V.G	80,567	6.3	193,067	3.5	11,521	17.7	38,207	10.4	323,362	1.9
Liguria	63,113	7.7	259,522	2.4	6964	22.6	13,194	16.3	342,793	1.7
Emilia Romagna	112,267	5.9	407,843	2.1	30,833	13.2	27,909	12.1	578,852	1.5
Toscana	67,764	8.4	813,811	1.4	43,697	9.4	103,392	5.8	1,028,665	1.1
Umbria	27,722	13.8	335,265	2.0	7373	22.2	13,568	18.5	383,928	1.6
Marche	40,568	10.0	198,079	3.2	16,237	16.6	30,020	10.7	284,904	2.1
Lazio	53,113	10.3	424,061	2.1	14,703	17.8	66,183	7.1	558,060	1.6
Abruzzo	138,276	5.1	218,970	3.3	17,293	14.2	34,077	10.0	408,616	1.8
Molise	7563	35.7	131,413	3.8	5308	35.0	6248	24.5	150,533	3.1
Campania	49,412	9.9	274,058	3.0	16,270	16.6	61,024	7.7	400,763	2.1
Puglia	52,676	8.9	69,408	6.4	18,162	15.6	2003	42.6	142,248	4.0
Basilicata	64,792	8.3	160,628	4.5	9679	19.4	51,399	8.0	286,498	2.7
Calabria	116,723	5.9	239,604	3.7	68,965	7.2	67,478	8.5	492,771	2.1
Sicilia	105,187	6.7	79,022	7.6	77,374	7.5	23,148	16.3	284,731	3.2
Sardegna	102,367	6.7	401,251	2.9	55,773	8.9	40,864	9.9	600,255	2.2
Italia	1,556,230	1.7	6,180,428	0.6	447,261	3.1	772,868	2.2	8,956,787	0.4

Table 9.6 Plantations area by stand origin / Estensione degli Impianti di arboricoltura da legno ripartiti per l'origine dei soprassuoli

Regione/Region	Plantations/Impianti di arboricoltura da legno					
	Planted		Not classified		Total	
	Origine artificiale		Non classificato		Totale	
	Area	ES	Area	ES	Area	ES
	(ha)	(%)	(ha)	(%)	(ha)	(%)
Piemonte	20,318	16.4	343	100.0	20,660	16.2
Valle d'Aosta	0	–	0	–	0	–
Lombardia	23,296	13.4	1836	83.7	25,132	12.5
Alto Adige	0	–	0	–	0	–
Trentino	0	–	0	–	0	–
Veneto	4933	35.5	344	99.9	5277	33.7
Friuli V.G	9194	17.0	0	–	9194	17.0
Liguria	367	99.2	0	–	367	99.2
Emilia Romagna	5680	33.0	368	100.0	6049	31.6
Toscana	6783	30.3	0	–	6783	30.3
Umbria	4974	33.7	1402	99.8	6377	33.7
Marche	6863	35.7	0	–	6863	35.7
Lazio	1756	45.4	420	100.0	2176	39.7
Abruzzo	2971	36.3	0	–	2971	36.3
Molise	2715	27.7	0	–	2715	27.7
Campania	3163	50.4	0	–	3163	50.4
Puglia	100	99.8	0	–	100	99.8
Basilicata	1522	93.6	0	–	1522	93.6
Calabria	2033	42.8	373	99.9	2406	39.3
Sicilia	758	70.7	0	–	758	70.7
Sardegna	25,138	12.0	746	70.7	25,885	11.8
Italia	122,566	6.3	5833	39.4	128,399	6.2

5.7 m^3 ha^{-1} on average. Stumps are 783.8 million in numbers (86.3 ha^{-1} on average) for a total volume of 14.0 million m^3 (1.5 m^3 ha^{-1}).

Per hectare values in the forest types of the Tall trees forest show that in general lying deadwood is the main component of conifer forests while standing dead tree deadwood is the most abundant in the broadleaved forests (Fig. 9.9); exceptions are the Mediterranean pines forests and the Hygrophilous forests.

Chestnut forests are marked by a very high value of trees per hectare (486.5 n ha^{-1} on average; Hornbeam and Hophornbeam forest type follows with 193.9 n ha^{-1}), and their timber volume of 28.6 m^3 ha^{-1} is more than twice that of any other forest type. Lying deadwood is also abundant, especially if compared to that of the broadleaved species forest types. The copious amount of deadwood in Chestnut forests had been

Table 9.7 Short trees forest, Sparse forest and Scrubland area by stand origin / Estensione di Boschi bassi, Boschi radi e Boscaglie ripartiti per l'origine dei soprassuoli

Region/Regione	Short trees forests, Sparse forest, Scrubland/Boschi bassi, Boschi radi, Boscaglie									
	Natural		Semi-natural		Planted		Not classified		Total	
	Origine naturale		Origine seminaturale		Origine artificiale		Non classificato		Totale	
	Area (ha)	ES (%)	Area (ha)	ES (%)	Area (ha)	ES (%)	Area (ha)	ES (%)	Area (ha)	ES (%)
Piemonte	21,200	19.6	5908	33.2	404	100.0	4900	44.7	32,412	15.2
Valle d'Aosta	2136	68.5	0	–	0	–	221	70.7	2357	62.5
Lombardia	6171	25.6	2535	41.0	0	–	3086	37.8	11,791	18.7
Alto Adige	2067	44.6	1767	42.9	0	–	1093	57.7	4928	26.0
Trentino	4216	29.5	960	56.0	0	–	1561	46.8	6737	22.8
Veneto	713	70.7	2243	40.7	0	–	748	70.6	3704	31.5
Friuli V.G	5859	24.7	743	70.5	0	–	3680	46.4	10,283	22.3
Liguria	8379	25.2	5682	32.8	1415	99.2	2515	61.2	17,992	18.5
Emilia Romagna	9514	23.4	9379	29.1	0	–	2583	37.8	21,475	16.7
Toscana	12,576	23.6	11,195	17.9	361	100.0	5420	25.8	29,552	13.0
Umbria	2877	55.0	3121	33.3	0	–	1106	57.6	7104	28.0
Marche	5360	40.5	1794	44.6	0	–	0	–	7153	32.2
Lazio	7633	35.1	9346	27.0	343	100.0	6134	42.0	23,456	18.4
Abruzzo	9011	26.8	1368	49.8	0	–	2158	69.8	12,537	22.9
Molise	125	99.8	6179	31.1	0	–	781	70.5	7085	28.2
Campania	6135	36.9	7304	27.9	0	–	2578	37.8	16,017	19.8
Puglia	12,998	21.3	3770	45.6	1165	57.5	388	99.8	18,321	17.9
Basilicata	4778	37.8	7830	21.6	0	–	6269	31.1	18,877	16.5
Calabria	9982	19.1	4851	27.6	3731	31.5	15,298	15.4	33,862	10.2
Sicilia	14,569	17.9	4396	40.3	2153	70.4	6703	29.9	27,822	14.2
Sardegna	34,384	12.5	41,133	10.3	746	70.7	10,417	18.8	86,680	7.2
Italia	180,680	5.8	131,506	6.2	10,320	25.3	77,640	8.5	400,146	3.7

detected in the past NFI (INFC2005) and explained by the devastating diseases infecting the species (Pignatti & De Natale, 2011).

Except the Chestnut forest type, there is more coarse woody debris in the Alpine coniferous forest; a similar amount is found for broadleaved species types only in the Hygrophilous forests. These are especially rich in lying deadwood (11.5 $m^3\ ha^{-1}$), but their value in standing dead trees is also relevant (8.3 $m^3\ ha^{-1}$) falling in second place only after that of Chestnut forest.

Stump volume is particularly abundant in the Alpine forest types of Larch and stone pine, Norway spruce and Fir. This is also due to the occurrence of such species

Table 9.8 Shrubs area by stand origin and Not accessible or not classified wooded area / Estensione degli Arbusteti ripartiti per l'origine dei soprassuoli e delle Aree boscate inaccessibili o non classificate

Region/Regione	Shrubs/Arbusteti										Not accessible or not classified wooded area / Aree boscate inaccessibili o non classificate	
	Natural / Origine naturale		Semi-natural / Origine seminaturale		Planted / Origine artificiale		Not classified / Non classificata		Total / Totale		Not classified / Non classificato	
	Area (ha)	ES (%)	Area (ha)	ES (%)	Area (ha)	ES (%)	Area (ha)	ES (%)	Area (ha)	ES (%)	Area (ha)	ES (%)
Piemonte	18,931	17.0	5253	27.7	0	–	6316	25.8	30,499	12.7	22,079	15.3
Valle d'Aosta	1927	44.3	0	–	0	–	771	70.3	2697	37.3	3680	31.2
Lombardia	28,314	15.0	2645	40.8	0	–	9504	27.5	40,462	12.3	17,998	17.8
Alto Adige	16,207	17.1	5671	25.6	0	–	5293	26.5	27,172	12.4	3981	30.1
Trentino	14,706	17.6	1441	50.0	0	–	6979	22.9	23,126	13.3	3963	29.2
Veneto	9004	20.0	1496	49.9	0	–	18,041	18.3	28,540	13.3	20,746	14.4
Friuli V.G	9503	26.3	743	70.5	0	–	10,297	26.8	20,543	17.1	10,232	18.5
Liguria	7216	22.1	1100	57.2	733	70.1	0	–	9048	19.7	17,045	17.2
Emilia Romagna	11,652	22.8	0	–	0	–	1473	50.0	13,126	21.0	19,314	13.2
Toscana	28,804	14.3	10,431	24.8	361	100.0	12,293	19.7	51,890	10.2	72,832	7.4
Umbria	8507	28.2	1475	49.9	0	–	737	70.6	10,719	23.9	5828	32.2
Marche	4619	45.6	0	–	0	–	2154	70.1	6772	37.1	7388	19.1
Lazio	33,997	13.6	6264	24.1	0	–	9545	23.5	49,806	10.6	14,649	17.4
Abruzzo	36,457	13.0	362	100.0	0	–	2172	40.8	38,991	12.4	11,483	14.9
Molise	2034	72.4	7891	26.5	0	–	391	99.8	10,315	24.6	2624	31.8
Campania	24,491	18.0	5590	33.5	0	–	20,315	18.4	50,397	11.2	20,918	13.6
Puglia	24,969	15.8	777	70.5	388	99.8	1265	53.6	27,399	14.7	3669	30.1
Basilicata	34,585	12.8	4947	27.1	373	99.9	26,005	12.6	65,910	8.3	19,605	13.4
Calabria	20,707	16.7	2985	35.3	0	–	13,122	21.2	36,814	12.2	84,768	6.5
Sicilia	40,888	11.7	4460	40.0	2185	69.8	18,220	19.4	65,753	9.4	8170	25.5
Sardegna	263,597	3.9	241,364	3.7	746	70.7	53,087	8.6	558,795	2.3	29,377	12.6
Italia	641,114	2.8	304,895	3.5	4787	37.9	217,980	4.9	1,168,776	1.9	400,350	3.2

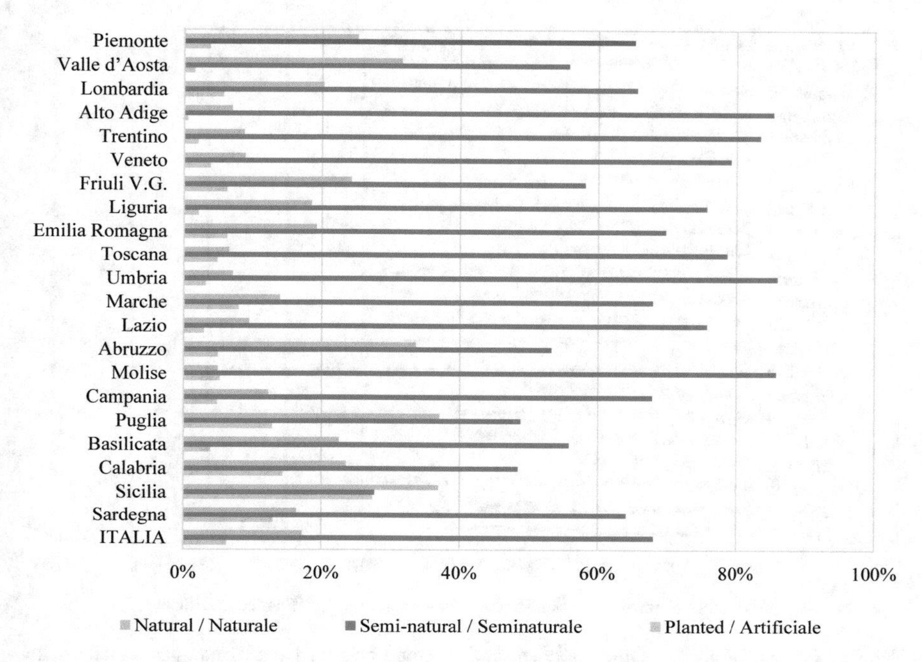

Fig. 9.5 Percent of area of Forest by stand origin / Ripartizione percentuale della superficie del Bosco per origine del soprassuolo

on steep slopes so that at ground level cut line in the uphill side may correspond to a very high stump height on the downhill side.

Abundance of the three components may also be compared through their relative deadwood biomass. Numerousness and volume are more closely linked to ecological functions, such as biodiversity and others, while deadwood biomass is more closely related to the carbon stock function, that is discussed in Chap. 12. It is noteworthy that the relative abundance of the three components may vary consistently if their biomass is used instead, because of the variable wood density values of the species and the decay class.

9.5 Wooded Lands in Protected Areas

During the inventory survey, when an inventory point was within a naturalistic protected area, the presence of naturalistic constraints was recorded. The main normative references for the definition of protected natural areas and the consequent naturalistic constraints are the Outline Law 394/91 at a national level, and the European Directives concerning 'Birds' (79/409/CEE) and 'Habitats' (92/43/CEE), adopted in Italy by Law 157/92 and Decree 357/97. The Convention on Wetlands of International Importance especially as Waterfowl Habitat, signed in Ramsar in 1971 and adopted

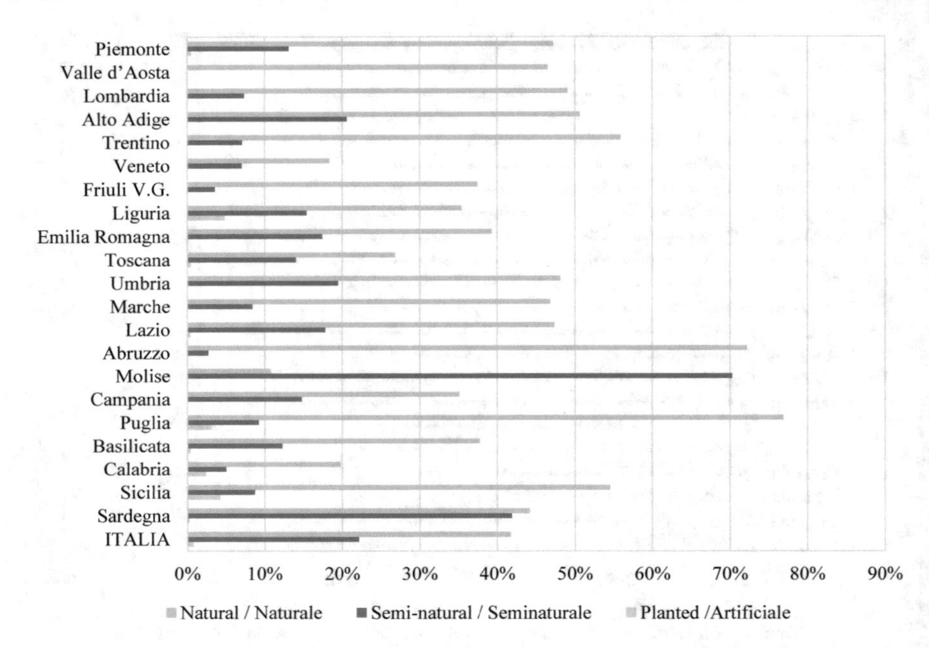

Fig. 9.6 Percent of area of Other wooded land by stand origin / Ripartizione percentuale della superficie delle Altre terre boscate per origine del soprassuolo

in Italy by Decree No. 448/76, must be considered in addition to the above-mentioned regulations. The different types of protected areas identified by these rules, and their definitions, are provided in Table 9.15, while in Table 9.16 the different levels of protection are defined for national parks and Natura2000 sites.

INFC2015 estimates 3.5 million hectares of wooded area are in protected areas, 31.8% of the national Total wooded area, of which 2.8 million hectares are Forest and almost 700,000 ha are Other wooded land. Figure 9.10 shows the area of Forest and Other wooded land included in the different types of protected areas. According to the national classification, different types of protection status may overlap on the same area. Natura2000 sites, for example, which spread over 3 million hectares of wooded lands (2.4 million hectares in Forest, about 600,000 hectares in Other wooded land), are often within national or regional parks and reserves.

The percentage of protected wooded land (Table 9.17, Fig. 9.11) is considerable in all Italian regions, but it is remarkable in some regions of central and southern Italy: more than 50% in Abruzzo, Campania, Puglia, and Sicilia. In the Alpine regions, which are marked by a higher forest cover, the percentage of protected wooded land is generally less than the national average, except for Veneto, which hosts 42.9% of its wooded lands in protected areas. The estimates on wooded lands in the different types of protected areas by inventory categories, as well as those on Tall trees forest types area by presence of naturalistic constraints, are available at inventarioforestale.org/statistiche_INFC.

Fig. 9.7 The deadwood component measured by INFC: **a** standing dead trees, **b** lying deadwood, and **c** stumps / Componenti di legno morto misurate da INFC: alberi morti in piedi (**a**), legno morto grosso a terra (**b**) e ceppaie residue (**c**)

a

b

c

Table 9.9 Total number and number per hectare of standing dead trees by Forest inventory category / Valori totali e per ettaro del numero di alberi morti in piedi per le categorie inventariali del Bosco

Region/Regione	Tall trees forest				Plantations				Total Forest			
	Boschi alti				Impianti di arboricoltura da legno				Totale Bosco			
	Number	ES	Number	ES	Number	ES	Number	ES	Number	ES	Number	ES
	Numero		Numero		Numero		Numero		Numero		Numero	
	(n)	(%)	(n ha^{-1})	(%)	(n)	(%)	(n ha^{-1})	(%)	(n)	(%)	(n ha^{-1})	(%)
Piemonte	204,350,828	7.4	234.9	7.3	1,367,107	93.5	66.2	93.2	205,717,935	7.4	231.0	7.3
Valle d'Aosta	7,645,974	19.4	77.0	18.9	0	–	0.0	–	7,645,974	19.4	77.0	18.9
Lombardia	93,226,186	10.1	156.2	9.9	561,232	59.8	22.3	56.7	93,787,418	10.0	150.8	9.9
Alto Adige	24,058,842	12.0	70.9	11.9	0	–	0.0	–	24,058,842	12.0	70.9	11.9
Trentino	55,888,345	12.0	149.7	11.9	0	–	0.0	–	55,888,345	12.0	149.7	11.9
Veneto	61,523,774	11.2	149.5	11.0	128,896	61.0	24.4	46.9	61,652,670	11.2	148.0	11.0
Friuli V.G	50,926,855	11.4	157.5	11.2	248,141	71.6	27.0	68.6	51,174,996	11.3	153.9	11.1
Liguria	102,238,399	10.9	298.3	10.8	0	–	0.0	–	102,238,399	10.9	297.9	10.8
Emilia Romagna	111,924,733	8.7	193.4	8.5	111,100	100.0	18.4	–	112,035,833	8.7	191.5	8.5
Toscana	217,056,118	7.3	211.0	7.2	48,007	71.3	7.1	70.6	217,104,124	7.3	209.7	7.2
Umbria	54,318,620	13.7	141.5	13.6	7120	100.0	1.1	–	54,325,740	13.7	139.2	13.6
Marche	31,272,777	12.4	109.8	12.2	134,768	78.2	19.6	82.2	31,407,545	12.4	107.6	12.2
Lazio	81,820,149	15.8	146.6	15.7	39,018	89.6	17.9	79.7	81,859,166	15.8	146.1	15.7
Abruzzo	42,540,600	11.7	104.1	11.5	103,341	97.0	34.8	88.9	42,643,941	11.7	103.6	11.5
Molise	13,416,111	18.4	89.1	18.1	66,067	100.0	24.3	–	13,482,178	18.3	88.0	18.0
Campania	75,044,408	15.1	187.3	14.9	125,106	71.8	39.5	75.8	75,169,513	15.1	186.1	14.9
Puglia	3,294,016	19.9	23.2	19.5	0	–	0.0	–	3,294,016	19.9	23.1	19.5
Basilicata	12,289,882	19.1	42.9	19.0	0	–	0.0	–	12,289,882	19.1	42.7	19.0
Calabria	64,702,933	20.4	131.3	20.3	258,610	100.0	107.5	–	64,961,544	20.3	131.2	20.2
Sicilia	9,333,253	16.1	32.8	15.8	0	–	0.0	–	9,333,253	16.1	32.7	15.8
Sardegna	44,421,355	18.4	74.0	18.3	609,344	84.8	23.5	84.0	45,030,700	18.1	71.9	18.1
Italia	1,361,294,158	2.9	152.0	2.9	3,807,858	38.7	29.7	38.4	1,365,102,015	2.9	150.3	2.9

Table 9.10 Total volume and volume per hectare of standing dead trees by Forest inventory category / Valori totali e per ettaro del volume degli alberi morti in piedi per le categorie inventariali del Bosco

Region/Regione	Tall trees forest				Plantations				Total Forest			
	Boschi alti				Impianti di arboricoltura da legno				Totale Bosco			
	Volume	ES	Volume	ES	Volume	ES	Volume	ES	Volume	ES	Volume	ES
	(m^3)	(%)	(m^3 ha^{-1})	(%)	(m^3)	(%)	(m^3 ha-$^{-1}$)	(%)	(m^3)	(%)	(m^3 ha^{-1})	(%)
Piemonte	11,142,686	7.1	12.8	7.0	22,958	64.4	1.1	63.0	11,165,644	7.1	12.5	7.0
Valle d'Aosta	554,936	17.1	5.6	16.6	0	–	0.0	–	554,936	17.1	5.6	16.6
Lombardia	6,203,788	9.5	10.4	9.3	75,477	72.2	3.0	69.4	6,279,265	9.4	10.1	9.2
Alto Adige	1,884,217	13.9	5.6	13.8	0	–	0.0	–	1,884,217	13.9	5.6	13.8
Trentino	3,221,304	18.0	8.6	17.9	0	–	0.0	–	3,221,304	18.0	8.6	17.9
Veneto	3,108,798	11.4	7.6	11.2	8632	58.8	1.6	44.3	3,117,430	11.4	7.5	11.2
Friuli V.G	3,774,248	12.1	11.7	11.9	5560	54.3	0.6	51.0	3,779,808	12.1	11.4	11.9
Liguria	5,496,419	8.7	16.0	8.6	0	–	0.0	–	5,496,419	8.7	16.0	8.6
Emilia Romagna	4,895,001	9.9	8.5	9.7	30,513	100.0	5.0	–	4,925,514	9.9	8.4	9.7
Toscana	12,080,801	7.3	11.7	7.2	8938	73.1	1.3	72.4	12,089,739	7.3	11.7	7.2
Umbria	1,010,893	14.0	2.6	13.9	761	100.0	0.1	–	1,011,654	14.0	2.6	13.9
Marche	1,141,673	16.0	4.0	15.8	9365	80.9	1.4	82.7	1,151,038	15.9	3.9	15.7
Lazio	2,298,127	13.6	4.1	13.5	10,974	96.5	5.0	86.9	2,309,101	13.5	4.1	13.4
Abruzzo	1,223,126	13.5	3.0	13.4	3346	97.5	1.1	89.3	1,226,471	13.5	3.0	13.3
Molise	361,983	21.2	2.4	20.9	8106	100.0	3.0	–	370,089	20.8	2.4	20.5
Campania	2,047,809	13.5	5.1	13.3	2309	72.1	0.7	76.0	2,050,118	13.5	5.1	13.3
Puglia	198,751	19.5	1.4	19.0	0	–	0.0	–	198,751	19.5	1.4	19.0
Basilicata	666,799	16.2	2.3	16.0	0	–	0.0	–	666,799	16.2	2.3	16.0
Calabria	3,969,876	14.4	8.1	14.3	21,865	100.0	9.1	–	3,991,741	14.4	8.1	14.2
Sicilia	830,353	24.1	2.9	24.0	0	–	0.0	–	830,353	24.1	2.9	24.0
Sardegna	1,010,193	18.9	1.7	18.9	12,146	62.7	0.5	61.7	1,022,339	18.7	1.6	18.7
Italia	67,121,780	2.8	7.5	2.8	220,950	32.0	1.7	31.3	67,342,730	2.8	7.4	2.8

Table 9.11 Total number and number per hectare of stumps by Forest inventory category / Valori totali e per ettaro del numero di ceppaie residue per le categorie inventariali del Bosco

Region/Regione	Tall trees forest				Plantations				Total Forest			
	Boschi alti				Impianti di arboricoltura da legno				Totale Bosco			
	Number	ES	Number	ES	Number	ES	Number	ES	Number	ES	Number	ES
	Numero		Numero		Numero		Numero		Numero		Numero	
	(n)	(%)	(n ha^{-1})	(%)	(n)	(%)	(n ha^{-1})	(%)	(n)	(%)	(n ha^{-1})	(%)
Piemonte	69,607,232	7.5	80.0	7.3	112,136	73.1	5.4	72.4	69,719,368	7.4	78.3	7.3
Valle d'Aosta	9,666,927	16.8	97.4	16.3	0	–	0.0	–	9,666,927	16.8	97.4	16.3
Lombardia	72,803,483	9.9	122.0	9.7	227,894	39.9	9.1	36.6	73,031,377	9.8	117.4	9.7
Alto Adige	99,702,713	6.8	293.9	6.6	0	–	0.0	–	99,702,713	6.8	293.9	6.6
Trentino	38,707,238	9.4	103.7	9.3	0	–	0.0	–	38,707,238	9.4	103.7	9.3
Veneto	56,256,145	8.9	136.7	8.6	84,102	92.4	15.9	91.9	56,340,247	8.9	135.2	8.6
Friuli V.G	32,053,286	9.5	99.1	9.2	188,283	52.2	20.5	49.1	32,241,568	9.4	97.0	9.2
Liguria	32,587,572	18.0	95.1	18.0	0	–	0.0	–	32,587,572	18.0	95.0	18.0
Emilia Romagna	26,724,295	17.8	46.2	17.8	6944	100.0	1.1	–	26,731,239	17.8	45.7	17.8
Toscana	50,593,378	8.7	49.2	8.6	785,040	68.8	115.7	68.3	51,378,418	8.6	49.6	8.6
Umbria	37,005,230	17.7	96.4	17.6	14,240	100.0	2.2	–	37,019,470	17.7	94.8	17.6
Marche	35,448,845	20.3	124.4	20.2	6997	100.0	1.0	–	35,455,842	20.3	121.5	20.2
Lazio	31,460,397	19.8	56.4	19.7	27,748	79.0	12.8	66.6	31,488,146	19.8	56.2	19.7
Abruzzo	38,358,170	23.1	93.9	23.0	0	–	0.0	–	38,358,170	23.1	93.2	23.0
Molise	5,289,086	22.7	35.1	22.5	161,497	100.0	59.5	–	5,450,584	22.2	35.6	22.0
Campania	43,477,379	15.1	108.5	14.9	304,534	70.0	96.3	67.4	43,781,913	15.0	108.4	14.8
Puglia	3,501,339	27.9	24.6	27.6	0	–	0.0	–	3,501,339	27.9	24.6	27.6
Basilicata	9,409,918	24.1	32.8	24.0	0	–	0.0	–	9,409,918	24.1	32.7	24.0
Calabria	59,306,777	11.5	120.4	11.4	336,917	70.7	140.0	58.8	59,643,695	11.5	120.4	11.4
Sicilia	4,777,575	22.5	16.8	22.3	0	–	0.0	–	4,777,575	22.5	16.7	22.3
Sardegna	22,555,258	14.6	37.6	14.5	2,275,730	24.8	87.9	23.2	24,830,988	13.4	39.7	13.3
Italia	779,292,242	3.1	87.0	3.0	4,532,062	19.4	35.3	19.0	783,824,304	3.0	86.3	3.0

Table 9.12 Total volume and volume per hectare of stumps by Forest inventory category / Valori totali e per ettaro del volume delle ceppaie residue per le categorie inventariali del Bosco

Region/Regione	Tall trees forest				Plantations				Total Forest			
	Boschi alti				Impianti di arboricoltura da legno				Totale Bosco			
	Volume	ES	Volume	ES	Volume	ES	Volume	ES	Volume	ES	Volume	ES
	(m^3)	(%)	(m^3 ha^{-1})	(%)	(m^3)	(%)	(m^3 ha^{-1})	(%)	(m^3)	(%)	(m^3 ha^{-1})	(%)
Piemonte	1,274,778	10.6	1.5	10.5	307	72.0	0.0	–	1,275,084	10.6	1.4	10.5
Valle d'Aosta	326,276	23.5	3.3	23.2	0	–	0.0	–	326,276	23.5	3.3	23.2
Lombardia	1,427,579	12.3	2.4	12.2	3776	42.8	0.2	40.4	1,431,355	12.3	2.3	12.2
Alto Adige	3,209,466	6.5	9.5	6.2	0	–	0.0	–	3,209,466	6.5	9.5	6.2
Trentino	1,383,672	9.8	3.7	9.7	0	–	0.0	–	1,383,672	9.8	3.7	9.7
Veneto	1,283,123	9.2	3.1	9.0	239	85.8	0.0	–	1,283,362	9.2	3.1	9.0
Friuli V.G	665,833	13.9	2.1	13.7	11,081	89.1	1.2	87.6	676,915	13.7	2.0	13.6
Liguria	369,079	14.6	1.1	14.5	0	–	0.0	–	369,079	14.6	1.1	14.5
Emilia Romagna	369,174	18.6	0.6	18.6	56	100.0	0.0	–	369,230	18.6	0.6	18.6
Toscana	738,091	13.5	0.7	13.4	13,985	79.3	2.1	78.4	752,076	13.3	0.7	13.3
Umbria	193,924	15.6	0.5	15.5	24	100.0	0.0	–	193,948	15.6	0.5	15.5
Marche	248,536	16.6	0.9	16.4	84	100.0	0.0	–	248,620	16.6	0.9	16.4
Lazio	364,024	18.8	0.7	18.7	572	87.5	0.3	76.5	364,596	18.7	0.7	18.7
Abruzzo	480,077	22.4	1.2	22.4	0	–	0.0	–	480,077	22.4	1.2	22.4
Molise	36,552	21.8	0.2	21.6	1460	100.0	0.5	–	38,012	21.3	0.2	21.1
Campania	347,513	15.9	0.9	15.8	2081	68.3	0.7	37.7	349,594	15.8	0.9	15.7
Puglia	16,376	25.5	0.1	25.1	0	–	0.0	–	16,376	25.5	0.1	25.1
Basilicata	87,741	19.0	0.3	18.8	0	–	0.0	–	87,741	19.0	0.3	18.8
Calabria	834,719	12.5	1.7	12.4	5860	70.7	2.4	58.8	840,579	12.4	1.7	12.3
Sicilia	84,504	27.8	0.3	27.6	0	–	0.0	–	84,504	27.8	0.3	27.6
Sardegna	208,348	14.8	0.3	14.7	14,661	28.1	0.6	25.9	223,009	13.9	0.4	13.8
Italia	13,949,385	3.1	1.6	3.1	54,185	29.9	0.4	29.6	14,003,571	3.1	1.5	3.1

Table 9.13 Total volume and volume per hectare of lying coarse woody debris by Forest inventory category / Valori totali e per ettaro del volume del legno morto grosso a terra per le categorie inventariali del Bosco

Region/Regione	Tall trees forest				Plantations				Total Forest			
	Boschi alti				Impianti di arboricoltura da legno				Totale Bosco			
	Volume	ES	Volume	ES	Volume	ES	Volume	ES	Volume	ES	Volume	ES
	(m^3)	(%)	(m^3 ha^{-1})	(%)	(m^3)	(%)	(m^3 ha^{-1})	(%)	(m^3)	(%)	(m^3 ha^{-1})	(%)
Piemonte	6,878,956	12.0	7.9	12.0	19,631	52.6	1.0	47.2	6,898,587	12.0	7.7	11.9
Valle d'Aosta	1,404,238	24.1	14.1	23.7	0	–	0.0	–	1,404,238	24.1	14.1	23.7
Lombardia	6,368,217	13.6	10.7	13.5	26,604	57.6	1.1	56.1	6,394,821	13.6	10.3	13.5
Alto Adige	3,400,468	16.8	10.0	16.7	0	–	0.0	–	3,400,468	16.8	10.0	16.7
Trentino	4,260,728	12.7	11.4	12.7	0	–	0.0	–	4,260,728	12.7	11.4	12.7
Veneto	2,503,037	10.9	6.1	10.7	0	–	0.0	–	2,503,037	10.9	6.0	10.7
Friuli V.G	4,315,704	13.8	13.3	13.6	641	100.0	0.1	98.6	4,316,346	13.8	13.0	13.6
Liguria	3,319,657	14.0	9.7	14.0	0	–	0.0	–	3,319,657	14.0	9.7	14.0
Emilia Romagna	2,724,392	12.7	4.7	12.5	7303	96.9	1.2	95.8	2,731,695	12.6	4.7	12.5
Toscana	5,863,173	12.9	5.7	12.8	8567	69.7	1.3	68.7	5,871,740	12.9	5.7	12.8
Umbria	1,026,263	25.5	2.7	25.5	45	100.0	0.0	–	1,026,308	25.5	2.6	25.5
Marche	1,161,743	17.6	4.1	17.4	558	73.8	0.1	76.1	1,162,301	17.6	4.0	17.4
Lazio	1,567,964	21.8	2.8	21.8	6203	100.0	2.9	–	1,574,167	21.7	2.8	21.7
Abruzzo	1,157,069	16.3	2.8	16.2	76	100.0	0.0	–	1,157,146	16.3	2.8	16.2
Molise	188,554	21.6	1.3	21.3	12,395	100.0	4.6	–	200,949	21.1	1.3	20.8
Campania	1,034,600	30.8	2.6	30.7	7459	100.0	2.4	–	1,042,059	30.6	2.6	30.5
Puglia	179,676	22.2	1.3	21.9	0	–	0.0	–	179,676	22.2	1.3	21.9
Basilicata	492,193	34.2	1.7	34.1	0	–	0.0	–	492,193	34.2	1.7	34.1
Calabria	2,517,446	26.2	5.1	26.1	3697	65.1	1.5	51.9	2,521,144	26.1	5.1	26.1
Sicilia	574,876	20.5	2.0	20.3	0	–	0.0	–	574,876	20.5	2.0	20.3
Sardegna	734,799	29.5	1.2	29.4	3928	51.3	0.2	50.3	738,727	29.3	1.2	29.3
Italia	51,673,754	4.1	5.8	4.0	97,109	27.0	0.8	26.1	51,770,863	4.0	5.7	4.0

Table 9.14 Total volume and volume per hectare of coarse woody debris by Forest inventory category / Valori totali e per ettaro del volume del legno morto grosso totale per le categorie inventariali del Bosco

Region/Regione	Tall trees forest				Plantations				Total Forest			
	Boschi alti				Impianti di arboricoltura da legno				Totale Bosco			
	Volume	ES	Volume	ES	Volume	ES	Volume	ES	Volume	ES	Volume	ES
	(m^3)	(%)	(m^3 ha^{-1})	(%)	(m^3)	(%)	(m^3 ha^{-1})	(%)	(m^3)	(%)	(m^3 ha^{-1})	(%)
Piemonte	19,296,419	6.3	22.2	6.1	42,896	43.7	2.1	39.7	19,339,315	6.2	21.7	6.1
Valle d'Aosta	2,285,451	17.0	23.0	16.5	0	–	0.0	–	2,285,451	17.0	23.0	16.5
Lombardia	13,999,584	8.1	23.5	7.8	105,857	60.9	4.2	58.0	14,105,441	8.0	22.7	7.8
Alto Adige	8,494,151	8.2	25.0	8.0	0	–	0.0	–	8,494,151	8.2	25.0	8.0
Trentino	8,865,704	10.8	23.8	10.7	0	–	0.0	–	8,865,704	10.8	23.8	10.7
Veneto	6,894,957	7.7	16.8	7.4	8871	57.2	1.7	42.6	6,903,829	7.7	16.6	7.4
Friuli V.G	8,755,786	9.2	27.1	9.0	17,283	63.4	1.9	61.0	8,773,069	9.2	26.4	9.0
Liguria	9,185,155	8.1	26.8	8.0	0	–	0.0	–	9,185,155	8.1	26.8	8.0
Emilia Romagna	7,988,567	8.1	13.8	7.9	37,873	99.4	6.3	98.3	8,026,439	8.1	13.7	7.9
Toscana	18,682,065	6.9	18.2	6.8	31,490	59.8	4.6	58.8	18,713,554	6.9	18.1	6.8
Umbria	2,231,080	14.2	5.8	14.1	830	91.2	0.1	94.0	2,231,910	14.2	5.7	14.1
Marche	2,551,952	12.7	9.0	12.4	10,006	80.4	1.5	82.2	2,561,959	12.6	8.8	12.4
Lazio	4,230,114	12.1	7.6	11.9	17,750	95.1	8.2	85.2	4,247,864	12.0	7.6	11.9
Abruzzo	2,860,273	10.9	7.0	10.8	3422	95.3	1.2	87.2	2,863,694	10.9	7.0	10.8
Molise	587,088	18.4	3.9	18.1	21,961	100.0	8.1	–	609,050	18.1	4.0	17.7
Campania	3,429,923	13.1	8.6	12.9	11,849	74.0	3.7	31.6	3,441,771	13.1	8.5	12.9
Puglia	394,804	14.3	2.8	13.7	0	–	0.0	–	394,804	14.3	2.8	13.7
Basilicata	1,246,733	17.3	4.4	17.1	0	–	0.0	–	1,246,733	17.3	4.3	17.1
Calabria	7,322,042	12.6	14.9	12.5	31,422	73.6	13.1	62.3	7,353,464	12.6	14.9	12.4
Sicilia	1,489,733	17.5	5.2	17.3	0	–	0.0	–	1,489,733	17.5	5.2	17.3
Sardegna	1,953,340	15.6	3.3	15.5	30,735	34.8	1.2	33.1	1,984,075	15.4	3.2	15.3
Italia	132,744,920	2.3	14.8	2.3	372,245	24.0	2.9	23.2	133,117,164	2.3	14.7	2.3

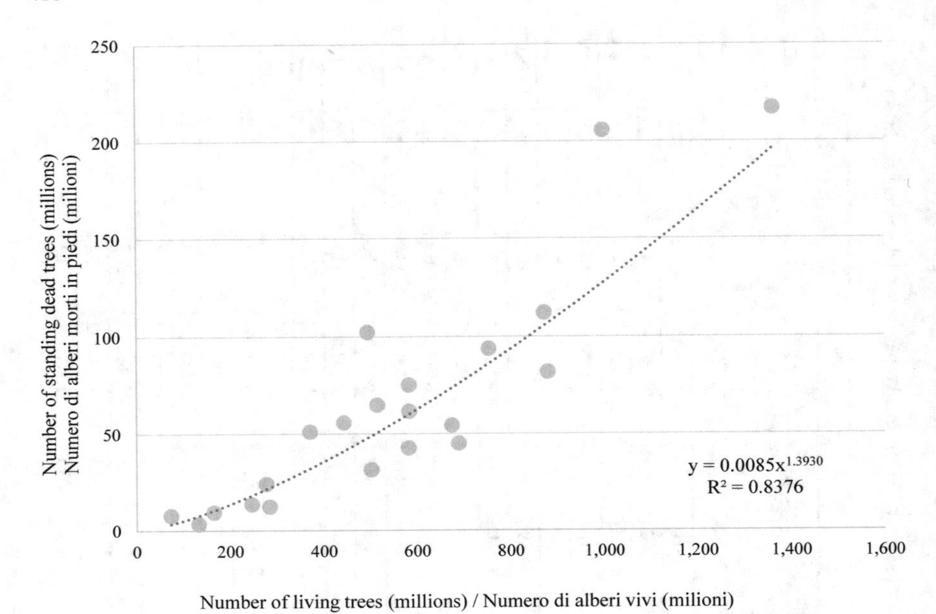

Fig. 9.8 Relationship between the number of standing dead trees and that of living trees in the Italian regions / Relazione tra il numero degli alberi morti in piedi e il numero degli alberi vivi nelle regioni italiane

Tables 9.18, 9.19 and 9.20 detail the area under naturalistic constraints at the inventory category level. All inventory categories have similar portions of protected areas, ranging between 30 and 35%, except from Plantations (23.4%).

Wooded lands within national parks are about 920,000 hectares: 725,000 ha in Forest (8.0%), 195,000 ha in Other wooded land (9.9%). At the regional level, this distribution varies considerably (Table 9.21), as it is conditioned by the distribution of the total area of the national parks. The rate of Forest in national parks is particularly high in three regions of central-southern Italy: Puglia (42.2%), Abruzzo (32.3%) and Calabria (26.2%). Table 9.22 (Forest) and Table 9.23 (Other wooded land) show the distribution of forest area in national parks by protection level. Considering Forest alone, Fig. 9.12 shows that the majority of protected area in national parks falls into Zone C (24.0%). Moreover, only 56.3% of such area is assigned to a specific class, while the remaining 43.7% is not classified to a specific protection level.

State nature reserves host 74,963 ha of the total wooded area, corresponding to 0.7% (Table 9.24). A similar rate (0.6%) is obtained when considering only the Forest. These Forest areas are concentrated in five regions: Tuscany, Lazio, Veneto, Abruzzo and Calabria, which constitute over 70% of the total.

Forests in regional nature parks (Table 9.25) cover a total of 713,048 ha of the total forest area, 592,554 ha of which is Forest (6.5%). The distribution at the regional level is rather variable: in Sicilia and Campania, for example, 28.6% and 24.6% of the Forest, respectively, is protected by regional parks.

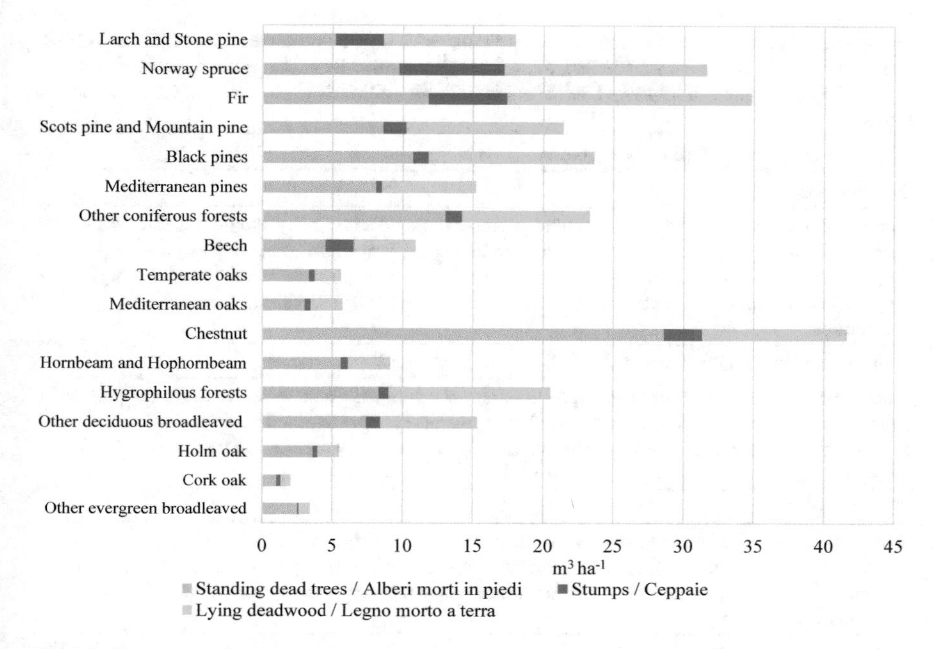

Fig. 9.9 Volume per hectare of coarse woody debris by component, in the Tall trees forest types / Volume per ettaro delle componenti di legno morto grosso, nelle categorie dei Boschi alti

The total forest area included in regional nature reserves is 1.1% (125,572 ha, Table 9.26). Very low values are observed for all regions, except for Sicilia, where 11.4% of Forest falls into this type of reserves.

A very small wooded area, 60,932 ha (Table 9.27, 0.6% of the total) is included in the other protected natural areas under the Law n.394/91.

Natura2000 sites (SCI and SPA) host 27.2% of wooded areas (3,011,119 ha), 2,408,882 ha of which are in Forest (Table 9.28). The regions with the largest rates, considering the regional wooded lands area, are Puglia (63.0%), Campania (51.0%), Abruzzo (50.7%), Sicilia (45.5%), and Veneto (45.2%). More detailed estimates on wooded land areas in SCI and SPA can be found at inventarioforestale.org/statistiche_INFC.

Table 9.29 shows Forest and Other wooded land area included in Wetlands of international interest (Ramsar Convention), which together account for about 5000 ha in the whole country.

Finally, although the national classification of protected areas is not fully comparable to that adopted for European reporting (MCPFE, 2003). Figure 9.13 shows the percentage of distribution of Forest area by mutually exclusive protection level classes, comparable to the MCPFE classes and used for Italy in the State of Europe's Forest 2011 (Forest Europe, UNECE and FAO, 2011), for indicator 4.9. For such classifications, more detailed inventory statistics are available at inventarioforestale.org/statistiche_INFC.

Table 9.15 Types of protected areas (*Source* Framework Law on protected areas n.394/91, Directive 79/409/CEE, Directive 92/43/CEE, Ramsar Convention) / Definizioni dei tipi di aree protette (fonte: Legge Quadro n.394/91, Direttiva 79/409/CEE, Direttiva 92/43/CEE, Convenzione di Ramsar)

Type of protected area/Tipo di area protetta
National parks
Terrestrial, river, lake, marine areas containing one or more intact or partly altered ecosystems and one or more physical, geological, geomorphologic or biological formations of international/national importance due to naturalistic, scientific, aesthetic, cultural, educative and recreation values, such as to require governmental intervention for their conservation for the benefit of current and future generations
Regional and interregional nature parks
Land, river, lake areas and sea stretch near the coast of naturalistic/environmental value, building up, within one or more neighbouring regions, a homogeneous system identified by the naturalistic conditions of the sites, by their landscape and artistic values and by the cultural traditions of local populations
Nature reserves
Land, river, lake areas and sea stretch with one or more naturalistically important plant and animal species, or one or more ecosystems important for the biological diversity or for the conservation of genetic resources. Natural reserves may be state or regional based on the importance of the naturalistic elements they contain
Other natural protected areas
Biotopes, environmental association oases, suburban parks, natural monuments, etc. not falling into the above classes. These are divided into public areas, instituted by regional law or similar measures, and privately managed areas, instituted by formal public provisions or by contractual deeds, such as concessions or equivalent forms
Wetland of international interest
Marshes, bogs, swamps, peat beds or natural/artificial watery zones, permanent or transitory, including sea water areas of depth at low tide not more than six metres, considered of international importance under the Ramsar convention
Special Protection Areas – (SPA)
Territories established under Directive 79/409/EEC suitable for conservation of the bird species listed in Annex I concerning protection of wild birds
Special Areas of Conservation (SAC) or Sites of Community Importance (SCI)
Natural areas, geographically defined and with delimited surface, established under Directive 92/43/EEC, which: (a) include land and water zones that are distinguished thanks to their geographic, abiotic and biotic characteristics, natural or semi natural (natural habitats) and that significantly contribute to conservation or recovery of a type of natural habitat or a species of wild flora and fauna listed in Annexes I and II to Directive 92/43/EEC on the conservation of natural habitats, and the conservation of wild fauna and flora at a favourable status in order to safeguard the biological diversity in the palearctic region by protecting the Alpine, Apennine and Mediterranean environments. Such areas are indicated as Special Areas of Conservation (SPA) (b) are designated by the state through a regulatory, administrative and/or contractual act, and in which the conservation measures necessary to maintain or recover to a satisfactory conservation status the natural habitats and/or the populations of the species for which the natural area is meant. Such areas are indicated as Sites of Community Importance (SCI)

(continued)

Table 9.15 (continued)

Type of protected area/Tipo di area protetta
Parchi nazionali
Aree terrestri, fluviali, lacuali o marine contenenti uno o più ecosistemi intatti o anche parzialmente alterati da interventi antropici, una o più formazioni fisiche, geologiche, geomorfologiche, biologiche, di rilievo internazionale o nazionale per valori naturalistici, scientifici, estetici, culturali, educativi e ricreativi tali da richiedere l'intervento dello Stato ai fini della loro conservazione per le generazioni presenti e future
Parchi naturali regionali e interregionali
Aree terrestri, fluviali, lacuali ed eventualmente tratti di mare prospicienti la costa, di valore naturalistico e ambientale, che costituiscono, nell'ambito di una o più regioni limitrofe, un sistema omogeneo, individuato dagli assetti naturalistici dei luoghi, dai valori paesaggistici e artistici e dalle tradizioni culturali delle popolazioni locali
Riserve naturali
Aree terrestri, fluviali, lacuali o marine che contengono una o più specie naturalisticamente rilevanti della flora e della fauna, ovvero presentino uno o più ecosistemi importanti per la diversità biologica o per la conservazione delle risorse genetiche. Le riserve naturali possono essere statali o regionali in base alla rilevanza degli elementi naturalistici in esse rappresentati
Altre aree naturali protette
Biotopi, oasi delle associazioni ambientaliste, parchi suburbani, monumenti naturali, ecc. che non rientrano nelle precedenti classi. Si dividono in aree a gestione pubblica, istituite cioè con leggi regionali o provvedimenti equivalenti, e aree a gestione privata, istituite con provvedimenti formali pubblici o con atti contrattuali quali concessioni o forme equivalenti
Zone umide di interesse internazionale
Aree acquitrinose, paludi, torbiere oppure zone naturali o artificiali d'acqua permanenti o transitorie, comprese zone di acqua marina la cui profondità con bassa marea non superi i sei metri che, per le loro caratteristiche, possono essere considerate di importanza internazionale ai sensi della convenzione di Ramsar
Zone di Protezione Speciale—(ZPS)
Territori designati ai sensi della direttiva 79/409/CEE, idonei alla conservazione delle specie di uccelli selvatici di cui all'allegato I della direttiva stessa
Zone Speciali di Conservazione (ZSC) o Siti di Importanza Comunitaria (SIC)
Designate ai sensi della direttiva 92/43/CEE, sono costituite da aree naturali, geograficamente definite e con superficie delimitata, che: (a) contengono zone terrestri o acquatiche che si distinguono grazie alle loro caratteristiche geografiche, abiotiche e biotiche, naturali o seminaturali (habitat naturali) e che contribuiscono in modo significativo a conservare, o ripristinare, un tipo di habitat naturale o una specie della flora e della fauna selvatiche di cui all'allegato I e II della direttiva 92/43/CEE, relativa alla conservazione degli habitat naturali e seminaturali e della flora e della fauna selvatiche in uno stato soddisfacente a tutelare la diversità biologica nella regione paleartica mediante la protezione degli ambienti alpino, appenninico e mediterraneo. Tali aree vengono indicate come Zone Speciali di Conservazione Comunitaria (ZSC); (b) sono designate dallo Stato mediante un atto regolamentare, amministrativo e/o contrattuale e nelle quali siano applicate le misure di conservazione necessarie al mantenimento o al ripristino, in uno stato di conservazione soddisfacente, degli habitat naturali e/o delle popolazioni delle specie per cui l'area naturale è designata. Tali aree vengono indicate come Siti di Importanza Comunitaria (SIC)

Table 9.16 Protected areas and their classification as to level of protection / Classi relative al grado di protezione in aree protette

Type of protected area	Protection level
National parks	Zone A—integral reserve
	Zone B—general oriented reserve
	Zone C—protection area
	Zone D—area of economic and social promotion
	Neighbouring area
	Protection level not ascertained
NATURA2000 (SCI and SPA) network sites	Sites of Community Importance—SCI—Habitat Directive 92/43/EEC
	Special Protection Areas—SPA Directive 79/409/EEC on the Conservation of Wild Birds
	SCI + SPA (overlay of the two site types)
Tipo di area protetta	**Grado di protezione**
Parchi nazionali	Zona A—riserva integrale
	Zona B—riserva generale orientata
	Zona C—area di protezione
	Zona D—area di promozione economica e sociale
	Area contigua
	Grado di protezione non accertato
Siti rete NATURA2000 (SIC e ZPS)	Siti di Importanza Comunitaria—SIC
	Zone di Protezione Speciale—ZPS
	SIC + ZPS (sovrapposizione dei due tipi di siti)

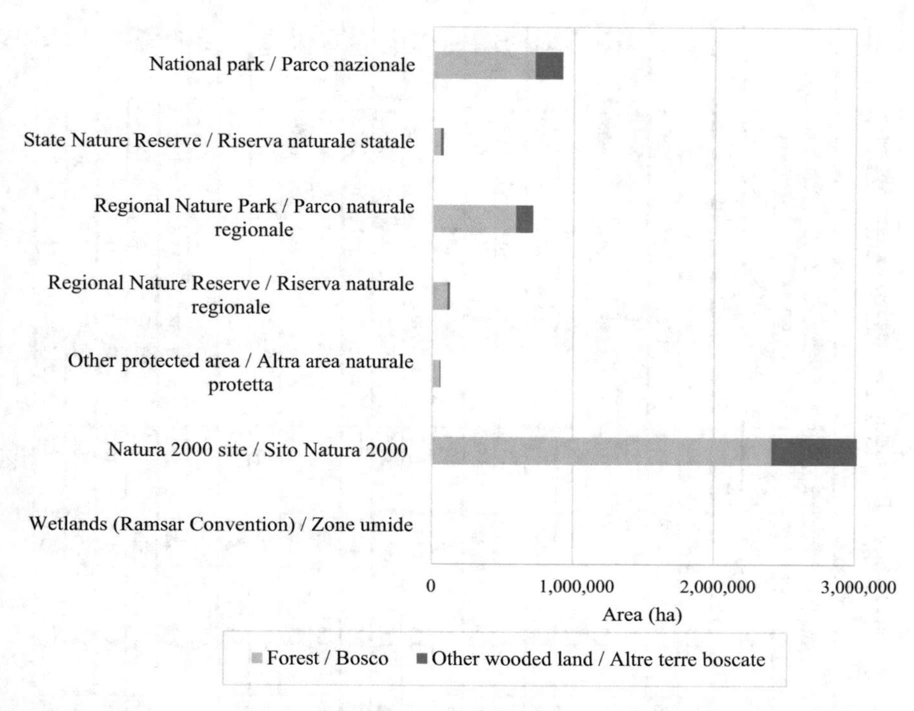

Fig. 9.10 Area of Forest and Other wooded land included in different types of protected areas, at national level / Estensione del Bosco e delle Altre terre boscate, a livello nazionale, ripartita per tipo di area protetta

Table 9.17 Forest and Other wooded land area included in protected areas / Superficie del Bosco e delle Altre terre boscate inclusa in aree protette

Region/Regione	Forest/Bosco						Other wooded land/Altre terre boscate					
	In protected areas		Not in protected areas		Total		In protected areas		Not in protected areas		Total	
	In aree protette		Non in aree protette		Totale		In aree protette		Non in aree protette		Totale	
	Area	ES	Area	ES	Area	ES	Area	ES	Area	ES	Area	ES
	(ha)	(%)	(ha)	(%)	(ha)	(%)	(ha)	(%)	(ha)	(%)	(ha)	(%)
Piemonte	135,247	5.5	755,186	1.6	890,433	1.3	20,927	16.9	64,064	9.5	84,991	8.0
Valle d'Aosta	12,715	16.5	86,527	4.4	99,243	3.6	1653	46.7	7081	27.7	8733	24.0
Lombardia	162,980	5.1	458,988	2.3	621,968	1.6	27,510	15.6	42,742	10.9	70,252	8.7
Alto Adige	57,648	7.6	281,622	2.3	339,270	1.7	14,415	16.0	21,666	14.2	36,081	10.4
Trentino	81,970	6.1	291,289	2.2	373,259	1.4	15,042	17.1	18,784	13.7	33,826	10.6
Veneto	178,935	4.0	237,769	3.4	416,704	1.9	38,870	10.9	14,121	18.3	52,991	9.1
Friuli V.G	74,866	6.6	257,690	2.7	332,556	1.9	18,176	18.1	22,882	14.4	41,058	10.6
Liguria	94,893	5.4	248,267	2.7	343,160	1.7	18,826	16.2	25,258	14.5	44,084	10.3
Emilia Romagna	122,996	5.2	461,905	2.1	584,901	1.5	9118	19.9	44,797	10.5	53,915	9.3
Toscana	208,303	4.0	827,145	1.5	1,035,448	1.1	41,334	11.2	112,940	6.2	154,275	5.2
Umbria	90,175	6.0	300,130	2.4	390,305	1.6	3687	31.5	19,965	17.1	23,651	15.2
Marche	100,238	5.5	191,529	3.5	291,767	2.1	4886	24.7	16,428	20.1	21,314	16.2
Lazio	237,998	3.4	322,238	2.9	560,236	1.6	37,337	11.6	50,574	11.2	87,912	7.7
Abruzzo	215,818	3.3	195,770	3.9	411,588	1.8	33,543	13.1	29,468	13.9	63,011	8.6
Molise	61,157	7.4	92,091	5.6	153,248	3.0	6935	28.5	13,090	20.4	20,025	16.0
Campania	261,918	3.2	142,008	5.0	403,927	2.1	59,986	9.5	27,347	16.1	87,332	7.6
Puglia	108,690	5.1	33,659	10.7	142,349	4.0	33,574	12.4	15,815	20.9	49,389	9.9
Basilicata	120,233	5.1	167,786	4.5	288,020	2.7	29,034	12.5	75,358	7.6	104,392	6.2
Calabria	171,827	4.4	323,349	3.1	495,177	2.0	63,245	7.6	92,199	6.7	155,443	4.8
Sicilia	162,830	4.7	122,659	6.3	285,489	3.2	51,896	10.1	49,849	11.1	101,745	7.1
Sardegna	167,196	4.7	458,944	2.7	626,140	2.1	159,406	5.0	515,445	2.5	674,851	2.0
Italia	2,828,635	1.1	6,256,551	0.6	9,085,186	0.4	689,398	2.6	1,279,874	1.8	1,969,272	1.4

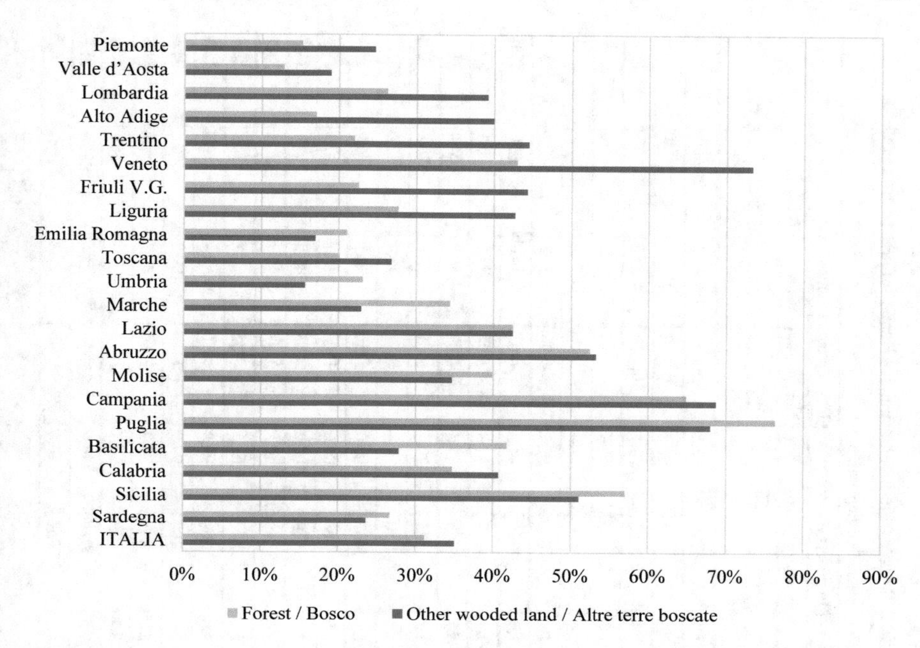

Fig. 9.11 Percent of area of Forest and Other wooded land included in protected areas / Ripartizione percentuale della superficie del Bosco e delle Altre terre boscate inclusa in aree protette

Table 9.18 Area of inventory categories of Forest included in protected areas / Superficie delle categorie inventariali del Bosco inclusa in aree protette

Region/Regione	Tall trees forest/Boschi alti						Plantations/Impianti					
	In protected areas		Not in protected areas		Tall trees forest		In protected areas		Not in protected areas		Plantations	
	In aree protette		Non in aree protette		Totale Boschi alti		In aree protette		Non in aree protette		Totale impianti	
	Area	ES	Area	ES	Area	ES	Area	ES	Area	ES	Area	ES
	(ha)	(%)	(ha)	(%)	(ha)	(%)	(ha)	(%)	(ha)	(%)	(ha)	(%)
Piemonte	131,468	5.5	738,305	1.6	869,773	1.3	3779	43.5	16,881	18.2	20,660	16.2
Valle d'Aosta	12,715	16.5	86,527	4.4	99,243	3.6	0	–	0	–	0	–
Lombardia	152,227	5.2	444,609	2.3	596,836	1.6	10,753	23.1	14,379	19.6	25,132	12.5
Alto Adige	57,648	7.6	281,622	2.3	339,270	1.7	0	–	0	–	0	–
Trentino	81,970	6.1	291,289	2.2	373,259	1.4	0	–	0	–	0	–
Veneto	178,248	4.0	233,179	3.5	411,427	1.9	687	69.5	4590	37.5	5277	33.7
Friuli V.G	73,134	6.7	250,228	2.7	323,362	1.9	1732	43.3	7462	19.6	9194	17.0
Liguria	94,893	5.4	247,900	2.7	342,793	1.7	0	–	367	99.2	367	99.2
Emilia Romagna	121,936	5.2	456,916	2.1	578,852	1.5	1060	57.5	4989	36.7	6049	31.6
Toscana	207,580	4.0	821,084	1.5	1,028,665	1.1	723	70.7	6061	32.8	6783	30.3
Umbria	90,175	6.0	293,753	2.5	383,928	1.6	0	–	6377	33.7	6377	33.7
Marche	98,825	5.4	186,079	3.5	284,904	2.1	1412	100.0	5450	38.5	6863	35.7
Lazio	237,630	3.4	320,431	2.9	558,060	1.6	368	100.0	1807	43.2	2176	39.7
Abruzzo	215,059	3.3	193,557	3.9	408,616	1.8	759	100.0	2213	47.2	2971	36.3
Molise	60,957	7.4	89,576	5.7	150,533	3.1	200	70.6	2515	29.4	2715	27.7
Campania	259,510	3.2	141,253	5.0	400,763	2.1	2408	63.5	755	62.1	3163	50.4
Puglia	108,589	5.1	33,659	10.7	142,248	4.0	100	99.8	0	–	100	99.8
Basilicata	120,133	5.1	166,365	4.5	286,498	2.7	100	99.9	1422	99.9	1522	93.6
Calabria	171,387	4.4	321,384	3.1	492,771	2.1	441	99.9	1966	42.7	2406	39.3
Sicilia	162,072	4.7	122,659	6.3	284,731	3.2	758	70.7	0	–	758	70.7
Sardegna	162,427	4.8	437,828	2.8	600,255	2.2	4769	37.9	21,116	12.0	25,885	11.8
Italia	2,798,584	1.1	6,158,203	0.6	8,956,787	0.4	30,051	14.6	98,348	7.2	128,399	6.2

Table 9.19 Area of Short trees forest, Sparse forests and Scrublands included in protected areas / Superficie di Boschi bassi, Boschi radi e Boscaglie inclusa in aree protette

Region/Regione	Short trees forest, Sparse forests and Scrublands Boschi bassi, Boschi radi e Boscaglie					
	In protected areas In aree protette		Not in protected areas Non in aree protette		Total Totale	
	Area (ha)	ES (%)	Area (ha)	ES (%)	Area (ha)	ES (%)
Piemonte	8472	26.1	23,940	18.8	32,412	15.2
Valle d'Aosta	0	–	2357	62.5	2357	62.5
Lombardia	2424	40.9	9367	21.2	11,791	18.7
Alto Adige	1430	49.2	3497	32.4	4928	26.0
Trentino	1802	44.7	4935	26.6	6737	22.8
Veneto	2209	40.8	1496	49.9	3704	31.5
Friuli V.G	4045	30.0	6238	31.4	10,283	22.3
Liguria	6896	28.7	11,095	25.2	17,992	18.5
Emilia Romagna	4052	30.1	17,423	19.4	21,475	16.7
Toscana	4697	27.7	24,855	14.6	29,552	13.0
Umbria	1475	49.9	5630	32.9	7104	28.0
Marche	1423	49.8	5730	38.4	7153	32.2
Lazio	8389	25.5	15,068	25.4	23,456	18.4
Abruzzo	6342	30.3	6195	35.8	12,537	22.9
Molise	1537	47.2	5548	33.7	7085	28.2
Campania	12,027	22.4	3990	43.0	16,017	19.8
Puglia	15,602	20.0	2719	37.5	18,321	17.9
Basilicata	5896	32.5	12,981	19.2	18,877	16.5
Calabria	11,194	18.1	22,668	12.6	33,862	10.2
Sicilia	16,754	18.0	11,067	24.0	27,822	14.2
Sardegna	22,194	14.0	64,486	8.5	86,680	7.2
Italia	138,861	6.0	261,285	4.7	400,146	3.7

Table 9.20 Shrubs area included in protected areas / Superficie degli Arbusteti inclusa in aree protette

Region/Regione	Shrubs Arbusteti					
	In protected areas In aree protette		Not in protected areas Non in aree protette		Total Totale	
	Area (ha)	ES (%)	Area (ha)	ES (%)	Area (ha)	ES (%)
Piemonte	4337	30.4	26,162	14.0	30,499	12.7
Valle d'Aosta	385	99.6	2312	40.4	2697	37.3
Lombardia	21,371	18.2	19,092	17.3	40,462	12.3
Alto Adige	12,228	17.4	14,944	17.9	27,172	12.4
Trentino	11,920	19.9	11,207	18.1	23,126	13.3
Veneto	21,845	16.0	6695	23.4	28,540	13.3
Friuli V.G	10,917	26.7	9626	23.4	20,543	17.1
Liguria	3917	29.9	5131	26.4	9048	19.7
Emilia Romagna	1473	50.0	11,652	22.8	13,126	21.0
Toscana	15,878	19.8	36,012	12.3	51,890	10.2
Umbria	1475	49.9	9244	26.6	10,719	23.9
Marche	371	100.0	6402	38.8	6772	37.1
Lazio	21,669	15.2	28,137	15.2	49,806	10.6
Abruzzo	21,419	17.9	17,572	19.3	38,991	12.4
Molise	4767	37.6	5548	33.7	10,315	24.6
Campania	31,980	14.7	18,417	20.6	50,397	11.2
Puglia	15,428	18.9	11,971	25.7	27,399	14.7
Basilicata	16,426	16.6	49,484	10.0	65,910	8.3
Calabria	11,101	18.1	25,713	15.7	36,814	12.2
Sicilia	30,609	13.4	35,144	13.8	65,753	9.4
Sardegna	132,868	5.5	425,926	2.8	558,795	2.3
Italia	392,385	3.6	776,391	2.4	1,168,776	1.9

Table 9.21 Forest and Other wooded land area included in National parks / Superficie di Bosco e delle Altre terre boscate inclusa in Parchi nazionali

Region/Regione	Forest/Bosco						Other wooded land/Altre terre boscate					
	In National parks		Not in National parks		Total		In National parks		Not in National parks		Total	
	In Parchi nazionali		Non in Parchi nazionali		Totale		In Parchi nazionali		Non in Parchi nazionali		Totale	
	Area	ES	Area	ES	Area	ES	Area	ES	Area	ES	Area	ES
	(ha)	(%)	(ha)	(%)	(ha)	(%)	(ha)	(%)	(ha)	(%)	(ha)	(%)
Piemonte	13,831	17.1	876,602	1.3	890,433	1.3	3033	53.8	81,958	8.1	84,991	8.0
Valle d'Aosta	5009	27.1	94,234	3.9	99,243	3.6	771	70.3	7963	25.6	8733	24.0
Lombardia	9698	21.2	612,270	1.6	621,968	1.6	7141	29.7	63,111	9.2	70,252	8.7
Alto Adige	20,256	13.4	319,014	1.9	339,270	1.7	756	70.5	35,324	10.5	36,081	10.4
Trentino	3736	31.8	369,524	1.5	373,259	1.4	360	100.2	33,466	10.6	33,826	10.6
Veneto	16,916	16.5	399,788	2.0	416,704	1.9	5200	26.6	47,791	9.8	52,991	9.1
Friuli V.G	0	–	332,556	1.9	332,556	1.9	0	–	41,058	10.6	41,058	10.6
Liguria	1466	49.6	341,693	1.7	343,160	1.7	0	–	44,084	10.3	44,084	10.3
Emilia Romagna	24,312	12.1	560,589	1.6	584,901	1.5	2947	35.3	50,968	9.7	53,915	9.3
Toscana	24,181	14.1	1011,267	1.1	1,035,448	1.1	5707	44.2	148,567	5.3	154,275	5.2
Umbria	7004	22.7	383,301	1.7	390,305	1.6	369	99.8	23,282	15.4	23,651	15.2
Marche	34,397	10.6	257,370	2.5	291,767	2.1	711	70.9	20,602	16.6	21,314	16.2
Lazio	16,212	14.9	544,024	1.7	560,236	1.6	1736	41.7	86,176	7.8	87,912	7.7
Abruzzo	132,779	4.7	278,808	2.9	411,588	1.8	19,670	16.6	43,341	11.3	63,011	8.6
Molise	6870	29.2	146,378	3.3	153,248	3.0	391	99.8	19,635	16.2	20,025	16.0
Campania	89,865	6.5	314,062	2.7	403,927	2.1	43,661	11.8	43,672	11.5	87,332	7.6
Puglia	60,091	7.9	82,258	6.1	142,349	4.0	11,268	23.3	38,121	11.8	49,389	9.9
Basilicata	87,432	6.4	200,588	3.8	288,020	2.7	18,504	16.7	85,888	6.9	104,392	6.2
Calabria	129,727	5.2	365,450	2.8	495,177	2.0	43,966	9.4	111,477	6.0	155,443	4.8
Sicilia	14,191	16.3	271,298	3.4	285,489	3.2	4339	40.5	97,407	7.3	101,745	7.1
Sardegna	27,232	11.6	598,908	2.2	626,140	2.1	24,325	12.3	650,526	2.0	674,851	2.0
Italia	725,204	2.2	8359,982	0.5	9,085,186	0.4	194,854	5.2	1,774,418	1.5	1,969,272	1.4

Table 9.22 Forest area included in National parks by protection level / Superficie del Bosco inclusa in Parchi nazionali, ripartita per grado di protezione

Region/Regione	Forest/Bosco															
	Zone A		Zone B		Zone C		Zone D		Neighbouring area		Not classified		Not in National parks		Total	
	Zona A—riserva integrale		Zona B—riserva generale orientata		Zona C—area di protezione		Zona D—area di promozione		Area contigua		Non classificato		Superficie esterna a Parchi nazionali		Totale	
	Area	ES	Area	ES	Area	ES	Area	ES	Area	ES	Area	ES	Area	ES	Area	ES
	(ha)	(%)	(ha)	(%)	(ha)	(%)	(ha)	(%)	(ha)	(%)	(ha)	(%)	(ha)	(%)	(ha)	(%)
Piemonte	808	70.7	3279	35.4	3232	35.3	404	100.0	0	–	6107	25.8	876,602	1.3	890,433	1.3
Valle d'Aosta	0	–	4624	28.3	0	–	0	–	0	–	385	99.6	94,234	3.9	99,243	3.6
Lombardia	0	–	441	100.0	0	–	441	100.0	0	–	8816	22.3	612,270	1.6	621,968	1.6
Alto Adige	0	–	378	99.8	756	70.5	0	–	0	–	19,122	13.9	319,014	1.9	339,270	1.7
Trentino	492	100.0	360	100.0	721	70.8	0	–	0	–	2162	40.8	369,524	1.5	373,259	1.4
Veneto	0	–	11,360	18.1	3365	33.2	0	–	0	–	2191	70.1	399,788	2.0	416,704	1.9
Friuli V.G	0	–	0	–	0	–	0	–	0	–	0	–	332,556	1.9	332,556	1.9
Liguria	0	–	733	70.1	0	–	0	–	0	–	733	70.1	341,693	1.7	343,160	1.7
Emilia Romagna	368	100.0	8104	21.2	13,998	16.1	368	100.0	368	100.0	1105	57.8	560,589	1.6	584,901	1.5
Toscana	361	100.0	7949	21.3	8672	20.3	0	–	0	–	7198	32.7	1,011,267	1.1	1,035,448	1.1
Umbria	0	–	0	–	0	–	0	–	0	–	7004	22.7	383,301	1.7	390,305	1.6
Marche	0	–	8825	24.6	4447	28.8	0	–	0	–	21,125	12.9	257,370	2.5	291,767	2.1
Lazio	5158	26.6	7001	22.8	0	–	0	–	0	–	4053	30.1	544,024	1.7	560,236	1.6
Abruzzo	23,165	12.3	24,530	13.8	4343	28.8	0	–	2172	40.8	78,570	6.3	278,808	2.9	411,588	1.8
Molise	391	99.8	3515	33.0	391	99.8	0	–	2574	60.3	0	–	146,378	3.3	153,248	3.0
Campania	29,461	10.9	25,272	12.8	34,396	11.7	737	70.8	0	–	0	–	314,062	2.7	403,927	2.1
Puglia	388	99.8	1827	81.4	1554	49.8	388	99.8	0	–	55,933	8.0	82,258	6.1	142,349	4.0
Basilicata	0	–	2186	40.5	57,654	8.4	0	–	373	99.9	27,219	11.3	200,588	3.8	288,020	2.7
Calabria	22,698	13.7	22,761	12.6	34,701	10.1	746	70.6	373	99.9	48,448	9.2	365,450	2.8	495,177	2.0
Sicilia	2217	40.8	4929	27.6	379	100.0	758	70.7	0	–	5908	26.0	271,298	3.4	285,489	3.2
Sardegna	373	100.0	373	100.0	5598	25.8	0	–	0	–	20,888	13.3	598,908	2.2	626,140	2.1
Italia	85,881	6.6	138,446	5.5	174,206	4.8	3843	31.7	5860	32.4	316,968	3.4	8,359,982	0.5	9,085,186	0.4

Table 9.23 Other wooded land area included in National parks by protection level / Superficie delle Altre terre boscate inclusa in Parchi nazionali, ripartita per grado di protezione

Region/Regione	Other wooded land/Altre terre boscate													
	Zone A		Zone B		Zone C		Neighbouring area		Not classified		Not in National parks		Total	
	Zona A—riserva integrale		Zona B—riserva generale orientata		Zona C—area di protezione		Area contigua		Non classificato		Superficie esterna a Parchi nazionali		Totale	
	Area	ES	Area	ES	Area	ES	Area	ES	Area	ES	Area	ES	Area	ES
	(ha)	(%)	(ha)	(%)	(ha)	(%)	(ha)	(%)	(ha)	(%)	(ha)	(%)	(ha)	(%)
Piemonte	0	–	0	–	404	100.0	0	–	2629	60.1	81,958	8.1	84,991	8.0
Valle d'Aosta	0	–	0	–	0	–	0	–	771	70.3	7963	25.6	8733	24.0
Lombardia	0	–	0	–	0	–	0	–	7141	29.7	63,111	9.2	70,252	8.7
Alto Adige	0	–	0	–	0	–	0	–	756	70.5	35,324	10.5	36,081	10.4
Trentino	0	–	360	100.0	0	–	0	–	0	–	33,466	10.6	33,826	10.6
Veneto	374	99.9	2243	40.7	0	–	0	–	2582	37.7	47,791	9.8	52,991	9.1
Friuli V.G	0	–	0	–	0	–	0	–	0	–	41,058	10.6	41,058	10.6
Liguria	0	–	0	–	0	–	0	–	0	–	44,084	10.3	44,084	10.3
Emilia Romagna	737	70.7	368	100.0	368	100.0	0	–	1473	50.0	50,968	9.7	53,915	9.3
Toscana	0	–	0	–	2843	70.0	0	–	2864	55.6	148,567	5.3	154,275	5.2
Umbria	0	–	0	–	0	–	0	–	369	99.8	23,282	15.4	23,651	15.2
Marche	0	–	341	100.0	0	–	0	–	371	100.0	20,602	16.6	21,314	16.2
Lazio	368	100.0	0	–	0	–	0	–	1367	45.6	86,176	7.8	87,912	7.7
Abruzzo	3137	51.2	1796	81.4	0	–	335	100.0	14,401	17.4	43,341	11.3	63,011	8.6
Molise	0	–	0	–	0	–	0	–	391	99.8	19,635	16.2	20,025	16.0
Campania	5888	24.9	7301	27.9	20,325	19.7	0	–	10,147	25.3	43,672	11.5	87,332	7.6
Puglia	0	–	0	–	1165	57.5	0	–	10,103	25.1	38,121	11.8	49,389	9.9
Basilicata	0	–	0	–	7015	28.8	0	–	11,489	20.7	85,888	6.9	104,392	6.2
Calabria	1119	57.6	4104	30.0	6343	24.1	0	–	32,399	11.2	111,477	6.0	155,443	4.8
Sicilia	1427	100.0	1459	50.1	0	–	0	–	1453	49.8	97,407	7.3	101,745	7.1
Sardegna	1866	44.7	2239	40.8	6707	23.5	0	–	13,513	16.5	650,526	2.0	674,851	2.0
Italia	14,917	19.5	20,212	16.0	45,170	12.1	335	100.0	114,219	6.6	1,774,418	1.5	1,969,272	1.4

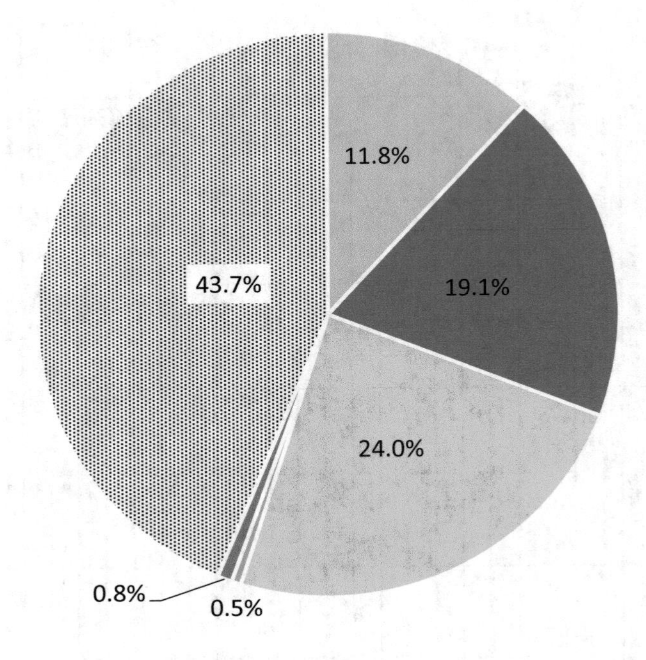

Fig. 9.12 Percentage area of Forest into national parks by different protection levels / Ripartizione percentuale del Bosco incluso in parchi nazionali, per grado di protezione

Table 9.24 Forest and Other wooded land area included in State nature reserves / Superficie di Bosco e Altre terre boscate inclusa in Riserve naturali statali

Region/Regione	Forest/Bosco						Other wooded land/Altre terre boscate					
	In State nature reserves		Not in State nature reserves		Total		In State nature reserves		Not in State nature reserves		Total	
	In riserve naturali statali		Non in riserve naturali statali		Totale		In Riserve naturali statali		Non in Riserve naturali statali		Totale	
	Area	ES	Area	ES	Area	ES	Area	ES	Area	ES	Area	ES
	(ha)	(%)	(ha)	(%)	(ha)	(%)	(ha)	(%)	(ha)	(%)	(ha)	(%)
Piemonte	2020	44.7	888,413	1.3	890,433	1.3	1417	100.0	83,574	8.0	84,991	8.0
Valle d'Aosta	0	–	99,243	3.6	99,243	3.6	0	–	8733	24.0	8733	24.0
Lombardia	441	100.0	621,527	1.6	621,968	1.6	0	–	70,252	8.7	70,252	8.7
Alto Adige	0	–	339,270	1.7	339,270	1.7	0	–	36,081	10.4	36,081	10.4
Trentino	0	–	373,259	1.4	373,259	1.4	0	–	33,826	10.6	33,826	10.6
Veneto	8690	25.3	408,014	1.9	416,704	1.9	5147	36.1	47,844	9.5	52,991	9.1
Friuli V.G	0	–	332,556	1.9	332,556	1.9	0	–	41,058	10.6	41,058	10.6
Liguria	0	–	343,160	1.7	343,160	1.7	0	–	44,084	10.3	44,084	10.3
Emilia Romagna	4789	27.7	580,112	1.6	584,901	1.5	368	100.0	53,546	9.4	53,915	9.3
Toscana	10,455	21.9	1,024,993	1.1	1,035,448	1.1	723	70.7	153,552	5.2	154,275	5.2
Umbria	0	–	390,305	1.6	390,305	1.6	0	–	23,651	15.2	23,651	15.2
Marche	3706	31.5	288,061	2.1	291,767	2.1	341	100.0	20,973	16.4	21,314	16.2
Lazio	9177	24.1	551,060	1.6	560,236	1.6	1842	44.7	86,069	7.8	87,912	7.7
Abruzzo	8380	20.7	403,208	1.8	411,588	1.8	2507	37.8	60,504	8.9	63,011	8.6
Molise	1172	57.5	152,076	3.1	153,248	3.0	0	–	20,025	16.0	20,025	16.0
Campania	737	70.8	403,190	2.1	403,927	2.1	0	–	87,332	7.6	87,332	7.6
Puglia	1165	57.5	141,183	4.1	142,349	4.0	0	–	49,389	9.9	49,389	9.9
Basilicata	746	70.6	287,274	2.7	288,020	2.7	0	–	104,392	6.2	104,392	6.2
Calabria	5224	26.6	489,953	2.1	495,177	2.0	5597	25.7	149,847	5.0	155,443	4.8
Sicilia	321	100.0	285,168	3.2	285,489	3.2	0	–	101,745	7.1	101,745	7.1
Sardegna	0	–	626,140	2.1	626,140	2.1	0	–	674,851	2.0	674,851	2.0
Italia	57,021	8.9	9,028,165	0.4	9,085,186	0.4	17,942	17.3	1,951,330	1.4	1,969,272	1.4

Table 9.25 Forest and Other wooded land area included in Regional nature parks / Superficie di Bosco e Altre terre boscate inclusa in Parchi naturali regionali

Region/Regione	Forest/Bosco						Other wooded land/Altre terre boscate					
	In Regional nature parks		Not in Regional nature parks		Total		In Regional nature parks		Not in Regional nature parks		Total	
	In Parchi naturali regionali		Non in Parchi naturali regionali		Totale		In Parchi naturali regionali		Non in Parchi naturali regionali		Totale	
	Area	ES	Area	ES	Area	ES	Area	ES	Area	ES	Area	ES
	(ha)	(%)	(ha)	(%)	(ha)	(%)	(ha)	(%)	(ha)	(%)	(ha)	(%)
Piemonte	36,147	10.4	854,286	1.4	890,433	1.3	6967	34.1	78,023	8.3	84,991	8.0
Valle d'Aosta	771	70.3	98,472	3.6	99,243	3.6	385	99.6	8348	24.8	8733	24.0
Lombardia	66,787	9.0	555,180	1.8	621,968	1.6	8799	33.6	61,454	9.0	70,252	8.7
Alto Adige	34,038	10.2	305,232	2.1	339,270	1.7	11,768	17.7	24,313	13.2	36,081	10.4
Trentino	31,387	10.5	341,872	1.7	373,259	1.4	5657	25.4	28,169	11.8	33,826	10.6
Veneto	14,956	15.6	401,748	2.0	416,704	1.9	6695	23.4	46,296	10.0	52,991	9.1
Friuli V.G	20,033	14.7	312,524	2.1	332,556	1.9	11,609	24.3	29,448	12.7	41,058	10.6
Liguria	16,126	14.7	327,033	1.8	343,160	1.7	2916	35.0	41,169	10.8	44,084	10.3
Emilia Romagna	23,580	12.3	561,321	1.6	584,901	1.5	1819	44.7	52,095	9.5	53,915	9.3
Toscana	26,016	11.7	1009,432	1.1	1035,448	1.1	8287	25.6	145,988	5.4	154,275	5.2
Umbria	15,043	17.4	375,263	1.7	390,305	1.6	1475	49.9	22,177	15.9	23,651	15.2
Marche	6601	30.0	285,166	2.1	291,767	2.1	0	–	21,314	16.2	21,314	16.2
Lazio	62,130	7.4	498,106	1.9	560,236	1.6	5098	36.2	82,813	8.0	87,912	7.7
Abruzzo	25,345	11.9	386,243	2.0	411,588	1.8	6742	34.0	56,269	9.3	63,011	8.6
Molise	0	–	153,248	3.0	153,248	3.0	0	–	20,025	16.0	20,025	16.0
Campania	99,553	5.8	304,374	2.8	403,927	2.1	7675	20.5	79,657	8.1	87,332	7.6
Puglia	6992	23.2	135,357	4.2	142,349	4.0	6489	30.7	42,900	10.9	49,389	9.9
Basilicata	12,299	17.2	275,720	2.8	288,020	2.7	6428	23.8	97,964	6.5	104,392	6.2
Calabria	9328	19.9	485,849	2.1	495,177	2.0	1835	44.7	153,609	4.9	155,443	4.8
Sicilia	81,691	7.0	203,798	4.3	285,489	3.2	16,491	16.6	85,254	8.1	101,745	7.1
Sardegna	3732	31.6	622,408	2.1	626,140	2.1	3359	33.3	671,493	2.0	674,851	2.0
Italia	592,554	2.6	8,492,632	0.4	9,085,186	0.4	120,494	6.7	1,848,778	1.5	1,969,272	1.4

Table 9.26 Forest and Other wooded land area included in Regional nature reserves / Superficie di Bosco e Altre terre boscate inclusa in Riserve naturali regionali

Region/Regione	Forest/Bosco						Other wooded land/Altre terre boscate					
	In Regional nature reserves		Not in Regional nature reserves		Total		In Regional nature reserves		Not in Regional nature reserves		Total	
	In riserve naturali regionali		Non in riserve naturali regionali		Totale		In Riserve naturali regionali		Non in Riserve naturali regionali		Totale	
	Area	ES	Area	ES	Area	ES	Area	ES	Area	ES	Area	ES
	(ha)	(%)	(ha)	(%)	(ha)	(%)	(ha)	(%)	(ha)	(%)	(ha)	(%)
Piemonte	10,802	24.7	879,631	1.3	890,433	1.3	451	100.0	84,540	8.0	84,991	8.0
Valle d'Aosta	0	–	99,243	3.6	99,243	3.6	0	–	8733	24.0	8733	24.0
Lombardia	2204	44.7	619,764	1.6	621,968	1.6	0	–	70,252	8.7	70,252	8.7
Alto Adige	100	99.8	339,170	1.7	339,270	1.7	0	–	36,081	10.4	36,081	10.4
Trentino	1441	50.0	371,818	1.5	373,259	1.4	0	–	33,826	10.6	33,826	10.6
Veneto	2243	40.7	414,461	1.9	416,704	1.9	374	99.9	52,617	9.2	52,991	9.1
Friuli V.G	2230	40.6	330,326	1.9	332,556	1.9	372	99.7	40,686	10.7	41,058	10.6
Liguria	367	99.2	342,793	1.7	343,160	1.7	0	–	44,084	10.3	44,084	10.3
Emilia Romagna	737	70.7	584,164	1.5	584,901	1.5	0	–	53,915	9.3	53,915	9.3
Toscana	19,512	13.5	1,015,936	1.1	1,035,448	1.1	2529	37.8	151,745	5.3	154,275	5.2
Umbria	369	99.8	389,936	1.6	390,305	1.6	0	–	23,651	15.2	23,651	15.2
Marche	741	70.8	291,026	2.1	291,767	2.1	0	–	21,314	16.2	21,314	16.2
Lazio	17,283	16.1	542,954	1.7	560,236	1.6	737	70.7	87,175	7.7	87,912	7.7
Abruzzo	6877	22.8	404,711	1.8	411,588	1.8	0	–	63,011	8.6	63,011	8.6
Molise	0	–	153,248	3.0	153,248	3.0	0	–	20,025	16.0	20,025	16.0
Campania	4419	28.8	399,508	2.1	403,927	2.1	1105	57.8	86,228	7.6	87,332	7.6
Puglia	1554	49.8	140,795	4.1	142,349	4.0	388	99.8	49,000	9.9	49,389	9.9
Basilicata	4096	30.0	283,923	2.7	288,020	2.7	0	–	104,392	6.2	104,392	6.2
Calabria	373	99.9	494,804	2.0	495,177	2.0	0	–	155,443	4.8	155,443	4.8
Sicilia	32,558	12.4	252,932	3.6	285,489	3.2	11,711	27.1	90,034	7.5	101,745	7.1
Sardegna	0	–	626,140	2.1	626,140	2.1	0	–	674,851	2.0	674,851	2.0
Italia	107,906	6.4	8,977,281	0.4	9,085,186	0.4	17,666	19.8	1,951,606	1.4	1,969,272	1.4

Table 9.27 Forest and Other wooded land area included in Other natural protected areas / Superficie di Bosco e Altre terre boscate inclusa in Altre aree naturali protette

Region/Regione	Forest/Bosco						Other wooded land/Altre terre boscate					
	In Other natural protected areas		Not in Other natural protected areas		Total		In Other natural protected areas		Not in Other natural protected areas		Total	
	In altre aree naturali protette		Non in altre aree naturali protette		Totale		In Altre aree naturali protette		Non in Altre aree naturali protette		Totale	
	Area	ES	Area	ES	Area	ES	Area	ES	Area	ES	Area	ES
	(ha)	(%)	(ha)	(%)	(ha)	(%)	(ha)	(%)	(ha)	(%)	(ha)	(%)
Piemonte	7273	23.5	883,160	1.3	890,433	1.3	404	100.0	84,587	8.0	84,991	8.0
Valle d'Aosta	0	–	99,243	3.6	99,243	3.6	0	–	8733	24.0	8733	24.0
Lombardia	4504	30.4	617,464	1.6	621,968	1.6	0	–	70,252	8.7	70,252	8.7
Alto Adige	856	63.4	338,414	1.7	339,270	1.7	1134	57.6	34,946	10.6	36,081	10.4
Trentino	0	–	373,259	1.4	373,259	1.4	0	–	33,826	10.6	33,826	10.6
Veneto	0	–	416,704	1.9	416,704	1.9	0	–	52,991	9.1	52,991	9.1
Friuli V.G	10,343	18.6	322,213	2.0	332,556	1.9	372	99.7	40,686	10.7	41,058	10.6
Liguria	1100	57.2	342,060	1.7	343,160	1.7	0	–	44,084	10.3	44,084	10.3
Emilia Romagna	1473	50.0	583,428	1.6	584,901	1.5	0	–	53,915	9.3	53,915	9.3
Toscana	10,327	18.6	1,025,120	1.1	1,035,448	1.1	361	100.0	153,913	5.2	154,275	5.2
Umbria	3687	31.5	386,618	1.6	390,305	1.6	369	99.8	23,282	15.4	23,651	15.2
Marche	2194	40.9	289,573	2.1	291,767	2.1	0	–	21,314	16.2	21,314	16.2
Lazio	2948	35.3	557,288	1.6	560,236	1.6	579	73.3	87,333	7.7	87,912	7.7
Abruzzo	2534	37.8	409,054	1.8	411,588	1.8	0	–	63,011	8.6	63,011	8.6
Molise	781	70.5	152,467	3.1	153,248	3.0	0	–	20,025	16.0	20,025	16.0
Campania	1841	44.7	402,086	2.1	403,927	2.1	0	–	87,332	7.6	87,332	7.6
Puglia	388	99.8	141,960	4.0	142,349	4.0	0	–	49,389	9.9	49,389	9.9
Basilicata	373	99.9	287,647	2.7	288,020	2.7	0	–	104,392	6.2	104,392	6.2
Calabria	373	99.9	494,804	2.0	495,177	2.0	0	–	155,443	4.8	155,443	4.8
Sicilia	0	–	285,489	3.2	285,489	3.2	0	–	101,745	7.1	101,745	7.1
Sardegna	4478	28.8	621,661	2.1	626,140	2.1	2239	40.8	672,612	2.0	674,851	2.0
Italia	55,474	8.2	9,029,712	0.4	9,085,186	0.4	5458	26.0	1,963,814	1.4	1,969,272	1.4

Table 9.28 Forest and Other wooded land area included in NATURA2000 sites (SCI and SPA) / Superficie di Bosco e Altre terre boscate inclusa in siti (SIC e ZPS) della rete NATURA2000

Region/Regione	Forest/Bosco						Other wooded land/Altre terre boscate					
	In NATURA2000 sites		Not in NATURA2000 sites		Total		In NATURA2000 sites		Not in NATURA2000 sites		Total	
	In siti rete NATURA2000		Non in siti rete NATURA2000		Totale		In siti rete NATURA2000		Non in siti rete NATURA2000		Totale	
	Area	ES	Area	ES	Area	ES	Area	ES	Area	ES	Area	ES
	(ha)	(%)	(ha)	(%)	(ha)	(%)	(ha)	(%)	(ha)	(%)	(ha)	(%)
Piemonte	121,540	5.7	768,893	1.6	890,433	1.3	19,106	16.9	65,885	9.4	84,991	8.0
Valle d'Aosta	12,715	16.5	86,527	4.4	99,243	3.6	1653	46.7	7081	27.7	8733	24.0
Lombardia	130,893	5.7	491,075	2.2	621,968	1.6	23,296	16.6	46,957	10.7	70,252	8.7
Alto Adige	40,783	9.2	298,487	2.1	339,270	1.7	11,809	17.7	24,272	13.2	36,081	10.4
Trentino	80,168	6.1	293,091	2.2	373,259	1.4	15,042	17.1	18,784	13.7	33,826	10.6
Veneto	173,327	4.1	243,377	3.3	416,704	1.9	38,870	10.9	14,121	18.3	52,991	9.1
Friuli V.G	65,638	7.2	266,918	2.6	332,556	1.9	17,432	18.7	23,626	14.1	41,058	10.6
Liguria	90,128	5.6	253,031	2.6	343,160	1.7	18,826	16.2	25,258	14.5	44,084	10.3
Emilia Romagna	116,734	5.3	468,167	2.0	584,901	1.5	8772	20.3	45,142	10.5	53,915	9.3
Toscana	181,787	4.2	853,661	1.4	1,035,448	1.1	35,265	11.1	119,009	6.1	154,275	5.2
Umbria	77,714	6.5	312,591	2.3	390,305	1.6	2949	35.2	20,702	16.7	23,651	15.2
Marche	78,141	6.1	213,626	3.1	291,767	2.1	4545	25.7	16,769	19.7	21,314	16.2
Lazio	208,925	3.6	351,311	2.7	560,236	1.6	35,495	12.0	52,417	10.9	87,912	7.7
Abruzzo	210,054	3.4	201,534	3.8	411,588	1.8	30,729	13.2	32,283	13.6	63,011	8.6
Molise	59,595	7.5	93,653	5.5	153,248	3.0	6935	28.5	13,090	20.4	20,025	16.0
Campania	215,396	3.6	188,531	4.2	403,927	2.1	34,912	11.9	52,420	10.9	87,332	7.6
Puglia	92,919	5.7	49,430	8.7	142,349	4.0	27,902	13.8	21,487	17.2	49,389	9.9
Basilicata	76,357	6.6	211,663	3.7	288,020	2.7	23,346	14.3	81,046	7.2	104,392	6.2
Calabria	89,860	6.2	405,316	2.5	495,177	2.0	50,963	8.6	104,481	6.2	155,443	4.8
Sicilia	135,433	5.3	150,056	5.4	285,489	3.2	40,944	11.0	60,801	10.0	101,745	7.1
Sardegna	150,776	5.1	475,364	2.6	626,140	2.1	153,446	5.1	521,405	2.4	674,851	2.0
Italia	2,408,882	1.2	6,676,304	0.6	9,085,186	0.4	602,237	2.8	1,367,035	1.8	1,969,272	1.4

Table 9.29 Forest and Other wooded land area included in Wetlands of international importance (Ramsar Convention) / Superficie di Bosco e Altre terre boscate inclusa in Zone umide di interesse internazionale (Convenzione di Ramsar)

Region/Regione	Forest/Bosco						Other wooded land/Altre terre boscate					
	In Wetlands		Not in Wetlands		Total		In Wetlands		Not in Wetlands		Total	
	In Zone umide		Non in Zone umide		Totale		In Zone umide		Non in Zone umide		Totale	
	Area	ES	Area	ES	Area	ES	Area	ES	Area	ES	Area	ES
	(ha)	(%)	(ha)	(%)	(ha)	(%)	(ha)	(%)	(ha)	(%)	(ha)	(%)
Piemonte	0	–	890,433	1.3	890,433	1.3	0	–	84,991	8.0	84,991	8.0
Valle d'Aosta	0	–	99,243	3.6	99,243	3.6	0	–	8733	24.0	8733	24.0
Lombardia	882	70.7	621,086	1.6	621,968	1.6	0	–	70,252	8.7	70,252	8.7
Alto Adige	0	–	339,270	1.7	339,270	1.7	0	–	36,081	10.4	36,081	10.4
Trentino	0	–	373,259	1.4	373,259	1.4	0	–	33,826	10.6	33,826	10.6
Veneto	0	–	416,704	1.9	416,704	1.9	0	–	52,991	9.1	52,991	9.1
Friuli V.G	0	–	332,556	1.9	332,556	1.9	0	–	41,058	10.6	41,058	10.6
Liguria	350	99.2	342,809	1.7	343,160	1.7	0	–	44,084	10.3	44,084	10.3
Emilia Romagna	714	70.8	584,187	1.5	584,901	1.5	0	–	53,915	9.3	53,915	9.3
Toscana	0	–	1,035,448	1.1	1,035,448	1.1	0	–	154,275	5.2	154,275	5.2
Umbria	0	–	390,305	1.6	390,305	1.6	0	–	23,651	15.2	23,651	15.2
Marche	0	–	291,767	2.1	291,767	2.1	0	–	21,314	16.2	21,314	16.2
Lazio	1105	57.7	559,131	1.6	560,236	1.6	368	100.0	87,543	7.7	87,912	7.7
Abruzzo	362	100.0	411,226	1.8	411,588	1.8	0	–	63,011	8.6	63,011	8.6
Molise	0	–	153,248	3.0	153,248	3.0	0	–	20,025	16.0	20,025	16.0
Campania	0	–	403,927	2.1	403,927	2.1	0	–	87,332	7.6	87,332	7.6
Puglia	0	–	142,349	4.0	142,349	4.0	0	–	49,389	9.9	49,389	9.9
Basilicata	373	99.9	287,647	2.7	288,020	2.7	0	–	104,392	6.2	104,392	6.2
Calabria	373	99.9	494,804	2.0	495,177	2.0	0	–	155,443	4.8	155,443	4.8
Sicilia	758	70.7	284,731	3.2	285,489	3.2	0	–	101,745	7.1	101,745	7.1
Sardegna	0	–	626,140	2.1	626,140	2.1	0	–	674,851	2.0	674,851	2.0
Italia	4917	27.8	9,080,269	0.4	9,085,186	0.4	368	100.0	1,968,904	1.4	1,969,272	1.4

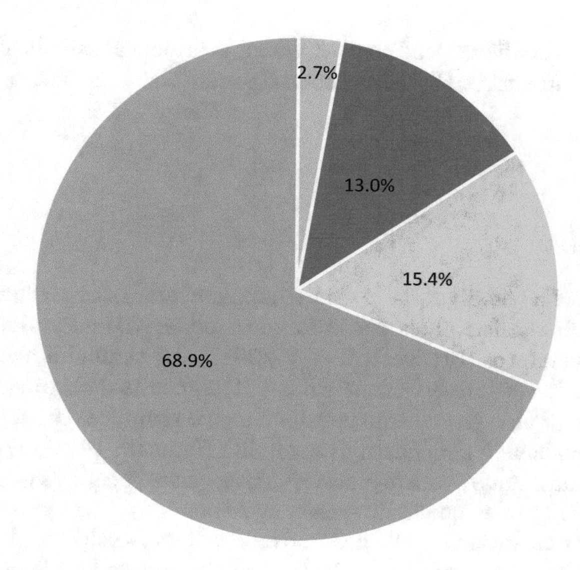

No active intervention / Nessun intervento
Minimum intervention / Interventi minimi
Conservation through active management / Conservazione tramite gestione attiva
Not protected / Non protetto

Fig. 9.13 Percentage area of Forest by protection level classes comparable to MCPFE classes / Ripartizione percentuale del Bosco secondo classi del livello di protezione confrontabili con la classificazione MCPFE

Appendix (Italian Version)

Riassunto Il valore delle foreste per gli aspetti diversi dalla produzione di legna e legname è cresciuto enormemente negli ultimi decenni con la consapevolezza dei rischi associati alla perdita delle foreste e con il riconoscimento dell'importanza della molteplicità di beni e servizi che da esse derivano. Sulla base di questa premessa, gli inventari forestali nazionali hanno ampliato il ventaglio delle variabili oggetto di stima e una parte considerevole di queste riguarda oggi aspetti di tipo naturalistico. Tra questi, la biodiversità riveste un ruolo importantissimo per le capacità di adattamento che essa offre agli ecosistemi forestali nell'ambito dei cambiamenti climatici. In questo capitolo sono presentate e commentate le stime relative alla diversità in specie rilevata da INFC2015, le stime sulle condizioni di naturalità nell'insediamento dei soprassuoli forestali e quelle sulla presenza e tipologia del legno morto. Alberi morti, ceppaie residue e legno morto a terra esprimono un potenziale molto elevato per la diversificazione delle specie, oltre al contributo offerto nell'immagazzinamento del carbonio organico. La consapevolezza in premessa è stata anche il fondamento per percorsi a tutela delle foreste di natura giuridica che hanno portato all'istituzione di aree protette diversificate per tipo e per grado; le principali statistiche relative alla protezione delle aree forestali italiane sono riportate e commentate in chiusura del capitolo.

Introduzione

L'importanza della biodiversità è stata ampiamente riconosciuta sia per la funzionalità degli ecosistemi sia per la fornitura dei servizi ad essi richiesti per il benessere dell'uomo (es. Brockerhoff et al., 2017). La comunità internazionale si è impegnata per la sua conservazione sin dal 1992, con la dichiarazione di Rio de Janeiro. Si tratta di un concetto solo apparentemente semplice ed intuitivo, neppure sempre corrispondente a condizioni di naturalità (Pignatti, 1995a). Sono comunemente riconosciuti cinque livelli di biodiversità, da quella genetica a quella di regione geografica, passando per quella di specie, ecosistemica e paesaggistica (Bernetti, 2005). Il rilievo di variabili utili a descrivere la biodiversità dei boschi è stato introdotto abbastanza recentemente dagli inventari forestali nazionali (Gasparini et al., 2013) e il contributo di INFC si esplica mediante la produzione di informazioni inerenti a quattro dei cinque livelli citati. Per i tre livelli più generali, nel Chap. 7 sono state presentate le stime sulle principali formazioni forestali presenti nei territori italiani e che concorrono a caratterizzarne anche i paesaggi. Informazioni ulteriori sono disponibili all'indirizzo inventarioforestale.org/statistiche_INFC, con particolare riferimento alle statistiche sulle sottocategorie forestali. In questo capitolo vengono illustrati i risultati di INFC2015 relativi alla diversità delle specie forestali (Sect. 9.2), all'origine del soprassuolo (Sect. 9.3), alla presenza del legno morto (Sect. 9.4) e alla superficie delle aree boscate incluse in aree protette (Sect. 9.5).

La vegetazione forestale italiana presenta un elevato grado di diversità (Pedrotti, 1995). Ad essa contribuisce la grande varietà di ambienti e climi del nostro Paese che è alla base della ricchezza floristica generale (Pignatti, 1994). Nonostante la biodiversità di una determinata area non possa essere esaurientemente espressa solo con il numero di specie presenti (es. Pignatti, 1995b), questo viene considerato un suo indicatore naturale, ed è alla base di molti modelli ecologici (Gotelli & Colwell, 2001). Il numero di specie rimane, inoltre, uno degli aspetti più importanti nella valutazione naturalistica dei boschi (Pignatti & De Natale, 2011).

La naturalità dei soprassuoli forestali viene anche valutata in base all'origine delle piante che sostituiranno quelle del vecchio ciclo. Con l'accezione adottata per l'origine del soprassuolo in ambito INFC, la condizione di massima naturalità si verifica in assenza di qualsiasi intervento umano volto a favorire l'insediamento di una nuova generazione di individui. All'estremo opposto, la rinnovazione artificiale prevede l'utilizzo di materiale vegetale anche di altra provenienza e/o specie. Dal punto di vista naturalistico, l'impatto può variare sensibilmente e valgano per questo gli esempi della rinnovazione artificiale posticipata con materiale vivaistico prodotto da seme precedentemente raccolto in loco o l'impianto di popolamenti per la produzione di legno con specie esotiche. La rinnovazione a seguito di interventi selvicolturali, che denota condizioni di semi-naturalità, garantisce la conservazione degli ecotipi locali, salvo i casi in cui perpetua ad esempio una specie precedentemente introdotta artificialmente.

Il legno morto rappresenta una componente di fondamentale importanza per gli ecosistemi forestali. Esso contribuisce alla conservazione della biodiversità, svolge un ruolo chiave nel ciclo degli elementi, influenza la dinamica del bosco e l'insediamento della rinnovazione naturale, concorre alla stabilità del suolo sui pendii e lungo i corsi d'acqua (es. Paletto et al., 2012) e ostacola la formazione di valanghe (Berretti et al., 2006). Come combustibile, oltre a rappresentare una risorsa di energia, la presenza di legno morto assume rilevanza anche per il rischio e le modalità di propagazione degli incendi. I Paesi che hanno sottoscritto gli accordi per la lotta ai cambiamenti climatici sono tenuti a stimare il carbonio immagazzinato nel legno morto, poiché esso costituisce una delle cinque riserve di carbonio forestale contabilizzate (IPCC, 2003). L'inclusione del legno morto tra le variabili rilevate dagli inventariali forestali nazionali ha portato anche ad una raccolta di dati senza precedenti e proprio la disponibilità di dati rilevati su larga scala ha mostrato l'inadeguatezza dell'uso di modelli per la stima di una variabile così complessa (es. Woodall et al., 2019) e avvalorato l'importanza di una misurazione diretta.

Le aree protette rivestono un ruolo cruciale nella conservazione della biodiversità e la salvaguardia di diversi ecosistemi, tra cui le cenosi forestali (Dudley & Phillips, 2006). Gli inventari forestali nazionali, insieme alla conoscenza dei confini delle aree protette, costituiscono una fonte importante di informazione sulle foreste in aree protette. La foresta è, in molti casi, una componente essenziale, sebbene non l'unica, delle aree protette presenti sul territorio italiano.

Le specie legnose

La biodiversità arborea è il presupposto all'edificazione di foreste in contesti ambientali diversi, con caratteristiche a valenza ecologica e funzionale molto diversificate, importante anche per soddisfare la molteplicità dei servizi ad esse richiesti. INFC2015 ha misurato individui ($d_{1.30} \geq 4.5$ cm) di circa 180 specie legnose.

La Table 9.1 riporta le stime per il Bosco a livello nazionale relative al numero degli alberi, al volume legnoso e alla fitomassa arborea epigea per le 45 specie maggiormente rappresentate in termini di volume. Le specie dell'elenco sono state selezionate in modo da assicurare che in ogni regione fosse rappresentato almeno il 90% del volume complessivo dei Boschi. In inventarioforestale.org/statistiche_INFC sono disponibili le statistiche sul numero di alberi, l'area basimetrica, il volume, l'incremento annuo, la fitomassa e il carbonio organico per categoria forestale, sia a livello nazionale sia per regione. In tutte le tabelle, le specie sono ordinate in maniera decrescente di volume, ma il gruppo "altre specie" è sempre riportato in fondo all'elenco. Questa voce rappresenta la classe residuale oltre la quarantacinquesima e può contenere un insieme di specie diverso nelle varie tabelle.

Le Figs. 9.1, 9.2 e 9.3 illustrano le frequenze cumulate, per il Bosco a livello nazionale, rispettivamente del numero di alberi, del volume e della fitomassa arborea epigea. Esse consentono di evidenziare che l'85% del volume e della biomassa arborea epigea è dovuto a 17 specie che, ad eccezione del pino laricio, sono anche tra le 27 specie che assicurano l'85% del numero di alberi. Le stesse specie sono le più importanti anche in termini di incremento di volume annuo, coerentemente con l'importanza del volume quale variabile predittiva principale nella stima dell'incremento (Bevilacqua, 1999; Gasparini et al., 2017); nel complesso, alle 17 specie principali si deve l'80% dell'incremento di volume annuo.

La specie più rappresentata nel paesaggio forestale italiano è il faggio; esso rappresenta il 10.8% del numero di alberi, il 15.2% dell'area basimetrica, il 18.1% del volume legnoso, il 20.7% della biomassa arborea epigea e il 14.2% dell'incremento annuo di volume. La seconda specie più rappresentata in termini di area basimetrica, volume, biomassa arborea epigea e incremento è l'abete rosso, che contribuisce con il 3.4% in numero di alberi, il 10.2% in area basimetrica, il 15.3% in volume, l'11.1% in biomassa e l'11.8% in incremento annuo di volume. Seguono il castagno e il cerro, anche se in ordine di importanza invertito per quanto riguarda il contributo in biomassa arborea epigea. Il castagno è presente con il 6.2% degli alberi, il 9.6% dell'area basimetrica, il 9.0% del volume, l'8,1% della biomassa arborea epigea e il 9.9% dell'incremento annuo di volume. Il cerro contribuisce per il 7.6% degli alberi totali, l'8.2% dell'area basimetrica, l'8.1% del volume, il 9.4% della biomassa e il 9.2% dell'incremento. Faggio, abete rosso, castagno e cerro assicurano nell'insieme poco più del 50% del volume legnoso e della biomassa arborea epigea del Bosco in Italia, nonché il 45.5% della produzione annua di volume. Le quattro specie caratterizzano ognuna una categoria forestale, nella quale rappresentano una quota dell'area basimetrica totale variabile ma sempre piuttosto elevata, dal 76.6% del cerro nella categoria delle Cerrete, boschi di farnetto, fragno e vallonea all'89.0% per il faggio nella categoria delle Faggete. La Fig. 9.4 mostra la ripartizione dell'area basimetrica delle quattro specie nelle diverse categorie del Bosco; il grafico presenta accorgimenti utili a dare risalto alle categorie dei Boschi alti e ad agevolare la lettura per i valori di percentuale bassi. L'insieme delle specie cui si deve l'85% del numero di alberi è naturalmente più ampio di quello ad esempio del volume, per il contributo di specie minori spesso allevate a ceduo o che si trovano in mescolanza in soprassuoli con quella forma di governo.

Fra le 45 specie elencate, ne compaiono alcune esotiche e tra queste è particolarmente degna di nota la robinia, non soltanto perché è ormai da tempo naturalizzata ma perché contribuisce in maniera non irrilevante al numero di alberi (2.5% del numero totale), al volume legnoso (1.6% del volume totale), alla fitomassa arborea epigea (1.9% del totale) e all'incremento annuo di volume (2.7% del totale); gli eucalipti (0.3% del volume totale e 0.6% dell'incremento totale annuo del Bosco) e la douglasia (0.6% del volume totale e 0.6% dell'incremento), impiegate nei rimboschimenti o negli impianti di arboricoltura da legno, hanno una presenza più contenuta.

Origine dei soprassuoli

Per origine del soprassuolo, INFC considera la naturalità della rinnovazione in relazione all'intensità degli interventi colturali. Sono distinte tre classi, descritte in Table 9.2. Le Tables 9.3 e 9.4 mostrano le stime di superficie per il Bosco e le Altre terre boscate secondo l'origine del soprassuolo. Le Tables 9.5 e 9.6 mostrano le corrispondenti statistiche per i Boschi alti e per gli Impianti di arboricoltura da legno. Le Tables 9.7 e 9.8 mostrano le statistiche per le categorie inventariali delle Altre terre boscate. Le statistiche sull'origine dei soprassuoli per le categorie forestali dei Boschi alti sono disponibili all'indirizzo inventarioforestale.org/statistiche_INFC.

La gran parte dei soprassuoli del Bosco italiano ha un'origine seminaturale; l'insediamento di una nuova generazione di alberi è, quindi, stato perseguito e/o favorito mediante interventi di natura selvicolturale. La proporzione di superficie del Bosco con queste caratteristiche ammonta al 68.0% a livello nazionale e la prevalenza è osservabile in tutte le regioni con l'unica eccezione della Sicilia, dove la classe di origine naturale interessa una aliquota maggiore di quella seminaturale (36.8% contro 27.7%). La Calabria e la Puglia sono le uniche altre due regioni dove la quota dei soprassuoli di origine seminaturale non supera il 50% del Bosco, anche se il dato se ne discosta di poco (rispettivamente 48.8% e 48.4%). La classe di origine naturale dei soprassuoli è stata riscontrata sul 17.1% della superficie del Bosco; anche per essa si registra una variabilità piuttosto elevata tra le regioni, ma rimane in genere la classe più estesa dopo quella dell'origine seminaturale, con l'unica eccezione del Molise, dove presenta valori di poco inferiori rispetto ai soprassuoli di origine artificiale. In questa classe rientrano i soprassuoli originatisi spontaneamente dopo eventi catastrofici, come ad esempio gli incendi, ma anche quelli legati alle successioni primarie e secondarie che sono importanti per l'espansione della superficie forestale.

Il confronto tra la Fig. 9.5, che mostra la ripartizione della superficie del Bosco secondo le tre classi di origine del soprassuolo adottate, e la Fig. 9.6, che mostra la stessa ripartizione per le Altre terre boscate, permette di evidenziare un'inversione di frequenza generalizzata tra le classi di origine seminaturale (prevalente nel Bosco) e naturale (prevalente nelle Altre terre boscate) sebbene si debba tener conto di un'incidenza non trascurabile della superficie non classificata per l'origine dei soprassuoli delle Altre terre boscate. In linea generale, però, sembra possibile ipotizzare un contributo cospicuo alla marcata naturalità dell'origine nelle Altre terre boscate dei processi di colonizzazione delle specie legnose su terreni divenuti marginali e non più coltivati, su cui sono ancora limitati gli interventi antropici (fanno eccezione il Molise e la Sardegna) per favorire la rinnovazione. Parte di questi terreni potranno evolvere verso l'edificazione di boschi veri e propri ed è proprio nell'ambito di questa dinamica che può essere spiegata anche una parte della superficie del Bosco nella classe di origine naturale del soprassuolo. Si tratta di una classe di transito determinata dal tempo più o meno lungo che può intercorrere tra il raggiungimento delle soglie adottate nella definizione di Bosco, sufficiente perché si passi dalle Altre terre boscate al Bosco, al momento di convenienza alle operazioni selvicolturali. La classe di origine artificiale è stata riscontrata su una superficie del 6.3% del Bosco a livello nazionale e sotto questa soglia si attesta la gran parte delle regioni; il dato della Sicilia si distingue nettamente sia perché presenta il valore massimo in percentuale rispetto alle altre regioni (27.4% della superficie del Bosco), sia perché arriva quasi ad equiparare quello dell'origine seminaturale (27.7%). Come atteso, nelle Altre terre boscate è stata riscontrata una quota molto limitata di soprassuoli con origine artificiale, quando non del tutto assente.

Il legno morto

INFC2015 ha stimato le tre componenti di legno morto degli alberi morti in piedi, del legno morto grosso a terra e delle ceppaie residue (Fig. 9.7) secondo soglie dimensionali genericamente in linea con quelle raccomandate per le rendicontazioni internazionali (es. FAO, 2018) e insieme afferenti al legno morto grosso o coarse woody debris. Per le tre componenti, INFC2015 ha stimato il volume, la necromassa e il contenuto di carbonio organico immagazzinato; degli alberi morti in piedi e delle ceppaie residue è stata stimata anche la numerosità. Di seguito vengono riportate e commentate le stime relative al numero e al volume degli alberi morti in piedi (Tables 9.9 e 9.10) e al numero e al volume delle ceppaie residue (Tables 9.11 e 9.12), al volume del legno morto grosso a terra (Table 9.13) e del legno morto grosso totale (Table 9.14), nella macrocategoria Bosco. In inventarioforestale.org/statistiche_INFC sono disponibili statistiche analoghe per le categorie forestali; inoltre, sono disponibili le statistiche relative alla necromassa e al contenuto di carbonio delle tre componenti di legno morto, sia per le categorie inventariali sia per le categorie forestali.

Nel Bosco italiano ci sono 1.37 miliardi di alberi morti in piedi, per un valore medio ad ettaro di 150.3. Si tratta di una quantità pari all'11.9% del numero degli alberi vivi; questa percentuale è piuttosto variabile tra le diverse regioni (dal 2.5% della Puglia al 20.7% della Liguria), ma se si rappresentano su un grafico i dati regionali del numero di alberi morti rispetto a quello degli alberi vivi rimane la possibilità di interpolare piuttosto efficacemente la relazione positiva secondo una funzione potenza (Fig. 9.8). In termini di volume, gli alberi morti in piedi corrispondono a 67.3 milioni di metri cubi di legno, con un valore per ettaro pari a 7.4 m^3. Il legno morto grosso a terra rappresenta la seconda componente per abbondanza del Bosco, con un valore complessivo di 51.8 milioni di m^3, corrispondenti a 5.7 m^3 ha^{-1}. La presenza di ceppaie residue è più contenuta, con 783.8 milioni di ceppaie totali (86.3 ceppaie per ettaro) ed un volume totale di 14.0 milioni di m^3 (1.5 m^3 ha^{-1}).

Per i valori ad ettaro nelle categorie forestali dei Boschi alti, si osserva che la componente di legno morto più abbondante è in genere quella del legno grosso a terra nei boschi di conifere, mentre prevale quella degli alberi morti in piedi nei boschi di latifoglie (Fig. 9.9); fanno eccezione le Pinete di pini mediterranei, gli Altri boschi di conifere e i Boschi igrofili. Dai dati sulle categorie forestali emerge il valore particolarmente elevato di legno morto grosso dei Castagneti. In essi è molto alto il numero di piante morte, pari a 486.5 per ettaro in media, rispetto, ad esempio, ai

193.9 degli Ostrieti e carpineti, che pure sono la seconda categoria nella classifica; il volume degli alberi morti in piedi è di 28.6 m^3 ha^{-1}, che corrisponde a più del doppio di quello di qualsiasi altra categoria, ma anche il volume del legno morto grosso a terra è considerevole soprattutto rispetto alle altre categorie tipicamente di latifoglie. La posizione dei Castagneti in merito al legno morto conferma una condizione già evidenziata da INFC2005 ed è stato osservato come essa possa dipendere dalle gravi patologie che affliggono la specie (Pignatti & De Natale, 2011). In linea generale, si può osservare che i boschi di conifere alpine sono i più ricchi di legno morto grosso e per quelli di latifoglie valori analoghi si riscontrano, oltre che per i Castagneti di cui si è già detto, anche per i Boschi igrofili; questi sono particolarmente ricchi di legno morto grosso a terra (11.5 m^3 ha^{-1}) ma presentano anche un valore elevato per gli alberi morti in piedi, secondo solo ai Castagneti nell'ambito delle categorie a latifoglie (8.3 m^3 ha^{-1}). Il volume delle ceppaie residue è particolarmente elevato nei boschi alpini delle categorie dei Boschi di Larice e cembro, Abete rosso e Abete bianco; a questa occorrenza contribuiscono le pendenze elevate dei versanti che determinano altezze delle ceppaie (ad esempio alla linea di taglio) basse a monte ma considerevoli a valle.

L'abbondanza relativa delle tre componenti di legno morto grosso può naturalmente essere esaminata anche sulla base delle rispettive necromasse. Numerosità e volumi hanno maggiore attinenza con le funzioni ecologiche legate alla biodiversità e ad altre funzioni naturalistiche mentre la biomassa consente una valutazione più pertinente sul contributo nell'accumulo di carbonio organico, cui è dedicato il capitolo 12. Naturalmente, le abbondanze relative delle tre componenti di legno morto commentate in questo paragrafo possono variare in maniera non trascurabile quando il confronto viene basato sui valori corrispondenti di necromassa, a causa delle variazioni di densità basale che intervengono con il gruppo di specie e con la classe di decadimento del legno.

Superficie forestale e aree protette

L'appartenenza di un punto inventariale a un'area protetta è stata declinata come presenza di vincolo naturalistico. I principali riferimenti normativi di definizione delle aree naturali protette e del conseguente vincolo naturalistico sono la Legge quadro sulle aree protette n. 394/91, a livello nazionale, e le Direttive 79/409/CEE ("Uccelli") e 92/43/CEE ("Habitat") a livello europeo. A questi va aggiunta la "Convenzione sulle zone umide di importanza internazionale come habitat degli uccelli acquatici e palustri", firmata a Ramsar nel 1971 e resa esecutiva dal DPR n. 448/76. I diversi tipi di aree protette individuati da tali norme, e le relative definizioni, sono riportate in Table 9.15, mentre in Table 9.16 vengono definite, per i parchi nazionali e i siti della rete Natura2000, le classi relative al grado di protezione.

INFC2015 stima in 3.5 milioni di ettari la Superficie forestale ricadente in aree protette, pari al 31.8% della Superficie forestale nazionale, di cui 2.8 milioni di ettari nel Bosco e quasi 700,000 ha nelle Altre terre boscate (Table 9.17). La Fig. 9.10

mostra l'area di Bosco e Altre terre boscate inclusa nei diversi tipi di aree protette. In base alla classificazione nazionale, diverse tipologie di tutela possono coesistere su una stessa superficie. Ad esempio, i siti della Rete Natura2000 ospitano complessivamente 3 milioni di ettari di superficie forestale (2.4 milioni di ettari in Bosco, circa 600,000 ha in Altre terre boscate) spesso inclusa anche in parchi e riserve nazionali o regionali. L'aliquota di foreste protetta (Fig. 9.11) è ragguardevole in tutte le regioni italiane, ma si osserva che la percentuale di Bosco interessata da vincoli naturalistici è più elevata in alcune regioni del centro e sud Italia, superando il 50% in Puglia, Campania, Sicilia e Abruzzo. Le regioni dell'Arco alpino, caratterizzate da un indice di boscosità più elevato, registrano percentuali più basse di Bosco in aree protette, con l'eccezione del Veneto che ne ospita il 42.9%.

La presenza di vincolo naturalistico viene ulteriormente dettagliata, per le categorie inventariali, nelle Tables 9.18, 9.19 e 9.20, dalle quali si evince che tutte le categorie inventariali presentano un'aliquota di superficie simile appartenente ad area protetta, variabile tra il 30% e il 35%, a eccezione degli Impianti, che presentano il 23.4% di superficie con vincolo naturalistico. Analoghe statistiche relative alle categorie forestali, ripartite per i diversi tipi di aree protette, sono disponibili in inventarioforestale.org/statistiche_INFC.

La Superficie forestale all'interno dei parchi nazionali è di circa 920,000 ha: 725,000 ha in Bosco (8.0%) e 195,000 ha in Altre terre boscate (9.9%) (Table 9.21). A livello di regioni, la distribuzione della Superficie forestale varia notevolmente, risentendo della ripartizione della superficie totale dei parchi nazionali. L'aliquota di Bosco in parchi nazionali è particolarmente elevata in tre regioni dell'Italia centro-meridionale, la Puglia (42.2%), l'Abruzzo (32.3%) e la Calabria (26.2%). Le Tables 9.22 (Bosco) e 9.23 (Altre terre boscate) mostrano la ripartizione della Superficie forestale ricadente in parchi nazionali per grado di protezione. La Fig. 9.12, che considera solo il Bosco, evidenzia che è la Zona C (aree di protezione) a contenerne la percentuale relativamente più alta, con il 24.0%. Sempre con riferimento al Bosco, si osserva che in tutto il territorio italiano soltanto il 56.3% della superficie risulta assegnato ad una zona specifica, mentre per il restante 43.7% manca l'attribuzione ad un grado di protezione specifico.

Le riserve naturali statali (Table 9.24) interessano appena lo 0.7% della Superficie forestale nazionale (74,963 ha) e lo 0.6% della superficie del Bosco (57,021 ha). Con riferimento al Bosco, tali superfici sono concentrate per oltre il 70% in cinque regioni: Toscana, Lazio, Veneto, Abruzzo e Calabria.

I parchi naturali regionali (Table 9.25) interessano complessivamente 713,048 ha di Superficie forestale totale, di cui 592,554 ha nel Bosco (6.5%). Anche in questo caso la ripartizione a livello di regioni è piuttosto variabile: in Sicilia e in Campania, ad esempio, è tutelato da parchi regionali rispettivamente il 28.6% ed il 24.6% del Bosco.

La Superficie forestale totale ricadente nelle riserve naturali regionali (Table 9.26) è l'1.1% (125,572 ha), con analoga percentuale per quanto riguarda il Bosco (107,906 ha). Si osservano per tutte le regioni valori molto contenuti, ad eccezione della Sicilia, dove l'11.4% del Bosco rientra in questo tipo di riserve.

Le altre aree naturali protette previste dalla Legge n.394/91 (Table 9.27) interessano soltanto lo 0.6% sia della Superficie forestale totale (60,932 ha), sia del Bosco (55,474 ha).

Le stime delle superfici incluse in siti della rete Natura 2000 (SIC e ZPS), spesso a loro volta inclusi in altri tipi di aree protette, sono riportate nella Table 9.28; complessivamente il 27.2% della Superficie forestale nazionale (3,011,119 ha) e il 26.5% del Bosco (2,408,882 ha) è localizzato all'interno di queste aree di tutela. Le regioni con le quote più consistenti, in termini relativi rispetto alla Superficie forestale, sono la Puglia (63.0%), la Campania (51.0%), l'Abruzzo (50.7%), la Sicilia (45.5%) e il Veneto (45.2%). In inventarioforestale.org/statistiche_INFC sono disponibili statistiche di maggiore dettaglio sulle Superfici forestali in SIC e ZPS.

Le aree umide di interesse internazionale accolgono circa 5000 ettari di Superficie forestale totale (Table 9.29). Nel Bosco di tredici regioni non sono state rilevate aree umide protette dalla Convenzione di Ramsar, mentre la superficie maggiore è stata stimata nel Lazio (1473 ha).

La classificazione nazionale delle aree protette non risulta perfettamente sovrapponibile a quella proposta da MCPFE (2003); tuttavia, nella Fig. 9.13 viene mostrata la ripartizione percentuale della superficie del Bosco secondo classi del livello di protezione mutualmente esclusive, equiparabili alle analoghe classi MCPFE ed utilizzate per l'Italia a partire dal report State of Europe's Forest 2011 (Forest Europe, UNECE and FAO, 2011), per l'indicatore 4.9. Per questo tipo di classificazione, statistiche più dettagliate sono disponibili in inventarioforestale.org/statis tiche_INFC.

References

Berretti, R., Caffo, L., Camerano, P., & Terzuolo, P. G. (2006). *Selvicoltura nelle foreste di protezione - esperienze e indirizzi gestionali in Piemonte e Valle d'Aosta*. Compagnia delle foreste, Arezzo (220 pp.). ISBN 978-88-901223-5-4.

Bernetti, G. (2005). *Atlante di selvicoltura – dizionario illustrato di alberi e foreste* (pp. 496). Edagricole.

Bevilacqua, E. (1999). *Growth responses in individual eastern white pine (*Pinus strobus* L.) trees following partial cutting treatments*. Ph.D. University of Toronto.

Brockerhoff, E. G., Barbaro, L., Castagneyrol, B., Forrester, D. I., Gardiner, B., Ramo´n Gonza´lez-Olabarria, J. R., O'B. Lyver, P., Meurisse, N., Oxbrough, A., Taki, H., Thompson, I. D., van der Plas, F., & Jactel, H. (2017). Forest biodiversity, ecosystem functioning and the provision of ecosystem services. *Biodiversity and Conservation, 26*, 3005–3035.https://doi.org/10.1007/s10 531-017-1453-2

Dudley, N., & Phillips, A. (2006). Forests and protected areas: *Guidance on the use of the IUCN protected area management categories*. IUCN, Gland, Switzerland and Cambridge, UK (58 pp.). ISBN 978-2-8317-0828-7.

FAO (2018). *Global Forest Resources Assessment 2020. FRA2020 terms and definitions*. Forest resources assessment working paper 188. Rome (32 pp.). Retrieved Jan 05, 2022, from https:// www.fao.org/3/I8661EN/i8661en.pdf.

Forest Europe, UNECE, FAO (2011). *State of Europe's forests 2011. Status and trends in sustainable forest management in Europe* (327 pp.). ISBN 978-82-92980-05-7.

Gasparini, P., Di Cosmo, L., Cenni, E., Pompei, E., & Ferretti, M. (2013). Towards the harmonization between National Forest Inventory and Forest Condition Monitoring. Consistency of plot allocation and effect of tree selection methods on sample statistics in Italy. *Environmental Monitoring and Assessment, 185*, 6155–6171. https://doi.org/10.1007/s10661-012-3014-1

Gasparini, P., Di Cosmo, L., Rizzo, M., & Giuliani, D. (2017). A stand-level model derived from National Forest Inventory data to predict periodic annual volume increment of forests in Italy. *Journal of Forest Research-JPN, 22*(4), 209–217. https://doi.org/10.1080/13416979.2017.1337260

Gotelli, N. J., & Colwell, R. K. (2001). Quantifying biodiversity: Procedures and pitfalls in the measurement and comparison of species richness. *Ecology Letters, 4,* 379–391.

IPCC (2003). *Good practice guidance for land use, land use change and forestry*. IPCC National Greenhouse Gas Inventories Programme. Institute for Global Environmental Strategies, Hayama, Japan (590 pp.). ISBN 4-88788-003-0. Retrieved Jan 04, 2022, from https://www.ipcc-nggip.iges.or.jp/public/gpglulucf/gpglulucf_files/GPG_LULUCF_FULL.pdf.

MCPFE (2003). *State of Europe's forests 2003. The MCPFE report on sustainable forest management in Europe. Ministerial conference on the protection of forests in Europe. Liaison Unit. Vienna* (pp. 107–110). ISBN 3-902073-09-08. Retrieved Jan 31, 2022, from https://unece.org/DAM/timber/docs/sfm/europe-2003.pdf.

Paletto, A., Ferretti, F., De Meo, I., Cantiani, P., & Focacci, M. (2012). Ecological and environmental role of deadwood in managed and unmanaged forests. In: J. J. Diez (Ed.), *Current research* (pp. 219–238). InTech, Rijeka. ISBN 978-953-51B-0621-0.

Pedrotti, F. (1995). La vegetazione forestale italiana. *Atti Dei Convegni Lincei, 115*, 39–78.

Pignatti, G., & De Natale, F. (2011). Biodiversità ed altri aspetti naturalistici. In: P. Gasparini & G. Tabacchi (Eds.), *L'Inventario Nazionale delle Foreste e dei serbatoi forestali di Carbonio INFC 2005. Secondo inventario forestale nazionale italiano. Metodi e risultati* (pp. 371–393). Edagricole-Il Sole 24 Ore, Milano. ISBN 978-88-506-5394-2.

Pignatti, S. (1994). La flora. In: S. Pignatti (Ed.), *Ecologia del paesaggio* (p. 15). UTET.

Pignatti, S. (1995a). Biodiversità della vegetazione mediterranea. *Atti Dei Convegni Lincei, 115*, 7–31.

Pignatti, S. (1995b). Diversità. In: S. Pignatti (Ed.), *Ecologia vegetale* (pp. 93–95). UTET.

Woodall, C. W., Monleon, V. J., Fraver, S., Russell, M. B., Hatfield, M. H., Campbell, J. L., & Domke, G. M. (2019). The downed and dead wood inventory of forests in the United States. *Scientific Data, 6*, 180303. https://doi.org/10.1038/sdata.2018.303.

Open Access This chapter is licensed under the terms of the Creative Commons Attribution 4.0 International License (http://creativecommons.org/licenses/by/4.0/), which permits use, sharing, adaptation, distribution and reproduction in any medium or format, as long as you give appropriate credit to the original author(s) and the source, provide a link to the Creative Commons license and indicate if changes were made.

The images or other third party material in this chapter are included in the chapter's Creative Commons license, unless indicated otherwise in a credit line to the material. If material is not included in the chapter's Creative Commons license and your intended use is not permitted by statutory regulation or exceeds the permitted use, you will need to obtain permission directly from the copyright holder.

Chapter 10
Forest Health

Lo stato di salute delle foreste

Maria Rizzo and Patrizia Gasparini

Abstract Forests mainly consist of long-lived trees or shrub species and are exposed to natural or human disturbances of different severities. They are essential components of the natural development of forest ecosystems, since by triggering natural selection and ecological succession processes, they can achieve the best status in terms of species composition and structure. Nevertheless, extreme events can cause serious economic or naturalistic losses and, in some cases, endanger specific forest ecosystems. Disturbance events that damage forests vary and include pests and diseases, fires, pollution, climate changes, overexploitation or inadequate silvicultural practices, excessive grazing and browsing, to name just a few. Health monitoring of forests is a necessary condition to provide useful information for the conservation of forest resources, as well as to support forest management practices aimed at increasing the resilience of forests and their adaptation capacity. Through ground surveys, the Italian national forest inventory INFC has classified the health condition of Italian forests and produced the estimates presented in the chapter. These concern the distribution of the Forest area by pathologies and damage presence, severity and cause. Estimates of the Forest area affected by defoliation, divided by defoliation class and localization of defoliation, are also presented.

Keywords Disturbances · Forest pathologies · Damages · Defoliation · Crown transparency

M. Rizzo (✉) · P. Gasparini
CREA Research Centre for Forestry and Wood, Trento, Italy
e-mail: maria.rizzo@crea.gov.it

P. Gasparini
e-mail: patrizia.gasparini@crea.gov.it

© The Author(s) 2022

P. Gasparini et al. (eds.), *Italian National Forest Inventory—Methods and Results of the Third Survey*, Springer Tracts in Civil Engineering,
https://doi.org/10.1007/978-3-030-98678-0_10

10.1 Introduction

Different disturbances of nature (biotic or abiotic) and intensity are an integral part of forest ecosystem dynamics (Seidl et al., 2017). The alteration of forest structure, composition and function allows young and old forests (trees) to turn over and create heterogeneity, promoting biodiversity and ecosystem renewal or reorganisation (Thom et al., 2017; White & Pickett, 1985). Nevertheless, beyond a certain limit, disturbances can cause serious ecological or economic damages, causing the interruption of some forest functions (protection, landscape, timber and wood production, etc.). A forest management policy, aimed at the prevention and mitigation of natural disturbances, is usually considered essential (Motta, 2018).

The main disturbances that can damage the health of forests, causing their degradation or limiting their functionality, can derive both from natural biotic or abiotic factors (pests, grazing and browsing, fires, windstorms, floods, severe snow, etc.) and from profound changes in the territory and the environment caused by human activities, due for example to pollution or to the excessive exploitation of the forest in terms of wood and timber utilisation or for recreational use.

Because of climate change, which triggers more and more extreme climate events (strong rainstorms, floods, windstorms, heatwaves, etc.), forest vulnerability is increasing. Today, the frequency of extreme climate events is three times higher than 50 years ago (FAO, 2021). This creates a reduction in plant defence mechanisms (Forzieri et al., 2021). Climate change can also be responsible for the spread of pathogenic species in environments where they were not present, causing worrisome epidemics of animal and plant parasites.

Forest health monitoring is a prerequisite for adequately defining forest management practices aimed at improving forest resilience and adaptation capacity. In the Italian national forest inventory, the health of forest stands is detected through the direct observation in the field of the presence of damage or pathologies on the trees included in a plot (AdS25, cf. Chap. 4) to estimate the areas of the Forest involved (Gasparini et al., 2013). Any signals are considered, both as an alteration resulting from the action of a damaging agent (e.g., lesions, necrosis, defoliation) and in the form of signs of the presence of a damage factor (nests of larva, insects, fungi, etc.). The survey refers to the presence of pathologies or damages occurring in the survey's year or previously, but whose effects are still observable.

According to INFC, pathologies and damages are considered "visible" when they affect at least 30% of the crown coverage of the analysed stand, with reference to the subjects (trees and shrubs) with DBH $\geq$ 4.5 cm. For these cases, in addition to the presence, the severity of the pathology or damage, the origin (which can be biotic or abiotic) and, if defoliation is present, its degree and localization on the affected trees are also assessed.

In this chapter, the estimate of the Forest area by pathology and damaging diffusion, severity and cause are presented. Estimates of Forest area affected by defoliation are also presented, by degree and localisation of defoliation. Area estimates of damage presence and diffusion for the inventory categories of Forest and the forest categories of the Tall trees forest, the latter at the national level, are also available at inventarioforestale.org/statistiche_INFC.

10.2 Diffusion and Severity of Damages

The third Italian national forest inventory INFC2015 classifies the presence of pathologies and damages according to classes of their degree of diffusion, measured as percentage of crown coverage affected: absence of damages, damages on less than 30% of the crown coverage, damages on 30–59% of the crown coverage and damages on at least 60% of the crown coverage.

Table 10.1 at the end of this chapter provides estimates on Forest area by degree of damage diffusion. Estimates show that almost 80% of the Forest area is not affected by pathologies or damage and 15.6% of the area is affected by minor damage on less than 30% of the crown coverage. Overall, the Forest area without damage or pathologies or with minor damage is equal to 94.9% of the total. The Forest area with visible (affected crown coverage 30–59%) and very visible (at least 60% crown coverage affected) damages is equal to 3.3% and 1.0% of the total, respectively.

Estimates at the regional level differ partially from national ones.[1] Umbria is the region least affected by damages and pathologies (99.2% of the Forest area not affected or with minor damages), followed by Molise (99.1%), Valle d'Aosta (98.6%), Basilicata (98.4%), Trentino (98.3%) and Abruzzo (98.1%). The regions most affected by visible or very visible damage are Piemonte (8.7 and 1.9% of the area, respectively), Emilia-Romagna (8.0 and 1.3%), Liguria (6.7 and 1.8%), Calabria (2.7 and 4.2%) and Sardegna (5.3 and 1.1%) (Fig. 10.1).

Damage severity was assessed by estimating the proportion of the damaged above-ground components (foliage, branches and stem) compared to the normal conditions that trees should present in that context (Table 10.2). Only permanent and lasting damage and not temporary damage (just on foliage, for example) were taken into account. The severity of the damage, as mentioned above, was assessed only in forests with visible or very visible damages (at least 30% of the crown coverage affected).

At the end of this chapter, estimates of Forest area by damage severity are given in Table 10.3. Excluding the not classified or not assessed area, the most frequent class is that of medium damage (52.8% of the assessed Forest area), followed by the class of severe damage (34.4%). At the regional level, in Molise and Basilicata the entire Forest area assessed for the severity of the damage is affected by severe damage. In Umbria and Valle d'Aosta, on the other hand, the entire assessed area is affected by medium severity damages (Fig. 10.2).

[1] It should be noted that the estimates presented are only partially affected by the effects of the Vaia storm in October 2018 which occurred during the survey campaign.

Table 10.1 Forest area by degree of damage diffusion / Estensione del Bosco ripartito per grado di diffusione di patologie e danni

Region/Regione	Absence of damages or pathologies		Damages or pathologies on less than 30% of the crown coverage		Damages on 30–59% of the crown coverage		Damages on at least 60% of the crown coverage		Not classified		Total Forest	
	Patologie o danni assenti		Patologie o danni su meno del 30% della copertura		Patologie o danni su 30–59% della copertura		Patologie o danni su almeno 60% della copertura		Non classificato		Totale Bosco	
	Area	ES	Area	ES	Area	ES	Area	ES	Area	ES	Area	ES
	(ha)	(%)	(ha)	(%)	(ha)	(%)	(ha)	(%)	(ha)	(%)	(ha)	(%)
Piemonte	626,695	2.9	166,147	8.4	77,351	13.3	16,900	30.0	3340	54.1	890,433	1.3
Valle d'Aosta	60,764	9.0	37,087	13.9	1272	56.8	0	–	120	99.6	99,243	3.6
Lombardia	529,304	2.7	67,982	15.1	12,091	35.3	8252	45.2	4338	41.7	621,968	1.6
Alto Adige	236,387	4.5	85,540	10.9	15,324	28.0	0	–	2020	45.0	339,270	1.7
Trentino	304,222	3.4	62,588	13.6	4525	50.8	1433	100.0	492	100.0	373,259	1.4
Veneto	391,814	2.3	12,028	30.7	5256	49.9	1147	99.9	6459	27.1	416,704	1.9
Friuli V.G	280,547	3.3	38,986	16.6	10,122	35.0	0	–	2901	68.1	332,556	1.9
Liguria	236,883	4.3	75,839	11.3	23,006	22.9	6306	44.9	1126	57.3	343,160	1.7
Emilia Romagna	284,385	5.3	231,937	6.4	46,776	16.5	7639	44.7	14,164	31.1	584,901	1.5
Toscana	821,026	2.3	189,725	8.3	22,457	25.5	0	–	2240	38.1	1,035,448	1.1
Umbria	375,351	2.2	11,770	32.9	379	99.8	0	–	2805	69.1	390,305	1.6
Marche	232,041	4.0	52,972	14.0	6754	44.1	0	–	0	–	291,767	2.1
Lazio	501,589	2.5	41,214	18.8	5505	52.2	9830	42.0	2099	41.4	560,236	1.6
Abruzzo	377,388	2.6	26,228	21.6	3034	63.5	3602	54.1	1336	57.4	411,588	1.8
Molise	143,525	3.9	8369	33.5	0	–	1354	72.2	0	–	153,248	3.0
Campania	369,746	2.8	13,951	33.0	15,922	29.0	3203	52.3	1105	57.8	403,927	2.1
Puglia	80,954	8.4	56,209	11.1	2635	63.3	1225	99.8	1326	50.1	142,349	4.0
Basilicata	257,186	3.6	26,141	20.9	0	–	1398	99.9	3294	33.1	288,020	2.7
Calabria	403,978	3.6	46,794	19.3	13,453	33.3	20,581	36.1	10,371	33.7	495,177	2.0
Sicilia	229,385	4.5	43,973	15.5	3870	58.5	4427	57.5	3833	55.5	285,489	3.2
Sardegna	459,536	3.5	125,433	9.3	33,433	18.3	7104	38.6	633	71.9	626,140	2.1
Italia	7,202,706	0.7	1,420,914	2.8	303,166	6.5	94,400	13.4	64,001	12.0	9,085,186	0.4

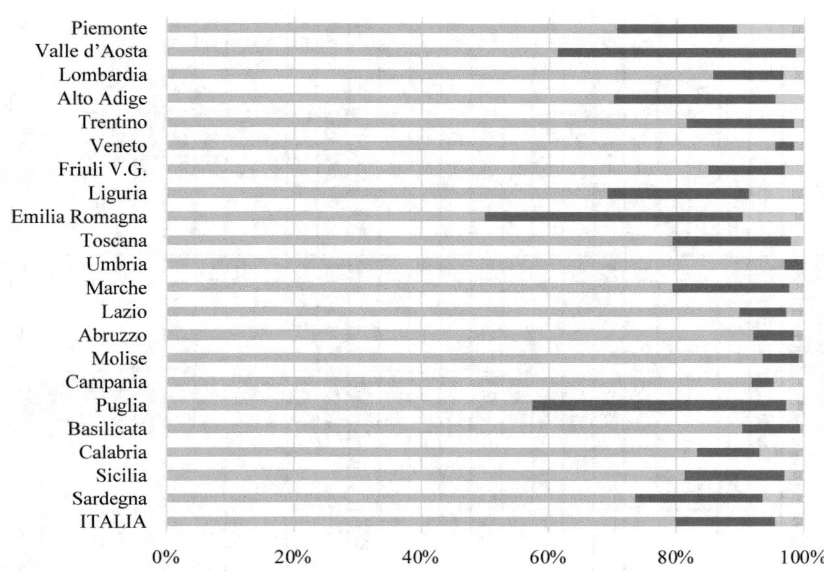

Fig. 10.1 Percentage of the Forest area by degree of diffusion of damages or pathologies / Ripartizione percentuale del Bosco per grado di diffusione di patologie o danni

Table 10.2 Classes of damage severity / Classi di intensità di patologie e danni

Severity classes	Description
Classi di intensità	Descrizione
Low damage Danno modesto	Damages affecting less than 30% of the aboveground tree components[1]
	Meno del 30% delle parti epigee dei soggetti arborei[1] è interessato da patologie o danni
Medium damage Danno medio	Damages affecting 30–59% of the aboveground tree components
	30–59% delle parti epigee dei soggetti arborei interessato da patologie o danni
Severe damage Danno intenso	Damages affecting at least 60% of the aboveground components
	60% o più delle parti epigee dei soggetti arborei interessato da patologie o danni

[1]Crown, trunk, branches / Chioma, fusto, rami

Table 10.3 Forest area by damage severity; it is assessed when more than 30% of the crown coverage is affected by damages or pathologies / Estensione del Bosco ripartito per intensità di patologie o danni; viene valutata quando questi interessano più del 30% della copertura delle chiome

Region/Regione	Low damage		Medium damage		Severe damage		Not assessed or not classified		Total Forest	
	Danno modesto		Danno medio		Danno intenso		Non valutato o non classificato		Totale Bosco	
	Area	ES	Area	ES	Area	ES	Area	ES	Area	ES
	(ha)	(%)	(ha)	(%)	(ha)	(%)	(ha)	(%)	(ha)	(%)
Piemonte	3489	63.5	65,532	14.6	25,231	24.6	796,182	1.9	890,433	1.3
Valle d'Aosta	0	–	1272	56.8	0	–	97,971	4.0	99,243	3.6
Lombardia	2915	73.7	12,770	35.2	4657	59.4	601,625	1.9	621,968	1.6
Alto Adige	8934	36.4	6390	44.5	0	–	323,946	2.4	339,270	1.7
Trentino	1372	100.0	1781	69.5	2805	70.8	367,301	2.0	373,259	1.4
Veneto	1565	99.9	1147	99.9	3691	57.1	410,301	2.1	416,704	1.9
Friuli V.G	720	99.7	7073	43.2	2329	73.6	322,434	2.4	332,556	1.9
Liguria	4270	56.7	13,044	31.5	11,999	32.5	313,848	2.8	343,160	1.7
Emilia Romagna	2864	70.8	32,255	20.0	19,296	27.4	530,486	2.3	584,901	1.5
Toscana	3723	63.6	12,226	35.3	6508	49.7	1,012,991	1.3	1,035,448	1.1
Umbria	0	–	379	99.8	0	–	389,926	1.9	390,305	1.6
Marche	1326	100.0	3272	60.8	2156	84.8	285,013	2.5	291,767	2.1
Lazio	2194	100.0	2873	62.9	10,268	40.4	544,901	2.0	560,236	1.6
Abruzzo	0	–	3034	63.5	3602	54.1	404,952	2.1	411,588	1.8
Molise	0	–	0	–	1354	72.2	151,893	3.4	153,248	3.0
Campania	8529	41.3	5981	44.9	4615	47.5	384,802	2.5	403,927	2.1
Puglia	0	–	2635	63.3	1225	99.8	138,488	4.2	142,349	4.0
Basilicata	0	–	0	–	1398	99.9	286,621	2.9	288,020	2.7
Calabria	3297	70.9	8655	41.4	22,081	34.3	461,143	2.8	495,177	2.0
Sicilia	1452	100.0	2419	71.8	4427	57.5	277,191	3.4	285,489	3.2
Sardegna	4112	58.6	27,201	20.1	9224	33.9	585,603	2.4	626,140	2.1
Italia	50,762	16.8	209,938	7.9	136,866	10.8	8,687,620	0.5	9,085,186	0.4

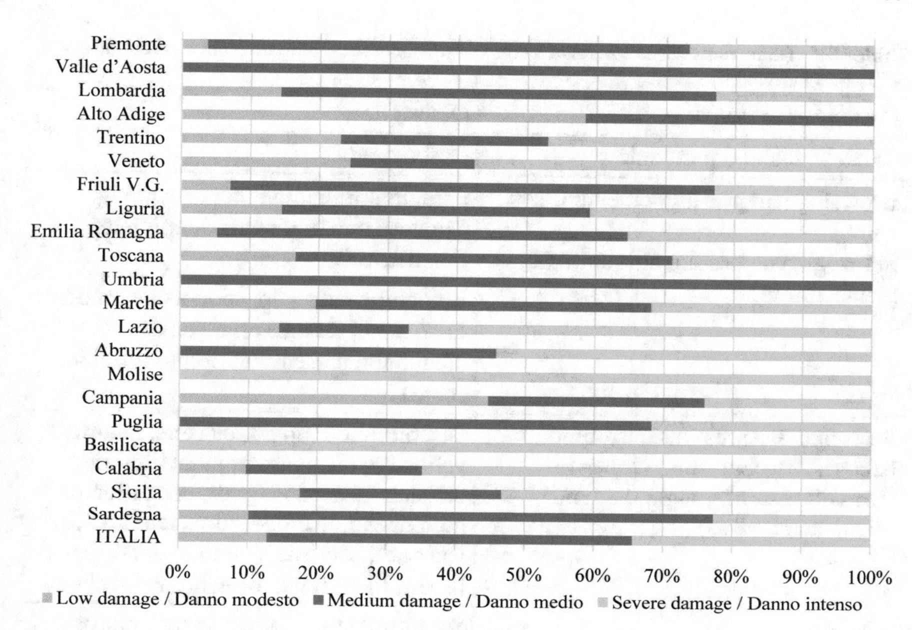

Fig. 10.2 Percentage of the assessed Forest area by classes of damage severity / Ripartizione percentuale della superficie del Bosco valutato per classi di intensità di patologie o danni

10.3 Damaging Causes

The causes of damages and pathologies are identified by observing traces or evident signs (cf. Table 10.4). If the cause is not clearly recognisable or there is a concomitant action of several factors, none of which prevails, the class 'unknown or complex causes' is assigned. If, among many, there is a prevalent or triggering cause, then this one is indicated.

Table 10.5 gives the estimates of Forest area by damaging cause. At the national level, the main disturbances are pests and diseases caused by insects, fungi, bacteria, mycoplasma, viruses (33.8% of the assessed Forest area), followed by extreme climate events (26.5%) and forest fires on crowns (20.7%) (Fig. 10.3).

The effect of the different causes of damage is variable at the regional level. Nevertheless, the main disturbances belong to the three most relevant causes at the national level.

Table 10.4 Damaging causes and related signs of recognition / Cause di danni o patologie e segni per la loro individuazione

Damaging causes	Description
Cause di danni o patologie	Descrizione
Browsing and grazing, other animals, epiphytes Selvaggina o pascolo, altri animali, piante epifite	Foraging, trampling, bark stripping, damages due to rodents or epiphytes (mistletoe, ivy)
	Brucatura, calpestio, scortecciatura dei fusti, danni da roditori e danni dovuti alla presenza di epifite (vischio, edera)
Pests and diseases (insects, fungi, bacteria, mycoplasma, viruses) Parassiti (insetti, funghi, batteri, micoplasmi, virus)	Discoloration, blight, cankers, chlorosis, yellowing, decrease in plant vigour, dieback, distortion, galls, exudation of gum or sap, leaf distortion, leaf scorch, leaf spot, necrosis, stunting, abnormal broom-like growth of many weak shoots
	Scolorimento, ruggine, cancri, clorosi, ingiallimento, diminuzione del vigore della pianta, deperimento, galle, emissione di essudati o resina, deformazione fogliare, bruciatura fogliare, macchie fogliari, necrosi, arresto della crescita, scopazzi
Extreme climate events Eventi meteorologici intensi	Crashes, truncations, uprooting of trees, leaf dieback/necrosis
	Schianti, troncature, sradicamenti di soggetti arborei; disseccamento/necrosi fogliare
Crown fires Incendio del soprassuolo	Blackening, combustion and carbonization of aboveground biomass
	Annerimento, bruciatura, combustione a carico del soprassuolo
Underwood fires Incendio del sottobosco	Blackening, combustion and carbonization of underwood biomass
	Annerimento, bruciatura, combustione a carico del sottobosco

(continued)

Table 10.4 (continued)

Damaging causes	Description
Cause di danni o patologie	Descrizione
Direct human actions Azione diretta dell'uomo	Damages to forest stands in form of crashes, debarking, truncations, crashing, and to regeneration due to incorrect silvicultural activities
	Danni al soprassuolo in forma di scortecciature, troncature, schianti di singoli soggetti e danni al novellame dovuti ad attività selvicolturali errate
Indirect human actions Azione indiretta dell'uomo	Pollution, recreational activities
	Inquinamento, danni da fruizione turistico-ricreativa
Complex or unknown causes Cause complesse o ignote	Presence of more than one cause or not clearly interpretable
	Presenza di più cause o non chiaramente interpretabili

10.4 Defoliation and Its Localization

If pathologies or damage effects are manifested as defoliation, then its degree and localization are observed on some representative sample trees (up to three), according to the INFC field protocol.

The degree of defoliation is defined as the level of transparency of the crown, compared to the normal conditions that trees should present in that context, for that species and in that period. It is classified according to four classes: defoliation absent or $\leq$10%, defoliation equal to 11–25%, 26–60% of greater than 60%. The evaluation of the localization concerns, instead, the indication of the areas of the crown affected by the defoliation (Table 10.6). It is assessed if the degree of defoliation is greater than 10%.

Tables 10.7 and 10.8 show the estimates of Forest area by degree and localization of defoliation, respectively. Forest area affected by defoliation $\leq$10% is equal to 34.3% of the assessed area, while the upper classes of defoliation respectively affect 24.4%, 24.6% and 16.8% of the assessed area (Fig. 10.4).

The Forest area characterized by a homogeneous distribution of defoliation over the entire canopy is equal to 66.7% of the assessed area, while 19.8% of the assessed Forest area is characterized by defoliation on the apical portion of the canopy (Fig. 10.5).

Table 10.5 Forest area by damaging causes; it is assessed when more than 30% of the crown coverage is affected by damages or pathologies / Estensione del Bosco ripartito per causa di danni o patologie; viene valutata quando questi interessano più del 30% della copertura delle chiome

Region/Regione	Browsing and grazing, other animals, epiphytes		Pests and diseases (insects, fungi, bacteria, mycoplasma, viruses)		Extreme climate events		Crown fires		Underwood fires		Direct human actions		Indirect human actions		Complex or unknown causes		Not assessed or not classified		Total Forest	
	Selvaggina o pascolo, altri animali, piante epifite		Parassiti (insetti, funghi, batteri, micoplasmi, virus)		Fenomeni meteorologici estremi		Incendio del soprassuolo		Incendio del sottobosco		Azione diretta dell'uomo		Azione indiretta dell'uomo		Cause complesse o ignote		Non valutato o non classificato		Totale Bosco	
	area	ES	area	ES	area	ES	area	ES	area	ES	area	ES	area	ES	area	ES	area	ES	area	ES
	(ha)	(%)	(ha)	(%)	(ha)	(%)	(ha)	(%)	(ha)	(%)	(ha)	(%)	(ha)	(%)	(ha)	(%)	(ha)	(%)	(ha)	(%)
Piemonte	19,021	27.4	30,932	21.7	31,809	21.3	4520	56.8	3077	70.7	0	–	0	–	4891	57.6	796,182	1.9	890,433	1.3
Valle d'Aosta	0	–	424	99.6	0	–	0	–	0	–	0	–	0	–	848	70.0	97,971	4.0	99,243	3.6
Lombardia	0	–	10,081	37.6	8341	45.2	1921	100.0	0	–	0	–	0	–	0	–	601,625	1.9	621,968	1.6
Alto Adige	1403	99.8	1520	79.0	10,978	33.1	0	–	0	–	0	–	0	–	1424	99.8	323,946	2.4	339,270	1.7
Trentino	1372	100.0	2263	72.4	2323	72.7	0	–	0	–	0	–	0	–	0	–	367,301	2.0	373,259	1.4
Veneto	0	–	2293	68.9	4110	58.1	0	–	0	–	0	–	0	–	0	–	410,301	2.1	416,704	1.9
Friuli V.G	0	–	5607	43.7	4515	57.7	0	–	0	–	0	–	0	–	0	–	322,434	2.4	332,556	1.9
Liguria	4270	56.7	15,540	28.8	5693	48.8	3809	54.4	0	–	0	–	0	–	0	–	313,848	2.8	343,160	1.7
Emilia Romagna	4398	57.8	14,689	30.0	24,949	23.4	0	–	1502	100.0	1637	100.0	1498	100.0	5742	51.2	530,486	2.3	584,901	1.5
Toscana	0	–	18,105	29.0	0	–	0	–	410	100.0	0	–	0	–	3943	60.3	1,012,991	1.3	1,035,448	1.1
Umbria	0	–	0	–	379	99.8	0	–	0	–	0	–	0	–	0	–	389,926	1.9	390,305	1.6
Marche	0	–	741	70.8	3112	71.6	2362	79.6	0	–	0	–	539	100.0	0	–	285,013	2.5	291,767	2.1

(continued)

Table 10.5 (continued)

Region/Regione	Browsing and grazing, other animals, epiphytes		Pests and diseases (insects, fungi, bacteria, mycoplasma, viruses)		Extreme climate events		Crown fires		Underwood fires		Direct human actions		Indirect human actions		Complex or unknown causes		Not assessed or not classified		Total Forest	
	Selvaggina o pascolo, altri animali, piante epifite		Parassiti (insetti, funghi, batteri, micoplasmi, virus)		Fenomeni meteorologici estremi		Incendio del soprassuolo		Incendio del sottobosco		Azione diretta dell'uomo		Azione indiretta dell'uomo		Cause complesse o ignote		Non valutato o non classificato		Totale Bosco	
	area	ES	area	ES	area	ES	area	ES	area	ES	area	ES	area	ES	area	ES	area	ES	area	ES
	(ha)	(%)	(ha)	(%)	(ha)	(%)	(ha)	(%)	(ha)	(%)	(ha)	(%)	(ha)	(%)	(ha)	(%)	(ha)	(%)	(ha)	(%)
Lazio	0	–	6452	51.6	2665	84.2	6218	49.6	0	–	0	–	0	–	0	–	544,901	2.0	560,236	1.6
Abruzzo	1374	100.0	0	–	362	100.0	4900	47.6	0	–	0	–	0	–	0	–	404,952	2.1	411,588	1.8
Molise	0	–	530	99.8	0	–	824	99.8	0	–	0	–	0	–	0	–	151,893	3.4	153,248	3.0
Campania	0	–	4928	51.8	1495	100.0	11,290	32.7	1412	100.0	0	–	0	–	0	–	384,802	2.5	403,927	2.1
Puglia	0	–	0	–	1962	77.9	1898	73.4	0	–	0	–	0	–	0	–	138,488	4.2	142,349	4.0
Basilicata	0	–	0	–	0	–	1398	99.9	0	–	0	–	0	–	0	–	286,621	2.9	288,020	2.7
Calabria	1797	99.9	3594	69.8	0	–	26,846	29.6	0	–	0	–	0	–	1797	99.9	461,143	2.8	495,177	2.0
Sicilia	0	–	992	100.0	0	–	7306	44.5	0	–	0	–	0	–	0	–	277,191	3.4	285,489	3.2
Sardegna	6284	40.6	15,552	27.4	2787	60.2	9014	36.6	1059	100.0	0	–	0	–	5841	44.2	585,603	2.4	626,140	2.1
Italia	39,918	18.6	134,243	10.1	105,478	11.5	82,307	14.2	7460	43.0	1637	100.0	2037	78.2	24,486	24.0	8,687,620	0.5	9,085,186	0.4

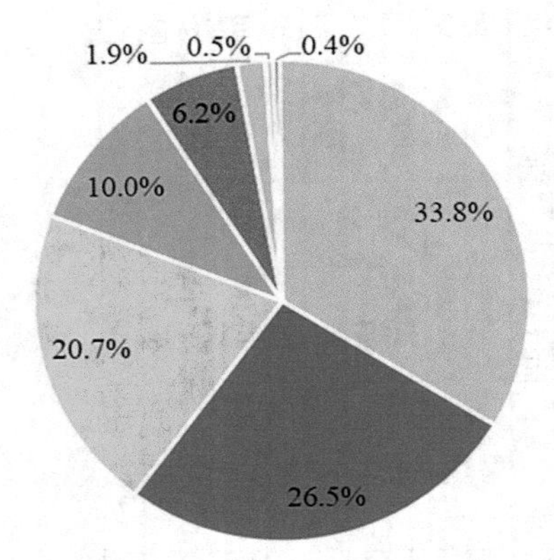

- Pests and diseases (insects, fungi, bacteria, mycoplasma, viruses) - Parassiti (insetti, funghi, batteri, micoplasmi, virus)
- Extreme climate events / Eventi meteorologici intensi
- Crown fires - Incendio del soprassuolo
- Browsing and grazing, other animals, epiphytes - Selvaggina o pascolo, altri animali, piante epifite
- Complex or unknown causes - Cause complesse o ignote
- Underwood fires - Incendio del sottobosco
- Indirect human actions - Azione indiretta dell'uomo
- Direct human actions - Azione diretta dell'uomo

Fig. 10.3 Percentage of the assessed Forest area by damaging causes at the national level / Ripartizione percentuale della superficie del Bosco valutato per causa di patologie o danni, a livello nazionale

Table 10.6 Classes of localization of defoliation / Classi di localizzazione della defogliazione

Localization of defoliation	Description
Localizzazione della defogliazione	Descrizione
Homogeneous Omogenea	Homogeneous defoliation over the entire crown
	Defogliazione omogenea su tutta la chioma
Basal Basale	Defoliation mainly on the basal portion
	Defogliazione principalmente sulla porzione basale
Localised Localizzata	Defoliation localized in single parts of the crown
	Defogliazione localizzata in singole parti della chioma
Apical Apicale	Defoliation mainly on the apical portion
	Defogliazione principalmente sulla porzione apicale

Table 10.7 Forest area by degree of defoliation; it is assessed when more than 30% of the crown coverage is affected by damages or pathologies / Estensione del Bosco ripartito per grado di defogliazione; viene valutato quando patologie o danni interessano più del 30% della copertura delle chiome

Region/Regione	Absence of defoliation or defoliation ≤ 10%		Defoliation 11–25%		Defoliation 26–60%		Defoliation > 60%		Not assessed or not classified		Total Forest	
	Defogliazione assente o defogliazione ≤ 10%		Defogliazione 11–25%		Defogliazione 26–60%		Defogliazione > 60%		Non valutato o non classificato		Totale Bosco	
	area	ES	area	ES	area	ES	area	ES	area	ES	area	ES
	(ha)	(%)	(ha)	(%)	(ha)	(%)	(ha)	(%)	(ha)	(%)	(ha)	(%)
Piemonte	63,153	14.8	15,164	31.9	12,753	35.0	3181	70.8	796,182	1.9	890,433	1.3
Valle d'Aosta	424	99.6	424	99.6	424	99.6	0	–	97,971	4.0	99,243	3.6
Lombardia	2327	83.3	5686	51.7	5875	51.8	6454	51.1	601,625	1.9	621,968	1.6
Alto Adige	6466	40.5	7056	44.3	1802	81.6	0	–	323,946	2.4	339,270	1.7
Trentino	3695	58.9	890	100.0	1372	100.0	0	–	367,301	2.0	373,259	1.4
Veneto	5256	49.9	0	–	0	–	1147	99.9	410,301	2.1	416,704	1.9
Friuli V.G	5235	51.6	2169	72.6	1901	81.9	818	99.7	322,434	2.4	332,556	1.9
Liguria	1423	99.2	11,736	32.9	12,694	32.2	3459	59.0	313,848	2.8	343,160	1.7
Emilia Romagna	19,868	26.0	16,042	30.2	18,505	27.8	0	–	530,486	2.3	584,901	1.5
Toscana	3604	63.7	6627	49.3	12,226	35.3	0	–	1,012,991	1.3	1,035,448	1.1
Umbria	0	–	379	99.8	0	–	0	–	389,926	1.9	390,305	1.6
Marche	1903	76.2	0	–	741	70.8	4110	62.2	285,013	2.5	291,767	2.1
Lazio	0	–	3402	67.7	2103	81.8	9830	42.0	544,901	2.0	560,236	1.6
Abruzzo	2959	63.0	1374	100.0	2303	63.1	0	–	404,952	2.1	411,588	1.8
Molise	0	–	0	–	824	99.8	530	99.8	151,893	3.4	153,248	3.0
Campania	3124	62.0	3819	56.7	4862	52.8	7320	42.0	384,802	2.5	403,927	2.1
Puglia	1439	99.8	1196	71.1	1225	99.8	0	–	138,488	4.2	142,349	4.0
Basilicata	0	–	0	–	0	–	1398	99.9	286,621	2.9	288,020	2.7
Calabria	5094	57.4	5062	51.7	4129	71.2	19,749	36.7	461,143	2.8	495,177	2.0
Sicilia	4108	58.9	1427	100.0	0	–	2762	69.7	277,191	3.4	285,489	3.2
Sardegna	6094	45.3	14,653	27.7	13,924	28.2	5865	43.9	585,603	2.4	626,140	2.1
Italia	136,174	10.0	97,105	11.8	97,664	11.9	66,624	16.5	8,687,620	0.5	9,085,186	0.4

Table 10.8 Forest area by classes of localization of defoliation; it is assessed when more than 30% of the crown coverage is affected by damages or pathologies and more than 10% of the foliage is missing / Estensione del Bosco ripartito per localizzazione della defogliazione; viene valutata quando patologie o danni interessano più del 30% della copertura delle chiome e quando risulta mancante più del 10% del fogliame

Region/Regione	Homogeneous defoliation		Basal defoliation		Localised defoliation		Apical defoliation		Not assessed or not classified		Total Forest	
	Defogliazione omogenea		Defogliazione basale		Defogliazione localizzata		Defogliazione apicale		Non valutato o non classificato		Totale Bosco	
	Area	ES	Area	ES	Area	ES	Area	ES	Area	ES	Area	ES
	(ha)	(%)	(ha)	(%)	(ha)	(%)	(ha)	(%)	(ha)	(%)	(ha)	(%)
Piemonte	24,867	24.7	0	–	1542	100.0	4688	57.5	859,336	1.6	890,433	1.3
Valle d'Aosta	424	99.6	0	–	0	–	424	99.6	98,395	4.0	99,243	3.6
Lombardia	15,066	32.4	0	–	0	–	2949	73.9	603,952	1.9	621,968	1.6
Alto Adige	1802	81.6	2827	70.5	1403	99.8	2827	70.5	330,412	2.3	339,270	1.7
Trentino	2263	72.4	0	–	0	–	0	–	370,997	1.9	373,259	1.4
Veneto	1147	99.9	0	–	0	–	0	–	415,557	2.0	416,704	1.9
Friuli V.G	3680	59.2	0	–	0	–	1207	74.8	327,669	2.3	332,556	1.9
Liguria	21,846	23.7	350	99.2	0	–	5693	48.8	315,271	2.8	343,160	1.7
Emilia Romagna	18,554	27.5	3109	70.8	3987	59.0	8897	40.5	550,354	2.0	584,901	1.5
Toscana	8284	43.9	0	–	0	–	10,569	38.0	1,016,595	1.3	1,035,448	1.1
Umbria	0	–	0	–	0	–	379	99.8	389,926	1.9	390,305	1.6
Marche	2324	80.4	741	70.8	0	–	1786	100.0	286,916	2.5	291,767	2.1
Lazio	11,864	37.6	471	100.0	368	100.0	2632	85.0	544,901	2.0	560,236	1.6
Abruzzo	2303	63.1	0	–	0	–	1374	100.0	407,911	2.1	411,588	1.8
Molise	1354	72.2	0	–	0	–	0	–	151,893	3.4	153,248	3.0
Campania	13,156	32.0	1350	57.5	1495	100.0	0	–	387,926	2.5	403,927	2.1
Puglia	523	99.8	1225	99.8	0	–	673	99.8	139,927	4.2	142,349	4.0
Basilicata	1398	99.9	0	–	0	–	0	–	286,621	2.9	288,020	2.7
Calabria	19,064	34.0	6237	76.7	3638	73.4	0	–	466,237	2.8	495,177	2.0
Sicilia	2762	69.7	0	–	1427	100.0	0	–	281,300	3.3	285,489	3.2
Sardegna	21,693	23.2	4162	50.3	866	100.0	7722	35.8	591,697	2.4	626,140	2.1
Italia	174,376	9.2	20,472	30.4	14,726	32.0	51,819	16.4	8,823,794	0.5	9,085,186	0.4

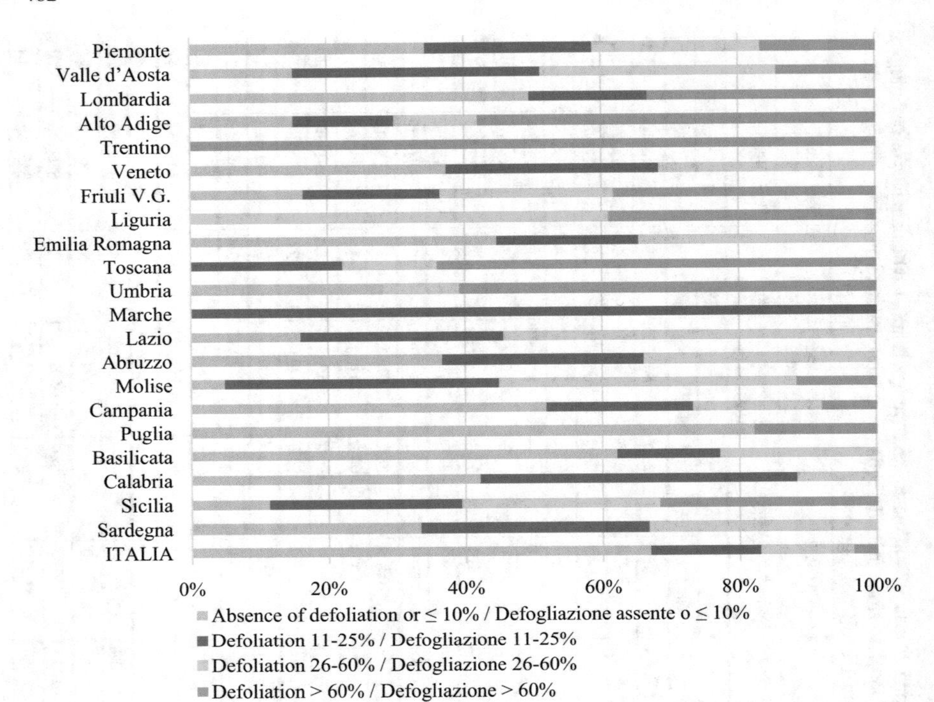

Fig. 10.4 Percentage of the assessed Forest area by degree of defoliation / Ripartizione percentuale della superficie del Bosco valutato per il grado di defogliazione

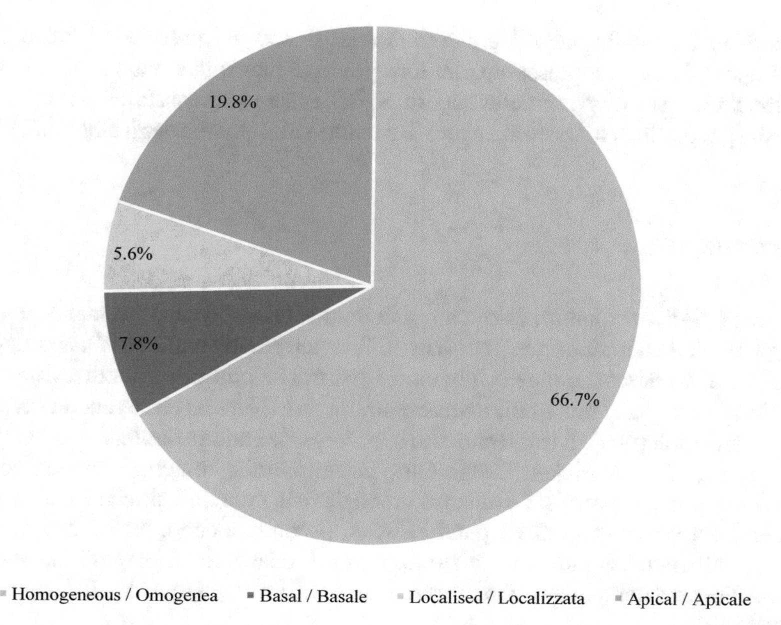

Fig. 10.5 Percentage of the assessed Forest area by localization of defoliation at the national level / Ripartizione percentuale della superficie del Bosco valutato per la localizzazione della defogliazione, a livello nazionale

Appendix (Italian Version)

Riassunto Le foreste sono prevalentemente costituite da specie arboree o arbustive caratterizzate da un lungo ciclo vitale, durante il quale possono subire gli effetti di fattori di disturbo di origine naturale o antropica di diversa intensità. Tali fattori sono fondamentali per l'evoluzione stessa delle foreste poiché, innescando processi di selezione naturale tra specie e individui, ne favoriscono la migliore composizione e struttura. Fattori di disturbo particolarmente intensi, tuttavia, possono causare perdita di valore naturalistico o economico e, in alcuni casi, avere conseguenze gravi e mettere a rischio la sopravvivenza di interi ecosistemi forestali.

L'elenco delle possibili cause di danno alle foreste è ampio e vario: parassiti, incendi, inquinamento, cambiamenti climatici, sovra-sfruttamento o pratiche di utilizzazione inadeguate, eccessivo pascolamento o impatto degli ungulati selvatici sono alcuni esempi.

Il monitoraggio dello stato di salute delle foreste rappresenta una condizione necessaria per fornire informazioni utili per la conservazione delle risorse forestali, come pure per supportare pratiche di gestione forestale finalizzate ad aumentare la resilienza delle foreste e la loro capacità di adattamento alle avversità. Attraverso i rilievi al suolo, l'inventario forestale nazionale italiano (INFC) ha classificato lo stato di salute delle foreste italiane e prodotto le stime presentate nel capitolo. Queste riguardano la ripartizione della superficie del Bosco per presenza e diffusione di patologie o danni al soprassuolo, del loro grado di intensità e della causa. Vengono inoltre presentate le stime della superficie del Bosco interessata da defogliazione, ripartita per grado di trasparenza e localizzazione sulla chioma degli alberi interessati.

Introduzione

I disturbi di diversa natura (biotica o abiotica) e intensità rappresentano una parte integrante della dinamica degli ecosistemi forestali (Seidl et al., 2017). Modificando la struttura, la composizione e la funzione, i disturbi favoriscono l'alternarsi nel tempo e nello spazio di fasi giovanili, mature e stramature e creano eterogeneità favorendo la diversità biologica e il rinnovamento o la riorganizzazione dell'ecosistema (Thom et al., 2017; White & Pickett, 1985). Oltre un certo limite, tuttavia, i disturbi possono provocare ingenti danni sia di natura ecologica sia economica, che incidono sulle funzioni del bosco (protettiva, paesaggistica, produttiva ecc.). Nella maggior parte dei casi si ritiene quindi necessario approntare una politica di gestione forestale indirizzata prevalentemente alla prevenzione ed alla mitigazione dei disturbi naturali (Motta, 2018).

Le principali minacce che possono alterare lo stato di salute delle foreste, provocandone la degradazione o limitando la loro funzionalità, sono riferibili sia a cause naturali di natura biotica o abiotica (parassiti vegetali o animali, presenza di selvaggina, incendi, alluvioni, neve abbondante, ecc.) sia a profondi cambiamenti del territorio e dell'ambiente ad opera dell'uomo, dovuti per esempio all'inquinamento o all'eccessivo sfruttamento del bosco in termini di utilizzazioni forestali o per uso ricreativo.

A causa dei cambiamenti climatici, che innescano sempre più spesso eventi meteorologici estremi (piogge intense, alluvioni, tempeste di vento, ondate di calore ecc.), la vulnerabilità delle foreste è in aumento. Gli eventi meteorologici estremi avvengono oggi con una frequenza tre volte superiore rispetto a cinquant'anni fa (FAO, 2021); questi, nel tempo, compromettono i meccanismi di difesa delle piante (Forzieri et al., 2021). I cambiamenti climatici, inoltre, possono innescare la diffusione di specie patogene in ambienti in cui non erano prima presenti, causando epidemie di parassiti animali e vegetali.

Monitorare lo stato di salute delle foreste rappresenta una condizione necessaria per definire pratiche di gestione forestale adeguate, volte a migliorare la resilienza e la capacità di adattamento alle avversità degli ecosistemi forestali. Nell'inventario forestale nazionale italiano, lo stato di salute dei soprassuoli forestali viene rilevato tramite l'osservazione diretta in campo della presenza di danni o patologie sui soggetti inclusi in un'area di saggio (AdS25, cfr. Chap. 4), al fine di stimare le superfici di Bosco interessate (Gasparini et al., 2013). Vengono considerate eventuali alterazioni conseguenti all'azione di un agente di danno (ad es. ferite, necrosi, defogliazione) o i segni della presenza di un fattore di danno o malattia (nidi di larve, insetti, corpi fruttiferi fungini, ecc.). L'osservazione si riferisce alla presenza di danni derivati da avversità verificatesi nell'anno del rilievo o precedentemente, ma i cui effetti sul soprassuolo siano ancora visibili.

Secondo INFC vengono considerati "evidenti" le patologie e i danni che riguardano almeno il 30% della copertura delle chiome all'interno dell'AdS25 (cfr. Chap. 4), con riferimento ai soggetti (alberi e arbusti) sopra la soglia di cavallettamento ($d_{1.30} \geq 4.5$ cm). Per questi casi, oltre alla presenza, viene valutata anche l'intensità della patologia o del danno, l'origine (che può essere di natura biotica o abiotica) e, se presente defogliazione, il suo grado e la localizzazione sui soggetti interessati.

Alla fine del capitolo vengono presentate le stime della superficie del Bosco ripartita per grado di diffusione di patologie o danni, intensità e causa. Vengono inoltre presentate le stime del Bosco interessato da defogliazione, per grado e localizzazione. All'indirizzo inventarioforestale.org/statistiche_INFC sono disponibili, inoltre, le stime di superficie relative alla ripartizione delle categorie inventariali del Bosco e delle categorie forestali dei Boschi alti, queste ultime a livello nazionale, per il grado di diffusione di patologie o danni.

Diffusione e intensità di patologie o danni

Il terzo inventario forestale italiano INFC2015 classifica la presenza di patologie o danni secondo classi del loro grado di diffusione, misurato in base alla percentuale di chiome interessate: assenza di danni e patologie, danni o patologie che interessano meno del 30%, oppure il 30–59% o almeno il 60% della copertura.

La Table 10.1 in fondo al capitolo riporta le stime della superficie del Bosco ripartita per il grado di diffusione di patologie o danni. Dalle stime emerge che quasi l'80% della superficie del Bosco non è interessato da danni e che il 15.6% della superficie è interessato da danni lievi, diffusi su meno del 30% della copertura. Complessivamente la superficie del Bosco senza danni o patologie o con danni lievi è pari al 94.9% del totale. L'aliquota di superficie con danni evidenti (copertura interessata pari a 30–59%) o molto evidenti (almeno il 60% della copertura interessata) è pari rispettivamente a 3.3 e 1.0%.

Le stime a livello regionale mostrano alcune variazioni rispetto alla media nazionale[2]; l'Umbria risulta la regione meno interessata da danni o patologie (99.2% della superficie del Bosco non interessata da patologie e danni o con danni lievi), seguita da Molise (99.1%), Valle d'Aosta (98.6%), Basilicata (98.4%), Trentino (98.3%) e Abruzzo (98.1%). Le regioni maggiormente interessate da danni evidenti o molto evidenti risultano essere il Piemonte (8.7 e 1.9% della superficie interessata), l'Emilia-Romagna (8.0 e 1.3%), la Liguria (6.7 e 1.8%), la Calabria (2.7 e 4.2%) e la Sardegna (5.3 e 1.1%) (Fig. 10.1).

[2] Le stime presentate risentono solo in parte degli effetti della tempesta Vaia dell'ottobre 2018, che si verificata nel corso della campagna di rilevo.

L'intensità del danno viene valutata stimando la proporzione delle parti epigee danneggiate (chioma, rami e fusto) rispetto alle condizioni di normalità per gli individui, nel contesto osservato (Table 10.2). Vengono presi in considerazione solo i danni duraturi e non quelli temporanei come, ad esempio, i danni all'apparato fogliare. L'intensità di patologie o danni, come detto in precedenza, viene valutata solo in caso di danni evidenti o molto evidenti (almeno il 30% della copertura interessata).

Le stime di superficie del Bosco per classi di intensità del danno sono riportate nella Table 10.3 in fondo al capitolo. Non considerando la classe "non classificato", corrispondente alla superficie non valutata per il grado di intensità del danno, la classe più frequente per il Bosco risulta quella di danno medio (52.8% della superficie del Bosco valutata), mentre la classe di danno intenso interessa il 34.4%. A livello regionale, in Molise e in Basilicata la totalità della superficie del Bosco valutata per l'intensità del danno risulta interessata da danni gravi. In Umbria e in Valle d'Aosta, invece, la totalità della superficie valutata è interessata da danni di media intensità (Fig. 10.2).

Cause di patologie o danni

La causa della patologia o del danno viene individuata tramite l'osservazione di tracce o segni evidenti secondo le descrizioni riportate nella Table 10.4. Qualora la causa non sia chiaramente riconoscibile, oppure in presenza dell'azione concomitante di più fattori, nessuno dei quali prevalente, viene assegnata la classe "cause ignote o complesse". In presenza di cause di danno o patologie diverse, si assegna la causa prevalente.

La Table 10.5 riporta le stime della superficie del Bosco per causa di danni o patologie. A livello nazionale le avversità più comuni sono rappresentate dagli attacchi parassitari, che interessano il 33.8% della superficie valutata, ossia della superficie con danni diffusi su almeno il 30% della copertura delle chiome; le cause di danno che seguono, in ordine di importanza per superficie interessata, sono gli eventi meteorici intensi (26.5%) e gli incendi del soprassuolo (20.7%) (Fig. 10.3).

A livello regionale l'incidenza delle diverse cause di danno o patologie è variabile, ma generalmente le principali cause di avversità appartengono ad una delle tre classi con maggiore rilevanza a livello nazionale.

Grado di defogliazione e sua localizzazione

Se gli effetti della patologia o del danno si manifestano come defogliazione, il protocollo di rilievo INFC prevede di rilevarne il grado e la localizzazione su alcuni soggetti arborei campione rappresentativi (fino a tre).

Il grado di defogliazione esprime l'aliquota di foglie mancanti e viene valutato osservando il livello di trasparenza delle chiome rispetto allo stato di fogliazione normale per quella specie, in quel contesto e in quel periodo. Il grado di defogliazione viene classificato secondo quattro classi: defogliazione assente o $\leq$ 10%, defogliazione pari a 11–25%, 26–60% e maggiore di 60%. La localizzazione riguarda, invece, l'indicazione delle zone della chioma interessate dalla defogliazione e viene valutata quando il grado di defogliazione supera il 10% (Table 10.6).

Tables 10.7 e 10.8 riportano le stime della superficie del Bosco rispettivamente per grado e localizzazione della defogliazione. La superficie del Bosco con un grado di defogliazione inferiore o uguale a 10% risulta pari al 34.3% della superficie valutata, mentre le classi superiori di defogliazione interessano rispettivamente il 24.4%, il 24.6% e il 16.8% della superficie valutata (Fig. 10.4).

Relativamente alla localizzazione della defogliazione, il 66.7% della superficie del Bosco con defogliazione superiore a 10% è caratterizzato da una distribuzione omogenea su tutta la chioma, mentre il 19.8% è caratterizzato da defogliazione nella porzione apicale della chioma (Fig. 10.5).

References

FAO (2021). *The impact of disasters and crises on agriculture and food security* (211 pp.). https://doi.org/10.4060/cb3673en.

Forzieri, G., Girardello, M., Ceccherini, G., Spinoni, J., Feyen, L., Hartmann, H., Beck, P. S. A., Camp-Valls, G., Chirici, G., Mauri, A., & Cescatti, A. (2021). Emergent vulnerability to climate-driven disturbances in European forests. *Nature Communications, 12*, 1081. https://doi.org/10.1038/s41467-021-21399-7

Gasparini, P., Di Cosmo, L., Cenni, E., Pompei, E., & Ferretti, M. (2013). Towards the harmonization between National Forest Inventory and Forest Condition Monitoring. Consistency of plot allocation and effect of tree selection methods on sample statistics in Italy. *Environmental Monitoring and Assessment, 185*, 6155–6171. https://doi.org/10.1007/s10661-012-3014-1

Motta, R. (2018). L'equilibrio della natura non esiste (e non è mai esistito!). *Forest@, 15*, 56–58. https://doi.org/10.3832/efor2839-015.

Seidl, R., Thom, D., Kautz, M., Martin-Benito, D., Peltoniemi, M., Vacchiano, G., Wild, J., Ascoli, D., Petr, M., Honkaniemi, J., Lexer, M. J., Trotsiuk, V., Mairota, P., Svoboda, M., Fabrika, M., Nagel, T. A., & Reyer, C. P. O. (2017). Forest disturbances under climate change. *Nature Climate Change, 7*, 395–402. https://doi.org/10.1038/nclimate3303

Thom, D., Rammer, W., & Seidl, R. (2017). Disturbances catalyze the adaptation of forest ecosystems to changing climate conditions. *Global Change Biology, 23*, 269–282. https://doi.org/10.1111/gcb.13506

White, P. S., & Pickett, S. T. A. (1985). Natural disturbance and patch dynamics: An introduction. In S. T. A. Pickett & P. S. White (Eds.), *The ecology of natural disturbance and patch dynamics* (pp. 3–13). Elsevier Inc. ISBN 9780125545204. https://doi.org/10.1016/B978-0-12-554520-4.50006-X.

Open Access This chapter is licensed under the terms of the Creative Commons Attribution 4.0 International License (http://creativecommons.org/licenses/by/4.0/), which permits use, sharing, adaptation, distribution and reproduction in any medium or format, as long as you give appropriate credit to the original author(s) and the source, provide a link to the Creative Commons license and indicate if changes were made.

The images or other third party material in this chapter are included in the chapter's Creative Commons license, unless indicated otherwise in a credit line to the material. If material is not included in the chapter's Creative Commons license and your intended use is not permitted by statutory regulation or exceeds the permitted use, you will need to obtain permission directly from the copyright holder.

Chapter 11
Protective Function and Primary Designated Management Objective

Funzione protettiva e funzione prioritaria

Antonio Floris and Lucio Di Cosmo

Abstract In a framework of multiple services supplied simultaneously by forests, the protection against natural hazards is one of the most important. Forests deliver conservation of natural resources, including soil and water, and other environmental services. They slow water dispersion and allow for infiltration and percolation of rainwater, which recharges soil and underground water storage. Forest cover, moreover, protects soil from wind and water erosion, avalanches and landslides. INFC collects a wide range of information related to the protective function of Italian wooded areas. This chapter shows estimates regarding such physical site characteristics, as slope, land position and aspect which, together with tree canopy coverage and terrain roughness, can condition the protective role of forests. Inventory statistics on terrain instability and hydrogeological constraint, as defined by national laws, are shown as well, the latter being a basis of most national and regional regulations on forest management. Finally, the presence of a primary designated management objective has been assessed with a particular focus on direct and indirect protection. Estimates on such attributes are shown in the last section of this chapter.

Keywords Terrain instability · Land position · Hydrogeological constraint · Soil protection · Forest recreation · Non-wood products · Ecosystem services

11.1 Introduction

One of the main reasons for protecting forests is that they protect us. Forests have many protective functions, from a local to global scale (FAO, 2005). Such purposes have traditionally been divided into two broad categories: direct protection and indirect protection. Direct protection occurs when a forest offers protection from specific

A. Floris (✉) · L. Di Cosmo
CREA Research Centre for Forestry and Wood, Trento, Italy
e-mail: antonio.floris@crea.gov.it

L. Di Cosmo
e-mail: lucio.dicosmo@crea.gov.it

© The Author(s) 2022

P. Gasparini et al. (eds.), *Italian National Forest Inventory—Methods and Results of the Third Survey*, Springer Tracts in Civil Engineering,
https://doi.org/10.1007/978-3-030-98678-0_11

risk factors in an area where people live or valuable objects are present (residential areas, buildings, roads, others). This is a conventional classification system, convenient in practice and broadly accepted, which classifies the forest function indirectly, through its value to the safeguarded territory and not by the function itself. Protection forests generally protect against several types of natural hazards (Dorren et al., 2004) but direct protection is local, since it is tied to the stand level (Schönenberger, 2001).

INFC assesses various functions of Italian forests. Chap. 9 provides inventory statistics on the wooded areas in protected areas, at the different levels of protection in line with national laws. This is particularly relevant for the biodiversity conservation, since forests represent the habitat of a considerable part of the flora and fauna (Fuhrer, 2000). Concerning the influence on climate, Chap. 12 shows the contribution of the aboveground woody vegetation to the carbon stock as well as estimates on the organic carbon stock annually stored by tree growth.

This chapter presents the estimates on some variables related to forest protective functions. Physical site characteristics as represented by slope, aspect and position of terrain may determine local climate conditions affecting distribution of phytocenosis on a topographical scale (De Philippis, 1961). These three variables also influence the composition and productivity of the forest, because they are closely related to soil features (depth, profile development, structure and texture) (Spurr & Barnes, 1980). Being correlated with climatic variables, slope and aspect, together with elevation, not only affect species composition, but also the productivity of stands (Stage & Salas, 2007). These variables are generally tested and included among the independent variables in growth models (e.g., Di Cosmo et al., 2020; Gasparini et al., 2017), separately or combined in complex indices (e.g., Roise & Betters, 1981; Stage, 1976).

The productive aspects are closely related to protection aspects of forests as well, because stands with different composition and productivity have different protection capabilities. This is not only due to their naturally diverse potential for protection, but also to the silvicultural techniques and the method of utilisation applied to obtain goods and ecosystem services that are themselves conditioned by composition and productivity.

It is simple to recognize the important protection effects of forests on steep slopes. Logging is generally limited or forbidden on terrains over 60% slope. However, Schönenberger and Brang (2004) reported that on steep terrain, pronounced variation in slope and aspect leads to a high variability in mesoclimate and small-scale microhabitat patterns, which increases the mutual influence between surface erosion and rockfalls on one side and forest on the other side. Viewing the relationship between terrain and stand from another perspective, we can observe that site features as represented by slope, aspect and land position originate different levels of vulnerability for soil stability to risk factors. For example, meso- and micro-climate features as affected by aspect may lead to different risk levels for causes of instability, such as snow movement (avalanches, creeping, etc.). Hence, estimates on wooded areas in the classes adopted for the three variables allow an indirect evaluation on the portion of the national territory for which forests have an important protective role (Sect. 11.2).

In Sect. 11.3, the relationship between stand and soil is observed in reverse to view the forest from a soil perspective. In fact, it is true that the forest protects the soil, but it is also true that soil stability is a prerequisite for the maintenance of forest cover. The main estimates on the wooded areas affected by different types of land instability are shown and discussed in this section. Italian national regulations have long recognised the fundamental importance of forests in the protection of soil and water. The estimates of wooded areas officially subject to hydrogeological constraint are shown and discussed in Sect. 11.4. Finally, in Sect. 11.5 estimates on the presence of forest management aimed at a primary objective are presented, including the functions of direct and indirect protection.

11.2 Physical Site Conditions

The physical characteristics considered to describe the site conditions are slope, land position and aspect.

Terrain slope, namely the ratio of difference in height between two points divided by the horizontal distance between the two points, has been calculated starting from inclinations measured in the field (cf. Chap. 4). Five slope classes were adopted (Table 11.1). Tables 11.2 and 11.3 show the estimates of Forest and Other wooded land area by terrain slope classes. A total of 45.1% of Forest area in Italy is characterised by steep terrains (slope above 40%, Fig. 11.1). In Other wooded land, the percentage of steep terrain area is lower (25.1%), but there is a high incidence of not classified area (35.4%). Considering only Forest, all Alpine regions (Piemonte, Valle d'Aosta, Lombardia, Alto Adige, Trentino, Friuli-Venezia Giulia and Liguria) have steep terrain rates higher than the national average, while Puglia (48.3%) and Sardegna (35.7%) have the highest rates of Forest area with a slope below 20%.

Land position was assessed in the field by sight classification of the extended and local area. The extended land position refers to a wide surface, variable from a few hectares to tens of hectares, surrounding the sample unit; the local area land position applies to AdS25 (about 2000 m^2). For both attributes five classes were adopted (Tables 11.4 and 11.5). Area estimates by extended area land position are shown in Table 11.6 (Forest) and in Table 11.7 (Other wooded land). Medium slope is the

Table 11.1 Slope classes (%) with respective inclinations (degrees) / Classi di pendenza percentuale e corrispondenti valori di inclinazione (gradi)

Slope classes (%)	Inclination angles (°)
Classi di pendenza (%)	Angoli di inclinazione (°)
0–20	0–11
21–40	12–22
41–60	23–31
61–80	32–39
>80	>39

Table 11.2 Forest area by ground slope class / Estensione del Bosco ripartito per classi di pendenza del terreno

Region/Regione	Forest/Bosco													
	0–20%		21–40%		41–60%		61–80%		>80%		Not classified Non classificata		Total Totale	
	Area (ha)	ES (%)	Area (ha)	ES (%)	Area (ha)	ES (%)	Area (ha)	ES (%)	Area (ha)	ES (%)	Area (ha)	ES (%)	Area (ha)	ES (%)
Piemonte	159,109	5.4	196,725	4.5	237,463	3.8	163,649	4.7	79,848	7.2	53,640	8.5	890,433	1.3
Valle d'Aosta	6165	24.3	11,286	17.6	30,979	10.8	22,469	11.9	17,195	13.9	11,150	17.6	99,243	3.6
Lombardia	98,917	6.8	91,865	6.8	140,619	5.4	141,571	5.4	95,215	6.4	53,782	9.4	621,968	1.6
Alto Adige	27,323	11.3	60,594b	7.3	86,739	6.1	101,971	5.6	39,053	9.4	23,590	12.5	339,270	1.7
Trentino	32,329	10.8	75,712	6.4	94,187	5.5	100,100	5.5	43,965	8.6	26,966	11.4	373,259	1.4
Veneto	53,990	9.5	95,898	6.1	110,106	5.6	71,413	6.8	50,244	8.3	35,053	11.0	416,704	1.9
Friuli V.G	66,275	7.5	50,879	8.4	78,709	6.4	62,436	7.1	36,050	9.7	38,207	10.4	332,556	1.9
Liguria	24,442	11.8	86,063	6.0	114,902	4.9	75,238	6.7	28,954	10.8	13,561	16.1	343,160	1.7
Emilia Romagna	128,301	5.6	201,102	3.8	132,526	4.9	62,991	7.3	31,311	10.6	28,669	11.9	584,901	1.5
Toscana	240,820	3.8	340,296	2.9	212,575	4.0	87,628	6.4	45,889	8.7	108,240	5.7	1,035,448	1.1
Umbria	71,855	7.4	131,121	5.0	106,002	5.3	51,980	7.9	12,903	16.6	16,445	17.8	390,305	1.6
Marche	30,106	13.3	74,619	7.2	82,558	5.9	52,257	7.8	22,207	12.6	30,020	10.7	291,767	2.1
Lazio	129,061	5.3	149,855	4.9	127,242	5.2	64,076	7.5	23,029	12.4	66,972	7.1	560,236	1.6
Abruzzo	65,951	8.2	130,898	5.1	123,649	5.0	42,830	8.8	14,182	15.7	34,077	10.0	411,588	1.8
Molise	41,185	10.6	58,782	8.1	28,838	11.5	13,899	18.3	4296	29.8	6248	24.5	153,248	3.0
Campania	62,860	8.2	105,434	5.8	102,997	6.1	48,472	8.8	22,771	13.6	61,392	7.6	403,927	2.1
Puglia	68,722	7.3	42,632	9.2	20,547	13.2	5049	27.4	2719	37.5	2679	37.6	142,349	4.0
Basilicata	68,233	8.0	88,170	6.6	48,202	9.4	21,008	16.3	8874	24.6	53,533	7.8	288,020	2.7
Calabria	93,167	6.5	133,856	5.4	118,030	5.6	49,065	9.5	29,104	11.1	71,955	8.1	495,177	2.0
Sicilia	60,376	8.7	105,318	6.4	58,133	9.2	25,802	15.1	12,332	22.3	23,528	16.1	285,489	3.2
Sardegna	223,481	4.3	203,117	4.3	109,377	6.2	36,188	10.0	11,880	20.5	42,097	9.7	626,140	2.1
Italia	1,752,668	1.5	2,434,222	1.2	2,164,378	1.3	1,300,094	1.7	632,021	2.4	801,804	2.2	9,085,186	0.4

Table 11.3 Other wooded land area by ground slope class / Estensione delle Altre terre boscate ripartite per classi di pendenza del terreno

Region/Regione	Other wooded land/Altre terre boscate													
	0–20%		21–40%		41–60%		61–80%		>80%		Not classified Non classificata		Total Totale	
	Area	ES	Area	ES	Area	ES	Area	ES	Area	ES	Area	ES	Area	ES
	(ha)	(%)	(ha)	(%)	(ha)	(%)	(ha)	(%)	(ha)	(%)	(ha)	(%)	(ha)	(%)
Piemonte	7172	37.0	10,492	22.8	12,766	22.0	11,254	21.8	10,011	28.6	33,295	12.9	84,991	8.0
Valle d'Aosta	0	–	385	99.6	2,906	53.7	385	99.6	385	99.6	4671	27.2	8733	24.0
Lombardia	2865	37.9	3086	37.8	7895	23.6	11,598	23.9	14,221	22.2	30,588	13.8	70,252	8.7
Alto Adige	2943	35.0	3799	30.6	7940	21.6	7489	28.0	3542	34.7	10,368	18.7	36,081	10.4
Trentino	0	–	3363	32.3	6030	25.0	7939	21.2	3992	42.0	12,503	16.7	33,826	10.6
Veneto	3704	31.5	2243	40.7	3704	31.5	2682	36.5	1122	57.7	39,535	11.1	52,991	9.1
Friuli V.G	4416	28.7	1487	49.8	2602	37.6	5458	40.3	2886	35.0	24,209	14.9	41,058	10.6
Liguria	3850	43.6	2932	35.0	10,945	21.1	6163	31.0	985	58.0	19,210	16.9	44,084	10.3
Emilia Romagna	8642	31.0	13,494	20.6	3684	31.6	2214	40.8	2511	61.5	23,370	12.0	53,915	9.3
Toscana	22,730	14.6	25,198	15.1	10,769	26.5	3951	43.3	720	70.7	90907	6.6	154275	5.2
Umbria	7769	30.2	3588	31.6	4253	40.6	737	70.6	0	–	7303	27.6	23,651	15.2
Marche	1052	57.3	5701	38.5	3236	50.5	1412	100.0	741	70.8	9171	22.2	21,314	16.2
Lazio	13,365	22.7	18,328	16.9	14,608	21.3	8739	31.5	2913	55.5	29,960	13.9	87,912	7.7
Abruzzo	12,346	25.1	18,820	19.1	8619	24.8	6354	35.5	1059	57.8	15,813	15.4	63,011	8.6
Molise	5905	37.2	6420	30.1	1953	44.4	1172	57.5	781	70.5	3796	28.2	20,025	16.0
Campania	13,998	25.6	14,303	21.4	8405	25.4	5402	40.8	1412	100.2	43,811	10.8	87,332	7.6
Puglia	25,920	15.0	13,485	23.5	3496	33.1	777	70.5	388	99.8	5323	25.3	49,389	9.9
Basilicata	13,402	16.4	26,920	14.4	10,331	27.3	1486	49.9	373	99.9	51,879	8.7	104,392	6.2
Calabria	8955	20.3	15,110	20.8	7743	21.7	5224	26.6	4104	30.0	114,307	5.7	155,443	4.8
Sicilia	15,464	20.4	22,795	15.6	19,488	18.5	8251	26.0	2654	37.7	33,093	13.6	101,745	7.1
Sardegna	197,992	4.6	193,770	4.6	131,007	5.5	46,886	9.5	11,195	18.2	94,001	6.5	674,851	2.0
Italia	372,490	3.7	405,719	3.5	282,380	4.2	145,575	6.2	65,996	9.5	697,112	2.5	1,969,272	1.4

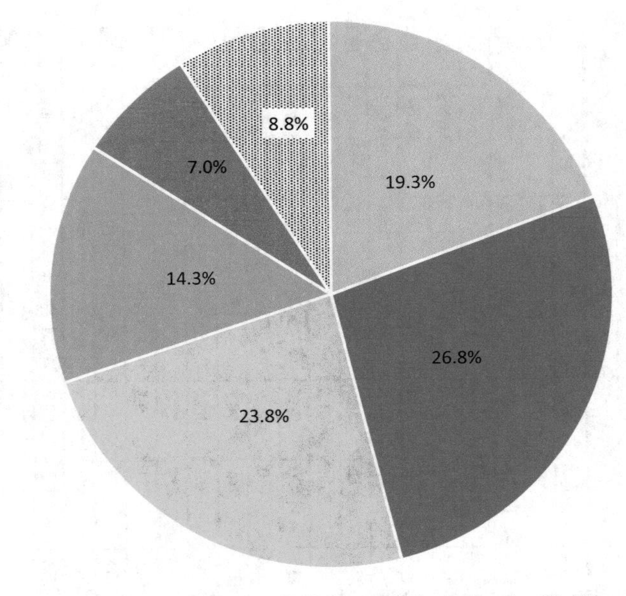

Fig. 11.1 Percent of forest area by ground slope class / Ripartizione percentuale dell'area del Bosco per classe di pendenza del terreno

Table 11.4 Classes of extended land position / Classi relative al tipo di giacitura dell'area estesa

Land position class Classe di giacitura	Description Descrizione
Flat Pianeggiante	Plains, wide valleys and terraces, plateaux Pianura, ampie vallate, ampi terrazzamenti, altopiani
High slope Alto versante	Crest, rises, saddles, slope summit where surface water outflow prevails Cresta, dosso, valico, parte sommitale di versante in cui prevalga il deflusso idrico superficiale
Medium slope Medio versante	Inclined surface where water in- and outflow balance Superficie inclinata in cui afflusso e deflusso idrico superficiale si equivalgono
Low slope Basso versante	Base portion, basin, valley, characterised by water accumulation Porzione basale, conca, avvallamento caratterizzati da prevalenza di accumulo idrico
Indeterminate Indeterminata	High variability of land position Condizioni di elevata variabilità della giacitura

Table 11.5 Classes of local land position / Classi relative al tipo di giacitura locale

Land position class Classe di giacitura	Description Descrizione
Flat Pianeggiante	Inclination not more than 5° Inclinazione non superiore a 5°
Convex Convessa	AdS25 of convex shape AdS25 di forma convessa
Sloped Su piano inclinato	AdS25 on slope AdS25 su piano inclinato
Concave Concava	AdS25 of concave shape AdS25 di forma concava
Indeterminate Indeterminata	High variability of land position Condizioni di elevata variabilità locale della giacitura

prevailing position type either in Forest (53.5%) and in Other wooded land (35.2%). Figure 11.2 shows the rates of all extended land position classes for Forest. National rates are also confirmed at the regional level. A high percentage of flat land position area can be noted in Puglia (20.6%) compared to the other regions.

In regard to local area land position, area estimates by position class are given for Forest in Tables 11.8 and in Fig. 11.3, and for Other wooded land in Table 11.9. In Forest, inclined plane position prevails both at the national level, with 68.7% of total area, and in all regions.

In the field, aspect was measured standing at the sample point and aiming a magnetic compass downward along the maximum slope direction. Nine aspect classes were considered (Table 11.10), each 45° wide. Aspect class rates at the national level are rather homogeneous both in Forest, ranging from 10 to 13%, and in Other wooded land (7–9%) as shown in Tables 11.11 and 11.12. The class 'no aspect' generally occurs for low percentage of areas (2–3%).

Additional estimates about ground slope, land position and terrain aspect for the inventory categories and for the Tall trees forest types are available at https://www.inventarioforestale.org/statistiche_INFC.

11.3 Terrain Instability

Terrain instability is conditioned by site morphological characteristics, especially slope, and by soil structure and geology. Instability may occur to such an extent as to affect the stability of stands.

INFC2015 has visibly determined four types of instability during the field surveys: shallow and deep-seated landslides, water erosion and flooding, rockfall and rolling, and snowslides and avalanches (Table 11.13). At least one type of instability was

Table 11.6 Forest area by extended land position class / Estensione del Bosco ripartito per classi di giacitura dell'area estesa

Region/Regione	Forest/Bosco													
	Flat		High slope		Medium slope		Low slope		Indeterminate		Not classified		Total	
	Pianeggiante		Alto versante		Medio versante		Basso versante		Indeterminata		Non classificata		Totale	
	Area	ES	Area	ES	Area	ES	Area	ES	Area	ES	Area	ES	Area	ES
	(ha)	(%)	(ha)	(%)	(ha)	(%)	(ha)	(%)	(ha)	(%)	(ha)	(%)	(ha)	(%)
Piemonte	89,908	7.7	112,274	6.2	532,323	2.2	103,500	6.5	0	–	52,428	8.6	890,433	1.3
Valle d'Aosta	1156	57.3	5042	38.2	77,271	4.5	4238	29.6	385	99.6	11,150	17.6	99,243	3.6
Lombardia	66,209	8.5	67,291	8.7	376,429	2.6	55,172	8.6	3967	33.3	52,900	9.4	621,968	1.6
Alto Adige	5393	26.1	43,927	9.3	220,630	3.0	43,839	8.8	1890	44.5	23,590	12.5	339,270	1.7
Trentino	7824	25.9	52,745	7.8	248,579	2.6	40,028	9.6	2162	40.8	21,921	12.7	373,259	1.4
Veneto	26,001	15.4	52,257	8.4	255,587	3.0	44,067	9.3	4861	27.6	33,931	11.3	416,704	1.9
Friuli V.G	40,333	9.5	51,251	8.3	136,686	4.4	57,569	7.8	8512	25.4	38,207	10.4	332,556	1.9
Liguria	3665	31.3	70,555	6.8	211,499	3.0	39,848	10.0	4398	28.5	13,194	16.3	343,160	1.7
Emilia Romagna	32,834	12.8	123,043	5.0	305,146	2.9	93,367	6.3	2947	35.3	27,564	12.2	584,901	1.5
Toscana	42,152	11.4	241,356	3.6	504,611	2.2	142,702	4.8	1445	50.0	103,181	5.8	1,035,448	1.1
Umbria	11,687	20.3	77,727	6.8	201,147	3.6	83,298	6.2	1475	49.9	14,970	19.0	390,305	1.6
Marche	2224	40.8	57,103	8.0	152,630	4.2	48,308	9.0	1482	50.0	30,020	10.7	291,767	2.1
Lazio	45,250	9.6	92,493	6.1	272,809	3.3	77,184	7.1	6264	24.1	66,235	7.1	560,236	1.6
Abruzzo	9495	26.3	80,343	6.4	222,605	3.5	62,534	7.7	2534	37.8	34,077	10.0	411,588	1.8
Molise	4130	29.8	36,417	9.3	74,649	6.6	30,632	11.7	1172	57.5	6248	24.5	153,248	3.0
Campania	9792	22.7	97,440	5.9	175,017	4.2	58,813	9.0	1841	44.7	61,024	7.7	403,927	2.1
Puglia	29,337	12.9	24,082	12.1	73,512	6.6	12,350	17.2	1554	49.8	1514	49.9	142,349	4.0
Basilicata	11,723	24.8	63,550	7.6	119,466	5.7	40,494	9.7	0	–	52,788	7.9	288,020	2.7
Calabria	13,843	18.5	121,109	5.3	233,728	3.7	56,469	9.2	2,549	61.5	67,478	8.5	495,177	2.0
Sicilia	10,890	26.9	55,059	10.3	155,368	4.8	40,646	10.6	379	100.0	23,148	16.3	285,489	3.2
Sardegna	69,996	7.8	104,672	6.3	306,385	3.5	97,026	6.6	6711	23.5	41,350	9.8	626,140	2.1
Italia	533,842	3.0	1,629,738	1.5	4,856,075	0.7	1,232,085	1.8	56,528	8.7	776,920	2.3	9,085,186	0.4

Table 11.7 Other wooded land area by extended land position class / Estensione delle Altre terre boscate ripartite per classi di giacitura dell'area estesa

Region/Regione	Other wooded land/Altre terre boscate													
	Flat		High slope		Medium slope		Low slope		Indeterminate		Not classified		Total	
	Pianeggiante		Alto versante		Medio versante		Basso versante		Indeterminata		Non classificata		Totale	
	Area	ES	Area	ES	Area	ES	Area	ES	Area	ES	Area	ES	Area	ES
	(ha)	(%)	(ha)	(%)	(ha)	(%)	(ha)	(%)	(ha)	(%)	(ha)	(%)	(ha)	(%)
Piemonte	2271	67.9	11,866	25.1	32,100	13.7	5457	35.0	0	–	33,295	12.9	84,991	8.0
Valle d'Aosta	0	–	1365	99.6	2697	37.3	0	–	0	–	4671	27.2	8733	24.0
Lombardia	331	100.1	8772	28.8	28,287	14.1	2275	44.8	0	–	30,588	13.8	70,252	8.7
Alto Adige	756	70.5	13,974	18.8	8277	21.1	2706	36.5	0	–	10,368	18.7	36,081	10.4
Trentino	0	–	10,857	18.4	9025	23.5	1441	50.0	0	–	12,503	16.7	33,826	10.6
Veneto	1835	44.7	2243	40.7	8256	20.9	748	70.6	374	99.9	39,535	11.1	52,991	9.1
Friuli V.G	1815	44.6	6945	33.4	5488	25.5	2230	40.6	0	–	24,581	14.8	41,058	10.6
Liguria	252	99.2	5813	32.4	16,627	17.6	1833	44.3	0	–	19,560	16.6	44,084	10.3
Emilia Romagna	368	100.0	6563	30.0	17,343	20.2	6270	24.2	0	–	23,370	12.0	53,915	9.3
Toscana	3970	43.2	16,574	18.0	35,597	13.1	7588	21.8	0	–	90,545	6.7	154,275	5.2
Umbria	1402	99.8	4524	38.5	8210	25.4	2212	40.7	0	–	7303	27.6	23,651	15.2
Marche	371	100.0	1082	57.9	8195	31.7	2495	62.1	0	–	9171	22.2	21,314	16.2
Lazio	2554	37.8	17,512	20.8	32,017	14.2	5501	25.7	368	100.0	29,960	13.9	87,912	7.7
Abruzzo	3941	43.2	11,006	23.6	22,319	17.3	9209	29.5	724	70.8	15,813	15.4	63,011	8.6
Molise	125	99.8	6089	31.7	6810	28.9	3205	50.5	0	–	3796	28.2	20,025	16.0
Campania	737	70.8	8345	29.2	26,520	16.5	7551	35.0	368	100.0	43,811	10.8	87,332	7.6
Puglia	13,171	22.7	10,592	26.7	15,930	19.8	3596	32.3	777	70.5	5323	25.3	49,389	9.9
Basilicata	4101	30.0	12,890	21.3	26,577	15.0	8944	20.2	0	–	51,879	8.7	104,392	6.2
Calabria	2612	37.7	11,847	17.5	21,143	15.4	6654	29.9	0	–	113,188	5.7	155,443	4.8
Sicilia	1832	44.6	8920	28.1	46,705	10.9	10,816	24.4	379	100.0	33,093	13.6	101,745	7.1
Sardegna	60,176	9.5	149,768	5.3	272,929	3.7	97,231	6.5	2985	35.3	91,762	6.6	674,851	2.0
Italia	102,619	7.4	327,549	4.0	651,053	2.8	187,961	5.1	5976	25.0	694,114	2.5	1,969,272	1.4

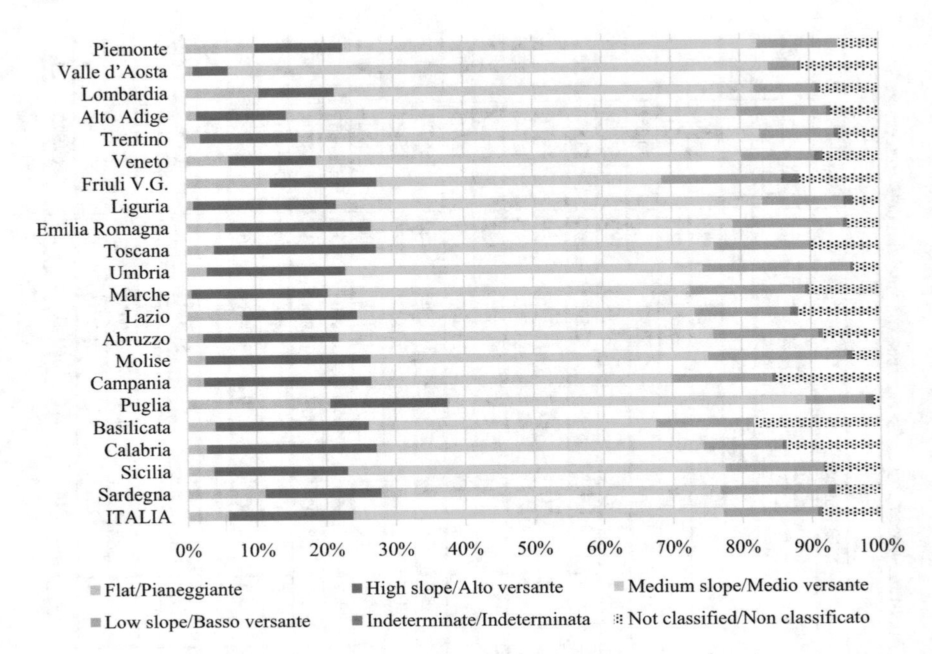

Fig. 11.2 Percent of Forest area by extended land position class / Ripartizione percentuale dell'area del Bosco per classe di giacitura estesa

observed on 15.2% of the national Forest area (Table 11.14 and Fig. 11.4), and at the regional level, the percentage ranges between 7.8% (Basilicata) and 26.3% (Valle d'Aosta). The most widespread instability type in Forest is rockfall and rolling (6.4% of total area).

Considering shallow and deep-seated landslides, a value much higher than the national average (3.6%) was observed in Emilia-Romagna (14.4%). The percentage of Forest area with instability due to snowslides and avalanches is very low at the national level (0.6%) and in most regions. It is noticeable in only three Alpine regions: Valle d'Aosta (5.4%), Alto Adige (5.6%) and Trentino (3.8%). In these three regions, the percent of area with the same instability causes is even higher in Other wooded land (17.6% in Valle d'Aosta, 16.9% in Alto Adige, 22.4% in Trentino; Table 11.15).

Additional estimates on hydrogeological instability concerning the inventory categories and the Tall trees forest types are available at https://www.inventarioforestale.org/statistiche_INFC.

11.4 Hydrogeological Constraints

The milestone Italian framework law for forests protection and management is the Royal Decree No. 3267/23, which contains the rules regarding the hydrogeological

Table 11.8 Forest area by local land position class / Estensione del Bosco ripartito per classi di giacitura dell'area locale

Region/Regione	Forest/Bosco													
	Flat		Convex		Sloped		Concave		Indeterminate		Not classified		Total	
	Pianeggiante		Convessa		Su piano inclinato		Concava		Indeterminata		Non classificata		Totale	
	Area	ES	Area	ES	Area	ES	Area	ES	Area	ES	Area	ES	Area	ES
	(ha)	(%)	(ha)	(%)	(ha)	(%)	(ha)	(%)	(ha)	(%)	(ha)	(%)	(ha)	(%)
Piemonte	99,887	7.1	75,163	7.3	548,771	2.1	88,121	6.8	23,640	13.9	54,852	8.4	890,433	1.3
Valle d'Aosta	1927	44.3	771	70.3	77,689	5.0	7321	22.2	385	99.6	11,150	17.6	99,243	3.6
Lombardia	68,413	8.3	19,055	15.0	440,887	2.3	31,456	12.2	6171	26.6	55,986	9.1	621,968	1.6
Alto Adige	9931	19.2	26,088	11.6	245,255	2.8	23,441	12.3	10,586	18.6	23,968	12.4	339,270	1.7
Trentino	8544	24.4	28,108	11.0	274,324	2.4	19,099	13.5	20,541	13.0	22,642	12.5	373,259	1.4
Veneto	27,870	14.6	20,564	13.2	290,967	2.7	29,537	10.9	13,460	16.5	34,305	11.2	416,704	1.9
Friuli V.G	45,871	9.2	5203	26.5	225,436	2.9	9663	19.3	7805	21.6	38,578	10.4	332,556	1.9
Liguria	7697	21.5	19,425	13.3	271,324	2.4	18,325	13.7	9163	19.6	17,226	14.2	343,160	1.7
Emilia Romagna	40,915	11.0	29,470	11.0	413,361	2.2	51,117	8.6	21,001	13.1	29,038	11.8	584,901	1.5
Toscana	73,274	7.6	80,144	7.0	663,607	1.8	77,662	6.8	24,570	12.0	116,189	5.4	1,035,448	1.1
Umbria	21,364	14.1	18,764	13.7	292,515	2.5	32,369	11.0	8110	21.1	17,182	17.3	390,305	1.6
Marche	5086	27.4	25,943	11.6	197,803	3.4	29,209	11.6	2224	40.8	31,502	10.4	291,767	2.1
Lazio	61,439	8.4	25,020	12.9	368,187	2.5	30,512	12.2	8106	21.2	66,972	7.1	560,236	1.6
Abruzzo	20,353	15.5	11,903	20.1	322,653	2.4	13,914	18.6	7601	21.7	35,163	9.8	411,588	1.8
Molise	10,769	18.3	5858	25.4	113,509	4.6	6639	23.8	8273	29.0	8201	21.3	153,248	3.0
Campania	22,582	14.6	11,724	20.4	280,978	3.0	13,257	16.5	12,153	17.3	63,234	7.5	403,927	2.1
Puglia	39,471	10.1	13,092	19.3	78,173	6.3	8545	20.9	0	–	3068	35.2	142,349	4.0
Basilicata	21,114	15.3	3723	31.5	199,955	3.9	7830	21.6	373	99.9	55,025	7.7	288,020	2.7
Calabria	33,556	11.5	37,125	11.3	290,074	3.2	37,281	9.7	25,931	14.3	71,209	8.2	495,177	2.0
Sicilia	15,149	19.5	22,158	15.0	200,011	4.3	18,577	14.0	4550	28.7	25,044	15.4	285,489	3.2
Sardegna	95,922	6.6	7464	22.3	445,976	2.7	29,084	11.2	5971	25.0	41,724	9.8	626,140	2.1
Italia	731,133	2.5	486,764	2.9	6,241,458	0.6	582,962	2.6	220,613	4.3	822,257	2.2	9,085,186	0.4

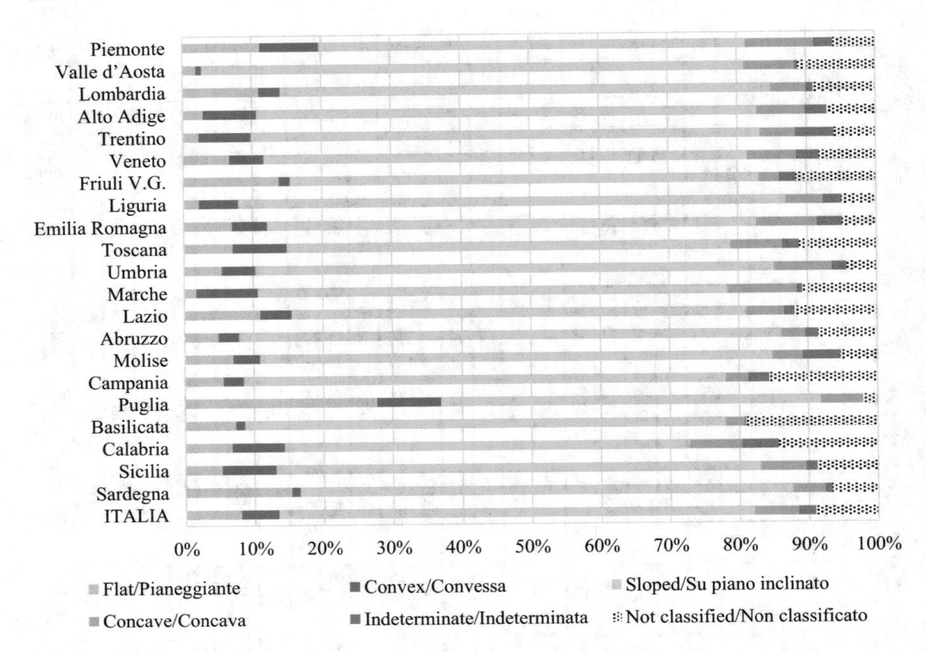

Fig. 11.3 Percent of Forest area by local land position class / Ripartizione percentuale dell'area del Bosco per classe di giacitura locale

constraints among its main objectives. This legal framework was later completed with other state and regional norms, to adapt to changing needs. The purpose of the hydrogeological constraint is to preserve the physical environment by avoiding forest utilisation practices that could bring about denudation, trigger erosion and cause loss of stability, disturb the water regimen or have the potential for other public damage. Nevertheless, the hydrogeological constraint does not preclude the possibility of intervention in forest areas, but places limits on owners, such as the necessity of authorisation for the transformation of forests into other types of cultivation and for the transformation of undisturbed terrain into terrain subjected to periodic tilling. Forest management on constrained terrains, moreover, must be conducted according to modalities defined by the so-called Overall and Forest Police Regulations.

Table 11.16 indicates the presence of hydrogeological constraints on the greater part of the total wooded area in Italy (80.1%). These constraints regard above all the Forest, where they are applicable to 86.6% of the area, while for Other wooded land the constrained area rate is 50.0%. The percentage of constrained Forest area is very high, with values above 95% in some regions of North and Central Italy (Fig. 11.5): Trentino, Alto Adige, Umbria, Toscana. Among the other regions, Sardegna is an exception, with only 51.6% of the Forest area subjected to hydrogeological constraints.

Table 11.9 Other wooded land area by local land position class / Estensione delle Altre terre boscate ripartite per classi di giacitura dell'area locale

Region/Regione	Other wooded land/Altre terre boscate													
	Flat		Convex		Sloped		Concave		Indeterminate		Not classified		Total	
	Pianeggiante		Convessa		Su piano inclinato		Concava		Indeterminata		Non classificata		Totale	
	Area	ES	Area	ES	Area	ES	Area	ES	Area	ES	Area	ES	Area	ES
	(ha)	(%)	(ha)	(%)	(ha)	(%)	(ha)	(%)	(ha)	(%)	(ha)	(%)	(ha)	(%)
Piemonte	3484	48.6	3232	35.3	39,576	12.9	4595	30.2	404	100.0	33,699	12.8	84,991	8.0
Valle d'Aosta	0	–	0	–	3677	44.8	385	99.6	0	–	4671	27.2	8733	24.0
Lombardia	1984	44.8	882	70.7	34,564	13.1	1283	58.6	952	70.9	30,588	13.8	70,252	8.7
Alto Adige	1471	49.9	4247	30.2	16,295	17.0	2187	40.3	1512	49.8	10,368	18.7	36,081	10.4
Trentino	0	–	2523	37.8	16,999	16.0	1441	50.0	360	100.0	12,503	16.7	33,826	10.6
Veneto	1835	44.7	748	70.6	8256	20.9	2 243	40.7	374	99.9	39,535	11.1	52,991	9.1
Friuli V.G	1815	44.6	372	99.7	12,061	22.2	1115	57.5	1115	57.5	24,581	14.8	41,058	10.6
Liguria	1351	50.1	1415	99.2	20,659	15.2	733	70.1	0	–	19,926	16.4	44,084	10.3
Emilia Romagna	2879	55.2	1774	82.0	20,434	16.5	3984	42.9	1105	57.8	23,738	11.9	53,915	9.3
Toscana	8667	24.8	5035	36.2	44,968	11.3	1807	44.7	1807	44.7	91,991	6.6	154,275	5.2
Umbria	3246	50.0	639	71.4	12,095	21.9	369	99.8	0	–	7303	27.6	23,651	15.2
Marche	371	100.0	711	70.9	8937	29.7	1753	83.1	371	100.0	9171	22.2	21,314	16.2
Lazio	5432	41.1	5860	32.8	44,106	11.7	1474	49.9	1080	57.7	29,960	13.9	87,912	7.7
Abruzzo	7452	34.9	2828	55.8	36,194	12.7	362	100.0	362	100.0	15,813	15.4	63,011	8.6
Molise	906	62.3	1953	44.4	12,740	22.6	631	72.5	0	–	3796	28.2	20,025	16.0
Campania	3622	46.3	1105	57.8	36,585	13.9	368	100.0	1841	44.7	43,811	10.8	87,332	7.6
Puglia	15,174	20.4	4895	37.5	19,840	17.7	3770	45.6	0	–	5711	24.5	49,389	9.9
Basilicata	6706	23.4	741	70.7	43,847	11.1	100	99.9	0	–	52,998	8.6	104,392	6.2
Calabria	4477	28.8	3358	33.2	29,943	13.0	2612	37.7	1492	49.9	113,561	5.7	155,443	4.8
Sicilia	4075	30.0	2975	35.3	50,818	10.8	6792	23.4	3233	63.1	33,851	13.3	101,745	7.1
Sardegna	85,769	7.9	26,050	12.8	432,522	2.7	29,481	11.1	4789	37.9	96,240	6.4	674,851	2.0
Italia	160,715	6.0	71,343	8.5	945,114	2.2	67,486	8.1	20,799	16.9	703,815	2.5	1,969,272	1.4

Table 11.10 Classes of terrain aspect / Classi di esposizione

Aspect class	Central azimuth of the class (°)
Classe di esposizione	Azimut centrale di classe (°)
North/Nord	0
North-East/Nord-Est	45
East/Est	90
South-East/Sud-Est	135
South/Sud	180
South-West/Sud-Ovest	225
West/Ovest	270
North-West/Nord-Ovest	315
No aspect/Nessuna esposizione	–

Table 11.11 Forest area by class of aspect / Estensione del Bosco ripartito per classi di esposizione

Region/Regione	Forest/Bosco											
	North		North-East		East		South-East		South		South-West	
	Nord		Nord-Est		Est		Sud-Est		Sud		Sud-Ovest	
	Area	ES	Area	ES	Area	ES	Area	ES	Area	ES	Area	ES
	(ha)	(%)	(ha)	(%)	(ha)	(%)	(ha)	(%)	(ha)	(%)	(ha)	(%)
Piemonte	124,763	5.8	99,993	6.3	107,112	6.5	84,955	7.2	91,379	7.1	87,106	6.9
Valle d'Aosta	11,174	17.7	11,174	17.7	13,101	16.2	7530	26.9	10,789	18.1	8027	25.7
Lombardia	75,242	8.0	53,867	9.0	62,860	8.7	62,342	8.3	78,004	7.3	72,465	7.8
Alto Adige	44,097	8.8	36,019	9.8	39,737	10.2	32,780	11.0	36,297	9.7	35,918	9.8
Trentino	43,135	8.7	38,919	9.2	36,529	9.6	46,383	8.7	37,118	9.5	36,653	10.1
Veneto	47,315	9.3	45,884	9.5	57,848	8.2	45,615	8.7	56,302	8.7	35,455	9.9
Friuli V.G	59,294	7.8	26,015	11.5	31,925	11.1	35,598	10.4	37,121	9.5	33,448	10.1
Liguria	44,331	8.5	39,102	9.2	38,467	9.2	36,634	9.5	44,613	9.3	40,201	9.0
Emilia Romagna	94,229	6.2	75,247	6.7	75,016	7.1	50,767	8.6	51,369	9.1	54,823	8.2
Toscana	129,950	5.4	116,443	5.6	113,081	5.6	102,209	6.0	123,744	5.3	95,939	6.1
Umbria	48,371	9.3	44,555	9.5	48,112	9.0	36,514	9.7	41,035	10.3	45,696	9.0
Marche	35,369	11.3	31,803	11.1	41,009	9.5	33,795	10.7	21,862	13.9	27,766	11.1
Lazio	76,684	7.3	64,778	7.7	50,898	8.2	40,886	10.6	52,251	8.8	55,821	7.8
Abruzzo	57,990	7.8	63,111	7.4	49,860	9.5	32,519	10.2	32,722	11.5	43,057	9.3
Molise	25,645	13.2	25,483	11.5	13,208	19.0	15,073	21.1	15,942	16.9	17,536	18.0
Campania	51,453	8.5	51,293	9.1	39,652	9.8	34,176	11.3	46,891	9.9	35,912	10.4
Puglia	22,883	13.5	16,861	17.8	12,390	17.2	11,110	21.4	10,448	18.8	17,753	15.9
Basilicata	34,307	11.7	32,264	11.7	27,149	13.0	28,953	15.0	23,092	12.4	19,736	13.4
Calabria	58,182	8.3	60,259	8.1	50,620	9.0	46,520	9.8	50,930	9.3	50,930	9.3
Sicilia	46,043	9.4	43,726	11.2	30,957	13.8	24,596	14.9	25,892	14.3	30,820	12.8
Sardegna	84,875	6.7	70,124	8.3	67,851	8.3	57,457	9.4	64,892	8.3	64,698	8.1
Italia	1,215,333	1.8	1,046,920	1.9	1,007,381	2.0	866,414	2.2	952,691	2.1	909,762	2.1

(continued)

Table 11.11 (continued)

Region / Regione	Forest / Bosco									
	West		North-West		No exposure		Not classified		Total	
	Ovest		Nord-Ovest		Nessuna esposizione		Non classificata		Totale	
	Area	ES	Area	ES	Area	ES	Area	ES	Area	ES
	(ha)	(%)	(ha)	(%)	(ha)	(%)	(ha)	(%)	(ha)	(%)
Piemonte	93 746	6.6	88 482	6.7	59 661	8.8	53 236	8.6	890 433	1.3
Valle d'Aosta	13 462	15.9	12 836	16.4	0	-	11 150	17.6	99 243	3.6
Lombardia	61 431	8.4	60 902	8.2	41 513	11.0	53 341	9.4	621 968	1.6
Alto Adige	42 953	8.9	45 988	8.6	1 890	44.5	23 590	12.5	339 270	1.7
Trentino	55 989	7.5	50 487	8.0	1 081	57.8	26 966	11.4	373 259	1.4
Veneto	51 118	7.5	34 024	10.2	8 091	30.1	35 053	11.0	416 704	1.9
Friuli V.G	24 529	11.9	25 236	12.8	21 184	12.9	38 207	10.4	332 556	1.9
Liguria	45 682	9.2	40 201	9.0	733	70.1	13 194	16.3	343 160	1.7
Emilia Romagna	64 430	7.3	71 809	6.8	18 909	17.5	28 301	12.0	584 901	1.5
Toscana	125 159	5.4	102 962	5.8	18 803	13.8	107 156	5.7	1 035 448	1.1
Umbria	61 026	7.5	48 553	8.6	369	99.8	16 076	18.1	390 305	1.6
Marche	32 983	11.4	36 419	10.2	741	70.8	30 020	10.7	291 767	2.1
Lazio	71 307	7.1	60 040	8.1	21 336	14.2	66 235	7.1	560 236	1.6
Abruzzo	51 282	9.0	42 627	9.7	4 343	28.8	34 077	10.0	411 588	1.8
Molise	12 953	19.1	18 816	16.3	2 343	40.5	6 248	24.5	153 248	3.0
Campania	42 510	9.4	39 651	10.3	1 364	50.5	61 024	7.7	403 927	2.1
Puglia	15 148	15.5	10 876	18.5	22 201	12.6	2 679	37.6	142 349	4.0
Basilicata	31 896	11.8	31 939	11.1	5 150	35.8	53 533	7.8	288 020	2.7
Calabria	50 620	9.0	49 999	8.3	5 161	35.8	71 955	8.1	495 177	2.0
Sicilia	30 372	12.3	25 912	12.8	3 644	46.4	23 528	16.1	285 489	3.2
Sardegna	71 070	7.4	74 035	7.5	29 041	12.3	42 097	9.7	626 140	2.1
Italia	1 049 668	1.9	971 792	2.0	267 558	4.2	797 667	2.2	9 085 186	0.4

Other estimates on hydrogeological constraint in wooded areas, for the inventory categories and for Tall trees forest types are available at https://www.inventarioforestale.org/statistiche_INFC.

11.5 Primary Designated Management Objective

During the field surveys, INFC2015 collected data on the presence of a primary designated management objective as a component of attributes related to sylviculture and management. Such data was obtained by direct assessment, planning documents and interviews with local forest technicians. Table 11.17 shows the eight classes adopted for primary management objectives and their definitions. Estimates about the presence of a primary function in Forest are shown in Table 11.18, while Fig. 11.6 shows the percentage of Forest area with a primary function, at regional and national levels.

Table 11.12 Other wooded land area by class of aspect / Estensione delle Altre terre boscate ripartite per classi di esposizione

Region/Regione	Other wooded land/Altre terre boscate											
	North		North-East		East		South-East		South		South-West	
	Nord		Nord-Est		Est		Sud-Est		Sud		Sud-Ovest	
	Area	ES	Area	ES	Area	ES	Area	ES	Area	ES	Area	ES
	(ha)	(%)	(ha)	(%)	(ha)	(%)	(ha)	(%)	(ha)	(%)	(ha)	(%)
Piemonte	8641	29.2	3683	33.3	2067	44.8	6921	34.3	4946	37.4	4895	28.9
Valle d'Aosta	1156	57.3	385	99.6	0	–	1365	99.6	0	–	0	–
Lombardia	3086	37.8	6788	34.9	3967	33.3	1763	50.0	6149	32.7	8164	26.8
Alto Adige	6138	24.9	2615	61.8	2328	39.2	2694	37.7	1512	49.8	2606	37.7
Trentino	853	71.7	3499	45.8	4685	27.7	1681	44.1	1802	44.7	1213	58.5
Veneto	1496	49.9	1122	57.7	1122	57.7	1869	44.6	4008	29.8	1122	57.7
Friuli V.G	5866	32.8	743	70.5	372	99.7	1115	57.5	2478	62.9	1858	44.5
Liguria	1083	57.3	2199	40.4	1833	44.3	5848	24.7	3184	33.1	6129	36.3
Emilia Romagna	4653	45.6	4357	40.2	3984	42.9	5094	35.9	1473	50.0	5526	25.8
Toscana	4312	40.6	6116	31.5	7564	27.2	10,428	24.8	11,512	23.1	6504	23.5
Umbria	1843	44.6	2410	63.0	2114	40.9	1106	57.6	2877	55.0	1475	49.9
Marche	2094	71.3	4278	48.6	0	–	741	70.8	1052	57.3	1453	50.1
Lazio	6056	37.6	5860	32.8	9142	27.6	3316	33.3	10,187	28.0	15,679	21.6
Abruzzo	7533	27.1	6354	35.5	2401	37.4	6433	35.3	7452	34.9	7118	28.1
Molise	2974	35.4	1953	44.4	2574	60.3	3055	52.0	0	–	1297	52.8
Campania	3990	43.0	1841	44.7	5710	44.0	4051	30.1	7916	33.7	9761	28.6
Puglia	4935	37.4	5697	39.2	3302	50.2	3770	45.6	8187	34.7	3884	31.3
Basilicata	5693	25.3	3356	33.2	9172	27.3	6269	31.1	7437	22.1	6328	24.1
Calabria	1866	44.6	5597	25.7	4477	28.8	7089	22.8	5224	26.6	7027	28.8
Sicilia	8978	24.5	6677	29.9	11453	25.7	7050	28.8	6671	29.9	8219	26.0
Sardegna	69,599	8.5	60,922	8.6	65,379	8.3	71,671	8.0	80,171	7.6	65,411	7.9
Italia	152,845	6.0	136,453	6.5	143,644	6.2	153,330	5.8	174,239	5.7	165,668	5.5

(continued)

Table 11.12 (continued)

Region / Regione	Other wooded land / Altre terre boscate									
	West		North-West		No exposure		Not classified		Total	
	Ovest		Nord-Ovest		Nessuna esposizione		Non classificata		Totale	
	Area	ES	Area	ES	Area	ES	Area	ES	Area	ES
	(ha)	(%)	(ha)	(%)	(ha)	(%)	(ha)	(%)	(ha)	(%)
Piemonte	13 084	26.2	6 604	24.9	855	70.9	33 295	12.9	84 991	8.0
Valle d'Aosta	0	-	1 156	57.3	0	-	4 671	27.2	8 733	24.0
Lombardia	4 937	38.1	4 479	31.6	331	100.0	30 588	13.8	70 252	8.7
Alto Adige	2 905	37.8	4 537	28.7	378	99.8	10 368	18.7	36 081	10.4
Trentino	2 523	37.8	5 068	27.3	0	-	12 503	16.7	33 826	10.6
Veneto	1 596	47.2	1 122	57.7	0	-	39 535	11.1	52 991	9.1
Friuli V.G	2 230	40.6	1 487	49.8	700	70.7	24 209	14.9	41 058	10.6
Liguria	2 199	40.4	2 148	69.6	252	99.2	19 210	16.9	44 084	10.3
Emilia Romagna	4 353	40.2	1 105	57.8	0	-	23 370	12.0	53 915	9.3
Toscana	5 758	32.9	7 902	30.4	3 271	33.3	90 907	6.6	154 275	5.2
Umbria	2 384	37.8	2 140	69.8	0	-	7 303	27.6	23 651	15.2
Marche	1 783	82.1	371	100.0	371	100.0	9 171	22.2	21 314	16.2
Lazio	4 053	30.1	1 817	44.7	1 842	44.7	29 960	13.9	87 912	7.7
Abruzzo	7 761	33.7	1 421	50.1	724	70.8	15 813	15.4	63 011	8.6
Molise	1 021	58.8	2 965	54.0	391	99.8	3 796	28.2	20 025	16.0
Campania	7 547	35.0	2 705	36.3	0	-	43 811	10.8	87 332	7.6
Puglia	4 273	29.9	2 331	40.6	7 689	22.0	5 323	25.3	49 389	9.9
Basilicata	6 572	35.1	3 962	52.6	3 723	31.5	51 879	8.7	104 392	6.2
Calabria	5 442	34.4	4 042	42.9	373	99.9	114 307	5.7	155 443	4.8
Sicilia	8 877	20.9	10 347	27.7	379	100.0	33 093	13.6	101 745	7.1
Sardegna	73 391	7.2	68 268	7.7	26 786	12.6	93 255	6.5	674 851	2.0
Italia	162 688	5.7	135 976	6.1	48 063	9.1	696 366	2.5	1 969 272	1.4

Table 11.13 Classes of terrain instability / Classi relative ai tipi di dissesto

Type of instability	Description
Tipo di dissesto	Descrizione
Shallow and deep-seated landslides Frane, smottamenti	Detachment and slip of terrain or rock along a slope, due to many factors, among which excessive erosion rate of the hillsides; includes both surface slips and massive, deeper slides. We can recognise: slides, with material that slides along an inclined plane, usually consisting of low-consistency material which behaves as a lubricant for the overlying material (this is frequent on layered slopes with alternating clay layers and layers of other nature); debris slides, with movement of loose or incoherent material along curved surfaces; shallow slides, i.e., surface movement of slope debris covers (soil flow, in the case of very slow movement); earth flow, similar to the above, but of small size debris such as clays weighed down and fluidised by water
	Fenomeni di distacco o scivolamento di terreno o di roccia lungo un pendio, dovuti a molti fattori tra i quali l'eccessiva velocità di erosione dei versanti; sono inclusi sia i movimenti franosi superficiali sia quelli di massa, più profondi. Si possono riconoscere: frane di scivolamento, con il materiale in frana che scivola lungo un piano inclinato, di solito costituito da materiale a bassa consistenza, che si comporta da lubrificante per il materiale soprastante (frequenti sui versanti a franapoggio con strati argillosi intercalati a strati di diversa natura); frane di scoscendimento, con movimento di materiale sciolto o poco cementato lungo superfici curve; smottamenti, in presenza di movimenti superficiali delle coperture detritiche dei versanti (soliflussi, se si tratta di movimenti molto lenti); colamenti, simili ai precedenti, ma che riguardano detriti di piccole dimensioni, come quelli delle argille, appesantiti e resi fluidi dall'acqua
Water erosion, flooding Erosione idrica, fenomeni alluvionali	Erosion phenomena: removal of mostly surface layers of terrain or total decapitation of the profile, with formation if incisions, channels, valleys due to water streams or periodic or occasional wash-away by rain and river water (bank erosion); the opposite case is represented by the accumulation of sediment transported by water, during intense rain/flooding/overflow caused by river spates

(continued)

Table 11.13 (continued)

Type of instability Tipo di dissesto	Description Descrizione
	Fenomeni di erosione: asportazione di strati, per lo più superficiali, del terreno o totale decapitazione del profilo, con formazione di incisioni, canali, valloncelli, per effetto del ruscellamento o del dilavamento periodico od occasionale delle acque meteoriche o fluviali (erosioni di ripa). Il caso opposto è rappresentato da fenomeni di accumulo di sedimenti trasportati dalle acque, in concomitanza di precipitazioni intense/inondazioni/esondazioni fluviali conseguenti a piene
Rock fall and rolling Caduta o rotolamento pietre	Detachment, collapse, sliding and rolling or arrest of stones, boulders and rock debris due to gravity, usually on very steep slopes; frequent phenomena in sites at the foot of unstable rocky cliffs; they can be due to natural causes, or triggered by earth movement when opening roads in uphill areas
	Fenomeni di distacco, crollo, scorrimento/rotolamento o arresto di pietre, massi o detriti lapidei per effetto della gravità, di solito su pendii molto inclinati; fenomeni frequenti in stazioni poste al piede di pareti rocciose instabili; frane di crollo e rotolamento di sassi, oltre che per cause naturali, possono essere innescate anche da movimenti di terra per l'apertura strade in zone a monte
Snowslides, avalanches Slavine, valanghe	Detachment and slide, or accumulation, of snow masses on slopes usually steeper than 30° at the detachment area (while the accumulation area can be flat)
	Fenomeni di distacco, scorrimento/scivolamento o accumulo di masse di neve su pendii di solito inclinati oltre 30° nella zona di distacco (mentre la zona di accumulo può anche essere piana)

In Italy, 86.8% of the Forest area has no primary designated management objective. The rate of Forest area with a primary objective exceeds 20% only in four regions: Valle d'Aosta, Lombardia, Umbria and Sardegna. Among the considered management objectives, wood and timber production has the highest rate nationwide, 6.7% of the total area, followed by nature conservation with 2.1%. The protective function, both direct and indirect, occurs only on 2.1% of Forest area at the national level, but the Forest area percentage is rather high in Valle d'Aosta (23.7%) and in Alto Adige (11.5%). Forest area primarily managed for recreation is negligible nationwide, while non-wood products (1.3% Forest area at the national level) mark a notable percent of Forest area only in Sardegna due to cork production (10.1%). Further estimates on the presence of a primary designated management objective,

Table 11.14 Forest area by class of terrain instability / Estensione del Bosco ripartito per tipo di dissesto

Region/Regione	Forest/Bosco													
	Absence		Shallow and deep-seated landslides		Water erosion, flooding		Rock fall and rolling		Snowslides, avalanches		Not classified		Total	
	Assenza di dissesto		Frane, smottamenti		Erosione idrica, fenomeni alluvionali		Caduta o rotolamento pietre		Slavine, valanghe		Non classificata		Totale	
	Area	ES	Area	ES	Area	ES	Area	ES	Area	ES	Area	ES	Area	ES
	(ha)	(%)	(ha)	(%)	(ha)	(%)	(ha)	(%)	(ha)	(%)	(ha)	(%)	(ha)	(%)
Piemonte	650,203	1.9	42,059	10.4	63,667	8.6	73,282	7.4	8794	25.5	52,428	8.6	890,433	1.3
Valle d'Aosta	61,947	6.1	2312	40.4	3083	34.8	15,413	14.8	5339	34.0	11,150	17.6	99,243	3.6
Lombardia	460,326	2.3	18,161	16.7	14,955	17.0	70,688	7.7	4937	38.1	52,900	9.4	621,968	1.6
Alto Adige	236,555	2.7	7313	22.8	8040	21.3	44,883	8.8	18,889	16.5	23,590	12.5	339,270	1.7
Trentino	294,012	2.2	5045	26.6	8905	23.8	29,190	10.8	14,186	15.8	21,921	12.7	373,259	1.4
Veneto	302,700	2.7	19,016	15.3	29,881	10.9	28,185	11.3	2991	35.3	33,931	11.3	416,704	1.9
Friuli V.G	241,947	2.8	11,521	17.7	21,184	12.9	18,954	13.7	743	70.5	38,207	10.4	332,556	1.9
Liguria	277,017	2.3	22,710	12.2	7949	21.0	22,290	13.6	0	–	13,194	16.3	343,160	1.7
Emilia Romagna	420,285	2.2	84,134	6.7	38,920	10.5	13,998	16.1	0	–	27,564	12.2	584,901	1.5
Toscana	831,550	1.4	38,277	10.1	44,373	10.0	18,067	14.0	0	–	103,181	5.8	1,035,448	1.1
Umbria	313,572	2.3	4055	30.0	33,844	10.7	23,864	12.1	0	–	14,970	19.0	390,305	1.6
Marche	219,967	3.0	7783	21.6	14,755	17.7	18,872	13.7	371	100.0	30,020	10.7	291,767	2.1
Lazio	420,135	2.3	4790	27.6	18,054	14.1	50,812	8.6	210	100.0	66,235	7.1	560,236	1.6
Abruzzo	329,608	2.4	10,207	22.1	23,151	13.4	12,735	16.6	1810	44.7	34,077	10.0	411,588	1.8
Molise	131,988	3.8	5798	32.7	2734	37.5	6479	30.4	0	–	6248	24.5	153,248	3.0
Campania	296,535	2.9	7365	22.3	12,241	17.1	26,762	13.2	0	–	61,024	7.7	403,927	2.1
Puglia	121,842	4.7	1165	57.5	2719	37.5	15,109	15.5	0	–	1,514	49.9	142,349	4.0
Basilicata	212,642	3.7	1864	44.6	11,101	21.2	9625	23.3	0	–	52,788	7.9	288,020	2.7
Calabria	350,652	2.7	16,976	19.2	28,853	13.9	31,217	12.1	0	–	67,478	8.5	495,177	2.0
Sicilia	220,615	4.0	9389	23.8	12,043	20.3	20,294	15.9	0	–	23,148	16.3	285,489	3.2
Sardegna	534,482	2.4	2612	37.8	14,543	15.9	33,151	11.1	0	–	41,350	9.8	626,140	2.1
Italia	6,928,582	0.6	322,554	3.6	414,992	3.2	583,869	2.6	58,270	9.2	776,920	2.3	9,085,186	0.4

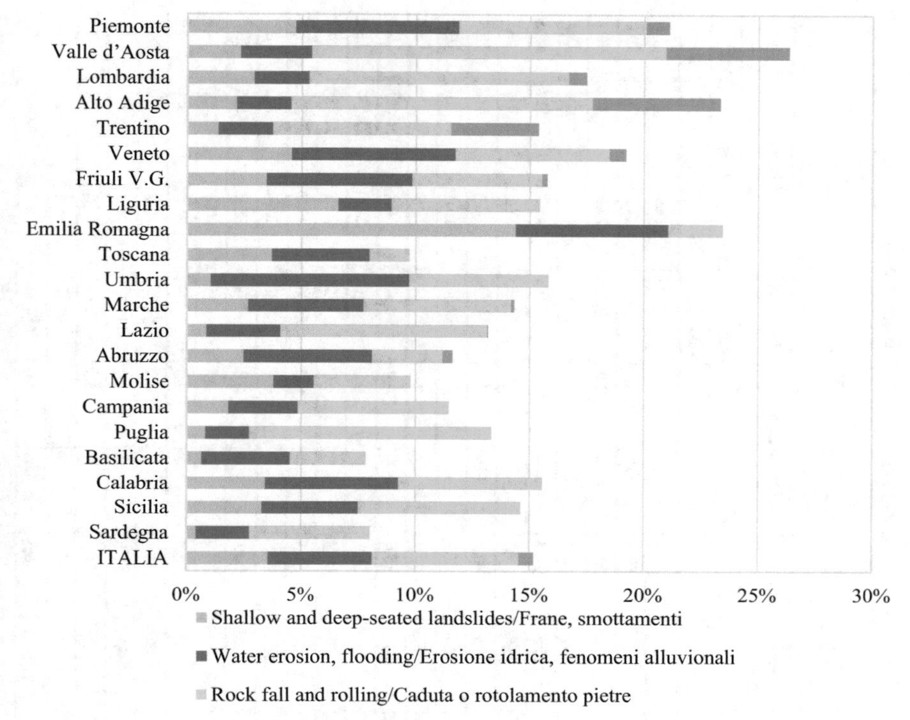

Fig. 11.4 Percent of Forest area by terrain instability class / Ripartizione percentuale dell'area del Bosco per classe di dissesto idrogeologico

for the inventory categories of Forest, are available at https://www.inventarioforestale.org/statistiche_INFC.

The percentage of Forest area where one management objective is predominant, is very low. This information, coupled with the wide presence of hydrogeological constraints, as discussed in Sect. 11.4, shows that a determinant role in protecting soil and water resources is recognized in forests regardless of a primary designated objective, which confirms how forest management in Italy is oriented towards ensuring the multifunctional role of forests.

Table 11.15 Other wooded land area by class of terrain instability / Estensione delle Altre terre boscate ripartite per tipo di dissesto

Region/Regione	Other wooded land/Altre terre boscate													
	Absence		Shallow and deep-seated landslides		Water erosion, flooding		Rock fall and rolling		Snowslides, avalanches		Not classified		Total	
	Assenza di dissesto		Frane, smottamenti		Erosione idrica, fenomeni alluvionali		Caduta o rotolamento pietre		Slavine, valanghe		Non classificata		Totale	
	Area	ES	Area	ES	Area	ES	Area	ES	Area	ES	Area	ES	Area	ES
	(ha)	(%)	(ha)	(%)	(ha)	(%)	(ha)	(%)	(ha)	(%)	(ha)	(%)	(ha)	(%)
Piemonte	34,779	13.3	3080	53.4	1212	57.8	9187	27.8	3437	48.9	33,295	12.9	84,991	8.0
Valle d'Aosta	771	70.3	0	0.0	0	0.0	1750	80.7	1541	49.6	4671	27.2	8733	24.0
Lombardia	28,710	13.4	2733	58.7	441	100.1	3196	35.1	4584	47.0	30,588	13.8	70,252	8.7
Alto Adige	12,313	17.0	3501	49.5	1393	57.5	2398	41.0	6109	24.4	10,368	18.7	36,081	10.4
Trentino	7807	21.1	360	100.2	360	100.2	5205	34.9	7591	22.0	12,503	16.7	33,826	10.6
Veneto	8291	21.0	748	70.6	1496	49.9	1461	50.0	1461	50.0	39,535	11.1	52,991	9.1
Friuli V.G	13,876	20.0	372	99.7	743	70.5	743	70.5	743	70.5	24,581	14.8	41,058	10.6
Liguria	19,508	16.7	1100	57.2	733	70.1	3184	33.1	0	0.0	19,560	16.6	44,084	10.3
Emilia Romagna	14,840	21.8	8405	25.3	4420	28.8	2879	55.2	0	0.0	23,370	12.0	53,915	9.3
Toscana	55,803	9.9	2891	35.3	2168	40.8	2867	55.6	0	0.0	90,545	6.7	154,275	5.2
Umbria	15,242	19.9	0	0.0	0	0.0	1106	57.6	0	0.0	7303	27.6	23,651	15.2
Marche	8596	30.7	1753	83.1	1423	49.8	371	100.2	0	0.0	9171	22.2	21,314	16.2
Lazio	48,108	11.0	5054	49.5	1105	57.7	3685	31.5	0	0.0	29,960	13.9	87,912	7.7
Abruzzo	41,474	11.8	724	70.8	2493	61.9	2507	37.8	0	0.0	15,813	15.4	63,011	8.6
Molise	13,255	21.9	1412	50.6	781	70.5	781	70.5	0	0.0	3796	28.2	20,025	16.0
Campania	34,193	13.9	3193	63.1	0	0.0	6135	36.9	0	0.0	43,811	10.8	87,332	7.6
Puglia	36,064	12.4	1165	57.5	1165	57.5	5672	33.8	0	0.0	5323	25.3	49,389	9.9
Basilicata	44,763	10.7	746	70.6	6259	31.1	746	70.6	0	0.0	51,879	8.7	104,392	6.2
Calabria	29,943	13.0	3358	33.2	4104	30.0	4851	27.6	0	0.0	113,188	5.7	155,443	4.8
Sicilia	55,762	10.4	1896	44.7	4929	27.6	6066	24.9	0	0.0	33,093	13.6	101,745	7.1
Sardegna	510,340	2.5	1866	44.7	21,271	13.2	49,612	8.5	0	0.0	91,762	6.6	674,851	2.0
Italia	1,034,439	2.1	44,355	12.6	56,497	8.7	114,401	6.5	25,466	14.6	694,114	2.5	1,969,272	1.4

Table 11.16 Forest and Other wooded land area by presence of hydrogeological constraints / Estensione del Bosco e delle Altre terre boscate ripartiti per presenza del vincolo idrogeologico

Region/Regione	Forest/Bosco								Other wooded land/Altre terre boscate							
	With constraint		Without constraint		Not classified		Total		With constraint		Without constraint		Not classified		Total	
	Con vincolo		Senza vincolo		Non classificato		Totale		Con vincolo		Senza vincolo		Non classificato		Totale	
	Area	ES	Area	ES	Area	ES	Area	ES	Area	ES	Area	ES	Area	ES	Area	ES
	(ha)	(%)	(ha)	(%)	(ha)	(%)	(ha)	(%)	(ha)	(%)	(ha)	(%)	(ha)	(%)	(ha)	(%)
Piemonte	750,576	1.6	139,453	5.7	404	100.0	890,433	1.3	51,286	10.3	8792	33.9	24,913	15.6	84,991	8.0
Valle d'Aosta	84,986	4.5	14,257	15.5	0	–	99,243	3.6	5054	35.2	0	–	3680	31.2	8733	24.0
Lombardia	510,265	1.9	108,016	6.7	3688	57.8	621,968	1.6	46,898	10.7	2535	41.0	20,819	17.8	70,252	8.7
Alto Adige	333,877	1.8	4637	28.1	756	70.5	339,270	1.7	32,099	11.2	0	–	3981	30.1	36,081	10.4
Trentino	370,016	1.5	3243	33.3	0	–	373,259	1.4	29,864	11.4	0	–	3963	29.2	33,826	10.6
Veneto	384,168	2.0	29,079	13.7	3457	58.0	416,704	1.9	26,045	11.7	374	99.9	26,571	14.5	52,991	9.1
Friuli V.G	281,883	2.2	47,773	8.9	2901	68.1	332,556	1.9	23,582	14.1	1443	49.9	16,033	20.0	41,058	10.6
Liguria	320,102	1.9	22,691	12.3	367	99.2	343,160	1.7	22,508	14.4	3117	51.2	18,460	17.3	44,084	10.3
Emilia Romagna	518,035	1.7	65,460	8.5	1406	100.0	584,901	1.5	29,948	13.3	4285	48.8	19,682	13.1	53,915	9.3
Toscana	989,940	1.1	43,725	9.4	1783	82.2	1,035,448	1.1	74,624	8.1	5035	36.2	74,615	7.5	154,275	5.2
Umbria	376,513	1.7	10,987	21.2	2805	69.1	390,305	1.6	17,823	17.8	0	–	5828	32.2	23,651	15.2
Marche	259,600	2.4	32,167	12.0	0	–	291,767	2.1	10,389	26.4	2124	70.8	8800	22.7	21,314	16.2
Lazio	515,879	1.8	43,338	10.6	1019	59.3	560,236	1.6	61,749	9.6	5757	25.0	20,405	18.3	87,912	7.7
Abruzzo	353,014	2.1	58,573	9.2	0	–	411,588	1.8	39,129	12.2	10,992	25.7	12,890	17.2	63,011	8.6
Molise	144,035	3.4	9213	24.0	0	–	153,248	3.0	15,367	18.5	2034	72.4	2624	31.8	20,025	16.0
Campania	340,671	2.5	61,475	8.1	1781	82.1	403,927	2.1	53,893	10.4	5459	34.2	27,980	14.7	87,332	7.6
Puglia	118,435	4.6	22,788	14.6	1125	57.7	142,349	4.0	30,995	13.5	14,725	20.9	3669	30.1	49,389	9.9
Basilicata	255,929	3.0	32,091	13.5	0	–	288,020	2.7	71,110	7.8	10,833	21.5	22,449	14.7	104,392	6.2
Calabria	398,280	2.5	84,368	6.6	12,529	29.9	495,177	2.0	51,864	9.1	14,925	15.6	88,654	6.7	155,443	4.8
Sicilia	242,562	3.6	37,218	12.6	5709	49.1	285,489	3.2	59,863	8.6	26,576	16.5	15,306	24.3	101,745	7.1
Sardegna	323,095	3.2	301,614	3.6	1431	100.0	626,140	2.1	230,015	3.9	414,029	2.9	30,808	12.8	674,851	2.0
Italia	7,871,861	0.5	1,172,167	1.9	41,158	16.8	9,085,186	0.4	984,106	2.2	533,034	2.8	452,132	3.3	1,969,272	1.4

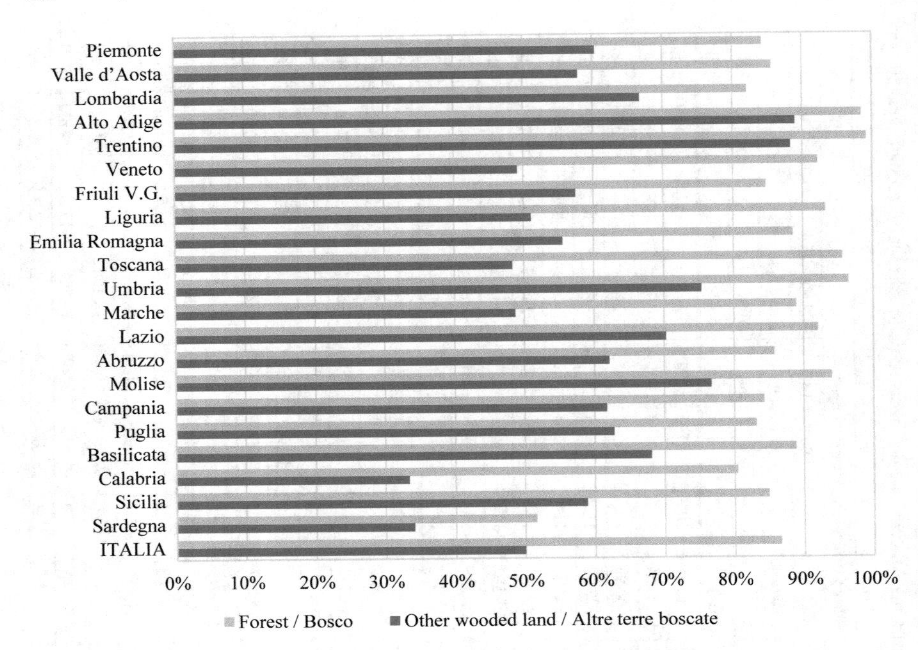

Fig. 11.5 Percent of Forest and Other wooded land area by presence of hydrogeological constraints / Ripartizione percentuale dell'area del Bosco e delle Altre terre boscate con presenza di vincolo idrogeologico

Table 11.17 Classes of primary designated management objective / Classi relative alla funzione prioritaria

Primary objective	Description
Funzione prioritaria	Descrizione
Timber/wood production Produzione legnosa	Management of forest is oriented mainly to wood or timber production. Plantations for timber/wood production are included in this class
	Comprende tutte le superfici destinate principalmente alla produzione e all'estrazione di prodotti legnosi, compresi tutti i popolamenti di arboricoltura da legno
Non-wood products Produzione non legnosa	Management of forest is oriented mainly to obtain non-woody products, such as cork, chestnuts, others. All stands classified into the "special cultural practices for secondary products" are included in this class

(continued)

Table 11.17 (continued)

Primary objective	Description
Funzione prioritaria	Descrizione
	Le superfici destinate principalmente alla produzione di prodotti non legnosi, come sughero, castagne, ecc. Vi rientrano i soprassuoli classificati nella sottoclasse "pratiche colturali speciali per produzioni secondarie"
Nature conservation Naturalistica	Forests specifically managed for biodiversity or naturalistic values conservation, and thus not used for timber or wood production. It includes integral nature reserves
	Le superfici destinate principalmente alla conservazione della biodiversità o di valori naturalistici e perciò sottratte a utilizzazioni di tipo produttivistico, gestite per esaltare la funzione naturalistica (es. riserve integrali quando è escluso qualsiasi tipo di intervento)
Recreation Ricreativa	Forests specifically managed for recreational and tourism services and activities, e.g., picnic areas or areas with equipped trails
	Le superfici destinate principalmente alla massimizzazione della funzione turistico-ricreativa, ad esempio aree pic-nic o fruibili mediante percorsi opportunamente indicati e attrezzati per offrire un servizio ai visitatori
Protection, indirect Protettiva indiretta	Forest stands on steep and loose soils where the protective function for soil and water is clearly predominant compared to other functions, based on visible oriented management interventions
	Vi rientrano i soprassuoli su terreni sciolti, acclivi dove è riconosciuta la preminente funzione di protezione del suolo dall'erosione, di regimazione delle acque e di tutela della qualità delle acque. Può comprendere i casi di popolamenti artificiali edificati su suoli particolarmente acclivi. Essendo comunemente noto che tutti i boschi esercitano un ruolo più o meno importante di protezione indiretta, l'attribuzione ha luogo solo qualora tale ruolo sia prioritario rispetto alle altre funzioni e reso evidente dagli interventi praticati localmente
Protection, direct Protettiva diretta	Forest stands protecting residential areas, settlements and human infrastructures in general (railways, bridges, roads etc.) against natural hazards (landslides, snowslides, etc.)

(continued)

Table 11.17 (continued)

Primary objective Funzione prioritaria	Description Descrizione
	Funzione riferita a soprassuoli il cui ruolo è specificamente riconosciuto per la protezione di abitati e altre infrastrutture artificiali (linee ferroviarie, strade, ecc) contro calamità naturali (valanghe, frane, ecc.) o laddove gli interventi praticati rendano evidente tale funzione prioritaria
Absence of a primary management objective Funzione prioritaria assente	Forests with several functions exerted at a similar level, and no one is clearly predominant above others
	Tutti i soprassuoli in cui diverse funzioni sono svolte contemporaneamente e nessuna sia chiaramente predominante sulle altre in misura tale da condizionare la selvicoltura e la gestione

Appendix (Italian Version)

Riassunto Nel contesto dei molteplici servizi erogati dalle foreste, la protezione dai rischi naturali è uno dei più importanti. Le foreste assicurano la conservazione delle risorse naturali, tra cui suolo e acqua, oltre ad altri servizi ambientali. Esse rallentano il deflusso superficiale dell'acqua, favoriscono l'infiltrazione e la percolazione delle piogge e la conseguente ricostituzione delle riserve di acqua nel sottosuolo. La copertura forestale, inoltre, protegge il suolo dall'erosione, da valanghe e smottamenti. INFC raccoglie molte informazioni sulla funzione protettiva delle aree boscate italiane. Questo capitolo illustra le statistiche inventariali su alcune caratteristiche fisico–stazionali, quali pendenza, giacitura ed esposizione che, insieme al grado di copertura delle chiome e all'accidentalità del terreno, possono condizionare il ruolo protettivo della foresta. Vengono inoltre presentati i risultati di INFC2015 relativi al dissesto idrogeologico e al vincolo idrogeologico come definito dalla normativa nazionale, nel cui rispetto sono state emanate altre norme nazionali e regionali inerenti alla gestione delle foreste. Nell'ultima sezione di questo capitolo, infine, sono presentate le stime di superficie relative alla presenza di funzioni prioritarie che condizionano la selvicoltura e la gestione forestale, con particolare riguardo alla funzione protettiva diretta e indiretta.

Introduzione

Una delle principali ragioni per le quali dobbiamo proteggere le foreste è che esse ci proteggono. Le foreste assolvono a molteplici funzioni protettive, dalla scala locale fino a quella globale (FAO, 2005). Tali funzioni sono state tradizionalmente ascritte

Table 11.18 Forest area by primary designated management objective / Estensione del Bosco ripartito per presenza e tipo di funzione prioritaria

Region/Regione	Forest/Bosco																	
	Timber/wood production		Non-wood products		Nature conservation		Recreation		Protective, indirect		Protective, direct		No primary management objective		Not classified		Total	
	Produzione legnosa		Produzione non legnosa		Naturalistica		Ricreativa		Protettiva indiretta		Protettiva diretta		Funzione prioritaria assente		Non classificata		Totale	
	Area	ES	Area	ES	Area	ES	Area	ES	Area	ES	Area	ES	Area	ES	Area	ES	Area	ES
	(ha)	(%)	(ha)	(%)	(ha)	(%)	(ha)	(%)	(ha)	(%)	(ha)	(%)	(ha)	(%)	(ha)	(%)	(ha)	(%)
Piemonte	63,781	14.0	15,545	31.4	7975	41.5	3489	63.5	4825	53.2	3885	57.9	789,159	1.9	1774	50.8	890,433	1.3
Valle d'Aosta	0	–	0	–	7395	39.9	0	–	2391	71.2	21,178	19.7	68,159	8.2	120	99.6	99,243	3.6
Lombardia	114,127	10.5	0	–	40,744	19.2	0	–	13,834	35.4	14,665	34.1	434,260	3.5	4338	41.7	621,968	1.6
Alto Adige	4308	55.9	4251	57.3	1802	81.6	3205	63.3	27,997	20.3	10,926	35.6	283,620	3.4	3161	45.7	339,270	1.7
Trentino	0	–	1372	100.0	360	100.0	3954	58.1	1372	100.0	4473	59.2	361,236	2.2	492	100.0	373,259	1.4
Veneto	5198	44.6	0	–	855	69.9	3022	70.9	0	–	0	–	401,171	2.1	6459	27.1	416,704	1.9
Friuli V.G	31,457	17.5	0	–	1652	99.7	2644	62.1	0	–	720	99.7	293,183	2.9	2901	68.1	332,556	1.9
Liguria	733	70.1	0	–	0	–	0	–	0	–	1808	80.6	339,493	2.1	1126	57.3	343,160	1.7
Emilia Romagna	44,854	17.1	8264	40.4	6779	34.1	5739	51.1	2327	66.9	5757	49.0	509,350	2.4	1831	43.8	584,901	1.5
Toscana	117,985	11.0	10,622	39.0	22,975	25.4	4620	52.8	1129	100.0	2175	84.7	873,702	2.1	2240	38.1	1,035,448	1.1
Umbria	90,286	10.8	1856	79.7	3066	69.8	1445	71.5	3715	48.4	1921	78.0	285,211	4.0	2805	69.1	390,305	1.6
Marche	18,164	26.1	741	70.8	1403	100.0	577	100.0	6699	34.2	2174	72.5	262,010	3.1	0	–	291,767	2.1
Lazio	52,380	16.7	5015	57.5	14,608	32.7	3473	64.2	3361	62.4	1275	74.1	478,026	2.6	2099	41.4	560,236	1.6
Abruzzo	3951	49.9	0	–	4063	57.9	2529	70.8	4683	50.8	3648	49.3	391,379	2.3	1336	57.4	411,588	1.8
Molise	1363	47.0	0	–	1931	77.4	0	–	0	–	0	–	149,954	3.5	0	–	153,248	3.0
Campania	2922	53.2	6985	43.0	3332	53.7	1982	81.4	1873	79.4	0	–	385,729	2.4	1105	57.8	403,927	2.1
Puglia	4524	45.8	505	99.8	11,666	25.8	2955	62.4	505	99.8	0	–	120,868	5.2	1326	50.1	142,349	4.0
Basilicata	9607	34.7	769	70.7	1771	81.6	0	–	1566	99.9	0	–	271,013	3.2	3294	33.1	288,020	2.7
Calabria	3918	55.6	0	–	1418	99.9	653	99.9	1306	99.9	6387	39.5	472,068	2.3	9427	35.7	495,177	2.0
Sicilia	5949	44.1	2881	55.1	25,922	20.0	1381	100.0	0	–	1381	100.0	244,742	4.1	3233	63.1	285,489	3.2
Sardegna	32,564	15.0	63,370	12.2	27,865	22.8	2119	70.0	14,124	30.2	14,350	29.7	471,116	3.3	633	71.9	626,140	2.1
Italia	608,070	4.5	122,177	9.8	187,581	8.4	43,786	17.1	91,706	11.7	96,721	11.3	7,885,446	0.6	49,699	12.4	9,085,186	0.4

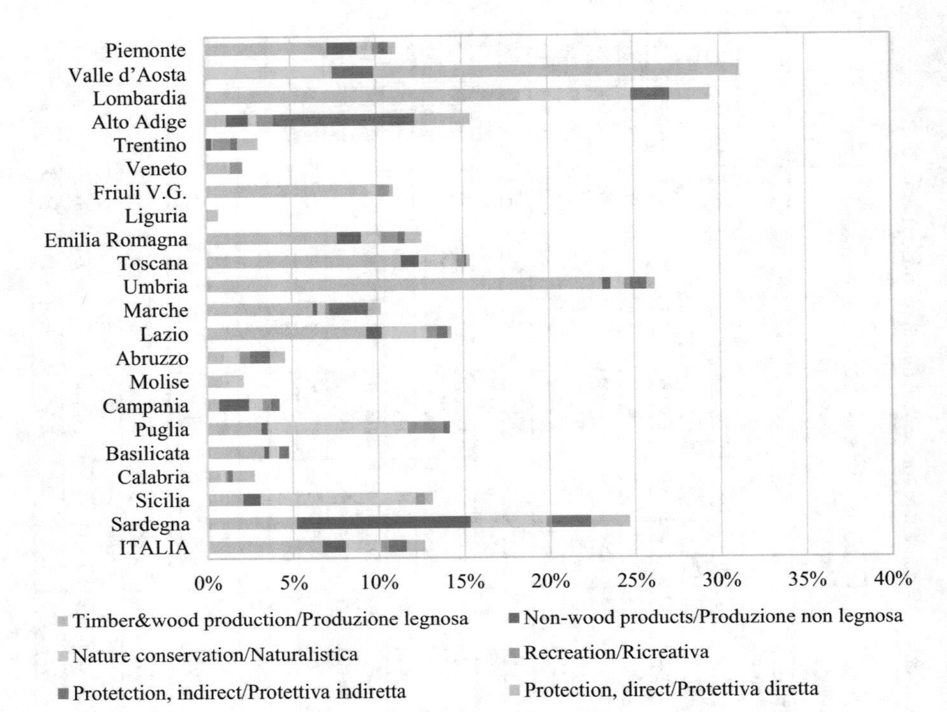

Fig. 11.6 Percent of Forest area by primary designated management objective / Ripartizione percentuale dell'area del Bosco per funzione prioritaria

alle due grandi categorie della protezione diretta e di quella indiretta. Si parla di funzione protettiva diretta in tutti quei casi in cui un bosco pone a riparo da specifici fattori di rischio una porzione di territorio caratterizzata da un certo grado di antropizzazione (centri abitati, edifici, strade, ecc.). Si tratta di un criterio di classificazione convenzionale, molto utile ai fini pratici e di generale accettazione, mediante il quale la funzione svolta dal bosco è classificata in maniera indiretta, attraverso l'importanza del territorio salvaguardato e non in sé stessa. I boschi di protezione di solito proteggono da più tipi di rischio (Dorren et al., 2004) ma la funzione protettiva diretta ha un rilievo locale, ancorato alla dimensione di un soprassuolo specifico (Schönenberghher, 2001).

INFC contribuisce alla valutazione dei vari servizi forniti dal bosco in Italia. Il Chap. 9 riporta le statistiche sulla superficie forestale in aree protette, ai vari gradi di protezione previsti dalla normativa nazionale, che ha una rilevanza prevalente per la conservazione della biodiversità, poiché le foreste costituiscono l'habitat di una parte considerevole della flora e della fauna (Führer, 2000). Per quanto riguarda lo stoccaggio del carbonio organico, il Chap. 12 riporta il contributo della componente legnosa epigea e la quantità di carbonio annualmente immagazzinata dagli alberi con la crescita.

In questo capitolo vengono illustrati i risultati relativi a diverse variabili correlate alle funzioni protettive del bosco. Le condizioni fisico-stazionali rappresentate dalla pendenza, dall'esposizione e dalla giacitura del terreno possono determinare caratteristiche del clima locale con un effetto distributivo delle fitocenosi a scala topografica (De Philippis, 1961). Queste tre variabili influenzano la composizione e la produttività dei soprassuoli anche in virtù della stretta relazione con le caratteristiche del suolo (profondità, profilo, struttura e tessitura) (Spurr & Barnes, 1980). Per la loro correlazione con le variabili climatiche, pendenza ed esposizione, insieme alla quota, influiscono non solo sulla composizione ma anche sulla produttività dei soprassuoli (Stage & Salas, 2007); tali variabili sono perciò generalmente incluse tra quelle esplicative nei modelli di incremento (es. Di Cosmo et al., 2020; Gasparini et al., 2017), singolarmente o combinate in indici complessi (es. Roise & Betters, 1981; Stage, 1976). Gli aspetti produttivi citati sono strettamente attinenti a quelli della protezione, perché è da considerare che i soprassuoli con caratteristiche compositive e produttive differenti esplicano protezione in misura diversa. Ciò è dovuto non solo al loro diverso potenziale intrinseco di capacità protettiva, ma anche alle forme di gestione e agli usi cui sono sottoposti concretamente per l'ottenimento dei beni e servizi ecosistemici, che dipendono ancora una volta dalla composizione e dalla produttività. L'azione positiva del bosco sui versanti acclivi è di facile intuizione. Le utilizzazioni sono di solito limitate o precluse su pendenze oltre il 60%. Tuttavia, Schönenberger and Brang (2004) hanno potuto evidenziare come, in contesti particolarmente difficili, la variabilità di mesoclima e di microhabitat indotta a scala locale dalle variazioni nella pendenza e nell'esposizione sono capaci di accrescere le reciproche influenze tra l'erosione superficiale e il rotolamento di massi da una parte e il bosco dall'altra. Analizzando la relazione suolo-soprassuolo da una prospettiva leggermente diversa, si può osservare che dalle condizioni stazionali legate a pendenza, esposizione e giacitura consegue un diverso grado di vulnerabilità del terreno ai fattori di rischio per la stabilità. Le condizioni meso- e micro-climatiche determinate dall'esposizione, per esempio, possono avere effetti anche su alcuni tipi di dissesto come le valanghe e le slavine. Le statistiche sulla superficie forestale nelle varie classi delle tre variabili considerate permettono, quindi, una valutazione indiretta sulla porzione di territorio nazionale per cui le aree boscate esplicano una importante funzione protettiva del suolo (Sect. 11.2). La Sect. 11.3 considera la relazione foresta-suolo a partire dal suolo. Se è vero, infatti, che le foreste proteggono il suolo è anche vero che la stabilità del suolo su cui insiste la foresta è a sua volta condizione necessaria per la conservazione della copertura forestale. Vengono presentate e commentate le stime relative ai fenomeni di dissesto osservati nelle aree forestali, che possono condizionare o pregiudicare la stabilità del soprassuolo. La normativa italiana ha riconosciuto da molto tempo il ruolo primario della funzione protettiva della foresta nei riguardi del suolo e delle acque. Le stime sulla superficie forestale sottoposta alla tutela di legge del vincolo idrogeologico sono illustrate e discusse nella Sect. 11.4. Nella Sect. 11.5, infine, vengono presentate le stime relative alla presenza di una gestione del Bosco finalizzata ad una funzione prioritaria; tra le possibili funzioni, in fase di rilievo al suolo INFC valuta in maniera esplicita anche quella protettiva diretta e indiretta.

Condizioni fisiche delle stazioni

I caratteri fisici considerati da INFC per descrivere le condizioni stazionali sono la pendenza, la giacitura e l'esposizione.

La pendenza del terreno, vale a dire il rapporto tra il dislivello e la distanza orizzontale tra due punti, espresso in percentuale, è stata calcolata a partire dalle misurazioni in campo dell'inclinazione (cfr. Chap. 4). La ripartizione delle superfici stimate è avvenuta secondo cinque classi di pendenza, descritte nella Table 11.1. Le Tables 11.2 e 11.3 riportano le stime della superficie di Bosco e Altre terre boscate ripartiti secondo le classi di pendenza del terreno. Il 45.1% della superficie del Bosco in Italia è caratterizzata da valori di pendenza elevati, superiori al 40% (Fig. 11.1). Nelle Altre terre boscate, l'aliquota di terreni in queste condizioni di pendenza è risultata più bassa (25.1%), ma si deve notare che l'aliquota di superficie priva di classificazione è piuttosto elevata (35.4%). Considerando solo il Bosco, tutte le regioni dell'Arco alpino (Piemonte, Valle d'Aosta, Lombardia, Alto Adige, Trentino, Friuli-Venezia Giulia e Liguria) hanno aliquote di terreno a pendenze maggiori del 40% più alte della media nazionale, mentre la Puglia, con il 48.3%, e la Sardegna, con il 35.7%, sono le regioni con le aliquote maggiori di superficie con pendenza inferiore al 20%. Altre tabelle con le stime di superficie per le classi di pendenza di tutte le categorie inventariali e delle categorie forestali dei Boschi alti sono disponibili all'indirizzo https://www.inventarioforestale.org/statistiche_INFC.

La giacitura è stata valutata durante i rilievi in campo tramite classificazione a vista, sia per l'area estesa sia per quella locale. La giacitura estesa si riferisce ad una superficie piuttosto ampia, variabile da qualche ettaro ad alcune decine di ettari circostanti l'unità di campionamento; la giacitura locale riguarda AdS25 (circa 2000 m^2). Per entrambi i livelli sono state distinte cinque classi (Tables 11.4 e 11.5). Le stime di superficie secondo le classi di giacitura dell'area estesa sono riportate in Table 11.6 (Bosco) e in Table 11.7 (Altre terre boscate). La giacitura di medio versante è quella prevalente sia per il Bosco (53.5%) sia per le Altre terre boscate (35.2%). La Fig. 11.2 mostra le aliquote delle classi di giacitura estesa per regione, per la superficie del Bosco. Quanto osservato per l'ambito nazionale trova analogia anche a livello regionale; si può notare per la Puglia una percentuale di superficie con giacitura pianeggiante (20.6%) elevata rispetto alle altre regioni. Riguardo alla giacitura locale, le stime per classi di giacitura sono fornite nelle Tables 11.8 e 11.9 rispettivamente per il Bosco e per le Altre terre boscate. Nella Fig. 11.3 si osserva come nel Bosco la forma di giacitura locale più diffusa sia quella su piano inclinato, che interessa il 68.7% della superficie. Tale forma è la prevalente, in misura analoga, in tutte le regioni.

L'esposizione è stata rilevata in campo mediante lettura dell'azimut magnetico, orientando la bussola dal punto di campionamento lungo la linea di massima pendenza del versante, verso valle. Sono state considerate nove classi di esposizione, ciascuna ampia 45° (Table 11.10). A livello nazionale le classi di esposizione sono ripartite omogeneamente sia per il Bosco, dove la superficie in ciascuna classe è compresa tra il 10% e il 13% di quella totale, sia per le Altre terre boscate, con variazione tra il 7% e il 9%, come riportato nelle Tables 11.11 e 11.12. La classe "Nessuna esposizione" rappresenta sempre un'aliquota molto bassa, attorno al 2–3% della superficie totale del Bosco o delle Altre terre boscate.

Stime di maggior dettaglio su pendenza, giacitura ed esposizione, relative alle categorie inventariali del Bosco e delle Altre terre boscate, e alle categorie forestali dei Boschi alti, sono disponibili all'indirizzo https://www.inventarioforestale.org/statistiche_INFC.

Dissesto idrogeologico

Pendenza e altre caratteristiche morfologiche del sito, struttura e composizione del suolo e degli strati geologici sottostanti sono tra gli aspetti più importanti per la vulnerabilità del territorio al dissesto idrogeologico. Il dissesto idrogeologico può manifestarsi in entità tale da costituire pregiudizio per la stabilità del soprassuolo arboreo. INFC2015 ha stimato la presenza di quattro tipi di dissesto, valutata a vista durante i rilievi in campo: frane e smottamenti, erosione idrica e fenomeni alluvionali, caduta o rotolamento di pietre, slavine e valanghe (Table 11.13).

Almeno un tipo di dissesto risulta essere presente sul 15.2% della superficie del Bosco a livello nazionale (Table 11.14 e Fig. 11.4), con ripartizione regionale compresa fra il 7.8% della Basilicata e il 26.3% della Valle d'Aosta. La causa di dissesto più diffusa è la caduta o rotolamento di pietre (6.4%). In Emilia-Romagna, frane e smottamenti interessano il 14.4% della superficie, valore molto più alto della media nazionale (3.6%). L'aliquota di superficie del Bosco interessata da valanghe e slavine è molto bassa a livello nazionale (0.6%) e nella maggior parte delle regioni, ma diventa apprezzabile in tre regioni alpine: Valle d'Aosta (5.4%), Alto Adige (5.6%) e Trentino (3.8%). Nelle stesse regioni, l'area è ancora più estesa nelle Altre terre boscate (17.6% in Valle d'Aosta, 16.9% in Alto Adige, 22.4% in Trentino; Table 11.15). In https://www.inventarioforestale.org/statistiche_INFC sono disponibili statistiche sulle forme di dissesto idrogeologico riguardanti tutte le categorie inventariali del Bosco e delle Altre terre boscate e le categorie forestali dei Boschi alti.

Vincolo idrogeologico

Il primo quadro normativo di riferimento a livello nazionale per la gestione delle risorse forestali è dato dalle disposizioni a tutela dei boschi per fini idrogeologici contenute nel R.D.L. n.3267/23, successivamente integrato da altre norme dello Stato e delle Regioni al fine di adeguare le disposizioni di legge alle mutate esigenze. Lo scopo della normativa sul vincolo idrogeologico è quello di impedire forme di utilizzazione dei terreni di qualsiasi natura e destinazione, ma con riferimento particolare ai boschi, che possano causare danno pubblico determinato da denudazioni, perdita di stabilità e turbamento del regime delle acque. Il vincolo idrogeologico, pur non precludendo la possibilità di realizzare interventi nelle aree forestali, impone ai proprietari l'ottenimento di autorizzazione per la trasformazione dei boschi in altri tipi di coltura e per la trasformazione di "terreni saldi in terreni sottoposti a periodica lavorazione". La gestione del bosco nei terreni vincolati, inoltre, deve rispettare le modalità definite dalle cosiddette Prescrizioni di Massima e di Polizia Forestale.

La Table 11.16 evidenzia che il vincolo idrogeologico è presente su gran parte della superficie forestale italiana (80.1%). Il vincolo riguarda soprattutto il Bosco, dove è presente sull'86.6% della superficie, mentre per le Altre terre boscate la superficie soggetta a vincolo è pari al 50.0%. La percentuale di Bosco soggetta a vincolo idrogeologico è molto alta, con valori superiori al 95% in alcune regioni del nord e centro Italia (Fig. 11.5): Trentino, Alto Adige, Umbria, Toscana. Un'eccezione è rappresentata dalla Sardegna, dove soltanto il 51.6% della superficie del Bosco è sottoposta a vincolo idrogeologico.

Altre statistiche sulla presenza di vincolo idrogeologico a livello di tutte le categorie inventariali e delle categorie forestali dei Boschi alti sono disponibili all'indirizzo https://www.inventarioforestale.org/statistiche_INFC.

Funzione prioritaria

Durante i rilievi in campo INFC2015 ha raccolto dati sulla eventuale presenza di una funzione prioritaria del Bosco, quale obiettivo principale di gestione e di pianificazione degli interventi. Tali dati sono derivati da osservazione diretta, da documenti della pianificazione forestale o da interviste con tecnici e gestori locali. La Table 11.17 riporta le classi per la funzione prioritaria e le relative definizioni. Le statistiche circa la presenza di una funzione prioritaria per il Bosco sono riportate nella Table 11.18; la Fig. 11.6 mostra la ripartizione della superficie del Bosco per funzione prioritaria, a livello regionale e nazionale. In Italia, l'86.8% della superficie del Bosco non è destinato a svolgere una funzione prioritaria. Solo in quattro regioni la superficie del Bosco gestito secondo un obiettivo prioritario supera il 20%: Valle d'Aosta, Lombardia, Umbria e Sardegna. Tra le funzioni prioritarie considerate, la produzione legnosa riguarda l'aliquota maggiore di superficie, con il 6.7%, seguita dalla funzione naturalistica, con il 2.1%. Sebbene contenuta a livello nazionale (2.1%), la percentuale di superficie con funzione prioritaria protettiva, diretta e indiretta, è rilevante in Valle d'Aosta (23.7%) e in Alto Adige (11.5%). La percentuale di superficie con gestione finalizzata alla funzione ricreativa è trascurabile in tutte le regioni, mentre le produzioni non legnose (1.3% a livello nazionale) assumono un ruolo consistente quale obiettivo prioritario di gestione solo in Sardegna, per la produzione di sughero (10.1%). Ulteriori statistiche sulla presenza di un obiettivo prioritario di gestione a livello delle categorie inventariali del Bosco sono disponibili all'indirizzo https://www.inventarioforestale.org/statistiche_INFC.

La proporzione di Bosco in cui sia possibile individuare chiaramente una funzione prevalente sulle altre, in misura tale da condizionare gli interventi selvicolturali e la gestione per incrementare o mantenere tale funzione, è piuttosto contenuta. Il confronto anche solo con quanto esposto nella precedente Sect. 11.4, che ha evidenziato come la capacità nel proteggere il suolo e le risorse idriche sia riconosciuta a percentuali molto elevate della superficie del Bosco in tutte le regioni, conferma come la gestione forestale in Italia sia orientata a valorizzare la multifunzionalità delle foreste, piuttosto che obiettivi specifici.

References

De Philippis, A. (1961). *Appunti dalle lezioni di ecologia forestale e selvicoltura generale* (pp. 56–82). Università degli studi di Firenze.

Di Cosmo, L., Giuliani, D., Dickson, M. M., & Gasparini, P. (2020). An individual-tree linear mixed-effect model for predicting the basal area increment of major forest species in Southern Europe. *Forest Systems, 29*(3), (2019), 13 pp. https://doi.org/10.5424/fs/2020293-15500.

Dorren, K. A., Berger, F., Imeson, A. C., Maier, B., & Rey, F. (2004). Integrity, stability and management of protection forests in the European Alps. *Forest Ecology and Management, 195*, 165–176. https://doi.org/10.1016/j.foreco.2004.02.057

Gasparini, P., Di Cosmo, L., Rizzo, M., & Giuliani, D. (2017). A stand level model derived from National Forest Inventory data to predict periodic annual volume increment of forests in Italy. *Journal of Forest Research, 22*(4), 209–217.

FAO (2005). Protective functions of forest resources. In: *Global forest resources assessment 2005* (pp. 95–106). Rome. Retrieved Jan 03, 2022, from https://www.fao.org/forest-resources-assessment/past-assessments/fra-2005/en/.

Führer, E. (2000). Forest functions, ecosystem stability and management. *Forest Ecology and Management, 132*, 29–38. https://doi.org/10.1016/S0378-1127(00)00377-7

Roise, J., & Betters, D. R. (1981). An aspect transformation with regard to elevation for site productivity models. *Forestry Sciences, 27*(3), 483–486.

Schönenberger, W. (2001). Trends in Mountain Forest Management in Switzerland. *Schweizerische Zeitschrift Für Forstwesen, 152*(4), 152–156.

Schönenberger, W., & Brang, P. (2004). Silviculture in mountain forests. In: J. Burley, J. Evans, & J. Youngquist (Eds.), *Encyclopedia of forest science* (pp. 1085–1094). Elsevier.

Spurr, S. H., & Barnes, B. V. (1980). *Forest ecology* (3rd ed., pp. 207–208). Wiley.

Stage, A. R. (1976). An expression for the effect of aspect, slope and habitat type on tree growth. *Forestry Sciences, 22*(4), 457–461.

Stage, A. R., & Salas, C. (2007). Interactions of elevation, aspect and slope in models of forest species composition and productivity. *Forestry Sciences, 53*(4), 486–492.

Open Access This chapter is licensed under the terms of the Creative Commons Attribution 4.0 International License (http://creativecommons.org/licenses/by/4.0/), which permits use, sharing, adaptation, distribution and reproduction in any medium or format, as long as you give appropriate credit to the original author(s) and the source, provide a link to the Creative Commons license and indicate if changes were made.

The images or other third party material in this chapter are included in the chapter's Creative Commons license, unless indicated otherwise in a credit line to the material. If material is not included in the chapter's Creative Commons license and your intended use is not permitted by statutory regulation or exceeds the permitted use, you will need to obtain permission directly from the copyright holder.

Chapter 12
Forest Carbon Stock

Il carbonio forestale in Italia

Lucio Di Cosmo, Patrizia Gasparini, and Antonio Floris

Abstract Forests affect climate globally and have an important role in the global carbon cycle. Countries that signed the United Nations Framework Convention on Climate Change (UNFCCC) and the agreements that followed, use national forest inventory data to estimate carbon sequestration related to land use, land use changes and forestry. Five terrestrial carbon pools are relevant for the estimation of carbon stocks and carbon stock changes under the UNFCCC and the Kyoto Protocol: soil, litter, belowground and aboveground biomass, and deadwood. The second Italian NFI (INFC2005) estimated the organic carbon stock stored within four out of the five pools and confirmed the major role of soil, which stored 57.6% of Italian forest organic carbon in the four pools studied. Estimating soil carbon change is challenging if the time elapsed between two assessment periods is not long, because the expected changes are small compared to the high carbon stock already present in soils. INFC2015 updated the estimates of the carbon stored in aboveground living biomass and in deadwood. The results are shown and commented on in Sects. 12.2 and 12.3. They allow for computation of the overall aboveground biomass carbon stock, i.e., the joined contribution of those two ecosystem components. Annual variation of carbon in growing stock is also due to carbon stored via growth and carbon removed with harvesting. These entries of the balance were estimated by INFC2015 and indicate that Italian forests act as carbon sinks; this is shown in Sect. 12.4.

Keywords Carbon pool · Carbon sequestration · Coarse woody debris · Removals · Carbon sink · Carbon balance

---

L. Di Cosmo (✉) · P. Gasparini · A. Floris
CREA Research Centre for Forestry and Wood, Trento, Italy
e-mail: lucio.dicosmo@crea.gov.it

P. Gasparini
e-mail: patrizia.gasparini@crea.gov.it

A. Floris
e-mail: antonio.floris@crea.gov.it

© The Author(s) 2022

P. Gasparini et al. (eds.), *Italian National Forest Inventory—Methods and Results of the Third Survey*, Springer Tracts in Civil Engineering,
https://doi.org/10.1007/978-3-030-98678-0_12

12.1 Introduction

Trees bind carbon dioxide (CO_2) from the atmosphere through photosynthesis and store carbon in their wood. The carbon content of a tree is approximated by half the dry weight of the tree itself. Therefore, trees are seen as individuals able to storage carbon, and for this reason, also trees outside forests (TOF) have gained attention (e.g., Guo et al., 2014; Russo et al., 2014; Schnell et al., 2015; Speak et al., 2020). The assumption that trees are a net CO_2 sink led to planting trees in cities as part of their climate protection plans. However, assessments on their real potential to act as carbon sinks are difficult to carry out, because they require, among others, knowing the total emissions caused by planting and maintenance during the entire life cycle. Some completed assessments have shown such plantations have limited effectiveness in carbon stocking (e.g., McPherson & Kendall, 2014). For stocking purposes, trees are more efficient by moving carbon from the atmosphere into the forest ecosystem.

The role of forest in climate change mitigation has been widely recognised in recent decades. Countries that report on greenhouse gas (GHG) emissions and removals, after signing the United Nations Framework Convention on Climate Change (UNFCCC) and following the subsequent agreements, need to estimate carbon sequestration related to land use, land use changes and forestry. Five terrestrial carbon pools are relevant for the estimation of carbon stocks and carbon stock changes under the UNFCCC and the Kyoto Protocol (IPCC, 2003): soil, litter, belowground and aboveground biomass, and deadwood. National forest inventories are important data sources for estimating forest carbon stock (e.g., Breidenbach et al., 2021; Brown, 2002; Mäkipää et al., 2008). The second Italian NFI (INFC2005) estimated the carbon stock stored within four out of the five pools, i.e., belowground biomass was not included. The comprehensive assessment confirmed the major role of soil in European forests, as soil stored 57.6% of Italian forest organic carbon in the four pools, followed by the aboveground living biomass (38.1%). Together, deadwood and litter stored 4.3% of carbon (Gasparini & Di Cosmo, 2015; Gasparini et al., 2013).

Extensive field campaigns for sampling soil and following laboratory analyses for carbon content measurements are very expensive. Furthermore, estimating the change requires enormous effort because the expected changes are small compared to the high carbon stock already present in soils (Schrumpf et al., 2008), especially when the period between the two assessments is not long (e.g., Conen et al., 2003). Considering the short period between the two forest inventories, INFC2015 updated the estimates of carbon stored within the aboveground living biomass and the deadwood biomass. This living biomass is represented by woody vegetation consisting of trees with DBH $\geq$ 4.5 cm, small trees (seedlings and saplings typical of regeneration processes) and shrubs taller than 50 cm. Deadwood is composed of standing dead trees (trees with DBH $\geq$ 4.5 cm), remaining stumps (diameter $\geq$ 9.5 cm) and deadwood lying on the ground (section diameter $\geq$ 9.5 cm, length between the end sections $\geq$ 9.5 cm); together standing dead trees, stumps and deadwood lying on the ground are considered coarse woody debris.

The amount of carbon annually sequestrated by trees from the atmosphere was assessed estimating Forest annual increment in biomass. The carbon content in forest stands, however, also varies annually due to natural mortality and the removal of utilised trees. For this reason, the carbon sink quantified as the annual increase of carbon in living trees minus the carbon content of removed trees cannot be considered a net annual accumulation of carbon (Tabacchi et al., 2010).

12.2 Carbon Stock in the Aboveground Living Biomass

Table 12.1 shows the carbon stock stored in the woody vegetation as represented by all individuals taller than 50 cm, for the inventory categories of Forest and Total. Tables 12.2 and 12.3 show similar data for the forest types. Trees, small trees (seedlings and saplings) and shrubs store 539.3 million carbon tonnes, 59.4 tonnes per hectare on average. Table 12.4 shows the estimates on the carbon stored by trees for the inventory categories of Forest and for Total Forest. Table 12.5 shows the carbon stored by small trees and Table 12.6 shows the amount stored by shrubs. At www.inventarioforestale.org/statistiche_INFC similar statistics for the forest types are available. The data in the cited tables indicate that 96.8% of the carbon in the living biomass is stored by trees; small trees and shrubs store 1.7% and 1.5% of the total, respectively. Only in a few regions do small trees or shrubs store at least 3% of total aboveground living biomass carbon. Carbon in small trees exceeds that threshold in Umbria (5.7%), Marche (4.7%) and Molise (4.6%). Carbon in shrubs exceeds it in Molise (3.9%) and Sardegna (5.3%). If we consider the joint contribution of small trees and shrubs, 3% of the overall living biomass is exceeded in five additional regions: Emilia-Romagna, Lazio, Campania, Puglia and Basilicata. Figure 12.1 shows the carbon stock in the regions differentiating the parts due to trees and to small trees and shrubs together.

Figure 12.2 shows the carbon stock in the aboveground living biomass in the Tall trees forest types.

Considering the forest types with more than 40 million carbon tonnes, we can note (cf. Chap. 7) that Beech forest, Temperate oaks and Mediterranean oaks forests and Other deciduous broadleaved forest cover wide areas (more than 1 million hectare each) and account for large timber volume (generally over 100 million cubic metres, except Temperate oaks forest with 94.4 million cubic metres). On the contrary, Norway spruce and Chestnut forest types store a great deal of carbon mainly because they are rich in biomass, as they do not spread over very large areas (586.7 thousand hectares for Norway spruce and 778.5 thousand hectares for Chestnut forest).

Table 12.1 Total value and value per hectare of carbon stock in the aboveground biomass by Forest inventory category / Valori totali e per ettaro di carbonio nella fitomassa epigea totale per le categorie inventariali del Bosco

Region/Regione	Tall trees forest				Plantations				Total Forest			
	Boschi alti				Impianti di arboricoltura da legno				Totale Bosco			
	Corg	ES	Corg	ES	Corg	ES	Corg	ES	Corg	ES	Corg	ES
	(Mg)	(%)	(Mg ha^{-1})	(%)	(Mg)	(%)	(Mg ha^{-1})	(%)	(Mg)	(%)	(Mg ha^{-1})	(%)
Piemonte	50,456,000	2.7	58.0	2.4	644,591	27.0	31.2	20.2	51,100,591	2.7	57.4	2.4
Valle d'Aosta	5,676,403	6.9	57.2	5.6	0	–	0.0	–	5,676,403	6.9	57.2	5.6
Lombardia	41,713,946	3.6	69.9	3.1	580,900	22.0	23.1	16.3	42,294,846	3.5	68.0	3.1
Alto Adige	30,943,327	4.2	91.2	3.8	0	–	0.0	–	30,943,327	4.2	91.2	3.8
Trentino	32,360,381	4.2	86.7	4.0	0	–	0.0	–	32,360,381	4.2	86.7	4.0
Veneto	31,611,548	3.8	76.8	3.2	182,362	37.7	34.6	13.2	31,793,911	3.8	76.3	3.2
Friuli V.G	26,261,933	4.7	81.2	4.2	284,871	26.1	31.0	19.0	26,546,804	4.6	79.8	4.2
Liguria	18,925,878	4.0	55.2	3.8	13,049	100.0	35.6	–	18,938,927	4.0	55.2	3.7
Emilia Romagna	32,311,088	3.9	55.8	3.4	195,144	33.1	32.3	30.0	32,506,232	3.9	55.6	3.4
Toscana	59,934,948	2.8	58.3	2.5	285,749	37.4	42.1	31.6	60,220,698	2.8	58.2	2.5
Umbria	16,267,006	4.4	42.4	4.0	75,528	33.2	11.8	9.9	16,342,534	4.3	41.9	4.0
Marche	12,527,268	5.3	44.0	4.6	83,184	37.1	12.1	26.6	12,610,452	5.2	43.2	4.6
Lazio	30,352,191	4.4	54.4	4.0	50,349	47.6	23.1	30.0	30,402,540	4.3	54.3	4.0
Abruzzo	25,597,365	4.2	62.6	3.8	21,599	45.6	7.3	33.7	25,618,964	4.2	62.2	3.8
Molise	8,701,630	6.9	57.8	6.1	50,460	73.3	18.6	60.9	8,752,090	6.8	57.1	6.1
Campania	23,117,688	5.4	57.7	4.9	28,186	65.5	8.9	24.6	23,145,874	5.4	57.3	4.9
Puglia	6,538,046	8.4	46.0	7.1	464	100.0	4.6	–	6,538,510	8.4	45.9	7.1
Basilicata	14,568,935	6.1	50.9	5.7	1110	97.7	0.7	100.0	14,570,045	6.1	50.6	5.7
Calabria	38,409,538	4.8	77.9	4.5	94,269	51.6	39.2	33.2	38,503,807	4.8	77.8	4.5
Sicilia	10,913,462	6.2	38.3	5.4	597	70.7	0.8	–	10,914,060	6.2	38.2	5.4
Sardegna	19,209,130	4.8	32.0	4.5	326,568	18.2	12.6	15.0	19,535,699	4.8	31.2	4.5
Italia	536,397,713	1.0	59.9	0.9	2,918,979	9.8	22.7	7.9	539,316,692	1.0	59.4	0.9

Table 12.2 Total value and value per hectare of carbon stock in the aboveground biomass by Tall trees forest type / Valori totali e per ettaro di carbonio nella fitomassa epigea totale per le categorie forestali dei Boschi alti

Region/Regione	Larch and Stone pine				Norway spruce				Fir			
	Boschi di larice e cembro				Boschi di abete rosso				Boschi di abete bianco			
	Corg	ES	Corg	ES	Corg	ES	Corg	ES	Corg	ES	Corg	ES
	(Mg)	(%)	(Mg ha^{-1})	(%)	(Mg)	(%)	(Mg ha^{-1})	(%)	(Mg)	(%)	(Mg ha^{-1})	(%)
Piemonte	3,848,553	11.8	48.1	–	2,020,544	19.2	111.1	–	1,569,779	19.3	111.0	–
Valle d'Aosta	2,426,382	12.8	50.9	–	1,510,180	16.5	82.8	–	99,097	60.0	85.7	–
Lombardia	3,805,675	13.8	71.2	–	9,014,212	10.2	102.6	–	1,195,068	24.3	126.9	–
Alto Adige	6,581,075	10.2	70.3	–	19,097,382	6.2	107.1	–	321,095	39.9	121.3	–
Trentino	3,504,675	13.6	56.4	–	15,817,873	7.5	115.6	–	2,196,440	17.8	135.4	–
Veneto	2,715,956	16.0	64.1	–	10,430,779	8.2	107.4	–	1,020,690	24.0	143.7	–
Friuli V.G	491,082	34.4	37.9	–	5,888,858	12.3	132.0	–	354,523	48.4	190.8	–
Liguria	51,067	64.7	46.4	–	25,548	100.0	69.7	–	379,357	37.0	129.4	–
Emilia Romagna	0	–	0.0	–	455,783	32.8	123.7	–	545,910	38.8	185.2	–
Toscana	0	–	0.0	–	63,765	100.0	176.5	–	801,701	30.7	158.5	–
Umbria	0	–	0.0	–	0	–	0.0	–	0	–	0.0	–
Marche	0	–	0.0	–	20,855	100.0	56.3	–	0	–	0.0	–
Lazio	0	–	0.0	–	89,978	100.0	244.2	–	0	–	0.0	–
Abruzzo	0	–	0.0	–	32,365	100.0	89.4	–	96,587	71.1	133.4	–
Molise	0	–	0.0	–	0	–	0.0	–	125,632	62.0	107.2	–
Campania	0	–	0.0	–	0	–	0.0	–	0	–	0.0	–
Puglia	0	–	0.0	–	0	–	0.0	–	0	–	0.0	–
Basilicata	0	–	0.0	–	0	–	0.0	–	90,255	75.4	121.0	–
Calabria	0	–	0.0	–	0	–	0.0	–	393,264	34.7	105.4	–
Sicilia	0	–	0.0	–	0	–	0.0	–	0	–	0.0	–
Sardegna	0	–	0.0	–	0	–	0.0	–	0	–	0.0	–
Italia	23,424,465	5.2	59.5	4.4	64,468,123	3.5	109.9	2.8	9,189,397	8.4	131.6	4.2

(continued)

Table 12.2 (continued)

Region/Regione	Scots pine and Mountain pine				Black pines				Mediterranean pines			
	Pinete di pino silvestre e montano				Pinete di pino nero, laricio e loricato				Pinete di pini mediterranei			
	Corg	ES	Corg	ES	Corg	ES	Corg	ES	Corg	ES	Corg	ES
	(Mg)	(%)	(Mg ha^{-1})	(%)	(Mg)	(%)	(Mg ha^{-1})	(%)	(Mg)	(%)	(Mg ha^{-1})	(%)
Piemonte	1,286,999	17.6	62.0	–	170,029	41.1	52.6	–	17,416	71.7	21.6	–
Valle d'Aosta	660,164	21.3	56.6	–	15,942	100.0	41.4	–	0	–	0.0	–
Lombardia	1,132,287	23.5	59.5	–	358,037	43.6	81.2	–	0	–	0.0	–
Alto Adige	3,274,643	13.4	88.4	–	2640	100.0	7.0	–	0	–	0.0	–
Trentino	1,352,202	16.5	64.7	–	466,912	29.4	81.0	–	0	–	0.0	–
Veneto	663,546	19.9	54.2	–	202,435	42.2	57.7	–	65,592	71.1	87.7	–
Friuli V.G	653,318	27.2	53.4	–	2,086,927	14.3	66.1	–	35,148	64.0	31.5	–
Liguria	462,749	23.3	45.1	–	466,249	36.0	84.8	–	656,500	25.6	26.7	–
Emilia Romagna	97,473	54.3	52.9	–	1,447,417	24.1	89.3	–	201,417	53.5	68.3	–
Toscana	115,894	59.2	106.9	–	2,349,829	24.0	130.1	–	3,273,895	15.9	72.5	–
Umbria	42,144	75.8	57.2	–	325,271	28.0	55.1	–	398,566	23.8	49.1	–
Marche	0	–	0.0	–	583,889	23.3	56.3	–	106,682	48.7	48.6	–
Lazio	0	–	0.0	–	588,594	25.3	69.5	–	416,002	28.4	56.6	–
Abruzzo	113,188	62.8	104.2	–	1,058,931	24.6	57.5	–	106,387	53.1	49.0	–
Molise	0	–	0.0	–	115,920	44.3	49.5	–	52,773	64.9	24.2	–
Campania	0	–	0.0	–	427,284	36.8	77.4	–	492,716	26.1	53.9	–
Puglia	0	–	0.0	–	177,724	53.4	114.4	–	899,358	20.5	32.4	–
Basilicata	0	–	0.0	–	130,709	43.6	50.1	–	310,346	24.6	34.7	–
Calabria	0	–	0.0	–	6,150,965	10.9	83.8	–	847,840	23.5	48.5	–
Sicilia	0	–	0.0	–	279,020	29.4	37.2	–	1,916,071	20.5	42.3	–
Sardegna	0	–	0.0	–	250,094	42.4	47.9	–	1,032,709	16.2	29.8	–
Italia	9,854,606	6.9	66.2	4.8	17,654,818	6.3	76.6	5.0	10,829,417	7.3	45.0	6.2

(continued)

Table 12.2 (continued)

Region/Regione	Other coniferous forests				Beech				Temperate oaks			
	Altri boschi di conifere, pure o miste				Faggete				Querceti di rovere, roverella e farnia			
	Corg	ES	Corg	ES	Corg	ES	Corg	ES	Corg	ES	Corg	ES
	(Mg)	(%)	(Mg ha^{-1})	(%)	(Mg)	(%)	(Mg ha^{-1})	(%)	(Mg)	(%)	(Mg ha^{-1})	(%)
Piemonte	368,566	34.4	57.0	–	9,560,696	7.9	81.6	–	3,198,425	14.1	42.0	–
Valle d'Aosta	0	–	0.0	–	133,446	67.5	115.4	–	203,711	37.6	48.1	–
Lombardia	428,248	38.9	67.5	–	5,741,284	12.2	86.1	–	2,289,803	15.0	55.7	–
Alto Adige	0	–	0.0	–	342,312	31.4	82.3	–	227,944	31.4	50.2	–
Trentino	0	–	0.0	–	5,435,815	10.4	86.8	–	308,267	33.1	53.5	–
Veneto	0	–	0.0	–	6,876,259	10.7	101.7	–	1,013,775	18.6	64.8	–
Friuli V.G	86,526	67.1	77.6	–	9,533,824	9.8	105.2	–	411,155	27.3	50.3	–
Liguria	49,166	100.0	134.1	–	3,724,112	11.3	100.6	–	1,794,215	14.8	40.9	–
Emilia Romagna	205,756	33.7	62.1	–	9,763,463	8.8	95.2	–	2,609,129	11.2	33.9	–
Toscana	1,224,625	23.5	102.9	–	8,655,497	9.5	118.0	–	6,186,088	7.8	40.7	–
Umbria	180,639	46.4	98.0	–	1,234,581	20.2	74.7	–	3,149,565	10.9	31.2	–
Marche	201,494	31.2	35.6	–	1,468,492	21.5	82.5	–	2,351,482	14.1	38.1	–
Lazio	153,585	58.3	104.2	–	8,675,828	10.4	116.6	–	2,573,718	13.2	31.5	–
Abruzzo	111,416	65.2	51.3	–	13,788,214	6.9	110.8	–	3,352,548	10.1	38.2	–
Molise	0	–	0.0	–	2,007,435	18.4	135.3	–	1,931,243	14.2	40.1	–
Campania	15,210	60.5	17.6	–	8,138,642	11.8	144.7	–	1,973,806	13.1	32.4	–
Puglia	149,309	49.5	64.1	–	721,068	31.4	154.7	–	948,734	21.5	36.7	–
Basilicata	234,632	37.2	45.6	–	3,484,412	16.3	129.9	–	1,191,718	15.6	27.7	–
Calabria	1,372,670	22.8	109.3	–	12,230,779	8.1	154.0	–	2,176,756	19.0	40.5	–
Sicilia	389,551	30.0	41.9	–	1,303,950	18.9	86.0	–	2,240,879	14.8	31.2	–
Sardegna	480,811	27.2	36.8	–	0	–	0.0	–	2,066,155	11.9	23.5	–
Italia	5,652,205	9.8	67.4	7.9	112,820,111	2.6	107.1	2.0	42,199,117	3.2	36.6	2.7

(continued)

Table 12.2 (continued)

Region/Regione	Mediterranean oaks				Chestnut				Hornbeam and Hophornbeam			
	Cerrete, boschi di farnetto, fragno e vallonea				Castagneti				Ostrieti, carpineti			
	Corg	ES	Corg	ES	Corg	ES	Corg	ES	Corg	ES	Corg	ES
	(Mg)	(%)	(Mg ha^{-1})	(%)	(Mg)	(%)	(Mg ha^{-1})	(%)	(Mg)	(%)	(Mg ha^{-1})	(%)
Piemonte	500,791	28.7	82.6	–	11,189,224	6.4	68.4	–	695,260	20.8	43.0	–
Valle d'Aosta	0	–	0.0	–	244,917	31.9	57.8	–	0	–	0.0	–
Lombardia	172,445	47.9	78.2	–	6,925,781	9.9	84.4	–	3,145,508	11.7	38.5	–
Alto Adige	0	–	0.0	–	129,064	53.7	68.3	–	541,928	22.6	61.4	–
Trentino	0	–	0.0	–	136,038	55.7	94.4	–	1,949,185	16.4	47.7	–
Veneto	0	–	0.0	–	1,024,070	19.1	59.5	–	4,687,051	10.0	57.1	–
Friuli V.G	0	–	0.0	–	1,373,424	18.6	112.0	–	2,151,724	15.6	46.0	–
Liguria	761,476	21.2	67.7	–	6,558,109	7.8	59.8	–	1,966,069	13.9	42.5	–
Emilia Romagna	5,081,539	11.3	50.2	–	2,728,451	12.3	67.3	–	4,787,875	9.5	43.0	–
Toscana	11,840,057	6.6	47.3	–	10,317,907	7.0	66.1	–	3,229,690	11.4	50.0	–
Umbria	5,618,524	8.8	43.7	–	108,327	48.9	49.0	–	2,005,481	11.8	34.3	–
Marche	1,555,610	20.8	62.9	–	247,668	34.6	66.8	–	2,837,799	10.0	37.2	–
Lazio	7,380,857	10.0	55.7	–	2,129,043	16.0	59.5	–	3,909,821	9.5	39.9	–
Abruzzo	1,896,550	16.4	59.0	–	342,834	28.8	72.9	–	1,926,053	14.1	40.2	–
Molise	3,182,574	14.2	61.2	–	23,208	100.0	59.4	–	557,472	22.8	62.1	–
Campania	3,994,914	15.2	53.5	–	2,072,712	13.6	37.0	–	2,391,579	15.4	45.1	–
Puglia	1,643,626	19.7	42.4	–	64,981	57.7	55.8	–	241,851	33.8	51.9	–
Basilicata	6,723,790	9.9	51.4	–	372,532	32.0	62.6	–	349,313	25.8	49.3	–
Calabria	3,578,279	14.8	82.1	–	3,326,072	13.8	48.2	–	63,927	83.3	14.3	–
Sicilia	1,395,043	17.0	45.0	–	425,552	27.2	48.8	–	30,564	66.4	17.6	–
Sardegna	0	–	0.0	–	84,838	46.7	45.5	–	0	–	0.0	–
Italia	55,326,077	3.4	52.2	2.9	49,824,753	3.1	64.0	2.4	37,468,150	3.3	43.6	2.7

(continued)

Table 12.2 (continued)

Region/Regione	Hygrophilous forests				Other deciduous broadleaved forests				Holm oak			
	Boschi igrofili				Altri boschi caducifogli				Leccete			
	Corg	ES	Corg	ES	Corg	ES	Corg	ES	Corg	ES	Corg	ES
	(Mg)	(%)	(Mg ha^{-1})	(%)	(Mg)	(%)	(Mg ha^{-1})	(%)	(Mg)	(%)	(Mg ha^{-1})	(%)
Piemonte	982,412	21.2	43.2	–	15,036,372	5.4	46.4	–	0	–	0.0	–
Valle d'Aosta	0	–	0.0	–	382,565	26.2	36.4	–	0	–	0.0	–
Lombardia	815,395	20.3	43.3	–	6,619,000	9.5	54.7	–	0	–	0.0	–
Alto Adige	139,094	39.1	50.6	–	250,446	38.4	70.9	–	0	–	0.0	–
Trentino	261,101	49.2	67.6	–	927,089	19.4	56.3	–	4783	100.0	13.3	–
Veneto	337,506	25.8	32.4	–	2,365,435	12.4	45.1	–	208,455	41.6	69.7	–
Friuli V.G	276,188	24.8	21.9	–	2,919,238	12.8	61.5	–	0	–	0.0	–
Liguria	207,923	43.3	63.0	–	1,209,535	15.9	38.0	–	606,417	23.0	43.6	–
Emilia Romagna	865,678	20.7	30.7	–	3,466,089	13.5	40.1	–	55,108	72.9	74.8	–
Toscana	1,333,484	17.6	48.1	–	3,987,368	10.9	47.2	–	6,172,326	8.1	47.6	–
Umbria	456,723	40.0	56.8	–	331,264	29.8	29.4	–	2,415,920	12.9	58.5	–
Marche	703,158	22.9	54.5	–	2,058,593	14.5	33.2	–	391,545	26.8	55.6	–
Lazio	319,659	28.0	34.7	–	1,830,067	19.7	33.9	–	2,049,721	18.1	42.5	–
Abruzzo	731,587	21.2	36.8	–	1,639,842	18.8	29.1	–	364,013	30.8	41.9	–
Molise	237,192	28.3	26.2	–	444,622	27.5	43.5	–	23,558	72.8	20.1	–
Campania	523,585	31.5	47.4	–	1,353,157	18.0	39.4	–	1,678,215	17.5	44.8	–
Puglia	14,806	100.0	38.1	–	828,662	30.4	71.8	–	656,511	27.0	37.8	–
Basilicata	253,183	28.2	19.5	–	1,040,049	18.6	34.8	–	302,446	31.9	29.2	–
Calabria	524,343	27.6	40.7	–	2,225,435	18.6	49.3	–	4,730,735	25.5	97.2	–
Sicilia	86,895	29.9	17.6	–	380,487	28.3	28.7	–	1,275,788	21.8	65.3	–
Sardegna	77,138	34.9	23.0	–	40,808	30.5	5.3	–	10,489,921	7.7	41.1	–
Italia	9,147,053	6.5	38.9	4.8	49,336,123	3.2	44.3	2.7	31,425,463	5.4	48.9	4.9

(continued)

Table 12.2 (continued)

Region/Regione	Coark oak				Other evergreen broadleaved forests				Not classified				Total Tall trees forest			
	Sugherete				Altri boschi di latifoglie sempreverdi				Non classificato				Totale Boschi alti			
	Corg	ES	Corg	ES	Corg	ES	Corg	ES	Corg	ES	Corg	ES	Corg	ES	Corg	ES
	(Mg)	(%)	(Mg ha^{-1})	(%)	(Mg)	(%)	(Mg ha^{-1})	(%)	(Mg)	(%)	(Mg ha^{-1})	(%)	(Mg)	(%)	(Mg ha^{-1})	(%)
Piemonte	0	–	0.0	–	0	–	0.0	–	10,934	100.0	21.5	–	50,456,000	2.7	58.0	2.4
Valle d'Aosta	0	–	0.0	–	0	–	0.0	–	0	–	0.0	–	5,676,403	6.9	57.2	5.6
Lombardia	0	–	0.0	–	60,009	62.0	26.2	–	11,193	100.0	25.4	–	41,713,946	3.6	69.9	3.1
Alto Adige	0	–	0.0	–	0	–	0.0	–	35,706	57.2	23.6	–	30,943,327	4.2	91.2	3.8
Trentino	0	–	0.0	–	0	–	0.0	–	0	–	0.0	–	32,360,381	4.2	86.7	4.0
Veneto	0	–	0.0	–	0	–	0.0	–	0	–	0.0	–	31,611,548	3.8	76.8	3.2
Friuli V.G	0	–	0.0	–	0	–	0.0	–	0	–	0.0	–	26,261,933	4.7	81.2	4.2
Liguria	0	–	0.0	–	0	–	0.0	–	7385	70.7	10.1	–	18,925,878	4.0	55.2	3.8
Emilia Romagna	0	–	0.0	–	0	–	0.0	–	0	–	0.0	–	32,311,088	3.9	55.8	3.4
Toscana	310,062	29.0	50.5	–	26,505	60.0	18.3	–	46,254	70.7	64.0	–	59,934,948	2.8	58.3	2.5
Umbria	0	–	0.0	–	0	–	0.0	–	0	–	0.0	–	16,267,006	4.4	42.4	4.0
Marche	0	–	0.0	–	0	–	0.0	–	0	–	0.0	–	12,527,268	5.3	44.0	4.6
Lazio	106,319	52.8	41.2	–	48,826	89.5	18.9	–	80,172	52.3	60.0	–	30,352,191	4.4	54.4	4.0
Abruzzo	0	–	0.0	–	32,571	100.0	23.1	–	4277	100.0	11.8	–	25,597,365	4.2	62.6	3.8
Molise	0	–	0.0	–	0	–	0.0	–	0	–	0.0	–	8,701,630	6.9	57.8	6.1
Campania	28,664	100.0	77.8	–	3588	70.7	4.9	–	23,617	100.0	64.1	–	23,117,688	5.4	57.7	4.9
Puglia	0	–	0.0	–	142,468	32.4	27.7	–	48,947	62.0	43.5	–	6,538,046	8.4	46.0	7.1
Basilicata	0	–	0.0	–	65,882	95.1	35.3	–	19,667	100.0	56.6	–	14,568,935	6.1	50.9	5.7
Calabria	305,856	41.6	58.6	–	362,759	30.8	17.2	–	119,857	41.1	53.5	–	38,409,538	4.8	77.9	4.5
Sicilia	296,795	23.7	17.2	–	892,866	16.4	22.8	–	0	–	0.0	–	10,913,462	6.2	38.3	5.4
Sardegna	4,162,460	8.6	27.2	–	520,406	20.5	13.7	–	3791	100.0	10.2	–	19,209,130	4.8	32.0	4.5
Italia	5,210,155	7.7	28.3	6.5	2,155,882	11.0	18.9	9.2	411,799	21.5	40.9	9.8	536,397,713	1.0	59.9	0.9

Table 12.3 Total value and value per hectare of carbon stock in the aboveground biomass by Plantations forest type / Valori totali e per ettaro di carbonio nella fitomassa epigea totale per le categorie forestali degli Impianti di arboricoltura da legno

Region/Regione	Poplar plantations				Other broadleaved plantations				Coniferous plantations				Total Plantations			
	Pioppeti artificiali				Piantagioni di altre latifoglie				Piantagioni di conifere				Totale Impianti di arboricoltura da legno			
	Corg	ES	Corg	ES	Corg	ES	Corg	ES	Corg	ES	Corg	ES	Corg	ES	Corg	ES
	(Mg)	(%)	(Mg ha^{-1})	(%)	(Mg)	(%)	(Mg ha^{-1})	(%)	(Mg)	(%)	(Mg ha^{-1})	(%)	(Mg)	(%)	(Mg ha^{-1})	(%)
Piemonte	384,085	36.2	24.8	–	152,914	54.1	38.2	–	107,591	61.2	93.5	–	644,591	27.0	31.2	20.2
Valle d'Aosta	0	–	0.0	–	0	–	0	–	0	–	0.0	–	0	–	0	–
Lombardia	327,662	25.7	17.7	–	204,436	47.1	33	–	48,803	100.0	110.7	–	580,900	22.0	23.1	16.3
Alto Adige	0	–	0.0	–	0	–	0	–	0	–	0.0	–	0	–	0.0	–
Trentino	0	–	0.0	–	0	–	0	–	0	–	0.0	–	0	–	0.0	–
Veneto	78,274	76.7	33.6	–	104,088	55.5	35	–	0	–	0.0	–	182,362	37.7	34.6	13.2
Friuli V.G	209,942	33.6	32.1	–	74,928	44.3	28.2	–	0	–	0.0	–	284,871	26.1	31.0	19.0
Liguria	13,049	100.0	35.6	–	0	–	0	–	0	–	0.0	–	13,049	100.0	35.6	–
Emilia Romagna	93,336	39.9	22.0	–	63,677	58.0	45	–	38,130	100.0	103.5	–	195,144	33.1	32.3	30.0
Toscana	36,616	71.0	49.7	–	73,831	42.9	16	–	175,302	56.7	120.7	–	285,749	37.4	42.1	31.6
Umbria	0	–	0.0	–	75,528	33.2	12	–	0	–	0.0	–	75,528	33.2	11.8	9.9
Marche	0	–	0.0	–	83,184	37.1	12.1	–	0	–	0.0	–	83,184	37.1	12.1	26.6
Lazio	9274	70.9	11.8	–	41,074	56.5	30	–	0	–	0.0	–	50,349	47.6	23.1	30.0
Abruzzo	5625	100.0	10.6	–	15,974	50.9	7	–	0	–	0.0	–	21,599	45.6	7.3	33.7
Molise	36,409	100.0	93.2	–	14,050	45.6	6	–	0	–	0.0	–	50,460	73.3	18.6	60.9
Campania	20,598	87.2	13.4	–	7588	56.6	5	–	0	–	0.0	–	28,186	65.5	8.9	24.6
Puglia	0	–	0.0	–	464	100.0	5	–	0	–	0.0	–	464	100.0	4.6	–
Basilicata	0	–	0.0	–	1110	97.7	1	–	0	–	0.0	–	1110	97.7	0.7	–
Calabria	970	100.0	9.7	–	28,316	56.5	18	–	64,982	70.7	87.1	–	94,269	51.6	39.2	33.2
Sicilia	0	–	0.0	–	597	70.7	1	–	0	–	0.0	–	597	70.7	0.8	–
Sardegna	0	–	0.0	–	297,291	19.4	12.2	–	29,277	50.1	19.6	–	326,568	18.2	12.6	15.0
Italia	1,215,841	16.2	23.6	12.9	1,239,052	13.7	17.4	9.8	464,086	30.7	82.1	16.9	2,918,979	9.8	22.7	7.9

Table 12.4 Total value and value per hectare of carbon stock in the aboveground tree biomass by Forest inventory category / Valori totali e per ettaro di carbonio nella fitomassa arborea epigea per le categorie inventariali del Bosco

Region/Regione	Tall trees forest				Plantations				Total Forest			
	Boschi alti				Impianti di arboricoltura da legno				Totale Bosco			
	Corg	ES	Corg	ES	Corg	ES	Corg	ES	Corg	ES	Corg	ES
	(Mg)	(%)	(Mg ha^{-1})	(%)	(Mg)	(%)	(Mg ha^{-1})	(%)	(Mg)	(%)	(Mg ha^{-1})	(%)
Piemonte	49,309,246	2.8	56.7	2.4	636,422	26.9	30.8	20.1	49,945,668	2.8	56.1	2.4
Valle d'Aosta	5,613,974	6.9	56.6	5.6	0	–	0.0	–	5,613,974	6.9	56.6	5.6
Lombardia	40,802,587	3.7	68.4	3.2	562,685	21.2	22.4	15.4	41,365,272	3.6	66.5	3.2
Alto Adige	30,753,986	4.3	90.6	3.9	0	–	0.0	–	30,753,986	4.3	90.6	3.9
Trentino	31,919,035	4.3	85.5	4.0	0	–	0.0	–	31,919,035	4.3	85.5	4.0
Veneto	30,725,695	4.0	74.7	3.3	173,653	38.2	32.9	12.4	30,899,348	4.0	74.2	3.3
Friuli V.G	25,835,420	4.8	79.9	4.3	273,612	26.3	29.8	19.3	26,109,033	4.7	78.5	4.3
Liguria	18,424,200	4.1	53.7	3.9	12,832	100.0	35.0	–	18,437,032	4.1	53.7	3.9
Emilia Romagna	31,431,271	4.0	54.3	3.6	194,234	33.2	32.1	30.1	31,625,505	4.0	54.1	3.6
Toscana	57,527,526	2.9	55.9	2.6	281,175	37.8	41.5	32.2	57,808,701	2.9	55.8	2.6
Umbria	14,901,993	4.8	38.8	4.4	68,359	33.1	10.7	10.4	14,970,352	4.8	38.4	4.4
Marche	11,584,899	5.7	40.7	5.1	70,242	40.0	10.2	29.8	11,655,141	5.7	39.9	5.1
Lazio	29,133,317	4.6	52.2	4.2	49,221	48.1	22.6	31.0	29,182,539	4.6	52.1	4.2
Abruzzo	24,925,479	4.3	61.0	3.9	18,506	48.5	6.2	37.3	24,943,985	4.3	60.6	3.9
Molise	7,957,709	7.5	52.9	6.8	48,102	74.5	17.7	62.1	8,005,810	7.5	52.2	6.8
Campania	22,194,010	5.6	55.4	5.2	21,195	66.3	6.7	26.8	22,215,205	5.6	55.0	5.2
Puglia	6,308,507	8.6	44.3	7.4	456	100.0	4.6	–	6,308,963	8.6	44.3	7.4
Basilicata	14,000,271	6.3	48.9	5.9	1019	100.0	0.7	–	14,001,290	6.3	48.6	5.9
Calabria	37,479,082	5.0	76.1	4.6	88,146	54.2	36.6	36.9	37,567,228	5.0	75.9	4.6
Sicilia	10,691,845	6.3	37.6	5.6	597	70.7	0.8	–	10,692,442	6.3	37.5	5.6
Sardegna	17,845,610	5.2	29.7	4.9	272,184	20.9	10.5	18.0	18,117,794	5.1	28.9	4.8
Italia	519,365,662	1.0	58.0	0.9	2,772,640	10.1	21.6	8.1	522,138,302	1.0	57.5	0.9

Table 12.5 Total value and value per hectare of carbon stock in small trees biomass by Forest inventory category / Valori totali e per ettaro di carbonio totale della rinnovazione per le categorie inventariali del Bosco

Region/Regione	Tall trees forest				Plantations				Total Forest			
	Boschi alti				Impianti di arboricoltura da legno				Totale Bosco			
	Corg	ES	Corg	ES	Corg	ES	Corg	ES	Corg	ES	Corg	ES
	(Mg)	(%)	(Mg ha^{-1})	(%)	(Mg)	(%)	(Mg ha^{-1})	(%)	(Mg)	(%)	(Mg ha^{-1})	(%)
Piemonte	572,243	8.3	0.7	8.2	1699	47.9	0.1	45.6	573,941	8.3	0.6	8.2
Valle d'Aosta	32,983	19.6	0.3	19.2	0	–	0	–	32,983	19.6	0.3	19.2
Lombardia	430,250	9.0	0.7	8.8	798	72.6	0	–	431,048	9.0	0.7	8.8
Alto Adige	136,830	13.7	0.4	13.5	0	–	0	–	136,830	13.7	0.4	13.5
Trentino	232,296	11.5	0.6	11.4	0	–	0	–	232,296	11.5	0.6	11.4
Veneto	393,125	9.7	1.0	9.3	8281	91.2	1.6	90.1	401,406	9.7	1.0	9.3
Friuli V.G	209,141	9.2	0.6	9.0	8313	89.8	0.9	88.2	217,454	9.5	0.7	9.3
Liguria	192,888	12.4	0.6	12.3	0	–	0	–	192,888	12.4	0.6	12.3
Emilia Romagna	530,585	11.3	0.9	11.1	53	83.7	0	–	530,637	11.3	0.9	11.1
Toscana	1,454,657	13.2	1.4	13.1	1263	47.6	0.2	42.0	1,455,921	13.2	1.4	13.1
Umbria	936,136	9.2	2.4	9.1	1549	52.8	0.2	57.1	937,685	9.2	2.4	9.0
Marche	585,637	9.5	2.1	9.2	6314	84.7	0.9	86.2	591,952	9.4	2.0	9.1
Lazio	762,310	14.6	1.4	14.5	316	100.0	0.1	–	762,626	14.6	1.4	14.4
Abruzzo	344,838	8.2	0.8	7.9	1919	58.8	0.6	45.4	346,757	8.2	0.8	7.9
Molise	405,796	11.9	2.7	11.5	509	62.2	0.2	54.9	406,306	11.9	2.7	11.5
Campania	537,175	18.7	1.3	18.6	1288	66.9	0.4	61.1	538,464	18.7	1.3	18.5
Puglia	108,069	14.5	0.8	14.0	8	100.0	0.1	–	108,077	14.5	0.8	14.0
Basilicata	318,959	16.4	1.1	16.3	27	74.1	0	–	318,986	16.4	1.1	16.3
Calabria	636,539	15.1	1.3	15.0	4747	93.0	2.0	84.3	641,286	15.0	1.3	14.9
Sicilia	80,141	18.5	0.3	18.2	0	–	0	–	80,141	18.5	0.3	18.2
Sardegna	356,580	13.8	0.6	13.8	27,775	33.8	1.1	32.4	384,355	13.1	0.6	13.0
Italia	9,257,179	3.4	1.0	3.4	64,859	24.5	0.5	24.3	9,322,038	3.4	1.0	3.3

Table 12.6 Total value and value per hectare of carbon stock in shrubs biomass by Forest inventory category / Valori totali e per ettaro di carbonio totale degli arbusti per le categorie inventariali del Bosco

Region/Regione	Tall trees forest				Plantations				Total Forest			
	Boschi alti				Impianti di arboricoltura da legno				Totale Bosco			
	Corg	ES	Corg	ES	Corg	ES	Corg	ES	Corg	ES	Corg	ES
	(Mg)	(%)	(Mg ha^{-1})	(%)	(Mg)	(%)	(Mg ha^{-1})	(%)	(Mg)	(%)	(Mg ha^{-1})	(%)
Piemonte	574,512	7.0	0.7	6.8	6470	58.6	0.3	54.6	580,981	6.9	0.7	6.7
Valle d'Aosta	29,446	21.2	0.3	20.9	0	–	0.0	–	29,446	21.2	0.3	20.9
Lombardia	481,109	8.9	0.8	8.8	17,417	89.2	0.7	86.7	498,526	9.2	0.8	9.0
Alto Adige	52,511	17.9	0.2	17.8	0	–	0.0	–	52,511	17.9	0.2	17.8
Trentino	209,051	12.5	0.6	12.3	0	–	0.0	–	209,051	12.5	0.6	12.3
Veneto	492,729	11.7	1.2	11.5	428	94.7	0.1	94.4	493,157	11.7	1.2	11.5
Friuli V.G	217,372	11.7	0.7	11.5	2946	42.9	0.3	38.1	220,317	11.6	0.7	11.4
Liguria	308,789	11.8	0.9	11.7	217	100.0	0.6	–	309,006	11.8	0.9	11.7
Emilia Romagna	349,232	7.8	0.6	7.5	857	53.9	0.1	53.8	350,089	7.7	0.6	7.5
Toscana	952,765	4.8	0.9	4.6	3311	54.6	0.5	43.5	956,076	4.8	0.9	4.6
Umbria	428,876	6.5	1.1	6.2	5621	44.3	0.9	18.1	434,497	6.4	1.1	6.1
Marche	356,732	6.9	1.3	6.4	6627	50.9	1.0	39.1	363,359	6.8	1.2	6.3
Lazio	456,564	7.3	0.8	7.0	812	99.7	0.4	90.6	457,375	7.3	0.8	7.0
Abruzzo	327,048	9.7	0.8	9.5	1174	93.1	0.4	94.1	328,222	9.7	0.8	9.4
Molise	338,125	9.1	2.2	8.4	1849	52.4	0.7	38.4	339,973	9.1	2.2	8.4
Campania	386,503	8.2	1.0	7.8	5702	73.2	1.8	31.0	392,206	8.1	1.0	7.8
Puglia	121,470	19.2	0.9	18.3	0	–	0.0	–	121,470	19.2	0.9	18.3
Basilicata	249,705	8.3	0.9	7.9	63	78.2	0.0	94.7	249,769	8.3	0.9	7.9
Calabria	293,917	13.4	0.6	13.1	1377	63.2	0.6	52.0	295,293	13.3	0.6	13.1
Sicilia	141,477	13.0	0.5	12.6	0	–	0.0	–	141,477	13.0	0.5	12.6
Sardegna	1,006,941	6.3	1.7	6.1	26,610	34.3	1.0	32.7	1,033,550	6.2	1.7	6.0
Italia	7,774,873	2.0	0.9	2.0	81,480	24.0	0.6	22.9	7,856,353	2.0	0.9	2.0

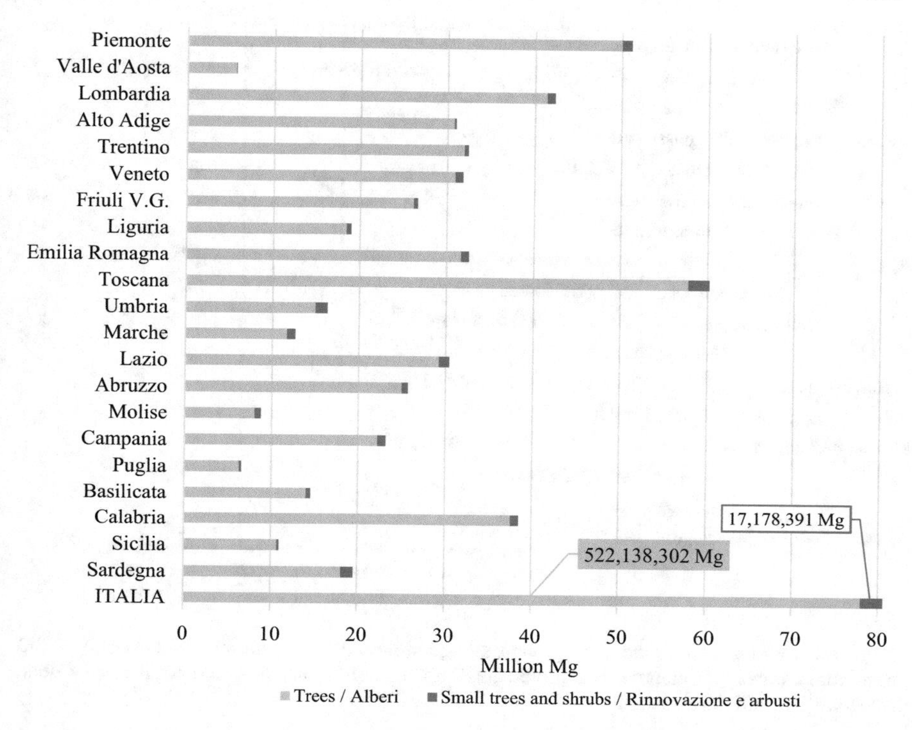

Fig. 12.1 Carbon in trees, small trees (regeneration) and shrubs in Italian Forest (X-axis was limited to improve readibility; exceeding values are given in numbers) / Contenuto di carbonio negli alberi, nella rinnovazione e negli arbusti nel Bosco italiano (l'asse delle ascisse è stato limitato per migliorare la leggibilità; i valori eccedenti sono mostrati in numero)

12.3 Carbon Stock in Deadwood

Table 12.7 shows the overall carbon stored in coarse woody debris in the forest inventory categories and for Total Forest. Tables 12.8 and 12.9 report similar information for the forest types. The three deadwood components considered by INFC2015 store altogether 29.8 million carbon tonnes, 3.3 tonnes per hectare on average. Table 12.10 shows the statistics for the standing dead trees, Table 12.11 shows those for stumps and Table 12.12 shows those for lying deadwood. The statistics for the three components in the forest types are available at www.inventarioforestale.org/statistiche_INFC.

Figure 12.3 reveals that 57.5% of coarse woody debris carbon stock is in standing dead trees; lying deadwood stores 33.2% of coarse woody debris carbon and stumps store 9.3%.

Figure 12.4 shows the carbon stock in the three deadwood components for the Tall trees forest types.

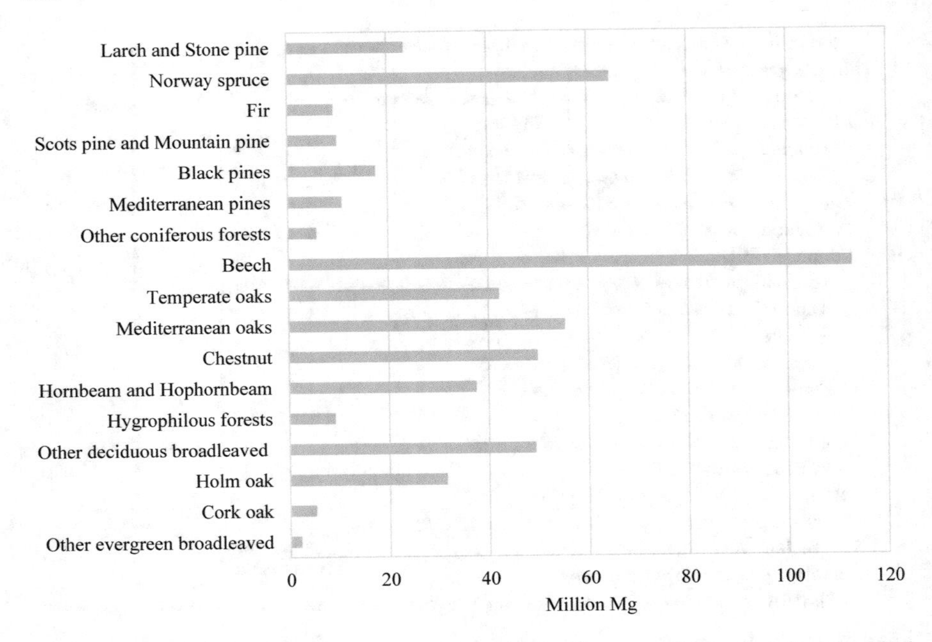

Fig. 12.2 Carbon stock in the aboveground living biomass (trees, small trees and shrubs) of Tall trees forest types / Contenuto di carbonio negli alberi, nella rinnovazione e negli arbusti delle categorie forestali dei Boschi alti

Lying deadwood is the main component in carbon storing in four forest types: Larch and Stone pine, Norway spruce, Fir and Hygrophilous forests. In these categories lying deadwood stores between 41.5% (Norway Spruce forest) and 49.4% (Hygrophilous forests) of total coarse woody debris carbon stock. In all other forest types, most carbon is stored in the standing dead trees component, which generally accounts for more than 50%, except Scots pine and Mountain pine forest (48.2%) and Beech forest (47.8%). The average national value for carbon stored by stumps is exceeded in five forest types: Larch and Stone pine, Norway spruce, Fir, Beech and Cork oak forests, whose values range from 12.8% (Cork oak forest) to 23.3% (Norway spruce forests). Figure 12.5 shows the carbon stock in the regions by deadwood components.

As a general picture, standing dead trees are the main source of carbon stock, with percentages ranging from 49.9% (Marche) to 70.7% (Toscana), but in Valle d'Aosta and in Trentino the main deadwood component for carbon stock is lying deadwood (55.1% and 43.3%, respectively) while in Alto Adige both stumps (36.7%) and lying deadwood (36.5%) store more carbon than standing dead trees (26.7%).

Table 12.7 Total value and value per hectare of carbon stock in coarse woody debris by Forest inventory category / Valori totali e per ettaro del carbonio nel legno morto grosso totale per le categorie inventariali del Bosco

Region/Regione	Tall trees forest				Plantations				Total Forest			
	Boschi alti				Impianti di arboricoltura da legno				Totale Bosco			
	Corg	ES	Corg	ES	Corg	ES	Corg	ES	Corg	ES	Corg	ES
	(Mg)	(%)	(Mg ha^{-1})	(%)	(Mg)	(%)	(Mg ha^{-1})	(%)	(Mg)	(%)	(Mg ha^{-1})	(%)
Piemonte	4,466,212	6.2	5.1	6.0	9800	47.7	0.5	44.1	4,476,013	6.2	5.0	6.0
Valle d'Aosta	441,332	15.5	4.4	14.9	0	–	0.0	–	441,332	15.5	4.4	14.9
Lombardia	3,049,024	7.8	5.1	7.6	26,696	63.8	1.1	61.0	3,075,720	7.7	4.9	7.5
Alto Adige	1,641,778	7.9	4.8	7.7	0	–	0.0	–	1,641,778	7.9	4.8	7.7
Trentino	1,800,674	10.5	4.8	10.4	0	–	0.0	–	1,800,674	10.5	4.8	10.4
Veneto	1,510,000	7.6	3.7	7.3	2453	57.4	0.5	42.8	1,512,453	7.6	3.6	7.3
Friuli V.G	1,829,989	9.3	5.7	9.0	3550	56.5	0.4	53.7	1,833,539	9.3	5.5	9.0
Liguria	2,102,955	8.0	6.1	7.9	0	–	0.0	–	2,102,955	8.0	6.1	7.9
Emilia Romagna	1,866,169	8.3	3.2	8.1	8171	99.4	1.4	98.3	1,874,340	8.2	3.2	8.0
Toscana	4,461,948	6.7	4.3	6.6	6625	59.9	1.0	58.8	4,468,573	6.7	4.3	6.6
Umbria	521,848	14.0	1.4	13.9	225	93.3	0.0	96.1	522,073	14.0	1.3	13.9
Marche	605,998	13.1	2.1	12.8	2398	77.6	0.3	79.6	608,396	13.1	2.1	12.8
Lazio	1,049,815	12.3	1.9	12.2	4592	96.1	2.1	86.3	1,054,407	12.3	1.9	12.1
Abruzzo	629,827	10.8	1.5	10.7	926	96.4	0.3	88.2	630,752	10.8	1.5	10.6
Molise	137,469	18.7	0.9	18.4	4942	100.0	1.8	–	142,410	18.4	0.9	18.0
Campania	821,633	11.7	2.1	11.4	2384	67.4	0.8	25.3	824,017	11.6	2.0	11.4
Puglia	91,662	14.3	0.6	13.7	0	–	0.0	–	91,662	14.3	0.6	13.7
Basilicata	294,365	18.0	1.0	17.8	0	–	0.0	–	294,365	18.0	1.0	17.8
Calabria	1,631,264	12.5	3.3	12.4	7664	80.5	3.2	70.2	1,638,929	12.5	3.3	12.3
Sicilia	342,230	17.7	1.2	17.5	0	–	0.0	–	342,230	17.7	1.2	17.5
Sardegna	445,905	14.9	0.7	14.8	7469	36.2	0.3	34.5	453,374	14.6	0.7	14.5
Italia	29,742,095	2.3	3.3	2.3	87,895	25.3	0.7	24.6	29,829,991	2.3	3.3	2.3

Table 12.8 Total value and value per hectare of carbon stock in coarse woody debris by Tall trees forest type / Valori totali e per ettaro del carbonio nel legno morto grosso totale per le categorie forestali dei Boschi alti

Region/Regione	Larch and Stone pine				Norway spruce				Fir			
	Boschi di larice e cembro				Boschi di abete rosso				Boschi di abete bianco			
	Corg	ES	Corg	ES	Corg	ES	Corg	ES	Corg	ES	Corg	ES
	(Mg)	(%)	(Mg ha^{-1})	(%)	(Mg)	(%)	(Mg ha^{-1})	(%)	(Mg)	(%)	(Mg ha^{-1})	(%)
Piemonte	168,210	18.9	2.1	–	208,693	40.2	11.5	–	62,021	37.5	4.4	–
Valle d'Aosta	107,789	30.8	2.3	–	84,132	22.1	4.6	–	35,938	87.1	31.1	–
Lombardia	249,379	30.1	4.7	–	478,621	17.9	5.4	–	53,337	35.8	5.7	–
Alto Adige	473,981	19.7	5.1	–	865,729	10.0	4.9	–	12,640	51.3	4.8	–
Trentino	214,579	22.2	3.5	–	1,024,121	17.2	7.5	–	83,638	22.9	5.2	–
Veneto	93,444	26.4	2.2	–	485,335	15.3	5.0	–	49,502	33.4	7.0	–
Friuli V.G	35,695	41.8	2.8	–	393,219	20.0	8.8	–	17,400	59.1	9.4	–
Liguria	10,744	83.3	9.8	–	23	100.0	0.1	–	20,526	78.9	7.0	–
Emilia Romagna	0	–	0.0	–	33,980	42.1	9.2	–	29,484	51.9	10.0	–
Toscana	0	–	0.0	–	8298	100.0	23.0	–	56,303	41.3	11.1	–
Umbria	0	–	0.0	–	0	–	0.0	–	0	–	0.0	–
Marche	0	–	0.0	–	412	100.0	1.1	–	0	–	0.0	–
Lazio	0	–	0.0	–	1214	100.0	3.3	–	0	–	0.0	–
Abruzzo	0	–	0.0	–	813	100.0	2.2	–	2569	90.0	3.5	–
Molise	0	–	0.0	–	0	–	0.0	–	2515	76.6	2.1	–
Campania	0	–	0.0	–	0	–	0.0	–	0	–	0.0	–
Puglia	0	–	0.0	–	0	–	0.0	–	0	–	0.0	–
Basilicata	0	–	0.0	–	0	–	0.0	–	3245	93.0	4.4	–
Calabria	0	–	0.0	–	0	–	0.0	–	17,130	57.5	4.6	–
Sicilia	0	–	0.0	–	0	–	0.0	–	0	–	0.0	–
Sardegna	0	–	0.0	–	0	–	0.0	–	0	–	0.0	–
Italia	1,353,821	10.3	3.4	10.0	3,584,591	7.1	6.1	6.8	446,249	13.9	6.4	11.6

(continued)

Table 12.8 (continued)

Region/Regione	Scots pine and Mountain pine				Black pines				Mediterranean pines			
	Pinete di pino silvestre e montano				Pinete di pino nero, laricio e loricato				Pinete di pini mediterranei			
	Corg	ES	Corg	ES	Corg	ES	Corg	ES	Corg	ES	Corg	ES
	(Mg)	(%)	(Mg ha^{-1})	(%)	(Mg)	(%)	(Mg ha^{-1})	(%)	(Mg)	(%)	(Mg ha^{-1})	(%)
Piemonte	120,091	26.0	5.8	–	8368	45.8	2.6	–	577	70.9	0.7	–
Valle d'Aosta	79,629	49.1	6.8	–	1073	100.0	2.8	–	0	–	0.0	–
Lombardia	57,132	32.9	3.0	–	29,236	48.8	6.6	–	0	–	0.0	–
Alto Adige	119,540	21.2	3.2	–	185	100.0	0.5	–	0	–	0.0	–
Trentino	53,091	25.0	2.5	–	14,409	40.1	2.5	–	0	–	0.0	–
Veneto	30,510	30.8	2.5	–	2100	72.3	0.6	–	5789	99.4	7.7	–
Friuli V.G	76,011	47.4	6.2	–	120,077	33.6	3.8	–	3085	89.7	2.8	–
Liguria	51,153	46.3	5.0	–	14,401	60.3	2.6	–	125,477	32.7	5.1	–
Emilia Romagna	5537	73.9	3.0	–	101,498	35.8	6.3	–	5668	66.5	1.9	–
Toscana	10,234	75.6	9.4	–	244,963	46.0	13.6	–	215,955	30.9	4.8	–
Umbria	1500	76.9	2.0	–	13,343	62.3	2.3	–	35,859	42.5	4.4	–
Marche	0	–	0.0	–	32,231	39.3	3.1	–	10,225	60.6	4.7	–
Lazio	0	–	0.0	–	24,509	44.1	2.9	–	22,596	40.9	3.1	–
Abruzzo	6804	63.7	6.3	–	25,795	40.1	1.4	–	6183	58.0	2.8	–
Molise	0	–	0.0	–	1296	57.3	0.6	–	14	70.7	0.0	–
Campania	0	–	0.0	–	41,134	38.0	7.4	–	49,601	47.5	5.4	–
Puglia	0	–	0.0	–	2277	68.7	1.5	–	18,051	36.4	0.7	–
Basilicata	0	–	0.0	–	2791	66.9	1.1	–	3349	39.2	0.4	–
Calabria	0	–	0.0	–	386,140	32.4	5.3	–	93,970	68.4	5.4	–
Sicilia	0	–	0.0	–	5225	50.6	0.7	–	107,141	39.1	2.4	–
Sardegna	0	–	0.0	–	5592	53.1	1.1	–	38,498	41.8	1.1	–
Italia	611,231	12.4	4.1	11.1	1,076,643	16.7	4.7	16.3	742,040	15.5	3.1	14.9

(continued)

Table 12.8 (continued)

Region/Regione	Other coniferous forests				Beech				Temperate oaks			
	Altri boschi di conifere, pure o miste				Faggete				Querceti di rovere, roverella e farnia			
	Corg	ES	Corg	ES	Corg	ES	Corg	ES	Corg	ES	Corg	ES
	(Mg)	(%)	(Mg ha^{-1})	(%)	(Mg)	(%)	(Mg ha^{-1})	(%)	(Mg)	(%)	(Mg ha^{-1})	(%)
Piemonte	33,820	40.6	5.2	–	350,808	15.0	3.0	–	215,658	20.6	2.8	–
Valle d'Aosta	0	–	0.0	–	7113	70.2	6.2	–	13,170	38.1	3.1	–
Lombardia	27,022	48.0	4.3	–	192,526	24.6	2.9	–	152,914	37.8	3.7	–
Alto Adige	0	–	0.0	–	32,846	35.8	7.9	–	18,692	39.9	4.1	–
Trentino	0	–	0.0	–	200,237	19.5	3.2	–	12,732	46.3	2.2	–
Veneto	0	–	0.0	–	233,977	17.2	3.5	–	38,534	26.1	2.5	–
Friuli V.G	2228	86.6	2.0	–	455,151	23.4	5.0	–	32,146	39.6	3.9	–
Liguria	3215	100.0	8.8	–	80,087	34.4	2.2	–	111,857	36.0	2.6	–
Emilia Romagna	19,436	38.2	5.9	–	183,503	16.1	1.8	–	108,474	18.6	1.4	–
Toscana	86,754	29.0	7.3	–	101,420	19.3	1.4	–	247,458	25.7	1.6	–
Umbria	10,137	49.4	5.5	–	27,835	29.8	1.7	–	55,286	21.2	0.5	–
Marche	4915	52.8	0.9	–	41,720	25.2	2.3	–	71,532	28.7	1.2	–
Lazio	3468	56.7	2.4	–	90,995	16.3	1.2	–	55,357	26.1	0.7	–
Abruzzo	2542	48.2	1.2	–	215,781	16.9	1.7	–	48,726	18.5	0.6	–
Molise	0	–	0.0	–	14,317	28.8	1.0	–	16,541	46.0	0.3	–
Campania	1964	64.9	2.3	–	126,214	21.2	2.2	–	67,667	27.3	1.1	–
Puglia	3600	66.3	1.5	–	5380	36.8	1.2	–	9804	33.4	0.4	–
Basilicata	15,953	69.4	3.1	–	48,407	27.4	1.8	–	10,942	31.4	0.3	–
Calabria	115,069	35.9	9.2	–	118,762	22.0	1.5	–	108,056	47.6	2.0	–
Sicilia	9559	45.1	1.0	–	8560	39.3	0.6	–	97,203	35.8	1.4	–
Sardegna	54,270	66.6	4.2	–	0	–	0.0	–	27,331	35.1	0.3	–
Italia	393,951	16.6	4.7	15.4	2,535,640	6.2	2.4	6.0	1,520,079	8.6	1.3	8.4

(continued)

Table 12.8 (continued)

Region/Regione	Mediterranean oaks				Chestnut				Hornbeam and Hophornbeam			
	Cerrete, boschi di farnetto, fragno e vallonea				Castagneti				Ostrieti, carpineti			
	Corg	ES	Corg	ES	Corg	ES	Corg	ES	Corg	ES	Corg	ES
	(Mg)	(%)	(Mg ha^{-1})	(%)	(Mg)	(%)	(Mg ha^{-1})	(%)	(Mg)	(%)	(Mg ha^{-1})	(%)
Piemonte	34,951	49.9	5.8	–	1,830,517	11.2	11.2	–	60,015	31.1	3.7	–
Valle d'Aosta	0	–	0.0	–	48,452	39.0	11.4	–	0	–	0.0	–
Lombardia	4888	66.5	2.2	–	963,674	15.5	11.7	–	150,917	22.1	1.8	–
Alto Adige	0	–	0.0	–	21,456	62.8	11.3	–	56,512	47.0	6.4	–
Trentino	0	–	0.0	–	7238	57.7	5.0	–	76,939	19.1	1.9	–
Veneto	0	–	0.0	–	184,723	28.2	10.7	–	237,390	22.9	2.9	–
Friuli V.G	0	–	0.0	–	189,889	26.2	15.5	–	160,642	30.6	3.4	–
Liguria	63,024	41.0	5.6	–	1,301,838	10.9	11.9	–	152,731	33.7	3.3	–
Emilia Romagna	236,481	24.1	2.3	–	399,019	21.1	9.8	–	325,115	19.4	2.9	–
Toscana	383,039	14.5	1.5	–	2,003,595	10.1	12.8	–	293,115	31.3	4.5	–
Umbria	171,860	32.5	1.3	–	8997	74.8	4.1	–	46,722	28.7	0.8	–
Marche	13,123	30.6	0.5	–	73,058	47.9	19.7	–	70,012	21.6	0.9	–
Lazio	221,558	30.3	1.7	–	183,549	34.0	5.1	–	102,983	19.7	1.1	–
Abruzzo	51,433	40.3	1.6	–	36,458	34.6	7.7	–	89,288	31.0	1.9	–
Molise	48,325	36.8	0.9	–	830	100.0	2.1	–	9122	39.6	1.0	–
Campania	33,513	25.5	0.4	–	257,008	23.8	4.6	–	49,746	21.1	0.9	–
Puglia	8165	30.3	0.2	–	2205	72.6	1.9	–	2422	50.0	0.5	–
Basilicata	47,803	19.4	0.4	–	67,636	62.1	11.4	–	8928	62.8	1.3	–
Calabria	148,558	33.8	3.4	–	337,896	25.9	4.9	–	0	–	0.0	–
Sicilia	10,729	37.8	0.3	–	42,425	55.9	4.9	–	1500	60.7	0.9	–
Sardegna	0	–	0.0	–	4537	56.8	2.4	–	0	–	0.0	–
Italia	1,477,449	9.2	1.4	9.0	7,964,999	5.0	10.2	4.6	1,894,098	8.3	2.2	8.0

(continued)

Table 12.8 (continued)

Region/Regione	Hygrophilous forests				Other deciduous broadleaved forests				Holm oak			
	Boschi igrofili				Altri boschi caducifogli				Leccete			
	Corg	ES	Corg	ES	Corg	ES	Corg	ES	Corg	ES	Corg	ES
	(Mg)	(%)	(Mg ha^{-1})	(%)	(Mg)	(%)	(Mg ha^{-1})	(%)	(Mg)	(%)	(Mg ha^{-1})	(%)
Piemonte	70,314	29.3	3.1	–	1,301,674	11.7	4.0	–	0	–	0.0	–
Valle d'Aosta	0	–	0.0	–	64,036	42.2	6.1	–	0	–	0.0	–
Lombardia	170,505	32.5	9.0	–	517,181	22.8	4.3	–	0	–	0.0	–
Alto Adige	10,908	46.9	4.0	–	12,107	40.9	3.4	–	0	–	0.0	–
Trentino	29,930	51.8	7.8	–	83,761	35.2	5.1	–	0	–	0.0	–
Veneto	36,455	33.3	3.5	–	106,811	15.6	2.0	–	5429	72.4	1.8	–
Friuli V.G	46,974	35.8	3.7	–	297,474	23.6	6.3	–	0	–	0.0	–
Liguria	25,643	51.5	7.8	–	102,910	27.4	3.2	–	39,155	46.6	2.8	–
Emilia Romagna	118,219	33.3	4.2	–	296,538	26.1	3.4	–	3218	71.7	4.4	–
Toscana	152,740	40.5	5.5	–	427,377	20.2	5.1	–	226,034	37.0	1.7	–
Umbria	50,543	54.6	6.3	–	14,199	54.5	1.3	–	85,567	32.6	2.1	–
Marche	117,546	43.3	9.1	–	166,583	25.2	2.7	–	4641	35.8	0.7	–
Lazio	64,068	50.3	7.0	–	192,782	31.6	3.6	–	72,394	73.3	1.5	–
Abruzzo	63,755	51.2	3.2	–	73,453	39.4	1.3	–	6156	54.2	0.7	–
Molise	20,139	43.3	2.2	–	23,486	61.5	2.3	–	882	85.9	0.8	–
Campania	42,923	49.3	3.9	–	98,387	53.2	2.9	–	52,537	41.8	1.4	–
Puglia	181	100.0	0.5	–	24,315	34.1	2.1	–	8490	52.7	0.5	–
Basilicata	37,914	43.4	2.9	–	34,244	42.6	1.1	–	340	100.0	0.0	–
Calabria	31,967	58.2	2.5	–	106,069	66.8	2.3	–	131,369	31.8	2.7	–
Sicilia	3237	87.0	0.7	–	6367	51.0	0.5	–	17,082	37.2	0.9	–
Sardegna	5723	53.1	1.7	–	814	60.0	0.1	–	230,172	21.3	0.9	–
Italia	1,099,684	11.6	4.7	10.9	3,950,571	6.8	3.5	6.6	883,466	14.2	1.4	14.0

(continued)

Table 12.8 (continued)

Region/Regione	Coark oak				Other evergreen broadleaved forests				Not classified				Total Tall trees forest			
	Sugherete				Altri boschi di latifoglie sempreverdi				Non classificato				Totale Boschi alti			
	Corg	ES	Corg	ES	Corg	ES	Corg	ES	Corg	ES	Corg	ES	Corg	ES	Corg	ES
	(Mg)	(%)	(Mg ha^{-1})	(%)	(Mg)	(%)	(Mg ha^{-1})	(%)	(Mg)	(%)	(Mg ha^{-1})	(%)	(Mg)	(%)	(Mg ha^{-1})	(%)
Piemonte	0	–	0.0	–	0	–	0.0	–	496	100.0	1.0	–	4,466,212	6.2	5.1	6.0
Valle d'Aosta	0	–	0.0	–	0	–	0.0	–	0	–	0.0	–	441,332	15.5	4.4	14.9
Lombardia	0	–	0.0	–	1006	71.8	0.4	–	688	100.0	1.6	–	3,049,024	7.8	5.1	7.6
Alto Adige	0	–	0.0	–	0	–	0.0	–	17,180	50.7	11.4	–	1,641,778	7.9	4.8	7.7
Trentino	0	–	0.0	–	0	–	0.0	–	0	–	0.0	–	1,800,674	10.5	4.8	10.4
Veneto	0	–	0.0	–	0	–	0.0	–	0	–	0.0	–	1,510,000	7.6	3.7	7.3
Friuli V.G	0	–	0.0	–	0	–	0.0	–	0	–	0.0	–	1,829,989	9.3	5.7	9.0
Liguria	0	–	0.0	–	0	–	0.0	–	171	70.7	0.2	–	2,102,955	8.0	6.1	7.9
Emilia Romagna	0	–	0.0	–	0	–	0.0	–	0	–	0.0	–	1,866,169	8.3	3.2	8.1
Toscana	3917	47.4	0.6	–	503	89.3	0.3	–	241	70.7	0.3	–	4,461,948	6.7	4.3	6.6
Umbria	0	–	0.0	–	0	–	0.0	–	0	–	0.0	–	521,848	14.0	1.4	13.9
Marche	0	–	0.0	–	0	–	0.0	–	0	–	0.0	–	605,998	13.1	2.1	12.8
Lazio	7197	58.7	2.8	–	6775	88.3	2.6	–	369	57.7	0.3	–	1,049,815	12.3	1.9	12.2
Abruzzo	0	–	0.0	–	0	–	0.0	–	71	100.0	0.2	–	629,827	10.8	1.5	10.7
Molise	0	–	0.0	–	0	–	0.0	–	0	–	0.0	–	137,469	18.7	0.9	18.4
Campania	816	100.0	2.2	–	0	–	0.0	–	123	100.0	0.3	–	821,633	11.7	2.1	11.4
Puglia	0	–	0.0	–	5959	70.1	1.2	–	813	67.6	0.7	–	91,662	14.3	0.6	13.7
Basilicata	0	–	0.0	–	12,522	100.0	6.7	–	291	100.0	0.8	–	294,365	18.0	1.0	17.8
Calabria	5765	44.2	1.1	–	28,302	61.7	1.3	–	2209	43.0	1.0	–	1,631,264	12.5	3.3	12.4
Sicilia	9859	51.5	0.6	–	23,344	35.6	0.6	–	0	–	0.0	–	342,230	17.7	1.2	17.5
Sardegna	61,149	28.1	0.4	–	17,731	41.3	0.5	–	88	100.0	0.2	–	445,905	14.9	0.7	14.8
Italia	88,704	21.1	0.5	20.7	96,141	26.3	0.8	26.0	22,741	38.8	2.3	33.6	29,742,095	2.3	3.3	2.3

Table 12.9 Total value and value per hectare of carbon stock in coarse woody debris by Plantations forest type / Valori totali e per ettaro di carbonio nel legno morto grosso totale per le categorie forestali degli Impianti di arboricoltura da legno

Region/Regione	Poplar plantations				Other broadleaved plantations				Coniferous plantations				Total Plantations			
	Pioppeti artificiali				Piantagioni di altre latifoglie				Piantagioni di conifere				Totale Impianti di arboricoltura da legno			
	Corg	ES	Corg	ES	Corg	ES	Corg	ES	Corg	ES	Corg	ES	Corg	ES	Corg	ES
	(Mg)	(%)	(Mg ha^{-1})	(%)	(Mg)	(%)	(Mg ha^{-1})	(%)	(Mg)	(%)	(Mg ha^{-1})	(%)	(Mg)	(%)	(Mg ha^{-1})	(%)
Piemonte	6213	67.4	0.4	–	1516	100.0	0.4	–	2071	68.2	1.8	–	9800	47.7	0.5	44.1
Valle d'Aosta	0	–	0.0	–	0	–	0.0	–	0	–	0.0	–	0	–	0.0	–
Lombardia	4183	64.9	0.2	–	19,266	85.7	3.1	–	3247	100.0	7.4	–	26,696	63.8	1.1	61.0
Alto Adige	0	–	0.0	–	0	–	0.0	–	0	–	0.0	–	0	–	0.0	–
Trentino	0	–	0.0	–	0	–	0.0	–	0	–	0.0	–	0	–	0.0	–
Veneto	156	94.1	0.1	–	2297	61.0	0.8	–	0	–	0.0	–	2453	57.4	0.5	42.8
Friuli V.G	349	96.6	0.1	–	3201	62.1	1.2	–	0	–	0.0	–	3550	56.5	0.4	53.7
Liguria	0	–	0.0	–	0	–	0.0	–	0	–	0.0	–	0	–	0.0	–
Emilia Romagna	51	100.0	0.0	–	0	–	0.0	–	8120	100.0	22.0	–	8171	99.4	1.4	98.3
Toscana	0	–	0.0	–	139	100.0	0.0	–	6487	61.1	4.5	–	6625	59.9	1.0	58.8
Umbria	0	–	0.0	–	225	93.3	0.0	–	0	–	0.0	–	225	93.3	0.0	96.1
Marche	0	–	0.0	–	2398	77.6	0.3	–	0	–	0.0	–	2398	77.6	0.3	79.6
Lazio	0	–	0.0	–	4592	96.1	3.3	–	0	–	0.0	–	4592	96.1	2.1	86.3
Abruzzo	13	100.0	0.0	–	913	97.8	0.4	–	0	–	0.0	–	926	96.4	0.3	88.2
Molise	4942	100.0	12.7	–	0	–	0.0	–	0	–	0.0	–	4942	100.0	1.8	88.9
Campania	1522	100.0	1.0	–	862	59.5	0.5	–	0	–	0.0	–	2384	67.4	0.8	25.3
Puglia	0	–	0.0	–	0	–	0.0	–	0	–	0.0	–	0	–	0.0	–
Basilicata	0	–	0.0	–	0	–	0.0	–	0	–	0.0	–	0	–	0.0	–
Calabria	0	–	0.0	–	6063	100.0	3.9	–	1601	70.7	2.1	–	7664	80.5	3.2	70.2
Sicilia	0	–	0.0	–	0	–	0.0	–	0	–	0.0	–	0	–	0.0	–
Sardegna	0	–	0.0	–	7378	36.6	0.3	–	91	100.0	0.1	–	7469	36.2	0.3	34.5
Italia	17,428	41.3	0.3	40.3	48,850	38.2	0.7	36.7	21,617	45.2	3.8	237.0	87,895	25.3	0.7	24.6

Table 12.10 Total value and value per hectare of carbon stock in standing dead trees by Forest inventory category / Valori totali e per ettaro di carbonio negli alberi morti in piedi per le categorie inventariali del Bosco

Region/Regione	Tall trees forest				Plantations				Total Forest			
	Boschi alti				Impianti di arboricoltura da legno				Totale Bosco			
	Corg	ES	Corg	ES	Corg	ES	Corg	ES	Corg	ES	Corg	ES
	(Mg)	(%)	(Mg ha^{-1})	(%)	(Mg)	(%)	(Mg ha^{-1})	(%)	(Mg)	(%)	(Mg ha^{-1})	(%)
Piemonte	2,888,189	7.2	3.3	7.1	6048	68.4	0.3	67.1	2,894,237	7.2	3.3	7.1
Valle d'Aosta	136,366	17.2	1.4	16.7	0	–	0.0	–	136,366	17.2	1.4	16.7
Lombardia	1,561,562	9.5	2.6	9.4	20,416	73.1	0.8	70.3	1,581,979	9.4	2.5	9.3
Alto Adige	439,157	13.6	1.3	13.5	0	–	0.0	–	439,157	13.6	1.3	13.5
Trentino	740,218	16.2	2.0	16.1	0	–	0.0	–	740,218	16.2	2.0	16.1
Veneto	760,674	10.9	1.8	10.7	2395	58.7	0.5	44.2	763,069	10.9	1.8	10.7
Friuli V.G	914,605	11.7	2.8	11.5	1529	54.7	0.2	51.4	916,134	11.7	2.8	11.5
Liguria	1,424,606	8.7	4.2	8.6	0	–	0.0	–	1,424,606	8.7	4.2	8.6
Emilia Romagna	1,245,199	10.0	2.2	9.9	6814	100.0	1.1	–	1,252,013	10.0	2.1	9.8
Toscana	3,156,601	7.3	3.1	7.2	1863	73.4	0.3	72.7	3,158,465	7.3	3.1	7.2
Umbria	269,476	14.1	0.7	14.0	210	100.0	0.0	–	269,687	14.1	0.7	14.0
Marche	301,379	16.0	1.1	15.8	2263	78.1	0.3	80.0	303,642	15.9	1.0	15.7
Lazio	624,347	13.8	1.1	13.7	2977	96.5	1.4	86.8	627,324	13.7	1.1	13.6
Abruzzo	315,278	13.7	0.8	13.5	913	97.8	0.3	89.6	316,191	13.6	0.8	13.5
Molise	93,702	21.2	0.6	20.9	2204	100.0	0.8	–	95,907	20.8	0.6	20.5
Campania	549,035	13.7	1.4	13.5	630	71.6	0.2	75.7	549,665	13.6	1.4	13.4
Puglia	53,672	19.8	0.4	19.3	0	–	0.0	–	53,672	19.8	0.4	19.3
Basilicata	169,111	15.8	0.6	15.6	0	–	0.0	–	169,111	15.8	0.6	15.6
Calabria	975,308	14.5	2.0	14.4	5991	100.0	2.5	–	981,299	14.4	2.0	14.3
Sicilia	207,152	23.2	0.7	23.0	0	–	0.0	–	207,152	23.2	0.7	23.0
Sardegna	268,509	19.3	0.4	19.2	3411	62.3	0.1	61.4	271,919	19.1	0.4	19.0
Italia	17,094,148	2.8	1.9	2.8	57,665	32.4	0.4	31.8	17,151,813	2.8	1.9	2.8

Table 12.11 Total value and value per hectare of carbon stock in stumps by Forest inventory category/Valori totali e per ettaro di carbonio nelle ceppaie residue per le categorie inventariali del Bosco

Region/Region	Tall trees forest				Plantations				Total Forest			
	Boschi alti				Impianti di arboricoltura da legno				Totale Bosco			
	Corg	ES	Corg	ES	Corg	ES	Corg	ES	Corg	ES	Corg	ES
	(Mg)	(%)	(Mg ha^{-1})	(%)	(Mg)	(%)	(Mg ha^{-1})	(%)	(Mg)	(%)	(Mg ha^{-1})	(%)
Piemonte	252,936	10.5	0.3	10.4	68	72.9	0.0	72.3	253,003	10.5	0.3	10.4
Valle d'Aosta	61,901	23.5	0.6	23.2	0	–	0.0	–	61,901	23.5	0.6	23.2
Lombardia	273,661	12.5	0.5	12.4	755	41.5	0.0	38.9	274,416	12.5	0.4	12.4
Alto Adige	603,173	6.9	1.8	6.6	0	–	0.0	–	603,173	6.9	1.8	6.6
Trentino	281,582	10.3	0.8	10.2	0	–	0.0	–	281,582	10.3	0.8	10.2
Veneto	257,941	9.4	0.6	9.2	57	85.0	0.0	84.5	257,998	9.4	0.6	9.2
Friuli V.G	122,677	14.9	0.4	14.8	1878	86.4	0.2	84.8	124,555	14.7	0.4	14.6
Liguria	73,656	15.1	0.2	15.0	0	–	0.0	–	73,656	15.1	0.2	15.0
Emilia Romagna	80,699	18.8	0.1	18.8	11	100.0	0.0	–	80,710	18.8	0.1	18.8
Toscana	149,059	13.0	0.1	13.0	3018	78.4	0.4	77.5	152,077	12.8	0.1	12.8
Umbria	44,595	16.6	0.1	16.5	5	100.0	0.0	–	44,600	16.6	0.1	16.5
Marche	55,634	17.1	0.2	16.9	19	100.0	0.0	–	55,654	17.1	0.2	16.9
Lazio	80,393	18.9	0.1	18.8	94	82.7	0.0	70.9	80,487	18.8	0.1	18.8
Abruzzo	94,443	22.4	0.2	22.4	0	–	0.0	–	94,443	22.4	0.2	22.4
Molise	8250	22.0	0.1	21.8	329	100.0	0.1	–	8579	21.5	0.1	21.3
Campania	79,786	16.3	0.2	16.2	456	67.7	0.1	39.6	80,242	16.2	0.2	16.1
Puglia	3509	25.4	0.0	25.1	0	–	0.0	–	3509	25.4	0.0	25.1
Basilicata	19,204	18.6	0.1	18.5	0	–	0.0	–	19,204	18.6	0.1	18.5
Calabria	168,146	11.9	0.3	11.7	1098	70.7	0.5	58.8	169,244	11.8	0.3	11.7
Sicilia	18,252	27.0	0.1	26.9	0	–	0.0	–	18,252	27.0	0.1	26.9
Sardegna	43,958	14.8	0.1	14.7	3156	27.1	0.1	24.6	47,114	13.9	0.1	13.8
Italia	2,773,455	3.2	0.3	3.2	10,945	28.7	0.1	28.4	2,784,401	3.2	0.3	3.1

Table 12.12 Total value and value per hectare of carbon stock in lying deadwood by Forest inventory category / Valori totali e per ettaro di carbonio nel legno morto grosso a terra per le categorie inventariali del Bosco

Region/Regione	Tall trees forest				Plantations				Total Forest			
	Boschi alti				Impianti di arboricoltura da legno				Totale Bosco			
	Corg	ES	Corg	ES	Corg	ES	Corg	ES	Corg	ES	Corg	ES
	(Mg)	(%)	(Mg ha^{-1})	(%)	(Mg)	(%)	(Mg ha^{-1})	(%)	(Mg)	(%)	(Mg ha^{-1})	(%)
Piemonte	1,325,087	12.0	1.5	11.9	3685	51.6	0.2	45.8	1,328,772	12.0	1.5	11.9
Valle d'Aosta	243,064	23.5	2.4	23.0	0	–	0.0	–	243,064	23.5	2.4	23.0
Lombardia	1,213,801	13.8	2.0	13.7	5525	56.8	0.2	55.1	1,219,326	13.7	2.0	13.6
Alto Adige	599,447	16.6	1.8	16.5	0	–	0.0	–	599,447	16.6	1.8	16.5
Trentino	778,875	12.9	2.1	12.8	0	–	0.0	–	778,875	12.9	2.1	12.8
Veneto	491,385	11.4	1.2	11.1	0	–	0.0	–	491,385	11.4	1.2	11.2
Friuli V.G	792,707	14.9	2.5	14.7	142	100.0	0.0	–	792,849	14.9	2.4	14.7
Liguria	604,693	14.9	1.8	14.9	0	–	0.0	–	604,693	14.9	1.8	14.9
Emilia Romagna	540,271	12.8	0.9	12.7	1346	96.3	0.2	95.2	541,617	12.8	0.9	12.6
Toscana	1,156,288	13.3	1.1	13.2	1744	70.2	0.3	69.3	1,158,031	13.2	1.1	13.2
Umbria	207,777	27.5	0.5	27.5	9	100.0	0.0	–	207,786	27.5	0.5	27.5
Marche	248,984	19.3	0.9	19.1	116	72.0	0.0	74.6	249,100	19.3	0.9	19.1
Lazio	345,075	23.7	0.6	23.7	1521	100.0	0.7	–	346,596	23.6	0.6	23.6
Abruzzo	220,105	16.5	0.5	16.4	13	100.0	0.0	–	220,118	16.5	0.5	16.4
Molise	35,516	21.2	0.2	20.9	2409	100.0	0.9	–	37,925	20.8	0.2	20.5
Campania	192,811	26.8	0.5	26.7	1298	100.0	0.4	–	194,110	26.6	0.5	26.5
Puglia	34,481	21.5	0.2	21.1	0	–	0.0	–	34,481	21.5	0.2	21.1
Basilicata	106,050	39.2	0.4	39.1	0	–	0.0	–	106,050	39.2	0.4	39.1
Calabria	487,810	27.3	1.0	27.2	575	63.1	0.2	49.3	488,385	27.3	1.0	27.2
Sicilia	116,825	22.7	0.4	22.4	0	–	0.0	–	116,825	22.7	0.4	22.4
Sardegna	133,438	28.0	0.2	27.9	903	51.7	0.0	50.8	134,341	27.8	0.2	27.7
Italia	9,874,492	4.1	1.1	4.1	19,286	26.8	0.2	26.0	9,893,777	4.1	1.1	4.1

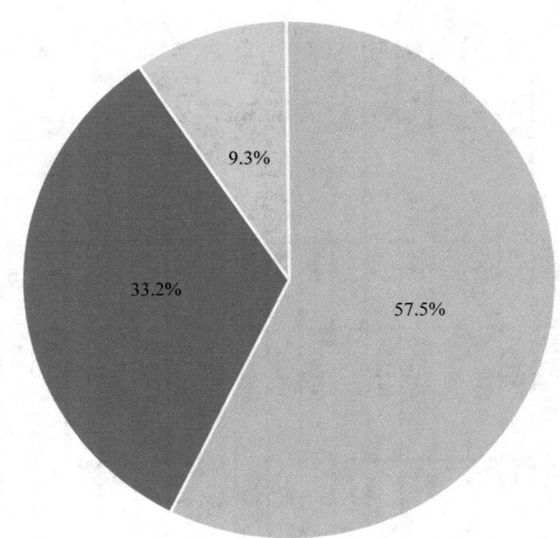

Fig. 12.3 Carbon stock in coarse woody debris in Forest / Contenuto di carbonio nelle componenti del legno morto grosso, nel Bosco

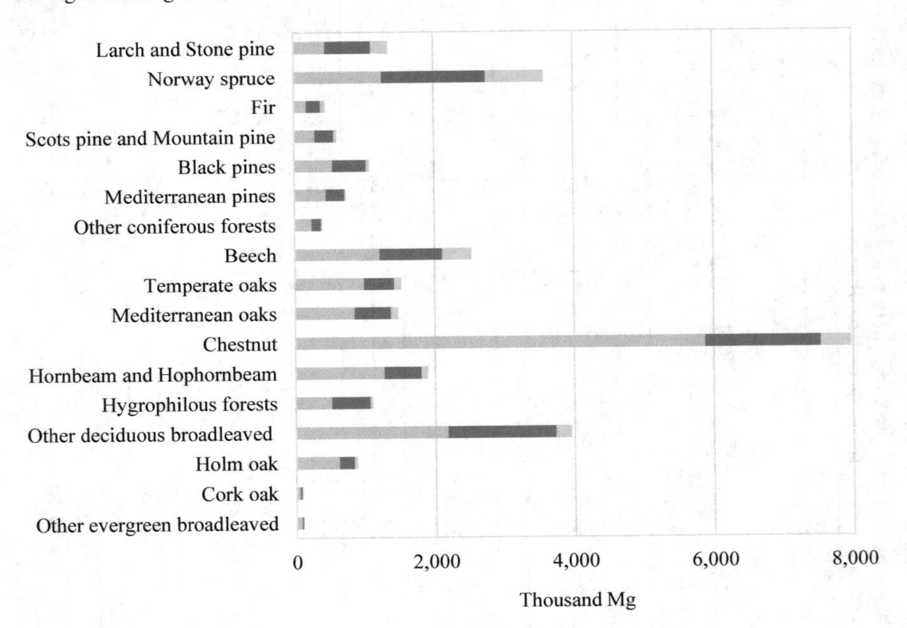

Fig. 12.4 Carbon stock in the Tall trees forest types by deadwood component / Contenuto di carbonio nelle categorie forestali dei Boschi alti per componente del legno morto

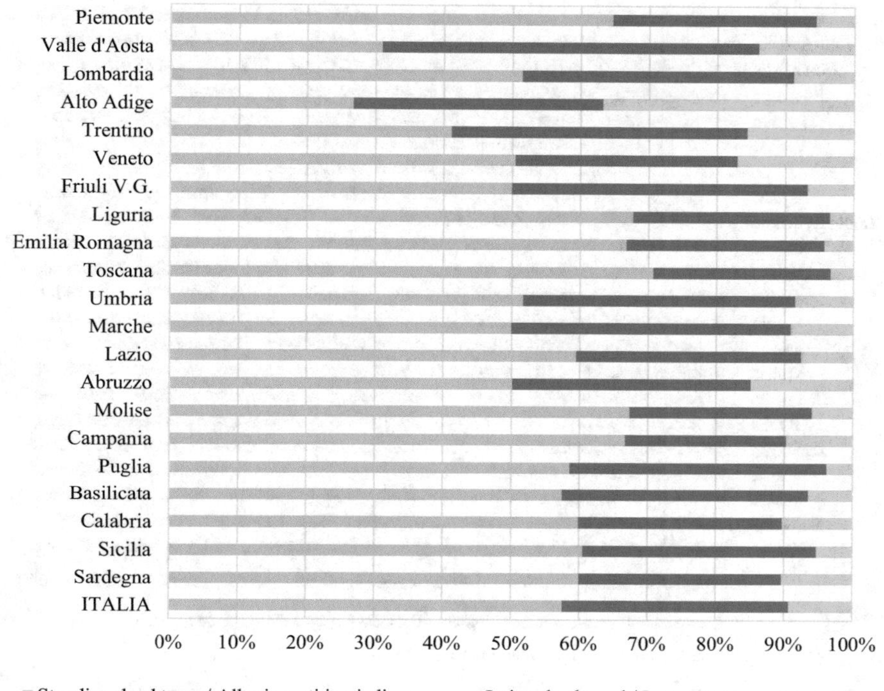

Fig. 12.5 Carbon stock in coarse woody debris by components, in the regions / Contenuto di carbonio nel legno morto per componente, nelle regioni

12.4 Overall Carbon Stock in the Aboveground Living and Deadwood Biomass

Together, woody vegetation taller than 50 cm and coarse woody debris store 569.1 million carbon tonnes. Carbon in trees and shrubs constitutes 94.8% of the total while deadwood comprises 5.2%. Figure 12.6 shows the cumulative percent of carbon in the aboveground living biomass and in coarse woody debris, in the regions.

The percentage of coarse woody debris is higher in the central-north regions. Its mean national value of 5.2% is exceeded in eight regions from the Alps to Toscana, except Alto Adige (5.0%) and Veneto (4.5%), which show values slightly lower. Marche and Calabria are the only two remaining regions whose coarse woody debris percent value is over 4%.

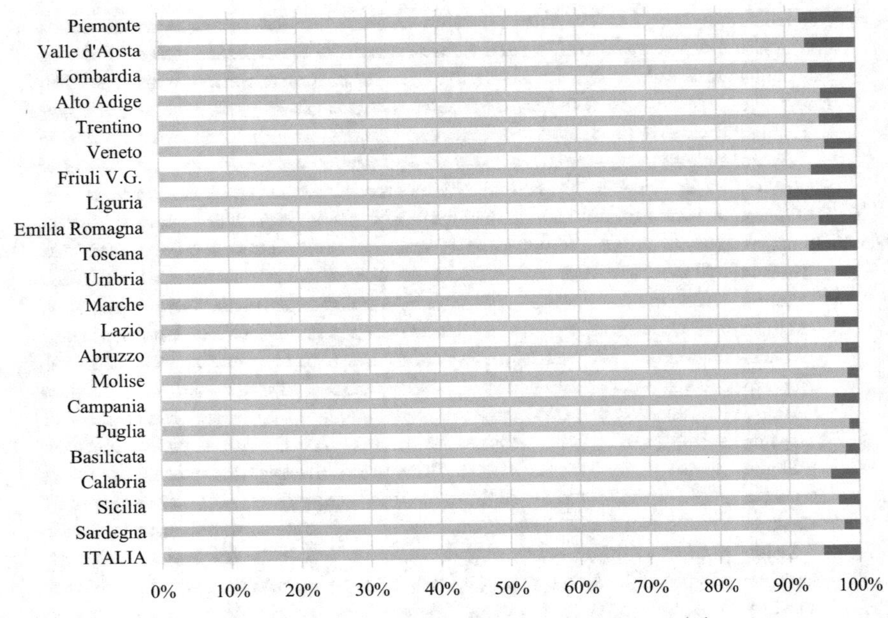

Fig. 12.6 Carbon stock in aboveground living biomass (trees and shrubs higher than 50 cm) and in coarse woody debris, in the regions / Contenuto di carbonio nella fitomassa epigea totale (alberi e arbusti più alti di 50 cm) e nel legno morto grosso totale, nelle regioni

12.5 Annual Variation of Tree Carbon Stock

The annual volume increment of Forest, estimated in 37.8 million cubic metres (cf. Chap. 7), represents approximately 13.5 million carbon tonnes, 49.5 million tonnes CO_2 equivalent. One hectare Forest stores 1.5 carbon tonnes per year and binds from the atmosphere 5.5 tonnes CO_2. Table 12.13 shows the estimates on carbon annually stored in wood by tree growth for the inventory categories of Forest. At www.inventarioforestale.org/statistiche_INFC the estimates for the forest types are available. Figure 12.7 shows the amount of carbon stored in the regions, total and per hectare.

A certain amount of carbon stock in the aboveground living biomass is annually removed from Forest due to felling. Table 12.14 shows the statistics on the carbon removed in the inventory categories of Forest. In Italy, almost 3.2 million carbon tonnes are annually removed from Forest, 11.7 million CO_2 tonnes equivalent. On average, 0.3 carbon tonnes are removed annually from one hectare Forest, 1.1 CO_2 tonnes equivalent. Figure 12.8 displays the data reported in the Table 12.14.

Table 12.13 Total value and value per hectare of carbon stock annual increase in aboveground tree biomass due to growth, by Forest inventory category / Valori totali e per ettaro dell'incremento annuo di carbonio nella fitomassa arborea epigea dovuto all'accrescimento, per le categorie inventariali del Bosco

Region/Regione	Tall trees forest				Plantations				Total Forest			
	Boschi alti				Impianti di arboricoltura da legno				Totale Bosco			
	Corg	ES	Corg	ES	Corg	ES	Corg	ES	Corg	ES	Corg	ES
	(Mg)	(%)	(Mg ha^{-1})	(%)	(Mg)	(%)	(Mg ha^{-1})	(%)	(Mg)	(%)	(Mg ha^{-1})	(%)
Piemonte	1,288,700	3.3	1.5	3.0	13,343	30.7	0.6	23.6	1,302,043	3.3	1.5	3.0
Valle d'Aosta	103,631	7.2	1.0	5.9	0	–	0.0	–	103,631	7.2	1.0	5.9
Lombardia	1,178,590	3.7	2.0	3.2	80,143	21.6	3.2	17.8	1,258,733	3.7	2.0	3.2
Alto Adige	536,507	4.0	1.6	3.6	0	–	0.0	–	536,507	4.0	1.6	3.6
Trentino	662,336	4.2	1.8	3.9	0	–	0.0	–	662,336	4.2	1.8	3.9
Veneto	753,090	3.7	1.8	2.9	28,317	65.9	5.4	53.9	781,407	4.2	1.9	3.6
Friuli V.G	570,200	4.4	1.8	3.9	9491	28.1	1.0	20.7	579,691	4.4	1.7	3.9
Liguria	418,669	4.2	1.2	4.0	341	100.0	0.9	–	419,010	4.2	1.2	4.0
Emilia Romagna	897,849	4.2	1.6	3.7	7185	40.1	1.2	31.2	905,034	4.2	1.5	3.7
Toscana	1,327,001	3.0	1.3	2.8	13,300	40.0	2.0	31.3	1,340,301	3.0	1.3	2.8
Umbria	429,948	4.4	1.1	4.0	3091	35.8	0.5	29.8	433,039	4.4	1.1	4.0
Marche	396,708	5.3	1.4	4.6	3389	38.0	0.5	34.8	400,097	5.2	1.4	4.6
Lazio	776,413	4.6	1.4	4.2	1343	44.2	0.6	20.2	777,756	4.6	1.4	4.2
Abruzzo	620,388	3.8	1.5	3.3	839	47.1	0.3	39.1	621,227	3.8	1.5	3.3
Molise	254,299	6.0	1.7	5.0	1792	71.6	0.7	59.1	256,092	6.0	1.7	5.0
Campania	790,676	4.9	2.0	4.3	1492	62.4	0.5	24.5	792,168	4.9	2.0	4.3
Puglia	156,149	6.9	1.1	5.0	18	100.0	0.2	–	156,168	6.9	1.1	5.0
Basilicata	422,760	5.5	1.5	4.9	59	100.0	0.0	–	422,819	5.5	1.5	4.9
Calabria	1,031,760	5.4	2.1	5.0	3572	48.8	1.5	32.4	1,035,331	5.4	2.1	5.0
Sicilia	285,825	6.7	1.0	5.8	28	70.7	0.0	–	285,854	6.7	1.0	5.8
Sardegna	457,938	5.8	0.8	5.5	21,638	21.1	0.8	18.9	479,576	5.6	0.8	5.3
Italia	13,359,437	1.0	1.5	0.9	189,381	14.3	1.5	13.2	13,548,819	1.0	1.5	0.9

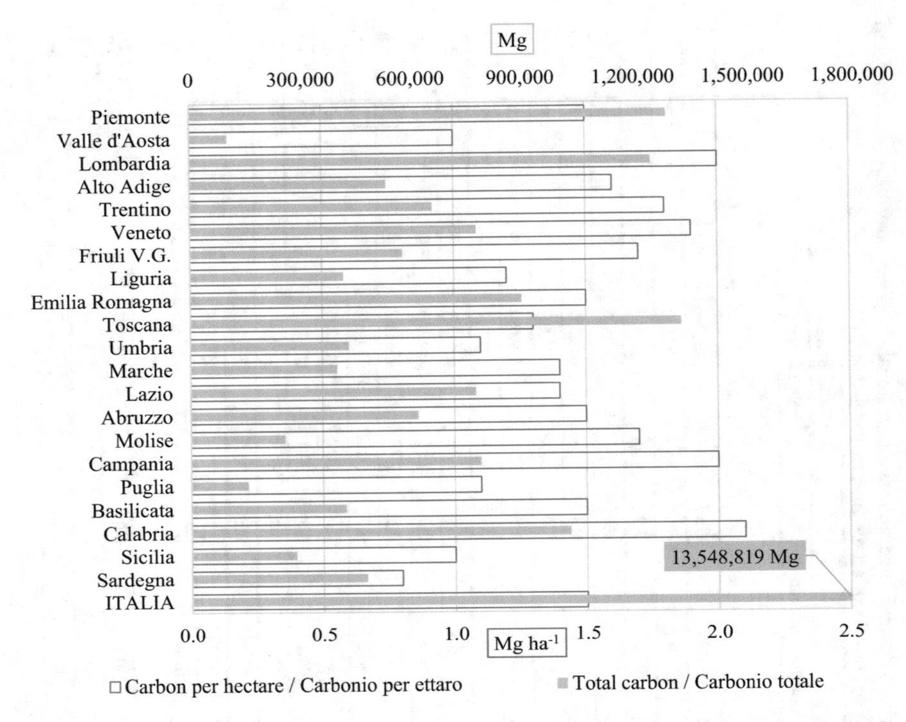

Fig. 12.7 Carbon annually stored by trees growth, total and per hectare (to improve readability, very long bars have been limited and true values are given in numbers) / Carbonio accumulato annualmente per accrescimento degli alberi, totale e per ettaro (per migliorare la lettura, è stata limitata la lunghezza di barre molto lunghe e i valori reali sono indicati con numero)

Figure 12.9 shows the amount of carbon stocked by trees every year and that extracted after felling.

For the whole country, 23.4% of carbon stock annually stored by trees growth is removed. Although the comparison was made not considering the natural mortality, which is not estimated by INFC, the strong positive balance confirms that Italian forests act as sinks. Among the regions, the balance is rather variable, ranging from 0.3% in Molise to 54.8% in Umbria.

Table 12.14 Total value and value per hectare of carbon stock in trees (stem, large and small branches) utilised in the 12 months before the survey / Valori totali e per ettaro del carbonio in fusto, rami grossi e ramaglia degli alberi utilizzati nei 12 mesi precedenti il rilievo, per le categorie inventariali del Bosco

Region/Regione	Tall trees forest				Plantations				Total Forest			
	Boschi alti				Impianti di arboricoltura da legno				Totale Bosco			
	Corg	ES	Corg	ES	Corg	ES	Corg	ES	Corg	ES	Corg	ES
	(Mg)	(%)	(Mg ha^{-1})	(%)	(Mg)	(%)	(Mg ha^{-1})	(%)	(Mg)	(%)	(Mg ha^{-1})	(%)
Piemonte	357,222	43.1	0.4	43.1	85	100.0	0.0	–	357,307	43.1	0.4	43.1
Valle d'Aosta	14,244	48.8	0.1	48.7	0	–	0	–	14,244	48.8	0.1	48.7
Lombardia	40,138	49.6	0.1	49.6	0	–	0.0	–	40,138	49.6	0.1	49.6
Alto Adige	145,346	39.6	0.4	39.6	0	–	0	–	145,346	39.6	0.4	39.6
Trentino	271,837	68.4	0.7	68.3	0	–	0	–	271,837	68.4	0.7	68.3
Veneto	381,327	40.5	0.9	40.4	412	88.5	0.1	88.2	381,740	40.5	0.9	40.4
Friuli V.G	62,814	55.5	0.2	55.4	755	81.1	0.1	79.6	63,569	54.8	0.2	54.8
Liguria	186,044	61.0	0.5	61.0	0	–	0.0	–	186,044	61.0	0.5	61.0
Emilia Romagna	376,423	68.6	0.7	68.6	0	–	0.0	–	376,423	68.6	0.6	68.6
Toscana	218,205	51.7	0.2	51.7	5598	100.0	0.8	–	223,803	50.5	0.2	50.5
Umbria	237,362	59.3	0.6	59.3	0	–	0.0	–	237,362	59.3	0.6	59.3
Marche	146,020	48.4	0.5	48.3	466	100.0	0.1	–	146,486	48.2	0.5	48.2
Lazio	176,084	75.2	0.3	75.2	0	–	0.0	–	176,084	75.2	0.3	75.2
Abruzzo	80,809	67.3	0.2	67.3	0	–	0.0	–	80,809	67.3	0.2	67.3
Molise	735	100.0	0.0	–	0	–	0.0	–	735	100.0	0.0	–
Campania	139,964	37.6	0.3	37.5	0	–	0.0	–	139,964	37.6	0.3	37.5
Puglia	864	100.0	0.0	–	0	–	0.0	–	864	100.0	0.0	–
Basilicata	64,208	61.8	0.2	61.8	0	–	0.0	–	64,208	61.8	0.2	61.8
Calabria	147,084	54.4	0.3	54.4	0	–	0.0	–	147,084	54.4	0.3	54.4
Sicilia	7545	76.4	0.0	76.4	0	–	0.0	–	7545	76.4	0.0	76.4
Sardegna	87,645	63.3	0.1	63.3	15,303	77.2	0.6	75.4	102,947	55.1	0.2	55.1
Italia	3,141,920	15.5	0.4	15.5	22,619	57.9	0.2	57.6	3,164,539	15.4	0.3	15.4

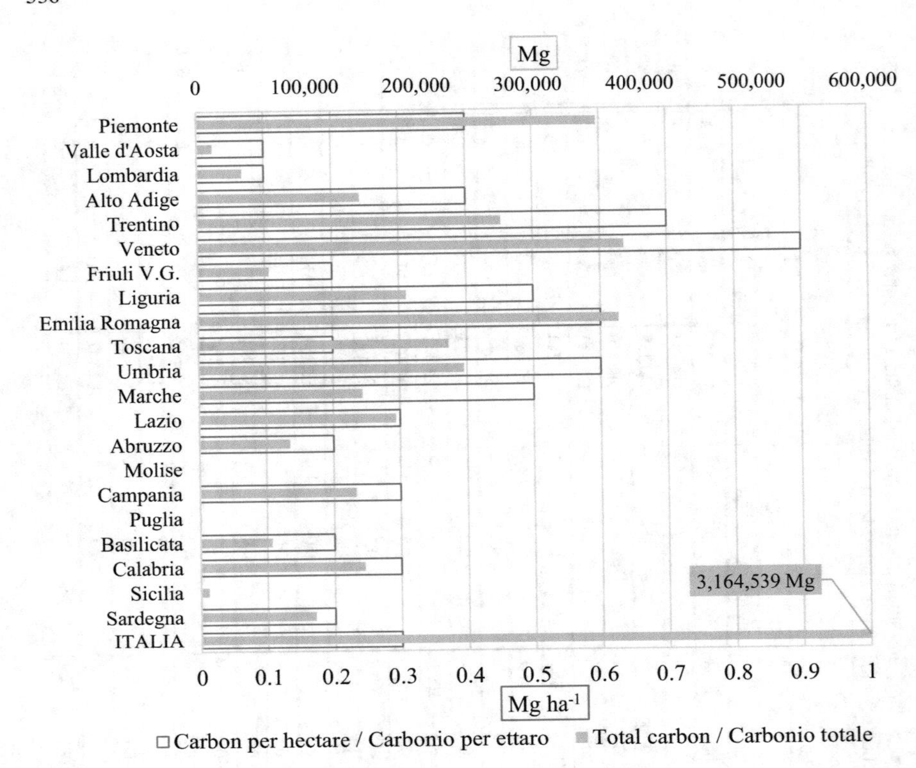

Fig. 12.8 Tree carbon removed annually from Forest with utilisations, total and per hectare (to improve readability, very long bars have been limited and true values are given in numbers) / Carbonio totale e per ettaro asportato annualmente dal Bosco con le utilizzazioni (per migliorare la lettura, è stata limitata la lunghezza di barre molto lunghe e i valori reali sono indicati con numero)

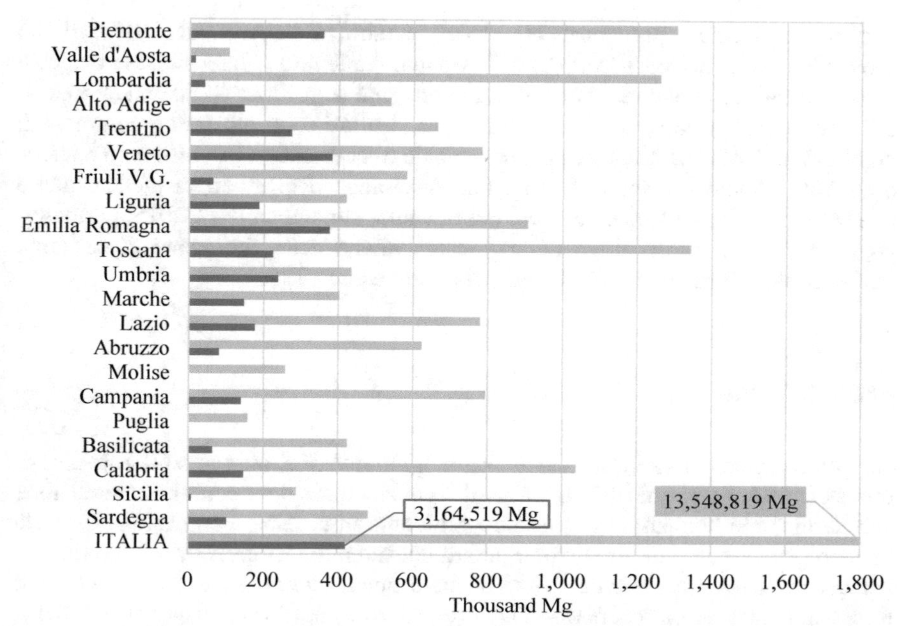

Fig. 12.9 Carbon annually stored by trees growth and removed with utilisations (to improve readability, very long bars have been limited and true values are given in numbers) / Carbonio fissato annualmente per accrescimento degli alberi e carbonio asportato con le utilizzazioni annue (per migliorare la lettura, è stata limitata la lunghezza di barre molto lunghe e i valori reali sono indicati con numero)

Appendix (Italian Version)

Riassunto Le foreste condizionano il clima a scala globale e rivestono un ruolo nel ciclo globale del carbonio. I Paesi che hanno sottoscritto la convenzione quadro sui cambiamenti climatici delle Nazioni Unite (UNFCCC) e i successivi impegni utilizzano i dati degli inventari forestali nazionali per stimare l'assorbimento di carbonio legato all'uso del suolo, al cambiamento d'uso e alle foreste. Cinque sono i serbatoi terrestri rilevanti nella stima del carbonio immagazzinato e della sua variazione per l'UNFCCC e il protocollo di Kyoto: il suolo, la lettiera, la fitomassa ipogea e quella epigea, il legno morto. Il secondo inventario forestale nazionale italiano (INFC2005) ha stimato il contenuto di carbonio organico in quattro dei cinque serbatoi e confermato il ruolo primario del suolo, che è risultato contenere il 57.6% del carbonio organico dei quattro serbatoi nelle foreste italiane. La stima delle variazioni di carbonio nei suoli forestali è particolarmente onerosa poiché le variazioni attese sono piccole se paragonate alle grandi quantità immagazzinate. INFC2015 ha provveduto ad aggiornare le stime del carbonio contenuto nella fitomassa epigea e nel legno morto, le due componenti maggiormente soggette a variazioni apprezzabili nel tempo trascorso tra i due inventari. I risultati sono mostrati e commentati nelle Sects. 12.2 e 12.3. La variazione annuale di carbonio nel soprassuolo è anche dovuta al carbonio immagazzinato con l'accrescimento degli alberi e a quello rimosso con le utilizzazioni. Queste due voci del bilancio, stimate da INFC2015, permettono di valutare il ruolo attivo delle foreste italiane nella rimozione di carbonio dall'atmosfera; questo contributo è discusso nella Sect. 12.4.

Introduzione

Gli alberi rimuovono anidride carbonica dall'atmosfera e, attraverso la fotosintesi, immagazzinano carbonio (C) nei tessuti legnosi. Si assume, con approssimazione accettabile, che la quantità di carbonio contenuta in un albero equivalga alla metà del suo peso secco. Gli alberi sono quindi considerati accumulatori di carbonio e, per questo, attenzione è stata rivolta anche a quelli che crescono in contesto non forestale (Trees outside forests—TOF) (es. Guo et al., 2014; Russo et al., 2014; Schnell et al., 2015; Speak et al., 2020). L'assunto che gli alberi siano un sink per il carbonio atmosferico ha portato ad inserire la piantumazione di alberi in contesto urbano nell'ambito delle strategie per la lotta ai cambiamenti climatici. Tuttavia, le valutazioni sulla loro reale capacità di comportarsi come accumulatori netti sono complesse perché richiedono di conoscere, tra l'altro, le emissioni dovute alle attività di impianto e cura nel corso dell'intero ciclo di vita, e alcune di quelle intraprese ne hanno ridimensionato il ruolo (es. McPherson & Kendall, 2014). Maggiore efficienza per lo stoccaggio del carbonio è garantita dalla capacità degli alberi di trasferire quello atmosferico nell'ecosistema forestale. Il contributo che le foreste possono dare alla lotta ai cambiamenti climatici è stato ampiamente riconosciuto negli ultimi decenni. I Paesi che compilano i report sulle emissioni e gli assorbimenti di gas serra, a seguito della sottoscrizione della convenzione sui cambiamenti climatici delle Nazioni Unite (UNFCCC) e gli accordi successivi, devono stimare l'assorbimento di carbonio legato all'uso del suolo, al cambiamento di uso del suolo e al settore forestale. Cinque sono i pool di carbonio considerati nelle stime per il carbonio immagazzinato e le sue variazioni ai fini del protocollo di Kyoto (IPCC, 2003): il suolo, la lettiera, la biomassa ipogea e quella epigea, il legno morto. Gli inventari forestali nazionali contribuiscono come un'importante fonte di dati per la stima del carbonio forestale (es. Breidenbach et al., 2021; Brown, 2002; Mäkipää et al., 2008). Il secondo inventario forestale nazionale italiano (INFC2005) ha stimato il carbonio immagazzinato in quattro dei cinque pool menzionati, non considerando la sola componente della biomassa ipogea. Le stime prodotte hanno confermato il ruolo principale che viene riconosciuto ai suoli nei contesti forestali europei, dacché il suolo del Bosco in Italia è risultato contenere il 57.6% del carbonio organico contenuto nei quattro pool indagati, seguito dalla fitomassa epigea (38.1%); lettiera e legno morto insieme sono risultati contribuire per il 4.3% del carbonio totale stimato (Gasparini & Di Cosmo, 2015; Gasparini et al., 2013).

Campagne di rilievo su larga scala per il campionamento dei suoli e le successive analisi di laboratorio per le determinazioni del contenuto di carbonio richiedono un impegno enorme perché le variazioni attese del contenuto di carbonio sono piccole se paragonate alle grandi quantità immagazzinate (Schrumpf et al., 2008), specialmente quando il periodo di tempo trascorso tra due campagne di rilievo non è lungo (es. Conen et al., 2003). Considerato il periodo relativamente breve trascorso dal precedente inventario, con INFC2015 si è proceduto ad aggiornare le stime della quantità di carbonio nella fitomassa epigea e nel legno morto grosso, le due componenti maggiormente soggette a variazioni apprezzabili. La fitomassa epigea è rappresentata dalla vegetazione legnosa costituita da individui (genericamente indicati come alberi) con diametro a 1.30 m dal suolo $\geq$ 4.5 cm, e dagli individui di rinnovazione e arbusti più alti di 50 cm. Il legno morto grosso si compone di soggetti morti in piedi (genericamente indicati come alberi, con diametro a 1.30 m dal suolo $\geq$ 4.5 cm), ceppaie residue (con diametro alla sezione di taglio $\geq$ 9.5 cm) e legno morto grosso a terra (con diametro alle sezioni estreme $\geq$ 9.5 cm e lunghezza $\geq$ 9.5 cm); in contesti internazionali, queste tre componenti sono spesso indicate come coarse woody debris.

La quantità di carbonio che gli alberi sequestrano annualmente dall'atmosfera per accrescimento è stata stimata attraverso l'incremento annuo in biomassa. Il contenuto di carbonio dei soprassuoli forestali varia annualmente anche a causa del taglio e dell'esbosco degli alberi e della mortalità naturale. Pertanto, l'assorbimento quantificato sottraendo al carbonio fissato annualmente per accrescimento degli alberi vivi quello rimosso con le utilizzazioni non può essere considerato un accumulo netto di carbonio annuo (Tabacchi et al., 2010).

Il contenuto di carbonio nella vegetazione

La Table 12.1 mostra la quantità complessiva di carbonio della vegetazione legnosa con altezza superiore a 50 cm, per le categorie inventariali del Bosco e per il suo totale. La Table 12.2 e la Table 12.3 riportano le stesse informazioni per le categorie forestali del Bosco. Alberi, rinnovazione e arbusti immagazzinano 539.3 milioni di tonnellate di carbonio, con un valore medio per ettaro di 59.4 tonnellate. Per le categorie inventariali del Bosco e per la sua totalità, la Table 12.4 riporta le stime della quantità di carbonio negli alberi, la Table 12.5 riporta stime analoghe per la rinnovazione e la Table 12.6 presenta quelle per gli arbusti. Le statistiche a livello di categoria forestale per il contenuto di carbonio negli alberi, nella rinnovazione e negli arbusti sono disponibili al link www.inventarioforestale.org/statistiche_INFC. Un confronto tra le tabelle citate consente di evidenziare che il 96.8% del carbonio nella componente viva è dovuto agli alberi, mentre rinnovazione e arbusti contribuiscono per percentuali rispettivamente dell'1.7% e dell'1.5%. Rinnovazione o arbusti solo in poche regioni contribuiscono ognuno per almeno il 3% del contenuto di carbonio della fitomassa epigea totale. La rinnovazione supera questa percentuale in Umbria (5.7%), Marche (4.7%) e Molise (4.6%); gli arbusti la superano in Molise (3.9%) e Sardegna (5.3%). Insieme, le due componenti superano il 3% in altre cinque regioni (Emilia-Romagna, Lazio, Campania, Puglia e Basilicata). La Fig. 12.1 mostra i valori del contenuto di carbonio della vegetazione legnosa nelle regioni, distinguendo quelli della componente arborea da quelli totali di rinnovazione e arbusti. La Fig. 12.2 mostra il contenuto di carbonio nella vegetazione legnosa delle varie categorie forestali dei Boschi alti. Per le categorie forestali che contribuiscono con più di 40 milioni di tonnellate, si può notare (cfr. Chap. 7) che le Faggete, i Querceti di rovere, roverella e farnia, le Cerrete, boschi di farnetto, fragno e vallonea, gli Altri boschi caducifogli sono caratterizzati sia da una superficie forestale elevata (sempre superiore al milione di ettari) sia da un volume legnoso generalmente sopra i 100 milioni di metri cubi (ad eccezione dei Querceti di rovere, roverella e farnia, con 94.4 milioni di metri cubi); i Boschi di abete rosso e i Castagneti, invece, contribuiscono cospicuamente allo stoccaggio del carbonio soprattutto perché ricchi in massa, poiché la superficie occupata è più limitata (586.7 migliaia di ettari per i Boschi di abete rosso e 778.5 migliaia di ettari per i Castagneti).

Il contenuto di carbonio nel legno morto

INFC2015 ha aggiornato le stime relativamente al carbonio organico delle componenti degli alberi morti in piedi, della necromassa grossa a terra e delle ceppaie residue. La Table 12.7 mostra il contenuto di carbonio complessivo per le tre componenti rilevate, per le categorie inventariali e per il Bosco in generale. Le Tables 12.8 e 12.9 riportano le stesse informazioni per le categorie forestali del Bosco. Nel legno morto sono presenti 29.8 milioni di tonnellate di carbonio, con un valore medio per ettaro di 3.3 tonnellate. La Table 12.10 riporta le stime sulla quantità di carbonio negli alberi morti in piedi, la Table 12.11 stime analoghe per le ceppaie residue e la Table 12.12 quelle per il legno morto grosso a terra, tutte per le categorie inventariali del Bosco. Le statistiche a livello di categoria forestale per il contenuto di carbonio negli alberi morti in piedi, nelle ceppaie residue e nel legno morto grosso a terra sono disponibili all'indirizzo www.inventarioforestale.org/statistiche_INFC.

La Fig. 12.3 mostra che il 57.5% del carbonio nel legno morto è contenuto negli alberi morti in piedi, seguiti dal legno morto grosso a terra, con il 33.2%, ed infine dalle ceppaie residue, che contribuiscono per il 9.3%. La Fig. 12.4 riporta il contenuto di carbonio nelle tre componenti del legno morto per le categorie forestali dei Boschi alti. Il legno morto grosso a terra costituisce la principale componente di necromassa per l'accumulo di carbonio in quattro categorie forestali, che sono i Boschi di larice e cembro, i Boschi di abete rosso, i Boschi di abete bianco e i Boschi igrofili. In queste categorie la percentuale del carbonio totale immagazzinato nel legno morto grosso a terra varia dal 41.5% (Boschi di abete rosso) al 49.4% (Boschi igrofili). In tutte le altre categorie, prevale sempre il contributo degli alberi morti in piedi che, ad eccezione delle Pinete di pino silvestre e montano (48.2%) e delle Faggete (47.8%), contribuiscono sempre con una quota superiore al 50% del carbonio accumulato dal legno morto grosso complessivo. Le ceppaie residue contribuiscono con valori superiori alla media nazionale nelle tre categorie a conifere già ricordate per il ruolo prevalente del legno morto a terra, nelle Faggete e nelle Sugherete. Ad eccezione di queste ultime (12.8%), il contributo percentuale supera sempre il 16% e arriva al 23.3% nel caso dei Boschi di abete rosso. La Fig. 12.5 mostra il contenuto di carbonio nelle varie componenti del legno morto per le regioni italiane. La componente che contribuisce maggiormente all'accumulo di carbonio è quasi sempre quella degli alberi morti in piedi, con quote dal 49.9% (nelle Marche) al 70.7% (Toscana), ma in Valle d'Aosta e in Trentino la componente principale è il legno morto a terra (55.1% e 43.3% rispettivamente) e in Alto Adige ceppaie residue (36.7%) e legno morto grosso a terra (36.5%) prevalgono sugli alberi morti in piedi, con valori molto simili tra loro.

Il contenuto complessivo di carbonio epigeo

Nel complesso, gli individui di specie legnose con altezza maggiore di 50 cm e il legno morto grosso immagazzinano 569.1 milioni di tonnellate di carbonio. Questa quantità è dovuta per il 94.8% agli alberi e agli arbusti e per il restante 5.2% al legno morto grosso. La Fig. 12.6 riporta le percentuali cumulate di carbonio nella vegetazione e nel legno morto per le diverse regioni. Si può notare che la quota dovuta al legno morto risulta più elevata nelle regioni del centro-nord. Il valore medio nazionale del 5.2% è superato in otto regioni, dall'arco alpino alla Toscana, con l'eccezione dell'Alto Adige (5.0%) e del Veneto (4.5%), che comunque non si discostano molto da quel valore e insieme alle Marche e alla Calabria sono le uniche regioni restanti con un contributo del legno morto sopra il 4%.

Le variazioni annuali dello stock di carbonio degli alberi

All'incremento annuo di volume del Bosco, stimato in 37.8 milioni di metri cubi (cfr. Chap. 7), corrisponde un accumulo di circa 13.5 milioni di tonnellate di carbonio, che equivalgono a 49.5 milioni di tonnellate di CO_2. Un ettaro di Bosco fissa annualmente 1.5 tonnellate di carbonio, rimuovendo dall'atmosfera 5.5 tonnellate di CO_2. La Table 12.13 mostra le stime del carbonio accumulato annualmente nel legno prodotto per le categorie inventariali del Bosco. All'indirizzo www.inventarioforestale.org/statistiche_INFC sono disponibili statistiche analoghe per le categorie forestali. La Fig. 12.7 mostra la quantità di carbonio totale e per ettaro assorbita annualmente nelle varie regioni.

A fronte della quantità di carbonio fissata annualmente, una parte del carbonio contenuto nei soprassuoli forestali viene rimossa con le utilizzazioni boschive. La

Table 12.14 mostra le stime della quantità di carbonio rimossa per le categorie inventariali del Bosco. A livello nazionale, tale quantità è di quasi 3.2 milioni di tonnellate, che corrispondono a 11.7 milioni di tonnellate di CO_2 equivalente. Da un ettaro di Bosco sono rimosse circa 0.3 tonnellate di carbonio, pari a 1.1 tonnellate equivalenti di CO_2. La Fig. 12.8 illustra i dati riportati nella tabella citata.

La Fig. 12.9 mette a confronto la quantità di carbonio fissato e rimossa annualmente dai soprassuoli forestali nelle diverse regioni. A livello nazionale, il carbonio rimosso dal Bosco con le utilizzazioni risulta pari al 23.4% del carbonio assorbito per accrescimento degli alberi. Al lordo delle perdite per mortalità naturale, la cui quantità non viene stimata direttamente da INFC, il confronto denota un bilancio fortemente positivo, a conferma del ruolo di sink dei soprassuoli forestali italiani. La percentuale di carbonio rimossa dal Bosco rispetto a quello fissato annualmente per accrescimento è piuttosto variabile tra le regioni, con valori che vanno dallo 0.3% del Molise al 54.8% dell'Umbria.

References

Breidenbach, J., McRoberts, R. E., Alberdi, I., Antón-Fernández, C., & Tomppo, E. (2021). A century of national forest inventories – informing past, present and future decisions. *Forest Ecosystems, 8*, 36. https://doi.org/10.1186/s40663-021-00315-x

Brown, S. (2002). Measuring carbon in forests: Current status and future challenges. *Environmental Pollution, 116*, 363–372.

Conen, F., Yakutin, M. V., & Sambuu, A. (2003). Potential for detecting changes in soil organic carbon concentration resulting from climate change. *Global Change Biology, 9*, 1515–1520.

Gasparini, P., Di Cosmo, L., & Pompei, E. (Eds.) (2013). *Il contenuto di carbonio delle foreste italiane. Inventario Nazionale delle Foreste e dei serbatoi forestali di Carbonio INFC2005. Metodi e risultati dell'indagine integrativa* (pp. 77–86). Consiglio per la Ricerca e la sperimentazione in Agricoltura, Roma. ISBN 978-88-97081-36-4.

Gasparini, P., & Di Cosmo, L. (2015). Forest carbon in Italian forests: Stocks, inherent variability and predictability using NFI data. *Forest Ecology and Management, 337*, 186–195.

Guo, Z. D., Hu, H. F., Pan, D., Birdsey, R. A., & Fang, J. Y. (2014). Increasing biomass carbon stocks in trees outside forests in China over the last three decades. *Biogeosciences, 11*(15), 4115–4122. https://doi.org/10.5194/bg-11-4115-2014

IPCC (2003). Good practice guidance for land use, land use change and forestry. IPCC National Greenhouse Gas Inventories Programme. Institute for Global Environmental Strategies, Hayama, Japan. ISBN 4-88788-003-0. Retrieved Jan 04, 2022, from https://www.ipcc-nggip.iges.or.jp/publ ic/gpglulucf/gpglulucf_files/GPG_LULUCF_FULL.pdf.

Mäkipää, R., Lehtonen, A., & Peltoniemi, M. (2008). Monitoring carbon stock changes in European forests using forest inventory data. In: A. J. Dolman, R. Valentini, & A. Freibauer (Eds.), *The continental-scale greenhouse gas balance of Europe* (pp. 191–214). Ecol. Stud., 203. Springer. https://doi.org/10.1007/978-0-387-76570-9_9.

McPherson, E. G., & Kendall, A. (2014). A life cycle carbon dioxide inventory of the Million Trees Los Angeles program. *International Journal of Life Cycle Assessment, 19*, 1653–1665. https://doi.org/10.1007/s11367-014-0772-8

Russo, A., Escobedo, F. J., Timilsina, N., Schmitt, A. O., Varela, S., & Zerbe, S. (2014). Assessing urban tree carbon storage and sequestration in Bolzano, Italy. *International Journal of Biodiversity Science, Ecosystem Services Management, 10*(1), 54–70. https://doi.org/10.1080/21513732.2013.873822

Schrumpf, M., Schumacher, J., Schöning, I., & Schulze, E. D. (2008). Monitoring carbon stock changes in European Soils: process understanding and sampling strategies. In: A. J. Dolman, R. Valentini, & A. Freibauer (Eds.), *The continental-scale greenhouse gas balance of Europe* (pp. 153–189). Ecol. Stud., 203. Springer. https://doi.org/10.1007/978-0-387-76570-9_9.

Schnell, S., Altrell, D., Ståhl, G., & Kleinn, C. (2015). The contribution of trees outside forests to national tree biomass and carbon stocks—A comparative study across three continents. *Environmental Monitoring and Assessment, 187*, 4197. https://doi.org/10.1007/s10661-014-4197-4

Speak, A., Escobedo, F. J., Russo, A., & Zerbe, S. (2020). Total urban tree carbon storage and waste management emissions estimated using a combination of LiDAR, field measurements and an end-of-life wood approach. *Journal of Cleaner Production, 256*, 120420. https://doi.org/10.1016/j.jclepro.2020.120420

Tabacchi, G., De Natale, F., & Gasparini, P. (2010). Coerenza ed entità delle statistiche forestali—stime degli assorbimenti netti di carbonio nelle foreste italiane. *Sherwood, 165*, 11–19.

Open Access This chapter is licensed under the terms of the Creative Commons Attribution 4.0 International License (http://creativecommons.org/licenses/by/4.0/), which permits use, sharing, adaptation, distribution and reproduction in any medium or format, as long as you give appropriate credit to the original author(s) and the source, provide a link to the Creative Commons license and indicate if changes were made.

The images or other third party material in this chapter are included in the chapter's Creative Commons license, unless indicated otherwise in a credit line to the material. If material is not included in the chapter's Creative Commons license and your intended use is not permitted by statutory regulation or exceeds the permitted use, you will need to obtain permission directly from the copyright holder.

Chapter 13
Changes of Italian Forests Over Time Captured by the National Forest Inventories

L'evoluzione nel tempo delle foreste italiane secondo gli inventari forestali nazionali

Patrizia Gasparini and Giovanni Tabacchi

Abstract In order to effectively compare the results of different surveys, they should share the study domain, and the definitions and the classification and measurement criteria of the observed variables should be consistent. The inventory domain in the two Italian forest inventories, INFC2005 and INFC2015, has remained unchanged. The two surveys adopt a common classification system and common stratification criteria, by region and forest category, as well as the same measurement thresholds for quantitative and qualitative variables. This chapter illustrates the results of the comparisons between the estimates obtained from the two surveys for some features considered relevant for the analysis of the current dynamics of the Italian forests, such as the extension of the areas, the growing stock volume, the aboveground tree biomass, the annual volume increment and removals, and the number of trees. The results of the comparisons are sufficiently clear at the national level, while the estimates disaggregated by region or forest category sometimes show very limited variations and are associated with high standard errors of estimates.

13.1 Introduction

Monitoring national forest resources requires reliable and comparable information, derived from repeated large-scale observations. In order to effectively compare the results of different surveys, it is necessary that they share the definition of the study domain, for example the definition of forest, and that the respective domains are defined according to common criteria, for example the crown coverage or the extension of the forest. It is also necessary that variable definitions and the classification and measurement criteria of the variables be consistent with each other.

P. Gasparini (✉) · G. Tabacchi
CREA Research Centre for Forestry and Wood, Trento, Italy
e-mail: patrizia.gasparini@crea.gov.it

© The Author(s) 2022

P. Gasparini et al. (eds.), *Italian National Forest Inventory—Methods and Results of the Third Survey*, Springer Tracts in Civil Engineering,
https://doi.org/10.1007/978-3-030-98678-0_13

In a chapter dedicated to illustrating some of the main changes that occurred in the period between the first Italian national forest inventory (IFNI85) and the second (INFC2005), De Natale and Gasparini (2011) highlighted the importance of having multiple surveys over time to make comparisons between the inventory statistics. This was useful in identifying trends in the Italian forests. Furthermore, the actions to harmonise the two forest inventories, which in their design diverge for some defining and classification aspects and for some measurement standards, were briefly mentioned. Those actions were undertaken to both allow for the desired comparisons and the need to respond with greater accuracy to requests from important supranational statistical processes (cf. Chap. 1).

Contrasted to past comparisons, the differences between INFC2005 and INFC2015 are definitely smaller and limited to the improvement of some operational procedures in the collection and recording of information in the field. The inventory domain, the classification system and the stratification criteria, by region and forest type, have remained unchanged (cf. Chap. 2). Similarly, the measurement thresholds of the quantitative variables and the classes used for the qualitative variables are the same in the two inventories (cf. Chap. 4), and the calculation and estimation procedures remained unchanged (cf. Chaps. 5 and 6). This favourable condition allows to compare the various homologous quantities detected in the two inventories more easily.

Another important aspect for the comparison between the INFC2005 and INFC2015 results concerns the complexity of inventory design and the methods of selecting the sampling units. The first phase sampling points remained the same in the two inventories, while the second and third phase samples of INFC2015 included additional sampling points selected from those affected by changes in land use and land cover (cf. Chaps. 2 and 5). The samples of the two surveys are therefore not independent. The literature devoted to comparisons between repeated multi-phase inventories and to the analysis of the distribution of the estimators used in such complex designs is unfortunately incomplete. However, for the purposes of this chapter, an analysis was carried out to capture the extent and the sign of the changes between the two inventories, taking into account the variability of the estimates, without statistical hypothesis testing.

In general, the analysis of the variations between quantities estimated by sampling must consider not only the values of the estimators, but also the corresponding standard errors (SE), which indicate the variability of the estimates for different samples of the same size, with the same sampling design (Gregoire & Valentine, 2008). If the extent of the variation between the two sample estimates is large and much greater than the relative variability, it is difficult to hypothesize that this variation is not really different from zero. For this reason, despite the limitations mentioned above, we conventionally compared the estimates considering the difference between two estimates "considerable" when their estimation intervals (equal to twice the SE) did not overlap or otherwise "negligible". In the comments, the percent difference between two estimates is shown together with their percent standard error, in order to immediately compare the size of the difference with the variability of the two estimates.

The results on the extent and signs of variations between the two inventories described in this chapter concern some inventory attributes considered more relevant for the analysis of the current dynamics of the Italian forests. They are the extension of wooded areas (Forest, Other wooded land and Total wooded area) and for the Tall trees forest, the growing stock, the aboveground biomass, the annual volume increment and removals, and the number of trees. The comparison was made by considering the estimates of the total values of each variable and the same estimates disaggregated by region and forest type. This was done to bring out any geographical differentiations in the variations that occurred in the period between the two surveys. Finally, the estimates of the new wooded areas and those of the areas affected by the change from wooded areas to other land uses resulting from the comparison of the sampling point classification in the two inventories are presented.

The results of the comparisons are sufficiently clear at the national level, while the estimates disaggregated by region or forest category sometimes show very limited variations and are associated with high standard errors of estimates.

13.2 Variations in the Wooded Area

13.2.1 Forest Area

The Forest area in Italy estimated by INFC2005 is equal to 8,759,202 ha, with a SE of 0.4% (Gasparini & Tabacchi, 2011). The analogous estimate from INFC2015 is equal to 9,085,186 ha, with a SE of 0.4%, and the variation between the two surveys is equal to 325,984 ha (+3.7%). Following the criteria described in the previous section, the variation between the estimates of the two surveys is considerable. At the regional level, it is observed that considerable area variations are reported only for Molise (20,686 ha, +15.6%), while the variations between the estimates for the other regions, all positive except for Trentino (−2143 ha, −0.6%) and Puglia (−3540 ha, −2.4%), are quantitatively negligible (Fig. 13.1). At the level of wider territories in the period between the two inventories, the extension of the Forest increased by 68,596 ha (+2.0%) in the Alpine regions, by 96,910 ha (+3.1%) in the regions of the northern and central Apennines and by 160,479 ha (+7.2%) in those of the southern Apennines and in the islands.[1]

Observing the area estimates for the Tall trees forest types, negative variations emerge for the Norway spruce (−0.4%), for the Scots pine and mountain pine (−2.3%), for the Black pines (−3.1%) and for the Chestnut (−1.8%); for the other types the area variation is always positive. However, it should be noted that only for the Other deciduous broadleaved (+11.4%) and the Other evergreen broadleaved

[1] Alpine regions: Piemonte, Valle d'Aosta, Lombardia, Alto Adige, Trentino, Veneto, Friuli-Venezia Giulia, Liguria; regions of the northern and central Apennines: Emilia-Romagna, Toscana, Umbria, Marche, Lazio, Abruzzo; regions of the southern Apennines and the islands: Molise, Campania, Puglia, Basilicata, Calabria, Sicilia, Sardegna.

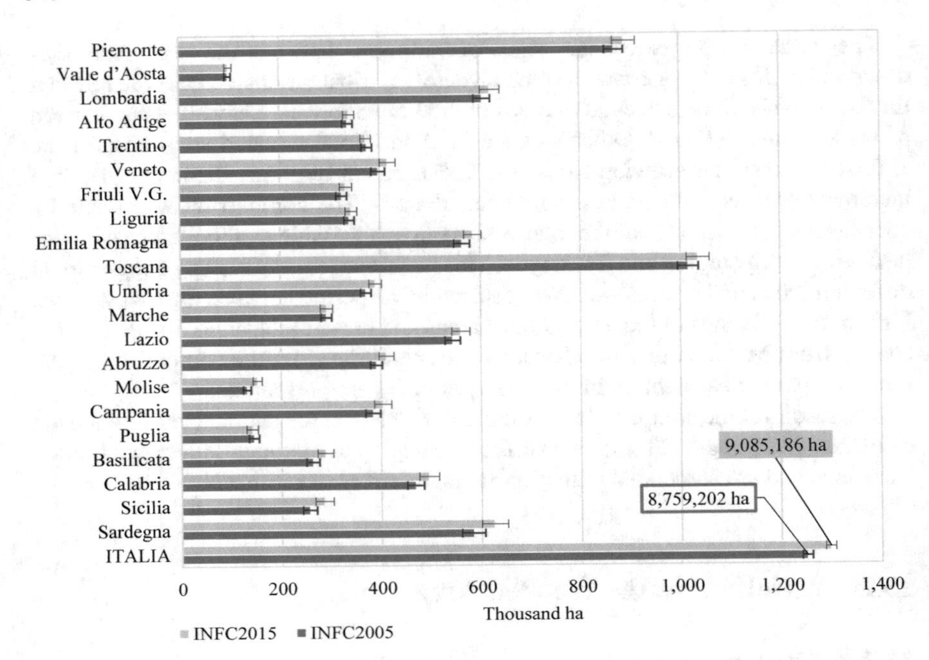

Fig. 13.1 Estimates of Forest area of INFC2005 and INFC2015, nationally and for individual regions; error bars are equal to twice the standard error; bars for Italy were shortened to improve readability; real values are given in numbers / Stime dell'estensione del Bosco per INFC2005 e INFC2015, a livello nazionale e per le singole regioni; barre di errore pari a due volte l'errore standard; le barre per l'Italia sono state ridotte per migliorare la lettura, i valori reali sono indicati in numero

(+33.6%) is the variation considerable (Fig. 13.2). These are two residual classes, in which newly formed forest stands, such as the Acero-lime groves and the *Robinia* and *Ailanthus* forests, converge. For the whole of the forest types of Tall trees forest, which occupy 98.6% of the area of the Forest in Italy according to INFC2015, the variation in extension is positive and considerable (319,837 ha, +3.7%).

13.2.2 Other Wooded Land Area

The area of the Other wooded land in Italy was estimated by INFC2005 to be 1,708,333 ha, with a SE of 1.3%; INFC2015 estimated for the same macro-category an area of 1,969,272 ha with a SE of 1.4%. The area variation between the two surveys is quantitatively considerable and equal to 260,939 ha (+15.3%). By disaggregating the area estimates of the Other wooded land by region, it can be observed that positive changes occur for all regions, but they are considerable just for Lazio (25,938 ha, +41,9%), Campania (26,453 ha, +43.5%) and Puglia (16,238 ha, +

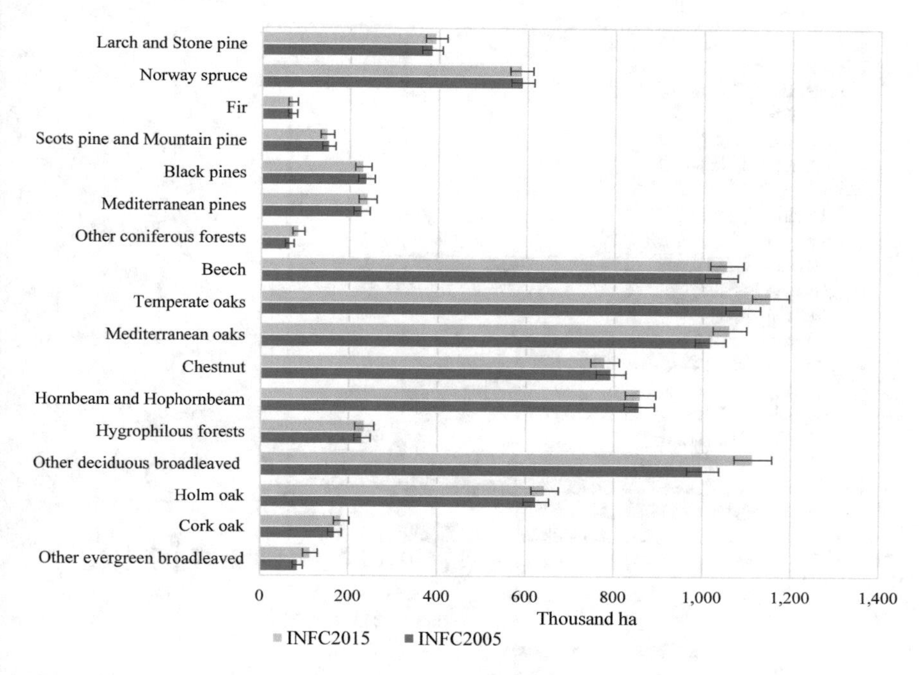

Fig. 13.2 Estimates of Tall trees forest types area of INFC2005 and INFC2015; error bars are equal to twice the standard error / Stime dell'estensione delle categorie forestali dei Boschi alti per INFC2005 e INFC2015; barre di errore pari a due volte l'errore standard

49,0%) (Fig. 13.3). At the level of wider territories, in the period between the two inventories the extension of the Other wooded land increased by 49,349 ha (+15.3%) in the Alpine regions, by 78,275 ha (+24.0%) in the regions of northern and central Apennines and 133,314 ha (+12.6%) in the regions of the southern Apennines and in the islands.

13.2.3 Total Wooded Area

In INFC2005 the Total wooded area (Forest plus Other wooded land) in Italy was estimated to be 10,467,533 ha with a SE of 0.3%; the similar INFC2015 estimate is equal to 11,054,458 ha, with a SE of 0.3%. The area variation between the two surveys is considerable and equal to 586,925 ha (+5.6%). By disaggregating the estimates of Total wooded area at the regional level, it can be observed that positive variations occur for all regions, with the sole exception of Trentino, by a very slight fraction (-0.1%). If we consider the uncertainty of the estimates, these variations appear considerable for many peninsular regions and for the islands: for Toscana (38,183 ha, + 3.3%), Umbria (23,701 ha, +6.1%), Lazio (42,289 ha, +7.0%), Abruzzo (36,009 ha,

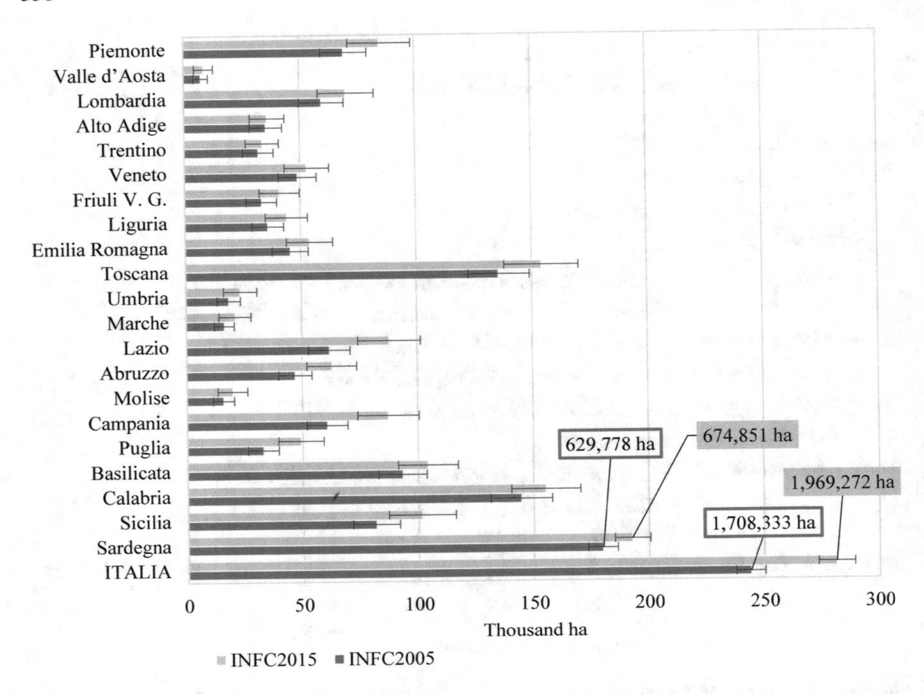

Fig. 13.3 Estimates of Other wooded land area of INFC2005 and INFC2015, nationally and for individual regions; error bars are equal to twice the standard error; bars for Italy and Sardegna were shortened to improve readability; real values are given in numbers / Stime dell'estensione delle Altre terre boscate per INFC2005 e INFC2015, a livello nazionale e per le singole regioni; barre di errore pari a due volte l'errore standard; le barre per Italia e Sardegna sono state ridotte per migliorare la lettura, i valori reali sono indicati in numero

8.2%), Molise (24,632 ha, +16.6%), Campania (45,985 ha, +10.3%), Basilicata (35,986 ha, +10.1%), Calabria (37,689 ha, +6.1%), Sicilia (49,063 ha, +14.5%) and Sardegna (87,741 ha, +7.2%) (Fig. 13.4). At the level of wider territories, in the period between the two inventories the Total wooded area increased by 117,944 ha (+3.2%) in the Alpine regions, by 175,184 ha (+5.0%) in the regions of the northern and central Apennines and 293,794 ha (+8.9%) in the regions of the southern Apennines and in the islands.

13.2.4 *Land Use and Land Cover Changes*

The variations in the Total wooded area and that of the Forest and Other wooded land resulting from the comparison between the relative estimates produced by the two most recent Italian forest inventories have been illustrated in the paragraphs above. Thanks to the repetition of the photointerpretation on all the points of the first phase

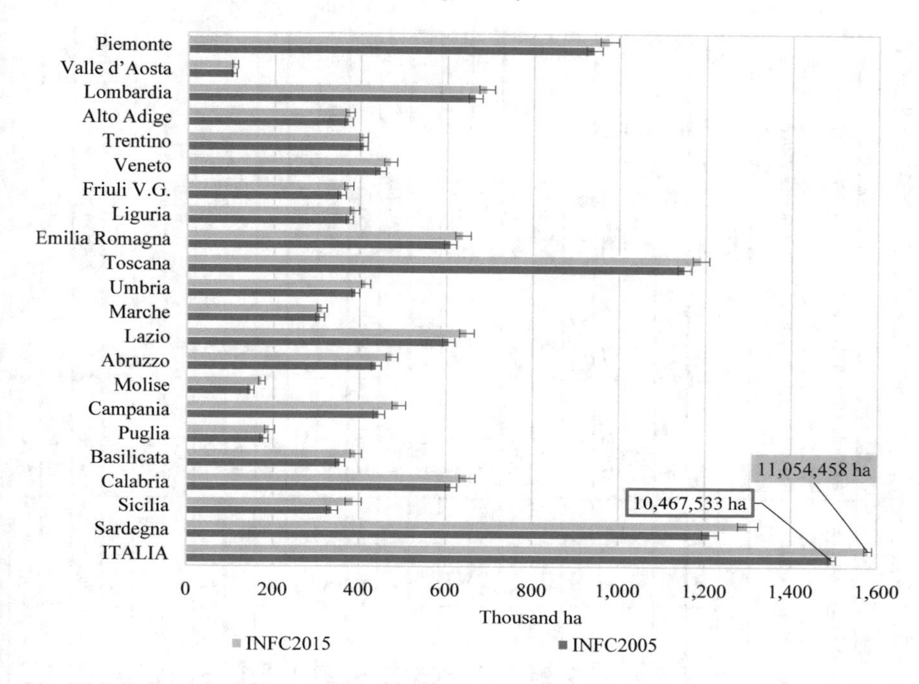

Fig. 13.4 Estimates of Total wooded area of INFC2005 and INFC2015, nationally and for individual regions; error bars are equal to twice the standard error; bars for Italy were shortened to improve readability; real values are given in numbers / Stime dell'estensione della Superficie forestale totale per INFC2005 e INFC2015, a livello nazionale e per le singole regioni; barre di errore pari a due volte l'errore standard; le barre per l'Italia sono state ridotte per migliorare la lettura, i valori reali sono indicati in numero

sample and to the subsequent ground check on the second phase sample with the INFC2015 surveys (cf. Chaps. 2 and 3), it was possible to identify the changes in land use and cover that are relevant for the purposes of the forest inventory, i.e., between forest use and cover and other land use and cover, and also between macro-categories and inventory categories.

Table 13.1 reports the INFC2015 estimates of areas not affected by relevant changes between the two Italian forest inventories, by region and national. It includes the estimates of new wooded areas and areas in other land uses additional to the previous inventory. It is noted that the Total wooded area confirmed by INFC2015 is estimated to be 10,304,862 ha with SE equal to 0.3%, while the area confirmed in Other land use and cover is estimated to be 18,914,517 ha (SE 0.2%). Nationally, the new Total wooded area is estimated at 749,596 ha with SE equal to 2.7% and the additional area in Other land use and cover is equal to 163,870 with SE equal to 4.7%. The difference between the two estimates, which expresses the balance between the wooded area gained and lost compared to the previous INFC2005 inventory, is equal to 585,726 ha, a value very close to that obtained by the difference between the estimates of the Total wooded area of the two inventories reported in Sect. 13.2.3. The

Table 13.1 Area estimates of land use and cover changes to and from the Total wooded area occurring between INFC2005 and INFC2015 / Stima dei cambiamenti di uso e copertura del suolo da e verso la Superficie forestale totale osservati tra INFC2005 e INFC2015

Region/Regione	Unchanged Other land use and cover		Unchanged Total wooded area		Additional Total wooded area		Additional Other land use and cover		Regional and country area
	Altro uso e copertura del suolo confermato		Superficie forestale confermata		Nuova Superficie forestale totale		Nuovo Altro uso e copertura del suolo		Superficie territoriale
	Area	ES	Area	ES	Area	ES	Area	ES	Area
	(ha)	(%)	(ha)	(%)	(ha)	(%)	(ha)	(%)	(ha)
Piemonte	1,533,346	0.7	908,976	1.0	66,448	8.3	31,213	10.6	2,539,983
Valle d'Aosta	216,299	1.5	103,881	2.8	4,095	42.1	2,047	42.1	326,322
Lombardia	1,676,412	0.5	648,049	1.2	44,171	10.7	17,653	14.3	2,386,285
Alto Adige	361,899	1.4	369,427	1.3	5,924	40.2	2,747	36.4	739,997
Trentino	209,148	2.2	403,074	1.1	4,012	42.4	4,456	29.0	620,690
Veneto	1,363,510	0.5	441,401	1.4	28,294	14.5	5,917	24.9	1,839,122
Friuli V.G	405,937	1.3	351,071	1.4	22,543	11.4	6,097	23.4	785,648
Liguria	152,597	3.5	373,091	1.2	14,153	24.3	2,183	40.5	542,024
Emilia Romagna	1,593,506	0.5	611,864	1.2	26,951	16.9	12,880	15.8	2,245,202
Toscana	1,100,597	0.9	1,142,890	0.8	46,832	11.9	8,699	20.2	2,299,018
Umbria	427,322	1.3	385,930	1.3	28,027	10.8	4,326	28.8	845,604
Marche	620,467	0.9	289,200	1.7	23,881	12.2	2,965	35.3	936,513
Lazio	1,067,600	0.8	600,839	1.2	47,309	11.4	5,020	26.8	1,720,768
Abruzzo	600,430	1.1	434,333	1.3	40,266	9.3	4,483	28.0	1,079,512
Molise	269,320	1.4	147,503	2.4	25,770	8.2	1,172	57.5	443,765
Campania	863,465	0.9	441,049	1.5	50,210	9.5	4,301	28.6	1,359,025
Puglia	1,737,193	0.3	171,391	2.7	20,347	17.2	7,649	22.0	1,936,580
Basilicata	604,445	1.1	353,822	1.5	38,590	11.2	2,605	37.7	999,461
Calabria	848,926	1.1	604,422	1.2	46,198	13.4	8,509	20.5	1,508,055
Sicilia	2,165,009	0.4	320,158	2.1	67,076	9.6	18,038	14.2	2,570,282
Sardegna	1,097,089	1.0	1,202,492	0.8	98,499	6.9	10,909	18.4	2,408,989
ITALIA	18,914,517	0.2	10,304,862	0.3	749,596	2.7	163,870	4.7	30,132,845

Table 13.2 Area estimates of land use and land cover changes to and from Forest and Other wooded land occurring between INFC2005 and INFC2015 / Stima dei cambiamenti verso e da Bosco e Altre terre boscate osservati tra INFC2005 e INFC2015

INFC2005	INFC2015					
	Forest		Other wooded land		Other land use and cover	
	Bosco		Altre terre boscate		Altro uso e copertura del suolo	
	Area	ES	Area	ES	Area	ES
	(ha)	(%)	(ha)	(%)	(ha)	(%)
Forest Bosco	8,599,194	0.4	33,194	10.5	127,930	5.3
Other wooded land Altre terre boscate	21,174	13.1	1,651,300	1.3	35,940	10.1
Other land use and cover Altro uso e copertura del suolo	464,818	4.3	284,778	6.1	18,914,517	0.2
Total/Totale	9,085,186	0.4	1,969,272	1.4	19,078,387	0.2

two estimates of the Total wooded area balance differ by a negligible and expected amount, due to the variability that characterises the estimates.

Table 13.2 reports the estimates of the areas affected by changes to and from the Forest and Other wooded land inventory categories at the national level. Similar estimates for the Italian regions are available at https://www.inventarioforestale.org/statistiche_INFC. It is observed that the area of the Forest estimated by INFC2015 that belonged to Other wooded land or to Other land use and cover according to INFC2005 is 21,174 ha and 464,818 ha, respectively. The increase in the Forest area due to the transformation of the areas previously under Other land use and cover is 5.3% of the Forest area according to INFC2005, while the loss of Forest area is 1.5%. The balance is positive for the Forest, which increased by 3.8%. For the Other wooded land, the transition to and from Other land use and cover is more marked, with an increase of 16.7%, a decrease of 2.1% and an overall balance of + 14.6%. There are also changes, for very small areas, even between the two inventory macro-categories, with a slightly positive but not quantitatively considerable balance in favour of the change from Forest to Other wooded land.

13.3 The Variations of Some Characteristics of the Forest Stands

13.3.1 The Growing Stock

INFC2005 has estimated the growing stock of the Forest in Italy to be 1,269,416,499 m^3 with a SE of 1.1%; with INFC2015 the same quantity is estimated to be 1,502,807,089 m^3 with a sample uncertainty of 1.0%. The variation between the two surveys is quantitatively considerable and equal to 233,390,590 m^3 (+18.4%).

Disaggregating the volume estimates by region, it can be observed that always positive and considerable variations occur for the following Italian regions: Piemonte (17,048,366 m^3, +13.1%), Lombardia (22,090,286 m^3, +20.4%), Veneto (17,269,061 m^3, +21.3%), Toscana (23,624,901 m^3, +17.9%), Umbria (6,470,477 m^3, +22.1%), Lazio (15,330,012 m^3, +26.7%), Abruzzo (12,947,702 m^3, +25.6%), Campania (14,259,671 m^3, +33.6%) and Calabria (23,666,353 m^3, +26.9%). For the other regions, the changes are still positive, generally more limited and, given the uncertainty of estimates, they are quantitatively negligible (Fig. 13.5).

Disaggregating the volume estimates by forest type, it is observed that, for the types of Tall trees forest, the variation is always positive and considerable for the Other coniferous forest (6,821,805 m^3, +54.5%), for the Beech (52,532,401 m^3, + 21.9%), for the Temperate oaks (16,963,720 m^3, +21.9%), for the Mediterranean oaks (25,549,227 m^3,+24.6%), for the Hornbeam and Hophornbeam (14,731,288 m^3, +21.3%), for Other deciduous broadleaved (33,626,655 m^3, +35.7%), and for the Holm oak forest (15,601,302 m^3, +34.2%). For the other forest types, the variation is still positive but negligible (Fig. 13.6). For the Tall trees forest as a whole, the overall variation is also quantitatively considerable (236,286,163 m^3, +18.8%).

Considering only the regions for which a considerable change in growing stock volume has occurred, the increase is largely due to the positive variation observed for the Apennine regions, representing over 41% of the overall increase. The geographical location of the positive volume variation is supported by the observation that it occurs especially in the Beech and Oaks forests, which are very present in these regions, constituting almost 47% of the overall volume increase.

13.3.2 The Aboveground Tree Biomass

The aboveground tree biomass of the Forest, expressed in terms of dry weight, was estimated at 874,443,096 Mg by INFC2005, with a SE equal to 1.0%. According to INFC2015, the same quantity is estimated at 1,044,276,604 Mg, with a SE of 1.0%. The variation between the two surveys is quantitatively considerable and equal to 169,833,508 Mg (+19.4%).

Considering the biomass estimates for the individual regions, it can be observed that always positive and considerable variations occur in Piemonte (12,341,412 Mg,

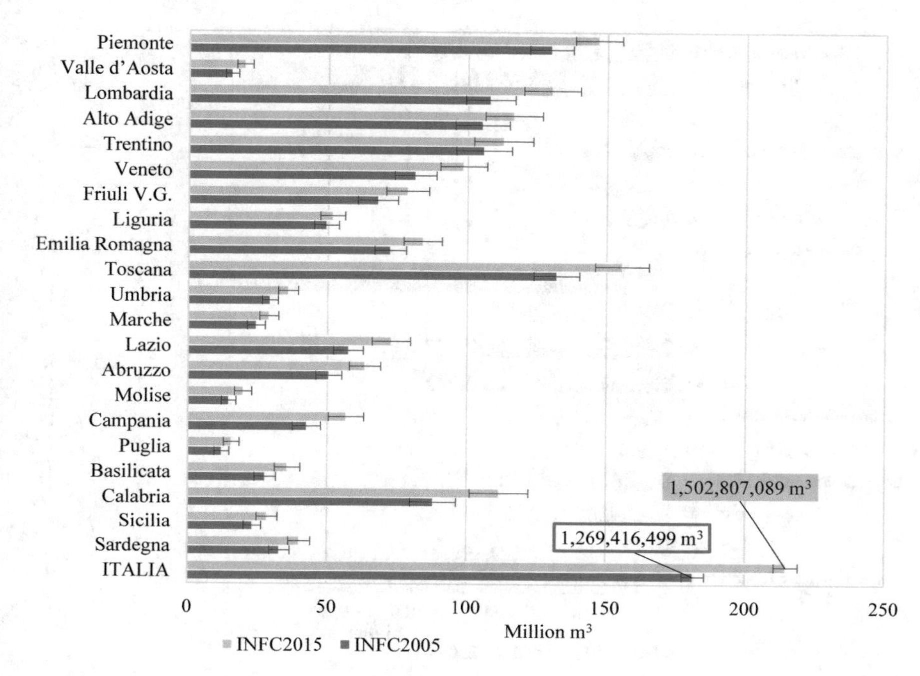

Fig. 13.5 Estimates of the total growing stock of Forest from INFC2005 and INFC2015, nationally and for individual regions; error bars equal to twice the standard error; bars for Italy were shortened to improve readability; real values are given in numbers / Stime del volume totale del Bosco per INFC2005 e INFC2015, a livello nazionale e per le singole regioni; barre di errore pari a due volte l'errore standard; le barre per l'Italia sono state ridotte per migliorare la lettura, i valori reali sono indicati in numero

+14.1%), Lombardia (14,265,600 Mg, +20.8%), Veneto (10,749,400 Mg, + 21.1%), Toscana (18,499,369 Mg, +19.0%), Umbria (5,146,881 Mg, +20.8%), Lazio (12,354,594 Mg, +26.9%), Abruzzo (9,828,520 Mg, +24.5%), Campania (11,079,508 Mg, +33.2%) and Calabria (16,766,828 Mg, +28.7%), while in the other regions the variations, still positive, are generally negligible (Fig. 13.7), in line with the results observed in the comparisons made for the total growing stock.

Disaggregating the estimates of aboveground tree biomass for the forest types of the Tall trees forest, it is observed that the variation is always positive and considerable for the Other coniferous forest (3,887.323 Mg, +53.7%), for the Beech (39,774,535 Mg, +21.6%), for the Temperate oaks (13,836,869 Mg, +21.3%), for the Mediterranean oaks (20,477,654 Mg, +24.1%), for the Hornbeam and Hophornbeam (11,292,435 Mg, +19.1%), for the Other deciduous broadleaved (24,100,050 Mg, +34.5%) and for the Holm oak forest (15,321,040 Mg, +35.0%). For the Tall trees forest as a whole, the variation is considerable and equal to 19.8%, and this is perfectly in line with what was observed in the comparisons made for the total volumes. For the increase of the aboveground tree biomass, the important contribution provided by the Apennine forests, especially Beech and Oak forests, is confirmed.

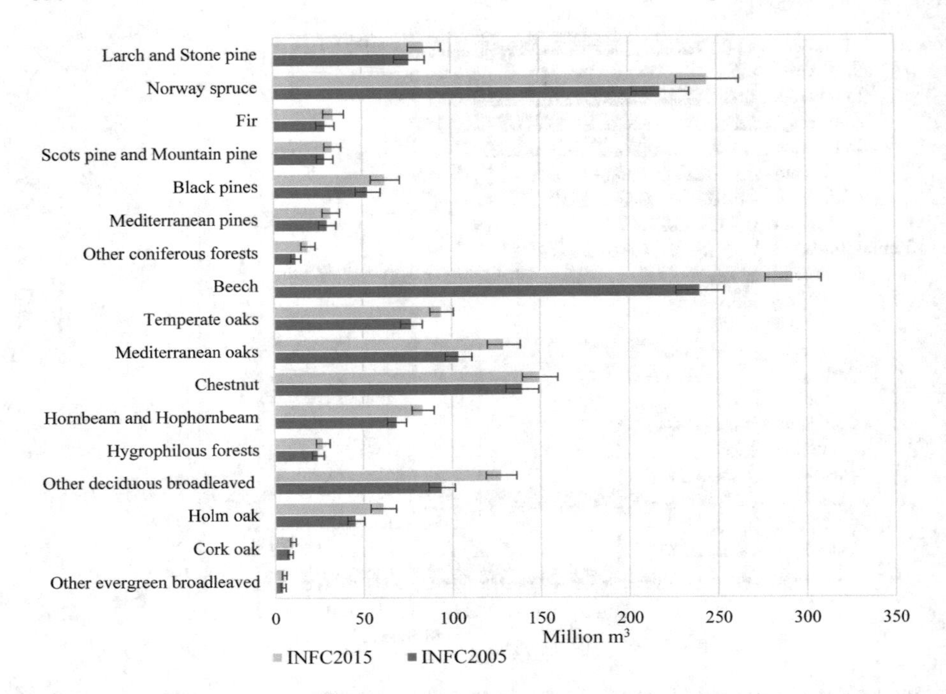

Fig. 13.6 Estimates of the total growing stock of Tall trees forest types from INFC2005 and INFC2015; error bars are equal to twice the standard error / Stime del volume totale delle categorie forestali dei Boschi alti secondo INFC2005 e INFC2015; barre di errore pari a due volte l'errore standard

13.3.3 The Annual Volume Increment

The annual volume increment of living trees standing in the Forest at the time of the INFC2005 inventory survey was estimated at 35,872,293 m^3, with an estimated SE equal to 1.1%. In INFC2015, the same quantity is estimated to be 37,787,784 m^3 with a SE of 1.0%. The positive variation between the two estimates is quantitatively considerable and equal to 1,915,491 m^3 (+5.3%).

By disaggregating the estimates of the annual volume increment according to the regions, it can be observed that positive and considerable variations occur for Lombardia (673.572 m^3, +21.4%), Umbria (208.009 m^3, +25.5%), Lazio (372.571 m^3, +24.1%), Abruzzo (253.181 m^3, +19.2%), Molise (198.589 m^3, + 47.0%), Campania (439.695 m^3, +28.1%) and Basilicata (330.271 m^3, +44.7%). A considerable negative change is observed for Liguria (−413.453 m^3, −26.2%) and for Toscana (−603.917 m^3, −14.5%). For the other regions the variations, positive and negative, are generally negligible according to the criteria adopted for the comparison (Fig. 13.9).

Disaggregating the same estimates of annual volume increment according to the forest type, it is observed that, for the types relating to the Tall trees forest, the

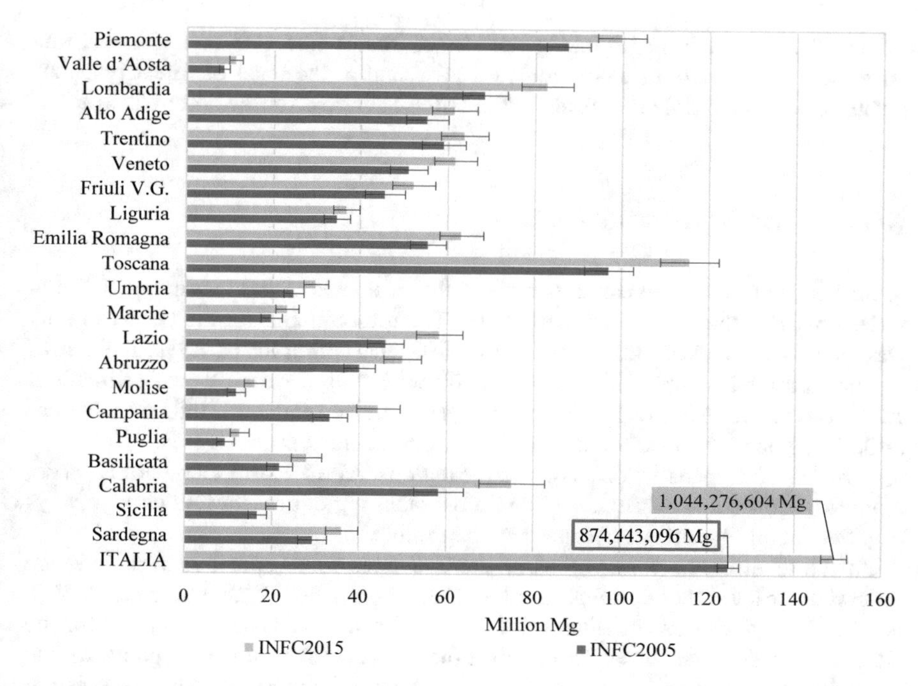

Fig. 13.7 Estimates of the total aboveground tree biomass of Forest from INFC2005 and INFC2015, nationally and for individual regions; error bars are equal to twice the standard error; bars for Italy were shortened to improve readability; real values are given in numbers / Stime della fitomassa arborea epigea del Bosco secondo INFC2005 e INFC2015, a livello nazionale e per le singole regioni; barre di errore pari a due volte l'errore standard; le barre per l'Italia sono state ridotte per migliorare la lettura, i valori reali sono indicati in numero

variation is positive and considerable for the Temperate oaks (586,013 m^3, +26.5%) and for the Other deciduous broadleaved (618.803 m^3, +15.4%), while it is negative and considerable for the Chestnut (−751.900 m^3, −15.2%) (Fig. 13.10). For the Tall trees forest as a whole, the variation is positive and considerable (2,219,885 m^3, + 6.4%).

13.3.4 *The Annual Removal of Growing Stock*

The growing stock removed through utilisations during the twelve months before the INFC2005 survey was estimated at 13,796,864 m^3, with a SE of 12.9%. In INFC2015 the same quantity is estimated to be 9,566,257 m^3 with a SE equal to 16.1%. The negative variation between the two estimates is equal to −4,230,607

m^3 (−30.7%) and is negligible given the estimates intervals considered. Similarly, observed variations at the regional level and at the level of forest types are quantitatively negligible according to the criteria adopted for the comparison.

13.3.5 The Number of Trees

A further interesting comparison between the estimates of the two national inventories concerns the number of living trees. The total number for the Forest in Italy was estimated by INFC2005 to be 11,949,630,797, with a SE of 1.3%; INFC2015 estimated a total number of 11,481,562,909 trees with a variability of estimate of 1.2%. The variation between the two surveys is therefore equal to - 468,067,888 trees (−3.9%), a negligible variation given the estimate intervals considered.

Positive differences are observed for the regions Valle d'Aosta, Lombardia, Lazio, Abruzzo, Molise and Sardegna, while for the other regions the variations are always negative. In all cases the differences are negligible.

At the level of the forest types of the Tall trees forest, positive variations are observed for the Other coniferous forest (33.2%), for the Temperate oaks (5.4%), for the Other deciduous broadleaved (10.4%), for the Cork oak (3.1%) and for the Other evergreen broadleaved forest, while the differences are always negative for the other forest types. For the type of Chestnut alone there is a considerable variation (−183,976,338 trees, −16.0%), while for all the other forest types the differences are quantitatively negligible.

Appendix (Italian Version)

Riassunto Per confrontare utilmente i risultati di indagini diverse è opportuno che esse condividano il dominio di studio, le definizioni e i criteri di classificazione e misura delle variabili osservate, o che questi siano coerenti tra loro. Il dominio inventariale nei due inventari forestali italiani INFC2005 e INFC2015 è rimasto invariato e le due indagini adottano un sistema di classificazione e criteri di stratificazione (regione e categoria forestale) comuni; anche le soglie di misura delle variabili quantitative e le classi delle variabili qualitative sono le stesse. Il capitolo illustra i risultati del confronto tra le stime ottenute dalle due indagini per alcune tra le grandezze più interessanti per valutazioni sulla dinamica delle foreste italiane, quali l'estensione delle aree, il volume e la fitomassa del bosco, l'incremento annuo di volume, il prelievo annuo e il numero degli alberi. I risultati del confronto sono sufficientemente esplicativi delle tendenze a livello nazionale, mentre le stime disaggregate per regione o per categoria forestale mostrano variazioni talvolta molto contenute e associate a errori standard delle stime elevati.

Introduzione

Il monitoraggio dello stato delle risorse forestali nazionali necessita di informazioni affidabili e confrontabili, derivanti da osservazioni ripetute su ampia scala. Al fine di poter confrontare i risultati di indagini diverse, è necessario che le stesse condividano i criteri per la definizione del dominio di studio, quali ad esempio la copertura arborea e la superficie per la definizione di foresta. È necessario, inoltre, che le definizioni e i criteri di classificazione e misura delle variabili osservate dalle diverse rilevazioni siano uguali o comunque coerenti tra loro.

Nell'introduzione al capitolo di De Natale and Gasparini (2011) dedicato all'illustrazione dei principali cambiamenti intervenuti tra il primo inventario forestale nazionale italiano (IFNI85) e il secondo (INFC2005), è stata sottolineata l'importanza di disporre di più rilevazioni nel tempo, al fine di operare utili confronti tra le statistiche risultanti e identificare così le tendenze in atto nel sistema forestale italiano. Sono state inoltre brevemente ricordate le azioni di armonizzazione tra i due inventari forestali citati - che nel loro disegno divergono per alcuni aspetti definitori, di classificazione e per alcuni standard di misura - in funzione sia del confronto auspicato sia della necessità di rispondere, con maggiore accuratezza rispetto al passato, alle richieste provenienti da importanti processi statistici sovranazionali (cfr. Chap. 1).

Rispetto al confronto precedente, le differenze metodologiche tra INFC2005 e INFC2015 sono decisamente più contenute, limitandosi al miglioramento di alcune procedure operative nella raccolta e nella registrazione delle informazioni in campo. Il dominio inventariale, il sistema di classificazione e i criteri di stratificazione, per regione e categoria forestale, sono rimasti invariati (cfr. Chap. 2); analogamente, le soglie di misura delle variabili quantitative e le classi utilizzate per le variabili qualitative sono le stesse (cfr. Chap. 4), così come sono rimaste invariate le procedure di calcolo e stima (cfr. Chaps. 5 and 6). Questa favorevole condizione permette di operare con maggiore tranquillità i vari confronti di interesse tra grandezze omologhe rilevate nelle due occasioni inventariali.

Altri aspetti da considerare per il confronto tra i risultati INFC2005 e INFC2015 riguardano la complessità del disegno inventariale e le modalità di selezione delle unità di campionamento. L'insieme dei punti di campionamento di prima fase è rimasto lo stesso nei due inventari, mentre i campioni di seconda e terza fase di INFC2015 includono punti aggiuntivi, selezionati tra quelli interessati da cambiamenti dell'uso e copertura del suolo (cfr. Chaps. 2 e 5). I campioni delle due indagini non sono quindi indipendenti. La letteratura dedicata ai confronti tra inventari ripetuti articolati in più fasi e all'analisi del comportamento probabilistico degli stimatori impiegati in disegni così complessi è purtroppo lacunosa. Tuttavia, per gli scopi del presente capitolo, si è proceduto ad un'analisi finalizzata a cogliere l'entità e il segno delle variazioni intervenute tra i due inventari, tenendo conto della variabilità delle stime nel valutare la rilevanza delle differenze, senza l'ausilio di un rigoroso test di ipotesi.

In generale, l'analisi delle variazioni tra grandezze stimate per via campionaria deve prendere in considerazione non solo i valori assunti dagli stimatori, ma anche i corrispondenti errori standard (ES), che indicano la variabilità che si osserva per campioni diversi di uguale dimensione, a parità di disegno campionario (Gregoire & Valentine, 2008). Se l'entità della variazione tra le due stime campionarie è ingente e molto più grande della relativa variabilità, risulta difficile ipotizzare che detta variazione non sia realmente diversa dallo zero. Per questo motivo, fermi restando i limiti sopra ricordati, si è convenzionalmente proceduto a confrontare le stime unitamente ai relativi errori campionari, considerando quantitativamente rilevanti le differenze con intervalli di stima, calcolati considerando due volte l'ES, non sovrapposti. Nei commenti che si riferiscono alle variabili prese in considerazione nel confronto tra INFC2005 e INFC2015 è riportata anche la corrispondente variabilità della stima, in termini di errore standard percentuale (ES%), in modo da cogliere con immediatezza i casi in cui variazioni e intervalli di stima possono confondersi.

I risultati del confronto presentati nel capitolo riguardano alcune grandezze inventariali ritenute interessanti per illustrare la dinamica attuale delle foreste italiane. In particolare, sono stati esaminati l'estensione delle aree boscate (Bosco, Altre terre boscate e Superficie forestale totale) e, tra le variabili che caratterizzano i Boschi alti, il volume degli alberi vivi, la fitomassa arborea epigea, l'incremento annuo di volume, il prelievo annuo e il numero di alberi. Il confronto riguarda le stime dei valori totali di ciascuna variabile e le stesse disaggregate per regione e per categoria forestale dei Boschi alti, per evidenziare eventuali differenze nelle variazioni intervenute tra le due rilevazioni nelle diverse aree geografiche. Infine vengono presentate le stime delle superfici interessate da cambiamenti di uso e copertura del suolo, da classi delle superfici forestali a classi di altro uso e copertura e viceversa, risultanti dal confronto della classificazione dei punti di campionamento nei due inventari.

I risultati del confronto sono sufficientemente esplicativi delle dinamiche in atto a livello nazionale. Per le stime disaggregate per regione o per categoria forestale, invece, le variazioni sono talvolta molto contenute e associate a errori campionari di stima elevati.

Le variazioni della superficie forestale

Il Bosco

La superficie totale del Bosco in Italia stimata da INFC2005 è pari a 8,759,202 ha, con un ES di 0.4% (Gasparini & Tabacchi, 2011); INFC2015 ha stimato la stessa grandezza in 9,085,186 ha, con ES pari a 0.4%. La variazione positiva di superficie intervenuta tra le due rilevazioni risulta quindi uguale a 325,984 ha (+3.7%). Secondo il criterio sopra descritto, il confronto tra le due stime evidenzia una variazione rilevante a livello nazionale tra le due indagini. Disaggregando le stime di superficie del Bosco a livello regionale, si osserva che variazioni di superficie rilevanti sono segnalate solo per il Molise (20,686 ha, +15.6%), mentre le variazioni per le altre regioni, tutte positive meno che per il Trentino (−2143 ha, −0.6%) e la Puglia (−3540 ha, −2.4%), non risultano quantitativamente rilevanti (Fig. 13.1). A livello di territori più ampi, nel periodo intercorso tra i due inventari l'estensione del Bosco è aumentata di 68,596 ha (+2.0%) nelle regioni dell'Arco alpino, di 96,910 ha (+3.1%) nelle regioni dell'Appennino settentrionale e centrale e di 160,479 ha (+7.2%) in quelle dell'Appennino meridionale e nelle isole.[2]

Osservando le stime di estensione del Bosco secondo le categorie forestali dei Boschi alti emergono variazioni negative per i Boschi di abete rosso (−0.4%), per le Pinete di pino silvestre e montano (−2.3%), per le Pinete di pino nero, laricio e loricato (−3.1%) e per i Castagneti (−1.8%); per le altre categorie la variazione di superficie è sempre positiva. Va però segnalato che solo per gli Altri boschi caducifogli (+11.4%) e per gli Altri boschi di latifoglie sempreverdi (+33.6%) la variazione è quantitativamente rilevante (Fig. 13.2). Si tratta di due classi residuali, in cui confluiscono soprassuoli forestali di recente formazione quali ad esempio gli Acerotiglieti di monte e i Robinieti e ailanteti. Per l'insieme delle categorie dei Boschi alti, che occupano il 98.6% della superficie del Bosco in Italia secondo INFC2015, la variazione di estensione risulta positiva e rilevante (319,837 ha, +3.7%).

Le Altre terre boscate

La superficie delle Altre terre boscate in Italia è stata stimata da INFC2005 in 1,708,333 ha, con un ES di 1.3%; INFC2015 ne ha stimato un'estensione pari a 1,969,272 ha, con ES di 1.4%. La variazione di superficie intervenuta tra le due rilevazioni risulta quantitativamente rilevante e pari a 260,939 ha (+15.3%). Disaggregando le stime di superficie delle Altre terre boscate per regione, si può osservare che le variazioni sono sempre positive, ma rilevanti solo per il Lazio (25,938 ha, +41.9%), la Campania (26,453 ha, +43.5%) e la Puglia (16,238 ha, +49.0%) (Fig. 13.3). A livello di territori più ampi, l'estensione delle Altre terre boscate è aumentata di 49,349 ha (+15.3%) nelle regioni dell'Arco alpino, di 78,275 ha (+24.0%) nelle regioni dell'Appennino settentrionale e centrale e di 133,314 ha (+12.6%) nelle regioni dell'Appennino meridionale e isole.

[2] Regioni dell'Arco alpino: Piemonte, Valle d'Aosta, Lombardia, Alto Adige, Trentino, Veneto, Friuli-Venezia Giulia, Liguria; regioni dell'Appennino settentrionale: Emilia-Romagna, Toscana, Umbria, Marche, Lazio, Abruzzo; regioni dell'Appennino meridionale e isole: Molise, Campania, Puglia, Basilicata, Calabria, Sicilia, Sardegna.

La Superficie forestale totale

Con INFC2005 la Superficie forestale totale (Bosco più Altre terre boscate) in Italia è stata stimata pari a 10,467,533 ha con un ES di 0.3%; la stima INFC2015 per la stessa grandezza è pari a 11,054,458 ha, con ES di 0.3%. La variazione di superficie intervenuta tra le due rilevazioni risulta quantitativamente rilevante e pari a 586,925 ha (+5.6%). Disaggregando le stime di Superficie forestale totale a livello regionale, si può osservare che variazioni positive si manifestano per tutte le regioni, con la sola eccezione del Trentino (-0.1%). Queste variazioni appaiono quantitativamente rilevanti per molte regioni peninsulari e per le isole: per Toscana (38,183 ha, + 3.3%), Umbria (23,701 ha, +6.1%), Lazio (42,289 ha, +7.0%), Abruzzo (36,009 ha, +8.2%), Molise (24,632 ha, +16.6%), Campania (45,985 ha, +10.3%), Basilicata (35,986 ha, + 10.1%), Calabria (37,689 ha, +6.1%), Sicilia (49,063 ha, + 14. 5%) e Sardegna (87,741 ha, +7.2%) (Fig. 13.4). A livello di territori più ampi, nel periodo intercorso tra i due inventari la Superficie forestale totale è aumentata di 117,944 ha (+3.2%) nelle regioni dell'Arco alpino, di 175,184 ha (+5.0%) nelle regioni dell'Appennino settentrionale e centrale e di 293,794 ha (+8.9%) nelle regioni dell'Appennino meridionale e isole.

I cambiamenti dell'uso e copertura del suolo

Le variazioni della Superficie forestale totale e di quella del Bosco e delle Altre terre boscate risultanti dal confronto tra le relative stime inventariali sono state illustrate nei paragrafi precedenti. Grazie alla ripetizione della fotointerpretazione su tutti i punti del campione di prima fase e al successivo rilievo al suolo sul campione di seconda fase INFC2015 (cfr. Chaps. 2 e 3), è stato possibile verificare i cambiamenti di uso e copertura avvenuti da e verso le classi di uso e copertura forestale, individuando i punti inventariali transitati da una classe all'altra, in particolare da uso e copertura forestale ad altro uso del suolo e viceversa, e le transizioni tra macrocategorie e categorie inventariali.

La Table 13.1 riporta le stime relative alle superfici non interessate da cambiamenti tra i due inventari forestali italiani, per regioni e nazionale, e le stime delle nuove superfici forestali e delle nuove superfici in altri usi del suolo rispetto a INFC2005. Si osserva che la Superficie forestale totale confermata da INFC2015 è stimata in 10,304,862 ha (ES 0.3%), mentre la superficie confermata in uso e copertura non forestale è stimata pari a 18,914,517 ha (ES 0.2%). A livello nazionale, la nuova Superficie forestale è stimata in 749,596 ha (ES 2.7%) e la nuova superficie in altri usi del suolo è pari a 163,870 (ES 4.7%). La differenza tra queste due stime di superficie esprime il bilancio tra Superficie forestale guadagnata e perduta rispetto al precedente inventario INFC2005. Essa è pari a 585,726 ha, un valore molto vicino a quello ottenuto per differenza tra le stime di Superficie forestale totale dei due inventari riportata nella Sect. 13.2.3 . Come atteso, i due valori relativi alle variazioni della Superficie forestale differiscono leggermente, per effetto dell'incertezza che caratterizza le stime campionarie, ma la differenza non è rilevante.

La Table 13.2 riporta le stime delle superfici interessate da cambiamenti di uso e copertura da e verso le categorie inventariali Bosco e Altre terre boscate, a livello nazionale. Stime analoghe per le regioni italiane sono disponibili all'indirizzo https://www.inventarioforestale.org/statistiche_INFC. Si osserva che la superficie del Bosco stimata da INFC2015 che apparteneva alla classe Altre terre boscate o ad Altro uso del suolo secondo INFC2005 è pari rispettivamente a 21,174 ha e 464,818 ha. L'aumento di superficie del Bosco dovuto a trasformazione di aree prima attribuite ad Altro uso del suolo è pari al 5.3% della superficie del Bosco secondo INFC2005, mentre la perdita di superficie è del 1.5%. Il bilancio è dunque positivo per la categoria Bosco, che aumenta del 3.8%. Per le Altre terre boscate il passaggio da e verso Altro uso e copertura del suolo è più marcato, con un aumento pari a 16.7%, una diminuzione pari a 2.1% e un bilancio complessivo di +14.6%. Si registrano, inoltre, passaggi di superfici molto piccoli anche tra le due categorie inventariali, con un bilancio di poco positivo ma quantitativamente non rilevante a favore del trasferimento da Bosco ad Altre terre boscate.

Le variazioni dei principali caratteri dei soprassuoli

Il volume totale

INFC2005 ha stimato il volume del Bosco in Italia pari a 1,269,416,499 m^3 con un ES di 1.1%; con INFC2015 la stessa variabile è stimata pari a 1,502,807,089 m^3 con un ES dell'1.0%. Ne risulta una variazione nelle stime quantitativamente rilevante pari a 233,390,590 m^3 (+18.4%).

Per le regioni, variazioni sempre positive e rilevanti si manifestano nei seguenti casi: Piemonte (17,048,366 m^3, +13.1%), Lombardia (22,090,286 m^3, +20.4%), Veneto (17,269,061 m^3, +21.3%), Toscana (23,624,901 m^3, +17.9%), Umbria (6,470,477 m^3, +22.1%), Lazio (15,330,012 m^3, +26.7%), Abruzzo (12,947,702 m^3, +25.6%), Campania (14,259,671 m^3, +33.6%) e Calabria (23,666,353 m^3, + 26.9%). Per le altre regioni, le variazioni risultano sempre positive, ma in genere sono anche contenute e non rilevanti (Fig. 13.5).

Disaggregando le stesse stime di volume per categoria forestale dei Boschi alti si osserva che la variazione, sempre positiva, è rilevante per gli Altri boschi di conifere, puri e misti (6,821,805 m^3, +54.5%), per le Faggete (52,532,401 m^3, +21.9%), per i Querceti di rovere, roverella e farnia (16,963,720 m^3, +21.9%), per le Cerrete, boschi di farnetto, fragno e vallonea (25,549,227 m^3, +24.6%), per gli Ostrieti, carpineti (14,731,288 m^3, +21.3%), per gli Altri boschi caducifogli (33,626,655 m^3, +35.7%), e per le Leccete (15,601,302 m^3, +34.2%). Per le altre categorie forestali la variazione, pur positiva, non è rilevante (Fig. 13.6). Anche per l'insieme dei Boschi alti la variazione complessiva risulta quantitativamente rilevante (236,286,163 m^3, +18.8%).

Considerando l'insieme delle regioni per cui si è manifestata una variazione rilevante nella stima del volume del Bosco, l'aumento è in buona parte dovuto alla variazione positiva osservata per le regioni appenniniche, pari a oltre il 41% dell'aumento complessivo. Questa considerazione è suffragata dall'osservazione che l'aumento di volume riguarda in particolare le Faggete e i Querceti, ben presenti in queste regioni, categorie che da sole coprono quasi il 47% dell'aumento complessivo di volume.

La fitomassa arborea epigea

La fitomassa arborea epigea del Bosco, espressa in termini di peso secco, è stata stimata pari a 874,443,096 Mg da INFC2005, con un ES pari a 1.0%; secondo INFC2015 la stessa grandezza ammonta a 1,044,276,604 Mg (ES di 1.0%). La variazione intervenuta tra le due rilevazioni risulta quantitativamente rilevante e pari a 169,833,508 Mg (+19.4%).

Considerando le stime per le singole regioni, si può osservare che variazioni sempre positive e rilevanti si manifestano in Piemonte (12,341,412 Mg, +14.1%), Lombardia (14,265,600 Mg, +20.8%), Veneto (10,749,400 Mg, + 21.1%), Toscana (18,499,369 Mg, +19.0%), Umbria (5,146,881 Mg, +20.8%), Lazio (12,354,594 Mg, +26.9%), Abruzzo (9,828,520 Mg, +24.5%), Campania (11,079,508 Mg, +33.2%) e Calabria (16,766,828 Mg, +28.7%), mentre nelle altre regioni le variazioni si mantengono positive ma non sono rilevanti (Fig. 13.7), in linea con le risultanze osservate nei confronti svolti per i volumi totali.

Disaggregando le stime di fitomassa arborea epigea per le categorie forestali dei Boschi alti si osserva una variazione positiva e rilevante per gli Altri boschi di conifere, puri e misti (3,887,323 Mg, +53.7%), per le Faggete (39,774,535 Mg, +21.6%), per i Querceti di rovere, roverella e farnia (13,836,869 Mg, +21.3%), per le Cerrete, boschi di farnetto, fragno e vallonea (20,477,654 Mg, +24.1%), per gli Ostrieti, carpineti (11,292,435 Mg, +19.1%), per gli Altri boschi caducifogli (24,100,050 Mg, +34.5%) e per le Leccete (15,321,040 Mg, +35.0%). Per le altre categorie forestali la variazione rimane positiva, ma più contenuta (Fig. 13.8).

Per l'insieme dei Boschi alti, la variazione è rilevante e pari al 19.8%, in linea con quanto osservato per la variazione del volume. Anche per la consistente variazione positiva della fitomassa arborea epigea si conferma l'importante contributo fornito dai boschi appenninici, soprattutto Faggete e Querceti.

L'incremento annuo di volume

L'incremento annuo di volume degli alberi vivi del Bosco al momento del rilevamento inventariale INFC2005 è stato stimato in 35,872,293 m^3, con un ES di stima pari

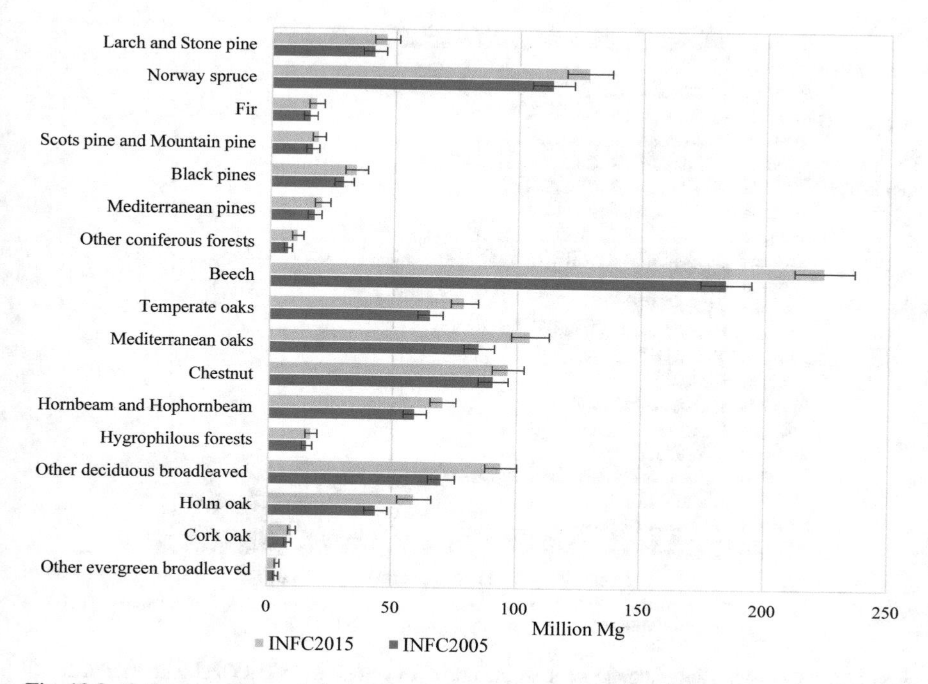

Fig. 13.8 Estimates of the total aboveground tree biomass of Tall trees forest types from INFC2005 and INFC2015; error bars are equal to twice the standard error / Stime della fitomassa arborea epigea delle categorie forestali dei Boschi alti secondo INFC2005 e INFC2015; barre di errore pari a due volte l'errore standard

a 1.1%, mentre con INFC2015 la stessa grandezza è stimata pari a 37,787,784 m^3 con un ES di 1.0%; la variazione positiva tra le due stime risulta quantitativamente rilevante e pari a 1,915,491 m^3 (+5.3%).

Disaggregando le stime per regione si possono osservare variazioni positive e rilevanti per Lombardia (673,572 m^3, +21.4%), Umbria (208,009 m^3, +25.5%), Lazio (372,571 m^3, +24.1%), Abruzzo (253,181 m^3, +19.2%), Molise (198,589 m^3, +47.0%), Campania (439,695 m^3, +28.1%) e Basilicata (330,271 m^3, + 44.7%). Variazioni negative rilevanti si osservano per la Liguria (−413,453 m^3, −26.2%) e per la Toscana (−603,917 m^3, −14.5%). Per le altre regioni le variazioni, positive o negative, non sono rilevanti secondo il criterio adottato per il confronto (Fig. 13.9).

Le stime di incremento annuo per categoria forestale dei Boschi alti permettono di osservare variazioni positive rilevanti per i Querceti di rovere, roverella e farnia (586,013 m^3, +26.5%) e per gli Altri boschi caducifogli (618,803 m^3, +15.4%), mentre è negativa e quantitativamente rilevante quella per i Castagneti (−751,900 m^3, −15,2%) (Fig. 13.10). Per l'insieme dei Boschi alti la variazione risulta positiva e rilevante (2,219,885 m^3, +6.4%).

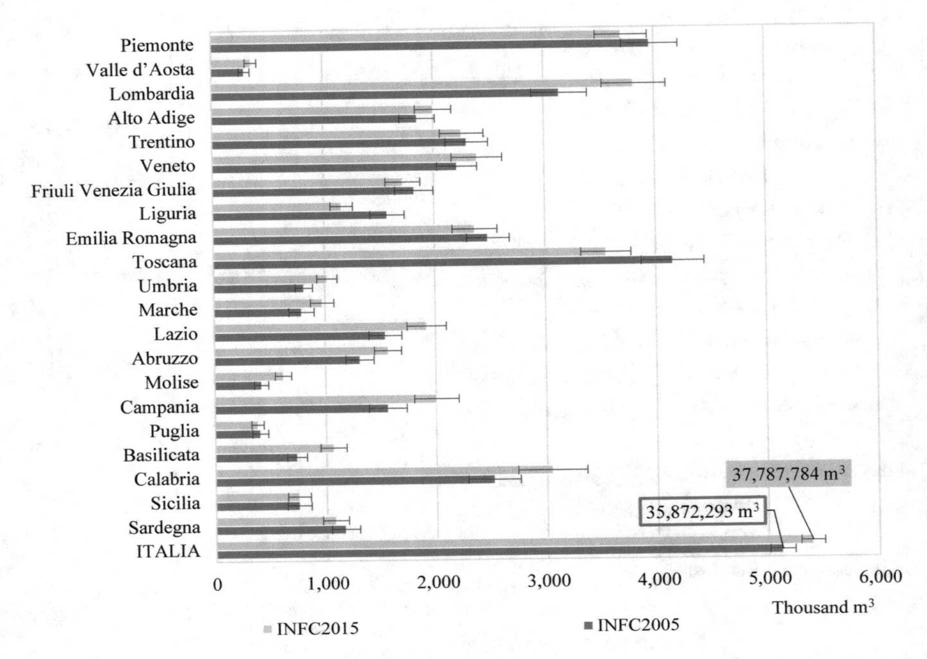

Fig. 13.9 Estimates of the annual volume increment of Forest from INFC2005 and INFC2015, nationally and for individual regions; error bars are equal to twice the standard error; bars for Italy were shortened to improve readability; real values are given in numbers / Stime dell'incremento annuo di volume del Bosco secondo INFC2005 e INFC2015, a livello nazionale e per le singole regioni; barre di errore pari a due volte l'errore standard; le barre per l'Italia sono state ridotte per migliorare la lettura, i valori reali sono indicati in numero

Il prelievo annuo

Il volume degli alberi utilizzati nei dodici mesi precedenti il rilevamento inventariale INFC2005 è stato stimato in 13,796,864 m^3, con ES pari a 12.9%; la stima INFC2015 della stessa grandezza è pari a 9,566,257 m^3 con ES uguale a 16.1%. La variazione negativa tra la due stime è pari a −4,230,607 m^3 (−30.7%), valore quantitativamente non rilevante sulla base delle convenzioni adottate per il confronto. Analogamente, le differenze osservate per le singole regioni e categorie forestali risultano non rilevanti.

Il numero di alberi

Un ulteriore confronto di interesse tra le stime dei due inventari nazionali riguarda il numero degli alberi vivi. Il loro numero totale per il Bosco in Italia è stato stimato da INFC2005 pari a 11,949,630,797, con un ES di 1.3%; INFC2015 ha stimato un numero totale di 11,481,562,909 alberi, con ES di 1.2%. La differenza tra le due stime

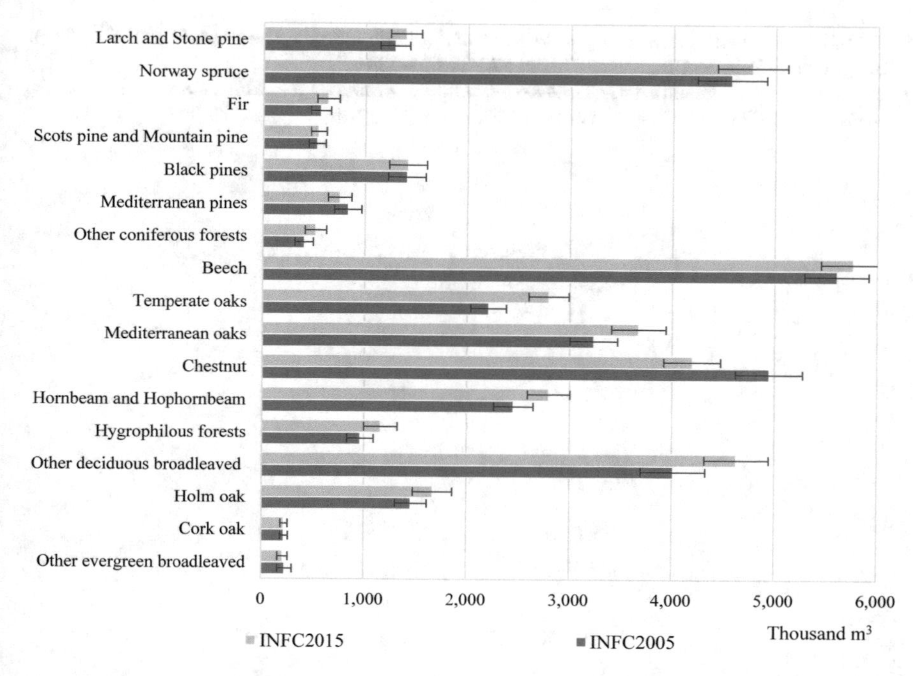

Fig. 13.10 Estimates of the annual volume increment of Tall trees forest types from INFC2005 and INFC2015; error bars are equal to twice the standard error / Stime dell'incremento annuo di volume delle categorie forestali dei Boschi alti secondo INFC2005 e INFC2015; barre di errore pari a due volte l'errore standard

risulta dunque uguale a −468,067,888 alberi (−3.9%), variazione non rilevante visti gli intervalli di stima calcolati. Differenze positive si osservano per la Valle d'Aosta, la Lombardia, il Lazio, l'Abruzzo, il Molise e la Sardegna, mentre per le altre regioni le variazioni sono sempre negative. A livello regionale le differenze non risultano mai quantitativamente rilevanti.

A livello delle categorie forestali dei Boschi alti, si osservano variazioni positive per gli Altri boschi di conifere puri e misti (+33.2%), per i Querceti di rovere, roverella e farnia (+5.4%), per gli Altri boschi caducifogli (+10.4%), per le Sugherete (+3.1%) e per gli Altri boschi di latifoglie sempreverdi (+27.1%), mentre le differenze sono sempre negative per le altre categorie forestali. Nel complesso, considerando gli intervalli di stima, le differenze non sono mai rilevanti, ad eccezione che per la categoria dei Castagneti (−183,976,338 alberi, −16.0%).

References

De Natale, F., & Gasparini, P. (2011). I cambiamenti nel periodo 1985–2005. In: P. Gasparini & G. Tabacchi (Eds.), *L'inventario Nazionale delle Foreste e dei serbatoi forestali di carbonio*

INFC2005. Secondo inventario forestale nazionale italiano. Metodi e risultati (pp. 585–589). Edagricole-Il Sole 24 Ore, Milano. ISBN 978-88-506-5394-2.

Gregoire, T. G., & Valentine, H. T. (2008). *Sampling strategies for natural resources and the environment* (474 pp.). Chapman & Hall/CRC.

Open Access This chapter is licensed under the terms of the Creative Commons Attribution 4.0 International License (http://creativecommons.org/licenses/by/4.0/), which permits use, sharing, adaptation, distribution and reproduction in any medium or format, as long as you give appropriate credit to the original author(s) and the source, provide a link to the Creative Commons license and indicate if changes were made.

The images or other third party material in this chapter are included in the chapter's Creative Commons license, unless indicated otherwise in a credit line to the material. If material is not included in the chapter's Creative Commons license and your intended use is not permitted by statutory regulation or exceeds the permitted use, you will need to obtain permission directly from the copyright holder.

Glossary/Glossario

This glossary provides definitions of technical terms as they were intended in INFC, which are relevant to the reading of this book. More rigorous definitions in other books might differ from those listed, especially for terms not commonly agreed upon in the existing research/Questo glossario riporta le definizioni di termini tecnici come adottate per INFC e che sono rilevanti per la lettura di questa pubblicazione. Definizioni più rigorose in altri libri potrebbero differire da quelle qui proposte, soprattutto per i termini con definizione non unanimemente accettata.

Aboveground tree biomass/Fitomassa arborea epigea Stand living biomass of callipered trees. This may refer to one tree or to a class (e.g., the aboveground biomass of all the trees in a forest type). It includes the biomass of stem, large branches and treetop up to 5 cm diameter section, bark; stump, small branches and treetop below 5 cm diameter section/Fitomassa della ceppaia, del fusto, dei rami grossi, della ramaglia e del cimale. Può essere riferita ad un albero o a un'intera classe come ad esempio una categoria forestale. Si parla di necromassa nel caso sia riferita ad alberi integri, morti in piedi.

Calliper/Cavalletto dendrometrico Tool for measuring DBH (also diameters of stumps and deadwood lying on the ground)/Strumento forestale per la misurazione dei diametri di alberi e tronchi.

Callipering/Cavallettamento Measurement of DBH using the calliper/Operazione di misurazione dei diametri con il cavalletto dendrometrico.

Clinometer/Clisimetro Optical or mechanical instrument used for measuring angles of slope/Strumento ottico o meccanico utilizzato per la misurazione della inclinazione del terreno.

Client Hardware or software component, connected to another computer hardware or software (Server) computer, using services provided by the latter/Computer, o componente software, collegato in rete ad un altro elaboratore o componente software (Server) dal quale fruisce di servizi software o dati.

Coordinates/Coordinate Values that are duplet or triplet, which uniquely identify a point's position on the Earth surface with respect to a reference parallel and

© The Editor(s) (if applicable) and The Author(s) 2022

P. Gasparini et al. (eds.), *Italian National Forest Inventory—Methods and Results of the Third Survey*, Springer Tracts in Civil Engineering,
https://doi.org/10.1007/978-3-030-98678-0

meridian. They can be geographic coordinates (angles, i.e., latitude and longitude) or plane coordinates (distances, e.g., N, E). The third coordinate, when present, is height above ellipsoid or altitude above sea level/Coppia o terna di valori che individuano univocamente la posizione di un punto sulla superficie terrestre rispetto a un parallelo e a un meridiano di riferimento. Esse possono essere c. geografiche (angoli, latitudine e longitudine) o c. piane (distanze, es. N, E). La terza coordinata, quando presente, è l'altezza sull'ellissoide o la quota sul livello del mare.

Datum A model for the Earth's shape and dimensions, used as a reference to compute coordinates. It usually consists of several components, such as an ellipsoid, an origin of coordinates and several control points precisely measured from the origin and marked; WGS84 is an example of datum/Modello di forma e dimensioni della Terra, usato come riferimento per calcolare coordinate. Solitamente costituito da diverse componenti, quali un ellissoide, una origine delle coordinate e numerosi punti di controllo misurati con precisione e materializzati. Un esempio di datum è WGS84.

Density (in statistics)/Densità (in statistica) Rate between the total value of a variable by the area to which it refers i.e., the mean value per unit area (generally hectare)/Rapporto statistico tra il totale di una determinata variabile e la superficie a cui fa riferimento, che ne misura l'ammontare per unità di superficie (generalmente l'ettaro).

Dilution of precision/Diluizione di precision In GNSS positioning, a multiplicative factor that increases the uncertainty range; it depends mainly on satellites and receiver geometry. Horizontal (HDOP) and Position (PDOP) are the indicators most used/Nel posizionamento con GNSS, fattore moltiplicativo che ne amplifica l'intervallo di incertezza. È causata soprattutto dalla geometria relativa che intercorre fra il ricevitore e i satelliti utilizzati. Gli indicatori più usati sono HDOP (Horizontal) e PDOP (Position).

Ellipsoid of rotation/Ellissoide di rotazione Solid figure obtained rotating an ellipse around one of its axes. In geodesy, it is the most used geometric figure to approximate the Earth's shape. Since two of its three axes are equal, it is often called a Spheroid/Figura solida ottenuta dalla rotazione di un'ellisse attorno a uno dei suoi assi. In geodesia è la figura geometrica più usata per descrivere la forma della Terra. Poiché due dei suoi tre assi sono uguali, si definisce anche Sferoide.

Forest/Bosco Land spanning more than 0.5 ha with trees higher than 5 m and a canopy cover of more than 10%, or trees able to reach these thresholds in situ. It does not include land that is predominantly under agricultural or urban land use. It includes young trees that have not yet reached but which are expected to reach a canopy cover of 10% and tree height of 5 m, temporarily unstocked areas due to clear-cutting or natural disasters, forest roads, firebreaks and small open areas, corridors of trees with an area of more than 0.5 ha and width of more than 20 m, cork-oak plantations/Area di superficie maggiore di 0.5 ha, caratterizzata da una copertura superiore a 10% di alberi in grado di superare un'altezza di 5 m a maturità, oppure in grado di superare tali soglie in situ. Sono escluse le aree con uso prevalente di carattere agricolo o urbano. Sono inclusi: soprassuoli giovani, in grado di raggiungere le soglie di copertura e altezza; le aree temporaneamente prive di soprassuolo per effetto di tagli a raso o disastri naturali; strade forestali, viali tagliafuoco e piccole aree prive di copertura; fasce di alberi con larghezza maggiore di 20 m; impianti di querce da sughero.

Forest type/Categoria forestale Each of the 23 classes adopted to classify forest area, based on the prevailing species in terms of crown coverage/Ciascuna delle 23 classi adottate per la classificazione della superficie forestale, definita sulla base della composizione specifica dei popolamenti.

Forest subtype/Sottocategoria forestale Each of the class distinguishing vegetation formations based on the dominant species or according to ecological criteria. Forest subtypes (91 in total) are the domains of greatest detail in INFC/Ciascuna delle 91 classi descrittive della vegetazione che costituiscono il livello di maggior dettaglio nella classificazione della superficie forestale adottata in INFC. Le sottocategorie forestali sono definite sulla base di un criterio ecologico o botanico.

Global Navigation Satellite System/Sistema di navigazione satellitare globale A constellation of satellites providing radio signals from space that transmit positioning and timing data, providing global coverage. Proper (GNSS) receivers then use this data to determine their position. Currently several systems are operating worldwide: GPS (USA), GLONASS (Russia), Beidou (China), Galileo (EU) and others/Una costellazione di satelliti emittenti segnali radio dallo spazio che veicolano dati di posizionamento e di tempo (cronologico). Opportuni ricevitori (GNSS) usano questi dati per determinare la loro posizione. Attualmente diversi sistemi sono operativi nel mondo: GPS (USA), GLONASS (Russia), Beidou (China), Galileo (EU) e altri.

Growing stock or growing stock volume The living woody species component of the standing volume. It is obtained by the stem volume of all subjects callipered. Stem volume includes large branches and tree top up to 5 cm diameter section, bark; stump, small branches and treetop below 5 cm diameter section are excluded/Il volume del soprassuolo ottenuto come somma dei volumi degli alberi cavallettati; comprende volume del fusto e dei rami grossi e cimale sopra la soglia di 5 cm alla sezione di svettamento.

Hypsometer/Ipsometro Optical instrument, with mechanical or electronic components, to measure tree height/Strumento, ottico, con componenti meccaniche oppure elettroniche, per la misurazione dell'altezza degli alberi.

Included polygon/Incluso Homogeneous area for land use and cover with an area between 500 and 5000 m^2 or, if elongated, with a width between 3 and 20 m, assigned to the land use and cover class of the nearest polygon that respects the minimum area and width thresholds of 0.5 ha and 20 m, respectively/Elemento territoriale omogeno per uso e copertura del suolo che non raggiunge le soglie minime dimensionali per essere classificato autonomamente e che viene classificato sulla base dell'uso del suolo più vicino.

Inventory attribute/Attributo inventariale The object of the estimate (also known as a variable)/Grandezza o caratteristica che viene osservata perché di interesse, in quanto oggetto di stima dell'inventario (sinonimo di variabile).

Inventory category/Categoria inventariale Each of the six classes adopted to classify Total wooded area, of which two refer to Forest and four to Other wooded land; classification is based on vegetation features assessable with field survey (potential tree height and tree or shrub crown cover) or cultivation practices (for Plantations)/Ciascuna delle 6 classi adottate per la classificazione della superficie forestale, al secondo livello gerarchico, in INFC. La categoria inventariale viene definita sulla base di caratteri fisionomici osservabili al suolo (in particolare l'altezza potenziale in situ e la copertura arborea distinta da quella arbustiva) e, in un unico caso, sulla base delle pratiche di coltivazione (impianti di arboricoltura da legno).

Inventory macro-category/Macrocategoria inventariale Each of the two classes that make Total wooded area (Forest and Other wooded land)/Ciascuna delle due classi di cui si compone la Superficie forestale totale (Bosco e Altre terre boscate).

Navigation/Navigazione Movement on the ground, even outside roads and trails, guided by a GNSS receiver and aimed at reaching a target point. From its own position, the device continuously updates distance and bearing to the target/Spostamento sul terreno, anche al di fuori di percorsi precostituiti come strade e sentieri, assistito da ricevitore GNSS e finalizzato al raggiungimento di un punto obiettivo. Dalla propria posizione istantanea il ricevitore aggiorna in continuo distanza e direzione al punto obiettivo.

Near Infrared/Infrarosso vicino Electromagnetic radiation with wavelength longer than that of visible light, between 750–1400 nm. In remote sensing, one of the typical bands/regions of multispectral images, particularly useful for vegetation monitoring/Radiazione elettromagnetica con lunghezza d'onda più lunga di quella della luce visibile, compresa fra 750 e 1400 nm. In telerilevamento, una delle tipiche bande delle immagini multispettrali, particolarmente utile per il monitoraggio della vegetazione.

Not assessed/Non valutato In the estimates produced by INFC2015, a quantity (e.g., an area) which was not classified according to a variable because the classification was not relevant/Nelle statistiche prodotte da INFC2015, una quantità (ad esempio una superficie) non classificata per la variabile perché la classificazione non era richiesta.

Not classified/Non classificato In the estimates produced by INFC, a quantity (e.g., an area) which was not classified according to the variable under estimation; it corresponds to a missing value/Nelle statistiche prodotte da INFC, una quantità (ad esempio una superficie) che manca della classificazione relativa alla variabile oggetto di stima; si tratta di un dato mancante.

Offset In GNSS positioning, the deviation between the real object/element to be positioned (usually in difficult conditions with respect to radio-signal reception) and one or more points where the measure is taken/Nel posizionamento satellitare, scarto fra la posizione reale dell'oggetto o elemento individuato, di solito in condizioni sfavorevoli per la ricezione del segnale radio, e quella di uno o più punti in cui avviene la misura.

Orthophoto/Ortofoto Aerial or satellite image with the same properties of an orthographic projection, derived by conventional perspective images, and with a uniform scale. The geometric rectification, resampling and georeferencing make it overlap other maps/Foto area o satellitare con le stesse proprietà di una proiezione ortografica, derivata da immagini con prospettiva convenzionale, con scala resa uniforme. La correzione geometrica, il ricampionamento e la georeferenziazione la rendono sovrapponibile alle mappe del territorio.

Other wooded land/Altre terre boscate Land not classified as 'Forest', spanning more than 0.5 ha, with trees higher than 5 m and a canopy cover of 5–10%, or trees able to reach these thresholds in situ; or with a combined cover of shrubs, bushes and trees above 10%. It does not include land that is predominantly under agricultural or urban land use. It includes areas with trees that will not reach a height of 5 m in situ and with a canopy cover of 10% or more, e.g., some Alpine tree vegetation types, or vegetation of arid zones/Area di superficie maggiore di 0.5 ha non classificata come "Bosco", caratterizzata da una copertura pari a 5–10% di alberi con altezza superiore a 5 m, o in grado di superare tali soglie in situ, oppure con una copertura complessiva di arbusti, cespugli e alberi superiore a 10%. Sono escluse le aree con uso prevalente di carattere agricolo o urbano. Include le aree con copertura superiore a 10% di alberi non in grado di raggiungere i 5 m a maturità, come alcune formazioni alpine o di zone aride.

Photointerpretation/Fotointerpretazione Land monitoring and classification technique by means of visual analysis of ground objects and patterns on aerial or satellite imagery/Tecnica di indagine e classificazione di elementi del territorio basata sull'osservazione e l'analisi visiva di immagini fotografiche aeree o satellitari.

Photopoints/Fotopunti Sample points overlapped on an aerial or satellite image to the purpose of classifying them by means of photointerpretation/Punti campione sovrapposti a riprese fotografiche o satellitari allo scopo di classificarli mediante fotointerpretazione.

Positioning/Posizionamento Determination of a point's coordinates on the ground. It is performed by triangulation (angles) or trilateration (distances) measurements with respect to known position points/Determinazione della posizione (coordinate) di un punto sul terreno. Viene eseguito con tecniche di triangolazione (misura di angoli) o trilaterazione (misura di distanze) a partire da altri punti con posizione nota.

Positioning accuracy/Accuratezza posizionale Euclidean distance between the measured position (coordinates) and the position assumed as true of a point; usually two components are considered, the planimetric and the altimetric/Differenza, intesa come distanza euclidea, fra la posizione misurata di un punto e la sua "vera" posizione (incognita); solitamente si distinguono un'accuratezza planimetrica e una altimetrica.

Quality Assurance The strategy adopted to guarantee good data quality; it involves decisions on both the design of the sampling strategy and the operational details of data collection, (crews' training, survey protocol, software for data recording, quality checks of stored data, assistance to surveyors on the implementation of the protocol, quality checks of surveyors' activities, final checks for data approval)/Strategia adottata allo scopo di assicurare la qualità dei risultati di un'indagine conoscitiva, a partire dalla definizione degli aspetti progettuali (scelta del piano di campionamento e delle modalità di rilevamento), fino ai dettagli organizzativi (formazione dei rilevatori, preparazione dei manuali) e operativi (realizzazione di applicativi per l'inserimento dei dati, controlli sul database, assistenza tecnica ai rilevatori, controlli di qualità, collaudi) della raccolta di dati.

Region/Regione By convention, the administrative units of Italy; Alto Adige and Trentino are actually autonomous provinces/Per convenzione, i distretti amministrativi in Italia anche nel caso delle Province autonome di Bolzano e di Trento.

Reference system/Sistema di riferimento In cartography, a system adopted to uniquely identify a point's coordinates on the Earth surface, in a plane projection. It consists of a Datum (e.g., WGS84), a projection (e.g., Universal Transverse Mercator) and some roto-translation parameters/In cartografia, sistema adottato per identificare univocamente le coordinate di un punto sulla superficie terrestre. Consiste in un Datum (es. WGS84), una proiezione (es. Universale Trasversa di Mercatore) e alcuni parametri di roto-traslazione.

Remote sensing/Telerilevamento In Earth observation disciplines, aircraft- or satellite-based sensor techniques to acquire information on Earth's objects by means of their reflected electromagnetic radiation/Nelle scienze della terra, tecniche di acquisizione a distanza di informazioni da sensori installati su piattaforma aerea o satellitare, mediante la riflessione di radiazione elettromagnetica da parte degli oggetti/elementi terrestri.

Sample plot/Area di saggio Reference area (circular in INFC) in which assessment of qualitative variables and measurement of quantitative variables are carried out/Superficie (sempre circolare in INFC) entro la quale condurre osservazioni o misure prestabilite (acronimo AdS); il termine inglese "plot" è entrato nell'uso nella letteratura tecnica anche per i testi italiani.

Sample point/Punto di sondaggio o punto di campionamento A point, defined by its geographic or cartographic coordinates, randomly selected. It is a reference sample unit for the variables surveyed/Singola unità campionaria, alla quale vengono riferite le classificazioni, le osservazioni e le misure previste dall'indagine campionaria. Tale unità corrisponde ad un punto del territorio in esame e viene individuato mediante una coppia di coordinate geografiche (o cartografiche).

Sampling error/Errore campionario The difference between the sample statistic and population parameter; it incurs when the statistical characteristics of a population are estimated from a sample because sample statistics generally differ from the statistics of the entire population/La parte di scostamento fra il valore di un parametro della popolazione e quello della sua stima campionaria, dovuta alle differenze fra le caratteristiche del campione e quelle complessive della popolazione da cui viene estratto.

Satellite based augmentation system A positioning precision improving system based on geostationary satellites and a ground control station network/Sistema di miglioramento della precisione del posizionamento basato su una rete di satelliti geostazionari e stazioni di controllo a terra.

Server In computing, a computer hardware or software component that provides services for other devices or software (clients) connected to it/In informatica è un computer o componente software che fornisce servizi ad altri componenti hardware o software (client) ad esso collegati.

Standard error/Errore standard The standard deviation of the sampling distribution of the estimates of a population parameter or an estimate of that standard deviation/È la deviazione standard della distribuzione delle stime di un parametro in tutti i possibili campioni casuali di numerosità n di una popolazione. Indica la precisione con cui una statistica permette di stimare un parametro della popolazione.

Statistical sampling/Campionamento statistico The selection of a subset (a statistical sample) of individuals from within a statistical population to estimate characteristics of the whole population (inference)/Metodo applicato in statistica per selezionare un determinato insieme (campione) di unità di una popolazione, in modo da ottenere risultati generalizzabili (inferenza statistica).

Stem biomass/Fitomassa del fusto Biomass of stem, large branches and treetop up to 5 cm section; it may be living biomass or dead biomass/Fitomassa (o necromassa, nel caso di alberi morti) del fusto, dei rami grossi e cimale fino alla sezione di taglio di 5 cm.

Stem volume/Volume del fusto The volume of stem, large branches and treetop up to 5 cm section. It may refer to volume of one tree (tree volume) or the volume of all trees in a class (e.g., a forest type); in the latter case, it is also known as growing stock/Volume del fusto, dei rami grossi e cimale fino alla sezione di taglio di 5 cm.; può essere riferito ad un albero (volume dell'albero) o a un'intera classe come per esempio una categoria forestale.

Stump/Ceppaia The base of a tree left in the ground after felling or truncated by natural events, less than 1.30 m high/La porzione basale del tronco che resta nel terreno dopo un'operazione di abbattimento, o per schianto naturale, minore di 1.30 m in altezza.

Texture/Tessitura In photointerpretation, characteristics of an image object due to the tone/colour differences between neighbouring pixels/In fotointerpretazione, caratteristiche peculiari di un oggetto raffigurato in un'immagine dovute alle variazioni di tono e di colore fra i pixel contigui.

Tone/Tono In photointerpretation, characteristics of an image object depending on intensity of colour/grey level/In fotointerpretazione, caratteristiche di un oggetto raffigurato in un'immagine dovuto all'intensità del colore o del livello di grigio.

Total tree height/Altezza dell'albero Length of the tree axis from the ground level to treetop/Altezza totale di un albero, dalla linea di terra alla cima più alta.

User interface/Interfaccia utente Software component, which allows all interactions, including data entry, between the user and the processing modules. It can be textual (TUI) or graphical (GUI)/Componente del software che si occupa di tutte le interazioni, compreso l'inserimento dei dati, fra l'utente e i moduli di elaborazione. Può essere testuale (TUI-Textual User Interface) o grafica (GUI-Graphical User Interface) a seconda della presenza o meno di elementi grafici atti a facilitarne la fruizione.

Variable/Variabile In INFC it is used as a synonym of attribute; the object of the estimate/In INFC è sinonimo di attributo, la grandezza oggetto della stima.

Printed in the United States
by Baker & Taylor Publisher Services